Interest, Principal, Rate, and Time Formulas

Interest	$I = PRT$	Time	Time in years $= \frac{I}{PR}$
Principal	$P = \frac{I}{RT}$		Time in days $= \frac{I}{PR} \times 360$
Rate	$R = \frac{I}{PT}$		

Maturity Value

The *maturity value*, M, of a principal of P dollars at a rate of interest R for T years is either

$$M = P + I$$

or, since $I = PRT$,

$$M = P(1 + RT).$$

Present Value at Simple Interest

The *present value at simple interest*, P, of a future value M at a rate of interest R for a time T is

$$P = \frac{M}{1 + RT}.$$

Simple Interest and Simple Discount

Variables Used for Simple Interest	Variables Used for Simple Discount
I = Interest	B = Discount
P = Principal (Face value)	P = Proceeds
R = Rate of interest	D = Discount rate
T = Time in years, or Fraction of a year	T = Time in years, or Fraction of a year
M = Maturity value	M = Maturity value

	Simple Interest	Simple Discount
Face value	Stated on note, or $P = \frac{M}{1 + RT}$	Same as maturity value, or $M = \frac{P}{1 - DT}$
Interest charge	$I = PRT$	$B = M \cdot D \cdot T$
Maturity value	$M = P + I$ or $M = P(1 + RT)$	Same as face value, or $M = \frac{P}{1 - DT}$
Amount received by borrower	Face value or principal	Proceeds: $P = M - B$ or $P = M(1 - DT)$
Identifying phrases	Interest at a certain rate Maturity value greater than face value	Discounted at a certain rate Proceeds Maturity value equal to face value
Annual interest rate	Same as stated rate, R	Greater than stated rate, D

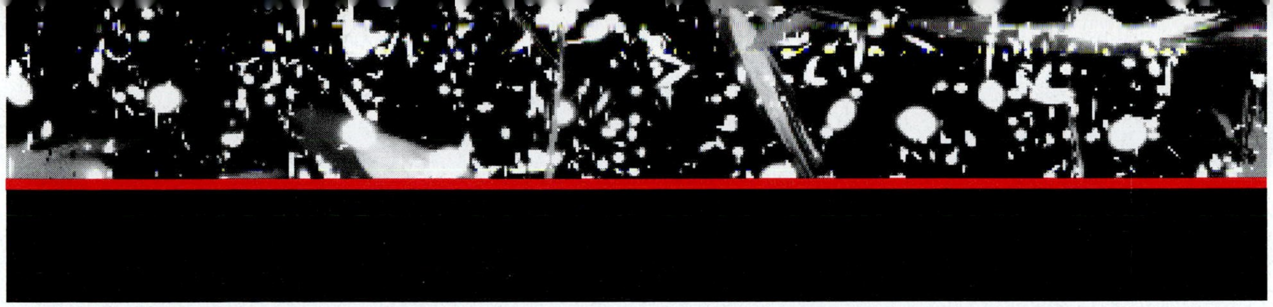

Mathematics *for* Business

Seventh Edition

Stanley A. Salzman

American River College

Charles D. Miller

Gary Clendenen

University of Texas—Tyler

Addison Wesley

Boston San Francisco New York
London Toronto Sydney Tokyo Singapore Madrid
Mexico City Munich Paris Cape Town Hong Kong Montreal

Publisher: *Jason A. Jordan*
Acquisitions Editor: *Jennifer Crum*
Editorial Project Manager: *Ruth Berry*
Managing Editor: *Ron Hampton*
Text Design Supervision: *Susan C. Raymond*
Cover Design: *Susan C. Raymond*
Production Coordinator: *Sheila Spinney*
Production Services: *UG / GGS Information Services, Inc.*
Editorial Assistant: *Sharon Smith*
Media Producer: *Lorie Reilly*
Marketing Manager: *Dona Kenly*
Marketing Coordinator: *Elan Hanson*
Prepress Services Buyer: *Caroline Fell*
First Print Buyer: *Evelyn Beaton*

Cover Image: © *Orion Press/Natural Selection*

Library of Congress Cataloging-in-Publication Data

Salzman, Stanley A.
Mathematics for business / Salzman, Miller, and Clendenen.—7th ed.
p. cm.
Includes index.
ISBN 0-321-06920-X (hc)
1. Business mathematics. I. Miller, Charles David, 1942– II. Clendenen, Gary. III. Title.

HF5691.S26 2000
650′.01′513—dc21 99-098182

ISBN: 0-321-06920-X

6 7 8 9 10 VHP 04

Contents

P4 Ch001

Test

Preface

The seventh edition of *Mathematics for Business* continues to provide solid, practical, and up-to-date coverage of those topics students must master to attain success in business today. More than ever, real-life examples from today's business world have been incorporated throughout the book. Many new graphs, tables, and news clippings have been added to increase the relevance of chapter content to the world students know. The globalization of our society is emphasized through examples and exercises that highlight foreign countries and international topics.

The new edition reflects the extensive business and teaching experience of the authors, as well as the suggestions of many reviewers nationwide. Providing solid, practical, and up-to-date coverage of business mathematics topics, the text begins with a brief review of basic mathematics and goes on to introduce key business topics, such as bank services, payroll, taxes, insurance, business discounts and markups, stocks and bonds, consumer loans, depreciation, financial statements, and business statistics. The text is accompanied by a greatly enhanced supplements package that provides many avenues—both print and media—for students to practice and explore further the concepts discussed in the chapters. (Please see pages xii to xiv of this Preface for full descriptions of the student and instructor supplements available.)

New Features

Chapter Openers Each chapter opens with a graph, table, or headline to capture students' interest. Chapter openers use an actual business example that applies to the specific content of that chapter.

Enhanced Treatment of Real-World Applications The seventh edition places a greater emphasis on real-world applications. The application problems have been updated throughout the text to be as relevant as possible to today's students, and they reference well-known companies such as Home Depot, McDonald's, Bank of America, Amazon.com, Nike, Starbucks, Borders, and Levi's. Applications are now preceded by headings that highlight particular topics such as sports, vacations, and financial planning.

New Art Program The art program of the seventh edition includes graphs and charts that utilize actual data from a variety of recognized sources. Rendered to draw student attention while emphasizing the data itself, the graphs and charts help students see that the mathematics of business is inherent to the world around them.

'Net Assets In each chapter, a one-page feature emphasizes the growing importance of the World Wide Web in business by showing an example of a company's Internet home page and providing questions that relate that company's on-line activities to the chapter concepts. Some of the corporations highlighted operate solely over the Internet. This feature, suitable even for students without access to the Web, is ideal for self-contained assignments that will illustrate the relevance of business math to actual corporate situations.

Financial and Estate Planning Foldout This innovative, six-page foldout focuses on key subject areas involved in personal finance and estate planning. Divided into nine sections, the foldout includes several guided activities to help students become actively involved in understanding their personal finances and how to invest wisely. Students are encouraged to look at maximizing lifetime earnings through career and education, and they are advised on how to save money, secure adequate insurance, build up a cash reserve, and avoid accumulating short-term debt. They are also asked to think about long-range planning related to home ownership, retirement, and wills. Finally, students can assess their current financial situation by completing a budget worksheet. The material in the foldout relates closely to the content in Chapters 1, 3, 7, 14, 15, 17, and 19 of the text, and optional questions at the end of selected exercise sets direct students to use the foldout.

Cumulative Reviews Five Cumulative Reviews, found after Chapters 3, 7, 10, 15, and 19, help students review groups of related chapter topics and reinforce their understanding of the material.

Metric System The metric system of measurement, found in Appendix B, gives students the information needed to understand and use the metric system.

Financial Calculator Solutions Several financial calculator solution boxes have been added, where applicable, to demonstrate to the student how this tool can be used to solve problems. Financial calculators are also illustrated in Appendix A.2, which includes exercises that students may solve using the financial calculator of their choice.

Quick Review with Chapter Terms The revised end-of-chapter Quick Review feature now begins with a list of key terms from the chapter and the pages on which they first appear. The Quick Review now uses a two-column format (Concepts and Examples) to help students review all the main points presented in the chapter.

Additional Features

Numerous Exercises Mastering business mathematics requires working many exercises, so we have included more than 3700 in the seventh edition. They range from simple drill problems to real-life application exercises that require several steps to solve. All problems have been independently checked to ensure accuracy.

Graded Application Exercises The application exercises in this text increase in difficulty level. Each even-numbered application exercise is the same type of problem as the preceding odd-numbered exercise. This allows the student to solve an odd-numbered exercise, check the answer in the answer section, and then solve the following even-numbered exercise with confidence.

Scientific Calculator Solutions Scientific Calculators are covered in depth in Appendix A.1, and scientific calculator solution boxes are shown throughout the text.

Supplementary Exercises Two sets of supplementary exercises, appearing in Chapter 3 help students review and synthesize difficult concepts. Answers to the odd-numbered supplementary exercises are located in the back of the text.

Writing Exercises Designed to help students better understand and relate the concepts within a section, these exercises require a short written answer. They often include references to a specific learning objective to help students formulate their answer.

Pretest A Pretest for business mathematics is included in the text's introduction. This tool helps students and instructors identify individual and class strengths and weaknesses.

Problem Solving Hints These hints provide helpful suggestions to the student and are located throughout the text.

Summary Exercises Every chapter ends with a Summary Exercise that has been designed to help students apply what they have learned in the chapter. These problems require students to synthesize most or all of the topics they have covered in the chapter in order to solve one cumulative exercise.

Glossary A glossary of key words, located at the back of the book, provides a quick reference for the main ideas of the course.

Summary of Formulas The inside covers of *Mathematics for Business* provide a handy summary of commonly used information and business formulas from the book.

New Content Highlights

Chapter 1, Problem Solving and Operations with Fractions, now begins with a review of problem solving. Chapters 1 and 2 then review the basics of fractions and algebra and contain numerous drill problems.

Chapter 3 contains over 350 exercises, most of which are application problems. The wide selection of application problems offers the instructor the chance to work on the reading and comprehension skills of students while reviewing the concepts of percent.

The material in Chapter 4, Banking Services, has been updated in keeping with the latest banking trends and practices. Banking charges and credit card deposit slips reflect the latest available materials. The reconciliation form has been simplified once again to reflect current industry changes.

In Chapter 5, Payroll, all wages and salaries have been updated along with FICA, Medicare, and tax-withholding rates. State withholding tax has been modified to more accurately represent state income taxes throughout the nation.

In Chapter 6, Taxes, the latest available tax forms and tables have been included in Section 6.3, Personal Income Tax. In Chapter 7, Risk Management, the insurance rates for motor vehicles and life insurance have been updated to more accurately reflect today's insurance costs.

Chapter 11, Simple Interest, has been updated to reflect current interest rates, and substantial new material on inflation and the consumer price index has been added. New examples show that a raise may not be of much help to an employee, depending on the increase in the cost of living that the employee experiences.

Chapter 13, Compound Interest, has been updated, and several application problems have been changed.

New coverage of Roth IRAs has been added to Section 14.1.

Chapter 15, Business and Consumer Loans, has been updated to reflect current interest rates, and the last section of the chapter has been modified to emphasize mortgages and real estate loans. Many students will be able to relate easily to this section since they may be planning to buy a home or may have bought one recently. Additional examples of the cost of home ownership are provided in the Financial and Estate Planning foldout—a new feature of the seventh edition (described in full on page x of this Preface).

Data from recent McDonald's Corporation financial statements are included in Chapter 17, Financial Statements and Ratios, so that students learn about financial statements using actual data from a company they know.

The sections on stocks and bonds in Chapter 18, Securities and Distribution of Profit and Overhead, have been updated to reflect more current price information. Emphasis is placed on using mutual funds to save for retirement and other long-term purposes, including the education of children.

Many of the problems in Chapter 19, Business Statistics, have been changed, and a number of graphs showing data from the business world have been added to increase the chapter's sense of realism.

Appendix A.1, Scientific Calculators, contains greatly expanded coverage of scientific calculators for professors who allow students to use calculators. Appendix A.2, Financial Calculators, reviews the basic functions of financial calculators. Appendix B, The Metric System, is new to this edition and gives students the information needed to understand and use the metric system.

Supplements

For the Instructor

Annotated Instructor's Edition

The Annotated Instructor's Edition provides answers directly on the page to all of the exercises in the text. In addition, an answer section at the back of both the Annotated Instructor's Edition and the student edition provides answers to the odd-numbered exercises in each section and to all chapter test exercises.
ISBN 0-321-08097-1

Instructor's Solutions Manual

This supplement contains solutions to all of the even-numbered exercises in each section of the text.
ISBN 0-321-06922-6

Printed Test Bank/Instructor's Resource Guide

This comprehensive manual contains teaching suggestions for each chapter, two pretests for the course (one for basic math and one for business math), four short-answer and two multiple-choice tests per chapter, two final exams, and 50 or more additional practice exercises for every chapter.
ISBN 0-321-06923-4

TestGen-EQ/QuizMaster-EQ

Available on a dual-platform, Windows/Macintosh CD-ROM, this fully networkable software enables instructors to create, edit, and administer tests using a computerized test bank of questions organized according to the chapter content of the text. Six question formats are available, and a built-in question editor allows the user to create graphs, import graphics, and insert mathematical symbols and templates, or text. An Export to HTML feature allows practice tests to be posted to the Internet, and instructors can use QuizMaster-EQ to post quizzes to a local computer network so that students can take them on-line. QuizMaster-EQ automatically grades the quizzes, stores results, and lets the instructor view or print a variety of reports for individual students or for an entire class or section.

ISBN 0-321-06925-0

InterAct Math® Plus

This networkable software provides course-management capabilities and on-line test administration for Addison Wesley Longman's InterAct Math® Tutorial Software (see *For the Student*). InterAct Math® Plus enables instructors to create and administer on-line tests, summarize students' results, and monitor students' progress in the tutorial software, providing an invaluable teaching and tracking resource.

InterAct Math® Instructor Package for Windows: ISBN 0-201-63555-0
InterAct Math® Instructor Package for Macintosh: ISBN 0-201-64805-9

Web Site: www.mathbusiness.com

The *Mathematics for Business* Web site contains InterAct Math® tutorial exercises for every chapter of the text, as well as other valuable resources for both instructors and students, such as chapter quizzes and downloadable figures and tables from the text.

***By the Numbers* Video Telecourse and Study Guide**

The Southern California Consortium has produced a videotape series called *By the Numbers: Practical Applications of Business Mathematics*, which has been aired on the Public Broadcasting System (PBS). Suitable for use in either a traditional or distance-learning classroom, these videotapes may be purchased from Intelecom by calling 626-796-7300. A study guide to accompany the tapes is available for purchase from Addison Wesley Longman. The study guide content is specific to the videos rather than the text, but a correlation insert is provided that lists relevant reading assignments and problem sets from *Mathematics for Business,* Seventh Edition. Instructors can integrate the videos, text, and study guide to create an effective telecourse in business mathematics, or provide supplemental video instruction for a lecture-based course.

Study Guide ISBN 0-321-71947-9

For the Student

Student's Solutions Manual

This supplement contains solutions to the odd-numbered exercises in every section as well as solutions to all chapter reviews, summary exercises, and cumulative review.

ISBN 0-321-06921-8

InterAct Math® Tutorial Software
Available on a dual-platform, Windows/Macintosh CD-ROM, this interactive tutorial software provides algorithmically-generated practice exercises that are correlated at the objective level to the content of the text. Every exercise in the program is accompanied by an example and a guided solution designed to involve students in the solution process. The software recognizes common student errors and provides appropriate feedback. It also tracks student activity and scores, and can generate printed summaries of students' progress. Instructors can use the InterAct Math® Plus course-management software (see *For the Instructor*) to create, administer, and track on-line tests and monitor student performance during their practice sessions in InterAct Math®.
ISBN 0-321-06924-2

InterAct MathXL: www.mathxl.com
InterAct MathXL is a Web-based tutorial system that helps students prepare for tests by allowing them to take practice tests and receive a personalized study plan based on their results. Practice tests are correlated directly to the section objectives in the text, and once a student has taken an on-line practice test, the software scores the test and generates a study plan that identifies strengths, pinpoints topics where more review is needed, and links directly to the appropriate section(s) of the InterAct Math® tutorial software for additional practice and review. Students gain access to the MathXL Web site through a password-protected subscription; subscriptions can either be bundled with the text or purchased separately.
InterAct MathXL subscription bundled with the text: ISBN 0-201-71739-5
InterAct MathXL subscription purchased separately: ISBN 0-201-71630-5

AWL Math Tutor Center
The AWL Math Tutor Center is staffed by qualified mathematics instructors who tutor students via toll-free telephone, fax, or e-mail on examples and exercises from their text. The Tutor Center is accessed through student registration number that may be bundled with new textbooks or purchased separately for those students who purchase used textbooks.
Tutor Center registration bundled with the text: ISBN 0-201-71741-7
Tutor Center registration purchased separately: ISBN 0-201-71049-8

Web Site: www.mathbusiness.com
The *Mathematics for Business* Web site contains InterAct Math® tutorial exercises for every chapter of the text, as well as other valuable resources for both instructors and students, such as chapter quizzes and downloadable figures and tables from the text.

Acknowledgments

We would like to thank the many users of the sixth edition for their insightful observations and suggestions for improving this book. We also wish to express our appreciation and thanks to the following reviewers of the seventh edition for their contributions:

Carl Ballard, *Central Piedmont Community College*
Viola Lee Bean, *Boise State University*
Linda Buchanan, *Howard College*
Judy Deme, *Lenoir Community College*

Jacqueline Dlatt, *College of DuPage*
Paul Franklin, *DeVry Institute of Technology, Kansas City*
Steven Johnson, *Kaskaskia College*
Merilyn Linney, *Guilford Technical Community College*
David G. Martin, *Milwaukee Area Technical College*
Cheryl McGahee, *Guilford Technical Community College*
Lawrence R. Melley, *Pennsylvania College of Technology*
A. Ally Mishal, *Stark State College of Technology*
Peter E. Moore, *Northern Kentucky University*
Alan Ransom, *Cypress College*
Ellen Sawyer, *College of DuPage*
Ned W. Schillow, *Lehigh Carbon Community College*
Sally Tarley, *Fairmont State College*
Sophia D. Waymyers, *Tri-County Technical College*

Our appreciation also goes to Linda Buchanan, Cheryl Davids, and Nelda Shelton for their careful accuracy checking all the exercises and examples in the book. We also would like to express our gratitude to our colleagues at American River College and the University of Texas at Tyler who have helped us immeasurably with their support and encouragement.

Also, special thanks and appreciation go to Sheri Minkner and Judy Martinez for their neat and accurate manuscript typing and to Larry and Cyndi Clendenen and Geri Wink for their help in checking the manuscript for errors.

At Addison Wesley Longman, we would like to thank all the talented and focused people who made sure that all the elements of this project came together in superb fashion.

Stanley A. Salzman
Gary Clendenen

Introduction for Students

Success in Business Mathematics

With our growing need for record keeping, establishing budgets, and understanding finance, taxation, and investment opportunities, mathematics has become a greater part of our daily lives. This text applies mathematics to daily business experience. Your success in future business courses and pursuits will be enhanced by the knowledge and skills you will learn in this course.

Studying business mathematics is different than studying subjects such as English or history. The key to success is regular practice. This should not be surprising. After all, can you learn to ski or to play a musical instrument without a lot of regular practice? The same is true for learning mathematics. Working problems nearly every day is the key to becoming successful. Here are some suggestions to help you succeed in business mathematics:

1. **Pay attention in class to what your instructor says and does, and take careful notes.** Note the problems the instructor works on the board and copy the complete solutions. Keep these notes separate from your homework to avoid confusion.
2. **Don't hesitate to ask questions in class.** Asking questions is not a sign of weakness, but of strength. There are always other students with the same question who are too shy to ask.
3. **Determine whether tutoring is available and know how to get help when needed.** Use the instructor's office hours to contact the instructor for suggestions and direction and call the AWL Math Tutor Center (see page xiv for details).
4. **Before you start on your homework assignment, rework the problems the instructor worked in class.** This will reinforce what you have learned. Many students say, "I understand it perfectly when you do it, but I get stuck when I try to work the problem."
5. **Read your text carefully.** Many students read only enough to get by, usually only the examples. Reading the complete section will help you to be successful with the homework problems. As you read the text, work the example problems and check the answers. This will test your understanding of what you have read. Pay special attention to highlighted statements and those labeled "Note" and "Problem-Solving Hint."
6. **Do your homework assignment only after reading the text and reviewing your notes from class.** Estimate the answer before you begin working the problem in the text. Check your work before looking at the answers in the back of the book. If you

get a problem wrong and are unable to see why, mark that problem and ask your instructor.

7. **Work as neatly as you can using a *pencil* and organize your work carefully.** Write your symbols clearly, and make sure the problems are clearly separated from each other.
8. **After you have completed a homework assignment, look over the text again.** Try to decide what the main ideas are in the lesson. Often they are clearly highlighted or boxed in the text.
9. **Keep any quizzes and tests that are returned to you for studying for future tests and the final exam.** These quizzes and tests indicate what your instructor considers most important. Be sure to correct any test problems that you missed. Write all quiz and test scores on the front page of your notebook.
10. **Don't worry if you do not understand a new topic right away.** As you read more about it and work through the problems, you will gain understanding. No one understands each topic completely right from the start.

Pretest in Business Mathematics

This pretest measures your business mathematics skills at the beginning of the course. The solutions to each of these problems are found in this book on page xxi.

(page 11) 1. Convert to an improper fraction: $1\frac{5}{8}$

(page 12) 2. Convert to a mixed number: $\frac{23}{3}$

(page 15) 3. Find the least common denominator of the fractions $\frac{5}{12}$, $\frac{7}{18}$, and $\frac{11}{20}$.

(page 17) 4. Mixed numbers—add: $34\frac{1}{2} + 23\frac{3}{4} + 34\frac{1}{2} + 23\frac{3}{4}$

(page 17) 5. Common fractions—subtract: $\frac{17}{18} - \frac{20}{27}$

(page 18) 6. Mixed numbers—subtract:

$$\begin{array}{r} 36\frac{2}{9} \\ -27\frac{5}{6} \\ \hline \end{array}$$

(page 23) 7. Mixed numbers—multiply: $5\frac{5}{8} \times 4\frac{1}{6}$

(page 24) 8. Common fractions—divide: $\frac{25}{36} \div \frac{15}{18}$

(page 26) 9. Convert the decimal 0.028 to a fraction.

(page 42) 10. Solve $y + 12.3 = 20.5$ for y.

(page 46) 11. Solve $5r - 2 = 2(r + 5)$ for r.

(page 55) 12. Solve for T in the formula $M = P(1 + RT)$

(page 62) 13. Find x in the proportion: $\frac{4}{9} = \frac{36}{x}$

(page 74) **14.** Express as a percent: 0.7

(page 75) **15.** Express as a decimal: 142%

(page 79) **16.** Solve for part: 1.2% of 180 is _______.

(page 86) **17.** Solve for base: 135 is 15% of _______.

(page 87) **18.** The 5% sales tax collected by Famous Footwear was $780. What was the amount of total sales?

(page 100) **19.** The price of a home sold by real estate agent Cas Shields this year is $121,000, which is 10% more than last year's value. Find the value of the home last year.

(page 103) **20.** After Sports About deducted 10% from the price of a pair of skis, Craig Bleyer paid $135. What was the original price of the skis?

(page 191) **21.** Suppose that during a certain quarter Leslie's Pool Supplies has collected $2765.42 from its employees for FICA tax, $638.17 for medicare tax, and $3572.86 in federal withholding tax. Compute the total amount due to the Internal Revenue Service from Leslie's Pool Supplies.

(page 214) **22.** Find the taxes on each of the following pieces of property. Assessed valuations and tax rates are given.

(a) $58,975; 8.4% **(b)** $875,400; $7.82 per $100

(c) $129,600; $64.21 per $1000 **(d)** $221,750; 94 mills

(page 224) **23.** Chris Kelly is single, has no dependents, and had an adjusted gross income of $26,735 last year. He had deductions of $1352 for other taxes, $3118 for mortgage interest, and $317 for charity. Find his taxable income and his income tax.

(page 242) **24.** Helen Dale owns an industrial building valued at $760,000. Her fire insurance policy (with an 80% coinsurance clause) has a face value of $570,000. The building suffers a fire loss of $144,000. Find the amount of the loss that the insurance company will pay and the amount that Dale must pay.

(page 288) **25.** Oaks Hardware is offered a series discount of 20/10 on a Porter-Cable cordless drill with a list price of $150. Find the net cost after the series discount.

(page 302) **26.** An invoice received by Oaks Hardware for $840 is dated July 1 and offers terms of 2/10, n/30. If the invoice is paid on July 8 and the shipping and insurance charges, which were "FOB shipping point," are $18.70, find the total amount due.

(page 325) **27.** The manager of Roseville Appliance purchased a coffee maker manufactured in Spain for $15 and will sell it for $18.75. Find the percent of markup based on cost.

(page 340) **28.** An athletic shoe manufacturer makes a walking shoe at a cost of $16.80 per pair. Based on past experience 10% of the shoes will be defective and must be sold as irregulars for $24 per pair. If the manufacturer produces 1000 pairs of the shoes and desires a markup of 100% on cost, find the selling price per pair.

(page 365) **29.** Suppose Olympic Sports and Leisure made the following purchases of the Explorer External Frame backpack during the year.

Beginning inventory	20 backpacks at $70
January	50 backpacks at $80
March	100 backpacks at $90
July	60 backpacks at $85
October	40 backpacks at $75

At the end of the year there are 75 backpacks in inventory. Use the weighted average method to find the inventory value.

(page 385) **30.** Jeff Guerrant took out a loan for $9000 for 9 months for a used truck and had an interest charge of $540. What was the interest rate?

(page 443) **31.** On February 27, Andrews Lincoln-Mercury receives a 150-day simple interest note with a face value of $3500 at 8% interest per year. On March 27, the firm discounts the note at the bank. Find the proceeds if the discount rate is 12%. (Use ordinary or banker's interest.)

(page 460) **32.** A savings account at Northstar Bank in Canada pays 7% per year compounded semiannually. If you initially deposit $2500, **(a)** find the compound amount after 3 years and **(b)** find the compound interest.

(page 515) **33.** KidsToys, Inc. sold $100,000 worth of bonds that must be paid off in 8 years. They now must set up a sinking fund to accumulate the necessary $100,000 to pay off their debt. Find the amount of each payment going into a sinking fund if the payments are made at the end of each year, and the fund earns 10% compounded annually.

(page 563) **34.** Bob Jones uses a $75,000 loan for 25 years at 8% to purchase a summer cabin. Annual insurance and taxes on the property are $654 and $1329 respectively. Find the monthly payment.

(page 593) **35.** City Saturn purchased an electronic smog analyzer for $9000. Using the sum-of-the-years'-digits method of depreciation, find the first and second years' depreciation if the analyzer has an estimated life of 4 years and no salvage value.

(page 601) **36.** A boat dock with a life of 10 years is installed on April 12 at a cost of $18,000. If the double-declining-balance method is used, find the depreciation for the first partial year and the next full year.

(page 629) **37.** Write each of the following items as a percent of net sales.

Gross sales	$209,000	Salaries and wages	$11,000
Returns	$9,000	Rent	$6,000
Cost of goods sold	$145,000	Advertising	$11,000

(page 665) **38.** Due to a steep drop in the price of oil, Alamo Energy paid no dividend last year. The company has done much better this year and the board of directors has set aside $175,000 for the payment of dividends. The company has outstanding 12,500 shares of cumulative preferred stock having par value of $50, with an 8% dividend. The company also has 40,000 shares of common stock. What dividend will be paid to the owners of each type of stock?

(page 688) **39.** Laura Cameron, Jay Davis, and Donna Friedman opened a tool rental business. Cameron contributed $250,000 to the opening of the business, which will be operated by Davis and Friedman. The partners agree that Cameron will first receive a 10% return on her investment before any further division of profits. Additional profits will be divided in the ratio 1:2:2. Find the amount that each partner would receive from a profit of $75,000.

(page 727) **40.** The diameter of a part coming out of a machining process is measured regularly. The diameters vary some as shown in the frequency table. Find the **(a)** mean, **(b)** median, and **(c)** mode.

Diameter (inches)	Frequency
0.720–0.729	3
0.730–0.739	12
0.740–0.749	8
0.750–0.759	9
0.760–0.769	2

Answers: 1. $\frac{13}{8}$ **2.** $7\frac{2}{3}$ **3.** 180 **4.** $116\frac{1}{2}$ **5.** $\frac{11}{54}$ **6.** $8\frac{7}{18}$ **7.** $23\frac{7}{16}$ **8.** $\frac{5}{6}$ **9.** $\frac{7}{250}$ **10.** 8.2 **11.** 4 **12.** $T = \frac{M - P}{PR}$ **13.** 81 **14.** 70% **15.** 1.42 **16.** 2.160 **17.** 900 **18.** \$15,600 **19.** \$110,000 **20.** \$150 **21.** \$10,380.04 **22. (a)** \$4953.90 **(b)** \$68,456.28 **(c)** \$8321.62 **(d)** \$20,844.50 **23.** \$19,198; \$2879.70 **24.** \$135,000; \$9000 **25.** \$108 **26.** \$841.90 **27.** 25% **28.** \$34.67 **29.** \$6249.75 **30.** 8% **31.** \$3469.59 **32. (a)** \$3073.14 **(b)** \$573.14 **33.** \$8744.40 **34.** \$744.25 **35.** \$3600; \$2700 **36.** \$2700; \$3060 **37.** 104.5%; 4.5%; 72.5%; 5.5%; 3%; 5.5% **38.** \$8; \$1.88 **39.** \$35,000; \$20,000; \$20,000 **40. (a)** 0.743 **(b)** 0.7445 **(c)** 0.7345

Part 1 Basic Mathematics

1 Problem Solving and Operations with Fractions

Mathematics is very much a part of our lives. We use mathematics when we calculate the amount of money we earn, the interest and finance charges on our loans, the interest earned on our investments, and the cost of those things most important to our future. For example, Figure 1.1 shows the projected cost of 4-year college degrees in both public and private colleges over a 14-year period. The projected cost includes tuition, fees, room and board, transportation, books, and other expenses.

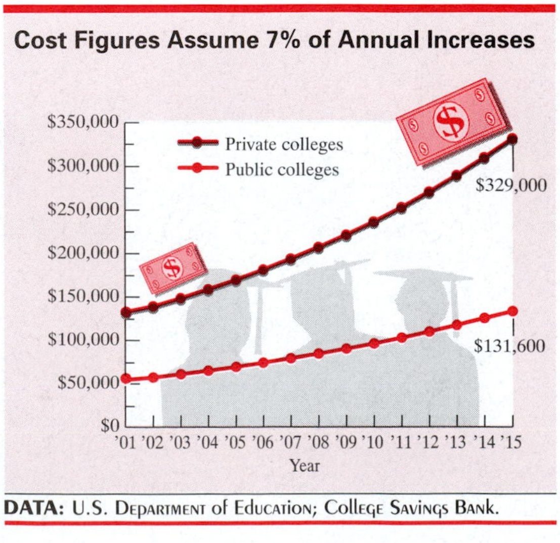

Figure 1.1 Projected cost of 4-year college degrees in public and private colleges.

What will the projected cost difference be between public and private colleges in 2015? How will students pay the costs of college and what are some possible sources of the needed funds?

Businesses use mathematics every day to prepare payrolls, find the interest on loans, determine the markups and markdowns on items to be sold, maintain the firm's financial records, or calculate the amount of taxes owed.

It is important for you to understand the fundamentals of mathematics, so that you can solve more advanced problems in business mathematics. The first chapter reviews problem solving and fractions, Chapter 2 looks briefly at basic equations and formulas, and Chapter 3 discusses percent. The remaining chapters apply these mathematical skills to actual business situations.

1.1 Problem Solving

Objectives

1 *Identify indicator words in application problems.*
2 *List the four steps for solving application problems.*
3 *Learn to estimate answers.*

Many business application problems require mathematics. We must read the words carefully to decide how to solve the problem.

Objective 1 *Identify Indicator Words in Application Problems.* Look for **indicators** in the application problem—words that indicate the necessary operations—either addition, subtraction, multiplication, or division. Some of these words appear below.

Addition	Subtraction	Multiplication	Division	Equals
plus	less	product	divided by	is
more	subtract	double	divided into	the same as
more than	subtracted from	triple	quotient	equals
added to	difference	times	goes into	equal to
increased by	less than	of	divide	yields
sum	fewer	twice	divided equally	results in
total	decreased by	twice as much	per	are
sum of	loss of			
increase of	minus			
gain of	take away			

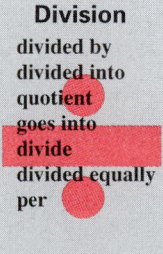

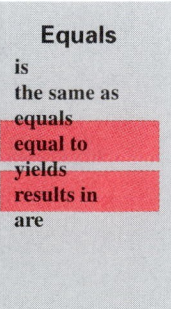

NOTE The word "and" does not indicate addition and does not appear as an indicator word above. Notice how the "and" shows the *location* of an operation sign.

The sum of 6 *and* 2 is $6 + 2$.

The difference of 6 *and* 2 is $6 - 2$.

The product of 6 *and* 2 is 6×2.

The quotient of 6 *and* 2 is $6 \div 2$. ▬

Objective 2 *List the Four Steps for Solving Application Problems.*

Solving Application Problems

Step 1. *Read* the problem carefully, and be certain that you understand what the problem is asking. It may be necessary to read the problem several times.

Step 2. Before doing any calculations, work out a plan and try to visualize the problem. Know which facts are given and which must be found. Use *indicator words* to help determine your plan.

Step 3. Estimate a *reasonable answer* using rounding.

Step 4. *Solve* the problem by using the facts given and your plan. Does the answer make sense? If the answer is reasonable, *check* your work. If the answer is not reasonable, begin again by rereading the problem.

NOTE Be careful not to make the mistake that some students do—they begin to solve a problem before they understand what the problem is asking. Be certain that you know what the problem is asking before you try to solve it. ■

Objective 3 *Learn to Estimate Answers.* These four steps give a systematic approach for solving application problems. Each of the steps is important, but special emphasis should be placed on Step 3, estimating a reasonable answer. Many times an answer just does not fit the problem.

What is a reasonable answer? Read the problem and try to determine the approximate size of the answer. Should the answer be part of a dollar, a few dollars, hundreds, thousands, or even millions of dollars? For example, if a problem asks for the cost of a man's shirt, would an answer of $20 be reasonable? $1000? $0.65? $65?

Always make an estimate of a reasonable answer; then check the answer you get to see if it is close to your estimate.

Example 1 Applying Division

At a recent garage sale, the total sales were $584.50. If the money was divided equally among Paul, Rachel, Maryangela, and Jose, how much did each person get? Round to the nearest cent.

Approach To find the amount received by each person, divide the total amount of sales by the number of people.

Solution

Step 1. A reading of the problem shows that the four members in the group divided $584.50 equally.

Step 2. The word indicators, ***divided equally***, show that the amount each received can be found by dividing $584.50 by 4.

Step 3. A reasonable answer would be a little less than $150 each, since $600 ÷ 4 = $150 ($584 rounded to $600).

Step 4. Find the actual answer by dividing \$584.50 by 4.

$$
\begin{array}{r}
146.125 \\
4\overline{)584.500} \\
\underline{4} \\
18 \\
\underline{16} \\
24 \\
\underline{24} \\
5 \\
\underline{4} \\
10 \\
\underline{8} \\
20 \\
\underline{20} \\
0
\end{array}
$$

The answer 146.125 rounds to \$146.13.
↑ 5 or larger, round up

This is a reasonable answer, as \$146.13 is close to the estimated answer of \$150. But is the answer \$146.13 correct? Check the work.

\$146.13		Amount received by each person
×	4	Number of people
\$584.52		Total sales

The check is not exact because of the rounding up. Actually, two people will get \$146.13 and two people will get \$146.12.

Example 2 Applying Addition

Matt Owens earns \$82.56 on Monday, \$72.23 on Tuesday, \$90.70 on Wednesday, \$94.38 on Thursday, and \$64.36 on Friday. Find his total earnings for the week. Round to the nearest dollar.

Approach To find the total for the week, add the earnings for each day.

Solution

Step 1. In this problem, the earnings for each day are given and the total earnings for the week must be found.

Step 2. The word indicator, ***total***, shows us to add the daily earnings to arrive at the weekly total.

Step 3. Because the earnings were about \$80 per day for a week of 5 days, a reasonable estimate would be around \$400 ($5 \times \$80 = \$400$).

Step 4 Find the actual answer by adding the earnings for the 5 days.

$$\begin{array}{r} \$\ 82.56 \\ 72.23 \\ 90.70 \\ 94.38 \\ +\ \ 64.36 \\ \hline \$404.23 \end{array} = \$404 \text{ rounded to the nearest dollar}$$

↑ 4 or less, round down

This answer is close to $400, and therefore reasonable.

Example 3 Applying Subtraction

The number of students enrolled in Chabot College this year is 4084 fewer than the number enrolled last year. Enrollment last year was 21,382. Find the enrollment this year.

Approach To find the number of students enrolled this year, subtract the enrollment decrease (fewer students) from last year.

Solution

Step 1. In this problem, the enrollment has decreased from last year to this year. The enrollment last year and the decrease in enrollment are given. This year's enrollment must be found.

Step 2. The indicator word "***fewer***" shows that subtraction must be used to find the number of students enrolled this year.

Step 3. Because the enrollment was about 21,000 students, and the decrease in enrollment is about 4000 students, a reasonable estimate would be 17,000 students (21,000 − 4000 = 17,000).

Step 4. Find the actual answer by subtracting 4084 from 21,382.

$$\begin{array}{r} 21{,}382 \\ -\ \ 4{,}084 \\ \hline 17{,}298 \end{array}$$

The enrollment this year is 17,298. The answer 17,298 is reasonable, as it is close to the estimate of 17,000. Check by adding.

$$\begin{array}{rl} 17{,}298 & \text{Enrollment this year} \\ +\ \ 4{,}084 & \text{Decrease in enrollment} \\ \hline 21{,}382 & \text{Enrollment last year} \end{array}$$

Example 4 Solving a Two-Step Problem

A landlord receives $680 from each of five tenants. After paying $1880 in expenses, how much rent does the landlord have left?

Approach To find the amount remaining, first find the total rent received. Next, subtract the expenses paid to find the amount remaining.

Solution

Step 1. There are five tenants and each pays the same rent.

Step 2. The wording *from each of five tenants* indicates that the five rents must be totaled. Since the rents are all the same, use multiplication to find the total rent received. Finally, subtract expenses.

Step 3. The amount of rent is about \$700, making the total rent received about \$3500 (\$700 × 5). The expenses are about \$2000. A reasonable estimate of the amount remaining is \$1500 (\$3500 − \$2000).

Step 4. Find the exact amount by first multiplying \$680 by 5 (the number of tenants).

$$\begin{array}{r} \$680 \\ \times \quad 5 \\ \hline \$3400 \end{array}$$

Finally, subtract the \$1880 in expenses from \$3400.

$$\$3400 - \$1880 = \$1520$$

The amount remaining is \$1520.

The answer \$1520 is reasonable, since it is close to the estimated answer of \$1500. Check the amount by adding the expenses and then dividing by 5.

$$\$1520 + \$1880 = \$3400$$

$$\begin{array}{r} \$680 \\ 5\overline{)3400} \end{array}$$

1.1 Exercises

Solve the following application problems.

1. COMPETITIVE CYCLIST TRAINING During a week of training, Beth Andrews rode her bike 80 miles on Monday, 75 miles on Tuesday, 135 miles on Wednesday, 40 miles on Thursday, and 52 miles on Friday. What is the total number of miles she rode in the five-day period?
2. COFFEE SALES During a recent week, Starbucks Coffee sold 325 pounds of Estate Java coffee, 75 pounds of Encanta Blend coffee, 137 pounds of Ethiopia Sidamo coffee, 495 pounds of Starbucks House-Blend Decaf coffee and 105 pounds of New Guinea Peaberry coffee. Find the total number of pounds of these coffees sold.
3. ATM CRIME According to *ATM Crime and Security* newsletter, in one region there were 70 ATM burglaries and attempted burglaries in 1992 and 200 in 1997. How many more of these crimes were there in 1997 than in 1992?
4. ATM FACTS The amount of cash in an ATM ranges from \$15,000 in small machines to \$250,000 in large bank machines. How much more money is there in the large machines than in the small machines?
5. VIETNAM VETERANS A group of American soldiers and nurses, veterans of the Vietnam War, rode bicycles from Hanoi to Saigon (Ho Chi Minh City). If they rode 75 miles each day and the trip took 16 days, what is the distance between these two cities in Vietnam?

6. **WORLD POPULATION GROWTH** The world population grows by 10,000 people each hour. Find the increase in world population in one year of 365 days.
7. **AUTOMOBILE WEIGHT** A car weighs 2425 pounds. If its 582-pound engine is removed and replaced with a 634-pound engine, how much will the car weigh after the engine change?
8. **PRESCHOOL MANAGER** Tiffany Connolly has $2324 in her preschool operating account. After spending $734 from this account, the class parents raise $568 in a rummage sale. Find the balance in the account after depositing the money from the rummage sale.
9. **BUSINESS ENTERPRISES** There are 24 million business enterprises in the United States. There are only 7000 of these businesses that are large businesses having 500 or more employees, while the rest are small and midsize businesses. Find the number of small and midsize businesses.
10. **WEIGHING FREIGHT** A truck weighs 9250 pounds when empty. After being loaded with firewood, the truck weighs 21,375 pounds. What is the weight of the firewood?
11. **EMPLOYER BUYOUTS** Last year 15,293 federal workers applied for buyouts (a bonus to retire early), which was the chance to get a $25,000 going-away present from Uncle Sam. If all of these employees received the bonus, what was the total cost of the buyouts?
12. **EARLY RETIREMENT** This year the government will offer a $25,000 early retirement bonus to 60,000 federal workers. If all of these employees receive this bonus, what is the total cost to the government?
13. **HOTEL ROOM COSTS** In a recent survey of high-priced hotels, the least expensive was Harrah's at a cost of $65 per night, while several of the hotels in the group were $90 per night. Find the amount saved on a 5-night stay at the least expensive hotel over the cost of staying at one of the more expensive hotels.
14. **LUXURY HOTELS** The most expensive hotel room in a recent study was the Ritz-Carlton at $375 per night, while the least expensive was Motel 6 at $32 per night. Find the amount saved on a 4-night stay at Motel 6 over the cost of staying at the Ritz-Carlton.
15. **PHYSICALLY IMPAIRED** The Enabling Supply House purchases 6 wheelchairs at $1256 each and 15 speech compression recorder-players at $895 each. Find the total cost of the equipment.
16. **COLLEGE BOOKSTORE** A college bookstore buys 17 computers at $506 each and 13 printers at $482 each. Find the total cost.
17. **THEATER RENOVATION** A theater owner is remodeling and wants to provide enough seating for 1250 people. The main floor has 30 rows of 25 seats in each row. If the balcony has 25 rows, how many seats must be in each row of the balcony to satisfy the owner's seating requirements?
18. **PACKING AND SHIPPING** Nancy Hart makes 24 grapevine wreaths per week to sell to gift shops. She works 30 weeks a year and packages six wreaths per box. If she ships equal quantities to each of five shops, find the number of boxes each shop will receive.
19. **MALL-SHOPPING FACTS** A recent study showed that the average trip to the mall lasts 4.4 hours, which is longer than a football game. How many hours would be needed for 8 of these trips to the mall?

20. **VENDING-MACHINE COST** The Generic Cold-Drink Vendor (Model CD-777G) is priced at $2299.99 at Price Costco. Find the cost of 14 of these machines.

21. **STOCKHOLDER LOSSES** British Imports announced a $38 million loss, or $0.58 a share. Find the number of shares of stock in the company. Round to the nearest tenth of a million.

22. **SHARES OF STOCK** Money Store officials said that they expect to post a loss of $42 million, or $0.65 a share. Find the number of shares of stock in the company. Round to the nearest tenth of a million.

23. **OLYMPIC GOLD COINS** If a five-dollar Olympic gold coin weighs 8.359 grams, find the number of coins that can be produced from 221 grams of gold. Round to the nearest whole number.

24. **MEDICINE DOSAGE** Each dosage of a medication contains 1.62 units of a certain ingredient. Find the number of dosages that can be made from 57.13 units of the ingredient. Round to the nearest whole number.

25. **STAMP COLLECTORS** The Elvis Presley stamp is the most popular commemorative stamp to be saved in collections. If 124 million of these 29¢ stamps were saved, find the amount paid for these stamps.

26. **STAMP SALES** There were 76.2 million commemorative wildflower stamps saved in collections, making it the second most popular stamp after the Elvis Presley stamp. Find the total amount paid for these 29¢ stamps.

27. **GERMAN MANAGEMENT EARNINGS** A department manager at Karstadt, Germany's largest department store, earns $2365 each month for working a 37-hour week. Find
    (a) the number of hours worked each month and
    (b) the manager's hourly earnings. (1 month = 4.3 weeks) Round to the nearest cent.

28. **WAL-MART MANAGER EARNINGS** A department manager at Wal-Mart earns $2528 each month for working a 48-hour week. Find
    (a) the number of hours worked each month and
    (b) the manager's hourly earnings. (1 month = 4.3 weeks) Round to the nearest cent.

29. Use Section 1 of the foldout to answer the following.
    (a) Find the difference in the average starting salaries of an accountant and a journalist.
    (b) How much higher is the average salary of a person with a bachelor's degree than that of a high school graduate?

1.2 Addition and Subtraction of Fractions

Objectives

1. *Recognize types of fractions.*
2. *Convert mixed numbers to improper fractions.*
3. *Write fractions in lowest terms.*
4. *Use the rules for divisibility.*
5. *Add or subtract like fractions.*
6. *Find the least common denominator.*
7. *Add unlike fractions.*
8. *Add mixed numbers.*
9. *Subtract unlike fractions.*
10. *Subtract mixed numbers.*

This section looks at **fractions**—numbers, like decimals, that can be used to represent parts of a whole. Fractions and decimals are two ways of representing the same quantity. Fractions are used in business and our personal lives. For example, the following newspaper clipping gives the size of the sleeping compartments (berths) for women on board the *USS Sullivans*.

Destroyer Fits Women, Too

ABOARD THE USS SULLIVANS—The *Sullivans* is equipped with the Navy's most advanced missiles and submarine-spotting sonar. But it wasn't only designed to have the latest in high-tech military equipment. It also was built with the idea that women would be among its crew. So what is it like for women on board?

Sleeping areas for enlisted personnel have doors. Visitors of the opposite gender must announce "Man on deck" or "Woman on deck" before entering.

Each of two berthing compartments for women holds 18 bunks, or racks, stacked three high. Each rack has a blue privacy curtain and is $6\frac{1}{2}$ feet long. There is $17\frac{3}{4}$ inches of space between the 3-inch-thick mattress and the bottom of the rack above.

Male racks are identical, except many are in much larger berthing areas. The largest of the male compartments holds 105 racks, another holds 70, a third 40. The other two compartments for men hold 24 and 18 racks.

Source: USA Today, January 12, 1998.

A **fraction** represents parts of a whole. Fractions are written as one number over another, with a line between the two numbers, as in the following.

$$\frac{5}{8} \qquad \frac{1}{4} \qquad \frac{9}{7} \qquad \frac{13}{10}$$

The number above the line is called the **numerator**, and the number below the line is called the **denominator**. In the fraction $\frac{2}{3}$, for example, the numerator is 2 and the denominator is 3. The denominator tells the number of equal parts into which the whole is divided and the numerator tells how many of these parts we are talking about. For example, $\frac{2}{3}$ is "2 parts out of 3 equal parts." (See Figure 1.2.)

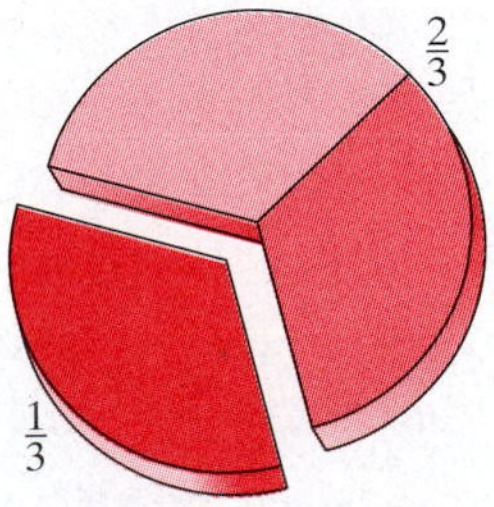

$\frac{2}{3}$ means 2 parts out of 3 equal parts

Figure 1.2

Objective 1 *Recognize Types of Fractions.* If the numerator of the fraction is smaller than the denominator, the fraction is a **proper fraction**. Examples of proper fractions include $\frac{2}{3}$, $\frac{3}{4}$, $\frac{15}{16}$, and $\frac{1}{8}$. A fraction with a numerator greater than or equal to the denominator is an **improper fraction**. Examples of improper fractions include $\frac{17}{8}$, $\frac{19}{12}$, $\frac{11}{2}$, and $\frac{5}{5}$. A proper fraction has a value less than 1, while an improper fraction has a value greater than or equal to 1.

To write a whole number as a fraction, place the whole number over 1; for example,

$$7 = \frac{7}{1} \qquad \text{and} \qquad 12 = \frac{12}{1}$$

Objective 2 *Convert Mixed Numbers to Improper Fractions.* The sum of a fraction and a whole number is called a **mixed number**. Examples of mixed numbers include $2\frac{2}{3}$, $3\frac{5}{8}$, and $9\frac{5}{6}$. The mixed number $2\frac{2}{3}$ is a short way of writing $2 + \frac{2}{3}$. A mixed number can be converted to an improper fraction. For example, to convert the mixed number $4\frac{5}{8}$ to an improper fraction, first multiply the denominator of the fraction part, 8, by the whole number part, 4. This gives $8 \times 4 = 32$. Then add the product (32) to the numerator (in this case, 5). This gives $32 + 5 = 37$. This sum is the numerator of the new improper fraction. The denominator stays the same.

$$4\frac{5}{8} = \frac{37}{8} \longleftarrow (8 \times 4) + 5$$

The following clipping notes that Home Depot stock jumped $1\frac{5}{8}$ points to $\$68\frac{1}{2}$. These are both mixed numbers. The Home Base stock was up $\frac{3}{8}$; this is a proper fraction.

Market View: Daily Stock Report

The Home Depot jumped $1\frac{5}{8}$ to $\$68\frac{1}{2}$. The home-improvement center reported increased quarterly earnings. Home Base, on a report of increased earnings, was up $\frac{3}{8}$ to $\$8\frac{3}{16}$.

Example 1 Converting Mixed Numbers to Improper Fractions

The increase in value and the closing price of Home Depot stock are both expressed as mixed numbers. Convert these mixed numbers to improper fractions.

(a) $1\frac{5}{8}$ **(b)** $68\frac{1}{2}$

Solution

(a) First multiply 8 (the denominator) and 1 (the whole number), and then add 5 (the numerator). This gives $(8 \times 1) + 5 = 8 + 5 = 13$. (The parentheses are used to show that 8 and 1 are multiplied first.) The denominator stays the same.

$$1\frac{5}{8} = \frac{13}{8} \longleftarrow (8 \times 1) + 5$$

(b) $68\frac{1}{2} = \frac{(2 \times 68) + 1}{2} = \frac{137}{2}$

Convert an improper fraction to a mixed number by dividing the numerator of the improper fraction by the denominator. The quotient is the whole number part of the mixed number, and the remainder is used as the numerator of the fraction part. The denominator stays the same.

Example 2 Converting Improper Fractions to Mixed Numbers

Convert $\frac{23}{3}$ to a mixed number.

Solution Divide 23 by 3.

$$\begin{array}{r} 7 \\ 3\overline{)23} \\ \underline{21} \\ 2 \end{array}$$

The whole number part is the quotient 7. The remainder 2 is used as the numerator of the fraction part. Keep 3 as the denominator.

$$\frac{23}{3} = 7\frac{2}{3}$$

NOTE A proper fraction has a value that is smaller than 1; an improper fraction has a value that is 1 or greater.

Objective 3 *Write Fractions in Lowest Terms.* If no number except 1 divides without remainder into both numerator and denominator of a fraction, the fraction is in **lowest terms**. For example, only 1 divides without remainder into both 2 and 3, so $\frac{2}{3}$ is in lowest terms. In the same way, $\frac{1}{9}$, $\frac{4}{11}$, $\frac{12}{17}$, $\frac{8}{9}$, and $\frac{11}{15}$ are all in lowest terms. The fraction $\frac{15}{25}$ is *not* in lowest terms, however, because both 15 and 25 may be divided by 5. Write $\frac{15}{25}$ in lowest terms by dividing numerator and denominator by 5.

$$\frac{15}{25} = \frac{15 \div 5}{25 \div 5} = \frac{3}{5}$$

Divide by 5.

Example 3 Writing Fractions in Lowest Terms

Write the following in lowest terms.

(a) $\frac{15}{40}$ **(b)** $\frac{33}{39}$

Solution Look for the largest number that divides both numerator and denominator without remainder.

(a) Both 15 and 40 can be divided by 5.

$$\frac{15}{40} = \frac{15 \div 5}{40 \div 5} = \frac{3}{8}$$

(b) Divide both numerator and denominator by 3.

$$\frac{33}{39} = \frac{33 \div 3}{39 \div 3} = \frac{11}{13}$$

Objective 4 *Use the Rules for Divisibility.* Deciding which numbers will divide into another number without remainder is sometimes difficult. The following **rules for divisibility** can help.

A number can be divided evenly by:

2	if the last digit is 0, 2, 4, 6, or 8
3	if the sum of the digits is divisible by 3
4	if the last two digits form a number divisible by 4
5	if the last digit is 0 or 5
6	if the number is even and the sum of the digits is divisible by 3
8	if the last three digits form a number divisible by 8
9	if the sum of the digits is divisible by 9
10	if the last digit is 0

Example 4 Using Rules for Divisibility

Determine whether the following statements are true.

(a) 3,746,892 is divisible by 4.
(b) 15,974,802 is divisible by 9.

Solution

(a) The number 3,746,892 is divisible by 4, since the last two digits form a number divisible by 4.

$$3,746,892$$

↑ **92** is divisible by 4.

(b) See if 15,974,802 is divisible by 9 by adding the digits of the number.

$$1 + 5 + 9 + 7 + 4 + 8 + 0 + 2 = 36$$

↑ **36** is divisible by 9.

Since 36 is divisible by 9, the given number is divisible by 9.

Problem-Solving Hint Testing for divisibility by adding the digits works only for 3 and 9.

NOTE The rules for divisibility only help to identify which single-digit numbers divide evenly into a larger number. The division must actually be done in order to find the number of times one number divides into another. ■

Objective 5 *Add or Subtract Like Fractions.* Fractions with the same denominator are called **like fractions**. Such fractions have a **common denominator**. For example, $\frac{3}{4}$ and $\frac{1}{4}$ are like fractions and 4 is the common denominator, but $\frac{4}{7}$ and $\frac{4}{9}$ are not like fractions. Add or subtract like fractions by adding or subtracting the numerators and then placing the result over the common denominator. The answer can then be written in lowest terms, if necessary.

Example 5 Adding and Subtracting Like Fractions

Add or subtract the following fractions.

(a) $\frac{3}{4} + \frac{1}{4} + \frac{5}{4}$ **(b)** $\frac{13}{25} - \frac{7}{25}$

Solution The fractions in both parts of this example are like fractions. Add or subtract the numerators and place the result over the common denominator. Write as a mixed number in lowest terms, as necessary.

(a) $\frac{3}{4} + \frac{1}{4} + \frac{5}{4} = \frac{3 + 1 + 5}{4} = \frac{9}{4} = 2\frac{1}{4}$ ← Add numerators. Write the answer as a mixed number. ← Write the common denominator.

(b) $\frac{13}{25} - \frac{7}{25} = \frac{13 - 7}{25} = \frac{6}{25}$

Objective 6 *Find the Least Common Denominator.* Fractions having different denominators are called **unlike fractions**. Add or subtract unlike fractions by first converting them to like fractions with a common denominator.

The **least common denominator (LCD)** for two or more fractions is the smallest whole number that can be divided, without remainder, by all the denominators of the fractions. For example, the least common denominator of the fractions $\frac{3}{4}$, $\frac{5}{6}$, and $\frac{1}{2}$ is 12, since 12 is the smallest number that can be divided by 4, 6, and 2.

In the cabinet specifications from American Landmark Cabinetry (shown below), the fractions in the Shelf-End Base drawing are **like fractions**, $23\frac{3}{16}$, $10\frac{9}{16}$, and $11\frac{3}{16}$. However, in the drawing of the Shelf-End Peninsula Base, the fractions are **unlike fractions**, $22\frac{7}{16}$, $11\frac{3}{32}$, and $11\frac{5}{8}$.

Shelf-End Base: Cross Section

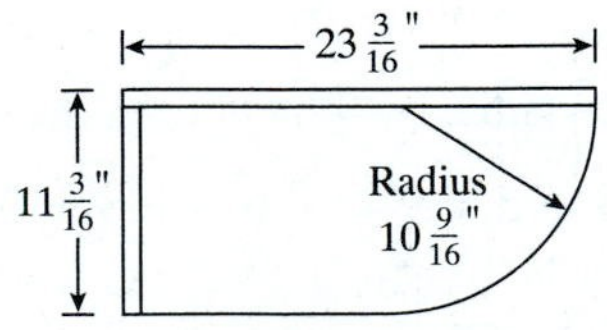

Detailed cross-section

Shelf-End Peninsula Base: Cross Section

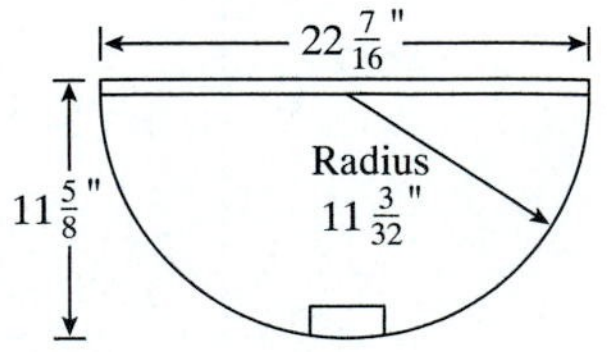

Detailed cross-section

There are two methods of finding the least common denominator; the *inspection method* and the *method of prime numbers*.

Inspection. Check to see if the least common denominator can be found by inspection. For example, the least common denominator of $\frac{1}{3}$ and $\frac{1}{4}$ is 12, since 12 is the smallest number into which 3 and 4 both divide with remainder zero. This method works best when the denominators involved are small.

Method of prime numbers. If you cannot find the least common denominator by inspection, use the method of prime numbers, as shown in the next example.

> A **prime number** is a number that can be divided without remainder only by itself and by 1. Prime numbers are 2, 3, 5, 7, 11, 13, 17, and so on.

NOTE All prime numbers other than 2 are odd numbers. However, not all odd numbers are prime numbers. For example, 27 is the product of 3 and 9, and it is not a prime number. ▬

Example 6 Finding the Least Common Denominator.

Use the method of prime numbers to find the least common denominator of the fractions $\frac{5}{12}$, $\frac{7}{18}$, and $\frac{11}{20}$.

Solution First write down the three denominators.

$$12 \qquad 18 \qquad 20$$

Begin by trying to divide the three denominators by the smallest prime number, 2. Write each quotient directly above the given denominator. (This way of writing the division process is just a handy way of writing the separate problems $2\overline{)12}$, $2\overline{)18}$, and $2\overline{)20}$.)

$$\begin{array}{r|rrr} & 6 & 9 & 10 \\ \hline 2 & 12 & 18 & 20 \end{array}$$

Two of the new quotients, 6 and 10, can still be divided by 2, so perform the division again. Since 9 cannot be divided by 2, just bring up the 9.

$$\begin{array}{r|rrr} & 3 & 9 & 5 \\ \hline 2 & 6 & 9 & 10 \\ \hline 2 & 12 & 18 & 20 \end{array} \quad \text{Just bring 9 up.}$$

None of the new quotients in the top row can be divided by 2, so try the next prime number, 3. The number 9 can be divided twice by 3 as shown below on the left.

$$\begin{array}{r|rrr} & 1 & 1 & 5 \\ \hline 3 & 1 & 3 & 5 \\ \hline 3 & 3 & 9 & 5 \\ \hline 2 & 6 & 9 & 10 \\ \hline 2 & 12 & 18 & 20 \end{array} \qquad \begin{array}{r|rrr} & 1 & 1 & 1 \\ \hline 5 & 1 & 1 & 5 \\ \hline 3 & 1 & 3 & 5 \\ \hline 3 & 3 & 9 & 5 \\ \hline 2 & 6 & 9 & 10 \\ \hline 2 & 12 & 18 & 20 \end{array}$$

Since none of the new quotients in the top row can be divided by 3, try the next prime number, 5. The number 5 can be used only once, as shown above on the right. Now that the top row contains only 1s, find the least common denominator by multiplying the prime numbers in the left column.

$$2 \times 2 \times 3 \times 3 \times 5 = 180$$

The least common denominator for $\frac{5}{12}$, $\frac{7}{18}$, and $\frac{11}{20}$ is 180.

NOTE It is not necessary to start with the smallest number as shown in Example 6. In fact, no matter which prime number we start with, we will still get the same least common denominator. ▬

Objective 7 *Add Unlike Fractions.* Add unlike fractions by rewriting the fractions with a common denominator. Since Example 6 shows that 180 is the least common denominator for $\frac{5}{12}$, $\frac{7}{18}$, and $\frac{11}{20}$, these three fractions can be added if each fraction is first written with a denominator of 180.

$$\frac{5}{12} = \frac{\quad}{180} \qquad \frac{7}{18} = \frac{\quad}{180} \qquad \frac{11}{20} = \frac{\quad}{180}$$

Rewrite these fractions with a common denominator by first dividing the common denominator by the denominator of the original fractions.

$$180 \div 12 = 15 \qquad 180 \div 18 = 10 \qquad 180 \div 20 = 9$$

Next, multiply each quotient by the original numerator.

$$15 \times 5 = 75 \qquad 10 \times 7 = 70 \qquad 9 \times 11 = 99$$

Finally, rewrite the fractions.

$$\frac{5}{12} = \frac{75}{180} \qquad \frac{7}{18} = \frac{70}{180} \qquad \frac{11}{20} = \frac{99}{180}$$

Now add the fractions.

$$\frac{5}{12} + \frac{7}{18} + \frac{11}{20} = \frac{75}{180} + \frac{70}{180} + \frac{99}{180} = \frac{75 + 70 + 99}{180} = \frac{244}{180} = 1\frac{64}{180} = 1\frac{16}{45}$$

Example 7 Adding Unlike Fractions

Add the following fractions.

(a) $\frac{3}{4} + \frac{1}{2} + \frac{5}{8}$ **(b)** $\frac{9}{10} + \frac{4}{5} + \frac{3}{8}$

Solution

(a) Inspection shows that the least common denominator is 8. Rewrite the fractions so they each have a denominator of 8. Then add.

$$\frac{3}{4} + \frac{1}{2} + \frac{5}{8} = \frac{6}{8} + \frac{4}{8} + \frac{5}{8} = \frac{6 + 4 + 5}{8} = \frac{15}{8} = 1\frac{7}{8}$$

(b) The method of prime numbers shows that the least common denominator is 40. Rewrite the fractions so they each have a denominator of 40. Then add.

$$\frac{9}{10} + \frac{4}{5} + \frac{3}{8} = \frac{36}{40} + \frac{32}{40} + \frac{15}{40} = \frac{36 + 32 + 15}{40} = \frac{83}{40} = 2\frac{3}{40}$$

Scientific *Calculator Approach*

All calculator solutions are shown using a scientific calculator. The calculator solution to Example 7(b) uses the fraction key on the scientific calculator.

9 [a b/c] 10 [+] 4 [a b/c] 5 [+] 3 [a b/c] 8 [=] $2\frac{3}{40}$

Objective 8 *Add Mixed Numbers.* Add two mixed numbers by first adding the whole number parts. Then add the fraction parts and combine the two sums.

Example 8 Adding Mixed Numbers

A rubber gasket must extend around all four edges (perimeter) of the dishwasher panel shown below before it is installed. Find the length of gasket material needed.

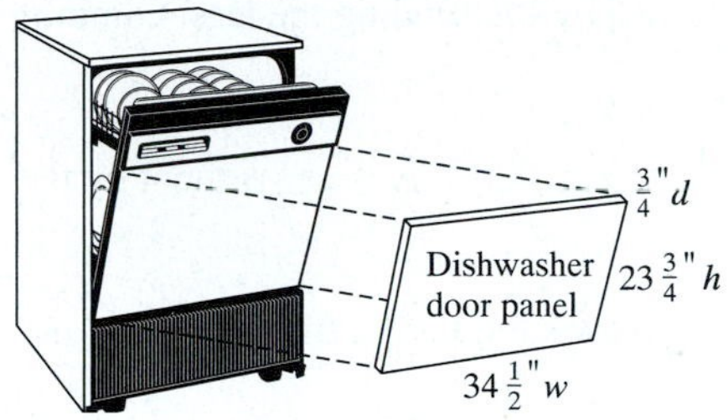

Add $34\frac{1}{2}$ inches and $23\frac{3}{4}$ inches and $34\frac{1}{2}$ inches and $23\frac{3}{4}$ inches.

Solution

$$\begin{aligned} 34\tfrac{1}{2} &= 34\tfrac{2}{4} \\ 23\tfrac{3}{4} &= 23\tfrac{3}{4} \\ 34\tfrac{1}{2} &= 34\tfrac{2}{4} \\ +\ 23\tfrac{3}{4} &= 23\tfrac{3}{4} \end{aligned} \qquad \frac{10}{4} = 2\frac{2}{4}$$

$$114\frac{10}{4} = 114 + 2\frac{2}{4} = 116\frac{2}{4} = 116\frac{1}{2} \text{ inches}$$

NOTE When adding mixed numbers, first add the fraction parts, then add the whole number parts. Then combine the two answers.

Objective 9 *Subtract Unlike Fractions.* If the fractions to be subtracted have different denominators, first find the least common denominator. For example, to subtract $\frac{1}{3}$ from $\frac{5}{8}$, first find the least common denominator, 24. Now write each fraction with a denominator of 24 and subtract.

$$\frac{5}{8} - \frac{1}{3} = \frac{15}{24} - \frac{8}{24} = \frac{15-8}{24} = \frac{7}{24}$$

Example 9 Subtracting Fractions

Subtract the following fractions.

(a) $\frac{3}{4} - \frac{5}{9}$ **(b)** $\frac{17}{18} - \frac{20}{27}$

Solution Find the common denominator and then subtract.

(a) $\frac{3}{4} - \frac{5}{9} = \frac{27}{36} - \frac{20}{36} = \frac{7}{36}$ **(b)** $\frac{17}{18} - \frac{20}{27} = \frac{51}{54} - \frac{40}{54} = \frac{11}{54}$

Objective 10 *Subtract Mixed Numbers.* Subtract two mixed numbers by changing the mixed numbers, if necessary, so that the fraction parts have a common denominator. Then subtract the fraction parts and the whole number parts separately. For example, subtract $3\frac{1}{12}$ from $8\frac{5}{8}$ by first finding the least common denominator, 24. Then rewrite the problem.

$$\begin{array}{rcl} 8\frac{5}{8} & = & 8\frac{15}{24} \\ -\ 3\frac{1}{12} & = & 3\frac{2}{24} \end{array}$$

Now subtract the fraction parts, and subtract the whole number parts.

$$\begin{array}{r} 8\frac{15}{24} \\ -\ 3\frac{2}{24} \\ \hline 5\frac{13}{24} \end{array}$$

Subtract fractions.
Subtract whole numbers.

The following example shows how to subtract when **borrowing** is needed.

Example 10 Subtracting with Borrowing

Subtract $27\frac{5}{6}$ from $36\frac{2}{9}$.

Solution Start by rewriting the problem with a common denominator.

$$\begin{array}{rcl} 36\frac{2}{9} & = & 36\frac{4}{18} \\ -\ 27\frac{5}{6} & = & 27\frac{15}{18} \end{array}$$

Subtracting $\frac{15}{18}$ from $\frac{4}{18}$ requires borrowing 1 from 36.

$$\begin{aligned} 36\frac{4}{18} &= 35 + 1 + \frac{4}{18} \\ &= 35 + \frac{18}{18} + \frac{4}{18} \qquad 1 = \frac{18}{18} \\ &= 35\frac{22}{18} \end{aligned}$$

$$\begin{array}{r} 35\frac{22}{18} \\ -\ 27\frac{15}{18} \\ \hline 8\frac{7}{18} \end{array}$$

Rewrite the problem as shown at the right. Check by adding $8\frac{7}{18}$ and $27\frac{5}{6}$. The answer should be $36\frac{2}{9}$.

Scientific *Calculator Approach*

The calculator solution to Example 10 uses the fraction key.

36 [a b/c] 2 [a b/c] 9 [−] 27 [a b/c] 5 [a b/c] 6 [=] $8\frac{7}{18}$

1.2 Exercises

Convert each of the following mixed numbers to an improper fraction.

1. $1\frac{3}{8}$ **2.** $2\frac{4}{5}$ **3.** $4\frac{1}{4}$ **4.** $2\frac{8}{11}$

5. $22\frac{7}{8}$ **6.** $15\frac{2}{3}$ **7.** $12\frac{5}{8}$ **8.** $17\frac{5}{8}$

Write each of the following fractions in lowest terms. Use the divisibility rules as needed.

9. $\frac{8}{16}$ **10.** $\frac{15}{20}$ **11.** $\frac{40}{75}$ **12.** $\frac{36}{42}$

13. $\frac{25}{40}$ **14.** $\frac{27}{45}$ **15.** $\frac{120}{150}$ **16.** $\frac{24}{64}$

17. $\frac{132}{144}$ **18.** $\frac{40}{96}$ **19.** $\frac{96}{180}$ **20.** $\frac{32}{128}$

Convert each of the following improper fractions to a mixed number and write it in lowest terms.

21. $\frac{7}{2}$ **22.** $\frac{9}{5}$ **23.** $\frac{76}{20}$ **24.** $\frac{42}{15}$

25. $\frac{14}{11}$ **26.** $\frac{55}{8}$ **27.** $\frac{21}{15}$ **28.** $\frac{85}{52}$

29. $\frac{124}{64}$ **30.** $\frac{190}{35}$ **31.** $\frac{81}{32}$ **32.** $\frac{360}{64}$

33. Your classmate asks you how to change a mixed number to an improper fraction. Write a couple of sentences explaining how this is done. (See Objective 2.)

34. Explain in a sentence or two how to change an improper fraction to a mixed number. (See Objective 2.)

Add each of the following and reduce to lowest terms.

35. $\frac{2}{5}+\frac{1}{5}$ **36.** $\frac{2}{9}+\frac{4}{9}$ **37.** $\frac{7}{10}+\frac{3}{20}$ **38.** $\frac{3}{8}+\frac{1}{4}$

39. $\frac{7}{12}+\frac{8}{15}$ **40.** $\frac{5}{8}+\frac{7}{12}$ **41.** $\frac{9}{11}+\frac{1}{22}$ **42.** $\frac{5}{6}+\frac{7}{9}$

43. $\frac{3}{4}+\frac{5}{9}+\frac{1}{3}$ **44.** $\frac{1}{4}+\frac{1}{8}+\frac{1}{12}$ **45.** $\frac{5}{6}+\frac{3}{4}+\frac{5}{8}$ **46.** $\frac{7}{10}+\frac{8}{15}+\frac{5}{6}$

47. $82\frac{3}{5} + 15\frac{1}{5}$ **48.** $25\frac{2}{7} + 14\frac{3}{7}$ **49.** $51\frac{1}{4} + 29\frac{1}{2}$ **50.** $38\frac{5}{6}$, $29\frac{1}{3}$, $+\ 47\frac{1}{2}$

51. $\begin{array}{r} 32\frac{3}{4} \\ 6\frac{1}{3} \\ +\ 14\frac{5}{8} \\ \hline \end{array}$

52. $\begin{array}{r} 16\frac{7}{10} \\ 26\frac{1}{5} \\ +\ \ 8\frac{3}{8} \\ \hline \end{array}$

53. $\begin{array}{r} 89\frac{5}{9} \\ 10\frac{1}{3} \\ +\ 87\frac{1}{9} \\ \hline \end{array}$

54. $\begin{array}{r} 74\frac{1}{5} \\ 58\frac{3}{7} \\ +\ 21\frac{3}{10} \\ \hline \end{array}$

Subtract each of the following and reduce to lowest terms.

55. $\frac{7}{8} - \frac{3}{8}$

56. $\frac{11}{12} - \frac{5}{12}$

57. $\frac{2}{3} - \frac{1}{6}$

58. $\frac{7}{8} - \frac{1}{2}$

59. $\frac{5}{12} - \frac{1}{16}$

60. $\frac{5}{6} - \frac{7}{9}$

61. $\frac{3}{4} - \frac{5}{12}$

62. $\frac{5}{7} - \frac{1}{3}$

63. $\begin{array}{r} 16\frac{3}{4} \\ -\ 12\frac{3}{8} \\ \hline \end{array}$

64. $\begin{array}{r} 25\frac{13}{24} \\ -\ 18\frac{5}{12} \\ \hline \end{array}$

65. $\begin{array}{r} 9\frac{7}{8} \\ -\ 6\frac{5}{12} \\ \hline \end{array}$

66. $\begin{array}{r} 24\frac{5}{6} \\ -\ 18\frac{5}{9} \\ \hline \end{array}$

67. $\begin{array}{r} 71\frac{3}{8} \\ -\ 62\frac{1}{3} \\ \hline \end{array}$

68. $\begin{array}{r} 19\frac{5}{6} \\ -\ 12\frac{3}{4} \\ \hline \end{array}$

69. $\begin{array}{r} 19 \\ -\ 12\frac{3}{4} \\ \hline \end{array}$

70. $\begin{array}{r} 374 \\ -\ 211\frac{5}{6} \\ \hline \end{array}$

71. Prime numbers are used to find the least common denominator. Give the definition of a prime number in your own words. (See Objective 6.)

72. Can you add or subtract fractions without using the least common denominator? Describe how you would do this.

73. Where are fractions used in everyday life? Think in terms of business applications, hobbies, and vacations. Give three examples.

74. When subtracting mixed numbers, explain when you need to borrow. Use an example to explain how to borrow.

Solve each of the following application problems.

75. CABINET INSTALLATION When installing cabinets, Kate Morgan must be certain that the proper type and size of mounting screw is used. Find the total length of the screw.

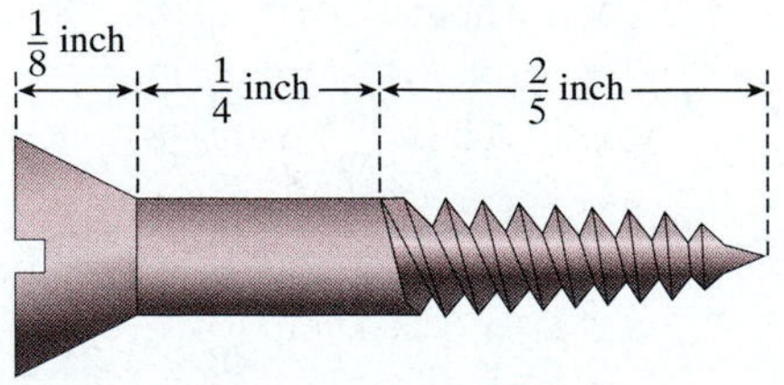

76. BULK PURCHASING The owner of Racy's Feed Store ordered $\frac{1}{3}$ cubic yard of corn, $\frac{3}{8}$ cubic yard of oats, and $\frac{1}{4}$ cubic yard of washed medium mesh gravel. Find the total cubic yards of products.

77. WETLANDS RESERVE A wetlands reserve has four sides, which measure $1\frac{7}{8}$ mile, $\frac{1}{2}$ mile, $1\frac{2}{3}$ mile, and $\frac{1}{3}$ mile. What is the total distance around the wetlands reserve?

78. MEASURING BRASS TRIM To complete a custom accessory order for a customer, Ruth Berry of Home Depot must find the number of inches of brass trim needed to go around the four sides of the lamp base plate shown below. Find the length of brass trim needed.

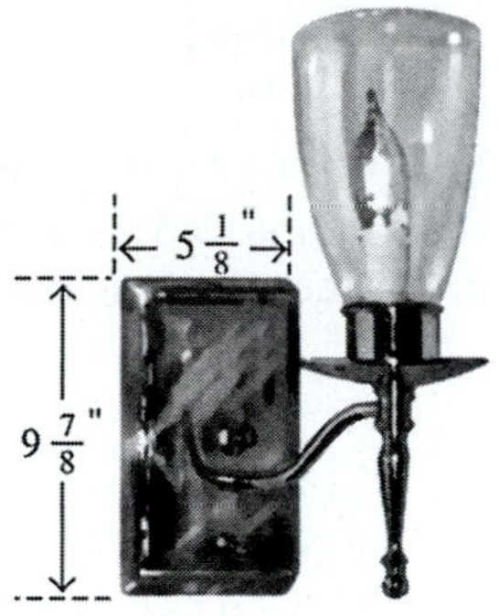

79. BRACKET INSTALLATION Find the diameter of the hole in the mounting bracket pictured below.

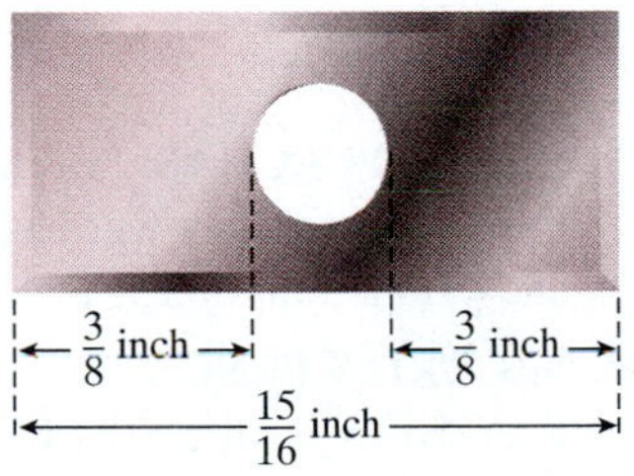

80. HYDRAULIC FLUID A hydraulic jack contains $\frac{7}{8}$ liter of hydraulic fluid. A cracked seal resulted in a loss of $\frac{1}{6}$ liter of fluid in the morning and another $\frac{1}{3}$ liter in the afternoon. Find the amount of fluid remaining.

81. HIGHWAY DRIVING On a recent vacation to Canada, Hernando Ramirez drove for $5\frac{1}{2}$ hours on the first day, $6\frac{1}{4}$ hours on the second day, $3\frac{3}{4}$ hours on the third day, and 7 hours on the fourth day. How many hours did he drive altogether?

82. WHOLESALE PRODUCE SALES Last month Lim's Wholesale Vegetable Market sold $3\frac{1}{4}$ tons of broccoli, $2\frac{3}{8}$ tons of spinach, $7\frac{1}{2}$ tons of corn, and $1\frac{5}{16}$ tons of turnips. Find the total number of tons of these vegetables sold by the firm last month

83. LAND ACREAGE Jenny Crum owns $83\frac{5}{8}$ acres of land in Mexico, $76\frac{3}{4}$ acres in the United States, and $182\frac{1}{3}$ acres in Canada. Find the total number of acres that she owns in these three countries.

84. MEASURING LAND Goldi's Resort decided to expand by buying a piece of property next to the resort. The property is irregularly shaped, with five sides. The lengths of the five sides are $146\frac{1}{2}$ feet, $98\frac{3}{4}$ feet, $196\frac{2}{3}$ feet, $76\frac{5}{8}$ feet, and $100\frac{7}{8}$ feet. Find the total distance around the piece of property.

85. AUTOMOTIVE SUPPLIES Comet Auto Supply sold $16\frac{1}{2}$ cases of generic brand oil last week, $12\frac{1}{8}$ cases of Havoline Oil, $8\frac{3}{4}$ cases of Valvoline Oil, and $12\frac{5}{8}$ cases of Castrol Oil. Find the number of cases of oil that Comet Auto Supply sold during the week.

86. GRAIN SALES In order to sample a shipment of grain, an inspector took $1\frac{5}{8}$ bushels from one part of a load, $3\frac{1}{4}$ bushels from a second part, $2\frac{3}{8}$ bushels from a third part, and $3\frac{1}{3}$ bushels from a fourth. Find the total number of bushels inspected.

87. WORK WEEK Julie Davis worked 40 hours during one week. She worked $8\frac{1}{4}$ hours on Monday, $6\frac{1}{6}$ hours on Tuesday, $7\frac{2}{3}$ hours on Wednesday, and $8\frac{3}{4}$ hours on Thursday. How many hours did she work on Friday?

88. LEADED GLASS FRAMING A craftsperson must attach a lead strip around all four sides of a leaded glass window before it is installed. The window measures $34\frac{1}{2}$ by $23\frac{3}{4}$ inches. Find the length of lead stripping needed.

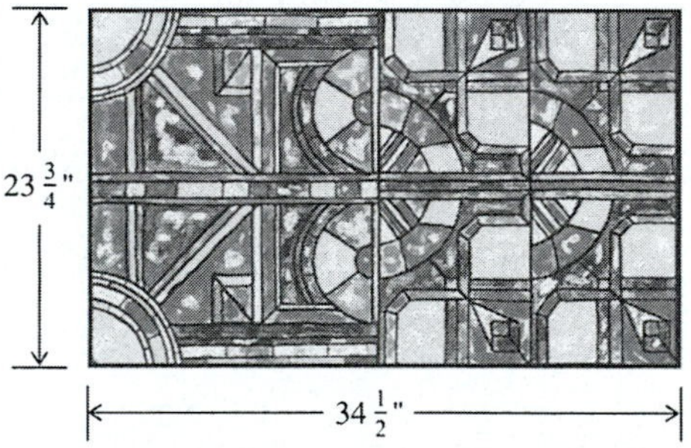

89. STOCK PURCHASES Eboni Perkins bought four shares of stock. The prices for three of the shares were $\$71\frac{3}{8}$, $\$18\frac{1}{2}$, and $\$143\frac{5}{8}$. Find the price of the fourth share if she paid a total of $\$352\frac{1}{8}$.

90. SECURITY FENCING The exercise yard at the correction center has four sides and is enclosed with $527\frac{1}{24}$ feet of security fencing around it. If three sides of the yard measure $107\frac{2}{3}$ feet, $150\frac{3}{4}$ feet, and $138\frac{5}{8}$ feet, find the length of the fourth side.

1.3 Multiplication and Division of Fractions

Objectives

1. *Multiply fractions.*
2. *Divide fractions.*
3. *Convert decimals to fractions.*
4. *Round decimals.*
5. *Convert fractions to decimals.*

Objective 1 *Multiply Fractions.* Multiply two fractions by first multiplying the numerators to form a new numerator and then multiplying the denominators to form a new denominator. Write the answer in lowest terms. For example, multiply $\frac{2}{3}$ and $\frac{5}{8}$.

Multiply numerators

$$\frac{2}{3} \times \frac{5}{8} = \frac{2 \times 5}{3 \times 8} = \frac{10}{24} = \frac{5}{12} \quad \text{(in lowest terms)}$$

Multiply denominators

This problem can be simplified by **cancellation**, a modification of the method of writing fractions in lowest terms. If a number divides evenly into a numerator and a denominator without remainder, then divide (cancel).

$$\frac{\overset{1}{\cancel{2}}}{3} \times \frac{5}{\underset{4}{\cancel{8}}} = \frac{1 \times 5}{3 \times 4} = \frac{5}{12}$$ Divide 2 into both 2 and 8.

NOTE It is not necessary to use a common denominator when multiplying or dividing fractions. ▬

Example 1 Multiplying Common Fractions

Multiply the following fractions.

(a) $\frac{6}{11} \times \frac{7}{8}$ **(b)** $\frac{35}{12} \times \frac{32}{25}$

Solution Use cancellation in both of these problems.

(a) $\frac{\overset{3}{\cancel{6}}}{11} \times \frac{7}{\underset{4}{\cancel{8}}} = \frac{3 \times 7}{11 \times 4} = \frac{21}{44}$ 2 was divided into both 6 and 8.

(b) $\frac{\overset{7}{\cancel{35}}}{\underset{3}{\cancel{12}}} \times \frac{\overset{8}{\cancel{32}}}{\underset{5}{\cancel{25}}} = \frac{7 \times 8}{3 \times 5} = \frac{56}{15} = 3\frac{11}{15}$ 4 was divided into both 12 and 32, while 5 was divided into both 35 and 25.

NOTE When cancelling, be certain that a numerator and a denominator are both divided by the same number. ▬

To multiply mixed numbers, change the mixed numbers to improper fractions, then multiply. For example, multiply $6\frac{1}{4}$ and $2\frac{2}{3}$.

$$6\frac{1}{4} \times 2\frac{2}{3} = \frac{25}{4} \times \frac{8}{3} = \frac{25}{\underset{1}{\cancel{4}}} \times \frac{\overset{2}{\cancel{8}}}{3} = \frac{25 \times 2}{1 \times 3} = \frac{50}{3} = 16\frac{2}{3}$$

Example 2 Multiplying Mixed Numbers

Multiply the following.

(a) $5\frac{5}{8} \times 4\frac{1}{6}$ **(b)** $1\frac{3}{5} \times 3\frac{1}{3} \times 1\frac{3}{4}$

Solution

(a) $\frac{\overset{15}{\cancel{45}}}{8} \times \frac{25}{\underset{2}{\cancel{6}}} = \frac{15 \times 25}{8 \times 2} = \frac{375}{16} = 23\frac{7}{16}$

(b) $\frac{\overset{2}{\cancel{8}}}{\underset{1}{\cancel{5}}} \times \frac{\overset{2}{\cancel{10}}}{3} \times \frac{7}{\underset{1}{\cancel{4}}} = \frac{2 \times 2 \times 7}{1 \times 3 \times 1} = \frac{28}{3} = 9\frac{1}{3}$

Scientific *Calculator Approach*

The calculator solution to Example 2(b) uses the fraction key.

1 [a b/c] 3 [a b/c] 5 [×] 3 [a b/c] 1 [a b/c] 3 [×] 1 [a b/c] 3 [a b/c] 4 [=] $9\frac{1}{3}$.

Objective 2 ***Divide Fractions.*** Divide two fractions by inverting the divisor and then multiplying. (Invert the second fraction by exchanging the numerator and denominator.)

Example 3 Dividing Common Fractions

Divide

(a) $\frac{25}{36} \div \frac{15}{18}$ **(b)** $\frac{21}{8} \div \frac{14}{16}$

Solution Invert the second fraction and multiply.

(a) $$\frac{25}{36} \div \frac{15}{18} = \frac{\overset{5}{\cancel{25}}}{\underset{2}{\cancel{36}}} \times \frac{\overset{1}{\cancel{18}}}{\underset{3}{\cancel{15}}} = \frac{5 \times 1}{2 \times 3} = \frac{5}{6}$$

(b) $$\frac{21}{8} \div \frac{14}{16} = \frac{\overset{3}{\cancel{21}}}{\underset{1}{\cancel{8}}} \times \frac{\overset{\overset{1}{\cancel{2}}}{\cancel{16}}}{\underset{\underset{1}{\cancel{2}}}{\cancel{14}}} = \frac{3 \times 1}{1 \times 1} = 3$$

NOTE The second fraction (divisor) is inverted when dividing by a fraction. Cancellation is done *only after inverting.* ■

Divide mixed numbers by changing all mixed numbers to improper fractions, as follows.

$$3\frac{5}{9} \div 2\frac{2}{5} = \frac{32}{9} \div \frac{12}{5} = \frac{\overset{8}{\cancel{32}}}{9} \times \frac{5}{\underset{3}{\cancel{12}}} = \frac{8 \times 5}{9 \times 3} = \frac{40}{27} = 1\frac{13}{27}$$

Multiply or divide by a whole number by first writing the whole number as a fraction over 1.

$$3\frac{3}{4} \times 16 = 3\frac{3}{4} \times \frac{16}{1} \quad \text{whole number over 1}$$

$$= \frac{15}{4} \times \frac{16}{1}$$

$$= \frac{15}{\underset{1}{\cancel{4}}} \times \frac{\overset{4}{\cancel{16}}}{1} = 15 \times 4 = 60$$

Also:

$$2\frac{2}{5} \div 3 = \frac{12}{5} \div \frac{3}{1} = \frac{\overset{4}{\cancel{12}}}{5} \times \frac{1}{\underset{1}{\cancel{3}}} = \frac{4 \times 1}{5 \times 1} = \frac{4}{5}$$

The newspaper article below reports that the prices of computer stocks have fallen. The stock of Gateway 2000 fell $1\frac{3}{8}$ (\$$1\frac{3}{8}$) to close at \$$37\frac{5}{8}$. Stock values are shown as mixed numbers representing the dollar value for one share. An investor must multiply the number of shares being purchased or sold by this mixed number to determine the total value of the stock.

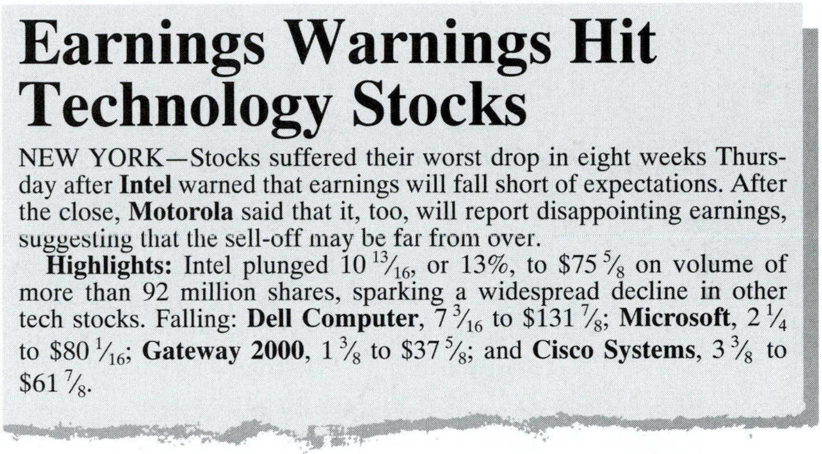

Earnings Warnings Hit Technology Stocks

NEW YORK—Stocks suffered their worst drop in eight weeks Thursday after **Intel** warned that earnings will fall short of expectations. After the close, **Motorola** said that it, too, will report disappointing earnings, suggesting that the sell-off may be far from over.

Highlights: Intel plunged $10\,{}^{13}/_{16}$, or 13%, to $\$75\,{}^{5}/_{8}$ on volume of more than 92 million shares, sparking a widespread decline in other tech stocks. Falling: **Dell Computer**, $7\,{}^{3}/_{16}$ to $\$131\,{}^{7}/_{8}$; **Microsoft**, $2\,{}^{1}/_{4}$ to $\$80\,{}^{1}/_{16}$; **Gateway 2000**, $1\,{}^{3}/_{8}$ to $\$37\,{}^{5}/_{8}$; and **Cisco Systems**, $3\,{}^{3}/_{8}$ to $\$61\,{}^{7}/_{8}$.

Example 4 Multiplying a Whole Number by a Mixed Number

Chia Ling purchased 80 shares of Gateway 2000 stock at a price of $\$37\frac{5}{8}$. Find her total cost for the stock.

Solution Multiply the number of shares purchased by the price per share, $37\frac{5}{8}$, or $\frac{301}{8}$. First, estimate a reasonable answer. A reasonable answer would be a little less than \$3200 since $80 \times \$40 = \$3200$ ($\$37\frac{5}{8}$ rounds to \$40). Now, find the actual answer.

$$80 \times \frac{301}{8} = \frac{\overset{10}{\cancel{80}}}{1} \times \frac{301}{\underset{1}{\cancel{8}}} = \frac{10 \times 301}{1 \times 1} = \frac{3010}{1} = \$3010$$

The answer is reasonable.

The total cost of the stock purchased by Ling is \$3010.

Objective 3 *Convert Decimals to Fractions.* A **decimal number** is really a fraction with a denominator that is a power of 10, such as 10, 100, or 1000. The digits written to the right of the decimal point have place values as shown in Figure 1.3.

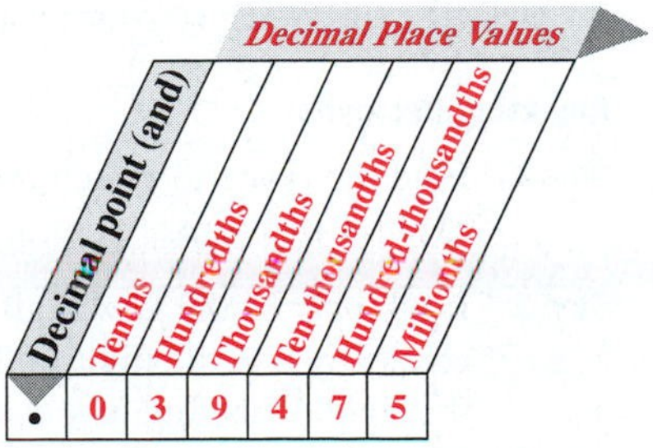

Figure 1.3

Convert a decimal to a fraction by thinking of the decimal as being written in words. For example, think of 0.47 as "forty-seven hundredths." Then write this in fraction form.

$$0.47 = \frac{47}{100}$$

In the same way, 0.3, read as "three tenths," is written in fraction form as follows.

$$0.3 = \frac{3}{10}$$

Also, 0.963, read as "nine hundred sixty-three thousandths," is written

$$\frac{963}{1000}$$

NOTE The word "and" when reading a number is used when crossing over the decimal point. The number 2.71 is read as "two and seventy-one hundredths" and written in fraction form as $2\frac{71}{100}$. ■

Example 5 Converting Decimals to Fractions

Convert the following decimals to fractions.

(a) 0.75 **(b)** 0.028

Solution

(a) 0.75 is read as "seventy-five hundredths."

$$0.75 = \frac{75}{100} = \frac{3}{4}$$

Here, $\frac{75}{100}$ is written in lowest terms as $\frac{3}{4}$.

(b) 0.028 is read as "twenty-eight thousandths," and the resulting fraction is written in lowest terms.

$$0.028 = \frac{28}{1000} = \frac{7}{250}$$

Objective 4 *Round Decimals.* It is important to be able to round decimals. For example, the 7-Eleven Store is selling two candy mints for \$0.75, but you only want to buy one mint. The price of one mint is \$0.75 ÷ 2, which is \$0.375. Since you cannot pay part of a cent, the store rounds the price to \$0.38 for one mint.

Use the following steps for rounding decimals.

Rounding Decimals

Step 1. Find the place to which the rounding is being done. Draw a line ***after*** that place to show that you are cutting off the rest of the digits.

Step 2. Look *only* at the *first* digit you are cutting off. If the first digit is *5 or more,* increase by one the digit in the place to which you are rounding. If the first digit to the right of the line is *4 or less,* do not change the digit in the place to which you are rounding.

Step 3. Drop all digits to the right of the place to which you have rounded.

NOTE Do not move the decimal point when rounding.

Example 6 Rounding to the Nearest Tenth

Round 98.5892 to the nearest tenth.

Solution

Step 1. Locate the tenths digit and draw a line.

98.5|892

tenths digit (5)

The tenths digit is 5.

Step 2. Locate the digit just to the right of the line.

98.5|892

just to the right of the line (8)

The digit just to the right of the line is 8.

Step 3. If the digit found in Step 2 is 4 or less, leave the digit of Step 1 alone. If the digit found in Step 2 is 5 or more, increase the digit of Step 1 by 1. The digit found in Step 2 is 8, so 98.5892 rounded to the nearest tenth is 98.6.

Example 7 Rounding to the Nearest Thousandth

Round 0.008572 to the nearest thousandth.

Solution Locate the thousandths digit and draw a line.

0.008|572

thousandths digit (8)

Since the digit to the right of the line is 5, increase the thousandths digit by 1; so that 0.008572 rounded to the nearest thousandth is 0.009.

Example 8 Rounding Decimals

When doing business with a foreign country, you must convert the foreign currency using the U.S. exchange rate. Round the following currencies to the indicated position.

(a) Germany (mark) 0.5319 to the nearest thousandth
(b) Peru (new sol) 0.2872 to the nearest tenth
(c) Britain (pound) 1.6238 to the nearest hundredth
(d) Italy (lira) 0.0005373 to the nearest hundred-thousandth

Solution Use the method described.

(a) 0.5319 to the nearest thousandth is 0.532
(b) 0.2872 to the nearest tenth is 0.3
(c) 1.6238 to the nearest hundredth is 1.62
(d) 0.0005373 to the nearest hundred-thousandth is 0.00054

Problem-Solving Hint — The answer to part (c) in Example 8 may be surprising because of the answer in (b). However, always round a number by going back to the *original number*, and not to some number that was rounded from the original number.

Objective 5 ***Convert Fractions to Decimals.*** To convert a fraction to a decimal, divide the numerator of the fraction by the denominator. Place a decimal point after the numerator and attach additional zeros, one at a time, to the right of the decimal point as the division is performed. Keep going until the division ends or until the desired degree of precision is reached. As a general rule, divide until the quotient has one more digit than the desired degree of precision, then round from the last digit. The result is the decimal equivalent to the fraction.

For example, to convert $\frac{1}{8}$ to a decimal, divide 1 by 8.

$$8\overline{)1.}$$

Since 8 will not divide into 1, place a 0 to the right of the decimal point. Now 8 divides into 10 once, with a remainder of 2.

$$\begin{array}{r} 0.1 \\ 8\overline{)1.0} \\ \underline{8} \\ 2 \end{array}$$

Be sure to move the decimal point up.

Continue placing 0s to the right of the decimal point and continue dividing. The division now gives a remainder of 0.

$$\begin{array}{r} 0.125 \\ 8\overline{)1.000} \\ \underline{8} \\ 20 \\ \underline{16} \\ 40 \\ \underline{40} \\ 0 \end{array}$$

Keep attaching zeros.

Remainder of 0.

Therefore, $\frac{1}{8} = 0.125$.

NOTE The decimal answer, 0.125, was not rounded; instead, the division was continued until there was no remainder. In most problems the answer will be rounded to the required accuracy. ▬

Example 9 Rounding of a Repeating Decimal

Convert $\frac{2}{3}$ to a decimal. Round to the nearest ten-thousandth.

Solution Divide 2 by 3.

$$
\begin{array}{r}
0.66666 \\
3\overline{)2.00000} \\
\underline{1\,8} \\
20 \\
\underline{18} \\
20 \\
\underline{18} \\
20 \\
\underline{18} \\
20 \\
\underline{18} \\
2
\end{array}
$$

Keep attaching zeros.

This division results in a **repeating decimal**, which is often indicated by placing a bar over the digit or digits that repeat. The answer could be written as follows.

$$0.\overline{6} \quad \text{or} \quad 0.6\overline{6} \quad \text{or} \quad 0.6666\overline{6}$$

However, rounded to the nearest ten-thousandth:

$$\frac{2}{3} = 0.6667$$

Scientific *Calculator Approach*

The calculator solution to Example 9 is

2 [÷] 3 [=] 0.666666667.

Decimal equivalents. Some of the more common decimal equivalents of fractions are listed below. These decimals appear from least to greatest value and are rounded to the nearest ten-thousandth. Sometimes decimals must be carried out further to give greater accuracy, while at other times they are not carried out as far and are rounded sooner.

Decimal Equivalents

$\frac{1}{16} = 0.0625$	$\frac{1}{4} = 0.25$	$\frac{5}{8} = 0.625$
$\frac{1}{9} = 0.1111$	$\frac{5}{16} = 0.3125$	$\frac{2}{3} = 0.6667$
$\frac{1}{8} = 0.125$	$\frac{1}{3} = 0.3333$	$\frac{11}{16} = 0.6875$
$\frac{1}{7} = 0.1429$	$\frac{3}{8} = 0.375$	$\frac{3}{4} = 0.75$
$\frac{1}{6} = 0.1667$	$\frac{7}{16} = 0.4375$	$\frac{13}{16} = 0.8125$
$\frac{3}{16} = 0.1875$	$\frac{1}{2} = 0.5$	$\frac{5}{6} = 0.8333$
$\frac{1}{5} = 0.2$	$\frac{9}{16} = 0.5625$	$\frac{7}{8} = 0.875$

1.3 Exercises

Multiply each of the following and write in lowest terms.

1. $\frac{5}{8} \times \frac{2}{3}$
2. $\frac{3}{8} \times \frac{1}{6}$
3. $\frac{9}{10} \times \frac{11}{16}$
4. $1\frac{1}{4} \times 3\frac{1}{2}$
5. $1\frac{2}{3} \times 2\frac{7}{10}$
6. $6 \times 4\frac{2}{3}$
7. $4\frac{3}{5} \times 15$
8. $\frac{3}{4} \times \frac{8}{9} \times 2\frac{1}{2}$
9. $\frac{5}{9} \times 2\frac{1}{4} \times 3\frac{2}{3}$
10. $\frac{2}{3} \times \frac{9}{8} \times 3\frac{1}{4}$
11. $12 \times 2\frac{1}{2} \times 3$
12. $18 \times 1\frac{2}{3} \times 2$

Divide each of the following and write in lowest terms.

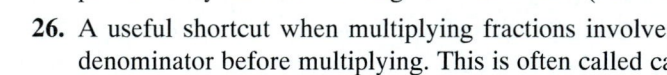

13. $\frac{1}{6} \div \frac{1}{3}$
14. $\frac{5}{8} \div \frac{3}{16}$
15. $\frac{13}{20} \div \frac{26}{30}$
16. $\frac{7}{8} \div \frac{3}{4}$
17. $\frac{15}{16} \div \frac{5}{8}$
18. $\frac{12}{11} \div \frac{3}{22}$
19. $2\frac{1}{2} \div 3\frac{3}{4}$
20. $6\frac{1}{2} \div \frac{1}{2}$
21. $3\frac{1}{8} \div \frac{15}{16}$
22. $5\frac{1}{2} \div 4$
23. $6 \div 1\frac{1}{4}$
24. $3 \div 1\frac{1}{4}$
25. Write in your own words the rule for multiplying fractions. Make up an example problem of your own showing how this works. (See Objective 1.)
26. A useful shortcut when multiplying fractions involves dividing a numerator and a denominator before multiplying. This is often called cancellation. Describe how this works and give an example. (See Objective 1.)

Find the total price for each of the following purchases of stock. It is common for the price of a share of stock to be given as a mixed number.

27. 80 shares of Mattel at $\$39\frac{5}{8}$ per share
28. 30 shares of Bank of America at $\$83\frac{1}{4}$ per share
29. 24 shares of Merck at $\$74\frac{3}{4}$ per share
30. 200 shares of Home Depot at $\$71\frac{5}{8}$ per share
31. 32 shares of Ford Motor at $\$57\frac{1}{8}$ per share
32. 56 shares of McDonald's at $\$47\frac{3}{8}$ per share

Convert each of the following decimals to fractions and reduce to lowest terms.

33. 0.8
34. 0.6
35. 0.24
36. 0.64
37. 0.73
38. 0.625
39. 0.875
40. 0.805
41. 0.0375
42. 0.8125
43. 0.1875
44. 0.3125

Round each of the following decimals to the nearest tenth and to the nearest hundredth.

45. 3.5218 **46.** 4.8361 **47.** 0.0837 **48.** 2.548

49. 8.643 **50.** 86.472 **51.** 58.956 **52.** 8.065

Convert each of the following fractions to decimals. Round the answer to the nearest thousandth if necessary.

53. $\frac{3}{4}$ **54.** $\frac{7}{8}$ **55.** $\frac{3}{8}$ **56.** $\frac{5}{6}$

57. $\frac{1}{6}$ **58.** $\frac{2}{3}$ **59.** $\frac{13}{16}$ **60.** $\frac{19}{50}$

61. $\frac{8}{25}$ **62.** $\frac{1}{3}$ **63.** $\frac{1}{99}$ **64.** $\frac{73}{93}$

65. A classmate of yours is confused about how to convert a decimal to a fraction. Write an explanation of this for your classmate including changing the fraction to lowest terms. (See Objective 3.)

66. Explain how to convert a fraction to a decimal. Be sure to mention rounding in your explanation. (See Objective 5.)

67. Explain in your own words the difference between hundreds and hundredths. (See Objective 3.)

68. Write the directions for rounding a money answer to the nearest cent.

Solve each of the following application problems.

69. PRODUCING CRAFTS Laura Griffin wants to make 16 holiday wreaths to sell at the craft fair. Each wreath needs $2\frac{1}{4}$ yard of ribbon. How many yards does she need?

70. EARNINGS CALCULATION Jason Hoffa worked $38\frac{1}{4}$ hours at \$10 per hour. How much money did he make?

71. STOCK PURCHASE Bobbi Kraham bought some stock in Telex Chile for $\$8\frac{3}{8}$ per share. If she paid \$5025 for the stock, how many shares did she buy?

72. TOTAL INVESTMENT Ms. Nishimoto bought 12 shares of stock at $\$18\frac{3}{4}$ per share, 24 shares at $\$36\frac{3}{8}$ per share, and 16 shares at $\$74\frac{1}{8}$ per share. Her broker charged her a commission of \$12. Find the total amount that she paid.

73. FINISH CARPENTRY Each home of a certain design needs $109\frac{1}{2}$ yards of prefinished baseboard. How many homes can be fitted with baseboards if there are 1314 yards of baseboard available?

74. COMMERCIAL FERTILIZER For one acre of crop, $7\frac{1}{2}$ gallons of fertilizer must be applied. How many acres can be fertilized with 1200 gallons of fertilizer?

75. INSECT CONTROL An insect spray manufactured by Dutch Chemicals Inc. is mixed with $1\frac{3}{4}$ ounces of chemical per gallon of water. How many ounces of chemical are needed for $12\frac{1}{2}$ gallons of water?

76. RESIDENTIAL ROOFING Each home in a new development requires $37\frac{3}{4}$ pounds of roofing nails. How many pounds of roofing nails are needed for 36 homes?

77. PROFESSIONAL PHOTOGRAPHY On average, a photographer uses $12\frac{3}{4}$ rolls of film at a wedding and $7\frac{1}{8}$ rolls of film at a retirement party. Find the total number of rolls needed for 28 weddings and 16 retirement parties.

78. HANDCRAFTED JEWELRY One necklace can be completed in $6\frac{1}{2}$ minutes, while a bracelet takes $3\frac{1}{8}$ minutes. Find the total time that it takes to complete 36 necklaces and 22 bracelets.

79. FIREWOOD SALES Amy Folsom had a small pickup truck that would carry $\frac{2}{3}$ cord of firewood. Find the number of trips needed to deliver 40 cords of wood.

80. WEATHER STRIPPING Bill Rhodes has a 200-yard roll of weather stripping material. Find the number of pieces of weather stripping $\frac{5}{8}$ yard in length that may be cut from the roll.

Chapter 1 Quick Review

Chapter Terms

Review the following terms to test your understanding of the chapter. For each term you do not know, refer to the page number found next to that term.

Concepts	Examples
1.1 Application problems Follow the steps. **1.** Read the problem carefully. **2.** Work out a plan using *indicator words* before starting. **3.** Estimate a reasonable answer. **4.** Solve the problem. If the answer is reasonable, check; if not, start over.	Manuel earns \$118 on Sunday, \$87 on Monday, and \$63 on Tuesday. Find his total earnings for the three days. *Total* means to add. \$268 Check 118 87 + 63 \$268 Total earnings
1.2 Types of fractions Proper: Numerator smaller than denominator. Improper: Numerator equal to or greater than denominator. Mixed: Whole number and proper fraction.	Proper $\frac{2}{3}, \frac{3}{4}, \frac{15}{16}, \frac{1}{8}$ Improper $\frac{17}{8}, \frac{19}{12}, \frac{11}{2}, \frac{5}{3}, \frac{7}{7}$ Mixed $2\frac{2}{3}, 3\frac{5}{8}, 9\frac{5}{6}$
1.2 Converting fractions Mixed to improper: Multiply denominator by whole number and add numerator. The denominator is unchanged. Improper to mixed: Divide numerator by denominator and place remainder over denominator.	$7\frac{2}{3} = \frac{23}{3}$ $\frac{17}{5} = 3\frac{2}{5}$
1.2 Writing fractions in lowest terms Divide the numerator and denominator by the same number.	$\frac{30}{42} = \frac{30 \div 6}{42 \div 6} = \frac{5}{7}$
1.2 Adding like fractions Add numerators and reduce to lowest terms. The denominator is unchanged.	$\frac{3}{4} + \frac{1}{4} + \frac{5}{4} = \frac{3+1+5}{4} = \frac{9}{4} = 2\frac{1}{4}$

1.2 Finding a least common denominator

Inspection method: Look to see if the least common denominator can be found.

$$\frac{1}{3}+\frac{1}{4}+\frac{1}{10}$$

Method of prime numbers: Use prime numbers to find the least common denominator.

$$\begin{array}{r|rrr} & 1 & 1 & 1 \\ \hline 5 & 1 & 1 & 5 \\ \hline 3 & 3 & 1 & 5 \\ \hline 2 & 3 & 2 & 5 \\ \hline 2 & 3 & 4 & 10 \end{array}$$

Multiply the prime numbers.

$$2 \times 2 \times 3 \times 5 = 60 \text{ LCD}$$

1.2 Adding unlike fractions

1. Find the least common denominator.
2. Rewrite fractions with the least common denominator.
3. Add numerators, placing answers over LCD and reduce to lowest terms.

$$\frac{1}{3}+\frac{1}{4}+\frac{1}{10} \quad \text{LCD} = 60$$

$$\frac{1}{3}=\frac{20}{60}, \frac{1}{4}=\frac{15}{60}, \frac{1}{10}=\frac{6}{60}$$

$$\frac{20+15+6}{60}=\frac{41}{60}$$

1.2 Adding mixed numbers

1. Add fractions.
2. Add whole numbers.
3. Combine the sums of whole numbers and fractions. Write answer in simplest terms.

$$\begin{array}{rcr} 9\frac{2}{3} & = & 9\frac{8}{12} \\ +\,6\frac{3}{4} & = & 6\frac{9}{12} \\ \hline & & 15\frac{17}{12} = 16\frac{5}{12} \end{array}$$

1.2 Subtracting fractions

1. Find the least common denominator.
2. Subtract numerator of subtrahend, borrowing if necessary.
3. Write difference over LCD and reduce to lowest terms.

$$\frac{5}{8}-\frac{1}{3}=\frac{15}{24}-\frac{8}{24}=\frac{7}{24}$$

1.2 Subtracting mixed numbers

1. Subtract fractions, borrowing if necessary.
2. Subtract whole numbers.
3. Combine the differences of whole numbers and fractions.

$$\begin{array}{rcr} 8\frac{5}{8} & = & 8\frac{15}{24} \\ -\,3\frac{1}{12} & = & 3\frac{2}{24} \\ \hline & & 5\frac{13}{24} \end{array}$$

1.3 Multiplying proper fractions

1. Cancel, then multiply numerators and denominators.
2. Reduce answer to lowest terms if canceling was not done.

$$\frac{6}{11}\times\frac{7}{8}=\frac{\overset{3}{\cancel{6}}}{11}\times\frac{7}{\underset{4}{\cancel{8}}}=\frac{21}{44}$$

1.3 Multiplying mixed numbers

1. Change mixed numbers to improper fractions.
2. Cancel if possible.
3. Multiply as proper fractions. Always reduce to lowest terms.

$$1\frac{3}{5} \times 3\frac{1}{3} = \frac{8}{\underset{1}{\cancel{5}}} \times \frac{\overset{2}{\cancel{10}}}{3} = \frac{8}{1} \times \frac{2}{3}$$

$$= \frac{16}{3} = 5\frac{1}{3}$$

1.3 Dividing proper fractions

Invert the divisor, cancel, then multiply as fractions.

$$\frac{25}{36} \div \frac{15}{18} = \frac{\overset{5}{\cancel{25}}}{\underset{2}{\cancel{36}}} \times \frac{\overset{1}{\cancel{18}}}{\underset{3}{\cancel{15}}} = \frac{5}{2} \times \frac{1}{3} = \frac{5}{6}$$

1.3 Dividing mixed numbers

Change mixed numbers to improper fractions. Invert the divisor, cancel if possible, and multiply as proper fractions.

$$3\frac{5}{9} \div 2\frac{2}{5} = \frac{32}{9} \div \frac{12}{5} = \frac{\overset{8}{\cancel{32}}}{9} \times \frac{5}{\underset{3}{\cancel{12}}}$$

$$= \frac{40}{27} = 1\frac{13}{27}$$

1.3 Converting decimals to fractions

Think of the decimal as being written in words, then write in fraction form. Reduce to lowest terms.

Convert 0.47 to a fraction. Think of 0.47 as "forty-seven hundredths," then write as $\frac{47}{100}$.

1.3 Rounding decimals

Decimal point
Tenths
Hundredths
Thousandths
Ten-thousandths
Hundred-thousandths
Millionths

0 . 0 7 3 2 6 5

Round 0.073265 to the nearest ten-thousandth.

0.0732|65

↑ ten-thousandths position

Since the digit to the right of the line is 6, increase the ten-thousandths digit by 1: 0.073265 rounds to 0.0733.

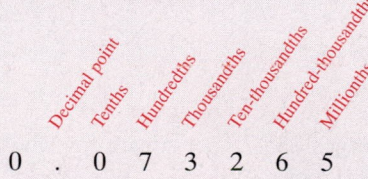

1.3 Converting fractions to decimals

Divide the numerator by the denominator. Round if necessary.

Convert $\frac{1}{8}$ to a decimal.

$$\begin{array}{r} 0.125 \\ 8\overline{)1.000} \\ \underline{8} \\ 20 \\ \underline{16} \\ 40 \\ \underline{40} \\ 0 \end{array}$$

$$\frac{1}{8} = 0.125$$

Chapter 1 Review Exercises

The answer section includes answers to all Review Exercises.

Write each of the following fractions in lowest terms. [1.2]

1. $\frac{24}{40}$ **2.** $\frac{32}{64}$ **3.** $\frac{27}{81}$ **4.** $\frac{147}{294}$

5. $\frac{63}{70}$ **6.** $\frac{84}{132}$ **7.** $\frac{24}{1200}$ **8.** $\frac{375}{1000}$

Convert each of the following improper fractions to mixed numbers and write in lowest terms. [1.2]

9. $\frac{65}{8}$ **10.** $\frac{56}{12}$ **11.** $\frac{38}{24}$ **12.** $\frac{55}{7}$

13. $\frac{120}{45}$ **14.** $\frac{196}{24}$ **15.** $\frac{258}{32}$ **16.** $\frac{194}{64}$

Solve each of the following problems and write in lowest terms. [1.2]

17. $\frac{5}{8} + \frac{7}{12}$ **18.** $\frac{1}{5} + \frac{3}{10} + \frac{3}{8}$ **19.** $\frac{5}{7} - \frac{1}{3}$ **20.** $\frac{3}{4} - \frac{2}{3}$

21. $25\frac{1}{6} + 46\frac{2}{3}$

22. $18\frac{3}{5} + 47\frac{7}{10} + 25\frac{8}{15}$

23. $6\frac{7}{12} - 2\frac{1}{3}$

24. $92\frac{5}{16} - 11\frac{1}{4}$

Solve each of the following application problems. [1.1 and 1.2]

25. Roofing material costs \$54.52 per square (10 ft × 10 ft). The roofer charges \$35.75 per square for labor, plus \$3.65 per square for supplies. Find the total cost for 26.3 squares of installed roof.

26. A federal law requires that all residential toilets sold in the United States use no more than 1.6 gallons of water per flush. Prior to this legislation, conventional toilets used 3.4 gallons of water per flush. Find the amount of water saved in one year by a family flushing the toilet 22 times each day.

27. Desiree Ramirez worked $5\frac{1}{2}$ hours on Wednesday, $6\frac{1}{4}$ hours on Thursday, $3\frac{3}{4}$ hours on Friday, and 7 hours on Saturday. How many hours did she work altogether?

28. A painting contractor arrived at a sixty-unit apartment complex with $147\frac{1}{2}$ gallons of exterior paint. His crew sprayed $68\frac{1}{2}$ gallons on the wood siding, rolled $37\frac{3}{8}$ gallons on the masonry exterior, and brushed $5\frac{3}{4}$ gallons on the trim. Find the number of gallons of paint remaining.

29. Three sides of Sheri Minkner's kiwi ranch are $202\frac{1}{8}$ feet, $370\frac{3}{4}$ feet, and $274\frac{1}{2}$ feet. If the distance around the ranch is $1166\frac{7}{8}$ feet, what is the length of the fourth side?

30. The Catering Crew served $12\frac{2}{3}$ pounds of American cheese, $16\frac{1}{8}$ pounds of jack cheese, $15\frac{1}{2}$ pounds of sharp cheddar cheese, and $10\frac{1}{6}$ pounds of muenster cheese at a catered event. Find the total weight of the cheese served.

Solve each problem and reduce to lowest terms. [1.2]

31. $\frac{5}{8} \times \frac{2}{3}$ **32.** $\frac{1}{3} \times \frac{7}{8} \times \frac{3}{5}$ **33.** $\frac{1}{6} \div \frac{1}{3}$ **34.** $10 \div \frac{5}{8}$

35. $2\frac{1}{2} \div 3\frac{3}{4}$ **36.** $3\frac{3}{4} \div \frac{27}{16}$ **37.** $12\frac{1}{2} \times 1\frac{2}{3}$ **38.** $12\frac{1}{3} \div 2$

Solve each of the following application problems. [1.1 and 1.3]

39. Barry bought 16.5 meters of rope at \$0.48 per meter and 3 meters of wire at \$1.05 per meter. How much change did he get from three \$5 bills? [1.4 and 1.5]

40. The earnings of Sierra West Bancorp last year were \$1.4 million, or \$0.39 per share of stock. Find the number of shares of stock in the company. Round to the nearest tenth of a million.

41. The area of a piece of land is $63\frac{3}{4}$ acres. One-third of the land is sold. What is the area of the land that is left?

42. Ellen Burke bought 25 shares of Korea Equity stock for $\$8\frac{3}{8}$ per share and 16 shares of Snyder Oil stock for $\$12\frac{1}{4}$ per share. How much did she pay altogether?

43. Find the number of window-blind pull cords that can be made from $157\frac{1}{2}$ yards of cord if $4\frac{3}{8}$ yards of cord are needed for each blind.

44. Play It Now Sports Center has decided to divide $\frac{2}{3}$ of the company's profit sharing funds evenly among the 8 store managers. What fraction of the total will each receive?

Convert each of the following decimals to a fraction and reduce to lowest terms. **[1.3]**

45. 0.25 **46.** 0.625 **47.** 0.93 **48.** 0.005

Round each of the following decimals to the nearest tenth and to the nearest hundredth. **[1.3]**

49. 68.433 **50.** 975.536

51. 0.3549 **52.** 8.025

53. 6.965 **54.** 0.428

55. 0.955 **56.** 71.249

Convert each of the following fractions to a decimal. Round to the nearest thousandth. **[1.3]**

57. $\frac{5}{8}$ **58.** $\frac{3}{4}$ **59.** $\frac{5}{6}$ **60.** $\frac{7}{16}$

Chapter 1 Summary Exercise
Fractions in Your Financial Future

Although retirement is many years off, David Perry has been investing his savings by buying stock in several companies. The daily changes in stock prices are shown in most major newspapers. The list here shows stock quotations in eight companies.

Stock Market News

52 Week High	52 Week Low		Close	Net Change
$68\frac{1}{2}$	33	AT&T	$56\frac{7}{16}$	$-\frac{1}{8}$
$81\frac{3}{8}$	$51\frac{5}{16}$	CocaCola	$78\frac{1}{16}$	$+\frac{11}{16}$
$38\frac{3}{16}$	$22\frac{9}{16}$	FruitLoom	$35\frac{7}{16}$	$-\frac{3}{8}$
$89\frac{7}{16}$	59	GenElec	$87\frac{7}{8}$	$+\frac{3}{8}$
$79\frac{3}{4}$	$54\frac{3}{8}$	ReynMetl	$60\frac{13}{16}$	$-1\frac{7}{8}$
$32\frac{3}{4}$	23	Rubbermaid	$32\frac{3}{8}$	$+\frac{1}{2}$
$37\frac{7}{8}$	$25\frac{3}{16}$	SherwinWill	$33\frac{1}{2}$	-1
$54\frac{3}{4}$	$29\frac{1}{2}$	Wal-Mart	$54\frac{3}{16}$	$-\frac{1}{16}$

(a) Stock prices use fractions to show parts of a dollar. For example, $\frac{1}{4} = \$0.25$; $\frac{1}{2} = \$0.50$ and $\frac{3}{4} = \$0.75$. Find the amounts represented by $\frac{3}{8}$, $\frac{3}{16}$, $\frac{7}{8}$ and $\frac{13}{16}$. Do not round.

(b) Find the difference between the highest price and the lowest price for Sherwin Williams in the last year.

(c) Find the difference between the highest and the lowest price for Fruit of the Loom during the last year.

(d) Which company had the greatest difference between the highest price and the lowest price for the year?

(e) David Perry bought 80 shares of General Electric stock at the closing price. Find his cost.

(f) David Perry bought 100 shares of WAL-MART stock at the closing price. Find his cost.

(g) With a $2000 investment, how many whole shares of Rubbermaid stock could be purchased at the closing price?

(h) During the past year David Perry has bought 50 shares of Coca Cola at $63\frac{1}{2}$ and 30 shares of Reynolds Metals at $76\frac{3}{4}$. Using the closing price, find the gain or loss on each of these stocks and the gain on their combined investment. Be sure to express a loss with a minus sign ($-$) before the number.

Net Assets Business on the Internet

Home Depot

Statistics

- 1978: Established
- 1999: $38.4 billion annual sales
- 170,000 employees
- 2002: 1600 stores (estimated)

The Home Depot, the country's largest home-improvement center, services exporters and importers worldwide. According to Arthur M. Blank, president and CEO, "The Home Depot is thriving in an industry that offers many opportunities for continued growth, and we are excited about our prospects for the future."

The company's progressive corporate culture includes a philanthropic budget that is directed back to the communities Home Depot serves and the interests of its employees through a matching-gift program. The major focuses are affordable housing, at-risk youth, and the environment. Team Depot, an organized volunteer force, was developed in 1992 to promote volunteer activities with the local communities the stores serve. For five consecutive years, the company has been ranked by *Fortune* magazine as America's Most-Admired Retailer.

1. A Home Depot customer needs a piece of 2-inch oak trim $22\frac{3}{8}$ inches long and another piece $15\frac{3}{16}$ inches long. Find the total length of oak trim needed.

2. A gutter downspout is 10 feet long. If a piece of gutter downspout 8 feet $8\frac{3}{8}$ inches is needed for a job, find the length of the piece remaining. (*Hint:* one foot equals 12 inches)

3. If the price of Home Depot stock is $\$58\frac{5}{8}$ per share. Find the cost of 130 shares of stock.

4. Home Depot stock is selling for $\$63\frac{3}{8}$ per share. Find the number of shares that can be purchased for $17,238.

2 Equations and Formulas

General Motors has been producing automobiles in the United States since 1908. The company manufactures Chevrolets, Buicks, Oldsmobiles, Cadillacs, GMCs, Pontiacs, Saturns, Isuzus, Opels, and Saabs. In 1998, they produced 5 million cars and trucks. In order to keep costs down and to meet demand, the company uses sophisticated algebra-based models to forecast demand for their products.

Why do you think managers use graphs such as Figure 2.1 and algebraic models in business?

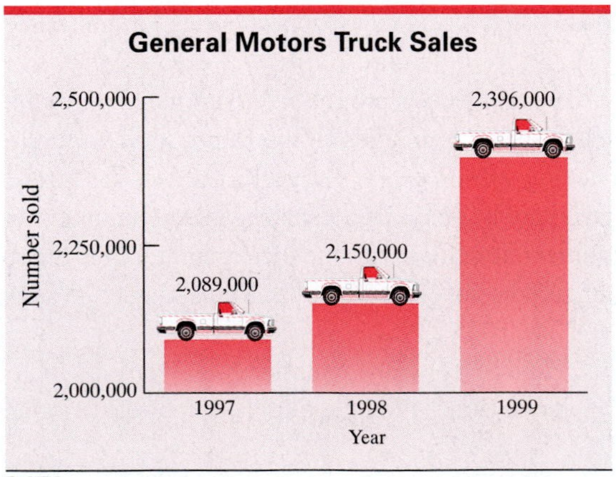

DATA: *www.gm.com*

Figure 2.1

Both equations and formulas occur again and again throughout business mathematics. For example, formulas are used for finding markup, interest, depreciation, and in other important areas of business. This chapter discusses various ways of solving basic equations and working with formulas.

2.1 Solving Equations

Objectives

1. *Learn the basic terminology of equations.*
2. *Use basic rules to solve equations.*
3. *Combine like terms in equations.*
4. *Use the distributive property to simplify equations.*

An **equation** is a statement that says two expressions are equal. For example, the equation

$$x + 5 = 9$$

says that the expression $x + 5$ and 9 are equal. In dealing with an equation certain terminology is used.

Objective 1 *Learn the Basic Terminology of Equations.* The letter x is called a **variable**—a letter that represents a number. The variable x, in the equation above, as well as the numbers 5 and 9, are called terms. A **term** is a single letter, a single number, or the product (or quotient) of a number and one or more letters. Different terms are separated from one another by + or − signs. The expression $x + 5$ is called the **left side** of the equation $x + 5 = 9$, while 9 is the **right side**. A **solution** to the equation is any number which can replace the variable and result in a true statement. The solution for this equation is the number 4, since the replacement of the variable x with the number 4 results in a true statement.

$$x + 5 = 9$$
$$4 + 5 = 9 \qquad \text{Let } x = 4.$$
$$9 = 9 \qquad \text{True.}$$

The preceding check is an example of **substitution**: the variable x was replaced with 4.

Objective 2 *Use Basic Rules to Solve Equations.* In solving equations, the object is to find numbers that can be used to replace the variable so that the equation is true. This is done by changing the equation so that all terms containing a variable are on one side of the equation and all numbers are on the other side. The two sides of an equation remain equal if the same change is made to both sides.

The rules for solving equations follow.

Addition Rule. The same number may be added or subtracted on both sides of an equation.
Multiplication Rule. Both sides of an equation may be multiplied or divided by the same nonzero number.

NOTE Remember, what you do to one side of an equation must also be done to the other side. ■

Example 1 Solving Equations Using Addition

Solve $x - 17 = 34$.

Solution To solve this equation, x must be alone on one side of the equal sign, and all numbers collected on the other side. To change the $x - 17$ to x, perform the opposite operation to "undo" what was done. The *opposite* of subtraction is addition, so add 17 to both sides.

$$\begin{aligned} x - 17 &= 34 \\ x - 17 + 17 &= 34 + 17 && \text{Add 17 to both sides.} \\ x + 0 &= 51 && \text{The sum of } -17 + 17 \text{ is 0.} \\ x &= 51 \end{aligned}$$

To check this answer, substitute 51 for x in the original equation.

$$\begin{aligned} x - 17 &= 34 && \text{Original equation.} \\ 51 - 17 &= 34 && \text{Let } x = 51. \\ 34 &= 34 && \text{True.} \end{aligned}$$

The answer checks, so $x = 51$.

Example 2 Solving Equations Using Subtraction

Solve $y + 12.3 = 20.5$.

Solution To isolate y on the left side, do the opposite of adding 12.3, which is *subtracting* 12.3 from both sides.

$$\begin{aligned} y + 12.3 &= 20.5 \\ y + 12.3 - 12.3 &= 20.5 - 12.3 && \text{Subtract 12.3 from both sides.} \\ y + 0 &= 8.2 \\ y &= 8.2 \end{aligned}$$

Check the answer by substituting 8.2 for y in the original equation.

$$\begin{aligned} y + 12.3 &= 8.2 + 12.3 \\ &= 20.5 \end{aligned}$$

The answer checks.

In formulas, the product of a number and a variable is often written without any special symbol for multiplication. As an example, the product of 5 and p could be written as $5p$, instead of $5 \times p$. The number 5 in the term $5p$ is the **coefficient** of p. Also, $\frac{1}{2}$ is the coefficient of z in the term $\frac{1}{2}z$, and the coefficient of the term s is 1 since $s = 1 \times s$.

Example 3 Solving Equations Using Division

Solve $5p = 60$.

Solution The term $5p$ indicates the product of 5 and p. Since the opposite of multiplication is division, solve the equation by *dividing* both sides by 5.

$$5p = 60$$

$$\frac{5p}{5} = \frac{60}{5} \quad \text{Divide both sides by 5.}$$

$$p = 12$$

Check by substituting 12 for p in the original equation.

Example 4 Solving Equations Using Multiplication

Solve $\frac{r}{8} = 13$.

Solution The bar $\frac{r}{8}$ means to divide ($r \div 8$), so solve the equation by multiplying both sides by 8 because 8 is the denominator. (The *opposite* of division is multiplication.) As in the following solution, it is common to use a dot to indicate multiplication.

$$\frac{r}{8} = 13$$

$$\frac{r}{8} \cdot 8 = 13 \cdot 8 \quad \text{Multiply both sides by 8.}$$

$$\frac{r}{\cancel{8}} \cdot \cancel{8} = 104 \quad \text{Cancel } \tfrac{8}{8}\text{, which is 1.}$$

$$r = 104$$

Example 5 shows how to solve an equation using a reciprocal. The product of a number and its **reciprocal** is 1. To get the reciprocal of a nonzero fraction, exchange the numerator and the denominator. For example, the reciprocal of $\frac{7}{9}$ is $\frac{9}{7}$.

$$\frac{\overset{1}{\cancel{7}}}{\underset{1}{\cancel{9}}} \cdot \frac{\overset{1}{\cancel{9}}}{\underset{1}{\cancel{7}}} = 1$$

Example 5 Solving Equations Using Reciprocals

Solve $\frac{4}{5}n = 20$.

Solution Solve this equation by multiplying both sides by $\frac{5}{4}$, the reciprocal of $\frac{4}{5}$. This process will give just $1n$, or n, on the left.

$$\frac{4}{5}n = 20$$

$$\frac{\overset{1}{\cancel{5}}}{\underset{1}{\cancel{4}}} \cdot \frac{\overset{1}{\cancel{4}}}{\underset{1}{\cancel{5}}}n = \frac{5}{4} \cdot 20 \quad \text{Multiply both sides by } \tfrac{5}{4}.$$

$$n = \frac{100}{4}$$

$$n = 25$$

The equation in Example 6 requires two steps to solve.

Example 6 Solving Equations Involving Several Steps

Solve $2m + 5 = 17$.

Solution To solve equations that require more than one step, first isolate the terms involving the unknown (or variable) on one side of the equation and constants (or numbers) on the other side by using addition and subtraction.

$$2m + 5 = 17$$
$$2m + 5 - 5 = 17 - 5 \quad \text{Subtract 5 from both sides.}$$
$$2m = 12$$

Now divide both sides by 2.

$$\frac{2m}{2} = \frac{12}{2} \quad \text{Divide by 2.}$$
$$m = 6$$

Check by substituting 6 for m in the original equation.

$$2m + 5 = 17$$
$$2(6) + 5 = 17$$
$$17 = 17$$

Thus, $m = 6$ is the solution.

NOTE The unknown can be on either side of the equal sign. The equation $6 = m$ is the same as $m = 6$. The number is the solution whether the equation has the variable by itself on either the left *or* the right side.

Objective 3 *Combine Like Terms in Equations.* Some equations have more than one term with the same variable. Terms with the same variables are called **like terms**, or **similar terms**. Like terms can be *combined* by adding or subtracting the coefficients, as shown.

$$8x + 2x = 10x$$
$$11y - 3y = 8y$$
$$12p - 5p + 2p = 9p$$
$$2z + z = 2z + 1z = 3z$$

NOTE Since multiplying by 1 does not change the value of a quantity, $1 \cdot z$ is the same as z.

Example 7 Combining Like Terms

Solve $9z - 3z + 2z = 50$.

Solution Start by combining terms on the left: $9z - 3z + 2z = 6z + 2z = 8z$. This gives the simplified equation:

$$8z = 50$$

$$\frac{8z}{8} = \frac{50}{8} \qquad \text{Divide by 8.}$$

$$z = 6\frac{1}{4} \text{ or } 6.25$$

Objective 4 *Use the Distributive Property to Simplify Equations.* Some of the more advanced formulas used later in this book involve a coefficient in front of parentheses. These formulas often require the use of the **distributive property**, by which a number on the outside of a parentheses can be multiplied by each term inside the parentheses, as shown here.

$$a(b + c) = ab + ac$$

The following diagram may help in remembering the distributive property.

$$a(b + c) = ab + ac$$

The term on the outside (a) is *distributed* over all terms in the parentheses:

$$2(m + 7) = 2m + 2 \cdot 7 = 2m + 14$$
$$8(k - 5) = 8k - 8 \cdot 5 = 8k - 40$$

Example 8 Solving Equations Using the Distributive Property

Solve $6(p - 2) = 30$.

Solution First use the distributive property on the left to remove the parentheses.

$$6(p - 2) = 30$$
$$6p - 12 = 30$$
$$6p - 12 + 12 = 30 + 12 \qquad \text{Add 12 to both sides.}$$
$$6p = 42$$
$$\frac{6p}{6} = \frac{42}{6} \qquad \text{Divide by 6.}$$
$$p = 7$$

Use the following steps to solve an equation.

Solving Linear Equations

Step 1. Remove all parentheses on both sides of the equation using the distributive property.

Step 2. Combine all similar terms on both sides of the equation.

Step 3. Add to or subtract from both sides whatever is needed to produce a term with the variable on one side and a number on the other side.

Step 4. Multiply or divide the variable term by whatever is needed to produce a term with a coefficient of 1. Multiply or divide the number term on the other side by the same quantity.

Example 9 Solving Equations Involving Several Steps

Solve $5r - 2 = 2(r + 5)$.

Solution

$$5r - 2 = 2(r + 5)$$

$$5r - 2 = 2r + 10 \quad \text{Use distributive property on the right.}$$

$$5r - 2 + 2 = 2r + 10 + 2 \quad \text{Add 2 to both sides to get all numbers on the right side.}$$

$$5r = 2r + 12$$

$$5r - 2r = 2r + 12 - 2r \quad \text{Subtract } 2r \text{ from both sides to get all variables on the left side.}$$

$$5r - 2r = 12$$

$$3r = 12 \quad \text{Combine like terms on the right.}$$

$$\frac{3r}{3} = \frac{12}{3} \quad \text{Divide both sides by 3 to get 1 as a coefficient.}$$

$$r = 4$$

Problem-Solving Hint Be sure to check the answer in the *original* equation and not in any other step.

2.1 Exercises

Solve each equation. Check each answer.

1. $b + 15 = 74$ **2.** $c + 21 = 146$ **3.** $r - 45 = 12$

4. $v - 29 = 17$ **5.** $25 = x + 12$ **6.** $312 = m - 40$

7. $10k = 42$ **8.** $7s = 84$ **9.** $12q = 144$

10. $8z = 136$ **11.** $60 = 30m$ **12.** $94 = 2z$

13. $5.9y = 17.7$ **14.** $16.5x = 39.6$ **15.** $1.54 = 0.7y$

16. $3.9a = 15.6$ **17.** $3.92w = 3.136$ **18.** $2.773m = 3.3276$

19. $0.0002x = 0.08$ **20.** $0.0324 = 0.0135y$ **21.** $\frac{s}{7} = 42$

22. $\frac{m}{5} = 6$ **23.** $\frac{r}{7} = 1$ **24.** $\frac{c}{7} = 2$

25. $\frac{2}{3}b = 8$ **26.** $22 = \frac{5}{4}s$ **27.** $35 = \frac{7}{5}t$

28. $\frac{7}{3}s = 21$ **29.** $2x = \frac{5}{3}$ **30.** $4y = \frac{1}{3}$

31. $3p = \frac{5}{12}$ **32.** $\frac{3}{4} = 9a$ **33.** $7b + 9 = 37$

34. $4x + 12 = 75$ **35.** $7y - 23 = 58$ **36.** $12r - 60 = 100$

37. $6p + 41.5 = 69.4$ **38.** $12.2s + 13.8 = 47.96$ **39.** $6c + \frac{3}{4} = 8$

40. $5z + \frac{2}{3} = 2$ **41.** $7q - \frac{2}{3} = 4$ **42.** $7a - \frac{5}{4} = \frac{9}{4}$

43. $5.2z - 4 = 1.2$ **44.** $3.6m + 2 = 6.32$ **45.** $27.85 = 3 + 7.1p$

46. $0.9 = 4t - 3.5$ **47.** $7m + 4m - 5m = 78$ **48.** $13r - 7r + 3r = 81$

49. $2s + s + 3s = 12$ **50.** $3.5k + k + k = 11.55$ **51.** $5y + 2 = 3(y + 4)$

52. $4z + 2 = 2(z + 2)$ **53.** $3(m - 4) = m + 2$ **54.** $s + 8 = 3(s - 6)$

55. $4(y + 8) = 3(y + 14)$ **56.** $7(z - 5) = 4(z + 8)$ **57.** $\frac{3}{4}s + \frac{1}{5}s = \frac{4}{5}$

58. $\frac{3}{4}q - \frac{1}{9} = \frac{1}{3} + \frac{1}{4}q$ **59.** $\frac{3}{8}y + \frac{1}{4} = \frac{9}{8}y - \frac{1}{4}$

60. $3(2p - 1) = 4(2.2 - p)$ **61.** $2(y + 1) = 4(4 - 2.5y)$

62. $9.1765y + 0.3284y = 6.65343$ **63.** $0.7452(3k - 1) = 3.94956$

64. $0.3255(1 + 7.5s) = 6.67275$ **65.** $1.2(2 + 3r) = 0.8(2r + 5)$

66. Explain why all terms with a variable should be placed on one side of the equation and all terms without a variable should be placed on the opposite side when solving an equation.

67. A student obtains the equation $6x = 5x$ after applying several steps correctly. The student then divides both sides by x and obtains the result $6 = 5$ and gives "no solution" as the answer. Is this correct? If not, state why not and give the correct solution. (See Objective 3.)

68. Explain the distributive property and give an example. (See Objective 4.)

2.2 Applications of Equations

Objectives

1. *Translate phrases into mathematical expressions.*
2. *Write equations from given information.*
3. *Solve application problems.*

Most problems in business are expressed in words. Before these problems can be solved, they must be converted into mathematical language.

Objective 1 *Translate Phrases into Mathematical Expressions.* Applied problems tend to have certain phrases that occur again and again. The key to solving such problems is to correctly translate these phrases into mathematical expressions. The next few examples illustrate this process.

Example 1 Translating Phrases into Expressions

Write the following verbal expressions as mathematical expressions. Use p to represent the unknown.

Verbal Expression	Mathematical Expression	Comments
(a) q plus a number	$q + p$	p represents the number and "plus" means addition
(b) Add 17 to a number	$p + 17$	p represents the number to which 17 is added
(c) The sum of a number and 12.7	$p + 12.7$	"sum" indicates addition
(d) 18 more than a number	$p + 18$	"more than" indicates addition

Example 2 Translating Phrases Involving Subtraction

Write each of the following verbal expressions as a mathematical expression. Use y as a variable.

Verbal Expression	Mathematical Expression	Comments
(a) 5.5 less than a number	$y - 5.5$	"less than" indicates subtraction
(b) A number decreased by $12\frac{1}{4}$	$y - 12\frac{1}{4}$	"decreased" indicates subtraction
(c) Eight fewer than a number	$y - 8$	"fewer than" indicates subtraction
(d) Eighteen minus a number	$18 - y$	"minus" indicates subtraction

Example 3 Translating Phrases Involving Multiplication and Division

Write the following verbal expressions as mathematical expressions. Use z as the variable.

Verbal Expression	Mathematical Expression	Comments
(a) The product of a number and 7.5	$7.5z$	"product" indicates multiplication
(b) Ten times a number	$10z$	"times" indicates multiplication
(c) One-fourth of a number	$\frac{1}{4}z$	"of" indicates multiplication
(d) The quotient of a number and 9	$\frac{z}{9}$	"quotient" indicates division
(e) The quotient of 8 and a number	$\frac{8}{z}$	"quotient" indicates division
(f) The sum of 6 and a number, multiplied by 8.3	$8.3(6 + z)$	multiplying a sum requires parentheses

NOTE If you are adding or multiplying, the order of the variable and the number doesn't matter. For example:

$$3 + x = x + 3 \quad \text{and} \quad 5 \cdot y = y \cdot 5$$

However, the problem changes when exchanging a variable and a number in a division or subtraction problem. ▬

Now that statements have been translated into mathematical expressions, you can use this knowledge to solve problems. The following steps provide an approach to solving application problems.

Solving Application Problems

Step 1. **First, read the problem very carefully.** Reread the problem to make sure that its meaning is clear.

Step 2. **Identify the unknown** and give it a variable name such as x. If possible, write other unknowns in terms of the same variable.

Step 3. Use the given information to write an equation describing the relationship given in the problem.

Step 4. **Solve the equation.**

Step 5. **Answer the question** asked in the problem.

Step 6. **Check the solution** using the *original* words of the problem.

Step 7. Be sure your answer is reasonable.

NOTE The third step is often the hardest. To write an equation from the information given in the problem, convert the facts stated in words into mathematical expressions, and then into equations. ■

Objective 2 *Write Equations from Given Information.* Any words that mean *equals* or *same* translate into an equal sign (=), producing an equation that can be solved for the unknown.

Example 4 Solving Number Problems

Translate the following statement into an equation: The product of 5, and a number decreased by 8, is 100. Use y as the variable. Solve the equation.

Solution Translate as follows.

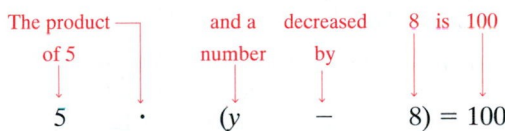

The product of 5		and a number	decreased by	8	is	100
5	$\cdot$	$(y$	$-$	$8)$	$=$	100

Simplify and complete the solution of the equation.

$$5 \cdot (y - 8) = 100$$
$$5y - 40 = 100 \quad \text{Apply the distributive property.}$$
$$5y = 140 \quad \text{Add 40 to both sides.}$$
$$y = 28 \quad \text{Divide by 5.}$$

NOTE If Example 4 had been written as "The product of 5 and a number, decreased by 8, is 100," then the corresponding equation would be $5y - 8 = 100$ rather than $5(y - 8) = 100$. These two equations have different solutions. ■

Objective 3 *Solve Application Problems.*

Example 5 Solving an Application Problem

A mattress is on sale for \$200, which is $\frac{4}{5}$ of its original price. Find the original price.

Solution The mattress is on sale for \$200, so it must have been priced at more than \$200 originally. Let p represent the original price; \$200 is the sale price, and the sale price is $\frac{4}{5}$ of the original price. Use all this information to write the equation.

sale price	is	$\frac{4}{5}$ of	original price
\$200	$=$	$\frac{4}{5} \times$	p

Solve the equation

$$200 = \frac{4}{5} \cdot p$$

$$\frac{5}{4} \cdot 200 = \frac{5}{4} \cdot \frac{4}{5} \cdot p \qquad \text{Multiply by reciprocal.}$$

$$\frac{1000}{4} = 1 \cdot p$$

$$250 = p$$

The original price is \$250. Check the solution.

$$\$200 = \frac{4}{5} \times p$$

$$\$200 = \frac{4}{5} \times \$250$$

$$\$200 = \$200$$

The answer checks.

Example 6 Solving an Inventory Problem

The Eastside Nursery ordered 27 trees. Some of the trees were elms, costing \$17 each, while the rest of the trees were maples at \$11 each. The total cost of the trees was \$375. Find the number of elms and the number of maples.

Solution Let x represent the number of elm trees, then $(27 - x)$ is the number of maples. The total cost of the elm trees is $17x$ and the total cost of the maples is $11(27 - x)$. A table can be very helpful in identifying the knowns and unknowns.

	Number of Trees	Cost per Tree	Total Cost
Elms	x	\$17	$17x$
Maples	$(27 - x)$	\$11	$11(27 - x)$
Totals	27		375

The information in the table is used to produce the following equation.

$$\text{Cost of elms} + \text{cost of maples} = \text{total cost}$$

$$17x + 11(27 - x) = 375$$

Now solve this equation. First use the distributive property.

$$17x + 297 - 11x = 375$$

$$6x + 297 = 375 \qquad \text{Combine terms.}$$

$$6x = 78 \qquad \text{Subtract 297 from each side.}$$

$$x = 13 \qquad \text{Divide each side by 6.}$$

We have only part of the answer; 13 elm trees. Now we need to find the number of maple trees, which is $27 - x$, or $27 - 13 = 14$. Check that the total number of trees is 27 and the total cost of the trees is $(13 \times \$17) + (14 \times \$11) = \$375$.

Example 7 Solving Investment Application Problems

Laurie Zimms has \$15,000 to invest. She places a portion of the funds in a passbook account and \$3000 more than twice this amount in a retirement account. How much is put into the passbook account? How much is placed in the retirement account?

Solution Let z represent the amount invested in the passbook account. To find the amount invested in the retirement account, translate as follows.

3000	more than	2 times the amount
↓	↓	↓
3000	+	$2z$

Since the sum of the two investments must be \$15,000, an equation can be formed as follows.

Amount invested in passbook	+	amount invested in retirement account	=	total amount invested
z	+	$(3000 + 2z)$	=	\$15,000

Now solve the equation.

$$z + (3000 + 2z) = 15{,}000$$

$$3z + 3000 = 15{,}000 \qquad \text{Combine like terms.}$$

$$3z = 12{,}000 \qquad \text{Subtract 3000.}$$

$$z = 4000 \qquad \text{Divide by 3.}$$

The amount invested in the passbook account is \$4000. The amount invested in the retirement account is $3000 + 2z$, or $3000 + 2(4000) = \$11{,}000$.

Check the solution as follows:

$$4000 + (3000 + 2 \cdot 4000) = 15{,}000$$

The answer checks.

2.2 Exercises

Write the following as mathematical expressions. Use x *as the variable.*

1. 27 plus a number

2. the sum of a number and $16\frac{1}{2}$

3. a number added to 22

4. 6.8 added to a number

5. 4 less than a number

6. 12 fewer than a number

7. subtract $3\frac{1}{2}$ from a number

8. subtract a number from 5.4

9. triple a number

10. the product of a number and 9

11. three-fifths of a number

12. four-thirds of a number

13. the quotient of 9 and a number

14. the quotient of a number and 11

15. 16 divided by a number

16. a number divided by 4

17. the product of 2.1 and the sum of 4 plus a number

18. the quantity of a number plus 4, divided by 9

19. 7 times the difference of a number and 3

20. the difference of a number and 2, multiplied by 7

Write mathematical expressions for each of the following.

21. PURCHASING CDs Find the cost of 12 CDs at y dollars each.

22. TUITION FEES Find the cost of x students paying tuition of $800 each.

23. LIVESTOCK FEED The demand forecast for next month is 472 tons of livestock feed. Find the amount that should be ordered if inventory is x tons.

24. Eighty-three of the x employees have computers. How many do not have computers?

25. UNION MEMBERSHIP A company has 73 employees of whom x are union. How many employees are nonunion?

26. CARD SALES The inventory of a small card shop is valued at $73,000. The value of the greeting cards is x. Find the value of the rest of the inventory.

27. PRODUCE BUYING A market paid $172 for x crates of berries. Find the cost of one crate of berries.

28. ADMISSION FEES A lodge paid $1853 for tickets for its x members to visit the state capitol. Find the cost of one ticket.

29. CHARITABLE DONATIONS Robin has 21 books on computers. She donates x of them to the school library. How many does she have left?

30. VIDEO STORE A video rental store is x years old. How old will it be in 8 years?

Solve the following application problems. Steps 1–7 are repeated here for your convenience—use them.

Solving Applied Problems

Step 1. Read the problem carefully.

Step 2. Identify the unknown and choose a variable to represent it. If possible, write any other unknowns in terms of the same variable.

Step 3. Translate the problem into an equation.

Step 4. Solve the equation.

Step 5. Answer the question asked in the problem.

Step 6. Check your solution using the original words of the problem.

Step 7. Be sure your answer is reasonable.

31. Three times 7, plus a number, is 36. Find the number.

32. Seventeen times a number, plus 5, equals 107. Find the number.

33. Six times the quantity of 4 minus a number is 15. Find the number.

34. Twelve times the quantity of a number less 1 is 72. Find the number.

35. When 6 is added to a number, the result is 7 times the number. Find the number.

36. If 6 is subtracted from three times a number, the result is 4 more than the number. Find the number.

37. When 5 times a number is added to twice the number, the result is 10. Find the number.

38. If 7 times a number is subtracted from 11 times the number, the result is 9. Find the number.

39. COMPUTER SALES Bill Thompson sold 18 more personal computers last month than did the other salesperson. Given that they sold 72 computers together, find the number sold by Thompson.

40. SODA SALES A grocery store sold 19 more cases of Coke than Sprite. Given that 43 cases were sold, find the number of cases of Coke sold.

41. At one company, 15 more people work in the production department than work in the packaging department. The total number of people in the two departments is 277. Find the number of people in the production department.

42. EXCHANGE PROGRAM TO MEXICO Twenty-one students went on a student exchange program to Guadalajara, Mexico. There were 11 more women than men. Find the number of women.

43. An automobile is on sale for \$18,450, which is $\frac{9}{10}$ of its original price. Find the original price.

44. INTERNATIONAL SHIPMENTS Because of handling and freight charges, Western Oil Equipment charges $\frac{5}{4}$ of the list price for an item shipped to Indonesia. Find the list price of an item that was charged at \$725.

45. HOME CONSTRUCTION A contractor built 105 homes last year, some economy models and some deluxe models. The number of economy models was $\frac{3}{2}$ the number of deluxe models. Find the number of each type of home that was built.

46. CULINARY SCHOOL The Regency Culinary School spent \$18,000 on advertising using radio and newspaper. The amount spent on newspaper advertising was $\frac{5}{4}$ that spent on radio advertising. Find the amount spent on each type of advertising.

47. RADIO STATION SALARIES Radio station KLRS spent \$10,500 on salaries one month. The amount spent on announcers was $\frac{3}{5}$ that amount spent on all other employees. Find the amount spent on announcers and on all other employees.

48. BUSY INTERSECTION An engineer is studying a busy intersection. In one hour, the number of cars going north-south was $\frac{3}{4}$ of those going east-west, and the total number of cars was 1400. Find the number of cars going north-south and east-west.

49. A building can be used either for retail stores or for offices. The owner wants to receive a total annual rent of \$67,500, with $3\frac{1}{2}$ times as much rent coming from retail stores as from offices. How much rent will come from offices? How much from retail stores?

50. Karen has a piece of material that is 106 inches long. She wishes to cut it into two pieces so that one piece is 12 inches longer than the other. What should be the length of each piece?

51. DEPARTMENT STORE There are 63 people employed at Dewey's Department Store. New workers receive \$6 per hour while experienced workers receive \$9 per hour. The company spends a total of \$483 per hour in wages. Find the number of each type of worker the company employs.

52. VEGETABLE SALES Jumbo Market makes \$0.05 on a head of lettuce and \$0.04 on a bunch of carrots. Last week, a total of 12,900 heads of lettuce and bunches of carrots were sold, with a total profit of \$587. How many heads of lettuce and how many bunches of carrots were sold?

53. NISSAN SALES Profits on Nissan Altimas and Sentras average \$1200 and \$850, respectively, at one dealership. The total profit in a month in which they sold 120 of these models was \$130,350. Find the number of each sold.

54. AUTO REPAIR One month, revenue from the 95 repairs at an auto repair shop totaled \$10,020. The average charges for repairs on personal and commercial vehicles were \$125 and \$90 respectively. Find the number of each type of vehicle repaired.

55. Are the problems in this section difficult for you? Explain why or why not. Explain two things you would recommend to a classmate who is having difficulty.

56. Write out the steps necessary to solve an applied problem. (See Objective 3.)

2.3 Formulas

Objectives

1. *Evaluate formulas for given values of the variables.*
2. *Solve formulas for a specific variable.*
3. *Use standard business formulas to solve word problems.*
4. *Evaluate formulas containing exponents.*

Many of the most useful rules and procedures in business are given as **formulas**, or equations, showing the relationship between different variables. For example, the formula for simple interest is:

$$\text{Interest} = \text{principal} \times \text{rate} \times \text{time}$$

or, using the first letter for each word:

$$I = P \times R \times T$$

Objective 1 *Evaluate Formulas for Given Values of the Variables.* By using letters to express the relationship between interest, principal, rate, and time we have shortened the process so that it can be written without words. Once any three values are substituted into the formula, we can then find the value of the remaining variable.

NOTE The variable T is always expressed as a fraction of a year in $I = PRT$. P and I are always in dollars and R is usually a percent.

Example 1 Finding Interest

Use the formula $I = PRT$ and find I if $P = 12{,}500$, $R = 0.08$, and $T = 1$ year.

Solution Substitute 12,500 for P, 0.08 for R, and 1 for T in the formula $I = PRT$. (Remember that writing P, R, and T together as PRT indicates the product of the three letters.)

$$I = PRT$$

$$I = 12{,}500 \times 0.08 \times 1$$

Multiply on the right to get the solution.

$$I = 1000 \qquad \text{\$1000 interest}$$

Example 2 Finding Time

Use the formula $I = PRT$ and find T if $P = \$4000$, $I = \$720$, and $R = 0.09$.

Solution Substitute the given numbers for the letters of the formula.

$$I = PRT$$
$$720 = 4000(0.09)T$$
$$720 = 360T$$
$$\frac{720}{360} = \frac{360T}{360} \qquad \text{Divide by 360.}$$
$$2 = T$$

Since T is in years, the answer is 2 years. Check by substituting values for P, R, and T.

$$I = PRT$$
$$= \$4000 \cdot 0.09 \cdot 2$$
$$= \$720$$

The interest checks.

Objective 2 *Solve Formulas for a Specific Variable.* In Example 2 we found the value of T when given the values of P, I, and R. If several problems of this type must be solved, it may be better to rewrite the formula $I = PRT$ so that T is alone on one side of the equation. Do this with the rules of equations given earlier. Since T is multiplied by PR, get T alone by dividing both sides of the equation by PR.

$$I = PRT$$
$$\frac{I}{PR} = \frac{PRT}{PR} \qquad \text{Divide by } PR.$$
$$\frac{I}{PR} = \frac{\cancel{PR}T}{\cancel{PR}}$$
$$\frac{I}{PR} = T \qquad or \qquad T = \frac{I}{PR}$$

This process of rearranging a formula is sometimes called *solving a formula for a specific variable.*

Example 3 Solving for a Specific Variable

Solve for T in the formula $M = P(1 + RT)$.

Solution This formula expresses the maturity value (M) of an initial amount of money (P) invested at a specified rate (R) for a certain period of time (T).

Start by using the distributive property on the right side.

$$M = P(1 + RT)$$
$$M = P + PRT$$

Now subtract P from both sides.

$$M - P = P + PRT - P$$
$$M - P = PRT$$

Divide each side by *PR*.

$$\frac{M-P}{PR}=\frac{PRT}{PR}$$

$$\frac{M-P}{PR}=T \quad \text{or} \quad T=\frac{M-P}{PR}$$

The original formula is now solved for *T*.

Example 4 Solving for a Specific Variable

An employee earns \$7.89 per hour plus \$0.45 for each circuit board she assembles. **(a)** Write an equation for her total income (*T*) in terms of hours (*H*) and number of circuit boards assembled (*C*). **(b)** Solve for the variable *C*. **(c)** Use this to find the number of circuit boards she must assemble in a 40 hour week in order to earn \$450.

Solution

(a) Her total earnings is the sum of her hourly wage times the number of hours plus the amount for each circuit board times the number of circuit boards assembled.

$$T = 7.89H + 0.45C$$

Subtract 7.89*H* from both sides.

(b)

$$T - 7.89H = 7.89H + 0.45C - 7.89H$$

$$T - 7.89H = 0.45C$$

Divide by 0.45.

$$\frac{T-7.89H}{0.45}=\frac{0.45C}{0.45}$$

$$C=\frac{T-7.89H}{0.45}$$

(c) Substitute \$450 in place of *T* and 40 in place of *H*.

$$C=\frac{450-7.89(40)}{0.45}$$

$$C=\frac{134.40}{0.45}$$

$$C = 298.666\ldots$$

Round to 299.

She must assemble 299 circuit boards.

Scientific *Calculator Approach*

The calculator solution to Example 4 uses chain calculations and the order of operations.

[( ] 450 [−] 7.89 [×] 40 [ )] [÷] .45 = 298.6666 . . .

Objective 3 ***Use Standard Business Formulas to Solve Word Problems.*** In the following two examples, application problems that use some common business formulas are solved. (These formulas are discussed in more detail later in the book.)

Example 5 Finding Gross Sales

Find the gross sales from selling 339 CDs at \$14.95 each.

Solution The formula for gross sales is

$$G = NP.$$

For this formula, N is the number of items sold and P is the price per item. To find the gross sales use the formula as shown.

$$G = NP$$
$$G = 339 \cdot 14.95$$
$$G = \$5068.05$$

The gross sales will be \$5068.05.

Example 6 Finding Selling Price

What is the selling price of a battle-simulator computer game if the cost is \$19.08 and the markup is \$10.87?

Solution The selling price of an item is found by adding the cost of the item and the markup.

$$S = C + M$$

The variable C is the cost of the item, and M is the markup (an amount added to cover expenses and profit).

$$S = 19.08 + 10.87$$
$$= 29.95$$

The selling price is \$29.95.

Objective 4 ***Evaluate Formulas Containing Exponents.*** Exponents are used to show repeated multiplication of some quantity. For example,

Exponent: number of times quantity is multiplied by itself

$$x \cdot x = x^2$$

Base: quantity being multiplied

Similarly, $z \cdot z \cdot z = z^3$ and $5 \cdot 5 \cdot 5 \cdot 5 = 5^4$, which is 625. Exponents are also referred to as "power." For example, x^5 is "x to the fifth power."

Example 7 Finding Monthly Sales

Trinity Sporting Goods has found that monthly sales can be approximated using

$$\text{Sales} = 40 + 1.6 \times (\text{advertising})^2$$

as long as advertising is less than \$4000. Sales and advertising in the equation above are in thousands of dollars. Estimate sales for a month with \$3500 in advertising.

Solution Place 3.5 (for \$3500) in the equation for the *number of thousands* of dollars of advertising and find sales.

$$\begin{aligned} \text{Sales} &= 40 + 1.6(3.5)^2 \\ &= 40 + 1.6(12.25) \quad \text{(in thousands)} \\ &= 40 + 19.6 \\ &= 59.6 \end{aligned}$$

Sales are projected to be \$59,600 for the month.

2.3 Exercises

In the following exercises a formula is given, along with the values of all but one of the variables in the formula. Find the value of the variable that is not given. Round to the nearest hundredth, if applicable.

1. $I = PRT$; $P = \$4600$, $R = 0.085$, $T = 1\frac{1}{2}$
2. $F = ma$; $m = 820$, $a = 12$
3. $P = \dfrac{nRT}{V}$; $n = 6$, $R = 0.0821$, $T = 315$, $V = 10$
4. $M = P(1 + RT)$; $P = 640$, $R = 0.10$, $T = 1\frac{3}{4}$
5. $R = \dfrac{D}{1 - DT}$; $D = 0.05$, $T = 4$
6. $\dfrac{I}{PR} = T$; $P = 100$, $R = 0.02$, $T = 500$
7. $P = 2L + 2W$; $P = 40$, $W = 6$
8. $P = 2L + 2W$; $P = 340$, $L = 70$
9. $P = \dfrac{I}{RT}$; $T = 3$, $I = 540$, $P = 2250$
10. $M = P(1 + RT)$; $R = 0.15$, $T = 2$, $M = 481$
11. $y = mx^2 + c$; $m = 3$, $x = 7$, $c = 4.2$
12. $F = \dfrac{GMm}{r^2}$; $^*G = 1$, $M = 3$, $m = 1$, $r = 2$
13. $M = P(1 + i)^n$; $P = 640$, $i = 0.02$, $n = 8$
14. $M = P(1 + i)^n$; $M = \$2400$, $i = 0.05$, $n = 4$
15. $E = mc^2$; $m = 7.5$, $c = 1$
16. $x = \frac{1}{2}at^2$; $t = 5$, $x = 150$
17. $A = \frac{1}{2}(b + B)h$; $A = 105$, $b = 19$, $B = 11$
18. $A = \frac{1}{2}(b + B)h$; $A = 70$, $b = 15$, $B = 20$
19. $P = \dfrac{S}{1 + RT}$; $S = 24{,}600$, $R = 0.06$, $T = \dfrac{5}{12}$
20. $P = \dfrac{S}{1 + RT}$; $S = 23{,}815$, $R = 0.09$, $T = \dfrac{11}{12}$

Solve each formula for the indicated variable.

21. $A = LW$; for L

22. $d = rt$; for t

23. $PV = nRT$; for V

24. $I = PRT$; for R

25. $M = P(1 + i)^n$; for P

26. $R(1 - DT) = D$; for R

27. $P = \dfrac{A}{1 + i}$; for i

28. $M = P(1 + RT)$; for R

29. $P = M(1 - DT)$; for D

30. $P = \dfrac{M}{1 + RT}$; for R

31. $A = \dfrac{1}{2}(b + B)h$; for h

32. $P = 2L + 2W$; for L

33. $M = Pe^{ni}$; for P

34. $S = R\left[\dfrac{(1 + i)^n - 1}{i}\right]$; for R

Solve the following application problems.

35. SHOE SALES A shoe store buys 82 pair of Nike tennis shoes for a total cost of \$3883.52. Find the average cost per pair.

36. WEB PAGES A retailer paid \$1305 to a student who built and added 15 pages to the firm's Web site. Find the cost per Web page.

37. MUSICAL INSTRUMENTS The Guitar Shoppe bought 6 sets of bongo drums and 7 Alvarez guitars for \$2445.80. The guitars averaged \$269 apiece. Find the cost for a set of bongo drums.

38. BICYCLE INVENTORY Bicycles Unlimited bought 8 racing bikes and 12 kid's bikes for \$3696. What is the cost of a racing bike if a kid's bike cost \$112?

39. An employee of Wilson's Department Store is paid a weekly salary given by the formula $S = 160 + 0.03x$, where x is the total sales of the employee for the week. Find the salary of employees having the following weekly sales: (a) \$1152, (b) \$1796, (c) \$2314.

40. A principal (P) of \$3500 invested over an unknown amount of time (T) at 9.5% ($R = 0.095$) yields \$748.13 in interest (I). Use $I = PRT$ to find the time rounded to the nearest hundredth of a year.

41. COMPUTER CHIPS A computer chip manufacturer had net sales of \$230 million with returns equal to $\frac{1}{40}$ of gross sales. Find gross sales rounded to the nearest million given that net sales equals gross sales less returns.

42. The Bridal Shop has net sales of \$33,000 and a return of $\frac{1}{12}$ of gross sales. If net sales equal gross sales minus returns, find gross sales.

43. One fashion store sets markup at $\frac{3}{4}$ of its cost for the item. What is the cost of an item if the selling price is \$84?

44. MARKUP ON BOOKS The bookstore at Ironwood Community College has a markup that is $\frac{1}{4}$ its cost on a book. Find the cost to the bookstore of a book selling for \$20.

45. MEXICAN FOOD Last year, the expenses at the Taco Mary Restaurant were $\frac{5}{6}$ of the revenue. The profit (the difference between revenue and expenses) was \$15,000. Find the revenue.

46. Computerworld has expenses that run $\frac{15}{16}$ of revenue. Profit (the difference between revenue and expenses) was $9000. Find the revenue.

47. Find the interest if principal of $5200 is invested at $7\frac{1}{2}$ % (or 0.075) for one year. ($I = PRT$)

48. LOAN TO AN UNCLE Ben Cross loaned $8000 to his uncle for 4 years and received $1920 in interest. Find the interest rate. ($I = PRT$)

49. AUTO PARTS STARTUP Terry Twitty made a $22,000 loan so that Melissa Graves could start an auto parts business. The loan was for 2 years, and interest was $5720. Find the rate of interest. ($I = PRT$)

50. Fred Tausz loaned $39,000 to Anne Topsy. The loan was at 7% (or 0.07), with interest of $13,650. Find the time for the loan. ($I = PRT$)

51. Jackie Williams loaned $18,200 to Colleen Sullivan. The two agreed on an interest rate of 11% (or 0.11), with interest of $8008. Find the time for the loan. ($I = PRT$)

52. Mary Scott invests $1000 at 8% (or 0.08) for 5 years. How much did Mary have in her account at the end of 5 years? [$M = P(1 + RT)$]

53. John Wood had $4560 in his account after 2 years. If the account paid 7% (or 0.07) interest, how much did John initially deposit in his account. [$M = P(1 + RT)$]

54. HAIR SALON STARTUP CAPITAL Jan Rice borrowed $2800 from her uncle to help her start a hair salon. She repayed $3472 exactly 3 years later. Find the interest rate. [$M = P(1 + RT)$]

55. Bill Abel paid a maturity value of $5989.50 on a 3-year note with annual interest of 10% (or 0.10) . Use $M = P(1 + i)^n$, where n is the number of years in this problem, to solve for the amount borrowed.

56. June Smith won $5000 in a lottery and placed it in a mutual fund yielding 12% (or 0.12) for 20 years. Use $M = P(1 + i)^n$, where n is the number of years to solve for the maturity value.

57. Write a step-by-step explanation of the procedure you would use to solve the equation $A = P + PRT$ for R. (See Objective 2.)

58. Formulas are used in business, physics, biology, chemistry, engineering, and many other places. Why are formulas so commonly used? (See Objective 1.)

2.4 Ratios and Proportions

Objectives

1. *Define a ratio.*
2. *Set up a proportion.*
3. *Solve a proportion for unknown values.*
4. *Use proportions to solve problems.*

Objective 1 *Define a Ratio.* A **ratio** is a quotient of two quantities that can be used to compare the quantities. The ratio of the number a to the number b is written in any of the following ways.

$$a \text{ to } b \qquad a:b \qquad \frac{a}{b}$$

This last way of writing a ratio is most common in algebra, while $a:b$ is perhaps most common in business. Both quantities in a ratio should be in *the same units if possible*, that is, cents to cents or dollars to dollars, not dollars to cents. However, many ratios will not have the same units in the numerator and denominator.

Example 1 Writing Ratios from Words

Write a ratio in the form $\frac{a}{b}$ for each phrase. (Notice in each phrase that the number mentioned first always becomes the numerator.)

Solution

(a) The ratio of 3 women to 5 children is $\frac{3}{5}$.

(b) To find the ratio of \$12 to 20 cents, first convert \$12 to cents (\$12 = 1200 cents), then write the ratio.

$$\frac{1200}{20} = \frac{60}{1} \textit{ or } 60:1$$

Converting 20 cents to \$0.20 results in the exact same ratio.

$$\frac{12}{0.20} = \frac{60}{1} \textit{ or } 60:1$$

(c) General Motors car and truck sales fell from 1,900,000 units in 1929 to 525,000 units during 1932, the worst year of the Depression. Write unit sales as a ratio.

$$\frac{1{,}900{,}\not{0}\not{0}\not{0}}{525{,}\not{0}\not{0}\not{0}} = \frac{1900}{525} = \frac{76}{21}$$

Cancel matching final zeros

(d) A dairy farmer received \$650 for 5000 pounds of milk. Write this as a ratio.

$$\frac{650}{5000} = \frac{13}{100} \textit{ or } \$13 \text{ per } 100 \text{ pounds}$$

NOTE In the ratios of Example 1(b), the quantities were first changed so that they had the same units (cents to cents or dollars to dollars).

Objective 2 *Set Up a Proportion.* A ratio is used to compare two numbers or amounts. A **proportion** says that two ratios are equal, as in the following example.

$$\frac{3}{4} = \frac{15}{20}$$

This proportion says that the ratios $\frac{3}{4}$ and $\frac{15}{20}$ are equal.

The ratios $\frac{a}{b}$ and $\frac{c}{d}$ are said to form a proportion if $\frac{a}{b} = \frac{c}{d}$. This can be determined by multiplying both sides of the equation by the product of the two denominators.

$$\frac{a}{b} = \frac{c}{d}$$

$$\not{b}d \cdot \frac{a}{\not{b}} = b\not{d} \cdot \frac{c}{\not{d}}$$

$$ad = bc$$

Therefore, $\frac{a}{b} = \frac{c}{d}$ if, and only if, $ad = bc$. This is referred to as the **method of cross products**.

$$\frac{a}{b} \nwarrow\!\!\!\!\times\!\!\!\!\nearrow \frac{c}{d}$$

is equivalent to $ad = bc$.

Example 2 Determining True Proportions

Decide if the following proportions are true.

(a) $\frac{3}{4} = \frac{25}{30}$ **(b)** $\frac{6.5}{\frac{3}{4}} = \frac{130}{15}$

Solution

(a) Find each cross product.

$$\frac{3}{4} \times \frac{25}{30}$$

$$3 \cdot 30 = 4 \cdot 25$$

$$90 \neq 100$$ The symbol $\neq$ means "not equal."

Since the cross products are not equal, the proportion is false.

(b) Find each cross product.

$$\frac{6.5}{\frac{3}{4}} \times \frac{130}{15}$$

$$6.5 \cdot 15 = \frac{3}{4} \cdot 130$$

$$97.5 = 0.75 \cdot 130$$

$$97.5 = 97.5$$

The proportion is true.

NOTE The numbers in a proportion need not be whole numbers. ■

Objective 3 *Solve a Proportion for Unknown Values.* Four numbers are used in a proportion. If *any* three of these numbers are known, *the fourth can be found.*

Example 3 Finding Unknown Values in a Proportion

Find the unknown that makes each proportion true.

(a) $\frac{4}{9} = \frac{36}{x}$ **(b)** $\frac{3.4}{12} = \frac{z}{96}$

Solution

(a) Set the cross products equal to one another, then solve the resulting equation.

$$4 \cdot x = 9 \cdot 36$$

$$4x = 324$$

$$\frac{4x}{4} = \frac{324}{4}$$ Divide by 4.

$$x = 81$$

Check by confirming that $\frac{4}{9}$ does equal $\frac{36}{81}$.

(b) Set the cross products equal to one another.

$$3.4 \cdot 96 = 12 \cdot z$$
$$326.4 = 12z$$
$$\frac{326.4}{12} = \frac{12z}{12} \quad \text{Divide by 12.}$$
$$27.2 = z$$

Example 4 Foreign Currency Exchange

While in Mexico on an exchange trip, Laura Axtell needed to exchange U.S. \$75 for Mexican pesos. If U.S. \$1 is equivalent to 9.8 Mexican pesos, how many pesos will she receive?

Solution Set the problem up as a proportion and solve for the unknown.

$$\frac{\text{U.S. \$1}}{\text{9.8 Mexican pesos}} = \frac{\text{U.S. \$75}}{x \text{ Mexican pesos}}$$
$$1 \cdot x = 9.8 \times 75$$
$$x = 735$$

Axtell will receive 735 Mexican pesos.

Objective 4 *Use Proportions to Solve Problems.* Proportions are used in many practical applications, as Example 4 shows.

Example 5 Solving Applications

A hospital charges a patient \$7.80 for 12 capsules. How much should it charge for 18 capsules?

Solution Let x be the cost of 18 capsules. Set up a proportion: one ratio in the proportion can involve the number of capsules, while the other ratio can use the costs. Make sure that corresponding numbers appear in the numerator and the denominator. Use this pattern.

$$\frac{\text{Capsules}}{\text{Capsules}} = \frac{\text{cost}}{\text{cost}}$$

Now substitute the given information.

$$\frac{12}{18} = \frac{7.80}{x}$$

Use cross products to solve the proportion.

$$12x = 18(7.80)$$
$$12x = 140.40$$
$$x = 11.70$$

The 18 capsules should cost \$11.70.

Example 6 Solving Applications

A firm in Hong Kong and one in Thailand agree to jointly develop a controller chip to be sold to North American auto manufacturers. They agree to split the development costs in a ratio of 8:3 (Hong Kong firm to Thailand firm), resulting in a cost of \$9,400,000 to the Hong Kong firm. Find the cost to the Thailand firm.

Solution Let x represent the cost to the Thailand firm, then

$$\frac{8}{3} = \frac{9{,}400{,}000}{x}$$

$$8x = 3 \cdot 9{,}400{,}000 \quad \text{Cross multiply.}$$

$$8x = 28{,}200{,}000$$

$$x = 3{,}525{,}000 \quad \text{Divide by 8.}$$

The Thailand firm's share of the costs is \$3,525,000.

Example 7 Solving Applications

Bill Thomas wishes to estimate the amount of timber on some forested land that he owns. One value he needs to estimate is the average height of the trees. One morning, Thomas notices that his own 6-foot body casts an 8-foot shadow at the same time that a typical tree casts a 34-foot shadow. Find the height of the tree.

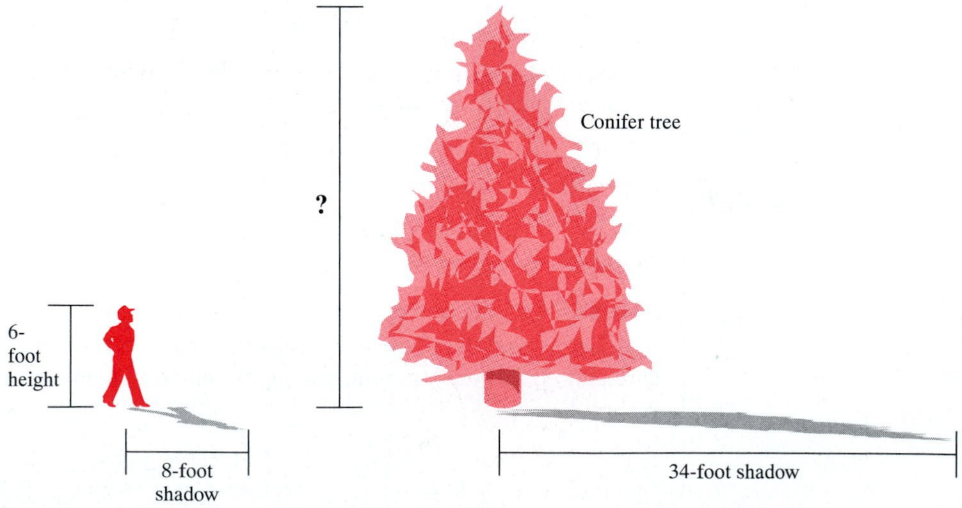

Solution Set up a proportion in which the height of the tree is given the variable name x.

$$\frac{6}{8} = \frac{x}{34}$$

$$6 \cdot 34 = 8 \cdot x \quad \text{Cross multiply.}$$

$$\frac{204}{8} = \frac{8 \cdot x}{8} \quad \text{Divide by 8.}$$

$$x = 25.5 \text{ feet}$$

The height of the tree is 25.5 feet.

2.4 Exercises

Write the following ratios. Write each ratio in lowest terms.

1. 18 kilometers to 64 kilometers

2. 18 defects out of 580 items

3. 216 students to 8 faculty

4. \$80 in returns to \$8360 in sales

5. 8 men to 6 women

6. 12 feet to 1 inch

7. 30 kilometers to 8 meters

8. 30 inches to 5 yards

9. 90 dollars to 40 cents

10. 148 minutes to 4 hours

11. 4 dollars to 10 quarters

12. 35 dimes to 6 dollars

13. 20 hours to 5 days

14. 6 days to 9 hours

15. \$0.80 to \$3

16. \$1.20 to \$0.75

17. \$3.24 to \$0.72

18. \$3.57 to \$0.42

Decide whether the following proportions are true or false.

19. $\frac{3}{5} = \frac{21}{35}$

20. $\frac{6}{13} = \frac{30}{65}$

21. $\frac{9}{7} = \frac{720}{480}$

22. $\frac{54}{14} = \frac{270}{70}$

23. $\frac{69}{320} = \frac{7}{102}$

24. $\frac{17}{19} = \frac{72}{84}$

25. $\frac{19}{32} = \frac{33}{77}$

26. $\frac{19}{30} = \frac{57}{90}$

27. $\frac{110}{18} = \frac{160}{27}$

28. $\frac{46}{17} = \frac{212}{95}$

29. $\frac{32}{75} = \frac{61}{108}$

30. $\frac{28}{75} = \frac{224}{600}$

31. $\frac{7.6}{10} = \frac{76}{100}$

32. $\frac{95}{64} = \frac{320}{217}$

33. $\frac{2\frac{1}{4}}{5} = \frac{9}{20}$

34. $\frac{\frac{3}{4}}{80} = \frac{\frac{9}{8}}{120}$

35. $\frac{4\frac{1}{5}}{6\frac{1}{8}} = \frac{27}{41}$

36. $\frac{1\frac{1}{2}}{12} = \frac{5\frac{1}{4}}{42}$

37. $\frac{8.15}{2.03} = \frac{61.125}{15.225}$

38. $\frac{423.88}{17.119} = \frac{330.6264}{13.35282}$

Solve each of the following proportions.

39. $\frac{x}{15} = \frac{49}{105}$

40. $\frac{y}{35} = \frac{27}{315}$

41. $\frac{6}{9} = \frac{r}{108}$

42. $\frac{16}{41} = \frac{112}{t}$

43. $\frac{63}{s} = \frac{3}{5}$

44. $\frac{260}{390} = \frac{x}{3}$

45. $\frac{1}{2} = \frac{r}{7}$

46. $\frac{2}{3} = \frac{5}{s}$

47. $\frac{\frac{3}{4}}{6} = \frac{3}{x}$

48. $\frac{3}{x} = \frac{11}{9}$

49. $\frac{12}{p} = \frac{23.571}{15.714}$

50. $\frac{86.112}{57.408} = \frac{k}{15}$

51. Explain the difference between ratio and proportion. (See Objective 2.)

52. Explain cross products using the rules of algebra. (See Objective 2.)

53. TICKET SALES One Ticketmaster outlet sold 350 rock-concert tickets in 2 days. At that rate, find the number of tickets it can expect to sell in 9 days.

54. BLOOD CELLS A 170-pound person has about 30 trillion blood cells. Estimate the number of blood cells in a 140-pound person to the nearest tenth of a trillion.

55. **REAL ESTATE DEVELOPMENT** Mike George paid \$172,000 for a 5-unit apartment house. Find the cost for a 12-unit apartment house.

56. **TIGER FOOD** A 450-pound circus tiger eats 15 pounds of meat per day. How many pounds of meat would you expect a 360-pound tiger to eat per day?

57. **SEWING** If 22 children's dresses cost \$176, what is the cost of 12 dresses?

58. **APARTMENT HOUSE** Jose paid \$199,500 for a 7-unit apartment house. Find the cost for a 16-unit apartment house.

59. **SEWING** Fifteen yards of material are needed for 5 dresses. How much material is needed for 12 dresses?

60. **FERTILIZER COVERAGE** Suppose that seven sacks of fertilizer cover 3325 square feet of lawn. Find the number of sacks needed for 7125 square feet.

61. **MAP READING** The distance between two cities on a road map is 2 inches. Actually, the cities are 120 miles apart. The distance between two other cities is 17 inches. How far apart are these cities?

62. **WOMAN'S CLOTHING SHOP** Jan Reus opened a woman's clothing shop and had sales of \$3720 during the first 3 weeks. At that rate, estimate sales for the first 4 weeks.

63. **SALES OF HEALTH FOOD** Natural Harvest had sales of \$274,312 for the first 20 weeks of the year. Estimate sales for the entire 52-week year.

64. **PARTNERSHIP PROFITS** Hite and Clark are partners who agree to divide any profits in the ratio 4:7. What is Clark's profit if Hite gets \$8000?

65. **DIVIDE PROFITS** Suppose two partners agree to divide their profits in the ratio 5:8. If the first partner receives \$15,000 in profits, how much does the second partner get?

66. **PRODUCTION EMPLOYEES** The owner of a factory has always kept the ratio of salespeople to production employees at 2:7. If she currently has 24 salespeople, how many production employees are there?

67. **SONGBIRD MIGRATION** Small songbirds migrate at 20 miles per hour whereas ducks migrate at 59 miles per hour. How far would ducks migrate in the amount of time it takes the songbirds to migrate 500 miles?

68. **ISLAND AREA** Indonesia has an area of 741,101 square miles and is made up of 13,677 islands. Assume the United States, with an area of 3,618,770 square miles, were similarly broken up into islands. How many islands would there be (to the nearest whole number)?

69. **ICEBERG VOLUME** Seven-eighths of an iceberg is below the water since icebergs are made up of freshwater, which is not as dense as seawater. Find the amount of an iceberg that is under water if the amount above water has a volume of 500,000 cubic meters.

70. **AUTO PRODUCTION** An auto plant produces 3 red sports models for every 7 blue family models. Find the number of red sports models produced if the plant produces 868 blue family models.

71. **JAPANESE YEN** Benjamin Lopez was in Japan for 2 months on a business trip. Find the number of U.S. dollars he will receive for 20,355 Japanese yen if U.S. \$1 can be exchanged for 105 yen. Round to the nearest cent.

72. **CANADIAN SALARY** Gina Harden was offered a job in Canada at an annual salary of \$54,700 in Canadian dollars. Find the salary in U.S. dollars if 1 Canadian dollar can be exchanged for U.S. \$0.68.

Chapter 2 Quick Review

Chapter Terms

Review the following terms to test your understanding of the chapter. For each term you do not know, refer to the page number found next to that term.

addition rule [p. 41]	equation [p. 41]	proportion [p. 61]	solution [p. 41]
coefficient [p. 42]	formulas [p. 54]	ratio [p. 60]	substitution [p. 41]
cross products [p. 62]	left side [p. 41]	reciprocal [p. 43]	term [p. 41]
distributive property [p. 45]	like terms [p. 44]	right side [p. 41]	variable [p. 41]
	multiplication rule [p. 41]	similar terms [p. 44]	

Concept	Example
2.1 Solving equations **1.** Remove all parentheses on both sides of the equation using the distributive property. **2.** Combine all like terms on both sides of the equation. **3.** ADDITION RULE Add to or subtract from both sides whatever is needed to produce a term with the variable on one side and a number on the other side. **4.** MULTIPLICATION RULE Multiply or divide the variable term by whatever is needed to produce a term with a coefficient of 1. Multiply or divide the number term on the other side by the same quantity.	$12(y + 2) = 84$ $12y + 24 = 84$ $12y + 24 - 24 = 84 - 24$ $12y = 60$ $y = 5$
2.2 Translating phrases Use mathematical symbols to represent verbal expressions.	6 times a number plus 3: $6x + 3$ 14 minus $2\frac{1}{2}$ times a number: $14 - 2\frac{1}{2}y$
2.3 Solving applied problems **1.** Read the problem carefully. **2.** Choose a variable to represent the unknown. If possible, write any other unknowns in terms of the same variable. **3.** Write an equation describing the relationship among the quantities. **4.** Solve the equation. **5.** Answer the problem using sentences. **6.** Check the solution.	A committee had 7 fewer men than women. The total number of people on the committee was 19. Find the number of women. Let x represent the number of women. $x + (x - 7) = 19$ $2x - 7 = 19$ $2x = 26$ $x = 13$ women Check the answer. $13 + (13 - 7) = 19$ $19 = 19$ The answer checks.

2.3 Evaluating formulas for given values of the variable

Substitute numerical values for variables and evaluate.

Use $I = PRT$ with $P = \$10{,}500$, $R = 0.09$, and $T = \frac{3}{4}$ to find interest (I).

$$I = PRT$$
$$I = 10{,}500 \cdot 0.09 \cdot \frac{3}{4}$$
$$= \$708.75$$

2.3 Solving formulas for a specific variable

Use the rules for solving equations.

Solve $M = P + PRT$ for T.

$$M - P = PRT$$
$$\frac{M - P}{PR} = T$$

2.3 Working with exponents

Use the definition of an exponent.

Find 6^4.

$$6^4 = 6 \cdot 6 \cdot 6 \cdot 6$$
$$= 1296$$

2.4 Solving a proportion for a missing part

Use the principle of cross products and solve the resulting equation.

$$\frac{a}{b} = \frac{c}{d} \text{ if } a \cdot d = b \cdot c$$

Find x in the proportion.

$$\frac{5}{x} \times \frac{35}{63}$$
$$5 \cdot 63 = 35x$$
$$315 = 35x$$
$$9 = x$$

2.4 Using proportions to solve problems

Set up the proportion, use the principle of cross products, and solve the resulting equation.

A video store charges \$12 to rent five tapes. How much does it charge for eight tapes?

$$\frac{12}{5} = \frac{x}{8}$$
$$12 \cdot 8 = 5 \cdot x$$
$$96 = 5x$$
$$19.20 = x$$

It charges \$19.20 for the tapes.

Chapter 2 Review Exercises

Solve each equation. [2.1]

1. $x + 45 = 96$

2. $r - 36 = 14.7$

3. $8t + 45 = 175.4$

4. $4t - 6 = 15$

5. $\frac{s}{6} = 42$

6. $\frac{5z}{8} = 85$

7. $\frac{m}{4} - 5 = 9$

8. $5(x - 3) = 3(x + 4)$

9. $6y = 2y + 28$

10. $3r - 7 = 2(4 - 3r)$

11. $0.15(2x - 3) = 5.85$

12. $0.6(y - 3) = 0.1y$

Write a mathematical expression. Use x *as the variable.* [2.2]

13. 94 times a number

14. $\frac{1}{2}$ times as number

15. 6 times a number is added to the number

16. 5 times a number is decreased by 11

17. The sum of 3 times a number and 7

Solve the following application problems. [2.2]

18. Molly Videtto wishes to purchase three CDs at $14.95 each and a Liz Claiborne sweater for $95. Given that she has $47.50, find the additional amount she needs to save. (Ignore taxes.)

19. A furniture store has found that profits (P) are related to advertising (A) according to $P = 2.775A + 4.5$ where all figures are in thousands of dollars. How much must they spend on advertising in order to obtain a quarterly profit of $60,000?

20. Phone and water bills together cost a company $540 for March. If the phone cost four times as much as the water, how much was each utility?

21. Five more than $\frac{1}{4}$ of the employees of a company have 25 years or more of service. If 24 employees have 25 or more years of service, how many employees does the company have?

22. The local movie theater sold 100 tickets for $390. If children's tickets cost $3 and adult tickets $6, how many of each were sold?

For each problem, use the formula to find the value of the variable that is not given. [2.3]

23. $I = PRT$; $I = \$960$, $R = 0.12$, $T = 2$

24. $M = P(1 + RT)$; $M = \$3770$, $R = 0.04$, $T = 4$

25. $M = P(1 + i)^n$; $M = \$14{,}526.80$, $i = 0.1$, $n = 6$

Solve each equation for the variable indicated. [2.3]

26. $I = PRT$; for R

27. $M = P(1 + RT)$; for T

28. $R = \frac{D}{1 - DT}$; for T

Write the following ratios and simplify. [2.4]

29. $17 to 50 cents

30. 9 days to 12 hours

31. $5000 to $250

32. 3 years to 15 months

33. $2 to 75 cents

Solve the following proportions. [2.4]

34. $\frac{v}{14} = \frac{27}{126}$

35. $\frac{5}{y} = \frac{20}{27}$

36. $\frac{3}{8} = \frac{z}{12}$

37. $\frac{6}{11} = \frac{90}{t}$

38. $\frac{20}{r} = \frac{60}{72}$

Solve the following application problems. [2.4]

39. Bass in a lake are sampled for a particular parasite; 14 of 60 bass have the parasite. Given that there are an estimated 18,400 bass in the lake, find the number with parasites. Round to the nearest whole number.

40. A college student majoring in petroleum engineering worked on an oil-drilling site. She noticed a down-hole pressure of 3220 pounds per square inch at 6700 feet below the surface. Estimate the pressure at the 9850-foot total depth of the well to the nearest pound per square inch. Assume that the ratio of pressure to depth does not change with different depths.

41. Gas costs $1.40 per gallon. If Joe fills his tank with 12.5 gallons, how much does the gas cost?

42. John proofreads seven pages in 12 minutes. How many pages does he proofread in 3 hours?

43. A company spends three times as much on training as it does on company cars. If $19,000 is spent on cars, how much is spent on training?

44. If eight shirts cost $223.20, how much would five shirts cost?

45. If four videocassettes cost $75, how many did Kim buy if she received $31.25 in change from her $500 check?

Chapter 2 Summary Exercise
Breakeven in Retail

The average selling price of a book at the Book Barn is $10.60. Typically, 70 percent of this amount goes to pay for the cost of the book, which includes shipping and handling. Monthly expenses at the Book Barn are:

Salaries (including owner's salary)	$2800
Rent	$2200
Utilities	$ 250
Supplies (other than books)	$ 200
Other	$ 650

These definitions may help you:
Gross revenue is the total of all revenue from all sales.
Breakeven is the point at which total revenue equals cost of goods sold plus total expenses.
Profit, or *net profit*, is the amount left over after all expenses have been paid.
Percents are parts out of 100, so that 70 percent is 70 parts out of 100 parts.

(a) Find the total monthly expenses.

(b) Write an equation for monthly net profit. Net profit is gross revenue from the sale of books (use N for the number of books sold in a month) less monthly expenses.

(c) How many books must they sell to break even (round up to next whole number)?

(d) What happens if they don't break even one month?

(e) How many books must they sell to reach a profit of $2500 in one month?

Net Assets Business on the Internet

General Motors

Statistics

- Founded 1908
- Operations in more than 50 countries
- 1999: 594,000 employees
- 1999: Produced 16.7 million cars and light trucks

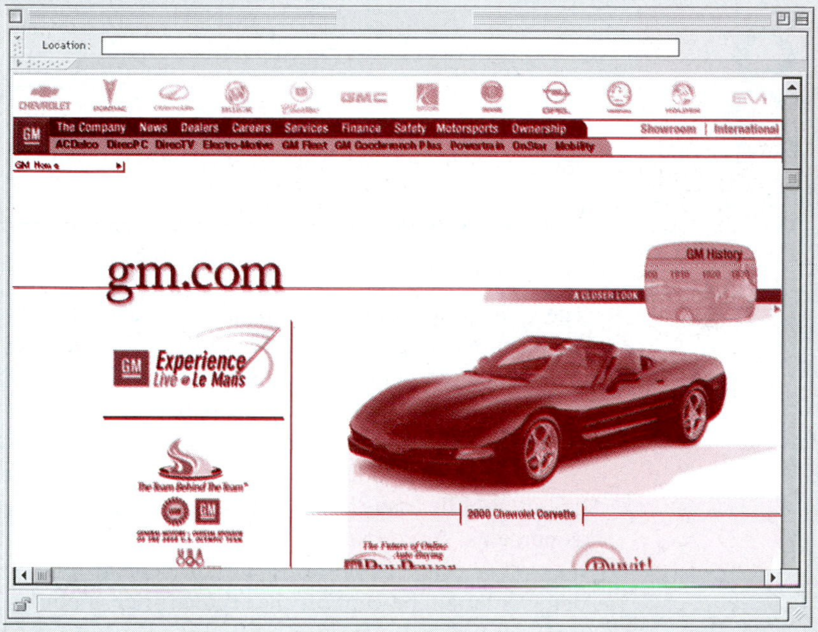

General Motors has been a primary producer of cars and trucks since it was founded in 1908. Their innovations include the first electric headlamp (1908), electric self-starter (1911), and the first all-steel body (1912). More recently, they have developed a nearly pollutant-free, gas-electric-powered vehicle (1990) and introduced the world's first pregnant crash test dummy (1993). General Motors produces vehicles under the familiar names of Chevrolet, Oldsmobile, Pontiac, Saturn, Buick, Cadillac, and GMC. They even produce an electric car called EVI that has no engine, no tailpipe, and no emissions.

A subsidiary of General Motors, General Motors Acceptance Corporation (GMAC) is one of the world's largest financial service companies with nearly \$100 billion in assets. GMAC provides a broad range of financial services, including consumer vehicle financing, full-service leasing and fleet leasing, dealer financing, car and truck extended-service contracts, residential and commercial mortgage services, and vehicle and homeowner's insurance. GMAC's business spans 33 markets worldwide.

1. The marketing department has found that the number of a particular model of Buick sold in one region of the country is related to dollars spent in advertising as follows: Cars sold $= 0.012 \times$ advertising dollars $+ 400$. Estimate the number of cars sold if advertising is \$78,000 during one quarter.

2. Estimate the total quarterly sales of General Motors during a quarter with sales of 2.16 million vehicles assuming that the average sales price per vehicle is \$19,400.

3. Managers predicted profits of 2.7% of gross revenue during one quarter. Write an algebraic equation for this relationship using P for profit and G for gross revenue.

4. Use the formula from question 3 to estimate profits for a quarter with sales found in question 2.

3 Percent

The word **percent** means "parts per one hundred." Percent is commonly used to compare different quantities or amounts. In Figure 3.1, the sales of fast-food hamburger chains are shown as a percent of total U.S. fast-food sales. This information is extremely important to the fast-food chains and their suppliers.

What types of strategies does a fast-food chain use to increase its market share?

Percents are widely used in business and everyday life. For example, interest rates on automobile loans, home loans, and other installment loans are almost always given as percents. Advertisers often claim that their products perform a certain percent better than other products or cost a certain percent less. Stores often advertise sale items as being a certain percent off the regular price. In business, marketing costs, damage, and theft may be expressed as a percent of sales; profit as a percent of investment; and labor as a percent of production costs. Current government figures for inflation, growth, and unemployment are also reported as percents. This chapter discusses the various types of percent problems that will be used throughout this text and in business situations.

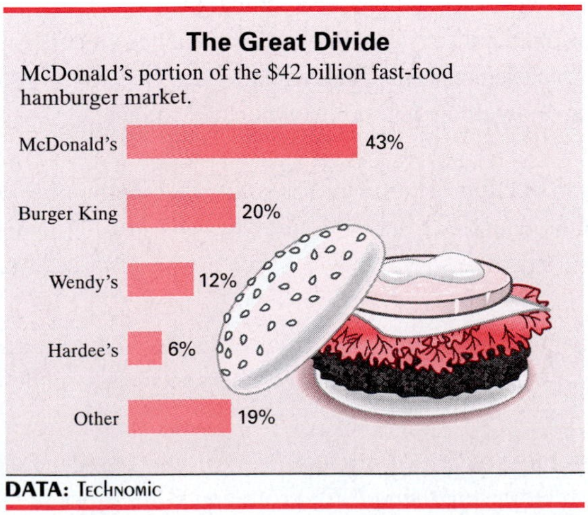

Figure 3.1

3.1 Writing Decimals and Fractions as Percents

Objectives

1. *Write a decimal as a percent.*
2. *Write a fraction as a percent.*
3. *Write a percent as a decimal.*
4. *Write a percent as a fraction.*
5. *Write a fractional percent as a decimal.*

Percents represent parts of a whole, just as fractions or decimals do. **Percents** are **hundredths**, or parts of a hundred. "One percent" means one of one hundred parts. Percents are written with a percent sign, %. For example, 25% refers to 25 parts out of 100 equal parts $\left(\frac{25}{100}\right)$, just as 50% refers to 50 out of 100 equal parts $\left(\frac{50}{100}\right)$, and 100% refers to all 100 of the 100 equal parts $\left(\frac{100}{100}\right)$. Therefore, 100% is equal to the whole item. If a percent is larger than 100% (for example, 150%), more than one item has been divided into 100 equal parts, and 150 of the parts are being considered $\left(\frac{150}{100}\right)$.

Objective 1 *Write a Decimal as a Percent.* Write a decimal as a percent by moving the decimal point two places to the right and attaching a % sign.

For example, write 0.75 as a percent by moving the decimal point two places to the right and attaching a % sign, giving 75% as the result.

Decimal	*Percent*
0.75 (0.75.) ← 2 places to the right	75% ← Attach a percent sign.

Example 1 Changing Decimals to Percents

When completing some market research you must express your findings as percents. Write the following decimals as percents.

(a) 0.25 **(b)** 0.38 **(c)** 0.65

Solution Move the decimal point two places to the right and attach a percent sign.

(a) 25% **(b)** 38% **(c)** 65%

If there is nothing in the hundredths position, place zeros to the right of the number to hold the hundredths position. For example, the decimal 0.5 is expressed as 50%, and the number 1.2 is 120%.

0.5 = 0.50. = 50% (Attach a zero.) 1.2 = 1.20. = 120% (Attach a zero.)

NOTE Another method of changing a decimal to a percent is to multiply the decimal by 100%. For example,

$$0.25 \times 100\% = 25\% \qquad 0.5 \times 100\% = 50\% \qquad 1.2 \times 100\% = 120\%$$

Example 2 Writing Decimals as Percents

Write the following decimals as percents.

(a) 0.7 (b) 1.3 (c) 0.1 (d) 3

Solution It is necessary to attach zeros to these decimals.

(a) 70% (b) 130% (c) 10% (d) 300%

If the decimal extends farther than the hundredths position, the resulting percent includes decimal parts of whole percents.

Example 3 Writing Decimals as Percents

When reading a newsletter, a hair salon owner sees the following decimals. Write these decimals as percents.

(a) 0.857 (b) 0.0057 (c) 0.0025

Solution

(a) 85.7% (b) 0.57% (c) 0.25%

NOTE In Example 3 both (b) and (c) are less than 1%; they are decimal parts of 1%.

Objective 2 *Write a Fraction as a Percent.* There are two ways to write a fraction as a percent. One way is to first convert the fraction to a decimal, as explained in Section 1.2. For example, to express the fraction $\frac{2}{5}$ as a pecent, write $\frac{2}{5}$ as a decimal by dividing 2 by 5. Then write the decimal as a percent.

Fraction	*Decimal*	*Percent*
$\frac{2}{5}$	0.4	40%

Example 4 Writing Fractions as Percents

An advertising account representative is given the following data in fraction form and must change the data to percent.

(a) $\frac{1}{4}$ (b) $\frac{3}{5}$ (c) $\frac{7}{8}$

Solution First write each fraction as a decimal. Then write the decimal as a percent.

(a) $\frac{1}{4} = 0.25 = 25\%$ (b) $\frac{3}{5} = 0.6 = 60\%$ (c) $\frac{7}{8} = 0.875 = 87.5\%$

A second way to write a fraction as a percent is to multiply the fraction by 100%. For example, write the fraction $\frac{2}{5}$ as a percent by multiplying $\frac{2}{5}$ by 100%.

$$\frac{2}{5} = \frac{2}{5} \times 100\% = \frac{200\%}{5} = 40\%$$

Example 5 Writing Fractions as Percents

Write the following fractions as percents.

(a) $\frac{3}{4}$ **(b)** $\frac{1}{3}$ **(c)** $\frac{5}{8}$

Solution Write these fractions as percents by multiplying each by 100%.

(a) $\frac{3}{4} \times 100\% = 75\%$ **(b)** $\frac{1}{3} \times 100\% = 33.3\%$ (rounded) **(c)** $\frac{5}{8} \times 100\% = 62.5\%$

NOTE When the fraction $\frac{1}{3}$ is written as a percent it is often expressed as $33\frac{1}{3}\%$. It is expressed this way because $\frac{1}{3} \times 100\% = 33\frac{1}{3}\%$. Also, $\frac{2}{3}$ is expressed as $66\frac{2}{3}\%$ ($\frac{2}{3} \times 100\% = 66\frac{2}{3}\%$).

Objective 3 ***Write a Percent as a Decimal.*** Write a percent as a decimal by moving the decimal point two places to the left and dropping the percent sign. For example, 50% becomes 0.50 or 0.5, 100% becomes 1, and 352% becomes 3.52.

Example 6 Writing Percents as Decimals

To calculate some insurance claims an insurance agent must change the following percents to decimals.

(a) 25% **(b)** 142% **(c)** $37\frac{1}{2}\%$ (*Hint:* $37\frac{1}{2}\% = 37.5\%$)

Solution Move the decimal point two places to the left and drop the percent sign.

(a) 0.25 **(b)** 1.42 **(c)** 0.375

NOTE Another method of changing a percent to a decimal number is to divide by 100%. For example,

$$25\% \div 100\% = 0.25 \qquad 142\% \div 100\% = 1.42 \qquad 37.5\% \div 100\% = 0.375.$$

Objective 4 ***Write a Percent as a Fraction.*** Write a percent as a fraction by first changing the percent to a decimal.

Example 7 Writing Percents as Fractions

The data in the following bar graph is from the National Association of Realtors and shows the technology tools used by star salespeople and the percent who use these tools. Write the percents seen in the bar graph as fractions.

(a) 72% **(b)** 63% **(c)** 36%

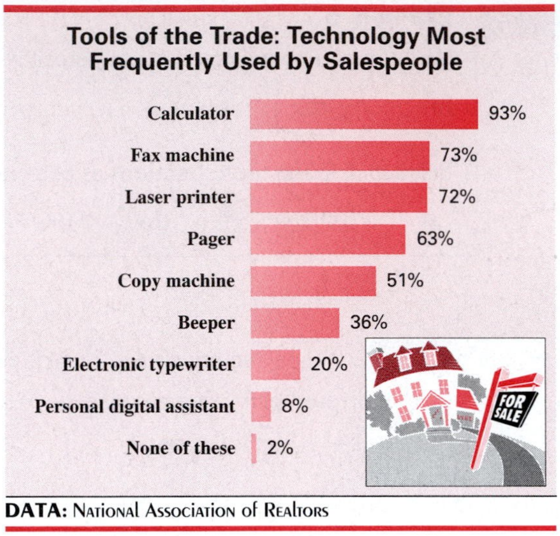

Solution First write each percent as a decimal and then write the decimal as a fraction in lowest terms.

(a) $72\% = 0.72 = \frac{72}{100} = \frac{18}{25}$ **(b)** $63\% = 0.63 = \frac{63}{100}$ **(c)** $36\% = 0.36 = \frac{36}{100} = \frac{9}{25}$

Objective 5 *Write a Fractional Percent as a Decimal.* A fractional percent such as $\frac{1}{2}\%$ has a value less than 1%. In fact, $\frac{1}{2}\%$ is equal to $\frac{1}{2}$ of 1%. Write a fractional percent as a decimal by first changing the fraction to a decimal, leaving the percent sign. For example, first write $\frac{1}{2}\%$ as 0.5%. Then write 0.5% as a decimal by moving the decimal point two places to the left and dropping the percent sign.

$$\frac{1}{2}\% = 0.5\% = 0.005$$

Written as decimal

Example 8 Writing Fractional Percents as Decimals

Write each of the following fractional percents as decimals.

(a) $\frac{1}{5}\%$ **(b)** $\frac{3}{4}\%$ **(c)** $1\frac{1}{8}\%$

Solution Begin by writing the fraction as a decimal.

(a) $\frac{1}{5}\% = 0.2\% = 0.002$ **(b)** $\frac{3}{4}\% = 0.75\% = 0.0075$

(c) $1\frac{1}{8}\% = 1.125\% = 0.01125$

The following box shows many fractions and their percent equivalents. It is often helpful to memorize those that are most commonly used.

Common Fractions and Their Equivalent Percents

$\frac{1}{100} = 1\%$	$\frac{1}{9} = 11\frac{1}{9}\%$	$\frac{1}{3} = 33\frac{1}{3}\%$	$\frac{4}{5} = 80\%$
$\frac{1}{50} = 2\%$	$\frac{1}{8} = 12\frac{1}{2}\%$	$\frac{3}{8} = 37\frac{1}{2}\%$	$\frac{5}{6} = 83\frac{1}{3}\%$
$\frac{1}{25} = 4\%$	$\frac{1}{7} = 14\frac{2}{7}\%$	$\frac{2}{5} = 40\%$	$\frac{7}{8} = 87\frac{1}{2}\%$
$\frac{1}{20} = 5\%$	$\frac{1}{6} = 16\frac{2}{3}\%$	$\frac{1}{2} = 50\%$	$1 = 100\%$
$\frac{1}{16} = 6\frac{1}{4}\%$	$\frac{3}{16} = 18\frac{3}{4}\%$	$\frac{3}{5} = 60\%$	$1\frac{1}{4} = 125\%$
$\frac{1}{12} = 8\frac{1}{3}\%$	$\frac{1}{5} = 20\%$	$\frac{5}{8} = 62\frac{1}{2}\%$	$1\frac{1}{2} = 150\%$
$\frac{1}{10} = 10\%$	$\frac{1}{4} = 25\%$	$\frac{2}{3} = 66\frac{2}{3}\%$	$1\frac{3}{4} = 175\%$
		$\frac{3}{4} = 75\%$	$2 = 200\%$

3.1 Exercises

Write the following decimals as percents.

1. 0.2 **2.** 0.5 **3.** 0.72 **4.** 0.86

5. 1.4 **6.** 3.017 **7.** 0.375 **8.** 0.875

9. 4.625 **10.** 7.8 **11.** 0.0025 **12.** 0.0008

13. 0.0015 **14.** 0.221 **15.** 3.45 **16.** 5.5

Write the following fractions as percents.

17. $\frac{1}{4}$ **18.** $\frac{5}{8}$ **19.** $\frac{1}{10}$ **20.** $\frac{1}{20}$

21. $\frac{1}{50}$ **22.** $\frac{1}{125}$ **23.** $\frac{3}{8}$ **24.** $\frac{4}{5}$

25. $\frac{1}{8}$ **26.** $\frac{13}{20}$ **27.** $\frac{1}{200}$ **28.** $\frac{1}{400}$

29. $\frac{7}{8}$ **30.** $\frac{1}{100}$ **31.** $\frac{3}{50}$ **32.** $\frac{4}{25}$

Write the following percents as decimals.

33. 65% **34.** 32% **35.** 75% **36.** 58%

37. 0.6% **38.** 0.5% **39.** 0.25% **40.** 0.125%

41. 315% **42.** 150% **43.** 200.6% **44.** 475.6%

45. 540.6% **46.** 135.6% **47.** 0.07% **48.** 0.05%

49. Fractions, decimals, and percents are all used to describe a part of something. The use of percent is much more common than fractions and decimals. Why do you suppose this is true?

50. List five uses of percent that are or will be part of your life. Consider the activities of working, shopping, saving, and planning for the future.

51. To change a fraction to a percent you must first change the fraction to a decimal. Why is this? (See Objective 2.)

52. The fractional percent $\frac{1}{2}\%$ is equal to 0.005. Explain each step as you change $\frac{1}{2}\%$ to its decimal equivalent. (See Objective 4.)

Determine the fraction, decimal, or percent equivalents for each of the following, as necessary. Write fractions in lowest terms.

	Fraction	Decimal	Percent
53.	$\frac{1}{2}$	______	______
54.	$\frac{3}{25}$	______	______
55.	______	______	15%
56.	______	______	87.5%
57.	______	0.25	______
58.	______	0.35	______
59.	$6\frac{1}{8}$	______	______
60.	$3\frac{1}{2}$	______	______
61.	______	7.25	______
62.	$1\frac{3}{4}$	______	______
63.	______	0.0025	______
64.	______	0.00125	______
65.	$\frac{1}{3}$	______	______
66.	______	______	$4\frac{1}{4}\%$
67.	______	______	$\frac{3}{4}\%$

68. Section 4 of the foldout shows that 40 percent of workers are not prepared financially for a potential layoff. In addition it shows how long savings would last. Change all of these percents to decimals and then to their equivalent fractions.

	Fraction	Decimal	Percent
69.	______	______	12.5%
70.	______	0.025	______
71.	______	2.5	______
72.	______	______	375%
73.	______	______	1038.35%
74.	______	23.82	______
75.	$4\frac{3}{8}$	______	______
76.	______	______	$37\frac{1}{2}\%$
77.	______	______	$6\frac{3}{4}\%$

3.2 Finding the Part

Objectives

1 *Identify the three elements of a percent problem.*
2 *Use the percent formula.*
3 *Apply the percent formula to a business problem.*
4 *Recognize the terms associated with base, rate, and part.*
5 *Use the basic percent equation.*

Objective 1 *Identify the Three Elements of a Percent Problem.* Problems involving percent have three main quantities. Usually two of these quantities are given and the third must be found. The three key quantities in a percent problem are as follows.

Base. The whole or total, the starting point, or that amount to which something is being compared.

Rate. A number followed by "%" or "percent."

Part. The result of multiplying the base and the rate. Part is always a portion of the base.

NOTE Percent and part are different quantities. The stated percent in a given problem is always the rate. The part is the product of the base and the rate. It is a portion of the base, as sales tax is a portion of the total sales and as the number of sports utility vehicles is a portion of the total number of motor vehicles. This makes part a quantity and it never appears with "percent" or "%" following it. ■

Objective 2 *Use the Percent Formula.* The above three quantities are related by the basic **percent formula**, which is expressed as

$$\text{Part} = \text{Base} \times \text{Rate} \quad or \quad P = B \times R \quad or \quad P = BR$$

A Century 21 real estate agent finds a \$120,000 home for his clients and will earn a 6% commission. Find 6% of \$120,000 by using $P = BR$, with $B = \$120{,}000$ and $R = 6\%$ (the rate). The rate must be changed to a decimal before multiplying.

$$P = BR$$
$$P = \$120{,}000 \times 6\%$$
$$P = \$120{,}000 \times 0.06$$
$$P = \$7200$$

Therefore, 6% of \$120,000 = \$7200.

Example 1 Solving for Part

Solve for part (P) using $P = B \times R$.

(a) 4% of 50 **(b)** 1.2% of 180

(c) 140% of 225 **(d)** $\frac{1}{4}\%$ of 560 (*Hint:* $\frac{1}{4}\% = 0.25\%$)

Solution

(a) $B \times R = P$
$50 \times 0.04 = 2$

(b) $B \times R = P$
$180 \times 0.012 = 2.16$

(c) $B \times R = P$
$225 \times 1.4 = 315$

(d) $B \times R = P$
$560 \times 0.0025 = 1.4$

Objective 3 *Apply the Percent Formula to a Business Problem.* Calculating **sales tax** is an excellent example of finding part. States, counties, and cities often collect taxes on sales to the consumer. The sales tax is a percent of the sale. This percent varies from as low as 3% in some states to 8% or more in other states. The percent formula is used for finding sales tax.

$$P = B \times R$$
$$\text{Sales tax} = \text{sales} \times \text{sales tax rate}$$

Example 2 Calculating Sales Tax

Racy Feed and Pet Supply sold $284.50 worth of merchandise. If the sales tax was 5%, what was the sales tax and the total sale, including the tax?

Solution The amount of sales, $284.50, is the starting point or base, and 5% is the rate. Since the tax is a *part* of total sales, use the formula $P = BR$ to find the part.

$$P = BR$$
$$P = \$284.50 \times 5\%$$
$$P = \$284.50 \times 0.05$$
$$P = 14.225 = \$14.23$$

The tax, or part, was $14.23.

To find the total amount of sales and tax, add the amount of sales, $284.50, to the sales tax, $14.23. The total is $298.73 ($284.50 + $14.23).

NOTE An alternative approach to finding the sales and tax would be to multiply $284.50 by 105% (100% sales + 5% sales tax) to get

$$\$284.50 \times 105\% = 284.5 \times 1.05 = \$298.73.$$

Example 3 Finding Part

The following bar graph shows that 27% of the office workers would like more storage space. If there are 14 million office workers, how many want more storage space?

Solution The number of office workers, 14 million, is the base. The rate, 27%, is the portion of the total number of office workers who want more storage space. Since the number of office workers who want more storage space is part of the total number of office workers, find the number of office workers who want more storage space by using the formula to find part.

$$P = B \times R$$
$$P = 14 \text{ million} \times 27\%$$
$$P = 14 \text{ million} \times 0.27$$
$$P = 3.78 \text{ million}$$

The number of office workers who want more storage space is 3.78 million, or 3,780,000.

Scientific *Calculator Approach*

The calculator solution to Example 3 is

14 [×] 27 [%] [=] 3.78 or 14 [×] .27 [=] 3.78

Objective 4 ***Recognize the Terms Associated with Base, Rate, and Part.*** Percent problems have certain similarities. For example, some phrases are associated with the base in the problem. Other phrases lead to the part, while "%" or "percent" following a number identifies the rate. Table 3.1 helps distinguish between the base and the part.

Most percent problems can be written in the following form.

_______% of _______ is _______

Table 3.1 DISTINGUISHING THE BASE FROM THE PART

Usually the Base	Usually the Part
Sales	Sales tax
Investment	Return
Savings	Interest
Value of bonds	Interest
Retail price	Amount of discount
Last year's anything	Increase or decrease
Value of real estate	Rents
Old salary	Raise
Total sales	Commission
Value of stocks	Dividends
Earnings	Expenditures
Original	Change

Objective 5 *Use the Basic Percent Equation.* This is known as the **basic percent equation**, several examples of which follow.

$$R \times B = P$$

5% of the automobiles are red.

4.2% of the workers are unemployed.

28% of the income is income tax.

75% of the students are full-time.

When expressed in this *standard form*, the elements in the percent problem appear in the following order.

R	$\times$	B		P
Rate	$\times$	Base	=	Part
↓	↓	↓	↓	↓
_______%	of	_______	is	_______

NOTE Rate is identified by "%" (the percent sign); the word "of" means "×" (multiplication); the *multiplicand*, or number being multiplied, is the base; the word "is" means "=" (equals); and the product, or answer, is a part of the base. ■

Example 4 Identifying the Elements in Percent Problems

Identify the elements given in the following percent problems and determine which element must be found.

(a) During a recent sale, Stockdale Marine offered a 15% discount on all new recreation equipment. Find the discount on a jet ski originally priced at $4895.

Solution First arrange this problem using the basic percent equation.

$$R \times B = P$$

_______% of _______ is _______

% of price is discount

15% of \$4895 = discount

$$R \times B = P$$

$$0.15 \times \$4895 = P$$

$$0.15 \times \$4895 = \$734.25 \text{ discount}$$

At this point, check that rate is given, base is given, and part must be found. To find the discount, multiply 0.15 by \$4895.

The discount is \$734.25.

(b) Round Table Pizza spends an amount equal to 5.8% of its sales on advertising. If sales for the month were \$12,500, find the amount spent on advertising.

Solution Use the basic percent equation.

$$R \times B = P$$

_______% of _______ is _______

% of sales is advertising

5.8% of \$12,500 = advertising

$$R \times B = P$$

Rate is given as 5.8%, base (sales) is \$12,500, and part (advertising) must be found. Find the amount spent on advertising by multiplying 0.058 and \$12,500.

$$0.058 \times \$12{,}500 = P$$

$$0.058 \times \$12{,}500 = \$725$$

The amount spent on advertising is \$725.

3.2 Exercises

Solve for part in each of the following.

1. 20% of 80 guests

2. 25% of 3500 salespeople

3. 22.5% of \$1086

4. 20.5% of \$1500

5. 4% of 120 feet

6. 125% of 2000 products

7. 175% of 5820 miles

8. 15% of 75 cases

9. 17.5% of 1040 homes

10. 52.5% of 1560 trucks

11. 118% of 125.8 yards

12. 500% of 142 units

13. $90\frac{1}{2}$% of \$5930

14. $7\frac{1}{2}$% of \$150

15. 0.5% of \$1300

16. 0.75% of 180,000 calls

17. Identify the three quantities in a percent problem and tell how you can identify each of these three quantities. (See Objective 1.)

18. There are words and phrases that are usually associated with base and part. Give three examples of words that usually identify base and the accompanying word for the part. (See Objective 4.)

Solve for part in each of the following application problems. Round to the nearest cent unless otherwise indicated.

19. PART-TIME EMPLOYMENT Aimee Toit works part-time, earns \$240 per week, and has 22% of this amount withheld for social security, Medicare, and taxes. Find the amount withheld.

20. GRADUATION REQUIREMENTS Kirsten Speed needs 124 credits to graduate. If Speed has already completed 75% of the credits, find the number of credits completed.

21. CHILD SUPPORT COLLECTION A collection agency specializing in collecting past-due child support charges \$25 as an application fee plus 20% of the amount collected. What is the total charge for collecting \$3100 of past-due child support?

22. WOMEN IN THE NAVY The Navy's guided-missile destroyer *USS Sullivans* has a 335-person crew of which 13% are female. Find the number of female crew members. Round to the nearest whole number.

23. SUPERMARKET SHOPPING The Point of Purchase Advertising Institute says that 55% of all supermarket shoppers have a written list of their needs. If there are 3680 shoppers per day entering the supermarket that you manage, what number of shoppers would you expect to have a written shopping list?

24. BAR SOAP A bar of Ivory Soap is $99\frac{44}{100}\%$ pure. If the bar of soap weighs 9 ounces, how many ounces are pure? (Round to the nearest hundredth.)

25. CANNED-MEAT SALES According to Hormel Foods Corporation, SPAM®, and SPAM® Lite together held 62.2% of the \$148 million canned lunchmeat category over a 52-week period (the entire year). Find the total annual sales of these Hormel products. Round to the nearest hundredth of a million.

26. IMPURITIES IN TUNA A U.S. Food and Drug Administration (FDA) biologist found that canned tuna is "relatively clean." Extraneous matter was found in 5% of the 1600 cans of tuna tested. How many cans of tuna contained extraneous matter?

27. CUBAN LABOR FORCE The size of Cuba's labor force is 3.8 million people. If 61% of the labor force is male, find (a) the percent who are female and (b) the number of workers who are male.

28. U.S. LABOR FORCE In the United States there are 132 million people in the labor force. If 54% of the labor force is male, find (a) the percent of the labor force who are female and (b) the number of workers who are male.

29. GLOBAL BUSINESS GROWTH A recent survey of 377 top executives found that 62% of them were upbeat about global business growth for the next three years. Find the number of executives that were upbeat about global business growth. Round to the nearest whole number.

30. BUSINESS GROWTH IN ASIA A survey of the same 377 executives in Exercise 29 found that 32% ranked China as the country with the best business opportunities in Asia over the next 3 years. How many executives are most favorable to business opportunities in China? Round to the nearest whole number.

31. NEW PRODUCT FAILURE Marketing Intelligence Service says that there were 15,401 new products introduced last year. If 86% of the products introduced last year failed to reach their business objectives, find the number of products that did reach their objectives. (Round to the nearest whole number.)

32. FAMILY BUDGET A family of four with a monthly income of \$2900 spends 90% of its earnings and saves the balance. Find (a) the monthly savings and (b) the annual savings of this family.

33. ORANGE JUICE IN CHINA This year the sales of Tropicana orange juice in China is a minuscule \$100 million. Seagram's Tropicana Beverage Group estimates that sales will increase by 35% next year. Find the amount of orange juice sales estimated for next year.

34. SUPER BOWL ADVERTISING The average cost of 1 minute of advertising during the Super Bowl last year was \$2.4 million. If the increase in cost this year is 8.3%, find the average cost of 1 minute of advertising during the Super Bowl this year. Round to the nearest tenth of a million.

35. SALES TAX COMPUTATION As the owner of a copy and print shop, you must collect $6\frac{1}{2}\%$ of the amount of each sale for sales tax. If sales for the month are \$48,680, what is the combined amount of sales and tax?

36. TOTAL COST A NuVac 3200 is priced at \$524 with an allowed trade-in of \$125 for an old unit. If sales tax of $7\frac{3}{4}\%$ is charged on the price of the NuVac unit, what is the total cost to the customer after receiving the trade-in? (*Hint:* Trade-in is subtracted last.)

37. AUTO-PARTS SALES J&K Mustang has increased its sales of auto parts by $32\frac{1}{2}\%$ over the last year. If the sale of parts last year amounted to \$385,200, what is the volume of parts sold this year?

38. FAMILY VACATIONS On average, a family of two adults and two children will pay \$95 per night for lodging and \$104 a day for food while on vacation. If this increases by 3% over the next year, find the total cost per day for lodging and meals for this family after the increase.

39. REAL ESTATE COMMISSIONS Thomas Dugally of Century 21 Realty sold a home for \$174,900. The commission was 6% of the sale price; however, Dugally receives only 60% of the commission while 40% remains with his broker. Find the amount of commission received by Dugally.

40. BUSINESS OWNERSHIP Rick Wilson has an 82% ownership in a company called Puppets and Clowns. The company has a value of \$49,200 and Wilson receives an income of 30% of the value of his ownership. Find the amount of his income.

41. Pamela and Peter Prentiss are saving for the future. They decide to save the suggested percent of earnings discussed in Section 2 of the foldout. If Pamela earns \$3150 per month and Peter earns \$28,400 per year, how much will they save in one year using the suggested percent?

3.3 Finding the Base

Objectives

1. *Use the basic percent equation to solve for base.*
2. *Find the amount of sales when taxes and rate of tax are known.*
3. *Find the amount of investment when expense and rate of expense are known.*
4. *Find base when rate and part are for different quantities.*

Objective 1 *Use the Basic Percent Equation to Solve for Base.* In some problems, the rate and part are given and the base must be found. For example, suppose that a couple, interested in purchasing a home, can make a monthly payment of \$770, which is 28% of their monthly income. To find their monthly income, use the rate (28%) and part (\$770) to find the base by using the basic percent equation, rate $\times$ base $=$ part. The key word here, indicating that their monthly income is the base, is "of."

$$R \times B = P$$

$$28\% \text{ of } \underline{\qquad\qquad} = \$770$$

$$0.28 \times B = 770$$

Now divide both sides by 0.28, as explained in Chapter 2.

$$\frac{\cancel{0.28}B}{\cancel{0.28}} = \frac{770}{0.28}$$

$$B = \frac{770}{0.28}$$

$$B = 2750$$

Their monthly income is \$2750.

Example 1 Solving for Base

Solve for base using the basic percent equation.

(a) 8 is 4% of ______ **(b)** 135 is 15% of ______

(c) 1.25 is 25% of ______

Solution

(a)

$$8 = 4\% \text{ of } \underline{\qquad\qquad}$$

$$8 = 4\% \times B$$

$$8 = 0.04B$$

$$\frac{8}{0.04} = \frac{\cancel{0.04}B}{\cancel{0.04}}$$

$$B = \frac{8}{0.04} = 200$$

(b)

$$135 = 15\% \text{ of } \underline{\qquad\qquad}$$

$$135 = 15\% \times B$$

$$135 = 0.15B$$

$$\frac{135}{0.15} = \frac{\cancel{0.15}B}{\cancel{0.15}}$$

$$B = \frac{135}{0.15} = 900$$

(c)

$$1.25 = 25\% \text{ of } \underline{\qquad\qquad}$$

$$1.25 = 25\% \times B$$

$$1.25 = 0.25B$$

$$\frac{1.25}{0.25} = \frac{\cancel{0.25}B}{\cancel{0.25}}$$

$$B = \frac{1.25}{0.25} = 5$$

Objective 2 *Find the Amount of Sales When Taxes and Rate of Tax Are Known.* A common business application of percent involves sales tax and the sales tax rate.

Example 2 Finding Sales When Sales Tax Is Given

The 5% sales tax collected by Famous Footwear was $780. What was the amount of total sales?

Solution Here the rate of tax collection is 5% and taxes collected are a part of total sales. The rate is 5% and the part is $780. Use the percent equation.

$$R \times B = P$$

$$5\% \text{ of } \underline{\hspace{2cm}} = \$780$$

$$0.05B = \$780$$

$$\frac{\cancel{0.05}B}{\cancel{0.05}} = \frac{\$780}{0.05}$$

$$B = \frac{\$780}{0.05} = \$15{,}600$$

The total sales of the company were $15,600.

Scientific *Calculator Approach*

In the calculator solution to Example 2, the percent key may be used when dividing.

780 [÷] 5 [%] [=] 15,600

Problem-Solving Hint

Consider whether your answer is reasonable. A common error in a base problem is to confuse the base and part. For example, if the taxes, $780, had been mistakenly used as the base, the resulting answer would have been $39 ($780 × 5%). Obviously, $39 is not a reasonable amount for total sales given $780 as sales tax.

> MOST PEOPLE INSIST they won't blow that refund check.
>
> In a survey to be released soon, most taxpayers who expect refunds this year said they plan to use the money to invest, save or pay off debts.
>
> Refunds approved by the IRS as of late March totaled $78.5 billion, up about 18% from a year earlier. The average refund rose 15% to $1589.

Source: Wall Street Journal, reprinted by permission of Dow Jones, Inc. via Copyright Clearance Center, Inc.

The newspaper clipping above reports that the number of income tax refunds and the average amount of a refund have both risen since last year and that this should help to strengthen the economy. In these calculations, the tax refund numbers from last year are the base, or starting point—the amounts to which this year is being compared.

Objective 3 *Find the Amount of Investment When Expense and Rate of Expense are Known.* The amount of an investment is the base. When the amount of expenses and the rate of expenses are known, the percent equation may be used to find the amount of the investment.

Example 3 Finding the Amount of an Investment

The yearly maintenance cost of an apartment is $3\frac{1}{2}\%$ of its value. If maintenance amounts to \$73,500 per year, find the value of the apartment complex.

Solution To find the total value of the complex, which is the base, use the percent equation.

$$R \times B = P$$

$$3\frac{1}{2}\% \text{ of } \underline{\qquad\qquad} = \$73{,}500$$

$$0.035B = \$73{,}500$$

$$\frac{0.035B}{0.035} = \frac{\$73{,}500}{0.035}$$

$$B = \frac{\$73{,}500}{0.035} = \$2{,}100{,}000$$

The total value of the complex is \$2,100,000.

NOTE When working with a fraction of a percent, it is best to change the fraction to a decimal. In Example 3, $3\frac{1}{2}\%$ was changed to 3.5%, which equals 0.035.

Objective 4 *Find the Base When Rate and Part Are for Different Quantities.* The rate used and the part given in a problem do not always refer to the same quantity. Always pay careful attention to reading and understanding a problem.

Example 4 Finding Base When Rate and Part Are for Different Quantities

United Hospital finds that 25% of its employees are men and 720 are women. Find the total number of employees.

Solution The rate given, 25%, refers to male employees, while 720, the part, is the number of women. A rate of 75%, which is the percent of women hospital employees (100% of all employees − 25% male employees) must be used to solve for the total number of employees, which is the base. Use the percent equation.

$$R \times B = P$$

$$75\% \text{ of } \underline{\qquad\qquad} = 720$$

$$0.75B = 720$$

$$\frac{0.75B}{0.75} = \frac{720}{0.75}$$

$$B = \frac{720}{0.75} = 960$$

The total number of employees is 960.

3.3 Exercises

Solve for base in each of the following. Round to the nearest hundredth.

1. 265 bowlers is 25% of _______ bowlers.
2. 240 letters is 80% of _______ letters.
3. 75 miles is 40% of _______ miles
4. 32 shipments is 8% of _______ shipments.
5. 55 packages is 5.5% of _______ packages.
6. \$850 is $4\frac{1}{4}$% of _______
7. 36 employees is 0.75% of _______ employees.
8. 23 workers is 0.5% of _______ workers.
9. 33 rolls is 0.15% of _______ rolls.
10. 54,600 boxes is 60% of _______ boxes.
11. 50 doors is 0.25% of _______ doors.
12. 39 bottles is 0.78% of _______ bottles.
13. \$33,870 is $37\frac{1}{2}$% of _______.
14. \$8,500 is $27\frac{1}{2}$% of _________.
15. 20% of _______ sacks is 350 sacks.
16. 16% of _______ is \$45.
17. 375 crates is 0.12% of _______ crates.
18. 3.5 quarts is 0.07% of _______ quarts.
19. 0.5% of _______ homes is 327 homes.
20. 6.5 barrels is 0.05% of _______ barrels.
21. 12 audits is 0.03% of _______ audits.
22. 8 banks is 0.04% of _______ banks.
23. The basic percent formula is $P = B \times R$. Show how to find the formula to solve for B (base). (See Objective 1.)
24. A problem includes amount of sales, sales tax, and a sales tax rate. Explain how you could identify the base, rate, and part in this problem. (See Objective 2.)

Solve for base in each of the following application problems.

25. **EMPLOYEE DOWNSIZING** Black and Decker plans to cut 3000 jobs from its workforce. If this represents 10% of Black and Decker's global workforce, find the total size of its global workforce.
26. **EMPLOYEE POPULATION BASE** In a large metropolitan area 81% of the employed population is enrolled in a health maintenance organization (HMO). If 700,650 employees are enrolled, find the total number of people in the employed population.

27. **COLLEGE ENROLLMENT** This semester there are 1785 married students on campus. If this figure represents 23% of the total enrollment, what is the total enrollment? (Round to the nearest whole number.)

28. **VOTER REGISTRATION** Registered voters make up 13.8% of the county population. If there are 345,000 registered voters in the county, what is the total population in the county?

29. **LOAN QUALIFICATION** Jenna DeMarco found a home for Stephen and Heather Hall that will require a monthly loan payment of $840. If the lender insists that the buyer's monthly payment not exceed 30% of the buyer's monthly income, find the minimum monthly income required by the lender.

30. **PERSONAL BUDGETING** Byron Hopkins spends 22% of his income on housing, 24% on food, 8% on clothing, 15% on transportation, 11% on education, and 7% on recreation and saves the balance. If his savings amount to $154 per month, what are his monthly earnings?

31. **CHILD RUNAWAYS** The Federal Administration on Children, Youth and Families estimates that 14% of adolescents age 12 to 17 have run away from home at least once. If the number of adolescents who have run away from home is 2.8 million, find the total number of adolescents.

32. **LICENSE TESTING** In analyzing the success of real estate license applicants, the state finds that 58.3% of those examined received a passing mark. If the records show that 8370 new licenses were issued, what was the number of applicants? (Round to the nearest whole number.)

33. **INCOME TAX RETURNS** According to the Internal Revenue Service, 5.2 million individual income tax returns reported medical and dental reductions last year. If this represents 4% of all individual returns filed, find the total number of returns filed.

34. **TOOL PRICE INCREASES** The price of a drill press manufactured by a Malaysian tool company was raised by $250, or 1.93%. Find the price of the drill press after the price increase. (Round to the nearest dollar.)

35. **GAMBLING PAYBACK** An Atlantic City casino advertises that it gives a 97.4% payback on slot machines, and the balance is retained by the casino. If the amount retained by the casino in one day is $4823, find the total amount played on the slot machines.

36. **SMOKING OR NONSMOKING** A casino hotel in Barbados states that 45% of its rooms are for nonsmokers. If the resort allows smoking in 484 rooms, find the total number of rooms.

Supplementary Exercises: Base and Part

Solve for base or part as indicated in the following application problems.

1. **FOREIGN INVESTMENTS** Foreign investments in China that went into agriculture amounted to $202 million last year. If this was only 1% of all foreign investments in China, find the total foreign investment in China last year.

2. **MINORITY LENDING** Chemical Banking Corporation made $338 million worth of mortgage loans to minorities last year. If this represented 18.6% of all their mortgages, find the total value of all mortgages that they originated last year. (Round to the nearest tenth of a million.)

3. **PROPERTY INSURANCE** A building is valued at \$423,750 and is insured for 68% of its value. Find the amount of insurance coverage.
4. **EMPLOYEE HEALTH PLANS** In a recent survey of 1100 employers, it was found that 84% offer only one health plan to employees. How many of these employers offer only one health plan?
5. **CAMAROS AND MUSTANGS** The Chevy Camaro was introduced in 1967. Sales that year were 220,917, which was 46.2% of the number of Ford Mustangs sold in the same year. Find the number of Mustangs sold in 1967. (Round to the nearest whole number.)
6. **AUTO-LOAN INDUSTRY** The Money Store Inc. announced that it has auto loans valued at \$776 million. If this is 5.1% of the company's total loan portfolio, what is the total amount of loans? (Round to the nearest tenth of a million.)
7. **FROZEN YOGURT SALES** The total sales of frozen yogurt were \$594 million in the past 12 months. If 15.8% of the sales were private label brands, what is the amount of sales that were private label brands? (Round to the nearest tenth of a million.)
8. **CALORIES FROM FAT** Häagen-Dazs vanilla ice cream has 270 calories per serving. If 60% of these calories come from fat, find the number of calories coming from fat.
9. **RETIREMENT ACCOUNTS** Nancy Barre has 9.5% of her earnings deposited into a retirement account. This amounts to \$308.75 per month. Find her annual earnings.
10. **SAVINGS ACCOUNT INTEREST** The Northridge PTA received \$79.75 in annual interest on its bank account. If the bank paid $5\frac{1}{2}\%$ interest per year, how much money was in the account?
11. **DRIVER SAFETY SURVEY** A survey at an intersection found that of 2200 drivers, 38% were wearing seat belts. How many drivers in the survey were wearing seat belts?
12. **BLOOD CHOLESTEROL LEVELS** At a recent health fair 32% of the people tested were found to have high blood cholesterol levels. If 350 people were tested, find the number having high blood cholesterol.
13. **CAT FOOD SALES** Sales of Whiskas canned cat food dropped 3%, or \$1.9 million in the past year. Find the sales of Whiskas cat food after the decrease. Round to the nearest tenth of a million.
14. **CANNED CAT FOOD** In the past year there has been a 14% increase in the sales of Tender Morsels canned cat food. If this increase amounts to \$1.41 million, find the sales of Tender Morsels canned cat food after the increase. Round to the nearest tenth of a million.

3.4 Finding the Rate

Objectives

1. *Use the percent equation to solve for rate.*
2. *Find the rate of return when the amount of the return and the investment are known.*
3. *Solve for the percent remaining when the total amount and amount used are given.*
4. *Find the percent of change.*

In the third type of percent problem, the part and base are given and the rate must be found. The rate is identified by the "%" sign, or "percent." For example, what percent of 32 is 8? Use the percent equation as shown next.

Objective 1 | ***Use the Percent Equation to Solve for Rate.***

$$R \times B = P$$

$$\_\_\_\_\_\_\_\_\% \text{ of } 32 = 8$$

$$R \times 32 = 8$$

$$32R = 8$$

Now divide both sides by 32.

$$\frac{\cancel{32}R}{\cancel{32}} = \frac{8}{32}$$

$$R = \frac{8}{32} - 0.25 = 25\%$$

Finally, 8 is 25% of 32, or 25% of 32 is 8.

NOTE When solving for rate, you *must* change the resulting decimal answer to a percent. ▬

Example 1 Solving for Rate

Solve for rate.

(a) 63 is what percent of 180?

(b) What percent of 500 is 100?

(c) 54 is what percent of 12?

Solution

(a) 63 is _______% of 180

$$\_\_\_\_\_\_\_\_\% \text{ of } 180 = 63$$

$$180R = 63$$

$$\frac{\cancel{180}R}{\cancel{180}} = \frac{63}{180}$$

$$R = \frac{63}{180} = 0.35 = 35\%$$

(b) 100 is _______% of 500

$$\_\_\_\_\_\_\_\_\% \text{ of } 500 = 100$$

$$500R = 100$$

$$\frac{\cancel{500}R}{\cancel{500}} = \frac{100}{500}$$

$$R = \frac{100}{500} = 0.2 = 20\%$$

(c) 54 is _______% of 12

$$\underline{\qquad\qquad}\% \text{ of } 12 = 54$$

$$12R = 54$$

$$\frac{\cancel{12}R}{\cancel{12}} = \frac{54}{12}$$

$$R = \frac{54}{12} = 4.5 = 450\%$$

Objective 2 ***Find the Rate of Return When the Amount of the Return and the Investment Are Known.*** It is often necessary to find the rate of return when the amount of the return and investment are known.

Example 2 Finding the Rate of Return

The accounting office of J. Susan Hessney and Associates invested \$1710 in a new computer. As a result of having the equipment, the company had additional income of \$1440. Find the rate of return.

Solution The amount of investment, \$1710, is the base, and the return, \$1440, is the part. The return is a part of the total investment. Start with $R \times B = P$.

$$\underline{\qquad\qquad}\% \text{ of } \$1710 = \$1440$$

$$R \times \$1710 = \$1440$$

$$\$1710R = \$1440$$

$$\frac{\cancel{1710}R}{\cancel{1710}} = \frac{1440}{1710}$$

$$R = \frac{1440}{1710} = 0.8421 = 84.2\%$$

Rounded to the nearest tenth of a percent.

NOTE The rate in Example 2 had to be rounded. The rules for rounding percents are identical to the rules for rounding discussed in Chapter 1. Change the decimal answer to percent and round as indicated. Here, 84.21% rounds to the nearest tenth as 84.2%.

Objective 3 ***Solve for the Percent Remaining When the Total Amount and Amount Used Are Given.*** When the total amount of something and the amount used are known, it is common to solve for the percent remaining.

Example 3 Solving for the Percent Remaining

A roof is expected to last 12 years before it needs replacement. If the roof is 10 years old, what percent of the roof's life remains? Round to the nearest tenth of a percent.

Solution The total life of the roof, 12 years, is the base. Subtract the amount of life already used, 10 years, from the total life, 12 years, to find the number of years remaining.

12 yrs. (total life) − 10 yrs. (life used) = 2 yrs. (life remaining)

Use the equation $R \times B = P$ to find the solution.

$$\_\_\_\_\_\_\% \text{ of } 12 \text{ is } 2$$

$$R \times 12 = 2$$

$$12R = 2$$

$$\frac{\cancel{12}R}{\cancel{12}} = \frac{2}{12}$$

$$R = \frac{2}{12} = \frac{1}{6} = 0.166\ldots = 16.7\%$$

Rounded to the nearest tenth of a percent.

If the age of the roof (10 years) had been used as part, the resulting answer, 83.3% (rounded), would be the percent of life used. Find the percent of remaining life by subtracting 83.3% from 100%. The result, 16.7% (the same answer), would be the percent of life remaining.

NOTE Remember that the base is always 100%.

Objective 4 ***Find the Percent of Change.*** A common business problem is to find the percent of change in amounts involved in operating a business, such as sales and returns, and to determine the percent of gain or loss of an investment.

Example 4 Finding the Percent of Increase

Sales of digital cameras at Circuit City climbed from \$36,600 last month to \$113,460 this month. Find the percent of increase.

Solution The sales last month, \$36,600 is the base. Subtract the sales last month \$36,600, from the sales this month, \$113,460, to find the increase in sales volume.

$$\$113{,}460 - \$36{,}600 = \$76{,}800 \text{ increase in sales volume (part)}$$

Use the basic percent equation as follows.

$$\_\_\_\_\_\_\% \text{ of } \$36{,}600 = \$76{,}800$$

$$R \times \$36{,}600 = \$76{,}800$$

$$\$36{,}600R = \$76{,}800$$

$$\frac{\cancel{\$36{,}600}R}{\cancel{\$36{,}600}} = \frac{76{,}800}{36{,}600}$$

$$R = \frac{76{,}800}{36{,}600} = 2.1 = 210\%$$

Problem-Solving Hint Remember, to find the percent of increase, the first step is to determine the amount of increase. The base is *always* the original amount, such as last year's or last month's amount, and the amount of increase is the part.

Example 5 Finding the Percent of Decrease

The newspaper clipping that follows shows that the sales of single-family homes declined in the month of April. The decline in sales was measured from the home sales in March and was found to be 2.5%.

Home Sales

The National Association of Realtors said sales of single-family homes fell 2.5% in April to 4.77 million units on an annualized basis from, 4.89 million in March. Analysts blamed the dip on foul weather.

Source: Money magazine. Reprinted by permission.

Sales of existing single-family homes fell to a seasonally adjusted annual number of 4.77 million units this month from 4.89 million units last month. Find the percent of decrease.

Solution The base is always the previous period, in this example, last month, which is 4.89 million units. Subtract the number of units sold this month, 4.77 million units, from the number of units sold last month, 4.89 million units, to find the decrease in the number of units sold.

$$4.89 \text{ million} - 4.77 \text{ million} = 0.12 \text{ million decrease in sales (part)}$$

Use the basic percent equation.

$$\underline{\qquad\qquad}\% \text{ of } 4.89 = 0.12$$

$$R \times 4.89 = 0.12$$

$$4.89R = 0.12$$

$$\frac{\cancel{4.89}R}{\cancel{4.89}} = \frac{0.12}{4.89}$$

$$R = \frac{0.12}{4.89} = 0.0245 = 2.5\%$$ Rounded to the nearest tenth of a percent.

Scientific *Calculator Approach*

The calculator solution to this example is to subtract to find the difference and then divide.

[( ] 4.89 [−] 4.77 [ )] [÷] 4.89 [=] 0.0245 = 0.025 (rounded)

Problem-Solving Hint To find the percent of decrease, first determine the amount of decrease. The amount of decrease is the part in the problem and the base is *always* the original amount or last year's, last month's, or last week's amount.

3.4 Exercises

Solve for rate in each of the following. Round to the nearest tenth of a percent.

1. _______% of 2760 listings is 276 listings.
2. _______% of 850 showings is 340 showings.
3. 310 phones is _______% of 248 phones.

4. 144 desks is _______% of 300 desks.

5. _______% of 78.57 ounces is 22.2 ounces.

6. _______% of 728 miles is 509.6 miles.

7. 73.1 quarts is _______% of 786.8 quarts.

8. $310.75 is _______% of $124.30.

9. _______% of $53.75 is $2.20.

10. _______% of 850 liters is 3.4 liters.

11. 46 shirts is _______% of 780 shirts.

12. 5.2 vats is _______% of 28.4 vats.

13. _______% of 2 acres is 2.05 acres.

14. _______% of $8 is $0.06.

15. 13,830 books is _______% of 78,400 books.

16. _______% of 73 cases is 350.4 cases.

17. _______% of $330 is $91.74.

18. _______% of 752 employees is 470 employees.

19. The basic percent formula is $P = B \times R$. Show how to use the formula to solve for R (rate). (See Objective 1.)

20. A problem includes last year's sales and this year's sales and asks for the percent of increase. Explain how you would identify the base, rate, and part in this problem. (See Objective 4.)

Solve for rate in each of the following application problems. Round to the nearest tenth of a percent.

21. ADVERTISING EXPENSES Stephanie Baldock of Wired Education Systems reports that sales were $132,900 while advertising expenses were $7442.40. What percent of last month's sales was spent on advertising?

22. BUSINESS EXPANSION Home Depot plans to add 144 more stores to the 616 stores that it now has. What percent of an addition is this?

23. WOMEN IN THE MILITARY A recent study by Rand's National Defense Research Institute examined 48,000 military jobs, such as Army attack-helicopter pilot or Navy gunner's mate. It was found that only 960 of these jobs are filled by women. What percent of these jobs are filled by women?

24. VOCABULARY KNOWLEDGE There are 55,000 words in Webster's dictionary, but most educated people can identify only 20,000 of these words. What percent of the words in the dictionary can these people identify?

25. ADVERTISING MEDIA Advertising expenditures for the Radisson Hotel are as follows.

Newspaper	$2250	Television	$1425
Radio	$954	Yellow Pages	$1605
Outdoor	$1950	Miscellaneous	$2775

What percent of the total advertising expenditures is spent on radio advertising?

26. **ANTIQUE SALES** Barbara's Antiquery says that of its 3800 items in inventory, 3344 are just plain junk, while the rest are antiques. What percent of the total inventory is antiques?

27. **HARLEY-DAVIDSON MOTORCYCLES** Harley-Davidson, the only major U.S.-based motorcycle maker, says that it expects to build 145,000 motorcycles this year, up from 131,000 last year. Find the percent of increase in production.

28. **RISING TUITION COSTS** Students were charged $1449 for tuition this quarter. If the tuition was $1228 last quarter, what is the percent of increase?

29. **THE MEXICAN PESO** Three years ago, the Mexican peso was valued at U.S. $0.135. Today, it has a value of U.S. $0.105. Find the percent of decrease in the peso's value.

30. **FRANCHISEE INCOME** Five years ago, McDonald's Corporation franchise owners averaged $124,290 in annual income per outlet. This year, the average income per outlet is $91,630. Find the percent of decrease.

Supplementary Exercises: Rate, Base, and Part

Solve for rate, base, or part as indicated in the following. Round rates to the nearest tenth of a percent.

1. **CHIROPRACTIC PATIENTS** Last year 20 million Americans visited chiropractors. If 3% of these chiropractic patients were referred by a medical doctor, how many patients were referred to chiropractors by medical doctors?

2. **EMPLOYEE HEALTH PLANS** When larger firms (over 1000 employees) were surveyed, it was found that 27% of the firms offered only one health plan to employees. If 1800 firms were surveyed, find the number offering only one health plan.

3. **RETAIL EXPANSION** General Nutrition Center now has 3200 stores and plans to add 450 more stores. Find the percent of additional stores that they have planned.

4. **DANGER OF EXTINCTION** Scientists tell us that there are 9600 species of birds and that 1000 of these species are in danger of extinction. What percent of the bird species are in danger of extinction?

5. **ECONOMY LODGING** According to industry figures there are 44,500 hotels and motels in America. Economy hotels and motels account for 38% of this total. Find the number of economy hotels and motels.

6. **AMERICAN CHIROPRACTIC ASSOCIATION** There are 50,000 licensed chiropractors in the nation. If 30% of these chiropractors belong to the American Chiropractic Association (ACA), find the number of chiropractors in the ACA.

7. **DRUNK-DRIVING ACCIDENTS** In the United States, 17,126 people were killed in alcohol-related driving accidents last year. This was 41% of all traffic deaths. Find the number of traffic deaths last year. (Round to the nearest whole number.)

8. **AUTOMOBILE DEALERSHIPS** There are 1200 new car dealers in the United States who are using some form of one-price selling—abandoning haggling in favor of a discounted but nonnegotiable price. These dealers represent 8% of the total number of new car dealers. Find the number of dealers in the United States.

9. **COST AFTER MARKDOWN** A fax machine priced at $398 is marked down 7% to promote the new model. If the sales tax is also 7%, what is the cost of the fax machine including sales tax?

10. BOOK PURCHASE College students are offered a 6% discount on a dictionary that sells for $18.50. If the sales tax is 6%, find the cost of the dictionary including the sales tax.

11. BLOOD-ALCOHOL LEVELS In the United States, 15 of the 50 states limit blood-alcohol levels for drivers to 0.08%. The remaining states limit these levels to 0.10%.

(a) What percent of the states have a blood alcohol limit of 0.08%?

(b) What percent have a limit of 0.10%?

12. WORLDWIDE BIRD SPECIES If there are 9600 species of birds worldwide and the populations of 6500 of these species are in decline, what percent of the bird species are declining in population?

13. AIRLINE LAYOFFS In the past year 51,156 airline workers have lost their jobs through layoffs. If this was a 9.8% reduction in the number of workers, how many workers were there after the layoffs.

14. AVERAGE HOME PRICE According to the National Association of Realtors the median national sales price of a house was down 1.4%, or $1390 from last month. Find the median national sales price this month. (Round to the nearest dollar.)

15. CREDIT CARD DEBT One person owes more than $30,000 to 15 different credit card companies. If his payment on this debt amounts to $1220 each month and $298 of this is interest, what percent of his payment is interest?

16. MACHINE TOOL ORDERS Machine tool orders rose $10.5 million last month from $359.8 million the month before. Find the percent of increase.

NATIONWIDE HOME SALES *The number of existing single-family homes sold in four regions of the country in the same month of two separate years are shown in the table below. Use the data in the table to answer Exercises 17–20. Round to the nearest tenth of a percent.*

EXISTING HOME SALES

Region	Last Year	This Year
Northeast	32,000	36,000
Midwest	65,000	66,300
South	82,000	77,500
West	54,000	49,600

17. Find the percent of increase in sales in the northeastern region.

18. Find the percent of increase in sales in the midwestern region.

19. What is the percent of decrease in sales in the southern region?

20. What is the percent of decrease in sales in the western region?

21. VENDING MACHINE SALES Of the total candy bars contained in a vending machine, 240 bars have been sold. If 25% of the bars have been sold, find the total number of candy bars that were in the machine.

22. TOTAL SALES If the sales tax rate is 6% and the sales tax collected is $478.20, what are the total sales?

23. FAMILY BUDGETING Lou and Sheri Minkner established a budget allowing 25% for rent, 30% for food, 8% for clothing, 20% for travel and recreation, and the remainder for savings. Lou takes home $1950 per month, and Sheri takes home $28,500 per year. How much will the couple save in a year?

24. CHICKEN NOODLE SOUP In one year there were 350 million cans of chicken noodle soup sold (all brands). If 60% of this soup is sold in the cold-and-flu season (October through March), how many cans were sold in the cold-and-flu season?

25. FLOOD INSURANCE According to the Federal Emergency Management Agency (FEMA), there are 11 million buildings at risk of flooding. The agency finds that only 2.6 million of these are currently insured for flooding. Find the percent that are insured.

26. REFRIGERATION CAPACITY A Hotpoint refrigerator has a capacity of 11.5 cubic feet in the refrigerator and 5.5 cubic feet in the freezer. What percent of the total capacity is the capacity of the freezer?

27. SIDE-IMPACT COLLISIONS Automobile accidents involving side-impact collision resulted in 9000 deaths last year. If automobiles were manufactured to meet a "side-impact standard" it is estimated that 63.8% of these deaths would have been prevented. How many deaths would have been prevented?

28. NEW HOME The average price of a new home rose 4.2%. If the average price of a new home was $131,500, what is the average price after the increase?

29. U.S. PATENT RECIPIENTS Among the 50 companies receiving the greatest number of U.S. patents last year, 18 were Japanese companies. What percent of the top 50 companies were Japanese companies?

30. LAYOFF ALTERNATIVE Instead of laying off workers, a company cut all employee hours from 40 hours a week to 30 hours a week. What was the percent cut in employee hours?

3.5 Increase and Decrease Problems

Objectives

1. *Learn to identify an increase or a decrease problem.*
2. *Apply the basic diagram for increase problems.*
3. *Use an equation to solve for base in increase problems.*
4. *Apply the basic diagram for decrease problems.*
5. *Use an equation to solve for base in decrease problems.*

Businesses commonly look at how amounts change, either up or down. For example, a manager might need to know the percent by which sales have increased, or the percent by which costs have decreased, while a consumer might need to know the percent by which the price of an item has changed. Identify these **increase and decrease problems** as follows.

Objective 1 | *Learn to Identify an Increase or a Decrease Problem.*

Increase Problem. The base (100%) plus some portion of the base, gives a new value, which is part. Phrases such as "after an increase of," "more than," or "greater than" often indicate an increase problem. The basic formula for an increase problem is

$$\underset{\text{base}}{\underset{\uparrow}{\text{Original}}} + \text{Increase} = \underset{\text{part}}{\underset{\uparrow}{\text{New value}}}.$$

Decrease Problem. The part equals the base (100%) minus some portion of the base, resulting in a new value. Phrases such as "after a decrease of," "less than," or "after a reduction of" often indicate a decrease problem. The basic formula for a decrease problem is

$$\underset{\text{base}}{\underset{\uparrow}{\text{Original}}} - \text{Decrease} = \underset{\text{part}}{\underset{\uparrow}{\text{New value}}}.$$

NOTE Base is always the original amount and we can solve for base in both increase and decrease problems. Base is always 100%.

Example 1 Using a Diagram to Understand an Increase Problem

The price of a home sold by real estate agent Cas Shields this year is $121,000, which is 10% more than last year's value. Find the value of the home last year.

Solution Use a diagram, such as Figure 3.2, to help solve this problem. Since base is the starting point, or that to which something is compared, the base here is last year's sales. Call base 100% and remember that

$$\text{Original} + \text{Increase} = \text{New value}.$$

Objective 2 *Apply the Basic Diagram for Increase Problems.*

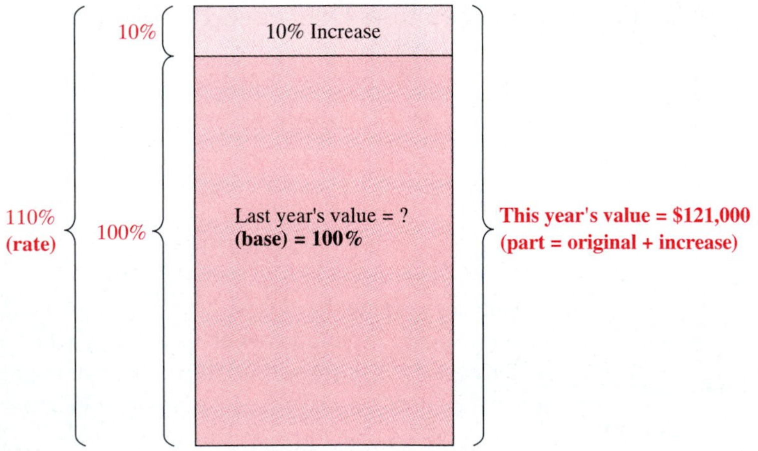

Figure 3.2

Objective 3 *Use an Equation to Solve for Base in Increase Problems.* As shown in Figure 3.2, the 10% increase is based on last year's value (which is unknown) and not on this year's value of $121,000. This year's value is the *result* of adding 10% of last year's value to the amount of last year's value. Therefore, this year's value is all of last year's value

(100%) plus 10% of last year's value (100% + 10% = 110%). Solve with the increase formula, using B to represent base.

$$\begin{aligned} \text{Original} + \text{Increase} &= \text{New value} \\ \text{Last year's value} + 10\% \text{ of last year's value} &= \text{this year's value} \\ 100\% \times B + 10\% \times B &= \$121{,}000 \\ 110\% \times B &= \$121{,}000 \\ 1.1B &= \$121{,}000 \\ \frac{\cancel{1.1}B}{\cancel{1.1}} &= \frac{\$121{,}000}{1.1} \\ B &= \frac{\$121{,}000}{1.1} \\ B &= \$110{,}000 \quad \text{Last year's value} \end{aligned}$$

Check the answer by taking 10% of last year's value and adding it to last year's value.

$$\begin{array}{rl} \$110{,}000 & \text{Last year's value} \\ +\quad 11{,}000 & (10\% \text{ of } \$110{,}000) \\ \hline \$121{,}000 & \text{This year's value} \end{array}$$

Problem-Solving Hint

The common error in solving for the base in an increase problem is thinking that the base is given and that the solution can be found by solving for part. Remember that the number given in Example 1, \$121,000, is the result of having added 10% of the base to the base (100% + 10% = 110%). In fact, the \$121,000 is the part, and base must be found.

The following graphic shows the rate of world population growth over many decades and the length of time that it takes for the population to double at various rates of growth. Population growth rates are an example of increase problems because each year's rate of growth (increase) is based on the previous year. Similarly, last year's rate of growth was based on the year prior to that.

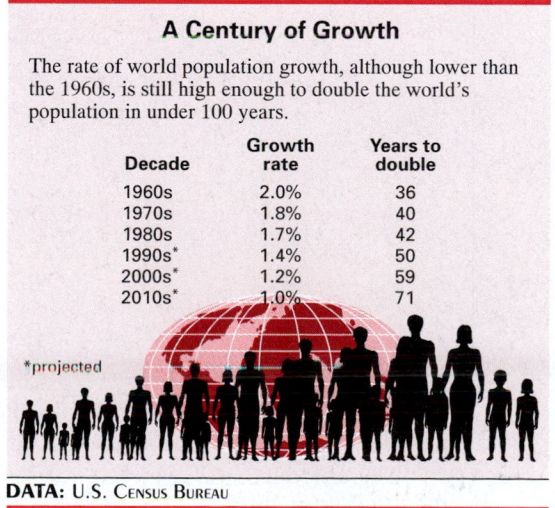

A Century of Growth

The rate of world population growth, although lower than the 1960s, is still high enough to double the world's population in under 100 years.

Decade	Growth rate	Years to double
1960s	2.0%	36
1970s	1.8%	40
1980s	1.7%	42
1990s*	1.4%	50
2000s*	1.2%	59
2010s*	1.0%	71

*projected

DATA: U.S. Census Bureau

Example 2 shows how to solve a problem with two increases.

Example 2 Finding Base after Two Increases

At Builder's Doors, production last year was 20% more than the year before. This year's production is 93,600 doors, which is 20% more than last year's. Find the number of doors produced two years ago.

Solution The two 20% increases cannot be added together because these increases are from two different years, with two separate bases. The problem must be solved in two steps. First, use a diagram to find last year's production.

From Figure 3.3, last year's production plus 20% of last year's production equals this year's production. Use the following formula:

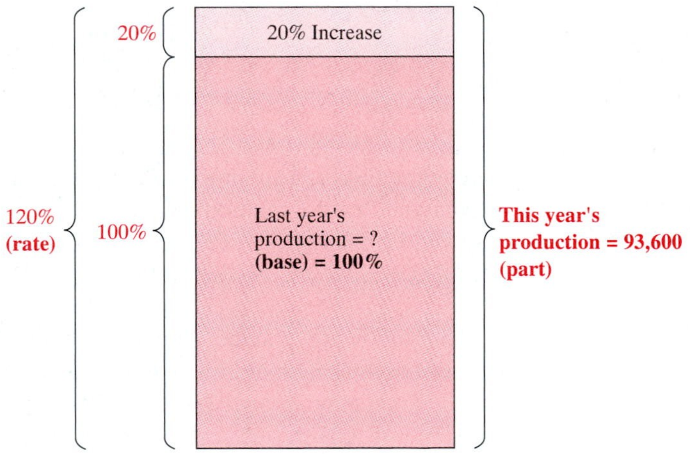

Figure 3.3

$$100\% \times B + 20\% \times B = 93{,}600$$
$$120\% \times B = 93{,}600$$
$$1.2B = 93{,}600$$
$$\frac{\cancel{1.2}B}{\cancel{1.2}} = \frac{93{,}600}{1.2}$$
$$B = \frac{93{,}600}{1.2}$$
$$B = 78{,}000 \quad \text{Last year's production}$$

Production last year was 78,000 units. Production for the preceding year (2 years ago) must now be found. Use another diagram (Figure 3.4).

Thus, production 2 years ago + 20% of production 2 years ago = last year's production. In the following solution b is used since B was used before.

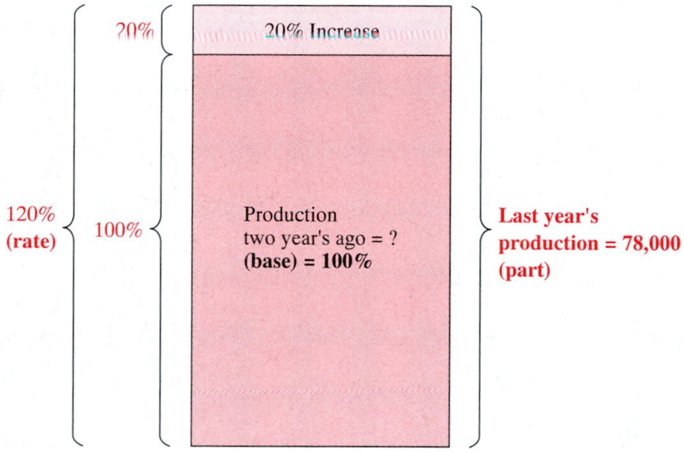

Figure 3.4

$$100\% \times b + 20\% \times b = 78{,}000$$
$$120\% \times b = 78{,}000$$
$$1.2b = 78{,}000$$
$$\frac{\cancel{1.2}b}{\cancel{1.2}} = \frac{78{,}000}{1.2}$$
$$b = \frac{78{,}000}{1.2}$$
$$b = 65{,}000 \quad \text{Production 2 years ago}$$

Check the answer.

65,000	Production 2 years ago
+ 13,000	20% increase
78,000	Production last year
+ 15,600	20% increase
93,600	Production this year

Scientific *Calculator Approach*

The calculator solution to this example divides in a series.

93,600 [÷] 1.2 [÷] 1.2 [=] 65,000

Problem-Solving Hint It is important to realize that the two 20% increases cannot be added together to equal one increase of 40%. Each 20% increase is calculated on a different base.

Objective 4 *Apply the Basic Diagram for Decrease Problems.*

Example 3 Using a Diagram to Understand a Decrease Problem

After Sports About deducted 10% from the price of a pair of skis, Craig Bleyer paid $135. What was the original price of the skis?

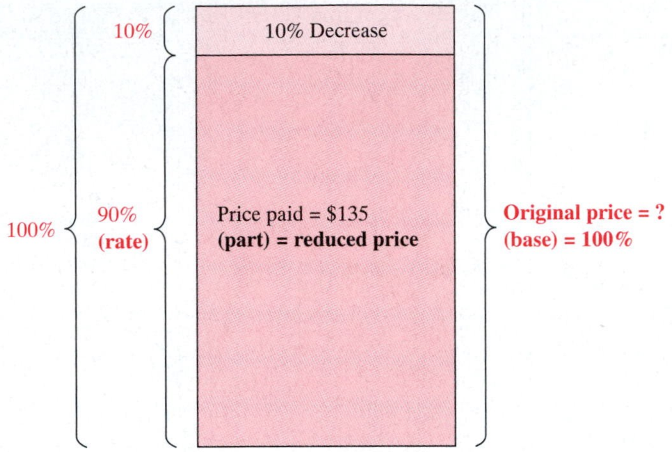

Figure 3.5

Solution Use a diagram (Figure 3.5) again and remember that the base is the starting point—in this case, the original price. As always, the base is 100%. Use the decrease formula because the price went down.

Objective 5 ***Use an Equation to Solve for Base in Decrease Problems.*** As Figure 3.5 shows, 10% was deducted from the original price. The result equals the price paid, which is 90% of the original price.

Be careful in finding the rate: 10% cannot be used because the original price on which 10% was calculated is not given. The rate 90% (the difference, 100% − 10% = 90%) must be used since 90% of the original price is the *resulting* \$135 price paid. Now find the original price.

$$\text{Original} - \text{Decrease} = \text{New value}$$
$$\text{Original price} - 10\% \text{ of the original price} = \text{Price paid}$$
$$100\% \times B - 10\% \times B = \$135$$
$$90\% \times B = \$135$$
$$0.9B = \$135$$
$$\frac{\cancel{0.9}B}{\cancel{0.9}} = \frac{\$135}{0.9}$$
$$B = \frac{\$135}{0.9}$$
$$B = \$150 \quad \text{Original price}$$

Check the answer.

\$150	Original price
− 15	10% discount
\$135	Price paid

Problem-Solving Hint The common mistake made in Example 3 is thinking that the reduced price, $135, is the base. The original price is the base, while the reduced price, $135, is a *result* of subtracting 10% of the base from the base. The reduced price is the part or 90% *of the base* (100% − 10% = 90%).

3.5 Exercises

Solve for base in each of the following. Round to the nearest cent.

Part (After Increase)	Rate of Increase
1. $450	20%
2. $800	25%
3. $30.70	10%
4. $10.09	5%

Solve for base in each of the following. Round to the nearest cent.

Part (After Decrease)	Rate of Decrease
5. $20	20%
6. $1530	15%
7. $598.15	30%
8. $98.38	15%

9. Certain words or word phrases help to identify an increase problem. Discuss how you will identify an increase problem. (See Objective 1.)

10. Certain words or word phrases help to identify a decrease problem. Discuss how you will identify a decrease problem. (See Objective 1.)

Solve each of the following application problems. Read each carefully to determine which are increase or decrease problems and work accordingly. Round to the nearest cent, when necessary.

11. HOME VALUE APPRECIATION Patricia Quinlin of Century 21 Realty just listed a home for $178,740. If this is 8% more than what the home sold for last year, what was last year's selling price?

12. DEALER'S COST John Chavez Auto Stereos sold an auto stereo for $337.92, a loss of 12% of the dealer's original cost. Find the original cost.

13. MOVIE-TICKET SALES The top ten films at the movie theaters had sales of $94 million over one holiday weekend. If this was an increase of 9% over last year's sales, how much were last year's sales? (Round to the nearest tenth of a million.)

14. STEERING WHEEL LOCKS Julie Ward bought a steering wheel lock for her pickup truck for $34.64 including $6\frac{1}{4}$% sales tax. (a) How much of the $34.64 was for the lock, and (b) how much was the sales tax?

15. ANTILOCK BRAKES In a recent test of an automobile antilock braking system (ABS) on wet pavement, the stopping distance was 114 feet. If this was 28.75% less than the distance needed to stop the same automobile without the ABS, find the distance needed to stop without the antilock braking system.

16. AUTO ALARMS Sandy Lindelof installed an electronic alarm and several simple auto-theft devices in her car. As a result her auto insurance policy premium was reduced by 5.2% to $1147 per year. Find the amount of her insurance premium before installing these devices. (Round to the nearest cent.)

17. WEDDING COSTS The average cost of a wedding today is $19,104, which is 26% more than the average cost 5 years ago. Find the average cost of a wedding 5 years ago. Round to the nearest cent.

18. POPULATION GROWTH In 2000 the population of Rio Linda was 10% more than it was in 1999. If the population was 26,620 in 2001, which was 10% more than in 2000, find the population in 1999.

19. WALLPAPER SALES Erika Guitierrez, owner of Wallpaper Plus, says that her sales have increased exactly 20% per year for the last 2 years. Her sales this year are $170,035.20. Find her sales 2 years ago.

20. VOLVO SALES In 5 years, Volvo plans to boost worldwide sales 25%, to 500,000 units a year. Find the annual Volvo sales this year.

21. NURSING HOME CARE The cost of nursing home care in the United States will jump nearly 12% to $66 billion next year. Find the cost of nursing home care this year. (Round to the nearest tenth of a billion.)

22. EXPENSIVE RESTAURANTS Among New York City's 20 most expensive restaurants, the average per-meal cost increased 6.5% to $69.33 in the last year. Find the price of this meal before the increase.

23. SURPLUS-EQUIPMENT SALES In a 3-day public sale of Jackson County surplus equipment, the first day brought $5750 in sales and the second day brought $4186 in sales, with 28% of the original equipment left to be sold on the third day. Find the value of equipment left to sell.

24. *COLLEGE EXPENSES* After spending $3450 for tuition and $4350 for dormitory fees, Donald Cole finds that 35% of his original savings remains. Find the amount of his savings that remains.

25. ROLL WITH IT This year's sales of Charmin toilet tissue are $1.052 billion. If this is a decrease in sales of 2.4% from last year, find last year's sales. (Round to the nearest thousandth of a billion.)

26. CONDOMINIUM SALES If an owner quickly sold her condominium for $86,330, which was a loss of 11% of the original purchase price, how much had the owner paid originally?

27. WINTER-WHEAT PLANTING Even though wheat prices are rising during the planting season, farmers have planted only 50.2 million acres of winter wheat varieties. If this is 2% fewer acres than last year, find the number of acres planted last year. (Round to the nearest tenth of a million.)

28. LEATHER-CLOTHING SALES Department stores and specialty chains sold 4.1 million leather jackets and coats this year. If this is a 38% drop in sales from last year, find last year's sales. (Round to the nearest tenth of a million.)

29. COMMUNITY-COLLEGE ENROLLMENT In 2000 the student enrollment at American River College was 8% more than it was in 1999. If the enrollment was 23,328 students in 2001, which was 8% more than it was in 2000, find the student enrollment in 1999.

30. UNIVERSITY FEES Students at the state universities are outraged. The annual university fees were 30% more last year than they were the year before. If the fees are \$2704 per year this year, which is 30% more than they were last year, find the annual student fees 2 years ago.

31. PERSONAL COMPUTERS Worldwide personal-computer shipments rose 15% this year to 79.81 million units. Find the number of personal computers shipped last year.

32. MINORITY LOANS A mortgage lender made 52% more loans to minorities this year than last year. If the number of loans to minorities this year is 2660, find the number of loans made to minorities last year.

33. NEW-HOME SALES New home sales this year in the Sacramento area were 14% fewer than last year. If the number of new homes sold this year was 5645, find the number of new homes sold last year. (Round to the nearest whole number.)

34. NEW-HOME PRICES The median price for a new home in the Sacramento area is \$148,950—down 2.5% from last year. Find the median price for a new home last year. (Round to the nearest dollar.)

Chapter 3 Quick Review

Chapter Terms

Review the following terms to test your understanding of the chapter. For each term you do not know, refer to the page number found next to that term.

Concepts	Examples
3.1 Writing a decimal as a percent Move the decimal point two places to the right and attach a % sign.	$0.75\ (0.75.) = 75\%$
3.1 Writing a fraction as a percent First change the fraction to a decimal, then write the decimal as a percent.	$\frac{2}{5} = 0.4$ $0.4\ (0.40.) = 40\%$
3.1 Writing a percent as a decimal Move the decimal point two places to the left and drop the % sign.	$50\%\ (0.50.\%) = 0.5$
3.1 Writing a percent as a fraction First change the percent to a decimal. Then write the decimal as a fraction in lowest terms.	$15\%\ (0.15.\%) = 0.15$ $= \frac{15}{100} = \frac{3}{20}$
3.1 Writing a fractional percent as a decimal First change the fraction to a decimal leaving the % sign, then move the decimal point two places to the left and drop the % sign.	$\frac{1}{2}\% = 0.5\%$ $0.5\% = 0.00.5$ $\frac{1}{2}\% = 0.005$
3.2 Solving for part using the percent formula Part = Base × Rate $P = B \times R$ $P = BR$ ______% of ______ is ______	A company offered a 15% discount on all sales. Find the discount on sales of \$1850. ______% of *sales* is discount 15% of \$1850 = discount $R \times B = P$ $0.15 \times \$1850 = P$ $0.15 \times \$1850 = \$277.50$ discount

3.3 Using the basic percent equation to solve for base

Remember that base is the starting point, reference point, all of something, or 100%.

Rate × Base = Part

______% × ______ is ______

If the sales tax rate is 4%, find the sales if the sales tax is $18.

$$R \times B = P$$
$$4\% \times \_\_\_\_\_\_ = \$18$$
$$0.04B = \$18$$
$$\frac{0.04B}{0.04} = \frac{\$18}{0.04}$$
$$B = \frac{18}{0.04}$$
$$= \$450 \text{ sales}$$

3.4 Using the basic percent equation to solve for rate

Remember that rate is a percent and is followed by a % sign.

Rate × Base = Part

______% × ______ is ______

The return is $307.80 on an investment of $3420. Find the rate of return.

$$R \times B = P$$
$$\_\_\_\_\_\_\% \text{ of } \$3420 \text{ is } \$307.80$$
$$R \times \$3420 = \$307.80$$
$$\$3420R = \$307.80$$
$$\frac{3420R}{3420} = \frac{307.80}{3420}$$
$$= 0.09$$
$$R = 9\%$$

3.4 Finding the percent of change

Calculate the change (increase or decrease), which is the part. Base is the amount before the change.

Use $R = \frac{P}{B}$

Production rose from 3820 units to 5157 units. Find the percent of increase.

$$5157 - 3820 = 1337 \text{ increase}$$
$$R = \frac{1337}{3820} = 0.35 = 35\%$$

3.5 Drawing a diagram and using an equation to solve an increase problem

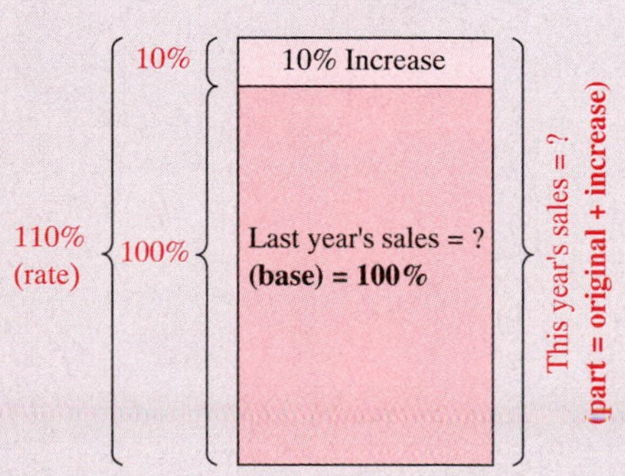

Solve for base given rate (110%) and part (after increase).

This year sales are $121,000, which is 10% more than last year's sales. Find last year's sales.

$$\text{Original} + \text{Increase} = \text{New value}$$
$$100\% \times B + 10\% \times B = \$121{,}000$$
$$1B + 0.1B = \$121{,}000$$
$$1.1B = \$121{,}000$$
$$B = \frac{\$121{,}000}{1.1}$$
$$= \$110{,}000 \text{ Last year's sales}$$

3.5 Drawing a diagram and using an equation to solve a decrease problem

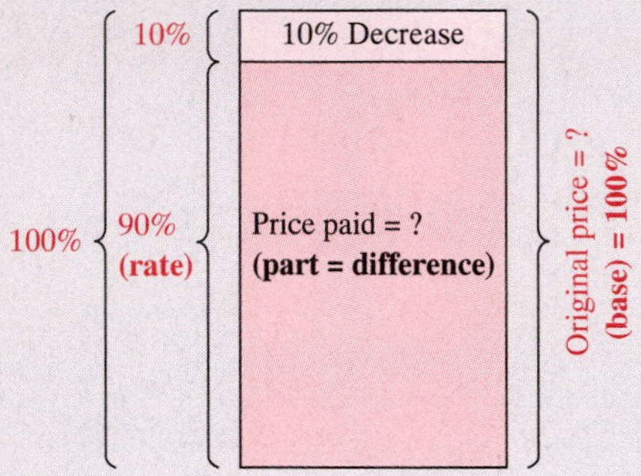

Solve for base given rate (90%) and part (after decrease).

After a deduction of 10% from the price, a customer paid $135. Find the original price.

$$\text{Original} - \text{Decrease} = \text{New value}$$
$$100\% \times B - 10\% \times B = \$135$$
$$1B - 0.1B = \$135$$
$$0.9B = \$135$$
$$B = \frac{\$135}{0.9}$$
$$B = \$150 \text{ original price}$$

Chapter 3 Review Exercises

The answer section includes answers to all Review Exercises.

Solve each of the following. [3.1–3.4]

1. 18 members is 12% of what number of members?

2. What is 5% of 480 vans?

3. 33 shippers is 3% of what number of shippers?

4. 36 accounts is what percent of 1440 accounts?

5. What is $\frac{1}{4}$% of \$1500?

6. Find the fractional equivalent of 24%.

7. 24 loads is $2\frac{1}{2}$% of how many loads?

8. Change 87.5% to its fractional equivalent.

9. \$70.55 is what percent of \$830?

10. What is the fractional equivalent of $\frac{1}{2}$%?

Solve each of the following application problems, reading each carefully to determine whether base, part, or rate is being asked for. Also, check to see which are increase or decrease problems, and work accordingly. (Round to the nearest cent or tenth of a percent, as necessary.)

11. One share of stock in Telefonos de Mexico sells for \$79.25 and pays a 2.1% dividend. Find the dividend per share. [3.2]

12. A supervisor at Barrett Manufacturing finds that rejects amount to 1120 units per month. If this amounts to 0.5% of total monthly production, find the total monthly production. [3.3]

13. Auto sales in eight Asian countries are expected to drop 29.2% this year from last year's sales of 3.83 million vehicles. Find the number of vehicles expected to be sold this year (after the drop). Round to the nearest hundredth of a million. [3.2]

14. It is estimated that 3 million people in the United States who are between the ages 55 and 64 have no health insurance. If this is 14% of the people in this age bracket, find the number of Americans in this age bracket. (Round to the nearest hundredth of a million.) [3.3]

15. Gabbert's Furniture, with a monthly advertising budget of \$3400, decides to set up a media budget. They plan to spend 22% for television, 38% for newspaper, 14% for outdoor signs, 15% for radio, and the remainder for bumper stickers. (a) What percent of the total budget do they plan to spend on bumper stickers? (b) How much do they plan to spend on bumper stickers for the entire year? [3.2]

16. The government is offering a \$25,000 bonus to federal employees for retiring early. After taxes and other deductions the employee will receive only \$17,000. What percent of the bonus will each employee actually receive? [3.4]

17. A digital camera is marked "reduced 25%, now only \$637.50." Find the original price of the digital camera. [3.5]

18. Last year's backpack sales were 10% more than they were the year before. This year's sales are 1452 units, which is 10% more than last year. Find the number of backpacks sold 2 years ago. [3.5]

19. One day on the London Stock Exchange, Unilever's stock shares increased 12.3 pence to 449.5 pence. Find the percent of increase. [3.4]

20. This year 8.4 million people paid income tax on their Social Security income. If this is an increase of 12.5% from last year, find the number of people who paid income tax on their Social Security income last year. Round to the nearest tenth of a million. [3.5]

21. Star-Kist is the leader in canned tuna sales with sales of \$553.7 million in the last 52 weeks. If total sales of canned tuna were \$1258.5 million, find the percent of sales that were Star-Kist. Round to the nearest whole percent. [3.4]

22. Most shampoos contain 75% to 90% water. If a 16-ounce bottle of shampoo contains 78% water, find the number of ounces of water in the 16-ounce bottle. (Round to the nearest tenth of an ounce.) [3.2]

23. Bookstore sales of the *Physicians Desk Reference*, which contains prescription drug information, rose 13.7% this year. If sales this year were 111,150 copies, find last year's sales. (Round to the nearest whole number.) [3.5]

24. After deducting 11.8% of total sales as her commission, George-Ann Hornor, a salesperson for Marx Toy Company, deposited \$35,138.88 to the company account. Find the total amount of her sales. [3.5]

25. The U.S. Patent Office received 230,000 patent applications last year and issued 112,091 patents. What percent of the patent applications resulted in patents? (Round to the nearest tenth of a percent.) [3.4]

26. Doctors in Argentina reported that 240,000 Argentines will have plastic surgery this year. If the population of the entire country is 33 million, what percent of the Argentines will have plastic surgery this year? [3.4]

27. The world population last year was 5.75 billion people. If 32% of the world population was under the age of 15, how many people in the world were under the age of 15? [3.2]

28. Tupperware, which was built on parties in peoples' homes where sellers demonstrated the use of plastic containers, is part of Premark International, Incorporated. Last year, Tupperware products accounted for 39% of Premark's $3.5 billion in sales. Find the amount of the Tupperware sales. [3.2]

29. The whooping crane, which is the tallest bird (5 feet) in North America, now has a population of 155 after dropping to near extinction in 1941, when there were only 15 birds. Find the percent of increase in the number of whooping cranes since 1941. [3.4]

30. The Small Business Administration (SBA) guaranteed 55,600 loans this year, compared with 36,000 loans last year. Find the percent of increase in the number of loans guaranteed. [3.4]

31. The number of business failures this year were 64,031, compared with 64,743 business failures last year. Find the percent of decrease. [3.4]

32. The number of Canadian tourists traveling to Florida this year has decreased 25% since 1990, when a record 2.4 million visited the "Sunshine State." Find the number of Canadian tourists visiting Florida this year. [3.2]

33. Volkswagen boosted production capacity in Mexico by 40%, to 350,000 cars this year. Find the production last year. [3.5]

34. General Motors light-truck sales were up 2% last month from the same month last year. If sales this year were 151,477 units, find the number of units sold last year. (Round to the nearest whole number.) [3.5]

35. Average hourly wages have risen 4.2% to $12.40 over the last year. Find the average hourly wages last year. [3.5]

36. Navistar International Corporation will cut daily production of class 8 trucks—those that can haul more than 33,000 pounds—by 26%, reducing daily production to 72 units. Find the daily production before the cutback. (Round to the nearest whole number.) [3.5]

Chapter 3 Summary Exercise
Mathematics for the Collector

Collecting baseball cards is an activity enjoyed by many that can also be very profitable. For instance, a 1952 Jackie Robinson rookie baseball card is currently valued at $1000. Listed to the right are some popular rookie cards along with price information for this year and last year. Find the card price last year, the percent of change from last year, or the card price this year, as necessary. Round dollar amounts to the nearest dollar, and percents to the nearest percent.

THE UPS AND DOWNS OF LAST YEAR

Rookie Card	Card Price Last Year	Card Price This Year	% Change From Last Year*
Chipper Jones	$40	______	0%
Cal Ripken, Jr.	$75	$70	______
Nomar Garciapara	______	$30	650%
Nolan Ryan	$1000	______	−10%
Ken Griffey, Jr.	______	$100	33%
Frank Thomas	$90	$80	______
Will Clark	$6	______	−17%
Mark McGwire	______	$150	650%
Sammy Sosa	$3	$150	______
Alex Rodriquez	$40	$30	______

*A minus sign (−) is used to indicate a drop in value.

Net Assets Business on the Internet

Century 21

Statistics

- 1999: Headquarters in Parsippany, New Jersey
- 110,000 brokers and agents
- 6300 offices in 25 countries

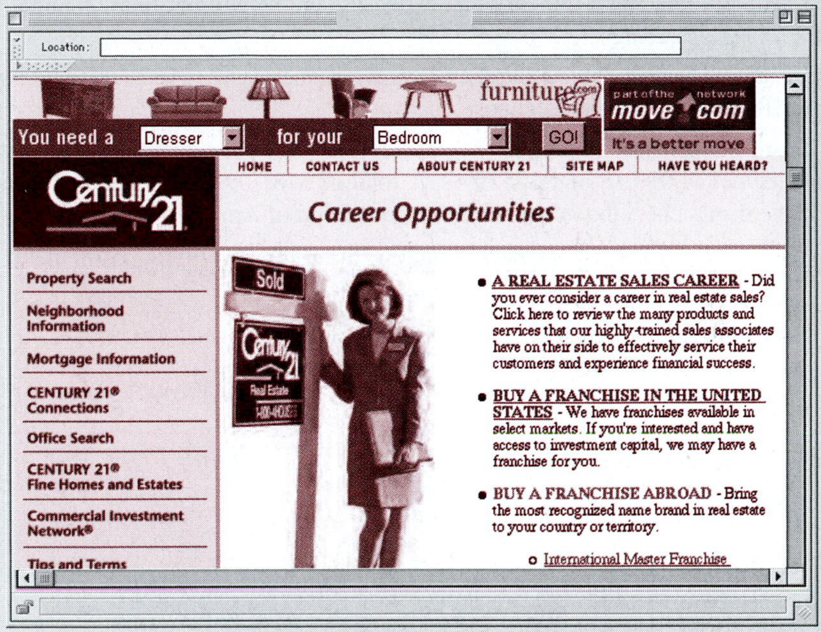

Century 21 is franchiser of the world's largest residential real estate sales organization. It provides comprehensive training, management, administrative, and marketing support for more than 6300 independently owned and operated offices with over 110,000 brokers and agents in more than 25 countries worldwide.

Century 21 is the number one consumer brand in the real estate industry and has the largest network and greatest global coverage of any competing brand. It is dedicated to continually providing buyers and sellers of real estate with the highest level of service possible.

1. A report by Century 21 says that home values in a certain area have increased by 8.2% since last year. Find the value of a home today that was valued at $125,000 last year.

2. A real estate office spent $1553.75 to sponsor a youth sports team. If this sponsorship equals 2.75% of the office's total advertising budget, find the total advertising budget.

3. A first-time home buyer made a $5960 down payment on a home selling for $74,500. What percent of the selling price is the down payment?

4. Total property sales in a Century 21 office this month were $4.76 million. If total sales last month were $4.25 million, find the percent of increase.

Part 1 Cumulative Review Chapters 1–3

To help you review, the numbers in brackets show the section in which the topic was introduced.

Solve the following application problems. [1.1]

1. Bryan Gripka decides to establish a budget. He will spend \$450 for rent, \$325 for food, \$320 for child care, \$182 for transportation, and \$150 for other expenses, and he will put the remainder in savings. If his take-home pay is \$1620, find his savings.

2. The Enabling Supply House purchases six wheelchairs at \$1256 each and 15 speech compression recorder-players at \$895 each. Find the total cost.

3. Software Supply had a bank balance of \$29,742.18 at the beginning of April. During the month, the firm made deposits of \$14,096.18 and \$6529.42. A total of \$18,709.51 in checks was paid by the bank during the month. Find the firm's checking account balance at the end of April.

4. Christine Grexa pays \$53.19 each month to the Bank of Bolivia. How many months will it take her to pay off \$1436.13?

Solve each of the following problems. [1.2–1.3]

5. Write $\frac{48}{54}$ in lowest terms.

6. Write $8\frac{1}{8}$ as an improper fraction.

7. Write $\frac{107}{15}$ as a mixed number.

8. $1\frac{2}{3} + 2\frac{3}{4} =$

9. $5\frac{7}{8} + 7\frac{2}{3} =$

10. $6\frac{1}{3} - 4\frac{7}{12} =$

11. $8\frac{1}{2} \times \frac{9}{17} \times \frac{2}{3} =$

12. $3\frac{3}{4} \div \frac{27}{16} =$

Solve each of the following application problems. [1.2–1.3]

13. The area of a piece of land is $63\frac{3}{4}$ acres. One-third of the land is sold. What is the area of the land that is left?

14. To prepare for the state real estate exam, Bonnie Maddison studied $5\frac{1}{2}$ hours on the first day, $6\frac{1}{4}$ hours on the second day, $3\frac{3}{4}$ hours on the third day, and 7 hours on the fourth day. How many hours did she study altogether?

15. The storage yard at American River Raft Rental has four sides and is enclosed with $527\frac{1}{24}$ feet of security fencing around it. If three sides of the yard measure $107\frac{2}{3}$ feet, $150\frac{3}{4}$ feet, and $138\frac{5}{8}$ feet, what is the length of the fourth side?

16. Play-It-Now Sports Center has decided to divide $\frac{2}{3}$ of the company's profit sharing funds evenly among the eight store managers. What fraction of the total amount will each receive?

Solve each of the following problems. [1.3]

17. Change 0.35 to a fraction.

18. Change $\frac{2}{3}$ to a decimal. Round to the nearest thousandth.

Round each of the following numbers as indicated. [1.3]

19. 78.572 to the nearest hundredth

20. 4732.489 to the nearest hundredth

21. 62.65 to the nearest tenth

22. 215.6749 to the nearest thousandth

Solve each equation. [2.1]

23. $x + 17 = 43$

24. $y - 33 = 52.4$

25. $\frac{z}{4} - 10 = 18$

26. $4(r - 2) = 2(r + 8)$

Write a mathematical expression. Use x *as the variable.* [2.2]

27. $\frac{3}{4}$ times a number

28. 5 times a number is added to the number

29. 8 times a number is decreased by 8

30. The sum of 6 times a number and 5

For each problem, use the formula to find the value of the variable that is not given. [2.3]

31. $I = PRT$; $I = \$2880$, $R = 0.08$, $P = \$12{,}000$

32. $M = P(1 + RT)$; $M = \$2035$, $R = 0.05$, $T = 2$

Write the following ratios. [2.4]

33. \$2000 to \$400

34. 21 feet to 5 yards

Solve the following proportions. [2.4]

35. $\frac{3}{x} = \frac{14}{42}$

36. $\frac{5}{8} = \frac{22}{y}$

Solve the following application problems. [2.4]

37. From a sample of 80 new home buyers, 24 were first-time home buyers. There were 2480 new homes sold in the county. Estimate the number of first-time home buyers.

38. A company spends four times as much on product development as it does on advertising. If \$38,500 is spent on advertising, how much is spent on product development?

Solve each of the following problems. [3.1–3.4]

39. Change $\frac{5}{8}$ to a percent.

40. Change 0.25% to a decimal.

41. Find 18% of 2500 prospects.

42. Find 134% of \$80.

43. 275 sales is what percent of 1100 sales?

44. 375 patients is what percent of 250 patients?

Solve each of the following application problems.

45. A mortgage company received 32,340 home loan applications from homebuyers last month and approved 20,860 of these loans. What percent of the applicants were approved? Round to the nearest tenth of a percent. [3.4]

46. The population of Elk Grove is 76,800 people, If 28.5% of the population is under the age of 18, find the number of people under the age of 18. [3.2]

47. The number of students enrolled in the Los Rios Community College District rose 9.5% this year. If the number of students enrolled this year is 64,040, find last year's enrollment. Round to the nearest whole number. [3.5]

48. As an antique dealer, Clarence Hanks charges a 25% selling fee. After deducting this fee, Hanks recently paid out \$12,570 to his sellers. Find the total amount of his sales. [3.5]

Part 2 Basic Business Applications

4 Banking Services

Modern banks and savings institutions today offer so many services that they have become more than places to deposit savings and take out loans. Today, many types of savings and checking accounts are offered, as well as services such as computerized home and business banking, automated teller machines (ATMs), credit cards, debit cards, investment securities services, and even payroll services for the business owner.

Most of the innovations in modern banking today are the result of electronic technology. While tradition keeps us writing and depositing checks, the growing use of direct deposits, ATM cards, debit cards, and home and business banking, all known as **electronic banking**, will become the common practice. Figure 4.1 shows past and projected use of home banking and the number of banks offering online banking service.

Aside from the increased customer benefits offered with online banking, how will the banks benefit from the greater use of online banking?

This chapter examines checking accounts and check registers. It also discusses business checking account services, the depositing of credit card transactions, and bank reconciliation (balancing the checking account).

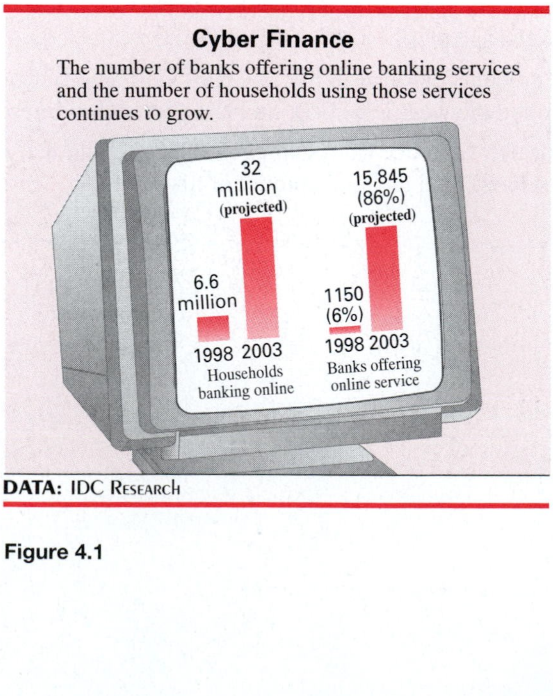

Figure 4.1

4.1 Checking Accounts and Check Registers

Objectives

1. *Identify the parts of a check.*
2. *Know the types of checking accounts.*
3. *Find the monthly service charges.*
4. *Identify the parts of a deposit slip.*
5. *Identify the parts of a check stub.*
6. *Complete the parts of a check register.*

Checking Facts

- Three of every four families have a checking account.
- The average adult writes over 100 checks each year.
- Checks became common after World War II.
- Today, 135 billion checks are processed each year.
- Checks represent about one-third of all consumer spending.

Even with the growth in **electronic commerce** (EC), where goods are purchased and sold electronically, the majority of business transactions today still involve checks. A small business may write several hundred checks each month and take in several thousand, while large businesses can take in several million checks in a month. This heavy reliance on checks makes it important for all people in business to have a good understanding of checks and checking accounts. The various parts of a check are explained in Figure 4.2.

Objective 1 | *Identify the Parts of a Check.*

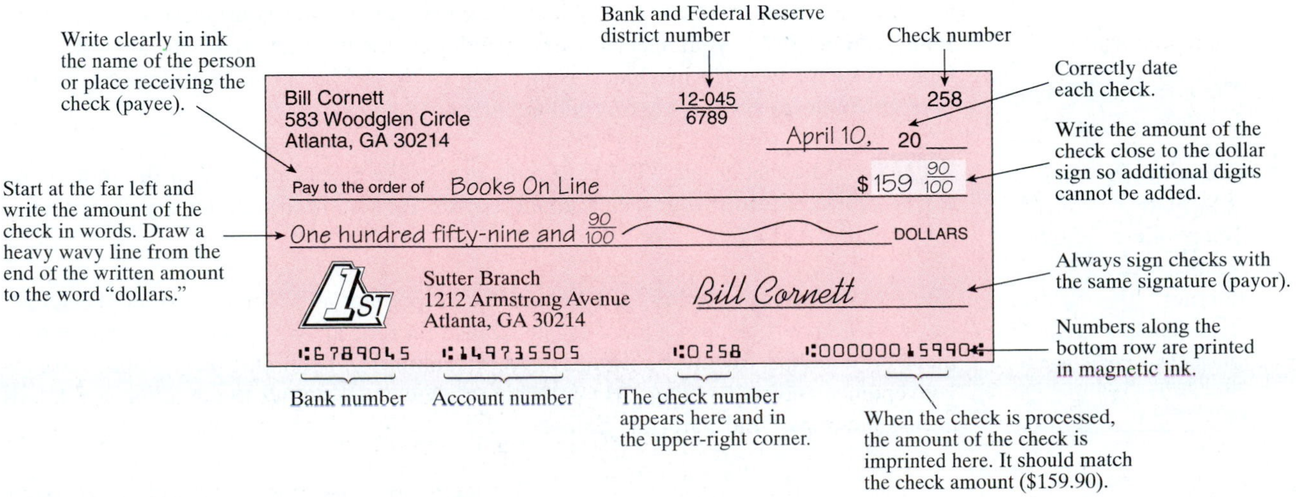
Bank and Federal Reserve district number
Check number
Write clearly in ink the name of the person or place receiving the check (payee).
Correctly date each check.
Bill Cornett
583 Woodglen Circle
Atlanta, GA 30214
12-045
6789
258
April 10, 20
Write the amount of the check close to the dollar sign so additional digits cannot be added.
Pay to the order of Books On Line
$ 159 90/100
Start at the far left and write the amount of the check in words. Draw a heavy wavy line from the end of the written amount to the word "dollars."
One hundred fifty-nine and 90/100
DOLLARS
Sutter Branch
1212 Armstrong Avenue
Atlanta, GA 30214
Bill Cornett
Always sign checks with the same signature (payor).
Numbers along the bottom row are printed in magnetic ink.
Bank number
Account number
The check number appears here and in the upper-right corner.
When the check is processed, the amount of the check is imprinted here. It should match the check amount ($159.90).

Figure 4.2

Objective 2 *Know the Types of Checking Accounts.* Two main types of checking accounts are available.

Personal checking accounts are used by individuals. The bank supplies printed checks (normally charging a check-printing fee) for the customer to use. Some banks offer the checking account at no charge to the customer, but most require that a minimum monthly balance remain in the checking account. If the minimum balance is not maintained during any month, a service charge is applied to the account. Today, the **flat-fee checking account** is common. For a fixed charge per month, the bank supplies the checking account, check printing, a bank credit card, an ATM card, a debit card, and a host of other services. Interest paid on checking account balances is common with personal checking accounts. These accounts are offered by savings and loan associations, credit unions, and banks and are available to individuals as well as a few business customers. These accounts often require much higher minimum balances than regular accounts.

Business checking accounts often receive more services than do personal accounts. For example, banks often arrange to receive payments on debts payable to business firms. The bank automatically credits the amount to the business account.

A popular service available to personal and business customers is the **automated teller machine (ATM)**. Offered by many banks, savings and loans, and credit unions, an ATM allows customers to perform a great number of transactions on a 24-hour basis. The customer can make cash withdrawals and deposits, transfer funds from one account to another, make payments on credit card accounts or other loans, and make account-balance inquiries. In addition, through several networking arrangements, the customer may make purchases and receive cash advances from hundreds, and in some cases thousands, of participating businesses nationally, and often worldwide.

NOTE Students traveling on exchange trips in foreign countries can often get cash in the local currency using their ATM cards. ■

These ATM cards are **debit cards**, *not* credit cards. When you use your debit card at a **point-of-sale terminal**, the amount of your purchase is instantly subtracted from your account and credit is given to the seller's account. When you use a credit card, you usually sign a receipt; when using a debit card, however, you enter your **personal identification number (PIN)** (your special code) to authorize the transaction. Cash can also be obtained from many ATM machines using credit cards such as VISA and MasterCard even in other countries such as Mexico and Germany.

Transaction Costs

The cost of different payment transactions

Bank branch	$1.07
U.S. Mail	$0.73
Telephone	$0.54
ATM	$0.27
Internet	$0.01

Data: Jupiter Communication 1997 Home Banking Report. Reprinted by permission.

NOTE When using an ATM card remember to keep receipts so that the transaction can be subtracted from your own account records. Be certain to subtract any charge made for using the ATM card. ■

The use of **electronic funds transfer (EFT)** is a popular alternative to paper checks because EFT saves businesses money. The table at the side shows the cost of different payment transactions. **Home banking** (Internet banking) is becoming very popular for its convenience and cost savings. Home banking allows a customer to pay bills, check balances, and move funds, all from home.

Objective 3 *Find the Monthly Service Charges.* Service charges for business checking accounts are based on the average balance for the period covered by the statement. This average

Table 4.1 TYPICAL BANK CHARGES FOR A BUSINESS CHECKING ACCOUNT

Average Balance	Maintenance Charge per Month	Per Check Charge
Less than $500	$12.00	$0.20
$500–$1999	$ 7.50	$0.20
$2000–$4999	$ 5.00	$0.10
$5000 or more	$ 0	$0

balance determines the **maintenance charge per month**, to which a **per debit charge** (per check charge) is added. The charges generally apply regardless of the amount of account activity. Some typical bank charges for a business checking account appear in Table 4.1.

Example 1 Finding the Checking Account Service Charge

Find the monthly service charge for the following business accounts.

(a) Pittsburgh Glass, 38 checks written, average balance $833

Solution From Table 4.1, an account having an average balance between $500 and $1999 will have a maintenance charge of $7.50 for the month. In addition, there is a per debit (check) charge of $0.20. Since 38 checks were written during the month, the monthly service charge is calculated as follows.

$$\$7.50 + 38(\$0.20) = \$7.50 + \$7.60 = \$15.10$$

(b) Fargo Western Auto, 87 checks written, average balance $2367

Solution Since the average balance is between $2000 and $4999, the maintenance charge for the month is $5.00, to which a $0.10 per debit (check) charge is added. The monthly service charge is $5.00 + 87($0.10) = $5.00 + $8.70 = $13.70.

Scientific *Calculator Approach*

The calculator solutions to Example 1 use chain calculations, with the calculator observing the order of operations.

(a) 7.5 [+] 38 [×] .2 [=] 15.1

(b) 5 [+] 87 [×] .1 [=] 13.7

Objective 4 *Identify the Parts of a Deposit Slip.* Money, either cash or checks, is placed into a checking account with a **deposit slip** or **deposit ticket** such as the one in Figure 4.3. The account number is written at the bottom of the slip in magnetic ink. The slip contains blanks in which are entered any cash (either currency or coins), as well as checks that are to be deposited.

When a check is deposited, it should have "for deposit only" and either the depositor's signature or the company stamp placed on the back within 1.5 inches of the vertical top edge. In this way, if a check is lost or stolen on the way to the bank, it will be worth-

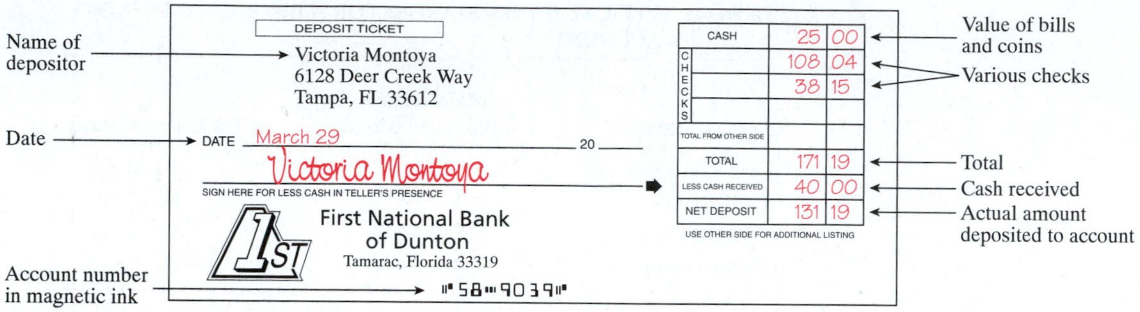

Figure 4.3

less to anyone finding it. Such an endorsement, which limits the ability to cash a check, is called a **restricted endorsement**. An example of a restricted endorsement is shown in Figure 4.4, along with two other types of endorsements. The most common endorsement by individuals is the **blank endorsement**, where only the name of the person being paid is signed. This endorsement should be used only at the moment of cashing a check. The

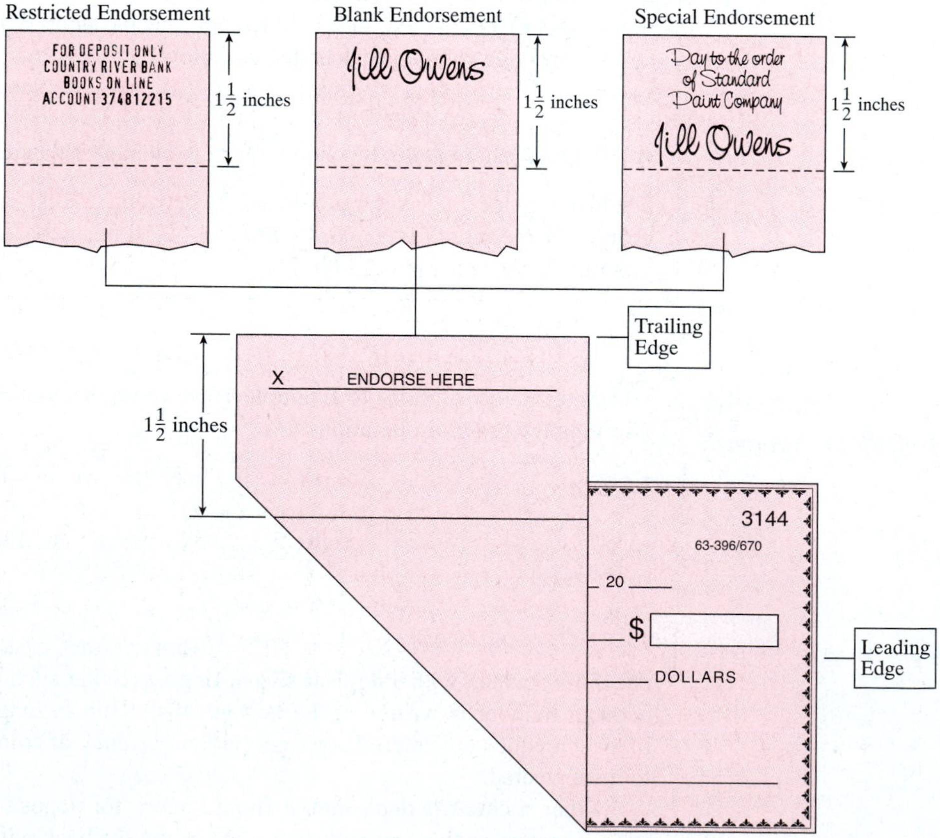

Figure 4.4

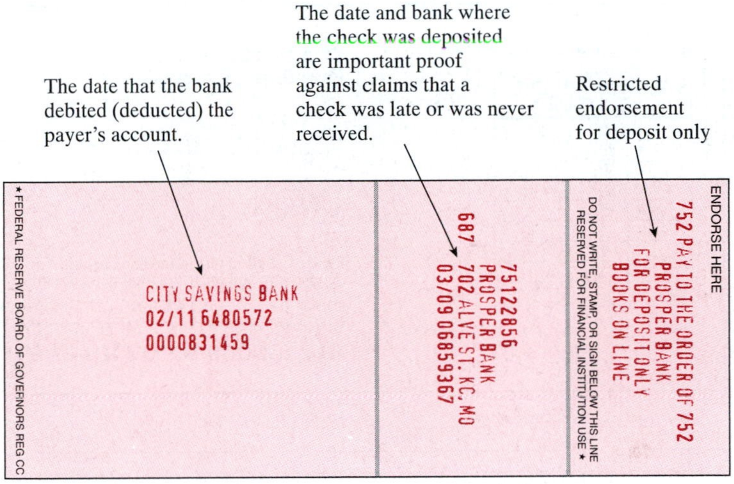

Figure 4.5

special endorsement, used to pass on the check to someone else, might be used to pay a bill on another account.

After the check is endorsed, it is normally cashed or deposited at a bank. The payee either is given cash or receives a credit in the account for the amount of the check. The check is then routed to a federal reserve bank, which forwards the check to the payer's bank. After going through this procedure, known as **processing**, the check is then **canceled** and returned to the payer. The check will now have additional processing information on its back, as shown above on Figure 4.5.

A two-sided commercial deposit slip is shown in Figure 4.6. Notice that much more space is given for an itemized list of customers' checks that are being deposited to the business account. Many financial institutions require that the bank and federal reserve district numbers be shown in the description column of the deposit slip. These numbers appear in the upper center portion of the check and are identified in the sample check on page 117.

Objective 5 *Identify the Parts of a Check Stub.* A record must be kept of every deposit made and every check written. Business firms normally do this with one **check stub** for each check. These stubs provide room to list the date, the person or firm to whom the check will be paid, and the purpose of the check. Also, the stub provides space to record the balance in the account after the last check was written (called the **balance brought forward**, abbreviated Bal. Bro't. For'd., on the stub), and any money deposited since the last check was written. The balance brought forward and amount deposited are added to provide the current balance in the checking account. The amount of the current check is then subtracted, and a new balance is found. This **balance forward** from the bottom of the stub should be written on the next stub. Figure 4.7 shows a typical check stub.

Example 2 Completing a Check Stub

Check number 2724 is made out on June 8 to Lillburn Utilities as payment for water and power. Assume that the check is for $182.15, that the balance brought forward is

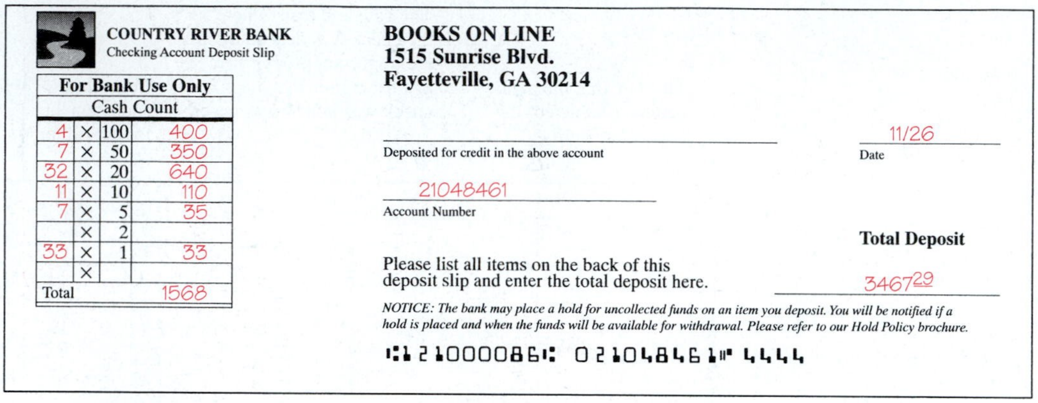

COUNTRY RIVER BANK
Checking Account Deposit Slip

BOOKS ON LINE
1515 Sunrise Blvd.
Fayetteville, GA 30214

For Bank Use Only			
Cash Count			
4	×	100	400
7	×	50	350
32	×	20	640
11	×	10	110
7	×	5	35
	×	2	
33	×	1	33
	×		
Total			1568

Deposited for credit in the above account

Date 11/26

Account Number 21048461

Total Deposit

Please list all items on the back of this deposit slip and enter the total deposit here. 3467.29

NOTICE: The bank may place a hold for uncollected funds on an item you deposit. You will be notified if a hold is placed and when the funds will be available for withdrawal. Please refer to our Hold Policy brochure.

⑆121000086⑆ 021048461⑈ 4444

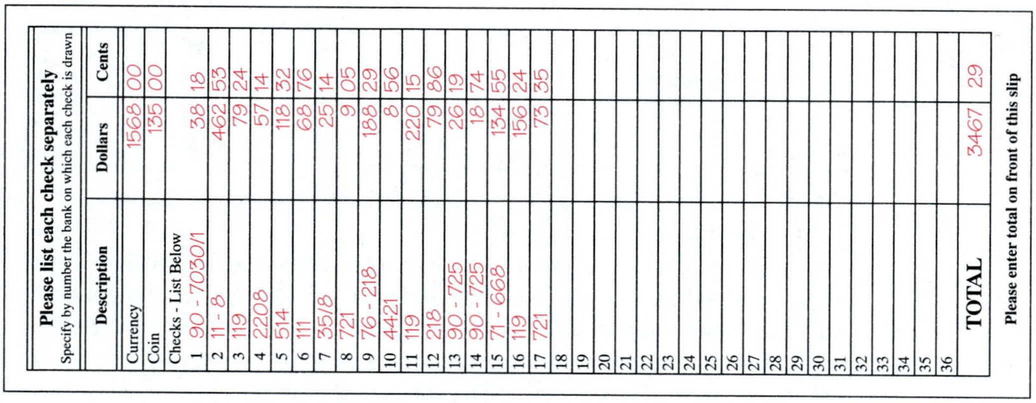

Please list each check separately
Specify by number the bank on which each check is drawn

Description	Dollars	Cents
Currency	1568	00
Coin	135	00
Checks - List Below		
1 90 - 7030/1	38	18
2 11 - 8	462	53
3 119	79	24
4 2208	57	14
5 514	118	32
6 111	68	76
7 35/8	25	14
8 721	9	05
9 76 - 218	188	29
10 4421	8	56
11 119	220	15
12 218	79	86
13 90 - 725	26	19
14 90 - 725	18	74
15 71 - 668	134	55
16 119	156	24
17 721	73	35
18		
19		
20		
21		
22		
23		
24		
25		
26		
27		
28		
29		
30		
31		
32		
33		
34		
35		
36		
TOTAL	3467	29

Please enter total on front of this slip

Figure 4.6

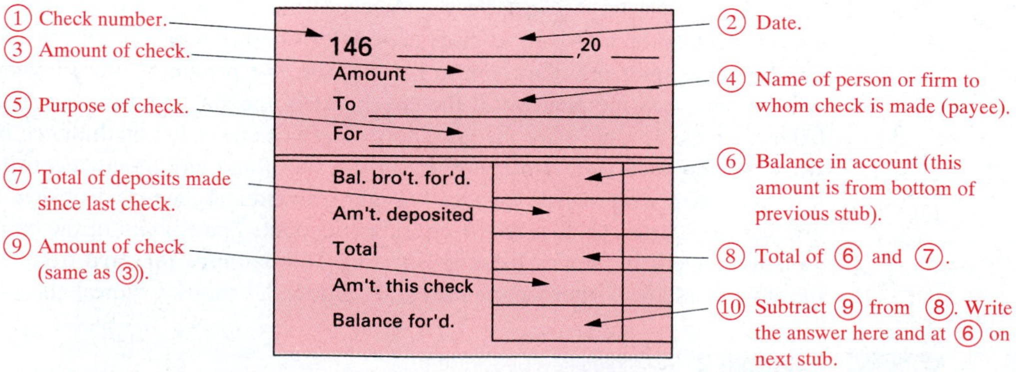

Figure 4.7

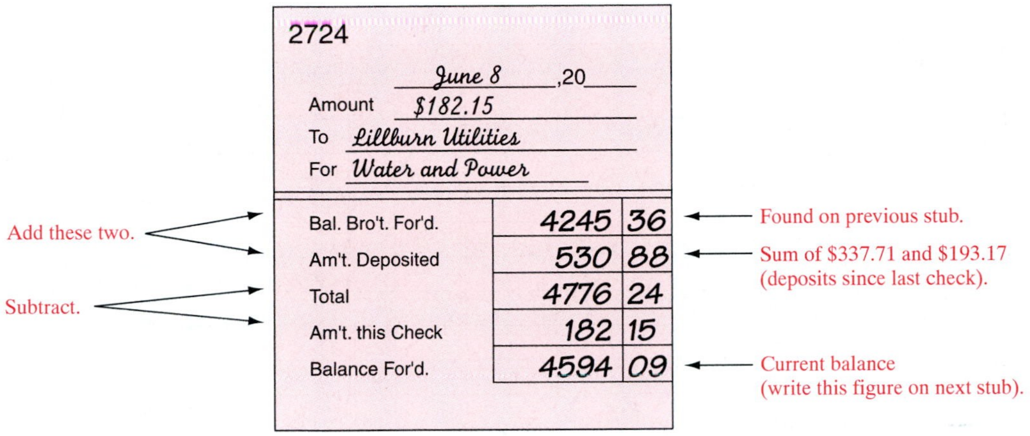

Figure 4.8

$4245.36, and that deposits of $337.71 and $193.17 have been made since the last check was written. Complete the check stub as shown in Figure 4.8.

Banks offer many styles of checkbooks. Notice that the two styles shown in Figure 4.9 offer two stubs and may be used for payrolls. The stub next to the check can be used as the employee's record of earnings and deductions. The second style provides space on the check itself for listing a group of invoices or bills that are being paid with that same check.

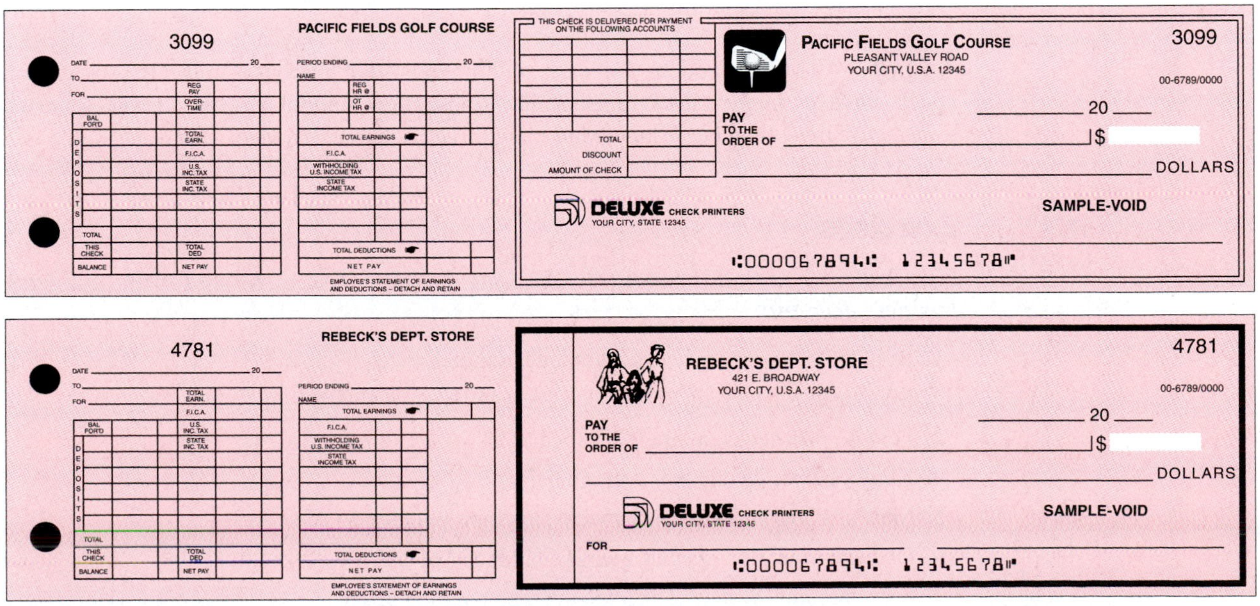

Figure 4.9

CHECK NO.	DATE	CHECK ISSUED TO	AMOUNT OF CHECK		✓	DATE OF DEP.	AMOUNT OF DEPOSIT		BALANCE	
		BALANCE BROUGHT FORWARD →							3518	72
1435	5/8	Swan Brothers	378	93					3139	79
1436	5/8	Class Acts	25	14					3114	65
1437	5/9	Mirror Lighting	519	65					2595	00
		Deposit				5/10	3821	17	6416	17
1438	5/10	Woodlake Auditorium	750	00					5666	17
		Deposit				5/12	500	00	6166	17
1439	5/12	Rick's Clowns	170	80					5995	37
1440	5/14	Y.M.C.A.	219	17					5776	20
	5/14	ATM	120	00					5656	20
		Deposit				5/15	326	15	5982	35
1441	5/16	Stage Door Playhouse	825	00					5157	35
1442	5/17	Gilbert Eckern	1785	00					3372	35
		Deposit				5/19	1580	25	4952	60

Figure 4.10

Objective 6 *Complete the Parts of a Check Register.* Some depositors prefer a check register to the check stubs, while others use both. The **check register** shows at a glance the checks written and deposits made, as seen in Figure 4.10. The column headed ✓ is used to check off each check after it is received back from the bank.

NOTE ATM transactions for cash withdrawals and purchases must be entered on check stubs or in the check register. The transaction amount and the charge for each transaction must then be subtracted to maintain an accurate balance. ▬

4.1 Exercises

Use Table 4.1 to find the monthly checking account service charge for the following business accounts.

1. Books On Line, 92 checks, average balance $4618
2. Mandarin Restaurant, 76 checks, average balance $3318
3. Pest-X, 40 checks, average balance $491
4. Kent's Keys, 23 checks, average balance $215
5. Mak's Smog and Tune, 48 checks, average balance $1763
6. Cellular One, 315 checks, average balance $6424
7. Software and More, 72 checks, average balance $516
8. Mart & Bottle, 74 checks, average balance $875

MAINTAINING BANK RECORDS *Use the following information to complete the check stubs that follow.*

	Date	To	For	Amount	Bal. Bro't. For'd.	Deposits
9.	Mar. 8	Patty Demko	Tutoring	$380.71	$3971.28	$79.26
10.	Oct. 15	Elizabeth Linton	Rent	$850.00	$2973.09	$1853.24
11.	Dec. 4	Paul's Pools	Chemicals	$37.52	$1126.73	—

9.

857

__________ 20___

Amount __________

To __________

For __________

Bal. Bro't. For'd.		
Am't. Deposited		
Total		
Am't. this Check		
Balance For'd.		

10.

1248

__________ 20___

Amount __________

To __________

For __________

Bal. Bro't. For'd.		
Am't. Deposited		
Total		
Am't. this Check		
Balance For'd.		

11.

735

__________ 20___

Amount __________

To __________

For __________

Bal. Bro't. For'd.		
Am't. Deposited		
Total		
Am't. this Check		
Balance For'd.		

12. List and explain at least six parts of a check. Draw a sketch showing where these parts appear on a check. (See Objective 1.)

13. Explain to a friend at least two advantages and two possible disadvantages of using an ATM card. Do this in writing. (See Objective 2.)

14. Write an explanation for a friend of two types of check endorsements. Describe where these endorsements must be placed. (See Objective 4.)

15. Explain in your own words the factors that determine the service charges on a business checking account. (See Objective 3.)

COMPLETING CHECK STUBS ***Using the information below, complete the following check stubs for Books On Line. The balance brought forward for check stub 5311 is $7223.69.***

CHECKS WRITTEN

Number	Date	To	For	Amount
5311	Oct. 7	Anessa Davis	Books	$1250.80
5312	Oct. 10	County Clerk	License	$39.12
5313	Oct. 15	United Parcel	Shipping	$356.28

DEPOSITS MADE

Date	Amount
Oct. 8	$752.18
Oct. 9	$23.32
Oct. 13	$1025.45

16.

5311

______ 20___

Amount ______

To ______

For ______

Bal. Bro't. For'd.		
Am't. Deposited		
Total		
Am't. this Check		
Balance For'd.		

17.

5312

______ 20___

Amount ______

To ______

For ______

Bal. Bro't. For'd.		
Am't. Deposited		
Total		
Am't. this Check		
Balance For'd.		

18.

5313

______ 20___

Amount ______

To ______

For ______

Bal. Bro't. For'd.		
Am't. Deposited		
Total		
Am't. this Check		
Balance For'd.		

BANK BALANCES ***For Exercises 19–22, complete the balance column in the following check registers after each check or deposit transaction.***

19.

CHECK NO.	DATE	CHECK ISSUED TO	AMOUNT OF CHECK		✓	DATE OF DEP.	AMOUNT OF DEPOSIT		BALANCE	
		BALANCE BROUGHT FORWARD ⟶							1629	86
861	7/3	Ahwahnee Hotel	250	45						
862	7/5	Willow Creek	149	00						
863	7/5	Void								
		Deposit				7/7	117	73		
864	7/9	Del Campo High School	69	80						
		Deposit				7/10	329	86		
		Deposit				7/12	418	30		
865	7/14	Big 5 Sporting Goods	109	76						
866	7/14	Dr. Yates	614	12						
867	7/16	Greyhound	32	18						
		Deposit				7/16	520	95		

20.

CHECK NO.	DATE	CHECK ISSUED TO	AMOUNT OF CHECK		✓	DATE OF DEP.	AMOUNT OF DEPOSIT		BALANCE	
		BALANCE BROUGHT FORWARD ⟶							832	15
1121	3/17	AirTouch Cellular	257	29						
1122	3/18	Curry Village	190	50						
		Deposit				3/19	78	29		
		Deposit				3/21	157	42		
1123	3/22	San Juan District	38	76						
1124	3/23	Macy's Gourmet	175	88						
		Deposit				3/23	379	28		
1125	3/24	Class Video	197	20						
1126	3/24	Water World	25	10						
1127	3/25	Bel Air Market	75	00						
		Deposit				3/28	722	35		

21.

CHECK NO.	DATE	CHECK ISSUED TO	AMOUNT OF CHECK		✓	DATE OF DEP.	AMOUNT OF DEPOSIT		BALANCE	
		BALANCE BROUGHT FORWARD ⟶							3852	48
2308	12/6	Web Masters	143	16						
2309	12/7	Water and Power	118	40						
		Deposit				12/8	286	32		
	12/10	ATM (cash)	80	00						
2310	12/11	Ann Kuick	986	22						
2311	12/11	Account Temps	375	50						
		Deposit				12/14	1201	82		
2312	12/14	Central Chevrolet	735	68						
2313	12/15	Miller Mining	223	94						
		Deposit				12/17	498	01		
2314	12/18	Federal Parcel	78	24						

22.

CHECK NO.	DATE	CHECK ISSUED TO	AMOUNT OF CHECK		✓	DATE OF DEP.	AMOUNT OF DEPOSIT		BALANCE	
		BALANCE BROUGHT FORWARD ⟶							8284	18
1917	6/4	Valley Electric	188	18						
1918	6/5	Harrold Ford	433	56						
1919	6/5	Paul Altier (photography)	138	17						
		Deposit				6/6	453	28		
		Deposit				6/8	1475	69		
1920	6/9	U.S. Rentals	335	82						
1921	6/11	Quick Turn Merchandise	573	27						
	6/11	ATM (gas)	16	35						
1922	6/14	Broadly Plumbing	195	15						
		Deposit				6/16	635	85		
1923	6/16	National Dues F.F.A.	317	20						

4.2 Checking Services and Depositing Credit Card Transactions

Objectives

1. ***Identify bank services available to customers.***
2. ***Understand interest-paying checking plans.***
3. ***Deposit credit card transactions.***
4. ***Calculate the discount fee on credit card deposits.***

Most business checking account charges are based on either the average balance or minimum balance in the account, together with specific charges for each service performed by the bank. Following are explanations of some of the more common services provided by banks, along with the typical charges. (These charges may vary from bank to bank.)

Objective 1 *Identify Bank Services Available to Customers.* An **overdraft** occurs when checks are written for which there are **nonsufficient funds (NSF)** in the checking account and when the customer has no overdraft protection. (This may also be referred to as "**bouncing a check**.") A typical charge is \$10 to \$30 per check. The same charges occur when a check is returned because it was improperly completed.

Some **ATM cards** can be used as debit cards when making point-of-sale purchases, although most can only be used to access savings and checking accounts. The fee for purchases varies from \$0.10 per transaction to \$1.00 per month for unlimited transactions. When used at the ATM machine there is usually no fee at your bank branch, a fee as high as \$2.00 at other banks, and an international fee as high as \$5.00.

Overdraft protection is given when an account balance is insufficient to cover the amount of a check and an overdraft occurs. Charges for this vary among banks.

A **returned-deposit item** is a check that was deposited and then returned to the bank, usually because of lack of funds in the account of the person or firm writing the check (\$15.00 per item).

A **stop-payment order** is a request to the bank that it not honor a check which the depositor has written ($15.00 per request).

A **cashier's check** is a check written by the financial institution itself. It therefore has the full faith and backing of the institution ($5.00).

A **money order** is an instrument that is purchased and is often used in place of cash. It is sometimes required by the payee instead of a personal or business check ($4.00).

Noncustomer check cashing is sometimes offered to an individual who does not have an account with the institution ($5.00).

A **notary service** (official certification of a signature on a document) is often given free to customers. This is a service that is required on certain business documents. There is usually a charge to a noncustomer ($10.00).

Objective 2 *Understand Interest-Paying Checking Plans.* Federal banking regulations now allow both personal and business **interest-paying checking plans**. Some of the plans combine two accounts—a savings account and a checking account—while others are simply checking accounts that pay interest on the average daily balance.

The following two pie charts show how consumers have changed the way they pay for their purchases.

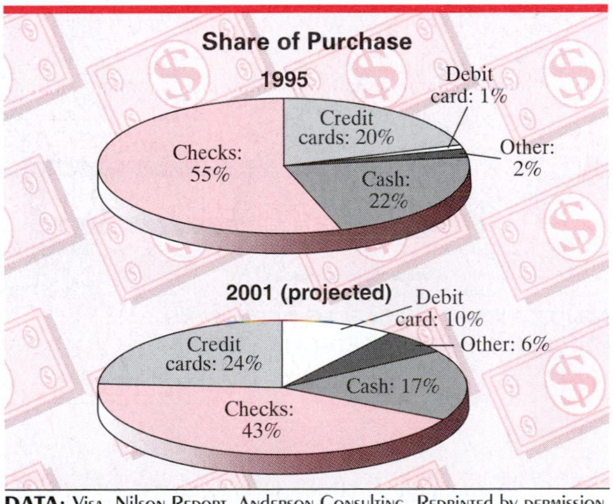

DATA: Visa, Nilson Report, Anderson Consulting. Reprinted by permission.

Objective 3 *Deposit Credit Card Transactions.* With the continuing growth in electronic commerce, more and more credit card transactions are being completed electronically by the retailer. However, a great number of retailers continue to process their credit card sales mechanically. These credit card sales are deposited into a business checking account with a **merchant batch header ticket**, such as the one shown in Example 1. This form is used with Visa or MasterCard credit card deposits. Notice that the form lists both sales slips and credit slips (refunds). Credit slips are used when merchandise purchased with a credit card is returned to the store. Entries in each of these categories are totaled, and the total credits are subtracted from the total sales to give the gross amount of deposit.

The merchant batch header ticket is a triplicate form and the bank copy along with the charge slips, credit slips, and printed calculator tape showing the itemized deposits and credits is deposited in the business checking account.

Example 1 Determining Deposits with Credit Card Transactions

Books On Line had the following credit card sales and refunds. Complete a merchant batch header ticket.

Sales		Refunds (Credit)
$82.31	$146.50	$13.83
$38.18	$78.80	$25.19
$65.29	$63.14	$78.56
$178.22	$208.67	

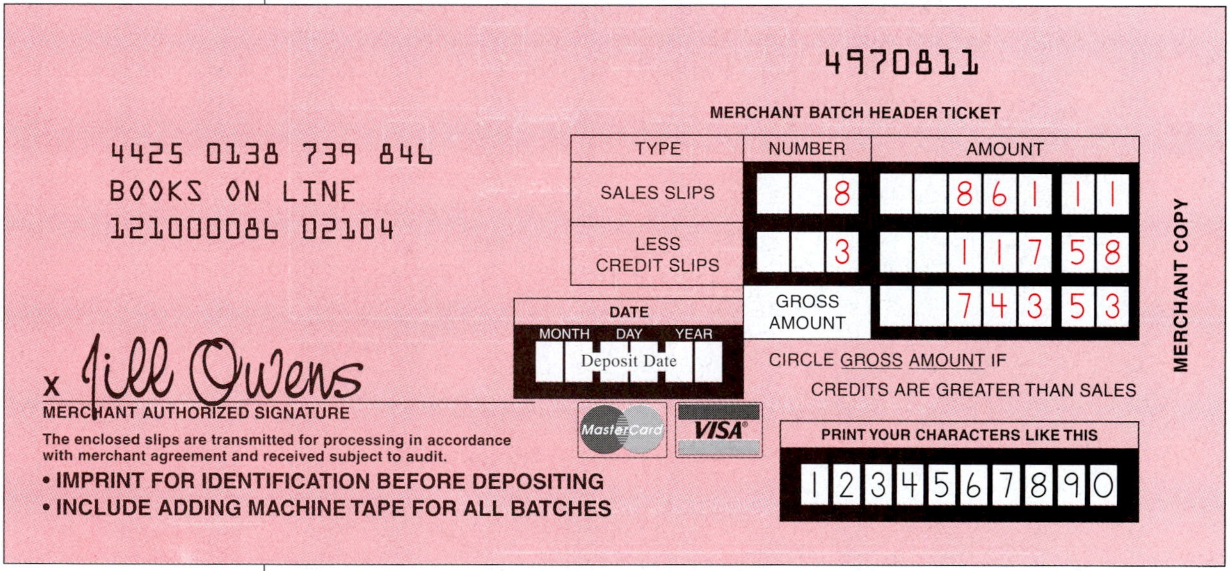

4970811

MERCHANT BATCH HEADER TICKET

4425 0138 739 846
BOOKS ON LINE
121000086 02104

TYPE	NUMBER	AMOUNT
SALES SLIPS	8	861 11
LESS CREDIT SLIPS	3	117 58
	GROSS AMOUNT	743 53

DATE
MONTH DAY YEAR
Deposit Date

CIRCLE GROSS AMOUNT IF CREDITS ARE GREATER THAN SALES

x Jill Owens
MERCHANT AUTHORIZED SIGNATURE

The enclosed slips are transmitted for processing in accordance with merchant agreement and received subject to audit.

- IMPRINT FOR IDENTIFICATION BEFORE DEPOSITING
- INCLUDE ADDING MACHINE TAPE FOR ALL BATCHES

MasterCard VISA

PRINT YOUR CHARACTERS LIKE THIS
1 2 3 4 5 6 7 8 9 0

MERCHANT COPY

Solution All credit slips and sales slips must be totaled. The number of each of these and the totals are written at the right on the form. The total sales slips are $861.11 and the credit slips total $117.58. The difference is the gross amount; here $743.53 is the gross deposit.

Objective 4 ***Calculate the Discount Fee on Credit Card Deposits.*** The bank collects a **discount fee** (a percent of sales) from the merchant and also an interest charge from the card user on all accounts not paid in full at the first billing. Although credit card transactions are deposited frequently by a business, the bank calculates the discount fee on the gross amount of the credit card deposits since the last bank statement date. The fee paid by the merchant varies from 2% to 5% of the sales slip amount and is determined by the type of processing used (electronic or manual), the dollar volume of credit card usage by the merchant, and the average amount of the sale at the merchant's store. All credit card deposits for the month are added, and the fee is subtracted from the total at the statement date.

Example 2 Finding the Discount and the Credit Given on a Credit Card Deposit

The deposit in Example 1 represents total credit card deposits for the month. Find the fee charged and the credit given to the merchant at the statement date if Books On Line pays a 3% fee.

Solution Since the total credit card deposit for Books On Line is \$743.53 and the fee is 3%, the discount charged is

$$\$743.53 \times 0.03\ (3\%) = \$22.31 \text{ discount charge (rounded)}$$

Out of a deposit of \$743.53, the merchant will receive a credit of \$743.53 − \$22.31 = \$721.22.

Scientific *Calculator Approach*

The calculator solution to Example 2 is

\$743.53 [−] 3 [%] [=] 721.2241.

4.2

Exercises

For each of the following businesses, find (a) the total charges, (b) the total credits, (c) the amount of the gross deposit, (d) the amount of the discount charged at the statement date, and (e) the amount of credit given after the fee is subtracted.

1. CREDIT CARD DEPOSITS Fry's Electronics does most of its business on a cash basis or through its own credit department, although it does honor major bank charge cards. In a recent period, the business had the following credit card charges and credits. The bank charges a 3% discount.

Sales		Credits
\$78.56	\$38.15	\$29.76
\$875.29	\$18.46	\$102.15
\$330.82	\$22.13	\$71.95
\$55.24	\$707.37	
\$47.83	\$245.91	

2. CREDIT CARD DEPOSITS Amber Tune Up and Brake accepts travel and leisure cards from customers for auto repair and the sale of parts. The following credit card transactions occurred during a recent period. The bank charges a 4% discount.

Sales		Credits
\$66.68	\$18.95	\$62.16
\$119.63	\$496.28	\$106.62
\$53.86	\$21.85	\$38.91
\$178.62	\$242.78	
\$219.78	\$176.93	

3. CREDIT CARD DEPOSITS Bayside Jeepers does most of its business on a cash basis or through its own credit department, although it does honor major bank charge cards. In a recent period, the business had the following credit card charges and credits. The bank charges a 4% discount.

	Sales		Credits
$25.18	$77.51	$14.73	$38.15
$15.73	$357.18	$106.78	$106.86
$138.97	$72.73	$88.34	$44.38
$58.73	$29.68	$72.21	
$255.18	$15.76	$262.73	

4. CREDIT CARD DEPOSITS Industrial Supply had the following credit card transactions during a recent period. The bank charges a 4.5% discount.

Sales		Credits
$42.60	$29.50	$22.10
$38.25	$72.85	$14.67
$16.60	$19.30	$30.30
$52.40	$6.75	
$14.38	$88.98	

5. CREDIT CARD DEPOSITS Maureen Tomlin Studios had the following credit card transactions during a recent period. The bank charges a 3% discount.

Sales		Credits
$7.84	$98.56	$13.86
$33.18	$318.72	$58.97
$50.76	$116.35	
$12.72	$23.78	
$9.36	$38.95	
$118.68	$235.82	

6. CREDIT CARD DEPOSITS David Fleming owns Campus Bicycle Shop near a college campus. The shop sells new and used bicycle parts, and does a major portion of its business in adjustments and repairs. The following credit card charges and credits took place during a recent period. The bank charges a 5% discount.

Sales		Credits
$16.40	$184.16	$23.17
$18.98	$137.61	$7.26
$6.76	$24.69	$14.53
$11.75	$86.17	
$29.63		

7. List and describe in your own words four services offered to business checking account customers. (See Objective 1.)

8. The merchant accepting a credit card from a customer must pay a fee of 2% to 5% of the transaction amount. Why is the merchant willing to do this? Who really pays this fee?

4.3 Reconciliation

Objectives

1. *Reconcile a bank statement with the checkbook.*
2. *List outstanding checks.*
3. *Find the adjusted bank balance or current balance.*
4. *Use the T-account form of reconciliation.*

Each month, banks send their checking account customers a **bank statement**. This bank statement shows all deposits made during the period covered by the statement, as well as all checks paid by the bank and any automated teller machine (ATM) and debit card transactions. Bank charges for the month covered by the statement are also listed. This is especially important with a business checking account because the bank charge normally varies from month to month. On occasion, a customer's check that was deposited must be returned due to nonsufficient funds (NSF) in the account. This is identified as a **returned check** and the amount of the check must be subtracted from the checkbook balance along with any other charges. The business must then resolve this matter with the writer of the bad check.

NOTE Reconciling the bank statement is an important step in maintaining accurate checking account records and in helping to avoid writing checks for which there are nonsufficient funds. In addition to nonsufficient funds charges, which can be costly, a certain amount of irresponsibility is associated with the person or business who writes "bad checks." ▬

Objective 1 *Reconcile a Bank Statement with the Checkbook.* Many businesses have automatic deposits from customers and other sources made to their accounts. These amounts must be added to the checkbook balance. When the bank statement is received, it is very important to verify its accuracy. In addition, it is a good time to check the accuracy of the check register, making certain that all checks written have been listed and subtracted and that all deposits have been added to the checking account balance. This process of checking the bank statement and the check register is called **reconciliation**.

Reconciliation is best done using the forms usually printed on the back of the bank statement. A sample bank statement is shown in Figure 4.11, and an example of the reconciliation process follows. Note the codes used by the bank, and their meaning, listed at the bottom of the statement. The codes on this bank statement indicate the following: RC means Returned Check, SC means Service Charge, IC means Interest Credit, ATM means Automated Teller Machine.

Figure 4.11

STATEMENT

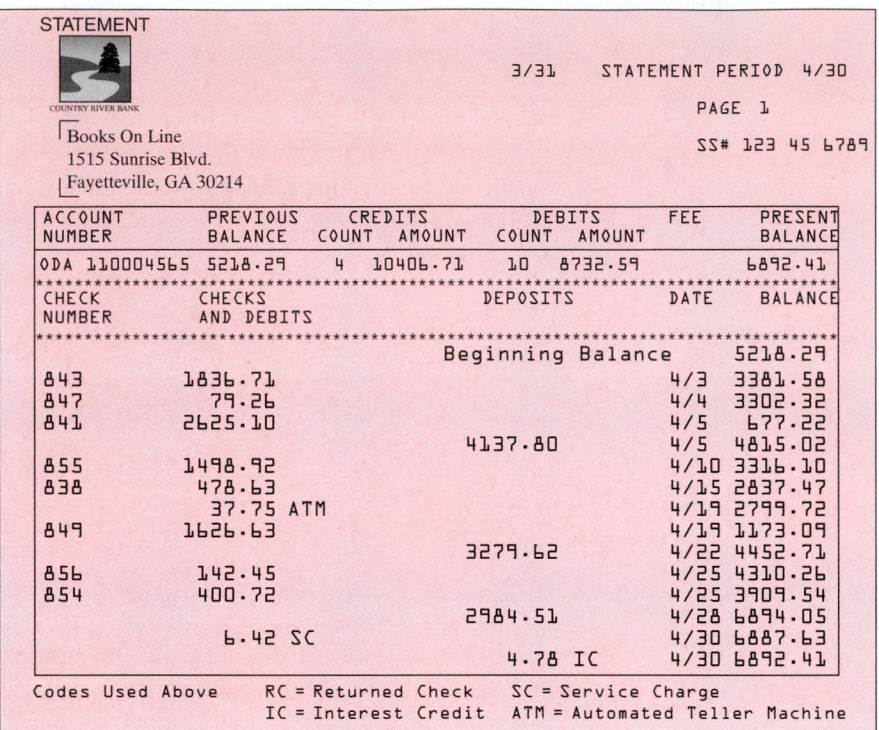

COUNTRY RIVER BANK

Books On Line
1515 Sunrise Blvd.
Fayetteville, GA 30214

3/31 STATEMENT PERIOD 4/30

PAGE 1

SS# 123 45 6789

ACCOUNT NUMBER	PREVIOUS BALANCE	CREDITS COUNT	CREDITS AMOUNT	DEBITS COUNT	DEBITS AMOUNT	FEE	PRESENT BALANCE
ODA 110004565	5218.29	4	10406.71	10	8732.59		6892.41

CHECK NUMBER	CHECKS AND DEBITS	DEPOSITS	DATE	BALANCE
		Beginning Balance		5218.29
843	1836.71		4/3	3381.58
847	79.26		4/4	3302.32
841	2625.10		4/5	677.22
		4137.80	4/5	4815.02
855	1498.92		4/10	3316.10
838	478.63		4/15	2837.47
	37.75 ATM		4/19	2799.72
849	1626.63		4/19	1173.09
		3279.62	4/22	4452.71
856	142.45		4/25	4310.26
854	400.72		4/25	3909.54
		2984.51	4/28	6894.05
	6.42 SC		4/30	6887.63
		4.78 IC	4/30	6892.41

Codes Used Above RC = Returned Check SC = Service Charge
IC = Interest Credit ATM = Automated Teller Machine

Example 1 Reconciling a Checking Account

Books On Line received its bank statement. The statement shows a balance of \$6892.41, after a bank service charge of \$6.42 and an interest credit of \$4.78. Books On Line checkbook now shows a balance of \$7576.38. Reconcile the account using the following steps (illustrated in Figure 4.12).

Solution

Figure 4.12

Checks Outstanding		
Number	Amount	
846	\$ 42	73
852	598	71
853	68	12
857	79	80
858	160	30
Total	\$ 949	66

Compare the list of checks paid by the bank with your records. List and total the checks not yet paid.

(1) Enter new balance from bank statement: \$ 6892.41

(2) List any deposits made by you and not yet recorded by the bank: + 892.41; + 739.58; + ; +

(3) Add all numbers from lines above. Total: 8524.40

(4) Write total of checks outstanding: − 949.66

(5) Subtract (4) from (3). This is adjusted bank balance: \$ 7574.74

To reconcile your records:

(6) List your checkbook balance: \$ 7576.38

(7) Write the total of any fees or charges deducted by the bank and not yet subtracted by you from your checkbook: − 6.42

(8) Subtract line (7) from line (6). 7569.96

(9) Enter interest credit: (Add to your checkbook) + 4.78

(10) Add line (9) to line (8). Adjusted checkbook balance. \$ 7574.74

New balance of your account; this number should be same as (5).

Objective 2 *List Outstanding Checks.* Compare the list of checks on the bank statement with the list of checks written by the firm. Checks that have been written by the firm but do not yet appear on the bank statement had not been paid by the bank as of the date of the statement. These unpaid checks are called **checks outstanding**. The following table shows those checks written by the firm that are outstanding.

Number	Amount	Number	Amount
846	$42.73	857	$79.80
852	$598.71	858	$160.30
853	$68.12		

After listing the outstanding checks in the space provided on the form, total them. The total is $949.66.

The following steps are used to reconcile the checking account of Books On Line.

Reconciling a Checking Account

Step 1. Enter the new balance from the front of the bank statement. As given, the new balance is $6892.41. Write this number in the space provided on the reconcilement form.

Step 2. List any deposits made that have not yet been recorded by the bank (deposits in transit). Suppose that Books On Line has deposits of $892.41 and $739.58 that are not yet recorded. These numbers are written at Step 2 on the form.

Step 3. All the numbers from Steps 1 and 2 are added. Here the total is $8524.40.

Step 4. Write down the total of outstanding checks. The total is $949.66.

Step 5. Subtract the total in Step 4 from the number in Step 3. The result here is $7574.74, called the **adjusted bank balance** or the **current balance**. This number should represent the current checking account balance.

Objective 3 *Find the Adjusted Bank Balance or Current Balance.* Now look at the firm's own records.

Step 6. List the firm's checkbook balance. As mentioned before, the checkbook balance for Books On Line is $7576.38. This number is entered on line 6.

Step 7. Enter any charges not yet deducted. The check charge here is $6.42. Since there are no other fees or charges, enter $6.42 on line 7.

Step 8. Subtract the charges on line 7 from the checkbook balance on line 6 to get $7569.96.

Step 9. Enter the interest credit on line 9. The interest credit here is $4.78. (This amount is interest paid on the money in the account.)

Step 10. Add the interest on line 9 to get $7574.74, the same result as in Step 5.

Since the result from Step 10 is the same as the result from Step 5, the account is balanced (reconciled). The correct current balance in the account is $7574.74.

Bank Reconciliation

Bank-statement balance	$ ______	**Checkbook balance**	$ ______
Add:		**Add:**	
1. Add all deposits not yet recorded.	$ ______	**1.** Add all miscellaneous credits, collections, and interest.	$ ______
Less:		**Less:**	
2. Subtract all outstanding checks.	$ ______	**2.** Subtract previously deposited overdrafts and bank charges.	$ ______
Adjusted balance	$ ======	**Adjusted balance**	$ ======

Figure 4.13

Objective 4 *Use the T-Account Form of Reconciliation.* Many business people and accountants prefer a **T-account form** for bank reconciliation. With this method, the bank statement balance is written on the left and the checkbook balance is written on the right. Adjustments are made to either the bank balance or the checkbook balance, depending on which side was unaware of the transaction or charge. T-account reconciliation uses the format in Figure 4.13. The adjusted balances must agree, with the result showing the actual amount remaining in the account.

Example 2 Using the T-Account Form

The bank statement of Hazel Nut Gifts shows a balance of \$4385.88. Checks outstanding are \$292.70, \$75.16, and \$636.55; deposits not yet recorded are \$483.11 and \$89.95. Also appearing are a service charge of \$7.90, a check-printing charge of \$9.20, a returned check (NSF) for \$94.25, and an interest credit of \$10.06. The checkbook now shows a balance of \$4055.82. Use the T-method to reconcile the checking account.

Solution The reconciliation is shown in Figure 4.14.

There are several typical reasons why checking accounts do not balance.

Why Checking Accounts Do Not Balance

- Forgetting to enter a check in the check register.
- Forgetting to enter a deposit in the check register.
- Transposing numbers (writing \$961.20 as \$916.20, for example).
- Addition or subtraction errors.
- Forgetting to subtract one of the bank service fees, such as those charged for using your debit card or for ATM use.
- The bank may have charged the customer an amount different from the check amount.
- A check may be altered or forged.

Figure 4.14

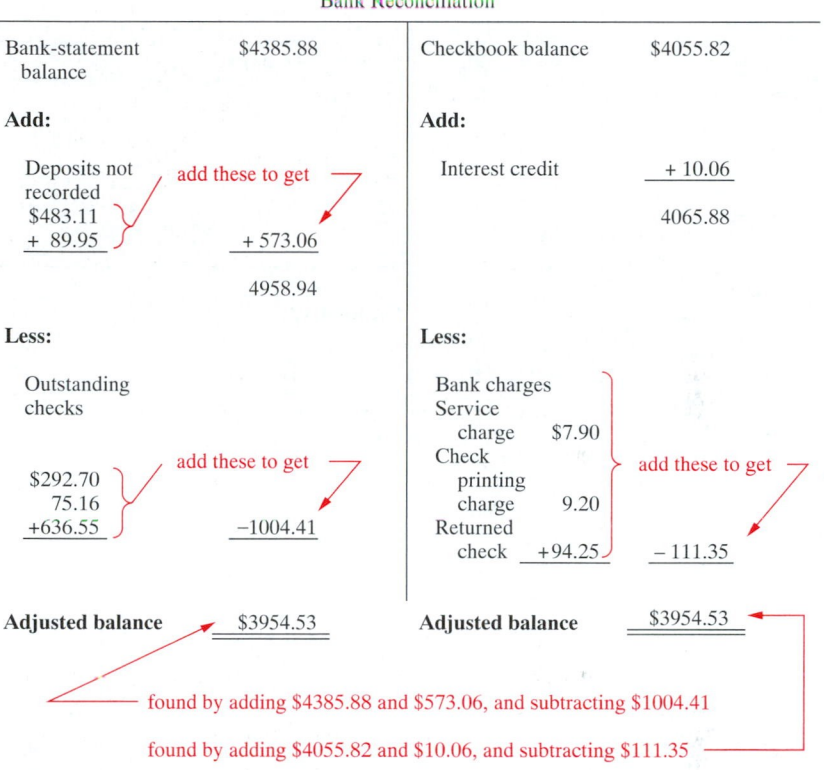

Bank Reconciliation

Bank-statement balance	$4385.88	Checkbook balance	$4055.82
Add:		**Add:**	
Deposits not recorded $483.11 + 89.95	+ 573.06	Interest credit	+ 10.06
	4958.94		4065.88
Less:		**Less:**	
Outstanding checks $292.70 75.16 +636.55	−1004.41	Bank charges Service charge $7.90 Check printing charge 9.20 Returned check +94.25	− 111.35
Adjusted balance	$3954.53	**Adjusted balance**	$3954.53

add these to get (deposits not recorded → + 573.06)

add these to get (outstanding checks → −1004.41)

add these to get (bank charges → − 111.35)

found by adding $4385.88 and $573.06, and subtracting $1004.41

found by adding $4055.82 and $10.06, and subtracting $111.35

Example 3 Reconciling a Checking Account

A checking account register is shown in Figure 4.15. A ✓on the register indicates that the check appeared on the previous month's bank statement. Reconcile the account with a bank statement shown in Figure 4.16. (Codes on the statement have the following meaning: RC, Returned Check; SC, Service Charge; IC, Interest Credit; ATM, Automated Teller Machine.)

Figure 4.15

CHECK NO.	DATE	CHECK ISSUED TO	AMOUNT OF CHECK		✓	DATE OF DEP.	AMOUNT OF DEPOSIT		BALANCE	
		BALANCE BROUGHT FORWARD →							2782	95
721	7/11	Miller's Outpost	138	50	✓				2644	45
722	7/12	Barber Advertising	73	08					2571	37
723	7/18	Wayside Lumber	318	62	✓				2252	75
		Deposit				7/20	1060	37	3313	12
724	7/25	I.R.S.	836	15					2476	97
725	7/26	John Lessor	450	00					2026	97
726	7/28	Sacramento Bee	67	80					1959	17
727	8/2	T.V.A.	59	25					1899	92
728	8/3	Carmichael Office	97	37					1802	55
		Deposit				8/6	795	45	2598	00
ATM	8/5	ATM Cash	80	00					2518	00

Bank Statement

CHECK NUMBER	CHECKS AND DEBITS		DEPOSITS	DATE	BALANCE
				7/20	2325.83
722	73.08			7/22	2252.75
			1060.37	7/24	3313.12
724	836.15			7/28	2476.97
725	450.00	49.07 RC		7/30	1977.90
727	59.25		3.22 IC	8/4	1921.87
		80.00 ATM		8/5	1841.87
		7.60 SC		8/5	1834.27

Figure 4.16

Bank Reconciliation

Bank statement balance		$1834.27	Checkbook balance		$2518.00
Add:			**Add:**		
Deposits not recorded		+ 795.45	Interest credit (IC)		+3.22
		2629.72			2521.22
Less:			**Less:**		
Outstanding checks			*Bank charges* Returned check (RC)	$49.07	
	$67.80 +97.37	−165.17	Service charge (SC)	+7.60	−56.67
Adjusted balance		$2464.55	**Adjusted balance**		$2464.55

Figure 4.17

Solution Follow the instructions on the form in Figure 4.13. The completed reconciliation is shown in T-form in Figure 4.17.

4.3 Exercises

Find the current balance for each of the following accounts.

	Balance from Bank Statement	Checks Outstanding		Deposits Not Yet Recorded
1.	$4,572.15	$ 97.68	$348.17	$816.14
		$418.25		$571.28
2.	$6,274.76	$381.40	$681.10	$346.65
		$875.14	$83.15	$198.96

3. $7,911.42	$52.38	$528.02	$492.80
	$95.42	$76.50	$38.72
4. $9,343.65	$840.71	$665.73	$971.64
	$78.68	$87.00	$3382.71
5. $19,523.20	$6853.60	$340.00	$6724.93
	$795.77	$22.85	$78.81
6. $32,489.50	$3589.70	$18,702.15	$7110.65
	$263.15	$7,269.78	$2218.63

RECONCILING CHECKING ACCOUNTS *Use the steps given in Example 1 and Figure 4.12 to reconcile the following accounts.*

	7.		**8.**	
Balance from bank statement		$6875.09		$14,928.42
Checks outstanding (check	421	$371.52	112	$84.76
number is given first)	424	$429.07	115	$109.38
	427	$883.69	117	$42.03
	429	$35.62	119	$1,429.12
Deposits not yet recorded		$701.56		$54.21
		$421.78		$394.76
		$689.35		$1,002.04
Bank charge		$8.75		$7.00
Interest credit		$10.71		$22.86
Checkbook balance		$6965.92		$14,698.28

RECONCILING CHECKING ACCOUNTS Use the T-account form, Figure 4.13, to reconcile the following accounts.

9. The checkbook of Dottie Fogel Furnishings shows a balance of $7779. When the bank statement was received, it showed a balance of $6237.44, a returned check amounting to $246.70, a service charge of $15.60, and a check-printing charge of $18.50. There were unrecorded deposits of $1442.44 and $479.50, and checks outstanding of $146.36, $91.52, $43.78, and $379.52.

10. Charles Hickman received a bank statement showing a balance of $1248.63, a returned check amounting to $35.17, a service charge of $7.70, and an interest credit of $2.51. Checks outstanding were $380, $36.66, $15.29, and $143.18; deposits not yet recorded were $478.18 and $359.12. The checkbook showed a balance of $1551.16. Reconcile the checking account.

11. The bank statement of William Poole Enterprises showed a bank balance of $4074.65, a returned check amounting to $168.40, a service charge of $7.08, and an interest credit of $10.18. There were unrecorded deposits of $907.82 and $1784.15 and checks outstanding of $642.55, $1082.98, $73.25, and $471.83. The checkbook showed a balance of $4661.31.

12. The bank statement of Laura Resca Decorating showed a balance of $1270.08. The checkbook balance showed $1626.63. There were unrecorded deposits of $370.64 and $219.38 and outstanding checks of $38.18, $185.10, $14.75, and $90.14. Check-printing charges were $8.50; the service charge was $3.80; there was a returned check of $83.85 and an interest credit of $1.45.

13. Explain in your own words the significance of writing a back check. What might the cost be in dollars? What are the other consequences? (See Objective 1.)

14. What happens when you as a businessperson receive a bad check? What are the financial costs to the business? What are you likely to do regarding this customer? (See Objective 1.)

15. Briefly describe the importance of reconciling a checking account. What are the benefits derived from keeping good checking records? (See Objective 2.)

16. Suppose your checking account will not balance. Name four types of errors that you will look for in trying to correct this problem.

Reconcile the following checking accounts. Compare the items appearing on the bank statement to the check register. A ✓ indicates that the check appeared on the previous month's statement. (Codes indicate the following: RC means Returned Check, SC means Service Charge, CP means Check Printing Charge, IC means Interest Credit, ATM means Automated Teller Machine.)

17.

CHECK NO.	DATE	CHECK ISSUED TO	AMOUNT OF CHECK	✓	DATE OF DEP.	AMOUNT OF DEPOSIT	BALANCE
		BALANCE BROUGHT FORWARD ⟶					7682 07
662	3/3	Action Packing Supplies	451 16				7230 91
663	3/3	Crown Paper	954 29	✓			6276 62
664	3/5	ATM Cash	80 00	✓			6196 62
		Deposit			3/7	913 28	7109 90
665	3/10	Fairless Water District	72 37				7037 53
666	3/12	Audia Temporary	340 88				6696 65
667	3/13	Lionel Toys	618 65				6078 00
668	3/14	Fairless Hills Power	100 50				5977 50
		Deposit			3/16	450 18	6427 68
		Deposit			3/18	163 55	6591 23
669	3/20	Hunt Roofing	238 50				6352 73
670	3/22	Standard Brands	315 62				6037 11
671	3/23	Penny-Saver Products	67 29				5969 82
		Deposit			3/24	830 75	6800 57

Bank Statement

CHECK NUMBER	CHECKS AND DEBITS		DEPOSITS	DATE	BALANCE
				3/5	6647.78
			913.28	3/7	7561.06
662	451.16			3/11	7109.90
666	340.88	82.15 RC	450.18	3/16	7137.05
665	72.37		22.48 IC	3/20	7087.16
667	618.65		163.55	3/22	6632.06
669	238.50	12.70 SC		3/26	6380.86

18.

CHECK NO.	DATE	CHECK ISSUED TO	AMOUNT OF CHECK	✓	DATE OF DEP.	AMOUNT OF DEPOSIT	BALANCE
		BALANCE BROUGHT FORWARD →					6669.34
760	2/8	Floors to Go	248.96				6420.38
762	2/9	Healthways Dist.	125.63				6294.75
		Deposit			2/11	618.34	6913.09
763	2/12	Franchise Tax	770.41	✓			6142.68
764	2/14	Foothill Repair	22.86	✓			6119.82
765	2/15	Yellow Pages	91.24				6028.58
		Deposit			2/17	826.03	6854.61
766	2/17	Morning Herald	71.59				6783.02
767	2/18	San Juan Electric	63.24				6719.78
ATM	2/22	ATM Gas	15.26				6704.52
769	2/23	West Construction	405.07				6299.45
770	2/24	Rent	525.00				5774.45
		Deposit			2/26	220.16	5994.61
771	2/28	Capital Alarm	135.76				5858.85

Bank Statement

CHECK NUMBER	CHECKS AND DEBITS		DEPOSITS	DATE	BALANCE
				2/14	5876.07
765	91.24		618.34	2/16	6403.17
760	248.96		826.03	2/17	6980.24
766	71.59			2/19	6908.65
762	125.63			2/21	6783.02
		198.17 RC		2/22	6584.85
		15.26 ATM	8.12 IC	2/24	6577.71
769	405.07	4.85 CP		2/26	6167.79
		6.28 SC		2/27	6161.51
770	525.00			2/28	5636.51

Chapter 4 Quick Review

Chapter Terms

Review the following terms to test your understanding of the chapter. For each term you do not know, refer to the page number found next to that term.

Concepts	Example
4.1 Checking account service charges There is usually a checking account maintenance charge and often a per check charge.	Find the monthly checking account service charge for a business with 36 checks and transactions, given a monthly maintenance charge of \$7.50 and a \$0.20 per check charge. \$7.50 + 36(\$0.20) = \$7.50 + \$7.20 = \$14.70 monthly service charge
4.2 Banking services offered The checking account customer must be aware of various banking services that are offered.	*Overdraft protection:* offered to protect the customer from bouncing a check (NSF) *ATM card:* used at automated teller machine to get cash or used as a debit card to make purchases *Stop-payment order:* stops payment on a check written in error *Cashier's check:* a check written by the financial institution itself *Money order:* an instrument used in place of cash *Notary service:* an official certification of a signature or document

4.2 Depositing credit card transactions

Subtract credit card refunds from total credit card sales to find the gross deposit. Then subtract the discount charge from this total.

The following are credit card charges and credits.

Sales		Credits
$28.15	$78.59	$21.86
$36.92	$63.82	$19.62

(a) Find total sales.

$$\$28.15 + \$36.92 + \$78.59 + \$63.82 = \$207.48$$

(b) Find total credits.

$$\$21.86 + \$19.62 = \$41.48$$

(c) Find gross amount.

$$\$207.48 - \mathbf{\$41.48} = \$166$$

(d) Given a 3% fee, find the amount of the charge.

$$\$166 \times \mathbf{0.03} = \$4.98$$

(e) Find the amount of credit given to the business.

$$\$166 - \mathbf{\$4.98} = \$161.02$$

4.3 Reconciliation of a checking account

A checking account customer must periodically verify checking account records with those of the bank or financial institution. The bank statement is used for this.

The accuracy of all checks written, deposits made, service charges incurred, and interest paid is checked and verified. The customer's checkbook balance and bank balance must be the same for the account to reconcile, or balance.

Chapter 4 Review Exercises

The answer section includes answers to all Review Exercises.

Use Table 4.1 on page 119 to find the monthly checking account service charge for the following accounts. **[4.1]**

1. The Sub Shop, 42 checks, average balance $1478
2. Sangi Market, 35 checks, average balance $485
3. Old English Chimney Sweep, 52 checks, average balance $3017

Complete the following three check stubs for Jack Armstrong International Trucking Company. The balance brought forward for stub 1561 is $16,409.82. Find the balance forward at the bottom of each stub. **[4.1]**

CHECKS WRITTEN

Number	Date	To	For	Amount
1561	Aug. 6	Fuel Depot	Fuel	$6892.12
1562	Aug. 8	First Bank	Payment	$1258.36
1563	Aug. 14	Security Service	Guard dogs	$416.14

Deposits made: $1572 on Aug. 7, $10,000 on Aug. 10.

4. 1561

Aug. 6, 20___
Amount $6,892.12
To Fuel Depot
For Fuel

Bal. Bro't. For'd.	$16,409	82
Am't. Deposited		
Total	16,409	82
Am't. this Check	6,892	12
Balance For'd.	9517	70

5. 1562

Aug. 8, 20___
Amount $1,258.36
To First Bank
For Payment

Bal. Bro't. For'd.	$9,517	70
Am't. Deposited	1,572	00
Total	11,089	70
Am't. this Check	1,258	36
Balance For'd.	9,831	34

6. 1563

Aug. 14, 20___
Amount $416.14
To Security Service
For Guard dogs

Bal. Bro't. For'd.	$9,831	34
Am't. Deposited	10,000	00
Total	19,831	34
Am't. this Check	416	14
Balance For'd.	19,415	20

Chuck Hickman owns Campus Bicycle Shop near campus. The shop sells new and used bicycles and does repairs as well. The following credit card transactions occurred during a recent period. **[4.2]**

Sales		Credits
$118.68	$235.82	$15.36
$7.84	$98.56	$57.47
$33.18	$318.72	
$50.76	$116.35	
$12.72	$23.78	
$9.36	$38.95	

7. Find the total charges for the store.
8. What is the total amount of the credits?
9. Find the amount of the gross deposit when these credit card transactions are deposited.

10. If the bank charges the retailer a 4% discount charge, what is the amount of the discount charge at the statement date?

11. Find the amount of credit given to Campus Bicycle Shop after the fee is subtracted.

Solve the following application problems.

12. Tracey Pittrof Antiques received a bank statement showing a balance of \$4964.52, a returned check amounting to \$140.68, a service charge of \$30.84, and an interest credit of \$10.04. Checks outstanding are \$1520, \$146.64, \$31.16, and \$572.76; deposits not yet recorded are \$1912.72 and \$1436.48. The checkbook shows a balance of \$6204.64. Use Figure 4.12 to reconcile the checking account. **[4.3]**

13. The bank statement of Home Page Services showed a bank balance of \$8149.30, a returned check amounting to \$336.80, a service charge of \$14.16, and an interest credit of \$20.36. There were unrecorded deposits of \$1815.64 and \$3568.30, and checks outstanding of \$1285.10, \$2165.96, \$146.50, and \$943.66. The checkbook shows a balance of \$9322.62. Use Figure 4.13 to reconcile the checking account. **[4.3]**

14. Use Figure 4.13 and the following check register and bank statement to reconcile the checking account. Compare the items appearing on the bank statement to the check register. A √ indicates that the check appeared on the previous month's statement. (Codes indicate the following: RC means Returned Check, SC means Service Charge, CP means Check Printing Charge, IC means Interest Credit, ATM means Automated Teller Machine.) **[4.3]**

CHECK NO.	DATE	CHECK ISSUED TO	AMOUNT OF CHECK	√	DATE OF DEP	AMOUNT OF DEPOSIT	BALANCE
						BALANCE BROUGHT FORWARD →	1876 93
318	9/6	MUIR TRAVEL	76 18	√			1800 75
319	9/6	NORTH COAST TOURS	322 40				1478 35
320	9/8	AMES PHOTO	41 12	√			1437 23
		DEPOSIT			9/10	851 62	2288 85
321	9/14	AMERICAN FLYERS	970 40				1318 45
322	9/15	REVERE INTER.	386 92				931 53
		DEPOSIT			9/18	995 20	1926 73
324	9/20	IDAHO EDISON	68 17				1858 56
325	9/20	WESSON SUPPLY	195 76				1662 80
326	9/22	PARKER PACKERS	348 33				1314 47
327	9/23	FREEZE DRY SUPPLY	215 84				1098 63
328	9/24	COUNTY WATER	169 56				929 07
		DEPOSIT			9/28	418 35	1347 42

BANK STATEMENT

Check Number	Checks and Debits		Deposits	Date	Balance
				9/9	1759.63
			851.62	9/10	2611.25
321	970.40			9/15	1640.85
319	322.40			9/18	1318.45
322	386.92	78.93 RC	995.20	9/20	1847.80
325	195.76		6.52 IC	9/23	1658.56
326	348.33	7.80 SC		9/25	1302.43

Chapter 4 Summary Exercise
The Banking Activities of a Retailer

Shafali Patel owns a retail store specializing in women's imported clothing. She sells authentic traditional fabrics and women's accessories from various parts of Europe, India, and other countries. Many of her customers use credit cards for their purchases and her credit card sales in a recent month amounted to $6438.50. During the same period she had $336.81 in credit slips and paid a credit card fee of 3.5 percent.

When she received her bank statement, the balance was $4228.34. The checks outstanding were found to be $758.14, $38.37, $1671.88, $120.13, $2264.75, $78.11, $3662.73, $816.25, and $400. There were also credit card deposits and bank deposits of $458.23, $771.18, $235.71, $1278.55, $663.52, and $1475.39 that were not recorded.

(a) Find the gross deposit when the credit card sales and credits are deposited.

(b) Find the amount of the credit given to Patel after the fee is subtracted.

(c) What is the total of the checks outstanding?

(d) Find the total of the deposits that were not recorded.

(e) Find the current balance in Patel's checking account.

Net Assets Business on the Internet

Amazon.com

Statistics

- 1995: Established
- 1995: 3 million World Wide Web buyers
- 1999: Offers 4.7 million book, video, music CD, and computer game titles
- 2002: 125 million Web buyers (estimated)

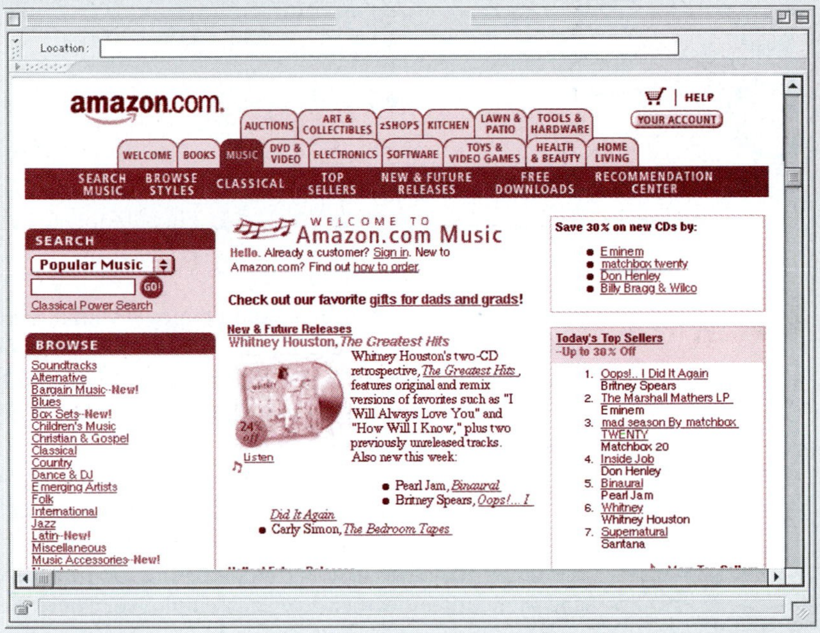

Amazon.com was started in 1995 by Jeff Bezos, who realized the potential for doing business on-line when he read of the 2300% annual growth in Web use. Using the business skills he acquired working on Wall Street, Bezos launched Amazon.com, the world's biggest bookstore. At first, he was selling books to a handful of customers. Today, Amazon.com is the world's e-commerce leader, offering millions of book, CD, and DVD titles, and even power tools, to millions of customers.

As the fastest-growing retailer in history, the Amazon.com corporate philosophy is simple: "If it's good for our customers, it's worth doing." The company mission is to leverage technology and expertise to provide the best buying experience on the Internet.

1. If Amazon.com pays a monthly checking-account fee of \$5.00 plus \$0.10 per check. Find the total checking account charge for a month when 836 checks were written.

2. The total credit card sales for Amazon.com during a certain period were \$837,422, while credit card returns for the same period were \$28,225. Find **(a)** the gross amount of the credit card deposit and **(b)** the amount of credit given to Amazon.com after a fee of 2% is subtracted.

Books On Line had a bank balance of \$9738 on April 1. During April, Jill Owens deposited \$36,282 received from sales, \$642 received as credits from suppliers, and \$63 as a public-utility refund. She paid out \$28,117 to suppliers; \$1476 for rent, utilities, and insurance; and \$3620 for salaries. Use this information to solve questions 3 and 4.

3. Find **(a)** the amount Jill Owens deposited in April and **(b)** the amount she paid out.

4. Find the balance in the account at the end of April.

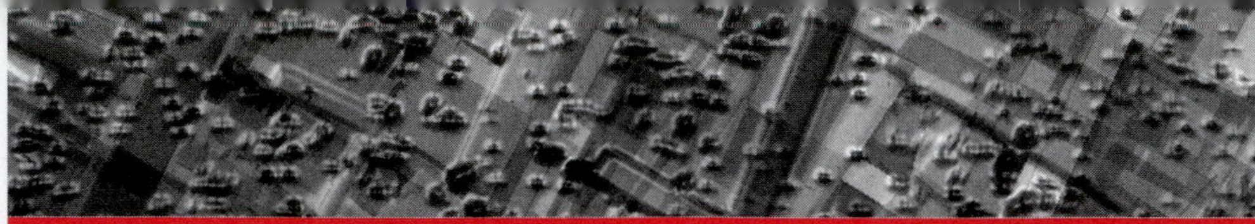

5 Payroll

Business owners, managers, and employees alike take a strong interest in payroll. The amount of pay earned by an employee is most often determined by the number of hours worked or by specific tasks accomplished. Figure 5.1 shows several reasons why full-time workers are spending more time on the job and gives useful information to both employees and employers.

Identify reasons that you have given for spending more time on the job.

Preparing the payroll is one of the most important jobs in any office. Payroll records must be accurate, and the payroll must be prepared on time so that the necessary checks can be written. The first step in preparing the payroll is to determine the **gross earnings** (the total amount earned) for each employee. There are many methods used to find gross earnings and several of these are discussed in this chapter. A number of **deductions** may be subtracted from gross earnings to find **net pay**, the amount actually received by the employee. These various deductions will also be discussed in this chapter. Finally, the employer must keep records to maintain an efficient business and to satisfy legal requirements.

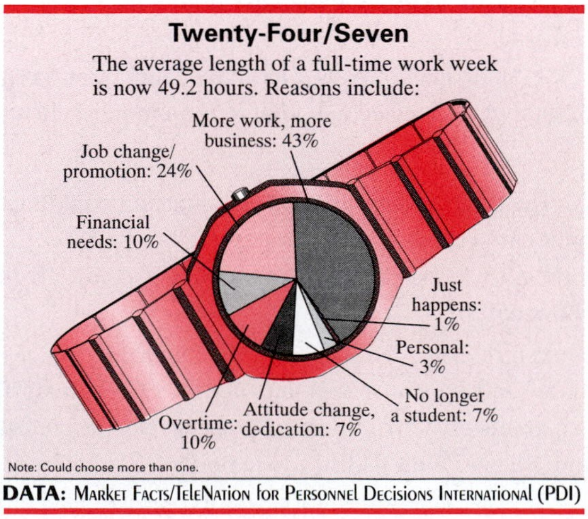

Figure 5.1

5.1 Gross Earnings (Wages and Salaries)

Objectives

1. *Use hourly rate to calculate gross earnings.*
2. *Find overtime earnings for over 40 hours of work per week.*
3. *Use overtime premium method of calculating gross earnings.*
4. *Find overtime earnings using time-and-a-half rate for over 8 hours of work per day.*
5. *Understand double time, shift differential, and split-shift premiums.*
6. *Find equivalent earnings for different pay periods.*
7. *Find gross earnings when overtime is paid to salaried employees.*

Several methods are used for finding an employee's pay. Two of these methods, salaries and wages, are discussed in this section; two additional methods, piecework and commission, will be discussed in the next two sections.

In many businesses, the first step in preparing the payroll is to look at the **time card** maintained for each employee. An example of a time card is shown in Figure 5.2. The card includes the dates of the pay period; the employee's name and other personal information; the days, times, and hours worked; the total number of hours worked; and a signature verification by the employee as to the accuracy of the card. While the card in Figure 5.2 is filled in by hand, many companies use a time clock that automatically stamps the days, dates, and times on the card. The information on these cards is then transferred to a **payroll register** (a chart showing all payroll information), as shown in Example 1.

Objective 1 *Use Hourly Rate to Calculate Gross Earnings.* Eboni Perkins, whose time card is shown in Figure 5.2, is paid an **hourly wage** of $9.80 (see the time card). Her gross earnings would be calculated with the formula

$$\text{Gross earnings} = \text{Number of hours worked} \times \text{Rate per hour}$$

For example, if Perkins works 7 hours at $9.80 per hour, her gross earnings would be

$$\text{Gross earnings} = 7 \times \$9.80 = \$68.60$$

Example 1 Completing a Payroll Register

Meg Holden is doing the payroll for two employees, S. Abruzzo and N. Williams. The first thing she must do is complete a payroll ledger.

	Hours Worked									
Employee	S	M	T	W	Th	F	S	Total Hours	Rate	Gross Earnings
Abruzzo, S.	—	2	4	8	6	3	—		$8.40	
Williams, N.	—	3.5	3	7	6.75	7	—		$7.12	

SEMI-MONTHLY OR TWO WEEKLY

PAYROLL CARD

NO TIME CLOCK REQUIRED

EMPL. NO. 1375 CARD NO. ______

FULL NAME EBONI PERKINS AGE (IF UNDER 18) ______

ADDRESS 1900 EAST LAKE SOCIAL SECURITY NO. 545-06-3189

DATE EMPLOYED ______ POSITION ______ RATE $ 9.80

PAY PERIOD STARTING 7/23 ENDING 7/27 19__

DATE	REGULAR TIME					OVER TIME		
	IN	OUT	IN	OUT	DAILY TOTALS	IN	OUT	DAILY TOTALS
7/23	8:00	11:50	12:20	4:30	8	4:30	6:30	2
7/24	7:58	12:00	12:30	4:30	8	5:00	7:30	2.5
7/25	8:00	12:00	12:30	4:32	8			
7/26	7:56	12:05	12:35	4:30	8	4:30	5:00	0.5
7/27	8:01	12:00	1:00	5:00	8			
APPROVED BY DCR FOREMAN				TOTAL REGULAR TIME	40		TOTAL OVER TIME	5

REGULAR DAYS WORKED 5 @ 8 HRS. @ 9.80 EARNINGS $ 392.00

ADDITIONAL COMPENSATION: VALUE OF MEALS, LODGING, GIFTS, ETC. AMOUNT $ ______

COMMISSIONS, FEES, BONUSES, GOODS, ETC. OT 5 @ $14.70 AMOUNT $ 73.50

OTHER REMUNERATIONS (KIND) $ ______

DEDUCTIONS: TOTAL EARNINGS $465.50

STATE DISAB. OR UNEMPL. TAX @____% $______

FEDERAL SOCIAL SECURITY TAX @____% $______

FEDERAL WITHHOLDING TAX ______ $______

STATE WITHHOLDING TAX ______ $______

OTHER DEDUCT. ______ $______

CASH ADVANCED ______ $______

TOTAL DEDUCTIONS (TO RIGHT) $______

NET PAY $______

I CERTIFY THE FOREGOING TO BE A CORRECT ACCOUNT OF THE TIME WORKED AND WAGES RECEIVED:

SIGNATURE ______ DATE PAID ______

4K402 REDIFORM®

Figure 5.2

Solution The first step is to find the total number of hours worked by each person.

Abruzzo: 2 + 4 + 8 + 6 + 3 = 23 hours

Williams: 3.5 + 3 + 7 + 6.75 + 7 = 27.25 hours

To find the gross earnings, multiply the number of hours worked and the rate per hour.

Abruzzo: 23 × $8.40 = $193.20

Williams: 27.25 × $7.12 = $194.02

The payroll ledger can now be completed.

Employee	Hours Worked							Total Hours	Rate	Gross Earnings
	S	M	T	W	Th	F	S			
Abruzzo, S.	—	2	4	8	6	3	—	23	$8.40	$193.20
Williams, N.	—	3.5	3	7	6.75	7	—	27.25	$7.12	$194.02

Objective 2 *Find Overtime Earnings for Over 40 Hours of Work Per Week.* The **Fair Labor Standards Act**, which covers the majority of full-time employees in the nation, establishes a workweek of 40 hours and sets the minimum hourly wage. The law states that an **overtime** wage (a higher than normal wage) must be paid for all hours worked over 40 hours per workweek. Also, a great number of companies not covered by the Fair Labor Standards Act have voluntarily followed the practice of paying a **time-and-a-half rate** (1.5 times the normal rate) for any work over 40 hours per week. With the time-and-a-half rate, gross earnings are found with the following formula.

Gross earnings = Earnings at regular rate + Earnings at time-and-a-half rate

Example 2 Completing a Payroll Register with Overtime

Complete the following payroll ledger.

Employee	Hours Worked							Total Hours		Reg. Rate	Gross Earnings		
	S	M	T	W	Th	F	S	Reg.	O.T.		Reg.	O.T.	Total
Chung, E.	6	9	8.25	8	9	4.5	—			$7.90			
Jenders, P.	—	10	6.75	9	6.25	10	4.25			$9.48			

Solution First find the total number of hours worked.

Chung: 6 + 9 + 8.25 + 8 + 9 + 4.5 = 44.75 hours

Jenders: 10 + 6.75 + 9 + 6.25 + 10 + 4.25 = 46.25 hours

Both employees worked more than 40 hours. Gross earnings at the regular rate can now be found as discussed previously. Chung earned 40 × $7.90 = $316 at the regular rate, and Jenders earned 40 × $9.48 = $379.20 at the regular rate. To find overtime earnings, first find the number of overtime hours worked by each employee.

Chung: 44.75 − 40 = 4.75 overtime hours

Jenders: 46.25 − 40 = 6.25 overtime hours

The regular rate given for each employee can be used to find the time-and-a-half rate.

Chung: $= 1\frac{1}{2} \times \$7.90 = \$11.85$ Per hour of overtime

Jenders: $= 1\frac{1}{2} \times \$9.48 = \$14.22$ Per hour of overtime

Now find the overtime earnings.

Chung: 4.75 hours × $11.85 per hour = $56.29 Rounded to the nearest cent

Jenders: 6.25 hours × $14.22 per hour = $88.88 Rounded

The ledger can now be completed.

Employee	Hours Worked							Total Hours		Reg. Rate	Gross Earnings		
	S	M	T	W	Th	F	S	Reg.	O.T.		Reg.	O.T.	Total
Chung, E.	6	9	8.25	8	9	4.5	—	40	4.75	$7.90	$316	$56.29	$372.29
Jenders, P.	—	10	6.75	9	6.25	10	4.25	40	6.25	$9.48	$379.20	$88.88	$468.08

Objective 3 *Use Overtime Premium Method of Calculating Gross Earnings.* Gross earnings with overtime is sometimes calculated with the **overtime premium method** (sometimes called the **overtime excess method**). With this method, which produces the same result as the method just described, the total hours at the regular rate are added to the overtime hours at one half of the regular rate to arrive at gross earnings.

The formula for the overtime premium method is

$$\begin{aligned} &\text{Total hours} \times \text{Regular rate} = \text{Straight-time earnings} \\ +\ &\text{Overtime hours} \times \tfrac{1}{2}\text{ regular rate} = \text{Overtime premium} \\ \hline &= \text{Gross earnings} \end{aligned}$$

Example 3 Using the Overtime Premium Method

This week Marcy Pleu worked 40 regular hours and 12 overtime hours. Her regular rate of pay is $12.38 per hour. Find her total gross pay using the overtime premium method.

Solution The total number of hours worked by Pleu is 52 (40 + 12) and her overtime premium rate is $6.19 ($\frac{1}{2}$ × $12.38).

52 hours × $12.38 = $643.76 Regular-rate earnings

12 overtime hours × $6.19 = $74.28 Overtime

$718.04 Gross earnings

Scientific *Calculator Approach*

Using chain calculations and the order of operations to solve Example 3, the regular earnings are calculated. Next the overtime earnings are calculated. Finally, they are added together.

52 [×] 12.38 [+] 12 [×] 12.38 [×] .5 [=] 718.04

NOTE Some companies prefer the overtime premium method since it readily identifies the extra cost of overtime labor and can be seen easily. Quite often, excessive use of overtime indicates inefficiencies in management. ▬

Objective 4 ***Find Overtime Earnings Using Time-and-a-Half Rate for Over 8 Hours of Work Per Day.*** Some companies pay the time-and-a-half rate for all time worked over 8 hours in any one day, no matter how many hours are worked in a week. This **daily overtime** is shown in the next example.

Example 4 Finding Overtime Each Day

Peter Harris worked 10 hours on Monday, 5 on Tuesday, 7 on Wednesday, and 12 on Thursday. His regular rate of pay is $10.10. Find his gross earnings for the week if everything over 8 hours in one day is overtime.

	S	M	T	W	Th	F	S	Total Hours
Reg.	—	8	5	7	8	—	—	28
O.T.	—	2	—	—	4	—	—	6

Solution Harris worked more than 8 hours on both Monday and Thursday. On Monday, he had $10 - 8 = 2$ hours of overtime, with $12 - 8 = 4$ hours of overtime on Thursday. For the week, he earned $2 + 4 = 6$ hours of overtime. His regular hours are 8 on Monday, 5 on Tuesday, 7 on Wednesday, and 8 on Thursday, or

$$8 + 5 + 7 + 8 = 28 \text{ hours at the regular rate.}$$

His hourly earnings are $10.10, giving

$$28 \times \$10.10 = \$282.80 \text{ at the regular rate.}$$

If the regular rate is $10.10, the time-and-a-half rate is

$$\$10.10 \times 1\frac{1}{2} = \$15.15$$

He earned time and a half for 6 hours.

$$6 \times \$15.15 = \$90.90$$

His gross earnings are found by adding regular earnings and overtime earnings.

$$\underbrace{\$282.80}_{\text{Total regular pay}} + \underbrace{\$90.90}_{\text{Total overtime}} = \underbrace{\$373.70}_{\text{Gross earnings}}$$

NOTE There are many careers that require unusual schedules and do not pay overtime for over 40 hours worked in one week or over 8 hours worked in one day. An obvious example is the work schedule of a firefighter where the employee may work 24 hours and then get 48 hours off. ■

Objective 5 *Understand Double Time, Shift Differential, and Split-Shift Premiums.* In addition to premiums paid for overtime, other **premium payment plans** include **double time** for holidays and, in some industries, Saturdays and Sundays. A **shift differential** is often given to compensate employees for working less desirable hours. For example, an additional amount per hour or per shift might be paid to swing shift (4 P.M. to midnight) and graveyard shift (midnight to 8:00 A.M.) employees.

Restaurant employees and telephone operators often receive a **split-shift premium**. The employees' hours are staggered so that the employees are on the job only during the busiest times. For example, an employee may work 4 hours, be off 4 hours, then work 4 hours. The employee is paid a premium because of this less desirable schedule.

Some employers offer **compensatory time**, or **comp time**, for overtime hours worked. Instead of receiving additional money, an employee is given time off from the regular work schedule as compensation for overtime hours already worked. Quite often, the compensatory time is given at $1\frac{1}{2}$ times the overtime hours worked. For example, 12 hours might be given as compensation for 8 hours of previously worked overtime. Occasionally an employee is given a choice of overtime pay or comp time. Many companies reserve the use of compensatory time for their supervisors or managerial employees. Also, compensatory time is very common in government agencies.

Objective 6 *Find Equivalent Earnings for Different Pay Periods.* The second common method of finding gross earnings uses a **salary**, an amount given as so much per **pay period** (time between pay checks). Common pay periods are weekly, biweekly, semimonthly, and monthly (Table 5.1).

NOTE One person's salary might be a certain amount per month, while another person might earn a certain amount every two weeks. Many people receive an annual salary, divided among shorter pay periods. ■

Table 5.1 COMMON PAY PERIODS

Monthly	12 paychecks each year
Semimonthly	Twice each month; 24 paychecks each year
Biweekly	Every two weeks; 26 paychecks each year
Weekly	52 paychecks each year

The table below shows average annual starting salaries for college graduates in various majors.

Average Starting Salaries for 1998 Graduates (Bachelor's Degrees)

Major	Average Starting Salary	% Change
Accounting	$30,154	3.6
Business administration	29,346	5.5
Economics/finance	31,333	3.8
MIS	35,133	4.8
Liberal Arts	28,875	6.5
Computer science	38,475	6.3
Computer engineering	39,593	6.3
Chemical engineering	42,802	3.9
Electrical engineering	39,456	4.2

Data: National Association of Colleges and Employers. Reprinted by permission.

Example 5 Determining Equivalent Salaries

You are a career counselor and want to compare the earnings of four clients for whom you have helped find jobs. John Cross receives a weekly salary of \$273, Melanie Goulet a biweekly salary of \$1686, Carla Lampsa a semimonthly salary of \$736, and Tom Shaffer a monthly salary of \$1818. For each worker, find the following: (a) earnings per year, (b) earnings per month, and (c) earnings per week.

Solution

John Cross

(a) $\$273 \times 52 = \$14{,}196$ per year
(b) $\$14{,}196 \div 12 = \$1183$ per month
(c) \$273 per week

Melanie Goulet

(a) $\$1686 \times 26 = \$43{,}836$ per year (biweekly = 26 per year)
(b) $\$43{,}836 \div 12 = \$3653$ per month
(c) $\$1686 \div 2 = \$843$ per week

Carla Lampsa

(a) $\$736 \times 24 = \$17{,}664$ per year
(b) $\$736 \times 2 = \$1472$ per month
(c) $\$17{,}664 \div 52 = \$339.69$ per week

Tom Shaffer

(a) $\$1818 \times 12 = \$21{,}816$ per year
(b) \$1818 per month
(c) $\$21{,}816 \div 52 = \$419.54$ per week

Objective 7 | ***Find Gross Earnings When Overtime Is Paid to Salaried Employees.*** A salary is paid for the performance of a certain job, without keeping track of the number of hours worked. However, the Fair Labor Standards Act requires that certain salaried positions receive additional compensation for overtime. Just as with wage earners, the salaried employee is often paid time and a half for all hours worked over the normal number of hours per week.

Example 6 Finding Overtime for Salaried Employees

Caralee Woods is paid $872 a week as an executive assistant. If her normal workweek is 40 hours, and she is paid time and a half for all overtime, find her gross earnings for a week in which she works 45 hours.

Solution The executive assistant's salary has an hourly equivalent of

$$\frac{\$872}{40 \text{ hours}} = \$21.80 \text{ per hour}$$

Since she must be paid overtime at the rate of $1\frac{1}{2}$ times her regular pay, she will get $32.70 per hour ($1\frac{1}{2}$ 3 $21.80) for overtime. Her gross earnings for the week are calculated as follows.

Salary for 40 hours =	$ 872.00	Regular-rate earnings
Overtime for 5 hours (5 × $32.70) =	163.50	Overtime
	$1035.50	Gross Earnings

Scientific *Calculator Approach*

The calculator solution to Example 6 is

872 [+] 872 [÷] 40 [×] 1.5 [×] 5 [=] 1035.5

Example 7 Finding Gross Earnings with Overtime

A conference coordinator is paid a salary of $432 per week. If his regular workweek is 36 hours, find his gross earnings for a week in which he works 46 hours. All overtime hours are paid at time and a half.

Solution The coordinator's salary has an hourly equivalent of

$$\frac{\$432}{36 \text{ hours}} = \$12 \text{ per hour}$$

Since he is paid $1\frac{1}{2}$ times the regular rate per hour, he will receive $18 per hour ($1\frac{1}{2} \times$ $12) for overtime. His gross earnings for the week are found as follows.

Salary for 36 hours =	$432	Regular-rate earnings
Overtime for 10 hours (10 × $18) =	180	Overtime
	$612	Gross Earnings

5.1 Exercises

THE PAYROLL LEDGER *Find the number of regular hours and overtime hours (any hours over 40) for each of the following employees. Then calculate the overtime rate (time and a half) for each employee.*

	Employee	S	M	T	W	Th	F	S	Reg. Hrs.	O.T. Hrs.	Reg. Rate	O.T. Rate
1.	Allen, K.	—	7	4	7	10	8	4			$8.10	
2.	Doran, C.	—	6.5	9	7.5	8	9.5	7			$8.24	
3.	Harris, T.	3	6	8.25	8	8.5	5	—			$7.80	
4.	Sheehan, A.	8.5	9	7.5	8	10	8.25	—			$9.50	
5.	Ulman, L.	—	9.5	7	9	9.25	10.5	—			$11.48	
6.	Fuqua, B.	—	8	8	9	7.25	6	7			$9.80	

GROSS EARNINGS *Find the earnings at the regular rate, the earnings at the overtime rate, and the gross earnings for each of the employees in problems 1–6.*

7. Allen, K.

8. Doran, C.

9. Harris, T.

10. Sheehan, A.

11. Ulman, L.

12. Fuqua, B.

Complete the following partial payroll register by finding the overtime rate at time and a half, the amount of earnings at regular pay, the amount at overtime pay, and the total gross wages for each employee.

		Total Hours				Gross Earnings		
	Employee	Reg.	O.T.	Reg. Rate	O.T. Rate	Reg.	O.T.	Total
13.	Fenton, C.	39.5	—	$8.80				
14.	Klein, A.	36.25	—	$10.20				
15.	Schultz, J.	40	4.5	$7.20				
16.	Stingley, J.	40	6.75	$6.06				
17.	Weisher, W.	40	4.25	$9.18				
18.	Tracey, N.	40	5	$7.10				

OVERTIME PREMIUM *Some companies use the overtime premium method to determine gross earnings. Use this method to complete the following partial payroll register. Overtime is paid at time-and-a-half rate for all hours over 40.*

	Employee	Hours Worked S	M	T	W	Th	F	S	Total Hours	Reg. Rate	O.T. Hours	O.T. Premium Rate	Gross Earnings Reg.	O.T.	Total
19.	Aragona, B.	10	9	8	5	12	7	—		$7.40					
20.	Biron, C.	7.75	10	5	9.75	8	10	—		$9.50					
21.	Cheever, P.	—	12	11	8	8.25	11	—		$8.60					
22.	Collins, C.	—	8.5	5.5	10	12	10.5	7		$7.50					
23.	Sherlock, F.	—	10	9.25	9.5	11.5	10	—		$10.20					
24.	Firavich, S.	8.5	7	9.75	—	10.5	12	—		$6.90					

DAILY OVERTIME PAYMENT *Some companies pay overtime for all time worked over 8 hours in a given day. Use this method to complete the following payroll register. Overtime is paid at time-and-a-half rate.*

	Employee	Hours Worked S	M	T	W	Th	F	S	Total Hours Reg.	O.T.	Reg. Rate	O.T. Rate	Gross Earnings Reg.	O.T.	Total
25.	Belinder, M.	—	10	9	11	6	5	—			$6.70				
26.	Cechvala, C.	—	9	8.75	7	8.5	10	—			$7.60				
27.	Deininger, M.	—	7.5	8	9	10.75	8	—			$6.70				
28.	Gingrich, D.	—	9	10	8	6	9.75	—			$8.60				
29.	Kaplan, L.	—	9.5	8.5	7.75	8	9.5	—			$10.20				
30.	Lerner, M.	6	8	6.5	8.75	—	10.25	—			$7.20				

31. Explain in your own words what premium payment plans are. Select a premium payment plan and describe it. (See Objective 5.)

32. If you were given a choice of overtime pay or compensatory time, which would you choose? Why? (See Objective 5.)

EQUIVALENT EARNINGS *Find the equivalent earnings for each of the following salaries as indicated.*

	Weekly Earnings	Biweekly Earnings	Semimonthly Earnings	Monthly Earnings	Annual Earnings
33.	$248	______	______	______	______
34.	______	______	$480	______	______
35.	______	$852	______	______	______
36.	______	______	______	$1150	______
37.	______	______	$1087.50	______	______
38.	$436	______	______	______	______
39.	______	______	______	$2680	______
40.	______	$768	______	______	______
41.	______	______	______	______	$21,580
42.	______	______	______	______	$26,100

SALARY WITH OVERTIME *Find the weekly gross earnings for the following people who are on salary and are paid time and a half for overtime. (Hint: Round hourly equivalents and overtime amounts to the nearest cent.)*

	Employee	Regular Hours per Week	Weekly Salary	Hours Worked	Weekly Gross Earnings
43.	Atkins, G.	40	$520	56	______
44.	Berry, M.	40	$360	42	______
45.	Bridges, C.	45	$418	50	______
46.	Kelley, R.	38	$340	40	______
47.	Magot, D.	32	$450	44	______
48.	Sypniewski, D.	30	$484	45	______

Solve each of the following application problems. (Hint: Round hourly equivalents, regular rates, and overtime rates to the nearest cent.)

49. RETAIL EMPLOYMENT Last week, Lori Merrill worked 48 hours at Blockbuster Video. Find her gross earnings for the week if she is paid $7.40 per hour and earns time and a half for all hours over 40.

50. ACCOUNTS PAYABLE CLERK Lisa Ventura is an accounts payable clerk and is paid $9.50 per hour for straight time, and time and a half for all hours over 40 worked in a week. Find her gross earnings for a week in which she worked 52 hours.

51. BOOKSTORE EMPLOYEE Byran Hopkins, a bookstore employee, earns $7.80 per hour and is paid time and a half for all time over 8 hours worked on a given day. Find his gross earnings for a week in which he worked the following hours: Monday, 9.5; Tuesday, 7; Wednesday, 10.75; Thursday, 4.5; and Friday, 8.75.

52. OFFICE ASSISTANT Tom Derungs is an office assistant and worked 10 hours on Monday, 9.75 hours on Tuesday, 5.5 hours on Wednesday, 12 hours on Thursday, and 7.25 hours on Friday. His regular rate of pay is $11.50 an hour, with time and a half paid for all hours over 8 worked in a given day. Find his gross earnings for the week.

53. ESCROW OFFICER Anne Felsted is paid $648 a week as an escrow officer at a bank. Her normal work week is 40 hours. She is paid time and a half for overtime. Find her gross earnings for a week in which she worked 46 hours.

54. FEEDSTORE SALES An employee at Valley Feed Stores is paid $298 for a normal work week of 35 hours. If she is paid time and a half for overtime, find her gross earnings for a week in which she worked 48 hours.

55. SALARY WITH OVERTIME Charles Dawkins, manager of the Cellular Phone Center, is paid a salary of $638 per week, has a normal workweek of 40 hours, and is paid time and a half for overtime. Find his gross earnings in a week in which he worked 52 hours.

56. HOME MORTGAGE BUSINESS Mike Carver, senior vice president of Countrywide Mortgage, worked 54 hours this week. If he is paid a weekly salary of $800, and has a normal workweek of 45 hours, find his gross earnings for the week. He is paid time and a half for all overtime.

57. EQUIVALENT EARNINGS An employee earns $630 weekly. Find the equivalent earnings if paid (a) biweekly, (b) semimonthly, (c) monthly, and (d) annually.

58. ANNUAL SALARY Angelica Canales is a plant supervisor and is paid $42,900 annually. Find the equivalent earnings if this amount is paid (a) weekly, (b) biweekly, (c) semimonthly, and (d) monthly.

59. Semimonthly pay periods result in 24 paychecks per year. Biweekly pay periods result in 26 paychecks per year. Which of these pay periods gives three checks in two months of the year? Will it always be the same two months? Explain.

60. How would you budget your money if you were paid just once a month instead of each week? Which would you prefer: a monthly pay period or a weekly pay period?

61. In Section 1 of the foldout, locate the earnings of an adult with a high school education and an adult with the bachelor's degree. Find the equivalent weekly, biweekly, semimonthly, and monthly earnings of (a) the high school graduate and (b) the college graduate.

5.2 Gross Earnings (Commission)

Objectives

1. *Find gross earnings using commission rate × sales (P = R × B).*
2. *Determine commission using a variable commission rate.*
3. *Use salary and commission rate to find gross earnings.*
4. *Use a drawing account and quota to find gross earnings.*
5. *Determine override as part of gross earnings.*

Many people in sales and marketing are paid on **commission**, usually a fixed percent of sales. This is an incentive system of compensation and the commissions are designed to produce maximum employee output, since pay is directly dependent on sales. This section discusses all of the common types of sales commissions.

Objective 1 *Find Gross Earnings Using Commission Rate × Sales* (**P = R × B**). With a **straight commission**, the salesperson is paid a fixed percent of sales. Gross earnings are found by the following formula.

$$\begin{array}{ccccc} P & = & R & \times & B \\ \text{Gross earnings} & = & \text{Commission rate} & \times & \text{Amount of sales} \end{array}$$

Example 1 Determining Earnings Using Commission

A real estate broker is paid a 6% commission. Find the commission on a house selling for $118,500.

Solution For selling the house the broker would receive

$$6\% \times \$118{,}500 = 0.06 \times \$118{,}500 = \$7110$$

Example 2 Subtracting Returns When Using Commissions

Arianne Weber, a textbook sales representative, had sales of $8295 with returns of $950. If her commission rate is 14%, find her gross earnings.

Solution The returns must first be subtracted from gross sales, then the difference, net sales, multiplied by the commission rate.

$$\begin{aligned} \text{Gross earnings} &= (\$8295 - \$950) \times 14\% \\ &= \$7345 \times 0.14 \\ &= \$1028.30 \end{aligned}$$

Problem-Solving Hint Before calculating the commission, all returned items are first subtracted from the amount of sales. The company will not pay a commission on sales that are not completed.

Objective 2 *Determine Commission Using a Variable Commission Rate.* The **sliding-scale** or **variable commission** plan is a method of pay designed to retain top-producing salespeople. With these plans, a higher rate of commission is paid as sales get larger and larger.

Example 3 Finding Earnings Using Variable Commission

Marika Colgan sells videotapes to video rental stores, such as Blockbuster Video, and is paid as follows.

Sales	Rate
Up to $10,000	6%
$10,001–$20,000	8%
$20,001 and up	9%

Find Colgan's earnings if she has video sales of $32,768.

Solution Use the three commission rates as follows.

(Total sales)	$32,768			
(First $10,000)	−10,000	at 6%	=	$600.00
	$22,768			
(Next $10,000)	−10,000	at 8%	=	$800.00
(Over $20,000)	$12,768	at 9%	=	$1149.12
Total commissions			=	$2549.12

Colgan had gross earnings of $2549.12.

Scientific *Calculator Approach*

The first thing to do is find the commission earned at the highest rate and *place* it in memory.

(32768 − 20000) × 9 % = STO

Next, find the commission at the second highest rate and *add* this to memory.

(20000 − 10000) × 8 % = + RCL = STO

Finally, find the commission at the lowest rate and *add* it to memory. The result is the total commission.

10000 × 6 % = + RCL = 2549.12

Objective 3 *Use Salary and Commission Rate to Find Gross Earnings.* With a **salary plus commission**, the salesperson is paid a fixed sum per pay period, plus a commission on all sales. This method of payment is commonly used by large retail stores. Gross earnings with salary plus commission are found by the following formula.

Gross earnings = Fixed amount per pay period + Amount earned on commission

Many salespeople favor this method of determining gross earnings. It is especially attractive to the beginning salesperson who lacks selling experience and personal self-confidence. While providing an incentive, it offers the security of a guaranteed income to cover basic living costs.

Example 4 Adding Commission to a Salary

Pat Quinlin is paid $225 per week by the Potters Exchange, plus 3% on all sales. Find her gross earnings for a week in which her sales were $7250.

Solution Use the formula in the box.

Gross earnings = Fixed earnings + Commission
= $225 + (0.03 × $7250)
= $225 + $217.50
= $442.50

Objective 4 *Use a Drawing Account and Quota to Find Gross Earnings.* The fixed amount of earnings is often a **draw**, or loan, against future commissions. A **drawing account** is set up with the amounts drawn repaid with future commissions. This is a loan against future commissions but offers the salesperson the assurance of a fixed sum per pay period. The salesperson must repay the drawing account as commissions are earned.

Example 5 Subtracting a Draw from Commission

Elizabeth Owens, a computer sales representative, has sales of $38,560 for the month and is paid a 7% commission rate. She had draws of $750 for the month. Find her gross earnings after repaying the drawing account.

Solution

$$\begin{aligned}\text{Gross earnings} &= \text{Commissions} - \text{Draw}\\ &= (0.07 \times \$38{,}560) - \$750\\ &= \$2699.20 - \$750\\ &= \$1949.20\end{aligned}$$

NOTE Commission earnings plans are a strong deterrent to attracting new salespeople. For this reason many companies offer salary-plus-commission and draw plans to help fill new sales positions.

A sales quota is often established for salespeople. The **quota** is the minimum amount of sales expected from the employee. If the salesperson continually falls short of the sales quota, termination may result. Normally, however, the salesperson is rewarded for passing the sales quota with a bonus or commission. This plan is called a **quota bonus system**.

Example 6 Using the Quota Bonus System

David Shea is a sales representative for a mountain bike manufacturer. During a recent week he had sales of $18,780 and was paid a commission of 8% after meeting the sales quota of $5000. Find his gross earnings.

Solution

$$\begin{aligned}\text{Gross earnings} &= \text{Commission rate} \times (\text{Sales} - \text{Quota})\\ &= 0.08 \times (\$18{,}780 - \$5000)\\ &= 0.08 \times \$13{,}780\\ &= \$1102.40\end{aligned}$$

No commission on the first $5000.

Objective 5 *Determine Override as Part of Gross Earnings.* Sales supervisors and department heads of retail stores are often paid a commission based on the total sales of their staff or department. This payment, for the efforts of others, rewards the supervisor or department head for doing a good job in training and maintaining a sales staff. This commission is called an **override**. Calculate it like any other commission, but use the total department sales.

Example 7 Finding Gross Earnings with Commission and Override

Outdoor Sports and More pays their managers a salary plus commission and override. Find the gross earnings for a manager given the following.

Personal sales	\$4,386	Personal returns	\$118
Store sales	\$11,865	Store returns	\$562
Salary	\$375	Personal quota	\$2500
Commission rate	4%	Override rate	1%

Solution First, find the manager's commission on personal sales.

Personal sales ↓ \$4386 − Returns ↓ \$118 − Quota ↓ \$2500 = Personal commission sales ↓ \$1768
= 0.04 × \$1768
= \$70.72

Now find the override commission on store sales.

Store sales ↓ \$11,865 − Returns ↓ \$562 = Override commission sales ↓ \$11,303
= 0.01 × \$11,303
= \$113.03

Calculate gross earnings as follows.

Gross earnings = Salary + Commission + Override
= \$375 + \$70.72 + \$113.03
= \$558.75

Scientific
Calculator Approach

In solving Example 7, the approach here is to first find the manager's commission on personal sales and *place* it in memory.

[( ] 4386 [−] 118 [−] 2500 [ )] [×] 4 [%] [=] [STO]

Next, find the override on store sales and *add* it to memory.

[( ] 11865 [−] 562 [ )] [×] 1 [%] [=] [+] [RCL] [=] [STO]

Finally, *add* the salary to the commission and override in memory. The result is the gross earnings.

375 [+] [RCL] [=] 558.75

5.2 Exercises

COMMISSION WITH RETURNS *Find the gross earnings for each of the following salespeople.*

	Employee	Total Sales	Returns and Allowances	Rate of Commission
1.	McKee, J.	$2,810	$208	8%
2.	Brown, D.	$5,734	$415	5%
3.	Pasnick, J.	$2,875	$64	15%
4.	Beckenstein, J.	$2,603	$76	18%
5.	Brown, K.	$25,658	$4083	9%
6.	Dramatinos, M.	$18,765	$386	8%
7.	Dobbins, G.	$45,618	$2281	1%
8.	Phares, H.	$34,183	$1169	2%

VARIABLE COMMISSION RATE *Bayside Janitorial Supply pays its salespeople the following commission.*

6% on first $7500 in sales
8% on next $7500 in sales
10% on any sales over $15,000

Find the gross earnings for each of the following employees.

	Employee	Total Sales		Employee	Total Sales
9.	Dean, D.	$18,550	**10.**	Davis, C.	$17,640
11.	Brueck, G.	$10,480	**12.**	Koch, R.	$16,250
13.	Sanchez, J.	$11,225	**14.**	Fisher, L.	$22,650
15.	Butter, M.	$25,860	**16.**	Manly, C.	$23,340

17. When you are paid a commission, there is always the possibility of higher earnings and also the uncertainty of a regular pay check. Explain in your own words the special budgetary planning you would have to do if you were paid on commission.

18. A variable commission plan is often referred to as an incentive within an incentive. Explain why this might be an accurate description of a variable commission plan. (See Objective 2.)

COMMISSION PAYROLL REGISTER *Complete the following commission payroll to find gross earnings.*

	Employee	Gross Sales	Sales Returns	Net Quota	Commission Sales	Commission Rate	Gross Commission	Salary	Gross Earnings
19.	Potter, D.	$5,250	$220	—		4%		$290	
20.	Schwartz, A.	$9,370	$840	$3000		10%		—	
21.	Reska, L.	$6,380	$295	$2000		6%		—	
22.	Wetherbee, S.	$3,270	$420	—		7%		$280	
23.	Chen, C.	$12,420	$390	$2500		3%		—	
24.	Ng, B.	$10,680	$490	$1500		6%		—	
25.	Jidobu, B.	$4,215	$318	$1000		5%		$210	
26.	Kroeger, T.	$3,850	$310	$1400		6%		$350	

Solve each of the following application problems.

27. PAINT SALES Helen Vasques is a sales representative for Watco Paints. She is paid an 8% commission rate, and has had a draw of $350 this week. If her sales are $9850 this week, find her gross earnings after repaying the drawing account.

28. COMMISSION DRAW Kim Craft has sales of $78,560 for the month and is paid a 6% commission rate. She has had draws totaling $1500 for the month. Find her gross earnings after repaying the drawing account.

29. SALARY PLUS COMMISSION Stacy Dwyer, a salesperson for Zapp Music, has sales of $194,800 this month and is paid a 2% commission by her company. She also receives a salary of $1750 each month. Find her gross earnings for the month.

30. ADVERTISING SALES Jim Snelling, an account representative for Ad-Art is paid a 3% commission rate and a salary of $300 each week. If his sales are $28,720 this week, find his gross earnings for the week.

31. FASTENER SALES John Chavez is a commission salesperson for Fastener Manufacturing which allows him to draw $800 per month. His commission is 6% of the first $6000 in sales, 8% of the next $16,000, and 15% of all sales over $22,000. If his sales for the month were $27,700, find (a) his total commission and (b) the earnings due at the end of the month after repaying the drawing account balance of $800.

32. GREETING CARD SALES Andrea Abriani is a sales representative for Hi Side Greeting Card Company. She is paid a monthly draw of $650. Her commission is 10% of the first $2000 in sales, 12% of the next $4000, and 20% of all sales over $6000. If her sales for the month were $8750, find (a) her total commission and (b) the gross earnings due at the end of the month after repaying the drawing account.

33. COMMISSION PLUS OVERRIDE The manager of Toys for Tots is paid a salary plus commission and an override. Find his gross earnings given the following.

Personal sales	$2825	Personal returns	$84
Department sales	$8656	Department returns	$317
Salary	$200	Personal quota	$1000
Commission rate	5%	Override	$1\frac{1}{2}\%$

34. APPLIANCE SALES MANAGER The sales manager for A & A Appliance is paid a salary plus commission and an override. Find her gross earnings given the following.

Personal sales	$5,856	Personal returns	$185
Store sales	$19,622	Store returns	$358
Salary	$250	Personal quota	$3000
Commission rate	3%	Override	2%

5.3 Gross Earnings (Piecework)

Objectives

1. *Find the gross earnings for piecework.*
2. *Find the gross earnings for differential piecework.*
3. *Determine chargebacks or dockings.*
4. *Find overtime earnings for piecework.*

The salaries and wages discussed in Section 5.1 are called **time rates**, since they depend only on the amount of time an employee was actually on the job. Commission earnings in Section 5.2 and the piecework methods discussed in this section are called **incentive rates**. These gross earnings are based on production and pay an employee for actual performance on the job. The ad for truck drivers lists piece rates of $0.46975 and $0.48475 per mile, while the automotive, insurance, loan officer, and industrial sales positions pay on a commission plan.

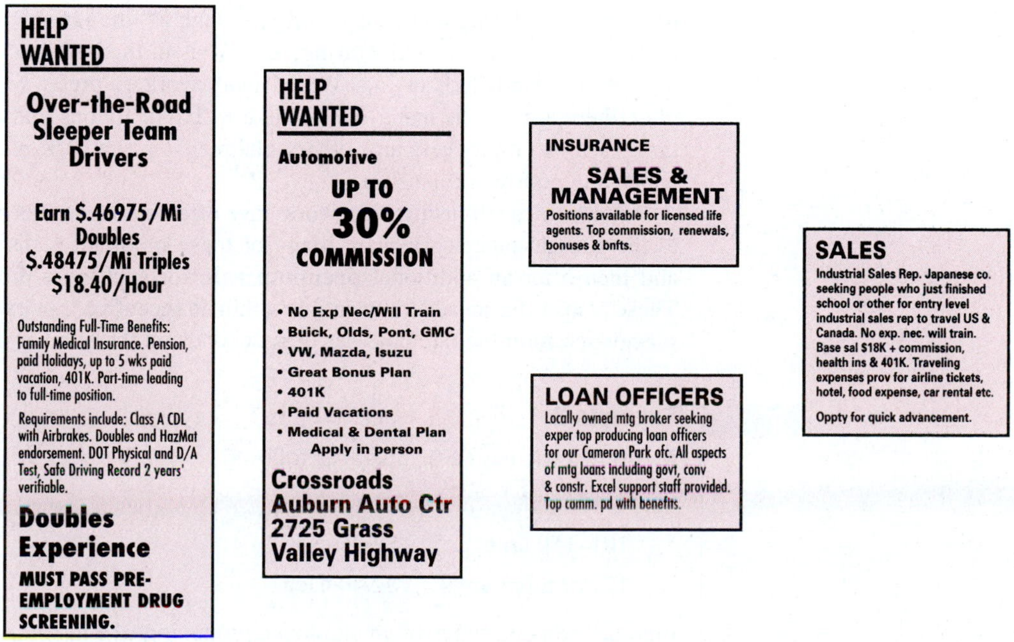

Objective 1 *Find the Gross Earnings for Piecework.* A **piecework rate** pays an employee so much per item produced. Gross earnings are found using the following formula.

$$\text{Gross earnings} = \text{Number of items} \times \text{Pay per item}$$

For example, a cabinet finisher who finishes 23 cabinets and is paid a piecework rate of \$4 per cabinet would have total gross earnings of

$$\text{Gross earnings} = \$4 \times 23 = \$92$$

Example 1 Finding Gross Earnings for Piecework

Dona Kenly was paid \$0.73 for sewing a jacket collar, \$0.87 for a sleeve with a cuff, and \$0.99 for a lapel. One week she sewed 318 jacket collars, 112 sleeves with cuffs, and 37 lapels. Find her gross earnings.

Solution Multiply the rate per item by the number of that type of item.

Item	Rate × Number	= Total
Jacket collars	\$0.73 × 318	= \$232.14
Sleeves with cuffs	\$0.87 × 112	= \$97.44
Lapels	\$0.99 × 37	= \$36.63

The gross earnings can be found by adding the three totals. \$232.14 + \$97.44 + \$36.63 = \$366.21.

Objective 2 *Find the Gross Earnings for Differential Piecework.* A **straight-piecework plan**, such as in Example 1, is perhaps the oldest of all incentive payment plans. It is used in many manufacturing and production jobs such as fine jewelry finishing, agricultural and farm work, garment manufacturing, as well as in the building trades for structural framing, roofing, and floor laying. While many workers prefer working under a piecework plan, there are just as many who dislike it. Labor unions and other employee organizations, senior employees, and others claim that piecework plans result in unsafe work habits and poor workmanship.

Companies still using piecework have often made various modifications and changes to the straight-piecework plan. Many of these modified plans incorporate initial quotas and then offer an additional **premium rate** for each item produced beyond the quota. These plans offer an added incentive within an incentive. For example, in the **differential-piecework plan** the rate paid per item depends on the number of items produced.

Example 2 Using Differential Piecework

Metro Electric pays assemblers as follows.

1–100 units	\$2.10 each
101–150 units	\$2.25 each
151 or more units	\$2.40 each

Find the gross earnings of an employee producing 214 items.

Solution The gross earnings of a worker producing 214 items would be found as follows.

(Total units)	214		
(First 100 units)	−100	at \$2.10 each =	\$210.00
	114		
(Next 50 units)	−50	at \$2.25 each =	\$112.50
(Number over 150)	64	at \$2.40 each =	+\$153.60
			\$476.10

The gross earnings are \$476.10.

NOTE With differential piecework, the highest amount paid only applies to the last units produced. In Example 2, \$2.10 is paid for units 1–100, \$2.25 is paid for units 101–150, and \$2.40 is only paid on those units beyond unit 150, which in this case is 64 units.

Objective 3 *Determine Chargebacks or Dockings.* While companies are often pleased to reward employees with premium rates for surpassing quotas, management is equally concerned with unacceptable quality and unusable production. Ruined items may produce a total loss of material and labor; correctable flaws require additional handling, resulting in added costs and decreased profits. To discourage carelessness and mistakes, many companies require the employee to share in the cost of the spoiled item. These penalties, called **chargebacks** or **dockings**, are normally at a lower rate than the employee receives for producing that piece. This lower rate is used because a small amount of production error is expected.

Example 3 Understanding and Using Chargebacks

In Example 2, suppose the company had a chargeback of \$1.50 per spoiled item and the employee had spoiled 14 items. Find the gross earnings after the chargeback.

Solution Gross earnings are found by subtracting the chargeback from piece rate earnings.

$$\begin{aligned}\text{Gross earnings} &= \text{Piecework earnings} - (\text{Spoiled items} \times \text{Chargeback rate})\\ &= \$476.10 - (14 \text{ items} \times \$1.50)\\ &= \$476.10 - \$21\\ &= \$455.10\end{aligned}$$

Piecework and differential-piecework rates are frequently modified to include some guaranteed hourly rate of pay. Often this is necessary to meet minimum wage laws. To satisfy the law, the employer may either pay minimum wage or piecework earnings, *whichever is higher*.

Example 4 Finding Earnings with a Guaranteed Hourly Wage

A tire installer at the Tire Center is paid \$8.40 per hour for an 8-hour day, or \$0.95 per tire installed—whichever is higher. Find the weekly earnings for an employee having the following rate of production.

Monday	86	installations
Tuesday	70	installations
Wednesday	88	installations
Thursday	68	installations
Friday	82	installations

Solution The hourly earnings for an 8-hour day are \$67.20 (8 × \$8.40). If the piecework earnings for the day are less than this amount, the hourly earnings will be paid.

Monday	86 × \$0.95 =	\$81.70	Piece rate
Tuesday	~~70 × \$0.95~~ =	\$67.20	Hourly (piece rate is \$66.50)
Wednesday	88 × \$0.95 =	\$83.60	Piece rate
Thursday	~~68 × \$0.95~~ =	\$67.20	Hourly (piece rate is \$64.60)
Friday	82 × \$0.95 =	\$77.90	Piece rate
		\$377.60	Weekly earnings

NOTE Since the piecework earnings on Tuesday and Thursday are below the minimum, the hourly rate is paid on those days.

Objective 4 ***Find Overtime Earnings for Piecework.*** Piecework employees, just as other workers, are paid time and a half for overtime. The overtime rate may be computed as $1\frac{1}{2}$ times the hourly rate, but most often the overtime rate is $1\frac{1}{2}$ times the regular rate per piece.

Example 5 Determining Earnings with Overtime Piecework

Tracy Light is paid \$0.84 per circuit board soldered. During one week she solders 480 circuit boards on regular time and 104 circuit boards during overtime hours. Find her gross earnings for the week if time and a half per panel is paid for overtime.

Solution

$$\begin{aligned} \text{Gross earnings} &= \text{Earnings at regular piece rate} + \text{Earnings at overtime piece rate} \\ &= 480 \times \$0.84 + (104 \times 1\tfrac{1}{2} \times \$0.84) \\ &= \$403.20 + \$131.04 \\ &= \$534.24 \end{aligned}$$

Scientific *Calculator Approach*

The calculator solution to Example 5 uses parentheses to first calculate the overtime piece rate.

480 [×] .84 + 104 [×] 1.5 [×] .84 [=] 534.24

5.3 Exercises

PET DIETARY PRODUCTS *Complete the following payroll register for Pet Salt Products. Employees are paid a straight piece rate. Rates per unit vary depending on worker skills involved.*

	Employee	Units Produced M	T	W	T	F	Total Pieces	Rate per Unit	Gross Earnings
1.	Campbell, C.	150	124	172	110	96		$0.39	
2.	Motton, T.	120	108	89	130	95		$0.87	
3.	Young, C.	98	86	79	108	80		$0.75	
4.	McIntosh, R.	67	54	72	83	59		$0.72	
5.	Todd, R.	118	124	143	132	148		$0.68	
6.	Eckern, G.	157	148	169	145	178		$0.59	
7.	Anderson, N.	125	118	115	132	98		$0.46	
8.	Demaree, D.	152	136	170	144	192		$0.43	
9.	Parker, R.	149	135	118	125	143		$0.78	
10.	Pearson, D.	96	84	115	102	96		$0.72	

CLASSIC RADIO SHOWS *Suppose that production workers at Classic Old Tyme Radio Shows are paid as follows for labeling and packaging CDs.*

1–500 CDs	$0.10 each
501–700 CDs	$0.12 each
701–1000 CDs	$0.14 each
Over 1000 CDs	$0.16 each

Find the gross earnings for each of the following employees.

	Employee	Number of CDs
11.	Dalton, S.	829
13.	Ngau, C.	1182
15.	Waipo, T.	1250

	Employee	Number of CDs
12.	Peterson, K.	926
14.	Farmber, P.	1380
16.	Roseborough, M.	1408

HOURLY/PIECEWORK RATES *Find the gross earnings for each of the following employees. Each has an 8-hour workday and is paid $0.75 for each unit of production or the hourly rate, whichever is greater.*

	Employee	Units Produced M	T	W	T	F	Hourly Rate	Gross Earnings
17.	Knab, C.	66	75	58	72	68	$6.18	
18.	Tracy, N.	62	78	79	80	81	$7.20	
19.	Wilson, M.	80	60	75	78	74	$6.80	

	Employee	Units Produced M	T	W	T	F	Hourly Rate	Gross Earnings
20.	Viale, D.	72	70	62	88	82	$6.50	
21.	Zurcher, S.	75	84	72	93	67	$6.75	
22.	Frase, E.	63	57	67	75	70	$5.70	
23.	Pantera, A.	73	62	78	64	81	$6.30	
24.	Enos, C.	90	77	89	102	99	$8.10	

PIECEWORK WITH OVERTIME *Find the gross earnings for each of the following employees. Overtime is $1\frac{1}{2}$ times the normal per piece rate. Rejected units are charged at the chargeback rate.*

	Employee	Units Produced Reg.	O.T.	Rejected Units	Rate per Unit	Chargeback per Unit	Gross Earnings
25.	Miller, J.	510	74	20	$0.72	$0.38	
26.	Kavanagh, M.	380	26	6	$0.69	$0.56	
27.	Boghoussian, A.	493	74	34	$0.86	$0.46	
28.	Carlson, K.	508	38	16	$0.59	$0.42	
29.	Balbi, G.	286	38	4	$0.95	$0.82	
30.	Fukano, H.	315	64	35	$0.74	$0.65	
31.	Hughes, G.	403	72	15	$0.68	$0.45	
32.	Dos Reis, A.	452	12	6	$0.59	$0.50	

33. Wages and salaries are known as time rates, while commissions are called incentive rates of pay. Explain in your own words the difference between these payment methods.

34. Describe what a chargeback or docking is for rejected units. Why do you think that the chargeback per unit is usually less than the rate paid per unit? (See Objective 3.)

Solve each of the following application problems.

35. KEYBOARD ASSEMBLY Greg Jackson is paid $4.75 for each keyboard assembled, charged $2.25 for each rejection, and paid time and a half for overtime production. Find his gross earnings for the week when he assembles 142 keyboards at the regular rate, 26 keyboards at the overtime rate, and has 7 chargebacks.

36. CAKE DECORATING John Davis decorates cakes at the French Meadow Bakery, for which he is paid $2.42 each. He is charged $1.40 per rejection and paid time and a half for all overtime production. Find his gross earnings when production for the week is 128 cakes at the regular rate and 19 cakes at the overtime rate. He has a total of 5 chargebacks.

37. ELECTRICAL INSTALLATION Robert Andrews is paid $1.35 per alternator installed, charged $0.85 for each rejection, and paid time and a half for overtime production. Find his gross earnings for the week when he installs 310 alternators at the regular rate, 110 alternators at the overtime rate, and has 20 chargebacks.

38. COSTUME PRODUCTION Kristen Clement sews lace on dance costumes, for which she is paid $1.18 each. She is charged $0.90 per rejection and is paid time and a half for all overtime production. Find her gross earnings when production for the week is 138 costumes at the regular rate and 28 costumes at the overtime rate. She has a total of 9 chargebacks.

39. GIFT PACKAGING Erica Gheen inspects and packages gift shipments. She is paid $0.85 per unit and is charged $0.35 for each rejection. Find her gross earnings for the week given the following production.

	M	T	W	T	F	Totals
Production	136	112	108	96	122	
Chargebacks	6	3	5	8	4	

40. PACKING AND SHIPPING George Parr checks and ships customer orders for Feathers and Stream Sportswear. He receives $0.35 per package and is charged $0.12 for each rejection. Find his gross earnings for the week given the following production.

	M	T	W	T	F	Totals
Production	178	165	186	171	174	
Chargebacks	5	3	2	7	4	

5.4 Social Security, Medicare, and Other Taxes

Objectives

1. *Understand FICA.*
2. *Find the maximum FICA tax paid by an employee in one year.*
3. *Understand Medicare tax.*
4. *Find FICA tax and Medicare tax.*
5. *Determine the FICA tax and the Medicare tax paid by a self-employed person.*
6. *Find state disability insurance deductions.*

Finding gross earnings is only the first step in preparing a payroll. The employer must then subtract all required deductions from gross earnings. For most employees, these deductions include Social Security tax, Medicare tax, federal income tax withholding, and state tax withholding. Other deductions may include state disability insurance, union dues, retirement, vacation pay, credit union savings or loan payments, purchase of bonds, uniform expenses, group insurance plans, and charitable contributions. Subtracting these deductions from gross earnings results in net pay, the amount the employee receives.

Objective 1 *Understand FICA.* The **Federal Insurance Contributions Act (FICA)** was passed into law in the 1930s during the middle of the Great Depression. This plan, now called **Social Security**, was originally designed to give monthly benefits to retired workers and their survivors. Also included today are death benefits and **Medicare** payments. As the number of people receiving benefits has increased along with the individual benefit amounts, people paying into Social Security have had to pay a larger amount of earnings into this fund each year. From 1937 through 1950 an employee paid 1% of income into Social Security, up to a maximum of $30 per year. This amount has increased over the years until an employee in 1999 paid 6.2% of income to FICA and 1.45% to Medicare, which together can total $6000 or more per year.

For many years both the Social Security tax rate and the Medicare tax rate were combined; however, since 1991 these tax rates have been expressed individually. Table 5.2 shows the tax rates and the maximum earnings on which Social Security and Medicare taxes are paid by the employee. The employer pays the same rate as the employee, matching all employee contributions dollar for dollar. Self-employed people pay double the rate paid by those who are employees. (They are paying for both employee and employer.)

Congress sets the tax rates and the maximum employee earnings subject to both Social Security tax and Medicare tax. Because these tax rates change, we will use 6.2% of the first $80,000 that the employee earns in a year for Social Security tax. For Medicare tax, we will use 1.45% of everything that the employee earns in a year. These figures are used in all examples and exercises in this chapter.

Each employee, whether a U.S. citizen or not, must have a Social Security card. Most post offices have application forms for the cards. All money set aside for an individual is credited to his or her account according to the Social Security number. Each year the Social Security Administration sends out a Social Security statement that shows workers how Social Security fits into their future. The statements are sent 3 months before the employees' birthday but only to workers who are 25 years of age and up. However, anyone may submit a **Request for Earnings and Benefit Estimate Statement** like

Table 5.2 MAXIMUM EARNINGS ON WHICH SOCIAL SECURITY AND MEDICARE TAXES ARE PAID

	Social Security Tax		Medicare Tax	
Year	Social Security Tax Rate	Employee Earnings Subject to the Tax	Medicare Tax Rate	Employee Earnings Subject to the Tax
1991	6.2%	$53,400	1.45%	$125,000
1992	6.2%	$55,500	1.45%	$130,200
1993	6.2%	$57,600	1.45%	$135,000
1994	6.2%	$59,600	1.45%	all
1995	6.2%	$61,200	1.45%	all
1996	6.2%	$62,700	1.45%	all
1997	6.2%	$65,400	1.45%	all
1998	6.2%	$68,400	1.45%	all
1999	6.2%	$72,600	1.45%	all
2000	6.2%	$76,200	1.45%	all
2001				
2002				

Request for Earnings and Benefit Estimate Statement

☐ Please check this box if you want to get your statement in Spanish instead of English.

Please print or type your answers. When you have completed the form, fold it and mail it to us. (If you prefer to send your request using the internet, contact us at http://www.ssa.gov)

1. Name shown on your Social Security card:

   First Name ______ Middle Initial ______

   Last Name Only ______

2. Your Social Security number as shown on your card:

   ☐☐☐-☐☐-☐☐☐☐

3. Your date of birth (Mo.-Day-Yr.)

   ☐☐-☐☐-☐☐☐☐

4. Other Social Security numbers you have used:

   ☐☐☐-☐☐-☐☐☐☐

   ☐☐☐-☐☐-☐☐☐☐

5. Your Sex: ☐ Male ☐ Female

For items 6 and 8 show only earnings covered by Social Security. Do NOT include wages from state, local, or federal government employment that are NOT covered for Social Security or that are covered ONLY by Medicare.

6. Show your actual earnings (wages and/or net self-employment income) for last year and your estimated earnings for this year.

   A. Last year s actual earnings: *(Dollars Only)*

   $☐☐☐,☐☐☐.00

   B. This year s estimated earnings: *(Dollars Only)*

   $☐☐☐,☐☐☐.00

7. Show the age at which you plan to stop working.

   ☐☐ *(Show only one age)*

8. Below, show the average yearly amount (not your total future lifetime earnings) that you think you will earn between now and when you plan to stop working. Include performance or scheduled pay increases or bonuses, but not cost-of-living increases.

   If you expect to earn significantly more or less in the future due to promotions, job changes, part-time work, or an absence from the work force, enter the amount that most closely reflects your future average yearly earnings.

   If you don t expect any significant changes, show the same amount you are earning now (the amount in 6B).

   Future average yearly earnings: *(Dollars Only)*

   $☐☐☐,☐☐☐.00

9. Do you want us to send the statement:
   ¥ To you? Enter your name and mailing address.
   ¥ To someone else (your accountant, pension plan, etc.)? Enter your name with c/o and the name and address of that person or organization.

   Name ______

   ______

   Street Address (Include Apt. No., P.O. Box, or Rural Route) ______

   City ______ State ______ Zip Code ______

NOTICE:
I am asking for information about my own Social Security record or the record of a person I am authorized to represent. I understand that if I deliberately request information under false pretenses, I may be guilty of a federal crime and could be fined and/or imprisoned. I authorize you to use a contractor to send the statement of earnings and benefit estimates to the person named in item 9.

▶

Please sign your name (Do Not Print)

Date ______ (Area Code) Daytime Telephone No. ______

Form SSA-7004-SM

Figure 5.3

the one in Figure 5.3. Since mistakes do occur, it is important to check the statements very carefully. There is a limit of about 3 years, after which errors may not be corrected. To obtain one of the forms and other information about Social Security you may phone 800-772-1213 or go on the World Wide Web to *www.ssa.gov*.

Objective 2 *Find the Maximum FICA Tax Paid by an Employee in One Year.* Remember that Social Security tax is paid on only the first \$80,000 of gross earnings in our examples. An employee earning \$80,000 during the first 10 months of a year would pay no more Social Security tax on additional earnings that year. The maximum Social Security tax to be paid by an employee is $\$80{,}000 \times 6.2\% = \$80{,}000 \times 0.062 = \$4960$ for that year.

NOTE Approximately 7% of income earners reach the Social Security maximum (cutoff point). ■

Objective 3 *Understand Medicare Tax.* Medicare tax is paid on all earnings in our examples. The total earnings are multiplied by 1.45%.

Objective 4 *Find FICA Tax and Medicare Tax.* When finding the amounts to be withheld for Social Security tax and Medicare tax, the employer must use the current rates.

Example 1 Finding FICA Tax and Medicare Tax

Find the Social Security tax and the Medicare tax for the following gross earnings.

(a) D. Horwitz; \$478.15 **(b)** C. Christensen; \$522.83

Solution

(a) The Social Security tax is found by multiplying gross earnings by 6.2%.

$$\$478.15 \times 6.2\% = \$478.15 \times 0.062 = \$29.65 \quad \text{Rounded}$$

Medicare tax is found by multiplying gross earnings by 1.45%.

$$\$478.15 \times 1.45\% = \$478.15 \times 0.0145 = \$6.93 \quad \text{Rounded}$$

(b) Social Security tax is

$$\$522.83 \times 6.2\% = \$522.83 \times 0.062 = \$32.42 \quad \text{Rounded}$$

Medicare tax is

$$\$522.83 \times 1.45\% = \$522.83 \times 0.0145 = \$7.58 \quad \text{Rounded}$$

Example 2 Finding FICA Tax

Cindy Herring has earned \$76,791.08 so far this year. Her gross earnings for the current pay period are \$4842.08. Find her Social Security tax.

Solution Social Security tax is paid on only the first \$80,000 earned in a year. Herring has already earned \$76,791.08. Subtract \$76,791.08 from \$80,000, to find that she has to pay Social Security tax on only \$3208.92 of her earnings for the rest of the year.

\$80,000.00	Maximum earnings subject to tax
− \$76,791.08	Earnings to date
\$3,208.92	Earnings on which tax is due

The Social Security tax on \$3208.92 is \$198.95 (\$3,208.92 × 6.2%). Therefore, Herring pays \$198.95 for the current pay period and no additional Social Security tax for the rest of the year.

The following graph shows what people of different age brackets expect from Social Security retirement income. Notice the difference in the retirement expectations of those in the older age brackets versus those in the younger age brackets?

Age	Most	Fair amount	Very little	None
18–29	4%	23%	50%	20%
30–49	6%	20%	51%	20%
50–59	13%	32%	40%	7%
Now retired	28%	37%	20%	12%

DATA: Putnam Investments survey

Objective 5 ***Determine the FICA Tax and the Medicare Tax Paid by a Self-Employed Person.*** People who are self-employed pay higher Social Security tax and higher Medicare tax than people who work for others. There is no employer to match the employee contribution so the self-employed person pays a rate that is double that of the employee. In our examples the self-employed person pays 12.4% (6.2% × 2) of self-employment earnings for Social Security tax and 2.9% (1.45% × 2) of self-employment earnings for Medicare tax.

Example 3 Finding FICA and Medicare Tax for the Self-Employed

Find the Social Security tax and the Medicare tax paid by Sashaya Davis, self-employed Web-page designer who earned $44,480 in self-employment income this year.

Solution

$$\text{Social Security tax} = \$44{,}480 \times 12.4\% = \$44{,}480 \times 0.124 = \$5515.52$$
$$\text{Medicare tax} = \$44{,}480 \times 2.9\% = \$44{,}480 \times 0.029 = \$1289.92$$

NOTE All employers and those who are self-employed should have the current tax rates for both Social Security and Medicare. These can always be found in Circular E, Employer's Tax Guide, which is available from the Internal Revenue Service. ■

Objective 6 ***Find State Disability Insurance Deductions.*** Many states have a state disability insurance program that is paid for by employees. If disabled, the employee would receive weekly benefits. A typical state program defines "disability" as "any illness or injury incurred on or off the job, either physical or mental, including pregnancy, childbirth, or related condition, that prevents you from doing your regular work." The employee processes the claim after obtaining certification from a doctor or other qualified examiner. Weekly disability benefits are determined by the highest quarter's earnings within the last year of employment.

A typical state program also requires the qualifying employee to pay a **state disability insurance (SDI)** deduction of 1% of the first $31,800 earned each year. There are no payments on earnings above this amount. Some states have similar programs, but their insurance is placed with private insurance companies rather than with the state.

Example 4 Finding State Disability Insurance Deductions

Find the state disability deduction for an employee at Comet Auto Parts with gross earnings of $418 this pay period. The SDI rate is 1%, and the employee has not earned $31,800 this year.

Solution The state disability deduction is $4.18 ($418 × 1%).

Example 5 Finding State Disability Insurance Deductions

Milo Lacy has earned $30,620 so far this year. Find the SDI deduction if gross earnings this pay period are $3096. Use an SDI rate of 1% on the first $31,800.

Solution The SDI deduction will be taken on $1180 of the current gross earnings.

$31,800	Maximum earnings subject to SDI
− $30,620	Earnings this year
$1,180	Earnings this year subject to SDI

The SDI deduction is $11.80 ($1180 × 1%).

NOTE Always be aware of the current rates and the maximum annual earning amounts against which FICA, Medicare, and SDI payroll deductions may be taken. Those involved in payroll work must always be up to date on federal and state laws and practices. ■

5.4 Exercises

SOCIAL SECURITY AND MEDICARE TAX *Find the Social Security tax and the Medicare tax for each of the following amounts of gross earnings. Assume a 6.2% FICA rate and a 1.45% Medicare tax rate.*

1. $324.72
2. $207.25
3. $463.24
4. $606.35
5. $854.71
6. $683.65
7. $1086.25
8. $1243.18

MAXIMUM SOCIAL SECURITY *Find the Social Security tax for each of the following employees for the current pay period. Assume a 6.2% FICA rate up to a maximum of $80,000.*

	Employee	Gross Earnings This Year (to date)	Earnings Current Pay Period
9.	Brown, M.	$77,871.24	$3218.36
10.	Cambell, K.	$78,818.93	$2700.00
11.	Carlson, K.	$75,721.59	$5780.00
12.	Floyd, D.	$78,018.67	$2162.34
13.	Johnson, R.	$79,819.75	$1915.38
14.	Levy, D.	$77,992.06	$3273.81

PAYROLL DEDUCTIONS *Find the regular earnings, overtime earnings, gross earnings, Social Security tax, Medicare tax, and state disability insurance deduction for each of the following employees. Assume time and a half is paid for any hours over a 40-hour week. Assume a 6.2% FICA rate, a 1.45% Medicare rate, a state disability rate of 1%; assume also that no one has earned more than the FICA or SDI maximum at the end of the current pay period.*

	Employee	Hours Worked	Regular Rate
15.	Thunstrom, P.	45.5	$9.22
16.	Lawler, J.	47.75	$7.52
17.	Odom, R.	44	$10.30
18.	Ruppart, A.	45	$6.58
19.	Leonard, B.	45	$8.18
20.	Taggart, G.	47	$11.68
21.	Yates, D.	46.75	$6.24
22.	Shotwell, E.	48.25	$7.40

Solve each of the following application problems. Assume that the FICA rate is 6.2%, the Medicare rate is 1.45%, the SDI rate is 1% and that earnings will not exceed $31,800.

23. SOCIAL SECURITY AND MEDICARE Adam Bryer worked 43.5 hours last week at Blockbuster Video. He is paid $8.58 per hour, plus time and a half for overtime (over 40 hours per week). Find his (a) Social Security tax and (b) Medicare tax for the week.

24. SOCIAL SECURITY AND MEDICARE Beth Kaufman receives 7% commission on all sales. Her sales on Monday of last week were $1412.20, with $1928.42 on Tuesday, $598.14 on Wednesday, $1051.12 on Thursday, and $958.72 on Friday. Find her (a) Social Security tax and (b) Medicare tax for the week.

25. STATE DISABILITY DEDUCTION Lynn Peterson is paid an 8% commission on sales. During a recent pay period, she had sales of $19,482 and returns and allowances of $193. Find the amount of (a) her Social Security tax, (b) her Medicare tax, and (c) her state disability for this pay period.

26. STATE DISABILITY DEDUCTIONS Peter Phelps is a representative for Delta International Machinery and is paid $350 per week plus a commission of 2% on sales. His sales last week were $17,240. Find the amount of (a) his Social Security tax, (b) his Medicare tax, and (c) his state disability for the pay period.

SELF-EMPLOYMENT DEDUCTIONS ***The following problems refer to self-employed individuals. These people pay Social Security tax of 12.4% and Medicare tax of 2.9%. Find the taxes on each of the following self-employment incomes. (See Example 3.)***

27. Jeremy Merz, owner of Sports Center, earned $36,852.80

28. Rachel Leach, an interior decorator, earned $28,286.20

29. Krystal McClellan, accountant, earned $34,817.16

30. Mike Viera, surveyor, earned $48,007.14

31. Lauren Midgley, shop owner, earned $26,843.60

32. Eric Lemmon, commission salesperson, earned $52,748.32

33. A young person who has just received her first paycheck is puzzled by the amounts that have been deducted from gross earnings. Briefly explain both the FICA and Medicare deductions to this person. (See Objectives 1–4.)

34. Describe the difference between the FICA paid by an employee and that paid by a self-employed person. (See Objective 5.)

5.5 Income Tax Withholding

Objectives

1. *Understand the Employee's Withholding Allowance Certificate (Form W-4).*
2. *Find the federal withholding tax from tables.*
3. *Find the federal tax using the percentage method.*
4. *Find the state withholding tax using the state income tax rate.*
5. *Find net pay when given gross wages, taxes, and other deductions.*

The **personal income tax** is the largest single source of money for the federal government. The law requires that the bulk of this tax owed by an individual be paid periodically, as the income is earned. For this reason, employers must deduct money from the gross earnings of almost every employee. These deductions, called **income tax withholdings**, are sent periodically to the Internal Revenue Service and credited to the accounts of the employees. The amount of money withheld depends on the employee's **marital status**, the number of **withholding allowances** claimed, and the amount of gross earnings. Generally, the withholding tax for a married person is less than the withholding tax for a single person making the same income.

Objective 1 *Understand the Employees Withholding Allowance Certificate (Form W-4).* Each employee must file with the employer a W-4 form as shown in Figure 5.4. On this form

Form **W-4**
Department of the Treasury
Internal Revenue Service

Employee's Withholding Allowance Certificate

▶ **For Privacy Act and Paperwork Reduction Act Notice, see page 2.**

OMB No. 1545-0010
200_

1 Type or print your first name and middle initial	Last name	**2** Your social security number
Home address (number and street or rural route)		**3** ☐ Single ☐ Married ☐ Married, but withhold at higher Single rate. **Note:** *If married, but legally separated, or spouse is a nonresident alien, check the Single box.*
City or town, state, and ZIP code		**4** If your last name differs from that on your social security card, check here. **You** must call 1-800-772-1213 for a new card . . . ▶ ☐

5 Total number of allowances you are claiming (from line H above or from the worksheets on page 2 if they apply) . **5**

6 Additional amount, if any, you want withheld from each paycheck . . . . **6** $

7 I claim exemption from withholding for 200_, and I certify that I meet **BOTH** of the following conditions for exemption:
- Last year I had a right to a refund of **ALL** Federal income tax withheld because I had **NO** tax liability **AND**
- This year I expect a refund of **ALL** Federal income tax withheld because I expect to have **NO** tax liability.

If you meet both conditions, write "EXEMPT" here . . . . ▶ **7**

Under penalties of perjury, I certify that I am entitled to the number of withholding allowances claimed on this certificate, or I am entitled to claim exempt status.
Employee's signature
(Form is not valid
unless you sign it) ▶ **Date** ▶

8 Employer's name and address (Employer: Complete 8 and 10 only if sending to the IRS)	**9** Office code (optional)	**10** Employer identification number

Cat. No. 10220Q

Figure 5.4

the employee states the number of withholding allowances being claimed along with additional information, so that the employer can withhold the proper amount for income tax.

A W-4 form is usually completed at the time of employment. A married person with three children will normally claim five allowances (one each for the employee and spouse, plus one for each child). However, when both spouses are employed, each may claim himself or herself. The number of allowances may be raised if an employee has been receiving a refund of withholding taxes, or the number may be lowered if the employee has had a balance due in previous tax years. The W-4 form has instructions to help determine the proper number of allowances. Some people enjoy receiving a tax refund when filing their income tax return, so they claim fewer allowances, having more withheld from each check. Other individuals would rather receive more of their income each pay period, so they claim the maximum number of allowances to which they are entitled. The exact number of allowances *must* be claimed when the income tax return is filed.

Objective 2 ***Find the Federal Withholding Tax from Tables.*** The withholding tax is found on the basis of the gross earnings per pay period. Income tax withholding is applied to all earnings, unlike Social Security tax. Generally, the higher a person's gross earnings, the more withholding tax is paid.

There are two methods that employers use to determine the amount of federal withholding tax to deduct from paychecks: the **wage bracket method** and the **percentage method**.

The Internal Revenue Service supplies withholding tax tables to be used with the wage bracket method. These tables are extensive and cover weekly, biweekly, monthly, and daily pay periods. Figures 5.5 and 5.6 show samples of the withholding tables. Figure 5.5 is for both single and married people who are paid weekly and Figure 5.6 is for both single and married people who are paid monthly.

Example 1 Finding Federal Withholding Using the Wage Bracket Method

Lisa Revies is single and claims no withholding allowances. (Some employees do this to receive a refund from the government or to avoid owing taxes at the end of the year. The proper number will be used when filing her income tax return.) Use the wage bracket method to find her withholding tax if her weekly gross earnings are $328.75.

Solution Use the table in Figure 5.5 for single persons—weekly payroll period. The given earnings are found in the row "at least $320 but less than $330." Go across this row to the column headed "0" (for no withholding allowances). From the table, the withholding is $41.

Example 2 Using the Wage Bracket Method for Federal Withholding

Larry Sifford is married, claims three withholding allowances, and has monthly gross earnings of $2947.35. Find his withholding tax using the wage bracket method.

Solution Use the table in Figure 5.6 for Married Persons—Monthly Payroll Period. Look down the two left columns, and find the range that includes Sifford's gross earnings: "at least $2920 but less than $2960." Read across the table to the column headed "3" (for the three withholding allowances). The withholding tax is $259. Had Sifford claimed six withholding allowances, his withholding tax would have been only $158.

SINGLE Persons—**WEEKLY** Payroll Period

(For Wages Paid in 20__)

If the wages are—		And the number of withholding allowances claimed is—										
At least	But less than	0	1	2	3	4	5	6	7	8	9	10
		The amount of income tax to be withheld is—										
125	130	11	4	0	0	0	0	0	0	0	0	0
130	135	12	4	0	0	0	0	0	0	0	0	0
135	140	13	5	0	0	0	0	0	0	0	0	0
140	145	14	6	0	0	0	0	0	0	0	0	0
145	150	14	7	0	0	0	0	0	0	0	0	0
150	155	15	7	0	0	0	0	0	0	0	0	0
155	160	16	8	0	0	0	0	0	0	0	0	0
160	165	17	9	1	0	0	0	0	0	0	0	0
165	170	17	10	2	0	0	0	0	0	0	0	0
170	175	18	10	3	0	0	0	0	0	0	0	0
175	180	19	11	3	0	0	0	0	0	0	0	0
180	185	20	12	4	0	0	0	0	0	0	0	0
185	190	20	13	5	0	0	0	0	0	0	0	0
190	195	21	13	6	0	0	0	0	0	0	0	0
195	200	22	14	6	0	0	0	0	0	0	0	0
200	210	23	15	8	0	0	0	0	0	0	0	0
210	220	25	17	9	1	0	0	0	0	0	0	0
220	230	26	18	11	3	0	0	0	0	0	0	0
230	240	28	20	12	4	0	0	0	0	0	0	0
240	250	29	21	14	6	0	0	0	0	0	0	0
250	260	31	23	15	7	0	0	0	0	0	0	0
260	270	32	24	17	9	1	0	0	0	0	0	0
270	280	34	26	18	10	2	0	0	0	0	0	0
280	290	35	27	20	12	4	0	0	0	0	0	0
290	300	37	29	21	13	5	0	0	0	0	0	0
300	310	38	30	23	15	7	0	0	0	0	0	0
310	320	40	32	24	16	8	1	0	0	0	0	0
320	330	41	33	26	18	10	2	0	0	0	0	0
330	340	43	35	27	19	11	4	0	0	0	0	0
340	350	44	36	29	21	13	5	0	0	0	0	0

MARRIED Persons—**WEEKLY** Payroll Period

(For Wages Paid in 20__)

If the wages are—		And the number of withholding allowances claimed is—										
At least	But less than	0	1	2	3	4	5	6	7	8	9	10
		The amount of income tax to be withheld is—										
440	450	48	40	33	25	17	9	1	0	0	0	0
450	460	50	42	34	26	18	11	3	0	0	0	0
460	470	51	43	36	28	20	12	4	0	0	0	0
470	480	53	45	37	29	21	14	6	0	0	0	0
480	490	54	46	39	31	23	15	7	0	0	0	0
490	500	56	48	40	32	24	17	9	1	0	0	0
500	510	57	49	42	34	26	18	10	3	0	0	0
510	520	59	51	43	35	27	20	12	4	0	0	0
520	530	60	52	45	37	29	21	13	6	0	0	0
530	540	62	54	46	38	30	23	15	7	0	0	0
540	550	63	55	48	40	32	24	16	9	1	0	0
550	560	65	57	49	41	33	26	18	10	2	0	0
560	570	66	58	51	43	35	27	19	12	4	0	0
570	580	68	60	52	44	36	29	21	13	5	0	0
580	590	69	61	54	46	38	30	22	15	7	0	0
590	600	71	63	55	47	39	32	24	16	8	1	0
600	610	72	64	57	49	41	33	25	18	10	2	0
610	620	74	66	58	50	42	35	27	19	11	4	0
620	630	75	67	60	52	44	36	28	21	13	5	0
630	640	77	69	61	53	45	38	30	22	14	7	0
640	650	78	70	63	55	47	39	31	24	16	8	0
650	660	80	72	64	56	48	41	33	25	17	10	2
660	670	81	73	66	58	50	42	34	27	19	11	3
670	680	83	75	67	59	51	44	36	28	20	13	5
680	690	84	76	69	61	53	45	37	30	22	14	6
690	700	86	78	70	62	54	47	39	31	23	16	8
700	710	87	79	72	64	56	48	40	33	25	17	9
710	720	89	81	73	65	57	50	42	34	26	19	11
720	730	90	82	75	67	59	51	43	36	28	20	12
730	740	92	84	76	68	60	53	45	37	29	22	14

Figure 5.5

SINGLE Persons—**MONTHLY** Payroll Period

(For Wages Paid in 20__)

If the wages are—		And the number of withholding allowances claimed is—										
At least	But less than	0	1	2	3	4	5	6	7	8	9	10
		The amount of income tax to be withheld is—										
840	880	96	62	28	0	0	0	0	0	0	0	0
880	920	102	68	34	1	0	0	0	0	0	0	0
920	960	108	74	40	7	0	0	0	0	0	0	0
960	1,000	114	80	46	13	0	0	0	0	0	0	0
1,000	1,040	120	86	52	19	0	0	0	0	0	0	0
1,040	1,080	126	92	58	25	0	0	0	0	0	0	0
1,080	1,120	132	98	64	31	0	0	0	0	0	0	0
1,120	1,160	138	104	70	37	3	0	0	0	0	0	0
1,160	1,200	144	110	76	43	9	0	0	0	0	0	0
1,200	1,240	150	116	82	49	15	0	0	0	0	0	0
1,240	1,280	156	122	88	55	21	0	0	0	0	0	0
1,280	1,320	162	128	94	61	27	0	0	0	0	0	0
1,320	1,360	168	134	100	67	33	0	0	0	0	0	0
1,360	1,400	174	140	106	73	39	5	0	0	0	0	0
1,400	1,440	180	146	112	79	45	11	0	0	0	0	0
1,440	1,480	186	152	118	85	51	17	0	0	0	0	0
1,480	1,520	192	158	124	91	57	23	0	0	0	0	0
1,520	1,560	198	164	130	97	63	29	0	0	0	0	0
1,560	1,600	204	170	136	103	69	35	1	0	0	0	0
1,600	1,640	210	176	142	109	75	41	7	0	0	0	0
1,640	1,680	216	182	148	115	81	47	13	0	0	0	0
1,680	1,720	222	188	154	121	87	53	19	0	0	0	0
1,720	1,760	228	194	160	127	93	59	25	0	0	0	0
1,760	1,800	234	200	166	133	99	65	31	0	0	0	0
1,800	1,840	240	206	172	139	105	71	37	4	0	0	0
1,840	1,880	246	212	178	145	111	77	43	10	0	0	0
1,880	1,920	252	218	184	151	117	83	49	16	0	0	0
1,920	1,960	258	224	190	157	123	89	55	22	0	0	0
1,960	2,000	264	230	196	163	129	95	61	28	0	0	0
2,000	2,040	270	236	202	169	135	101	67	34	0	0	0

MARRIED Persons—**MONTHLY** Payroll Period

(For Wages Paid in 20__)

If the wages are—		And the number of withholding allowances claimed is—										
At least	But less than	0	1	2	3	4	5	6	7	8	9	10
		The amount of income tax to be withheld is—										
2,040	2,080	228	195	161	127	93	60	26	0	0	0	0
2,080	2,120	234	201	167	133	99	66	32	0	0	0	0
2,120	2,160	240	207	173	139	105	72	38	4	0	0	0
2,160	2,200	246	213	179	145	111	78	44	10	0	0	0
2,200	2,240	252	219	185	151	117	84	50	16	0	0	0
2,240	2,280	258	225	191	157	123	90	56	22	0	0	0
2,280	2,320	264	231	197	163	129	96	62	28	0	0	0
2,320	2,360	270	237	203	169	135	102	68	34	0	0	0
2,360	2,400	276	243	209	175	141	108	74	40	6	0	0
2,400	2,440	282	249	215	181	147	114	80	46	12	0	0
2,440	2,480	288	255	221	187	153	120	86	52	18	0	0
2,480	2,520	294	261	227	193	159	126	92	58	24	0	0
2,520	2,560	300	267	233	199	165	132	98	64	30	0	0
2,560	2,600	306	273	239	205	171	138	104	70	36	3	0
2,600	2,640	312	279	245	211	177	144	110	76	42	9	0
2,640	2,680	318	285	251	217	183	150	116	82	48	15	0
2,680	2,720	324	291	257	223	189	156	122	88	54	21	0
2,720	2,760	330	297	263	229	195	162	128	94	60	27	0
2,760	2,800	336	303	269	235	201	168	134	100	66	33	0
2,800	2,840	342	309	275	241	207	174	140	106	72	39	5
2,840	2,880	348	315	281	247	213	180	146	112	78	45	11
2,880	2,920	354	321	287	253	219	186	152	118	84	51	17
2,920	2,960	360	327	293	259	225	192	158	124	90	57	23
2,960	3,000	366	333	299	265	231	198	164	130	96	63	29
3,000	3,040	372	339	305	271	237	204	170	136	102	69	35
3,040	3,080	378	345	311	277	243	210	176	142	108	75	41
3,080	3,120	384	351	317	283	249	216	182	148	114	81	47
3,120	3,160	390	357	323	289	255	222	188	154	120	87	53
3,160	3,200	396	363	329	295	261	228	194	160	126	93	59
3,200	3,240	402	369	335	301	267	234	200	166	132	99	65

Figure 5.6

Objective 3 *Find the Federal Tax Using the Percentage Method.* Many companies today prefer to use the **percentage method** to determine federal withholding tax. The percentage method does not require the several pages of tables needed with the wage bracket method and is more easily adapted to computer applications in the processing of payrolls. Instead, the table shown in Figure 5.7 is used.

Example 3 Finding Federal Withholding Using the Percentage Method

Steve Tomlin is married, claims four withholding allowances, and has weekly gross earnings of $1150. Use the percentage method to find his withholding tax.

Solution

Step 1. Find the amount of one withholding allowance *for one* on the weekly payroll period in Figure 5.7. The amount is $52.88. Since Tomlin claims four allowances, multiply the one withholding allowance, $52.88, by the number of withholding allowances, 4.

$$\$52.88 \times 4 = \$211.52$$

Step 2. Subtract the amount in Step 1 from gross earnings.

$$\$1150 - \$211.52 = \$938.48$$

Step 3. Find the "married person weekly" section of the percentage method withholding table. Since $938.48 is over $913 but not over $1894, an amount of $118.35 is added to 28% of the excess over $913.

$$\$938.48 - \$913 = \$25.48 \quad \text{Excess over \$913}$$

$$\$25.48 \times 28\% = \$25.48 \times 0.28 = \$7.13$$

$$\$118.35 + \$7.13 = \$125.48 \quad \text{Withholding tax}$$

NOTE In Step 1, for an employee who is paid monthly, the number of withholding allowances would be multiplied by $229.17, the amount from the table for one monthly withholding allowance. ■

Scientific *Calculator Approach*

The calculator solution to Example 3 is as follows.

1150 [−] 52.88 [×] 4 [−] 913 [=]

[×] .28 [+] 118.35 [=] 125.4844

NOTE The amount of withholding tax found using the wage bracket method can vary slightly from the amount of withholding tax found using the percentage method. Any differences would be eliminated when the income tax return is filed. ■

Objective 4 *Find the State Withholding Tax Using the State Income Tax Rate.* **State income taxes** vary, with no income tax in the states of Alaska, Florida, Nevada, South Dakota, Texas, Washington, and Wyoming. A few states have a flat tax rate (percent of income);

Payroll Period	One Withholding Allowance
Weekly	$ 52.88
Biweekly	105.77
Semimonthly	114.58
Monthly	229.17
Quarterly	687.50
Semiannually	1375.00
Annually	2750.00
Daily or miscellaneous (each day of the payroll period)	10.58

Tables for Percentage Method of Withholding

TABLE 1—WEEKLY Payroll Period

(a) SINGLE person (including head of household)—

If the amount of wages after subtracting withholding allowances is: / The amount of income tax to withhold is:

Not over $51 . . . . . $0

Over—	But not over—		of excess over—
$51	—$525	15%	—$51
$525	—$1,125	$71.10 plus 28%	—$525
$1,125	—$2,535	$239.10 plus 31%	—$1,125
$2,535	—$5,475	$676.20 plus 36%	—$2,535
$5,475		$1,734.60 plus 39.6%	—$5,475

(b) MARRIED person—

If the amount of wages after subtracting withholding allowances is: / The amount of income tax to withhold is:

Not over $124 . . . . $0

Over—	But not over—		of excess over—
$124	—$913	15%	—$124
$913	—$1,894	$118.35 plus 28%	—$913
$1,894	—$3,135	$393.03 plus 31%	—$1,894
$3,135	—$5,531	$777.74 plus 36%	—$3,135
$5,531		$1,640.30 plus 39.6%	—$5,531

TABLE 2—BIWEEKLY Payroll Period

(a) SINGLE person (including head of household)—

If the amount of wages after subtracting withholding allowances is: / The amount of income tax to withhold is:

Not over $102 . . . . $0

Over—	But not over—		of excess over—
$102	—$1,050	15%	—$102
$1,050	—$2,250	$142.20 plus 28%	—$1,050
$2,250	—$5,069	$478.20 plus 31%	—$2,250
$5,069	—$10,950	$1,352.09 plus 36%	—$5,069
$10,950		$3,469.25 plus 39.6%	—$10,950

(b) MARRIED person—

If the amount of wages after subtracting withholding allowances is: / The amount of income tax to withhold is:

Not over $248 . . . . $0

Over—	But not over—		of excess over—
$248	—$1,827	15%	—$248
$1,827	—$3,788	$236.85 plus 28%	—$1,827
$3,788	—$6,269	$785.93 plus 31%	—$3,788
$6,269	—$11,062	$1,555.04 plus 36%	—$6,269
$11,062		$3,280.52 plus 39.6%	—$11,062

TABLE 3—SEMIMONTHLY Payroll Period

(a) SINGLE person (including head of household)—

If the amount of wages after subtracting withholding allowances is: / The amount of income tax to withhold is:

Not over $110 . . . . $0

Over—	But not over—		of excess over—
$110	—$1,138	15%	—$110
$1,138	—$2,438	$154.20 plus 28%	—$1,138
$2,438	—$5,492	$518.20 plus 31%	—$2,438
$5,492	—$11,863	$1,464.94 plus 36%	—$5,492
$11,863		$3,758.50 plus 39.6%	—$11,863

(b) MARRIED person—

If the amount of wages after subtracting withholding allowances is: / The amount of income tax to withhold is:

Not over $269 . . . . $0

Over—	But not over—		of excess over—
$269	—$1,979	15%	—$269
$1,979	—$4,104	$256.50 plus 28%	—$1,979
$4,104	—$6,792	$851.50 plus 31%	—$4,104
$6,792	—$11,983	$1,684.78 plus 36%	—$6,792
$11,983		$3,553.54 plus 39.6%	—$11,983

TABLE 4—MONTHLY Payroll Period

(a) SINGLE person (including head of household)—

If the amount of wages after subtracting withholding allowances is: / The amount of income tax to withhold is:

Not over $221 . . . . $0

Over—	But not over—		of excess over—
$221	—$2,275	15%	—$221
$2,275	—$4,875	$308.10 plus 28%	—$2,275
$4,875	—$10,983	$1,036.10 plus 31%	—$4,875
$10,983	—$23,725	$2,929.58 plus 36%	—$10,983
$23,725		$7,516.70 plus 39.6%	—$23,725

(b) MARRIED person—

If the amount of wages after subtracting withholding allowances is: / The amount of income tax to withhold is:

Not over $538 . . . . $0

Over—	But not over—		of excess over—
$538	—$3,958	15%	—$538
$3,958	—$8,208	$513.00 plus 28%	—$3,958
$8,208	—$13,583	$1,703.00 plus 31%	—$8,208
$13,583	—$23,967	$3,369.25 plus 36%	—$13,583
$23,967		$7,107.49 plus 39.6%	—$23,967

Figure 5.7

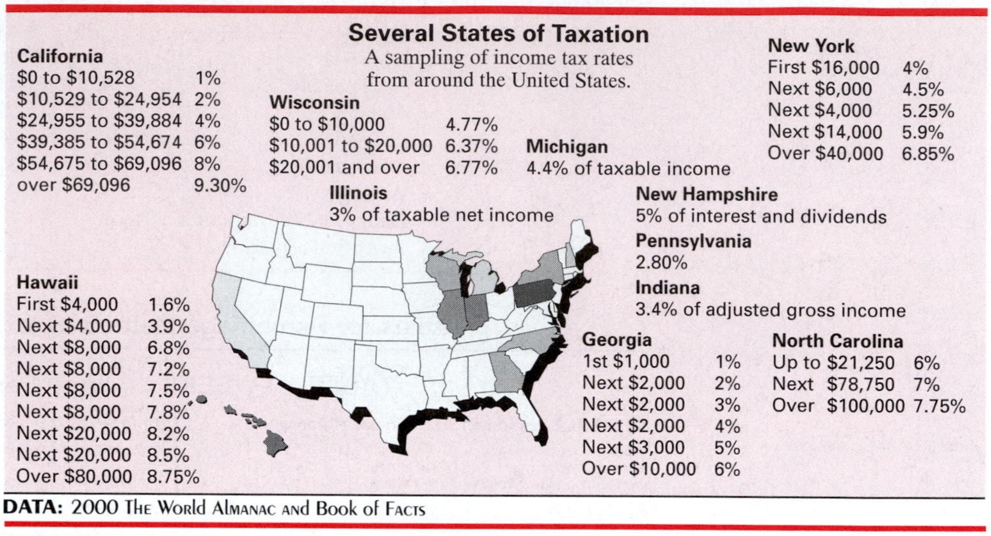

Figure 5.8

the majority of the states issue tax tables, with taxes going as high as 9% and 10%. A few of the state income tax rates are shown in Figure 5.8.

Example 4 Finding State Withholding Tax

Andrea Novak has gross earnings for the month of $3642. If her state has a 4.4% income tax rate, find the state withholding tax.

Solution State withholding tax can be found by multiplying 4.4% by the amount of earnings.

$$4.4\% \times \$3642 = 0.044 \times \$3642 = \$160.25 \quad \text{Rounded.}$$

The amount of state withholding tax is $160.25.

Objective 5 ***Find Net Pay When Given Gross Wages, Taxes, and Other Deductions.*** It is common for employees to request additional deductions, such as union dues and credit union payments. The final amount of pay received by the employee, called the net pay, is given by the following formula.

$$\text{Net pay} = \text{Gross earnings} - \text{FICA tax (Social Security)} - \text{Medicare tax} - \text{Federal withholding tax} - \text{State withholding tax} - \text{Other deductions}$$

Example 5 Determining Net Pay after Deductions

Ann Stypuloski is married and claims three withholding allowances. Her weekly gross earnings are $538.25. Her state withholding is 2.5% and her union dues are $20. Find her net pay using the percentage method of withholding.

Solution First find FICA (Social Security) tax, which is \$33.37, then Medicare, \$7.80. Federal withholding tax is \$38.34 and state withholding is \$13.46. Total deductions are

\$33.37	FICA tax (6.2% × \$538.25)
\$7.80	Medicare tax (1.45% × \$538.25)
\$38.34	Federal withholding
\$13.46	State withholding (2.5% × \$538.25)
+ \$20.00	Union dues
\$112.97	Total deductions

NOTE To find the federal withholding tax, use these steps:

Step 1. \$52.88 × 3 = \$158.64

Step 2. \$538.25 − \$158.64 = \$379.61

Step 3. (\$379.61 − \$124) × .15 = \$38.34 ■

Find net pay by subtracting total deductions from gross earnings.

\$538.25	Gross earnings
−\$112.97	Total deductions
\$425.28	Net pay

Stypuloski will receive a check for \$425.28.

5.5 Exercises

FEDERAL WITHHOLDING TAX *Find the federal withholding tax for each of the following employees. Use the wage bracket method.*

	Employee	Gross Earnings	Marital Status	Withholding Allowances
1.	MacDonald, T.	\$2817.33 monthly	Married	3
2.	Martin, L.	\$257.98 weekly	Single	2
3.	Brownlee, K.	\$485.61 weekly	Married	0
4.	Caven, A.	\$2745.73 monthly	Married	1
5.	Miller, J.	\$1829.83 monthly	Single	2
6.	Radcliff, J.	\$1810.42 monthly	Single	1
7.	Remington, J.	\$2387.92 monthly	Married	6
8.	Moran, K.	\$332.14 weekly	Single	1
9.	Seither, A.	\$1598.14 monthly	Single	4
10.	Thunstrom, P.	\$479.08 weekly	Married	0

STATE WITHHOLDING TAX *Use the state income tax rate given to find the state withholding tax for each of the following employees.*

	Employee	Gross Weekly Earnings	State Income Tax Rate	
11.	Owens, E.	$188.60	4.4%	______
12.	Childs, M.	$235.68	5.0%	______
13.	Christensen, C.	$317.43	2.8%	______
14.	deJamaer, K.	$547.54	3.4%	______
15.	Galvin, M.	$1476.32	6.0%	______
16.	Hampson, N.	$2720.85	5.95%	______

EMPLOYEE NET PAY *Use a 6.2% FICA rate to find FICA tax, 1.45% to find Medicare tax, and the percentage method of withholding to find federal withholding tax for each of the following employees. Then find the net pay for each employee. The number of withholding allowances and the marital status are listed after each employee's name. Assume that no employee has earned over $80,000 so far this year. (See Examples 3 and 5.)*

	Employee	Gross Earnings				
17.	Kolasa; 4, M	$417.58 weekly	______	______	______	______
18.	Lagerstrom; 3, S	$2356.21 monthly	______	______	______	______
19.	Moore; 1, S	$1532.18 monthly	______	______	______	______
20.	Scandone; 2, M	$625 weekly	______	______	______	______
21.	Williams; 3, M	$1938.76 semimonthly	______	______	______	______
22.	Wagnon; 1, S	$382.46 weekly	______	______	______	______
23.	Monroe; 6, M	$1971.06 semimonthly	______	______	______	______
24.	Watnick; 2, M	$1450 weekly	______	______	______	______
25.	Stephenson; 3, S	$710.56 biweekly	______	______	______	______
26.	Schimmenti; 2, S	$3998.17 monthly	______	______	______	______
27.	Young; 1, S	$915.34 weekly	______	______	______	______
28.	Weber; 4, M	$2705.20 biweekly	______	______	______	______
29.	Wilson; 3, M	$5312.59 monthly	______	______	______	______
30.	Deal; 2, M	$948.75 semimonthly	______	______	______	______
31.	Coop; 2, S	$431.25 weekly	______	______	______	______
32.	Beltram; 1, S	$3285.20 monthly	______	______	______	______

33. Write an explanation of how to determine the federal withholding tax using the wage bracket (tax tables) method.

34. Write an explanation of how to find the federal withholding tax using the percentage method.

Use the percentage method of withholding, a FICA rate of 6.2%, a Medicare rate of 1.45%, an SDI rate of 1%, and a state withholding tax of 3.4% in the following problems.

35. MARKETING REPRESENTATIVE Doug Gilbert, marketing representative, has weekly earnings of $783. He is married and claims four withholding allowances. His deductions include FICA, Medicare, federal withholding, state disability insurance, state withholding, union dues of $15.50, and credit union savings of $100. Find his net pay for a week in February.

36. WEEKLY NET PAY Cindy Herring has earnings of $588 in one week of March. She is single and claims four withholding allowances. Her deductions include FICA, Medicare, federal withholding, state disability insurance, state withholding, a United Way contribution of $10, and a savings bond of $50. Find her net pay for the week.

37. EDUCATIONAL SALES The top salesperson for Pearson Education is paid a salary of $410 per week plus 7% of all sales over $5000. She is single and claims two withholding allowances. Her deductions include FICA, Medicare, federal withholding, state disability insurance, state withholding, credit union savings of $50, a Salvation Army contribution of $10, and dues of $15 to the National Association of Professional Saleswomen. Find her net pay for a week in April during which she has sales of $11,284 with returns and allowances of $424.50.

38. TRAVEL-AGENCY SALES Elane Hanson, an account executive for a travel agency specializing in travel to South America, is paid on a variable commission, is married, and claims four withholding allowances. She receives 2% of the first $20,000 in sales, 3% of the next $10,000 in sales, and 5% of all sales over $30,000. This week she has sales of $38,740 and the following deductions: FICA, Medicare, federal withholding, state disability insurance, state withholding, a retirement contribution of $23.83, a savings bond of $37.50, and charitable contributions of $14. Find her net pay after subtracting all of her deductions.

39. HEATING-COMPANY REPRESENTATIVE Sheri Minkner, a commission sales representative for Alternative Heating Company, is paid a monthly salary of $4200 plus a bonus of 1.5% on monthly sales. She is married and claims three withholding allowances. Her deductions include FICA, Medicare, federal withholding, state disability insurance, credit union savings of $150, charitable contributions of $25, and a savings bond of $50. Find her net pay for a month in which her sales were $42,618. The state in which Minkner works has no state income tax.

40. RIVER-RAFTING MANAGER River Raft Adventures pays its manager Susan Estey a monthly salary of $2880 plus a commission of 0.8% based on total monthly sales volume. In the month of May, River Raft Adventures has total sales of $86,280. Estey is married and claims five withholding allowances. Her deductions include FICA, Medicare, federal withholding, state disability insurance, state withholding of $159.30, credit union payment of $300, March of Dimes contributions of $20, and savings bonds of $250. Find her net pay for May.

5.6 Payroll Records and Quarterly Returns

Objectives

1. *Understand payroll records kept by employers.*
2. *Calculate employer's matching Social Security and Medicare contributions.*
3. *Find the quarterly amount due the Internal Revenue Service.*
4. *Identify the form used by employers to file a quarterly tax return.*
5. *Find the amount of federal unemployment tax due.*

Employers keep payroll records for many reasons. Individual payroll records for each employee are used to keep track of Social Security tax, Medicare tax, federal and state withholding, and many other items. The amounts withheld from employee earnings are sent periodically to the proper agency; most are paid entirely by the employee and others are matched by the employer. Usually these records are filed quarterly. These quarters and the filing dates are shown in Table 5.3.

Objective 1 *Understand Payroll Records Kept by Employers.* The payroll register, discussed in Section 5.1, was a record of the number of hours worked by all of the employees for a certain time period. The Employee's Earnings Record shown in Figure 5.9 details quarterly totals for an individual employee. This record shows the gross earnings, deduction amounts, and net pay for each pay period during the quarter.

Objective 2 *Calculate Employer's Matching Social Security and Medicare Contributions.* The employer must check the earnings of each employee to make sure that the FICA, Medicare, and the federal unemployment tax cutoff points are not passed. Since the employer must also give an end-of-year wage and tax statement (Form W-2) to each employee, the records are also used as the source of this information. In addition, accurate payroll records are important because the employer is required by law to match the employee's Social Security and Medicare contributions.

Objective 3 *Find the Quarterly Amount Due the Internal Revenue Service.* An employee's contribution to Social Security and Medicare must be matched by the employer.

Table 5.3 FILING SCHEDULES FOR EMPLOYEE WITHHOLDING

Quarter	Ending	Due Date
January, February, March	March 31	April 30
April, May, June	June 30	July 31
July, August, September	September 30	October 31
October, November, December	December 31	January 31

NAME *Lisa Kamins* CLOCK NUMBER *114* DEPT *Production* *M* (MARITAL STATUS) *3* (NO. OF EXEMPT)

STREET *407 Glen Oak Dr.* SOC.SEC. NUMBER *123-45-6789* ☐ M. ☒ F.

CITY *Forth Worth, Texas* PHONE NO. *482-6319* DATE STARTED DATE LEFT

RECORD OF PAY CHANGES: *1/1/20* — *7.20*

YEAR *200*__ ENCIRCLE QUARTERS (1) 2 3 4

TIME WORKED	DATE PAY PERIOD ENDING	SUN	M	TU	W	TH	F	SAT	GROSS PAYROLL	F.W.T.	SOC. SEC.	MEDI	S.W.T.	U.D.	D. INS.	CHECK NO.	NET PAY	
	BROUGHT FORWARD →																	
	1/7		8	8	7	8	10		298.80	18.00	19.42	4.48	13.80	5.00	18.00	1186	220.10	1
	1/14		8	8	8	8	8		288.00	16.00	18.72	4.32	13.40	5.00	18.00	1295	212.56	2
	1/21		8	10	8	9	8	4	363.60	28.00	23.63	5.45	15.06	5.00	18.00	1378	268.46	3
	1/28		8	8	8	8	8		288.00	16.00	18.72	4.32	13.40	5.00	18.00	1498	212.56	4
	2/4		8	8	6	8	8	2	288.00	16.00	18.72	4.32	13.40	5.00	18.00	1601	212.56	5
	2/11		8	8	8	8	8		288.00	16.00	18.72	4.32	13.40	5.00	18.00	1738	212.56	6
	2/18		8	8	8	8	8		288.00	16.00	18.72	4.32	13.40	5.00	18.00	1856	212.56	7
	2/25		10	8	10	6	6	5	342.00	25.00	22.23	5.13	14.70	5.00	18.00	2023	251.94	8
	3/3		8	8	8	8	8		288.00	16.00	18.72	4.32	13.40	5.00	18.00	2186	212.56	9
	3/10		8	7	8	9	10		309.60	19.00	20.12	4.64	13.90	5.00	18.00	2316	228.94	10
	3/17		8	8	8	8	8		288.00	16.00	18.72	4.32	13.40	5.00	18.00	2479	212.56	11
	3/24		10	8	10	8	8		391.00	33.00	25.42	5.87	16.60	5.00	18.00	2632	287.11	12
	3/31		8	8	8	9	8	4	403.75	34.00	26.24	6.06	17.06	5.00	18.00	2801	297.39	13
																		14
																		15
																		16
	QTR.								4124.75	269.00	268.10	61.87	184.92	65.00	234.80		3041.86	
	TO DATE																	

(F.W.T. through D. INS. are DEDUCTIONS; ← DEDUCTION AMOUNTS)

Figure 5.9

Example 1 Finding the Amount of FICA and Medicare Tax Due

If the employees at Fair Oaks Automotive Repair pay a total of \$789.10 in Social Security tax, and \$182.10 in Medicare tax, how much must the employer send to the Internal Revenue Service?

Solution The employer must match this and send a total of \$971.20 (\$789.10 + \$182.10 from employees) + \$971.20 (from employer) = \$1942.40 to the government.

NOTE In addition to the employee's Social Security tax and a matching amount paid by the employer, the employer must also send the amount withheld for income tax to the Internal Revenue Service on a quarterly basis. ▬

Example 2 Finding the Employer's Amount Due the IRS

Suppose that during a certain quarter Leslie's Pool Supplies has collected \$2765.42 from its employees for FICA tax, \$638.71 for Medicare tax, and \$3572.86 in federal withholding tax. Compute the total amount due to the Internal Revenue Service from Leslie's Pool Supplies.

Solution

$2,765.42	Collected from employees for FICA tax.
$2,765.42	Equal amount paid by employer.
$638.17	Collected from employees for medicare tax.
$638.17	Equal amount paid by employer.
$3,572.86	Federal withholding tax.
$10,380.04	Total due to Internal Revenue Service.

The firm must send $10,380.04 to the Internal Revenue Service.

Scientific *Calculator Approach*

The calculator solution for Example 2 is as follows.

2765.42 [×] 2 [+] 638.17 [×] 2 [+] 3572.86 [=] 10380.04

Objective 4 ***Identify the Form Used by Employers to File a Quarterly Tax Return.*** At the end of each quarter, employers must file the **Employer's Quarterly Federal Tax Return (Form 941)**. This form reports the amount of income tax withheld by the employer, and the amount of Social Security taxes and Medicare taxes due.

A copy of Form 941 is shown in Figure 5.10. The right column (lines 1 through 16) itemizes total employee wages and earnings, and the income, FICA, and Medicare taxes due.

The "Net taxes" (line 13) must equal line (17(d)) "Total liability for quarter." This lower portion of Form 941 (line 17) divides the quarter into its 3 months and the amount of the tax liability (employee and employer) are entered on the line of the proper month in that quarter. The withheld income tax and employee and employer FICA and medicare taxes must be deposited with an authorized financial institution or a federal reserve bank or branch. The amount of taxes owed determines how often these deposits must be made. If the amount of taxes owed by an employer is less than $1000 at the end of a quarter, the money does not have to be deposited, but is sent directly to the IRS. Funds can be deposited electronically or by using the **Federal Tax Deposit Coupon (Form 8109)**, shown in Figure 5.11. This form is filed, along with the proper amount of money, at an authorized financial institution, which then forwards the money to the U.S. Treasury.

Objective 5 ***Find the Amount of Federal Unemployment Tax Due.*** The **Federal Unemployment Tax Act (FUTA)** requires employers to pay an additional tax. This unemployment insurance tax, paid entirely by employers, is used to pay unemployment benefits to an individual who has become unemployed and is unable to find work.

In general, all employers who paid wages of $1000 or more in a calendar quarter or had one or more employees for some part of a day in any 20 different weeks in a calendar year must file an employer's annual *Federal Unemployment Tax Return (FUTA)*. This federal return must be filed each January for the preceding year. The FUTA tax that must be paid by the employer is 6.2% of the first $7000 in earnings for that year for each employee. Most states have unemployment taxes (SUTA) paid by the employer for which the employer is given credit on the FUTA return.

The credit given cannot exceed 5.4%, so the minimum amount due for FUTA is 0.8% (6.2% − 5.4%). States normally require quarterly payment of the unemployment

Form **941** (Rev. January 1999) Department of the Treasury Internal Revenue Service (O)

Employer's Quarterly Federal Tax Return

▶ **See separate instructions for information on completing this return.**

Please type or print.

Enter state code for state in which deposits were made ONLY if different from state in address to the right ▶ (see page 2 of instructions).

Name (as distinguished from trade name): Shirley Cicero	Date quarter ended	OMB No. 1545-0029
Trade name, if any: Desktop Publishing Company	Employer identification number	T / FF / FD / FP / I / T
Address (number and street): P.O. Box 505	City, state, and ZIP code	
Carmicheal, MO 93834		

IRS Use: 1 1 1 1 1 1 1 1 1 1 2 3 3 3 3 3 3 3 3 4 4 4 5 5 5 / 6 7 8 8 8 8 8 8 8 8 9 9 9 9 9 10 10 10 10 10 10 10 10 10 10

If address is different from prior return, check here ▶ ☐

If you do not have to file returns in the future, check here ▶ ☐ and enter date final wages paid ▶

If you are a seasonal employer, see **Seasonal employers** on page 1 of the instructions and check here ▶ ☐

Line	Description			Line	Amount	Cents
1	Number of employees in the pay period that includes March 12th . ▶ 1				9	
2	Total wages and tips, plus other compensation			2	46,228	
3	Total income tax withheld from wages, tips, and sick pay			3	5815	
4	Adjustment of withheld income tax for preceding quarters of calendar year			4		
5	Adjusted total of income tax withheld (line 3 as adjusted by line 4—see instructions)			5	5815	
6	Taxable social security wages	6a 46,228	× 12.4% (.124) =	6b	5732	27
	Taxable social security tips	6c	× 12.4% (.124) =	6d		
7	Taxable Medicare wages and tips	7a 46,228	× 2.9% (.029) =	7b	1340	61
8	Total social security and Medicare taxes (add lines 6b, 6d, and 7b). Check here if wages are not subject to social security and/or Medicare tax ▶ ☐			8	7072	88
9	Adjustment of social security and Medicare taxes (see instructions for required explanation) Sick Pay $ ____ ± Fractions of Cents $ ____ ± Other $ ____ =			9		
10	Adjusted total of social security and Medicare taxes (line 8 as adjusted by line 9—see instructions)			10	7072	88
11	**Total taxes** (add lines 5 and 10)			11	12,887	88
12	Advance earned income credit (EIC) payments made to employees			12		
13	Net taxes (subtract line 12 from line 11). **If $1,000 or more, this must equal line 17, column (d) below (or line D of Schedule B (Form 941))**			13	12,887	88
14	Total deposits for quarter, including overpayment applied from a prior quarter			14	12,887	88
15	**Balance due** (subtract line 14 from line 13). See instructions			15	Ø	

16 **Overpayment.** If line 14 is more than line 13, enter excess here ▶ $ ________ and check if to be: ☐ Applied to next return **OR** ☐ Refunded.

- **All filers:** If line 13 is less than $1,000, you need not complete line 17 or Schedule B (Form 941).
- **Semiweekly schedule depositors:** Complete Schedule B (Form 941) and check here ▶ ☐
- **Monthly schedule depositors:** Complete line 17, columns (a) through (d), and check here ▶ ☐

17 Monthly Summary of Federal Tax Liability. Do not complete if you were a semiweekly schedule depositor.

(a) First month liability	(b) Second month liability	(c) Third month liability	(d) Total liability for quarter
$3810.30	$4707.10	$4370.48	$12,887.88

Sign Here Under penalties of perjury, I declare that I have examined this return, including accompanying schedules and statements, and to the best of my knowledge and belief, it is true, correct, and complete.

Signature ▶ S. Cicero Print Your Name and Title ▶ Cicero-Pres Date ▶ 4/15

For Privacy Act and Paperwork Reduction Act Notice, see back of form. Cat. No. 17001Z Form **941** (Rev. 1-99)

Figure 5.10

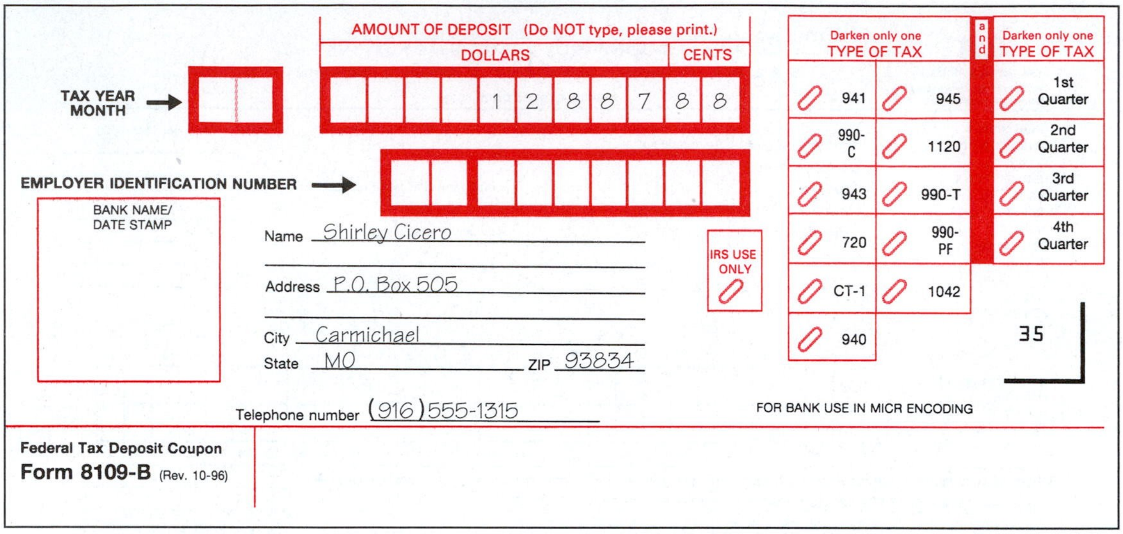

AMOUNT OF DEPOSIT (Do NOT type, please print.)
DOLLARS | CENTS
1 2 8 8 7 | 8 8
TAX YEAR MONTH
EMPLOYER IDENTIFICATION NUMBER
BANK NAME/ DATE STAMP
Name Shirley Cicero
Address P.O. Box 505
City Carmichael
State MO ZIP 93834
Telephone number (916)555-1315
IRS USE ONLY
Darken only one TYPE OF TAX
941 945
990-C 1120
943 990-T
720 990-PF
CT-1 1042
940
and
Darken only one TYPE OF TAX
1st Quarter
2nd Quarter
3rd Quarter
4th Quarter
35
FOR BANK USE IN MICR ENCODING
Federal Tax Deposit Coupon
Form 8109-B (Rev. 10-96)

Figure 5.11

tax. As soon as an employee reaches earnings of \$7000, no additional unemployment tax must be paid.

If an employer has a history of few layoffs, an individual state may drop the unemployment tax rate below 5.4%. On the other hand, an employer with high labor turnover may have to pay the higher unemployment tax rate.

Example 3 Finding the Amount of Unemployment Tax Due

Jennifer Eddy earns \$2400 each quarter. Find the following amounts: (a) Eddy's earnings in the third quarter of the year that are subject to unemployment tax and (b) the amount of unemployment tax due on third-quarter earnings. (Assume a tax rate of 6.2%.)

Solution

(a) Since earnings in the first two quarters are \$4800 (2 × \$2400), the earnings subject to unemployment taxes in the third quarter are \$2200 (\$7000 − \$4800).

(b) The federal unemployment tax due is \$136.40 (0.062 × \$2200).

5.6 Exercises

EMPLOYER PAYROLL RECORDS *Find the total combined amount of Social Security and Medicare tax (employee's contribution plus employer's contribution) for each of the following firms. Use a FICA rate of 6.2% and a Medicare tax rate of 1.45%, and assume no employee has earned over \$80,000 so far this year.*

	Firm	**Total Employee Earnings**
1.	Atlasta Ranch	$15,634.18
2.	Norm's Cel Phones	$10,046.53
3.	Adin Feed and Fuel	$17,462.10
4.	Leisure Outdoor Furniture	$15,324.15
5.	Computers Plus	$14,131.59
6.	Owl Drugstore	$21,281.60
7.	Plescia Produce	$62,475.80
8.	Cottonwood Nursery	$122,819.50

TOTAL EMPLOYER PAYMENT ***Calculate the total amount due from each of the following firms. Use a FICA rate of 6.2% and a Medicare tax rate of 1.45%.***

	Firm	**Total Employee Earnings**	**Total Withholding for Income Tax**
9.	Web Design	$6,150.82	$629.18
10.	McIntosh Meats	$8,714.55	$2,008.76
11.	Childlike Publishers	$32,121.85	$8,215.08
12.	Mart & Bottle	$20,255.60	$4,436.80
13.	Todd Consultants	$37,271.39	$7,128.64
14.	Mail Boxes Etc.	$10,158.24	$2,768.62
15.	Tony Balony's	$34,547.86	$12,628.19
16.	Oaks Hardware	$21,394.77	$5,671.30

17. An employer must file Form 941, Employer's Quarterly Federal Tax Return each quarter. What kinds of information are included on this form? (See Objective 4.)

18. Describe in your own words four differences between Social Security and FUTA. Consider the purpose, amount paid, and who pays.

Solve the following application problems. Assume no employee has earned over $80,000 so far this year. Use a FICA rate of 6.2% and a Medicare tax rate of 1.45%.

19. HAIR SALON OWNER Total employee earnings at City Hair Care are $21,928.10. Find the combined total amount of FICA and Medicare taxes sent to the IRS by the employer.

20. PAVING CONTRACTOR Biondi Paving has a payroll of $13,650.30. How much FICA and Medicare taxes should be sent to the IRS?

21. AM-PM OWNER The AM-PM Mart has an employee payroll of $7622.84. During the same time period, $1625.68 was withheld as income tax. Find the total amount due to the IRS.

22. FLORIST SHOP OWNER The payroll at the Oak Mill Florist is $22,607.72. During the same pay period, $3898.14 was withheld as income tax. Find the total amount that must be sent to the IRS.

FEDERAL UNEMPLOYMENT TAX ***Assume a FUTA rate of 6.2% of the first $7000 in earnings in each of the following.***

23. An employee earns $3280 in the first quarter of the year, $2600 in the second quarter, and $3156 in the third quarter. Find (a) the amount of earnings subject to unemployment tax in the third quarter and (b) the amount of unemployment tax due on third quarter earnings.
24. Terry White earns $910 per month. Find (a) the amount of earnings subject to unemployment tax in the third quarter and (b) the amount of unemployment tax due on third quarter earnings.
25. Emma Price is paid $1800 per month. Find (a) the amount of earnings subject to unemployment tax in the second quarter and (b) the amount of unemployment tax due on second quarter earnings.
26. Michael Booth received $4820 in earnings in the first quarter of the year. His earnings in the second quarter are $3815. Find (a) the amount of his earnings subject to FUTA in the second quarter and (b) the amount of tax due in the second quarter.

Chapter 5 Quick Review

Chapter Terms

Review the following terms to test your understanding of the chapter. For each term you do not know, refer to the page number found next to that term.

charge backs [p. 169]
commission rate [p. 160]
compensatory (comp) time [p. 154]
daily overtime [p. 153]
deductions [p. 148]
differential-piecework plan [p. 168]
dockings [p. 169]
double time [p. 154]
drawing account [p. 163]
Employer's Quarterly Federal Tax Return (Form 941) [p. 192]
Fair Labor Standards Act [p. 151]
Federal Insurance Contributions Act (FICA) [p. 174]
Federal Tax Deposit Coupon (Form 8109) [p. 192]
Federal Unemployment Tax Act (FUTA) [p. 192]
gross earnings [p. 148]
hourly wage [p. 149]
incentive rates [p. 167]
income tax withholdings [p. 180]
marital status [p. 180]
Medicare [p. 174]
net pay [p. 148]
override [p. 163]
overtime [p. 151]
overtime premium (excess) method [p. 152]
pay period [p.154]
payroll register [p. 149]
percentage method [p. 181]
personal income tax [p. 180]
piecework rate [p. 168]
premium payment plan [p. 154]
premium rate [p. 168]
quota bonus system [p. 163]
quotas [p. 163]
Request for Earnings and Benefit Estimate Statement [p. 174]
returns [p. 161]
salary [p. 154]
salary plus commission [p. 162]
shift differential [p. 154]
sliding-scale commission [p. 160]
Social Security [p. 174]
split-shift premium [p. 154]
state disability insurance (SDI) [p. 177]
state income tax [p. 184]
straight commission [p. 160]
straight-piecework plan [p. 168]
time card [p. 149]
time rates [p. 167]
time-and-a-half rate [p. 151]
variable commission [p. 161]
wage bracket method [p. 181]
withholding allowances [p. 180]

Concepts / Examples

5.1 Gross earnings

Gross earnings = Hours worked × Rate per hour

40 hours at \$7.10 per hour

$$\text{Gross earnings} = 40 \times \$7.10 = \$284$$

5.1 Gross earnings with overtime

First, find regular earnings. Then, determine overtime pay at overtime rate. Finally, add regular and overtime earnings.

40 hours at \$7.10 per hour; 10 hours at time and a half.

$$\begin{aligned}\text{Gross earnings} &= (40 \times \$7.10) + (10 \times \$7.10 \times 1\tfrac{1}{2})\\ &= \$284 + \$106.50\\ &= \$390.50\end{aligned}$$

5.1 Common pay periods

Pay Period	Paychecks per Year
Monthly	12
Semimonthly	24
Biweekly	26
Weekly	52

Find the earnings equivalent to \$1400 per month for other pay periods.

$$\text{Semimonthly} = \frac{1400}{2} = \$700$$

$$\text{Biweekly} = \frac{1400 \times 12}{26} = \$646.15$$

$$\text{Weekly} = \frac{1400 \times 12}{52} = \$323.08$$

5.1 Overtime for salaried employees

First, find the equivalent hourly rate. Next, multiply the rate by the overtime hours by $1\frac{1}{2}$. Finally, add overtime earnings to salary.

Salary is \$324 per week for 36 hours. Find earnings for 36 hours.

$$\$324 \div 36 = \$9.00 \text{ per hour}$$
$$\$9.00 \times 10 \times 1\tfrac{1}{2} = \$135 \text{ overtime}$$
$$\$324 + \$135 = \$459$$

5.2 Straight commission

Gross earnings = Commission rate × Amount of sales

Sales of \$25,800; commission rate is 5%.

$$0.05 \times \$25{,}800 = \$1290$$

5.2 Variable commission

Commission rate varies at different sales levels.

Up to \$10,000, 6%; \$10,001–\$20,000, 8%; \$20,001 and up, 9%. Find the commission on sales of \$32,768.

0.06 × \$10,000 =	\$600.00	(First \$10,000)
0.08 × \$10,000 =	\$800.00	(Next \$10,000)
0.09 × \$12,768 =	\$1149.12	(Over \$20,000)
Totals \$32,768	\$2549.12	

5.2 Salary and commission

Gross earnings = Fixed earnings + Commission

Salary, \$250 per week; commission rate, 3%. Find gross earnings on sales of \$6848.

$$\begin{aligned}\text{Gross earnings} &= \$250 + (0.03 \times \$6848)\\ &= \$250 + \$205.44\\ &= \$455.44\end{aligned}$$

5.2 Commissions with a drawing account

Gross earnings = Commission − Draw

Sales for month, \$28,560; commission rate, 7%; draw, \$750 for month. Find gross earnings.

$$\begin{aligned}\text{Gross earnings} &= (0.07 \times \$28{,}560) - \$750\\ &= \$1999.20 - \$750\\ &= \$1249.20\end{aligned}$$

5.2 Commissions with a quota bonus

Gross earnings = Commission rate × (Sales − Quota)

Sales for week, \$14,370; commission rate, 10% after meeting the sales quota of \$4000. Find gross earnings.

$$\begin{aligned}\text{Gross earnings} &= 0.1 \times (\$14{,}370 - \$4000)\\ &= 0.1 \times \$10{,}370\\ &= \$1037\end{aligned}$$

5.3 Gross earnings for piecework

Gross earnings = Number of items × Payment per item

Items produced, 175; payment per item, \$0.65. Find gross earnings.

$$175 \times \$0.65 = \$113.75$$

5.3 Gross earnings for differential piecework

The rate paid per item produced varies with level of production.

1–100 items \$0.75 each; 101–150 items \$0.90 each; 151+ items \$1.04 each. Find gross earnings for producing 214 items.

100 × \$0.75 =	75.00	(First 100 units)
50 × \$0.90 =	\$45.00	(Next 50 units)
64 × \$1.04 =	\$66.56	(More than 150)
214	\$186.56	

5.3 Gross earnings with piece rate and chargebacks

Find piece rate earnings; then calculate chargebacks and subtract them from piece rate earnings to find gross earnings.

Items produced, 318; piece rate, \$0.75; spoiled items, 26; chargeback, \$0.60. Find gross earnings.

$$318 \times \$0.75 = \$238.50$$
$$26 \times \$0.60 = \$15.60$$
$$\$238.50 - \$15.60 = \$222.90$$

5.3 Overtime earnings on piecework

Gross earnings = Earnings at regular rate + Earnings at overtime rate

Items produced on regular time, 530; items produced on overtime, 110; piece rate, \$0.34. Find gross earnings

$$\text{Gross earnings} = 530 \times \$0.34 + 110\left(1\frac{1}{2} \times \$0.34\right)$$
$$= \$180.20 + \$56.10$$
$$= \$236.30$$

5.4 FICA; Social Security tax

The gross earnings are multiplied by the rate. When the maximum earnings are reached, no additional FICA is withheld that year.

Gross earnings, \$458; social security tax rate, 6.2%. Find Social Security tax.

$$\$458 \times 0.062 = \$28.40$$

5.4 Medicare tax

The gross earnings are multiplied by the tax rate. There is no maximum earnings on medicare.

Gross earnings, \$458; Medicare tax rate, 1.45%. Find Medicare tax.

$$\$458 \times 0.0145 = \$6.64$$

5.4 State disability insurance deductions

Multiply the gross earnings by the SDI tax rate. When maximum earnings are reached, no additional taxes are paid in that year.

Gross earnings, \$3210; SDI tax rate 1%. Find SDI tax.

$$\$3210 \times 0.01 = \$32.10$$

5.5 Federal withholding tax

Tax is paid on total earnings. No maximum as with FICA.

Single employee with 3 allowances; weekly earnings of \$326; find the federal withholding tax.

Using wage bracket amount "at least \$320, but less than \$330," withholding is \$18. With the percentage method withholding is

$$\$326 - (\$52.88 \times 3) = \$167.36$$
$$\$167.36 - \$51 = \$116.36$$
$$\$116.36 \times 0.15 = \$17.45$$

5.5 State withholding tax

Tax is paid on total earnings. No maximum as with FICA.

Married employee with weekly earnings of \$392; find the state withholding tax given a state withholding tax rate of 4.5%.

$$4.5\% \times \$392 = 0.045 \times \$392 = \$17.64$$

5.6 Quarterly report, Form 941

Filed each quarter; FICA and federal withholding are sent to the IRS.

(FICA + Medicare) × 2 (employer) + federal withholding tax

Quarterly FICA withheld from employees is \$5269, Medicare tax is \$1581, and federal withholding tax is \$14,780. Find the total due the IRS.

$$(\$5269 + \$1581) \times 2 + \$14{,}780 = \$28{,}480$$

5.6 Federal Unemployment Tax (FUTA)

FUTA is paid by the employer on the first $7000 of earnings each year for each employee.

An employee earned $3850 in the first quarter of a year. Find the amount of unemployment tax using a 6.2% tax rate.

$$\$3850 \times 6.2\% = \$238.70$$

Chapter 5 Review Exercises

The answer section includes answers to all Review Exercises.

Complete the following partial payroll register. Find the total gross earnings for each employee. Time and a half is paid on all hours over 40 in one week. [5.1]

	Employee	Hours Worked	Reg. Hours	O.T. Hours	Reg. Rate	Gross Earnings
1.	Darasz, B.	48.5			$9.14	
2.	Davidson, D.	48			$8.50	
3.	Mandler, S.	38.25			$7.40	
4.	Rosenthal, L.	57.25			$6.80	

Find the equivalent earnings for each of the following salaries as indicated. [5.1]

	Weekly	Biweekly	Semimonthly	Monthly	Annually
5.	$410.80	______	______	______	______
6.	______	$1060	______	______	______
7.	______	______	______	______	$18,000
8.	______	______	$875	______	______

Find the weekly gross earnings for the following people who are on salary and are paid time and a half for overtime. [5.1]

	Employee	Regular Hours per Week	Weekly Salary	Hours Worked	Weekly Gross Earnings
9.	Uldall, E.	40	$640	45	______
10.	Donovan-Dickerson, K.	36	$342	42	______

Find the gross earnings for each of the following salespeople. **[5.2]**

	Employee	Total Sales	Returns	Rate of Commission
11.	Gonsalves, R.	$48,620	$3106	8%
12.	Kaufman, B.	$38,740	$1245	9%

Solve each of the following application problems. **[5.3]**

13. Twenty Minute Lube and Oil pays its employees $1.25 for an oil change, $1.50 for each car lubed, and $2.50 if a car gets both an oil change and a lube. This week Andre Herrebout changed the oil in 63 cars, lubed 46 cars, and gave an oil change and lube to 38 cars. Find his gross pay for the week.

14. At a Jalisco Electronics in Mexicali, Mexico, assemblers are paid as follows: 1–20 units in a week, $4.50 each; 21–30 units, $5.50 each; and more than 30 units, $7 each. Adrian Ortega assembled 28 units in one week. Find his gross pay.

15. Samantha Walker receives a commission of 6% for selling a $135,000 house. Half the commission goes to the broker, and half of the remainder to another salesperson. Walker gets the rest. Find the amount she receives.

Employees at Appliance Giant are paid a commission on the following schedule: first $2000 in sales, 6%; next $2000 in sales, 8%; sales over $4000, 10%. Use this information for Exercises 16 and 17. **[5.2]**

16. Find the gross earnings for an employee with total sales of $5850.

17. Find the gross earnings for an employee with total sales of $7200.

18. Pat Rowell ties bows on Christmas wreaths and is paid $0.12 for each bow tied. She is charged $0.09 for each rejection and is paid time and a half for overtime production. Find her gross earnings for a week when she produces 1850 at the regular rate, 285 at the overtime rate, and has 92 rejections. **[5.3]**

An employee is paid a salary of $7855 per month. If the FICA rate is 6.2% on the first $80,000 of earnings and the Medicare tax rate is 1.45% on all earnings, how much will the employee pay in (a) FICA tax and (b) Medicare tax during the following months? **[5.4]**

19. October

20. November

Find the federal withholding tax using the wage bracket method for each of the following employees. **[5.5]**

21. Flahive: 2 withholding allowances, single, $278.65 weekly earnings.

22. Hoffa: 2 withholding allowances, married, $705.91 weekly earnings.

23. Howard: 3 withholding allowances, married, $2208.79 monthly earnings.

24. Kluesner: 4 withholding allowances, single, $1757.23 monthly earnings.

25. Lawrence: 6 withholding allowances, married, $2580.76 monthly earnings.

26. Tewell: 2 withholding allowances, single, $335.75 weekly earnings.

Find the net pay for each of the following employees after FICA, Medicare, federal withholding tax, state disability, and other deductions have been made. Assume that no one has earned over $80,000 so far this year. Assume a FICA rate of 6.2%, Medicare rate of 1.45%, and a state disability rate of 1%. Use the percentage method of withholding. **[5.5]**

27. Precilo: $1852.75 monthly earnings, 1 withholding allowances, single, $37.80 in other deductions.

28. Colley: $522.11 weekly earnings, 4 withholding allowances, married, state withholding of $15.34, credit union savings of $20, educational television contribution of $7.50.

29. Harper: $677.92 weekly earnings, 6 withholding allowances, married, state withholding of $22.18, union dues of $14, charitable contribution of $15.

Solve the following application problems. Assume no employee has earned over $80,000 so far this year. Use a FICA rate of 6.2% and a rate of 1.45% for Medicare. **[5.6]**

30. Total employee earnings for Round Table Pizza are $12,720.15. Find the total amount of FICA and Medicare taxes sent to the IRS by the employer.

31. San Juan Electric has an employee payroll of $29,185.17. During the same period $4921 was withheld as income tax. Find the total amount due the IRS.

Solve the following application problems.

32. A salesperson is paid $452 per week plus a commission of 2% on all sales. The salesperson sold $712 worth of goods on Monday, $523 on Tuesday, $1002 on Wednesday, $391 on Thursday, and $609 on Friday. Returns and allowances for the week were $114. Find the employee's (a) Social Security tax (6.2%), (b) Medicare tax (1.45%), and (c) state disability insurance (1%) for the week. **[5.3 and 5.4]**

33. Kara Gourley earned $78,325.18 so far this year until last week. Last week she earned $1956.44. Find her (a) FICA tax and (b) Medicare tax for last week's earnings.

34. The employees of Miracle Floor Covering paid a total of $1496.11 in Social Security tax last month, $345.30 in Medicare tax, and $1768.43 in federal withholding tax. Find the total amount that the employer must send to the Internal Revenue Service.

For Exercises 35 and 36, find (a) the Social Security tax and (b) the Medicare tax for each of the following self-employed people. Use a FICA tax rate of 12.4% and a Medicare tax rate of 2.9%. **[5.4]**

35. Kula, S.: $38,795.22

36. Biondi, E.: $27,618.53

Assume a FUTA rate of 6.2% on the first $7000 in earnings in each of the following. **[5.6]**

37. An employee earns $2875 in the first quarter of the year, $3212 in the second quarter, and $2942 in the third quarter. Find (a) the amount of earnings subject to unemployment tax in the third quarter and (b) the amount of unemployment tax due on third quarter earnings.

38. Amy Berk earns $810 per month. Find (a) the amount of earnings subject to unemployment tax in the third quarter and (b) the amount of unemployment tax due on third quarter earnings.

Chapter 5 Summary Exercise
Payroll: Finding Your Take-Home Pay

Paige Dunbar, the manager of a Blockbuster Video receives an annual salary of $32,240, which is paid weekly. Her normal workweek is 40 hours and she is paid time and a half for all overtime. She is single and claims one withholding allowance. Her deductions include FICA, Medicare, federal withholding, state disability insurance, state withholding, credit union payments of $125, retirement deductions of $75, association dues of $12, and a Diabetes Association contribution of $15. Find each of the following for a week in which she works 52 hours.

(a) Regular weekly earnings
(b) Overtime earnings
(c) Total gross earnings
(d) FICA
(e) Medicare
(f) Federal withholding
(g) State disability
(h) State withholding (state income tax rate is 4.4%)
(i) Net pay

Net Assets Business on the Internet

Social Security Administration

Statistics

- 1935: Social Security Act passed
- 1937: First Social Security taxes collected
- 1965: Medicare bill signed into law
- 1998: 6 million families received $49 billion in disability payments.

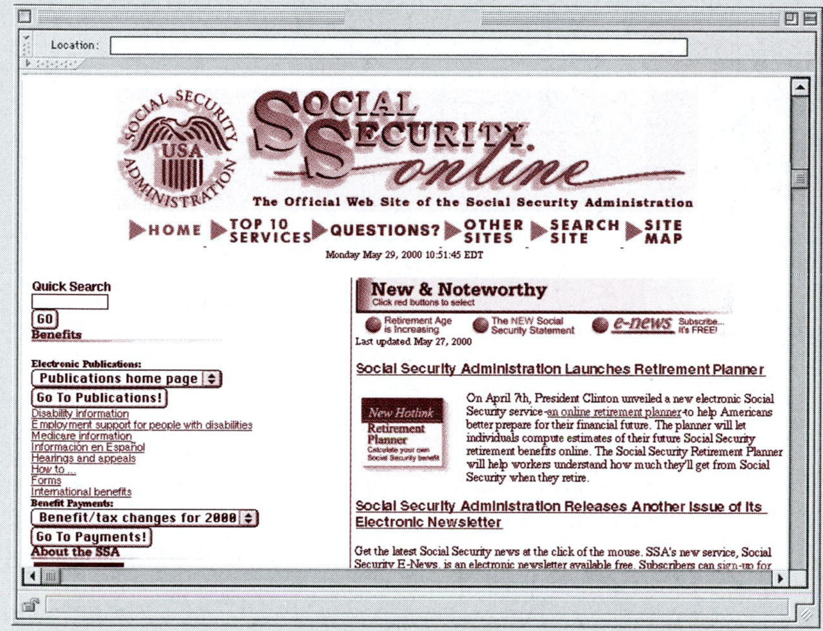

In the 1930s, during the Great Depression, the United States economy continued to move from agricultural production toward industrial production. The uncertainties associated with disability and old age, which in the past were the responsibility of family and local community, were becoming a much greater concern to many Americans. This concern resulted in the Social Security Act of 1935. As a part of this act, both employee and employer contribute to a fund that would provide benefits to retiring workers.

On January 1, 1940, the first monthly retirement check was issued to Ida May Fuller of Ludlow, Vermont, in the amount of $22.54. Miss Fuller received retirement checks for 35 years until her death in 1975 at the age of 100. Today, on the Internet, we can get electronic publications from the Social Security Administration in addition to information on benefits and many more on-line direct services.

1. A part-time employee works 4 hours on Monday, 6.5 hours on Wednesday, 5.25 hours on Friday, and 7.5 hours on both Saturday and Sunday. If the employee is paid $7.20 per hour, find the gross earnings for the week.

2. If a store manager is paid an annual salary of $31,200, find her equivalent earnings for monthly, semimonthly, and biweekly pay periods.

3. Renee Hampton worked 44.5 hours last week at Roadway Package Systems. She is paid $8.70 per hour, plus time and a half for all hours over 40 per week. Find her **(a)** Social Security tax and **(b)** Medicare tax for the week.

4. Last month, the employees at a Blockbuster Video paid a total of $792 in FICA, $237 in Medicare tax, and $2217 in federal withholding tax. Find the total amount the employer must send to the Internal Revenue Service.

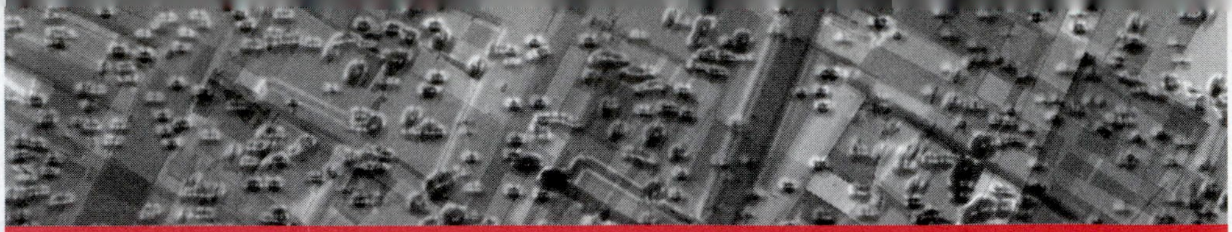

6 Taxes

There is one thing on which almost everyone agrees. "Taxes are too high." But taxes are a fact of life. In the early part of last century, Justice Oliver Wendell Holmes, Jr., of the U.S. Supreme Court said, "Taxes are the price we pay to live in a civilized society." Tax dollars pay for education, health services, national defense, streets and highways, parks and recreation facilities, police and fire protection, public assistance for the poor, libraries, and even street lights. As government provides more services, taxes go up. Figure 6.1 shows how long each year Americans must work just to pay all federal, state, and local taxes.

For which additional services, if any, would Americans be willing to pay more taxes (health care, education, libraries, police and fire protection, etc.)?

There are basically three forms of taxation: taxes on sales, property, and income. This chapter examines the basics of taxation and discusses the calculations necessary to work with each type of tax.

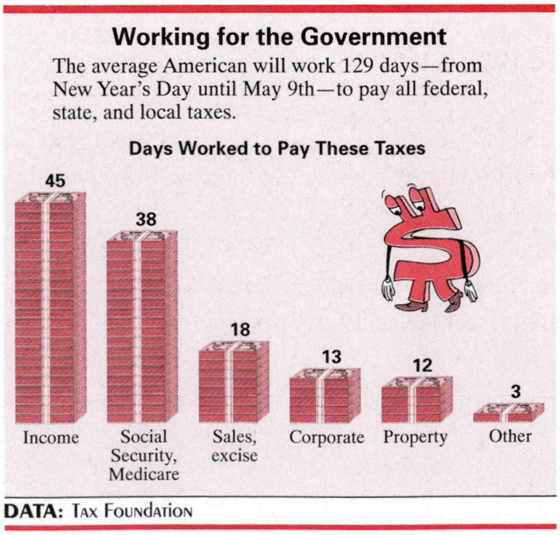

Figure 6.1

6.1 Sales Tax

Objectives

1. *Understand how sales tax is determined.*
2. *Find the amount of sales tax and the total sale.*
3. *Find the selling price when the sales tax is known.*
4. *Find the amount of the sale when the total price is known.*
5. *Define excise tax.*
6. *Find the total cost, including sales tax and excise tax.*

Most states and a large number of counties and cities have a tax on sales. This tax, called **sales tax**, is paid on the amount of retail sales. Normally calculated as a percent of the retail price, this tax varies from area to area. For instance, since food is considered a basic necessity, most states do not tax the majority of food items purchased in a grocery store. The same food items, however, may be taxed when purchased in a restaurant.

Objective 1 *Understand How Sales Tax Is Determined.* Many small retail stores use a tax table to compute sales tax. The sales clerk looks up the tax on the sales tax table and enters the amount into the cash register.

Today many stores have cash register systems that keep track of taxable items and automatically calculate the necessary tax. Sales tax tables from a typical state appear in Table 6.1. Table 6.2 shows the 1999 **sales tax rate** in each state, the District of Columbia, and New York City.

The basic percent equation is used to solve sales tax problems. The amount of sales tax is the part, the amount of the sale is the base, and the sales tax rate is the rate.

$$\begin{array}{ccccc} \text{Sales tax} & = & \text{Amount of Sale} & \times & \text{Sales tax rate} \\ P & = & B & \times & R \end{array}$$

Objective 2 *Find the Amount of Sales Tax and the Total Sale.* A common calculation in business is to find the amount of sales tax and the total amount of the sale, including tax.

Example 1 Finding Sales Tax and the Total Sale

A customer at Round Table Pizza purchases several pizzas for \$49.95. Sales tax in the state is 5%. Find (a) the amount of sales tax and (b) the total amount collected from the customer.

Solution

(a) The amount of sales tax (part) is found by multiplying the sales tax rate (rate) by the amount of the sale (base). Use the percent equation to find the sales tax.

$$\begin{aligned} P &= BR \\ &= \$49.95 \times 5\% \\ &= \$49.95 \times 0.05 = \$2.50 \quad \text{Rounded.} \end{aligned}$$

The sales tax is \$2.50.

Table 6.1 $7\frac{3}{4}\%$ SALES TAX COLLECTION SCHEDULE

Up to and Including	Tax
$0.06	$0.00
0.19	0.01
0.32	0.02
0.45	0.03
0.58	0.04
0.70	0.05
0.83	0.06
0.96	0.07
1.09	0.08
1.22	0.09
1.35	0.10
1.48	0.11
1.61	0.12
1.74	0.13
1.87	0.14
1.99	0.15
2.12	0.16
2.25	0.17
2.38	0.18
2.51	0.19

Table 6.2 1999 SALES TAX RATES

Alabama (plus city and/or county tax where applicable)	4%
Alaska (local tax where applicable)	0%
Arizona	5%
Arkansas (plus local tax where applicable)	4.625%
California (plus $\frac{1}{2}$% or more local tax where applicable)	7.25%
Colorado	3%
Connecticut	6%
Delaware	0%
District of Columbia	5.75%
Florida (plus local tax where applicable)	6%
Georgia (plus local tax where applicable)	4%
Hawaii	4%
Idaho	5%
Illinois	6.25%
Indiana	5%
Iowa (plus local tax where applicable)	5%
Kansas	4.9%
Kentucky	6%
Louisiana	4%
Maine	5.5%
Maryland	5%
Massachusetts	5%
Michigan	6%
Minnesota	6.5%
Mississippi	7%
Missouri (plus local tax where applicable)	4.225%
Montana	0%
Nebraska (plus local tax where applicable)	4.5%
Nevada (plus local tax where applicable)	6.5%
New Hampshire	0%
New Jersey	6%
New Mexico	5%
New York City	4.25%
New York State (plus city or county tax where applicable)	4%
North Carolina (plus local tax where applicable)	4%
North Dakota (plus local tax where applicable)	5%
Ohio (plus local tax where applicable)	5%
Oklahoma (plus local tax where applicable)	4.5%
Oregon	0%
Pennsylvania (plus local tax where applicable)	6%
Rhode Island	7%
South Carolina	5%
South Dakota (plus local tax where applicable)	4%
Tennessee (plus city or county tax where applicable)	6%
Texas (plus local tax where applicable)	6.25%
Utah (plus county and local taxes where applicable)	4.75%
Vermont	5%
Virginia	3.5%
Washington (plus local tax where applicable)	6.5%
West Virginia	6%
Wisconsin (plus local tax where applicable)	5%
Wyoming (plus local tax where applicable)	4%

(b) Add to find the total amount collected from the customer.

$49.95	Amount of sale
+ 2.50	Sales tax
$52.45	Total sale including tax

NOTE Several states do not have any sales tax. Many counties and cities impose a tax on sales to raise additional revenue which is then used locally.

Objective 3 ***Find the Selling Price When the Sales Tax Is Known.*** If the amount of sales tax and the sales tax rate are known, it is possible to determine the selling price.

Example 2 Finding the Price When Sales Tax Is Known

Sales tax on a Murray 5-horsepower Ultra Push Mower is $16.14. If the sales tax rate is 6%, find the price of the mower.

Solution Since sales tax (part) is found by multiplying the sales tax rate (rate) by the amount of the sale (base), the amount of sale is found using the percent equation and solving for base.

$$P = BR$$
$$\$16.14 = B \times 6\%$$
$$\$16.14 = 0.06B$$
$$\frac{\$16.14}{0.06} = \frac{0.06B}{0.06}$$
$$\$269 = B$$

The mower sells for $269.

Objective 4 ***Find the Amount of the Sale When the Total Price Is Known.*** When the total price including sales tax and the sales tax rate are known, the sale amount can be found as shown in the next example.

Example 3 Determining the Sale Amount When Total Price Is Known

Coleman Headstart Preschool purchased some playground equipment at a total cost of $3663.36, including sales tax of 6%. Find the price of the equipment before the sales tax.

Solution First, remember that the amount of sale + sales tax = total. Then, use the percent equation to find the amount of the sale (base).

$$B + (6\%)B = \$3663.36$$
$$1B + 0.06B = \$3663.36$$
$$1.06B = \$3663.36$$
$$B = \frac{\$3663.36}{1.06}$$
$$B = \$3456$$

The amount of the sale is $3456.

To check this answer, multiply the amount of the sale ($3456) by the tax rate (6%) to find the tax. Then, add the tax to the amount of the sale to find the total amount.

$$BR = P$$

$$\$3456 \times 6\% = P$$

$$\$3456 \times 0.06 = \$207.36 \quad \text{Amount of sales tax.}$$

Now add.

$$\$3456 \text{ (sale)} + \$207.36 \text{ (tax)} = \$3663.36 \quad \text{Total sale}$$

Objective 5 *Define Excise Tax.* An **excise tax**, or **luxury tax**, is charged on certain items by the federal, state, or local government. The tax is similar to sales tax since it is paid by or passed on to the consumer of goods and services. Excise taxes are charged on gasoline, tires, luxury cars, and services such as telephone, entertainment, air transportation, and business licenses.

Excise taxes are either a percent of the sale price of an item or a fixed amount for each unit sold. Table 6.3 shows the current federal excise taxes charged on several items.

Objective 6 *Find the Total Cost Including Sales Tax and Excise Tax.*

Example 4 Finding the Total Cost with Sales and Excise Taxes

A tire for a John Deere backhoe weighs 76 pounds and sells for $680. Find the total cost of the tire including 7% sales tax and excise tax of $4.50 plus 30¢ for each pound over 70 pounds.

Solution First, find the sales tax using the percent equation.

$$BR = P$$

$$\$680 \times 7\% = P$$

$$\$680 \times 0.07 = \$47.60 \quad \text{Amount of sales tax.}$$

Next, find the excise tax.

$$\begin{aligned}\text{Excise tax} &= \$4.50 + (76 - 70)(\$0.30)\\ &= \$4.50 + (6 \times \$0.30)\\ &= \$4.50 + \$1.80\\ &= \$6.30\end{aligned}$$

Now find the total cost.

$$\text{Amount of sale} + \text{Sales tax} + \text{Excise tax} = \text{Total cost}$$

$$\$680 + \$47.60 + \$6.30 = \$733.90$$

The total cost of the tire is $733.90

Table 6.3 FEDERAL EXCISE TAXES*

Product or Service	Rate	Product or Service	Rate
Telephone service	3%	Tires (by weight)	
Teletypewriter service	3%	Under 40 lb	No tax
Air transportation	7.5%	40–69 lb	15¢/lb over 40 lb
International air travel	$12.20/person	70–89 lb	$4.50 plus 30¢/lb over 70 lb
Air freight	6.25%	90 lb and more	$10.50 plus 50¢/lb over 90 lb
Coal		Truck and trailer, chassis and bodies	12%
Underground (lower amount)	$1.10/ton or 4.4%	Inland waterways fuel	24.4¢/gal.
Surface (lower amount)	55¢/ton or 4.4%	Ship passenger tax	$3/passenger
Fishing rods	10%	Luxury cars (amount over $36,000)	6%
Bows and arrows	12.4%		
Gasoline	18.4¢/gal.		
Diesel fuel	24.4¢/gal.		
Aviation fuel	21.9¢/gal.		

Source: Publication 510, I.R.S., Excise Taxes for 1999.

*In addition to the federal excise taxes shown here, there are a number of additional excise taxes that apply to alcoholic beverages, tobacco products, and firearms. Special ATF forms are used by businesses when selling these products.

Scientific
Calculator Approach

To solve Example 4 using a calculator, use parentheses to set aside the excise tax calculation and allow excise tax to be added in the chain calculation.

680 [+] 680 [×] 7 [%] [=] [+] 4.5 [+] [(] 76 [−] 70 [)] [×] .3 [=] 733.9

NOTE Excise taxes are added to the price in addition to sales tax. The excise tax is calculated on the amount of the sale before sales tax is added. ■

6.1 Exercises

SALES AND EXCISE TAX *Find (a) the amount of sales tax, (b) the amount of excise tax, and (c) the total sale price including taxes in each of the following problems.*

	Sale Price	Sales Tax Rate	Excise Tax Rate
1.	$76.20	3%	11%
2.	$59.80	6%	10%
3.	$47.70	4.5%	3%
4.	$21.15	8.25%	12%
5.	$173.50	5%	9¢/gal.; 165 gal.

	Sale Price	Sales Tax Rate	Excise Tax Rate
6.	$216.75	3%	14¢ gal.; 190 gal.
7.	$822.18	7%	12%
8.	$648.52	4%	10%
9.	$29,400	6.25%	$3/person; 168 people
10.	$57,552	4.5%	$3/person; 218 people

SALES TAX COMPUTATIONS *Find the sale price and total price when given the amount of sales tax and the sales tax rate. Round to the nearest cent.*

	Amount of Sales Tax	Sales Tax Rate
11.	$9.60	6%
12.	$4.58	5%
13.	$6.30	4%
14.	$21.84	8%
15.	$21.45	6.5%
16.	$58.00	7.25%
17.	$63.84	5%
18.	$22.32	4%

SALES TAX APPLICATIONS *Find the amount of the sale before sales tax was added and the amount of sales tax in each of the following. Round to the nearest cent.*

	Total Sale	Sales Tax Rate
19.	$107.31	5%
20.	$273.92	7%
21.	$551.52	6%
22.	$312.66	5.5%
23.	$20.60	4.25%
24.	$85.28	4%
25.	$333.90	6%
26.	$1352.01	3%
27.	$2945.76	7%
28.	$4469.64	5%

29. What is the sales tax where you live? Is there a different tax rate in a county or city near you? Explain why this difference exists.

30. List three items that you, your family, or your employer purchased within the last year on which an excise tax was paid. (See Objective 5.)

Solve each of the following application problems.

31. ARCHERY EQUIPMENT Open Range Archery sells an archery set (bow and arrows) for \$119.80. If sales tax if 6% and the excise tax is 12.4%, find (a) the amount of the sales tax, (b) the amount of the excise tax, and (c) the total sale including sales tax and excise tax.

32. TRUCK CHASSIS PURCHASE Cross Town Trucking pays \$43,135.20 for a truck chassis. If sales tax is 7% and excise tax is 12%, find (a) the amount of the sales tax, (b) the amount of the excise tax, and (c) the total price including sales tax and excise tax.

33. MOUNTAIN BICYCLE PURCHASE The sales tax on a mountain bike is \$17.10. If the sales tax rate is $4\frac{1}{2}\%$, find the sale price of the mountain bike.

34. MOTORCYCLE PURCHASE Sales tax on a motorcycle was \$456.75. If the tax rate was 7%, find the sales price of the motorcycle.

35. LAPTOP COMPUTER PURCHASE The price of a Compaq laptop computer is \$1979.55, including a 6% sales tax. Find the amount of the sale price.

36. HOBBY SALES K-Ron's Hobbies charges \$321.36 including 3% sales tax for a radio-controlled glider. Find the price of the glider without tax.

37. COCKTAIL LOUNGE Total sales for one day at Club Sunset were \$1285.44, including the 4% sales tax charged on all purchases. Find the amount that is sales tax.

38. TROPIC FISH SALES The Tropical Fish Place has total receipts for the day of \$875.43. This includes 6.5% sales tax on all sales. Find the amount that is sales tax.

39. EMPLOYEE MATH ERROR At the close of business one day, Mike Roche, a new employee, totaled the amount in the cash register and found \$1908. He multiplied this sum by the sales tax rate of 6% to find the amount of sales tax due. This procedure is incorrect. Find (a) the correct amount of sales tax and (b) the amount of error made by Roche.

40. TAX OVERPAYMENT Santiago Rowland has total daily receipts of \$1856. His lounge manager multiplies this total by 5.5%, the tax rate in the area. This procedure is incorrect. Find (a) the correct amount of sales tax and (b) the amount of error made by the manager.

41. TRUCK TIRE SALES Auburn Tire offers a 74-pound truck tire for \$182 plus tax. If sales tax is 7.5% and excise tax is \$4.50 plus 30¢ per pound over 70 pounds, find the total cost including tax.

42. TIRE EXCISE TAX Jack Anderson purchased a set of eight truck tires at a cost of \$390 each. Each tire weighs 52 pounds, sales tax is 6%, and excise tax is 15¢ per pound over 40 pounds for each tire. Find the total cost to Anderson.

43. CHARTERED TRAVEL Bob Towers Travels pays \$33,850 for a chartered international flight. Sales tax is 7.5% and excise tax is \$12.20 per person. If 240 people make the flight, what is the total cost including tax?

44. INTERNATIONAL AIR TRAVEL Ticket sales for an international flight are \$26,970 before taxes. Sales tax is 5.5% and excise tax is \$12.20 per person. Find the total sales including tax if there are 128 passengers.

6.2 Property Tax

Objectives

1. *Understand fair market value and find assessed valuation.*
2. *Use the tax rate formula.*
3. *Use the formula for property tax.*
4. *Express tax rates in percent, dollars per \$100, dollars per \$1000, and mills.*
5. *Find taxes given the assessed valuation and the tax rate.*
6. *Find the assessed valuation given the tax rate and the tax.*
7. *Find the tax rate given the assessed valuation and the tax.*

In virtually every area of the nation, the owners of **real property** (such as buildings and land) must pay a property tax on their property. In many areas **personal property** (such as mobile homes, furnishings, appliances, motor homes, trailers, boats, and other non–real estate items) is also taxed. Some areas handle these two taxes separately, while others combine them. The money raised by this property tax is used to provide services in the local community, such as police and fire protection, roads, schools, and parks.

Objective 1 *Understand Fair Market Value and Find Assessed Valuation.* To find the amount of this tax, each piece of real property in an area must be assessed. In this process, a local official, called the assessor, makes an estimate of the **fair market value** of the property, the price for which the property could reasonably be expected to be sold. The **assessed valuation** of the property is then found by multiplying the fair market value by a certain percent called the **assessment rate**. The percent that is used varies drastically from state to state, but normally remains constant within a state.

In some states, assessed valuation is 25% of fair market value, while in other states, the assessed valuation is 40% to 60% or even 100% of fair market value. Occasionally, different rates will be used for homes and for businesses. In theory, this step is unnecessary in calculating property tax. However, using an assessed valuation that is a percent of fair market value has become an accepted practice over the years.

Example 1 Finding the Assessed Value of Property

Find the assessed valuation for the following pieces of property owned by Lynn Colgin.

(a) Fair market value, \$112,000; assessment rate (percent), 25%
(b) Fair market value, \$1,382,500; assessment rate (percent), 60%

Solution Multiply the fair market value by the assessment rate.

(a) \$112,000 × 0.25 = \$28,000 assessed valuation
(b) \$1,382,500 × 0.60 = \$829,500 assessed valuation

Objective 2 *Use the Tax Rate Formula.* After calculating the assessed valuation of all the taxable property in an area and determining the amount of money needed to provide the necessary services (the budget), the agency responsible for levying the tax announces the annual **property tax rate**.

This tax rate is determined by the following formula.

$$R = \frac{P}{B}$$

$$\text{Tax rate} = \frac{\text{Total tax amount needed}}{\text{Total assessed value}}$$

Example 2 Finding the Tax Rate

Find the tax rate for the following park districts in River County.

(a) Total tax amount needed, $368,400; total assessed value, $7,368,000
(b) Total tax amount needed, $633,750; total assessed value, $28,800,000

Solution Divide the total tax amount needed by the total assessed value.

(a) $368,400 ÷ $7,368,000 = 0.05 = 5% tax rate
(b) $633,750 ÷ $28,800,000= 0.022 = 2.2% tax rate (rounded)

Objective 3 *Use the Formula for Property Tax.* Property tax rates are expressed in different ways in different parts of the country. However, property tax is always found with the formula

$$P = R \times B$$
$$\text{Tax} = \text{Tax rate} \times \text{Assessed valuation}$$

Objective 4 *Express Tax Rates in Percent, Dollars per $100, Dollars per $1000, and Mills.*

Percent. Some areas express tax rates as a percent of assessed valuation. The tax on a piece of property with an assessed valuation of $74,000 at a tax rate of 9.42% (9.42% = 0.0942) would be

Tax = 0.0942 × $74,000 = $6970.80.

Dollars per $100. In other areas, the rate is expressed as a number of dollars per $100 of assessed valuation. For example, the rate might be expressed as $11.42 per $100 of assessed valuation. Assuming a tax rate of $11.42 per $100, find the tax on a piece of land having an assessed valuation of $42,000 as follows. First divide by 100 by moving the decimal point two places to the left to find the number of hundreds in $42,000.

$42,000 = 420 hundreds   Move the decimal two places to the left.

Then, find the tax.

Tax = Tax rate × Number of hundreds of valuation
= $11.42 × 420 = $4796.40

Dollars per $1000. In other areas, the tax rate is expressed as a number of dollars per $1000 of assessed valuation. If the tax rate is $98.12 per $1000, a piece of property having an assessed valuation of $197,000 would be taxed as follows.

$197,000 = 197 thousands   Move the decimal three places to the left.
Tax = $98.12 × 197 = $19,329.64

Table 6.4 WRITING TAX RATES IN FOUR SYSTEMS

Percent	Per $100	Per $1000	In Mills
12.52	$12.52	$125.20	125.2
3.2	$3.20	$32.00	32
9.87	$9.87	$98.70	98.7

Mills. Finally, some areas express tax rates in mills (a **mill** is one-tenth of a cent, or one-thousandth of a dollar). For example, a tax rate might be expressed as 46 mills per dollar (or $0.046 per dollar). The tax on a property having an assessed valuation of $81,000, at a tax rate of 46 mills, is

$$\text{Tax} = 0.046 \times \$81{,}000 \qquad \text{46 mills} = \$0.046$$
$$= \$3726.$$

Table 6.4 shows the same tax rates written in the four different systems. Although expressed differently, they are equivalent tax rates.

NOTE The number of decimal places used and rounding practices in tax rates vary among taxing jurisdictions. A common practice is to round *up* the last digit used. In Table 6.4 rounding 125.2 mills and 98.7 mills to whole mills would result in 126 mills and 99 mills, respectively.

Objective 5 ***Find Taxes Given the Assessed Valuation and the Tax Rate.*** Property taxes are found by multiplying the tax rate by the assessed valuation, as shown in the following example.

Example 3 Finding the Property Tax

Find the taxes on each of the following pieces of property. Assessed valuations and tax rates are given.

(a) $58,975; 8.4% **(b)** $875,400; $7.82 per $100

(c) $129,600; $64.21 per $1000 **(d)** $221,750; 94 mills

Solution Multiply tax rate by the assessed valuation.

(a) 8.4% = 0.084

$$\text{Tax} = \text{Tax rate} \times \text{Assessed valuation}$$
$$\text{Tax} = 0.084 \times \$58{,}975 = \$4953.90$$

(b) $875,400 = 8754 hundreds

$$\text{Tax} = \$7.82 \times 8754 = \$68{,}456.28$$

(c) $129,600 = 129.6 thousands

$$\text{Tax} = \$64.21 \times 129.6 = \$8321.62$$

(d) 94 mills = 0.094

$$\text{Tax} = 0.094 \times \$221{,}750 = \$20{,}844.50$$

NOTE Some states offer certain tax exemptions that reduce the amount of property tax due. One type of exemption is the homeowner's tax exemption, which allows a specific amount of tax exemption to a person who owns and occupies a home or condominium as a personal residence. ▬

The following figure shows the sales tax, state income tax, and the property tax rates in five locations across the United States. These towns were identified by the *Wall Street Journal* as "Five Hometowns for the Future." Notice the variations in the tax rates.

Someplace to Be

Location	Sales Tax	State Income Tax	Property Tax
San Juan Island, Washington	7.7%	None	$8.92 per $1000
Destin/South Walton Beach, Florida	7%	None	$1.55 per $1000 City of Destin; $7.59 per $1000 Walton County
Kailua-Kona, Hawaii	None	2%–10% based on income	$8.50 per $1000
Petoskey/Harbor Springs, Michigan	6%	4.4%	$34.16 per $1000
Corolla, North Carolina	6%	6%–7%	$.64 per $100

DATA: Wall Street Journal

Objective 6 *Find the Assessed Valuation Given the Tax Rate and the Tax.* The tax formula can also be used to find the assessed valuation when given the amount of tax and the tax rate.

Example 4 Finding the Assessed Valuation

The property tax on a car wash in Boden County is $1024. If the tax rate is $1.65 per $100, find the assessed valuation.

Solution Use the formula for finding tax.

$$\text{Tax} = \text{Tax rate} \times \text{Assessed valuation}$$
$$\$1024 = \$1.65 \times \text{Assessed valuation}$$
$$\frac{1024}{1.65} = \text{Assessed valuation}$$
$$620.61 \text{ hundreds} = \text{Assessed valuation}$$

The assessed valuation is $62,061 (620.61 × 100).

Scientific *Calculator Approach*

The calculator solution to Example 4 is as follows.

1024 [÷] 1.65 [×] 100 [=] 62061 Rounded to the nearest dollar.

Objective 7 *Find the Tax Rate Given the Assessed Valuation and the Tax.* The tax rate may be found by using the tax formula when the assessed valuation and the amount of tax are given.

Example 5 Finding the Tax Rate Given Assessed Valuation and Tax

A commercial property in Hampton County has an assessed valuation of \$186,800 and an annual property tax of \$6771.50. Find the tax rate per \$1000.

Solution Use the formula for finding tax.

$$\text{Tax} = \text{Tax rate} \times \text{Assessed valuation}$$

The assessed valuation is 186.8 thousands (\$186,800 ÷ 1000).

$$\$6771.50 = \text{Tax rate} \times 186.8$$

$$\frac{6771.5}{186.8} = \text{Tax rate}$$

$$\$36.25 = \text{Tax rate per } \$1000$$

The tax rate per \$1000 is \$36.25.

6.2 Exercises

ASSESSED VALUATION *Find the assessed valuation for each of the following pieces of property.*

	Fair Market Value	Rate of Assessment
1.	\$64,000	40%
2.	\$136,500	50%
3.	\$173,800	35%
4.	\$98,200	42%
5.	\$1,300,500	25%
6.	\$2,450,000	80%

PROPERTY TAX RATES *Find the tax rate for the following. Write the tax rate as a percent rounded to the nearest tenth.*

	Total Tax Amount Needed	Total Assessed Value
7.	\$625,000	\$5,200,000
8.	\$322,500	\$4,300,000
9.	\$1,580,000	\$19,750,000
10.	\$2,175,000	\$54,375,000
11.	\$1,224,000	\$40,800,000
12.	\$2,941,500	\$81,700,000

TAX RATE COMPARISON *Complete the following list comparing tax rates.*

	Percent	Per $100	Per $1000	In Mills
13.	(a) ______%	(b) ______	(c) ______	28
14.	(a) ______%	$6.75	(b) ______	(c) ______
15.	2.41%	(a) ______	(b) ______	(c) ______
16.	7.42%	(a) ______	(b) ______	(c) ______
17.	(a) ______%	$7.08	(b) ______	(c) ______
18.	(a) ______%	(b) ______	$35	(c) ______

19. What is the difference between fair market value and assessed value? How is the assessment rate used when finding the assessed value? (See Objective 1.)

20. Select any tax rate and express it as a percent. Write this tax rate in three additional equivalent forms and explain what each form means. (See Objective 4.)

FINDING PROPERTY TAX *Find the tax for each of the following.*

	Assessed Valuation	Tax Rate
21.	$86,200	$6.80 per $100
22.	$41,300	$46.40 per $1000
23.	$685,400	6.93%
24.	$128,200	42 mills
25.	$58,200	$1.80 per $100
26.	$38,250	$89.70 per $1000

ASSESSED VALUE, TAX RATE, AND TAX *Find the missing quantity.*

	Assessed Valuation	Tax Rate	Tax
27.	$49,250	________%	$2,856.50
28.	________	$7.18 per $100	$15,652.40
29.	$73,800	85 mills	________
30.	________	$48.18 per $1000	$1,903.11
31.	$152,680	________ per $100	$8,015.70
32.	$435,500	37.6 mills	________
33.	________	4.3%	$10,182.40
34.	$96,200	________ per $1000	$3,367

Solve each of the following application problems.

35. PIZZA RESTAURANT OWNER Julie Maxey owns the real estate on which she operates her business, Round Table Pizza. The property has a fair market value of

$378,000, property in the area is assessed at 30% of market value, and the tax rate is 4.28%. Find the amount of the property tax.

36. PET CARE Lilly Bolton owns the Pet Hotel with a fair market value of $218,000. Property in the area is assessed at 30% of market value and the property tax is $3597. Find the tax rate as a percent.

37. FM RADIO BROADCASTING A new FM radio station broadcasts from a building having a fair market value of $334,400. The building is in an area where property is assessed at 25% of market value and the tax rate is $75.30 per $1000 of assessed value. Find the property tax.

38. INDUSTRIAL PROPERTY The Consumer's Cooperative owns property with a fair market value of $785,200. The property is located in a county that assesses at 80% of market value. Find the property tax if the tax rate is $14.30 per $1000 of assessed value.

39. COMMERCIAL OFFICE PROPERTY Downtown Office Park has a fair market value of $5,700,000. Property is assessed in the area at 25% of market value. The tax rate is $14.10 per $100 of assessed valuation. Find the property tax.

40. HARLEY-DAVIDSON SHOP Harley-Davidson of Lincoln has property with a fair market value of $518,600. The property is located in an area that is assessed at 35% of market value. The tax rate is $7.35 per $100. Find the property tax.

41. COUNTY PROPERTY TAX In one county, property is assessed at 40% of market value, with a tax rate of 32.1 mills. In a second county, property is assessed at 24% of market value with a tax rate of 50.2 mills. Feathers Custom Wood Products is trying to decide where to place a building with a fair market value of $95,000. (a) Which county would charge the lower property tax? (b) Find the annual amount saved.

42. PROPERTY TAX COMPARISON Property taxes vary from one county to the next. In one county, property is assessed at 30% of market value, with a tax rate of 45.6 mills. In a second county, property is assessed at 48% of market value, with a tax rate of 29.3 mills. Misty Arce is trying to decide where to build her $140,000 dream house. (a) Which county would charge the lower property tax? (b) Find the annual amount saved.

43. RESIDENTIAL PROPERTY TAX Last year the property tax on an executive's estate was $4625, and the tax rate was $12.50 per $1000. After a reassessment this year, the assessed value was increased by $25,000 and the property tax due is $5350. Find the percent of increase in the tax rate. Round to the nearest tenth of a percent. Hint: Do not round until the final answer.

44. INDUSTRIAL PROPERTY TAX Last year the property tax on a warehouse was $3042, and the tax rate was $3.60 per $100. After a reassessment this year, the assessed value was increased by $10,500 and the property tax due is $3705. Find the percent of increase in the tax rate. Round to the nearest tenth of a percent.

45. UNDEVELOPED LAND TAX A prime commercial corner lot was assessed at $45,000, and the tax was $1327.50. The following year the property tax increased to $1353.75 while the tax rate decreased by $0.10 per $100. Find the amount of increase in the assessed value of the commercial lot.

46. INVESTMENT PROPERTY TAX An investment property was assessed at $240,000, and the tax was $5400. The following year the property tax increased to

$5805, while the tax rate decreased by $1.00 per $1000. Find the amount of increase in the assessed value of the property.

6.3 Personal Income Tax

Objectives

1. *Know the four steps that determine tax liability.*
2. *Identify information needed to find adjusted gross income.*
3. *Calculate adjusted gross income.*
4. *Know the standard deduction amounts.*
5. *Know the tax rates.*
6. *List possible itemized deductions to find taxable income.*
7. *Calculate income tax.*
8. *Determine a balance due or a refund from the Internal Revenue Service.*
9. *Prepare a 1040A and a Schedule 1 federal tax form.*

The federal government, most states, many local governments, and some cities use income tax as a source of revenue. However, for most people the federal income tax is the largest tax expense.

As shown in Figure 6.2, the individual income tax provides the largest single source of income to the federal government.

Instructions provided with the tax forms have to cover the situation of every possible taxpayer, from students who earn very little money to professional people, such as lawyers and doctors, who often have complicated financial affairs. For this reason, over half (56.2% in 1999) of the people take their tax returns to a professional tax preparer. But even tax preparers do not solve all the problems of the taxpayer—the taxpayer still

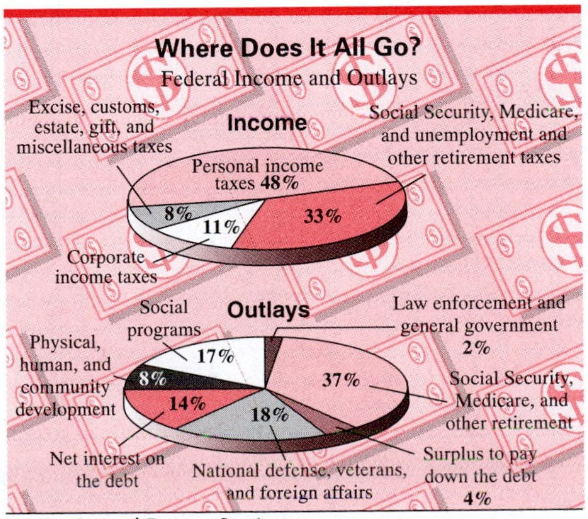

DATA: Internal Revenue Service.

Figure 6.2

must supply all the necessary information. Tax preparers only insert the figures in the correct places on the correct forms and then do the necessary calculations.

Objective 1 | *Know the Four Steps That Determine Tax Liability.* There are four basic steps in finding a person's total tax liability.

Preparing Your Income Tax Return

Step 1. Find the adjusted gross income (AGI) for the year.

Step 2. Find the taxable income.

Step 3. Find the tax.

Step 4. Check to see if a refund is due or if more money is owed to the government.

Objective 2 | *Identify Information Needed to Find Adjusted Gross Income.* These steps are explained in order. The first step is to find the **adjusted gross income** for the year by collecting all the **W-2 forms** that were provided by employers during the year. The form shows the total amount of money paid to the employee by the employer, and also shows the total amount that was withheld from the employee's paycheck and sent, in his or her name, to the IRS. Add all the amounts paid to the employee.

Next, collect any **1099 forms** that may have been received. These informational forms, copies of which the employer sends to the IRS, show miscellaneous income re-

a Control number: 22222	Void ☐	**For Official Use Only ▶ OMB No. 1545-0008**
b Employer identification number 94–1287319	**1** Wages, tips, other compensation $ 24,738.41	**2** Federal income tax withheld $ 3275.60
c Employer's name, address, and ZIP code Class Printing 1568 Liberty Heights Avenue Baltimore, MD 21230	**3** Social security wages $ 24,738.41	**4** Social security tax withheld $ 1533.78
	5 Medicare wages and tips $ 24,738.41	**6** Medicare tax withheld $ 358.71
	7 Social security tips	**8** Allocated tips
d Employee's social security number 123–45–6789	**9** Advance EIC payment	**10** Dependent care benefits
e Employee's name (first, middle initial, last)	**11** Nonqualified plans	**12** Benefits included in box 1
Jennifer Crum 2136 Old Road Towson, MD 21285	**13** See instrs. for box 13	**14** Other
f Employee's address and ZIP code	**15** Statutory employee ☐ Deceased ☐ Pension plan ☐	Legal rep. ☐ Deferred compensation ☐

16 State	Employer's state I.D. no.	**17** State wages, tips, etc.	**18** State income tax	**19** Locality name	**20** Local wages, tips, etc.	**21** Local income tax
MD	600–5076					

Form **W-2** **Wage and Tax Statement**

Department of the Treasury–Internal Revenue Service

9292 ☐ VOID ☐ CORRECTED

PAYER'S name, street address, city, state, ZIP code, and telephone no.		Payer's RTN (optional)	OMB No. 1545-0112	
Employees Credit Union 2572 Brookhaven Drive Dundalk, MD 21222			200_	**Interest Income**
PAYER'S Federal identification number 94-1287319	RECIPIENT'S identification number 123-45-6789	1 Interest income not included in box 3 $ 427.82		**Copy A For Internal Revenue Service Center**
RECIPIENT'S name Jennifer Crum		2 Early withdrawal penalty $	3 Interest on U.S. Savings Bonds and Treas. obligations $	**File with Form 1096.**
Street address (including apt. no.) 2136 Old Road		4 Federal income tax withheld $		For Paperwork Reduction Act Notice and instructions for completing this form, see the **1998 Instructions for Forms 1099, 1098, 5498, and W-2G.**
City, state, and ZIP code Towson, MD 21285		5 Foreign tax paid	6 Foreign country or U.S. possession	
Account number (optional)	2nd TIN Not. ☐	$	$	

Form **1099-INT** Cat. No. 14410K Department of the Treasury - Internal Revenue Service

ceived, such as interest on checking or savings accounts, as well as income received as stock dividends. Also, include any tips or other employee compensation, and enter the total on the correct line of the income tax form.

Objective 3 *Calculate Adjusted Gross Income.* Using the information from the W-2 and 1099 forms, and adding any dividends, capital gains, unemployment compensation, tips, or other employee compensation, enter the total on the correct line of the income tax form. Subtract any adjustments to income, such as an **individual retirement account (IRA)** or alimony payments. An IRA is a retirement plan that allows contributions to be deducted from income. The result is the adjusted gross income.

Adjusted gross income is found with the following formula.

Adjusted gross income = Total income − Total adjustments

Example 1 Finding Adjusted Gross Income (AGI)

As an assistant manager at Class Printing, Jennifer Crum earned $24,738.41 last year and $427.82 in interest from her credit union (see her W-2 and 1099 forms). She had $1500 in IRA contributions. Find her adjusted gross income.

Solution Add the income from her job ($24,738.41) and the interest ($427.82). Then subtract the IRA contributions.

$24,738.41 + $427.82 − $1500 = $23,666.23

NOTE A copy of all W-2 forms is sent to the Internal Revenue Service along with the completed tax forms. However, the IRS does not require the taxpayer to send copies of 1099 forms to them.

Objective 4 *Know the Standard Deduction Amounts.* Most people are almost finished at this point. If deductions for medical expenses, interest, and so on are not to be itemized, and if there are no further adjustments, then the **standard deduction** amount must be determined and subtracted from the adjusted gross income. Although these amounts change each year, the most current standard deduction amounts are shown as follows.

- $4300 for single people
- $7200 for married people filing jointly and qualifying widow(er)s
- $3600 for married people filing separately
- $6350 for head of a household

Additional standard deductions are given for taxpayers and dependents who are blind or 65 years of age or older.

Now, only one step remains before the tax owed is found: determine the number of **personal exemptions**. An exemption is taken for the head of the household and for each of his or her dependents, including spouse and children. For example, a married person with a spouse and three children would be allowed to claim five exemptions. The taxpayer is allowed a $2750 reduction in gross income for each exemption. After subtracting $2750 per exemption from the adjusted gross income, the result, **taxable income**, is multiplied by the proper tax to determine taxes due.

NOTE Both the standard deduction amounts and the personal exemption amounts change each year. It is always important to use current tax information. ▬

Objective 5 *Know the Tax Rates.* Most recently, individual income tax rates have been either 15%, 28%, 31%, 36%, or 39.6%, depending on the amount of taxable income and the taxpayer's filing status. Table 6.5, the tax rate schedule, shows the individual tax rates for each filing status.

Table 6.5 TAX RATE SCHEDULE

Single				Married Filing Jointly or Qualifying Widow(er)			
If taxable income is:				If taxable income is:			
Over—	But Not Over—	Tax Is—	of the Amount Over—	Over—	But Not Over—	Tax Is—	of the Amount Over—
$0	$25,750	...... 15%	$0	$0	$43,050	...... 15%	$0
25,750	62,450	$3,862.50 + 28%	25,750	43,050	104,050	$6,457.50 + 28%	43,050
62,450	130,250	14,138.50 + 31%	62,450	104,050	158,550	23,537.50 + 31%	104,050
130,250	283,150	35,156.50 + 36%	130,250	158,550	283,150	40,432.50 + 36%	158,550
283,150	......	90,200.50 + 39.6%	283,150	283,150	......	85,288.50 + 39.6%	283,150
Married Filing Separately				**Head of Household**			
If taxable income is:				If taxable income is:			
Over—	But Not Over—	Tax Is—	of the Amount Over—	Over—	But Not Over—	Tax Is—	of the Amount Over—
$0	$21,525	...... 15%	$0	$0	$34,450	..... 15%	$0
21,525	52,025	$3,228.75 + 28%	21,525	34,450	89,150	$5,167.50 + 28%	34,450
52,025	79,275	11,768.75 + 31%	52,025	89,150	144,400	20,637.50 + 31%	89,150
79,275	141,575	20,216.25 + 36%	79,275	144,400	283,150	37,765.00 + 36%	144,400
141,575	......	42,644.25 + 39.6%	141,575	283,150	......	87,715.00 + 39.6%	283,150

Example 2 Finding Taxable Income and the Income Tax Amount

Find the taxable income and the tax for each of the following people.

(a) David Shea, single, 1 exemption; adjusted gross income, \$21,835
(b) The Zagorins, married, filing jointly, 5 exemptions; adjusted gross income, \$74,308

Solution

(a) Taxable income is \$14,785 (\$21,835 − \$4300 standard deduction − \$2750 for one exemption). Since the total adjusted gross income is below \$25,750, the 15% tax rate applies. Using the rate in Table 6.5, the tax is calculated as follows:

$$15\% \times \$14{,}785 = \$2217.75$$

The tax is \$2217.75.

(b) Taxable income is \$53,358 (\$74,308 − \$7200 standard deduction − \$13,750 for five exemptions). "Married filing jointly" tax is:

$$15\% \times \$43{,}050 = \$6457.50$$
$$28\% \times \$10{,}308\ (\$53{,}358 - \$43{,}050) = \underline{\$2886.24}$$
$$\text{Total } \$9343.74$$

The tax is \$9343.74.

Problem-Solving Hint When taxable income goes beyond the 15% tax rate amount, do not make the mistake of using the 28% tax rate on the entire amount of taxable income. For example, for a single person having a taxable income greater than \$25,750, a tax rate of 15% is used for the first \$25,750 and a tax rate of 28% is used *only* on the amount *over* \$25,750, up to the next bracket.

Objective 6 *List Possible Itemized Deductions to Find Taxable Income.* An **itemized tax deduction** is any expense that the IRS will allow the taxpayer to subtract from adjusted gross income. To be of benefit, itemized deductions must exceed the automatic standard deduction allowed by the IRS. Usually, taxpayers will benefit from itemized deductions when they take out a loan in order to purchase a home and are allowed to deduct the interest on the loan.

The 20% of the American population who do itemize all their deductions must go through one additional step before subtracting exemptions to determine taxable income: all deductions must be listed. The most common deductions are given here.

Medical and dental expenses. Not all such expenses may be deductible. In general, only amounts in excess of 7.5% of the adjusted gross income may be deducted. For most people, however, this restriction limits medical deductions to catastrophic illnesses.

The medical and dental payments you may possibly deduct include all visits to medical doctors, dentists, chiropractors, and therapists, all medical examinations and treatments, nursing help, hospital care, ambulance service, a mileage deduction for travel to and from medical services and those medical and dental insurance premiums paid by the taxpayer. Expenses reimbursed by insurance companies cannot be deducted.

Taxes. State and local income taxes, real estate taxes, and personal property taxes may be deducted. You may not deduct federal income tax, gasoline taxes, Social Security or Medicare taxes, or any sales taxes.

Interest. Home mortgage interest on the taxpayer's principal residence and a qualified second home is deductible. Other interest charges (including credit card interest) may not be deducted.

Contributions. Contributions to qualified charities may be deducted.

Miscellaneous deductions. Miscellaneous expenses are only deductible to the extent that the total amount of such deductions exceeds 2% of the taxpayer's adjusted gross income. These deductions include union dues, qualified education expenses, income tax preparation fees, tax preparation books or computer software, appraisal fees for tax purposes, legal fees for tax planning or tax litigation, and safe deposit box rental fees.

NOTE The taxpayer gets to take whichever is higher—the standard deduction *or* the itemized deductions. When itemizing deductions, the gain in deductions is not the total of all itemized deductions but the difference between the standard deduction and the total itemized deductions. The taxpayer must be able to document these deductions with carefully kept records and receipts to satisfy any questions by the Internal Revenue Service. ▬

Objective 7 *Calculate Income Tax.* After the taxable income is determined, the amount of income tax must be found.

The amount of income tax is found using the following formula.

$$\text{Income tax} = \text{Taxable income} \times \text{Tax rate}$$

Example 3 Using Itemized Deductions to Find Taxable Income and Income Tax

Chris Kelly is single, has no dependents, and had an adjusted gross income of $26,735 last year. He had deductions of $1352 for other taxes, $3118 for mortgage interest, and $317 for charity. Find his taxable income and his income tax.

Solution First find the total of all deductions.

$$\text{Deductions} = \$1352 + \$3118 + \$317 = \$4787$$

Since Kelly is single, and the standard deduction is $4300, the larger itemized deduction amount, $4787, is taken as well as one personal exemption. Now find his taxable income.

$$\text{Taxable income} = \$26{,}735 - \$4787 - \$2750 = \$19{,}198$$

Finally, income tax is determined.

$$15\% \times \$19{,}198 = \$2879.70$$

His tax is $2879.70.

NOTE In preparing personal income tax, refer to current Internal Revenue Service publications, and always use the current tax rates. Check to see if you qualify for any tax credits. You may be entitled to receive the dependent care credit, the credit for the elderly, the adoption credit, the child tax credit, the mortgage interest credit, or the education credit. ▬

Objective 8 *Determine a Balance Due or a Refund From the Internal Revenue Service.* After calculating the proper tax, determine whether a refund is due from the government. Look again at the W-2 forms to find out how much already has been paid toward the tax bill. These forms show the total amount the employer has withheld and sent to the government. (Usually, no money is withheld for amounts on 1099 forms.) If the amount withheld is greater than the tax owed, the taxpayer is entitled to a refund. If the amount withheld is less than the tax owed, then the taxpayer must send the difference along with the tax return.

Electronic Update

Many more people are filing electronically, which typically means speedier refunds and fewer mistakes. The IRS received 20.6 million returns electronically as of March 27, up 25% from a year earlier. This includes filings by computer or by phone.

Source: Wall Street Journal, reprinted by permission of Dow Jones, Inc. via Copyright Clearance Center, Inc. © Dow Jones & Co., Inc. All rights reserved.

Example 4 Determining Tax Due or a Tax Refund

Gale Klein had \$375.20 per month withheld for federal income tax from her checks last year. She is single and has taxable income of \$24,056 for the year. Does she get a refund? If so, how much?

Solution Klein had $\$375.20 \times 12 = \$4502.40$ withheld from her checks last year. The tax due on taxable income of \$24,056 is \$3608.40 ($15\% \times \$24,056$); therefore, she will receive a refund.

$$\$4502.40 - \$3608.40 = \$894 \quad \text{Refund.}$$

Problem-Solving Hint Be certain that the proper income tax form is used when filing your individual income tax. The 1040 EZ form is used by many students. If you have over \$400 in interest or any adjustments to income, a 1040A form is used. In order to itemize deductions the taxpayer must use a 1040 form. There are additional considerations and restrictions that determine whether the 1040 EZ, 1040A, or 1040 forms should be used.

Objective 9 *Prepare a 1040A and a Schedule 1 Federal Tax Form.* The next example shows how to complete an income tax return using **Form 1040A** and a **Schedule 1 (Form 1040)**. When completing income tax forms and calculations, notice that all amounts may be rounded to the nearest dollar.

Example 5 Preparing a 1040A and a Schedule 1

Jennifer Crum is single and claims one exemption. Her income appears on the W-2 and 1099 forms on pages 220 and 221. Crum contributes $1500 to an IRA. Since she has interest income over $400 but does not itemize her deductions she may use Form 1040A and must also file a Schedule 1 (Form 1040A). Complete her income tax return.

Form **1040A** Department of the Treasury—Internal Revenue Service
U.S. Individual Income Tax Return (H) IRS Use Only—Do not write or staple in this space.
OMB No. 1545-0085

Label (See page 19.)
Use the IRS label. Otherwise, please print or type.
LABEL HERE

Your first name and initial	Last name	Your social security number
Jennifer	Crum	123 : 45 : 6789
If a joint return, spouse's first name and initial	Last name	Spouse's social security number
		: :

Home address (number and street). If you have a P.O. box, see page 20.	Apt. no.
2136 Old Road	
City, town or post office, state, and ZIP code. If you have a foreign address, see page 20.	
Towson, MD 21285	

▲ **IMPORTANT!** ▲ You **must** enter your SSN(s) above.

Presidential Election Campaign Fund (See page 20.)

	Yes	No
Do you want $3 to go to this fund?		☒
If a joint return, does your spouse want $3 to go to this fund?		

Note. Checking "Yes" will not change your tax or reduce your refund.

Filing status (Check only one box.)

1 ☒ Single
2 ☐ Married filing joint return (even if only one had income)
3 ☐ Married filing separate return. Enter spouse's social security number above and full name here. ▶
4 ☐ Head of household (with qualifying person). (See page 21.) If the qualifying person is a child but not your dependent, enter this child's name here. ▶
5 ☐ Qualifying widow(er) with dependent child (year spouse died ▶ 19). (See page 22.)

Exemptions

6a ☒ **Yourself.** If your parent (or someone else) can claim you as a dependent on his or her tax return, **do not** check box 6a.
b ☐ **Spouse**

No. of boxes checked on 6a and 6b: 1

c **Dependents:** (If more than seven dependents, see page 22.)

(1) First name Last name	(2) Dependent's social security number	(3) Dependent's relationship to you	(4) ✓ if qualifying child for child tax credit (see page 23)
			☐
			☐
			☐
			☐
			☐
			☐
			☐

No. of your children on 6c who: • lived with you ___ • did not live with you due to divorce or separation (see page 24) ___
Dependents on 6c not entered above ___
Add numbers entered on lines above ___

d Total number of exemptions claimed.

Income

Attach Copy B of your Form(s) W-2 here. Also attach Form(s) 1099-R if tax was withheld.
If you did not get a W-2, see page 25.
Enclose, but do not staple, any payment.

Line	Description		Amount
7	Wages, salaries, tips, etc. Attach Form(s) W-2.	7	24,738
8a	**Taxable** interest. Attach Schedule 1 if required.	8a	428
b	**Tax-exempt** interest. DO NOT include on line 8a. 8b		
9	Ordinary dividends. Attach Schedule 1 if required.	9	
10a	Total IRA distributions. 10a — **10b** Taxable amount (see page 25).	10b	
11a	Total pensions and annuities. 11a — **11b** Taxable amount (see page 26).	11b	
12	Unemployment compensation, qualified state tuition program earnings, and Alaska Permanent Fund dividends.	12	
13a	Social security benefits. 13a — **13b** Taxable amount (see page 28).	13b	
14	Add lines 7 through 13b (far right column). This is your **total income.** ▶	14	25,166

Adjusted gross income

Line	Description		Amount
15	IRA deduction (see page 30). 15 — 1500		
16	Student loan interest deduction (see page 30). 16		
17	Add lines 15 and 16. These are your **total adjustments.**	17	1500
18	Subtract line 17 from line 14. This is your **adjusted gross income.** ▶	18	23,666

For Disclosure, Privacy Act, and Paperwork Reduction Act Notice, see page 53. Cat. No. 11327A Form **1040A**

Form 1040A Page 2

Section	Line	Description	Line	Amount
Taxable income	19	Enter the amount from line 18.	19	23,666
	20a	Check if: ☐ **You** were 65 or older ☐ Blind; ☐ **Spouse** was 65 or older ☐ Blind. **Enter number of boxes checked** ▶	20a	☐
	b	If you are married filing separately and your spouse itemizes deductions, see page 32 and check here . . . . ▶	20b	☐
	21	Enter the **standard deduction** for your filing status. **But** see page 33 if you checked any box on line 20a or 20b **OR** if someone can claim you as a dependent. • Single—$4,300 • Married filing jointly or Qualifying widow(er)—$7,200 • Head of household—$6,350 • Married filing separately—$3,600	21	4300
	22	Subtract line 21 from line 19. If line 21 is more than line 19, enter -0-.	22	19,366
	23	Multiply $2,750 by the total number of exemptions claimed on line 6d.	23	2750
	24	Subtract line 23 from line 22. If line 23 is more than line 22, enter -0-. This is your **taxable income.** ▶	24	16,616
Tax, credits, and payments	25	Find the tax on the amount on line 24 (see page 34).	25	2492
	26	Credit for child and dependent care expenses. Attach Schedule 2. (26)		
	27	Credit for the elderly or the disabled. Attach Schedule 3. (27)		
	28	Child tax credit (see page 35). (28)		
	29	Education credits. Attach Form 8863. (29)		
	30	Adoption credit. Attach Form 8839. (30)		
	31	Add lines 26 through 30. These are your **total credits.**	31	
	32	Subtract line 31 from line 25. If line 31 is more than line 25, enter -0-.	32	2492
	33	Advance earned income credit payments from Form(s) W-2.	33	
	34	Add lines 32 and 33. This is your **total tax.** ▶	34	2492
	35	Total Federal income tax withheld from Forms W-2 and 1099. (35) 3276		
	36	1999 estimated tax payments and amount applied from 1998 return. (36)		
	37a	**Earned income credit.** Attach Schedule EIC if you have a qualifying child. (37a)		
	b	Nontaxable earned income: amount ▶ and type ▶		
	38	Additional child tax credit. Attach Form 8812. (38)		
	39	Add lines 35, 36, 37a, and 38. These are your **total payments.** ▶	39	3276
Refund	40	If line 39 is more than line 34, subtract line 34 from line 39. This is the amount you **overpaid.**	40	784
Have it directly deposited! See page 47 and fill in 41b, 41c, and 41d.	41a	Amount of line 40 you want **refunded to you.**	41a	784
	▶ b	Routing number ▶ c Type: ☐ Checking ☐ Savings		
	▶ d	Account number		
	42	Amount of line 40 you want **applied to your 2000 estimated tax.** (42)		
Amount you owe	43	If line 34 is more than line 39, subtract line 39 from line 34. This is the **amount you owe.** For details on how to pay, see page 48.	43	
	44	Estimated tax penalty (see page 48). (44)		

Sign here

Joint return? See page 20. Keep a copy for your records.

Under penalties of perjury, I declare that I have examined this return and accompanying schedules and statements, and to the best of my knowledge and belief, they are true, correct, and accurately list all amounts and sources of income I received during the tax year. Declaration of preparer (other than the taxpayer) is based on all information of which the preparer has any knowledge.

Your signature	Date	Your occupation	Daytime telephone number (optional)
Jennifer Crum	3/08/	Asst. Manager	( )
Spouse's signature. If joint return, BOTH must sign.	Date	Spouse's occupation	

Paid preparer's use only

Preparer's signature ▶	Date	Check if self-employed ☐	Preparer's SSN or PTIN
Firm's name (or yours if self-employed) and address ▶			EIN
			ZIP code

Form **1040A**

Schedule 1
(Form 1040A)

Department of the Treasury—Internal Revenue Service

Interest and Ordinary Dividends for Form 1040A Filers (H)

OMB No. 1545-0085

Name(s) shown on Form 1040A: Jennifer Crum

Your social security number: 123 : 45 : 6789

Part I

Interest

(See page 60 and the instructions for Form 1040A, line 8a.)

Note. If you received a Form 1099-INT, Form 1099-OID, or substitute statement from a brokerage firm, enter the firm's name and the total interest shown on that form.

			Amount
1	List name of payer. If any interest is from a seller-financed mortgage and the buyer used the property as a personal residence, see page 60 and list this interest first. Also, show that buyer's social security number and address.		
	Employee Credit union	1	428
2	Add the amounts on line 1.	2	428
3	Excludable interest on series EE and I U.S. savings bonds issued after 1989 from Form 8815, line 14. You **must** attach Form 8815.	3	
4	Subtract line 3 from line 2. Enter the result here and on Form 1040A, line 8a.	4	428

Part II

Ordinary dividends

(See page 60 and the instructions for Form 1040A, line 9.)

Note. If you received a Form 1099-DIV or substitute statement from a brokerage firm, enter the firm's name and the ordinary dividends shown on that form.

			Amount
5	List name of payer	5	
6	Add the amounts on line 5. Enter the total here and on Form 1040A, line 9.	6	

For Paperwork Reduction Act Notice, see Form 1040A instructions. Cat. No. 12075R **Schedule 1 (Form 1040A)**

NOTE The annual IRA contribution reduces taxable income for that year. In Example 5, the resulting tax saving is the amount of the IRA contribution multiplied by the tax rate or $\$1500 \times 15\% = \$225$. ▬

6.3 Exercises

ADJUSTED GROSS INCOME *Find the adjusted gross income for each of the following.*

	Name	Income from Jobs	Interest	Misc. Income	Dividend Income	Adjustments to Income
1.	R. Garrett	$18,610	$74	$1936	$115	$135
2.	C. Manly	$38,156	$285	$73	$542	$317
3.	The Hanks	$21,380	$625	$139	$184	$618
4.	The Jazwinskis	$33,650	$722	$375	$218	$473
5.	The Brashers	$38,643	$95	$188	$105	$0
6.	The Ameens	$41,379	$1147	$536	$186	$2258

TAXABLE INCOME AND TAX *Find the amount of taxable income and the tax for each of the following. Use the tax rate schedule (Table 6.5). The letter following the name indicates marital status, and all married people are filing jointly.*

	Name	Number of Exemptions	Adjusted Gross Income	Total Deductions
7.	E. Gragg, S	1	$24,200	$1,795
8.	P. Phelps, S	1	$15,615	$3,182
9.	The Jordans, M	3	$38,751	$5,968
10.	The Loveridges, M	7	$52,532	$6,972
11.	The Lanes, M	5	$71,800	$8,851
12.	G. Clark, S	1	$32,322	$4,318
13.	R. Bowtell, S	1	$40,350	$3,885
14.	D. Collins, S	1	$39,502	$4,365
15.	N. Weggener, S	1	$68,574	$2,793
16.	The Printices, M	5	$119,378	$17,382
17.	The Albers, M	2	$62,613	$7,681
18.	The Reents, M	8	$98,544	$7,053

TAX REFUND OR TAX DUE *Find the amount of any refund or tax due for the following people. The letter following the name indicates marital status. Assume a 52-week year and that married people are filing jointly.*

	Name	Taxable Income	Federal Income Tax Withheld from Checks
19.	Karecki, L., S	\$12,947	\$243.10 monthly
20.	Woo, C., S	\$28,072	\$347.80 monthly
21.	Hunziker, B., S	\$22,988	\$72.18 weekly
22.	The Fungs, M	\$49,238	\$149.27 weekly
23.	The Todds, M	\$31,786	\$358.44 monthly
24.	The Fords, M	\$45,436	\$128.35 weekly

25. List four sources of income for which an individual might receive W-2 and 1099 forms. Which form would commonly be received for each? (See Objective 3.)

26. List four possible tax deductions and explain the effect that a tax deduction will have on taxable income and on income tax due. (See Objective 6.)

Find the tax in each of the following application problems.

27. MARRIED—INCOME TAX The Tobins had an adjusted gross income of \$54,378 last year. They had deductions of \$682 for state income tax, \$187 for city income tax, \$472 for property tax, \$4260 in mortgage interest, and \$785 in contributions. They file a joint return and claim four exemptions.

28. SINGLE—INCOME TAX Diane Bolton had an adjusted gross income of \$40,502 last year. She had deductions of \$971 for state income tax, \$564 for property tax, \$2747 in mortgage interest, and \$235 in contributions. Bolton claims one exemption and files as a single person.

29. SINGLE—INCOME TAX Susan Winslow, filing as a single person and claiming one exemption, had an adjusted gross income of \$36,998. Her deductions amounted to \$3255.

30. MARRIED—INCOME TAX The Slausons had an adjusted gross income of \$36,116 last year. They had deductions of \$1078 for state income tax, \$253 for city income tax, \$879 for property tax, \$5218 in mortgage interest, and \$386 in contributions. They claim three exemptions and file a joint return.

31. STUDENT—INCOME TAX Sue Brasher, a full-time college student and a single person, had wages from three part-time jobs amounting to \$4108, \$2653, and \$1838. She had interest of \$137, no adjustments to income and no itemized deductions. Since her parents claim her as an exemption, Brasher must claim zero exemptions. Brasher is allowed the standard deduction.

32. SINGLE—INCOME TAX Tyrone Goodwin, a single person, had wages from three part-time jobs while attending college full-time. Wages amounted to \$974, \$2793, and \$3210. He had interest of \$96, no adjustments to income, and no itemized deductions. Goodwin must claim zero exemptions since his parents still claim him as a tax exemption. Goodwin is allowed the standard deduction.

33. MARRIED—INCOME TAX The Rusks had wages of $68,645, dividends of $385, interest of $672, and adjustments to income of $1058 last year. They had deductions of $877 for state income tax, $342 for city income tax, $786 for property tax, $8180 in mortgage interest, and $186 in contributions. They claim five exemptions and file a joint return.

34. SINGLE—INCOME TAX John Walker had wages of $34,218, other income of $2892, dividends of $240, interest of $315, and an IRA contribution of $750 last year. He had deductions of $1163 for state income tax, $1268 for property tax, $2826 in mortgage interest, and $850 in contributions. Walker claims one exemption and files as a single person.

35. MARRIED—INCOME TAX John and Vicki Karsten had combined wages and salaries of $45,428, other income of $5283, dividend income of $324, and interest income of $668. They have adjustments to income of $2484. Their itemized deductions are $7615 in mortgage interest, $729 in state income tax, $1185 in real estate taxes, and $1219 in charitable contributions. The Karstens filed a joint return and claimed six exemptions.

36. SINGLE—INCOME TAX Colleen Mannel had wages and salaries of $43,846, other income of $1682, dividend income of $478, and interest income of $986. She has an adjustment to income of $1452. Her itemized deductions are $4615 in mortgage interest, $1136 in state income tax, $856 in real estate taxes, and $835 in charitable contributions. Mannel claims one exemption and is a single person.

Chapter 6 Quick Review

Chapter Terms

Review the following terms to test your understanding of the chapter. For each term you do not know, refer to the page number found next to that term.

1099 forms [p. 220]
adjusted gross income [p. 220
assessed valuation [p. 212]
assessment rate [p. 212]
contributions [p. 224]
dollars per \$100 [p. 213]
dollars per \$1000 [p. 213]
excise (luxury) tax [p. 208]
fair market value [p. 212]
Form 1040A [p. 225]
1040 EZ [p. 225]
individual retirement account (IRA) [p. 221]
interest [p. 224]
itemized tax deduction [p. 223]
medical and dental expenses [p. 223]
mills [p. 214]
miscellaneous deductions [p. 224]
personal exemptions [p. 222]
personal property [p. 212]
property tax rate [p. 212]
real property [p. 212]
sales tax [p. 205]
sales tax rates [p. 205]
Schedule 1 (Form 1040) [p. 225]
standard deduction [p. 221]
taxable income [p. 222]
W-2 forms [p.220]

Concepts	Examples
6.1 Finding sales tax Collected by most states, and some counties and cities. Use $P = BR$, where P is the sales tax, B is the amount of the sale, and R is the sales tax rate.	Sales tax of 5% is charged on a sale of \$173.15. Find the amount of sales tax and the total sale including tax. $P = \$173.15 \times 0.05 = \$8.66$ \$173.15 sale + \$8.66 tax = \$181.81 total sale
6.1 Finding selling price when the sales tax is known Use the basic percent equation. $P = B \times R$ Sales tax = Selling price × Tax rate	Sales tax is \$4.59; sales tax rate is 6%. Find the amount of the sale. $\$4.59 = B \times 0.06$ $\frac{\$4.59}{0.06} = B$ $\$76.50 = B$ selling price
6.1 Finding the amount of the sale when the total price is known Use the percent equation and remember Amount of sale + Sales tax = Total	Total price including tax, \$128.96; sales tax, 4%. Find the amount of the sale. $B + (0.04)B = \$128.96$ $1.04B = \$128.96$ $B = \frac{\$128.96}{1.04} = \$124$
6.1 Excise tax (luxury tax) A tax charged on certain items by the federal, state, or local government. It may be either a percent of the sale price or a certain amount per item.	A tire weighs 50 pounds and sells for \$118; sales tax rate is 7%; excise tax is 15¢ per pound over 40 pounds. Find the total cost. \$118.00 — Tire 8.26 — Sales tax: 0.07 × \$118 + 1.50 — Excise tax: 0.15(50 − 40) \$127.76 — Total cost

6.2 Fair market value and assessed valuation Multiply the market value of the property by the assessment rate (a local assessed percent) to arrive at the assessed valuation.	The assessment rate is 30%; fair market value is \$115,000. Find the assessed valuation. $0.3 \times \$115{,}000 = \$34{,}500$
6.2 Tax rate The tax rate formula follows. $$\text{Tax rate} = \frac{\text{Total tax amount needed}}{\text{Total assessed value}}$$	Tax amount needed \$245,664; total assessed value, \$3,070,800. Find the tax rate. $$\frac{245{,}664}{3{,}070{,}800} = 0.08 = 8\%$$
6.2 Expressing tax rates in different forms and finding tax	
1. Percent: multiply assessed valuation by rate.	Assessed value, \$90,000; tax rate, 2.5%. $\$90{,}000 \times 0.025 = \$2250$
2. Dollars per \$100: move decimal 2 places to left in assessed valuation and multiply.	Tax rate, \$2.50 per \$100 $900 \times \$2.50 = \$2250$
3. Dollars per \$1000: move decimal 3 places to left in assessed valuation and multiply.	Tax rate, \$25 per \$1000 $90 \times \$25 = \$2250$
4. Mills: move decimal 3 places to the left in rate and multiply by assessed valuation.	Tax rate, 25 mills $\$90{,}000 \times 0.025 = \$2250$
6.3 Adjusted gross income Adjusted gross income includes wages, salaries, tips, dividends, and interest. Any IRA contributions or alimony payments are subtracted.	Salary, \$32,540; interest income, \$875; dividends, \$315. Find adjusted gross income. $\$32{,}540 + \$875 + \$315 = \$33{,}730$
6.3 Standard deduction amounts The majority of taxpayers use the standard deduction allowed by the IRS.	\$4300 for single people; \$7200 for married people filing jointly; \$3600 for married people filing separately; \$6350 for head of household
6.3 Taxable income The larger of either the total of itemized deductions or the standard deduction is subtracted from adjusted gross income along with \$2750 for each personal exemption.	Adjusted gross income, \$18,200; single taxpayer; itemized deductions total \$2830; find taxable income. Standard deduction is \$4300; larger than \$2850 itemized deduction. $\text{Taxable income} = \$18{,}200 - \$4300 - \$2750 = \$11{,}150$
6.3 Tax rates There are five tax rates: 15%, 28%, 31%, 36%, and 39.6%.	Single: 15%; over \$25,750, 28%; over \$62,450, 31%; over \$130,250, 36%; over \$283,150, 39.6%. Married filing jointly or qualifying widow(er)s: 15%; over \$43,050, 28%; over \$104,050, 31%; over \$158,550, 36%; over \$283,150, 39.6%. Married filing separately: 15%; over \$21,525, 28%; over \$52,025, 31%; over \$79,275, 36%; over \$141,575, 39.6%.

	Head of household: 15%; over $34,450, 28%; over $89,150, 31%; over $144,400, 36%; over $283,150, 39.6%.
6.3 Income tax The amount of income tax is found by using the correct tax table and the proper tax rates.	Taxable income, $52,058; married filing jointly; find income tax. 15% × $43,050 = $6457.50 28% × $9008 ($52,058 − $43,050) = $2522.24 Total $8979.74
6.3 Tax due or refund If the total amount withheld by employers is greater than the tax owed, a refund results. If the tax owed is the greater amount, a balance is due.	Tax owed, $2506; tax withheld, $226 per month for 12 months. Find balance due or refund. $226 × 12 = $2712 withheld $2712 withheld − $2506 owed = $206 refund

Chapter 6 Review Exercises

Find the amount of the sales tax, the excise tax, and the total sale price including taxes in each of the following problems. **[6.1]**

	Sale Price	Sales Tax Rate	Excise Tax Rate
1.	$852.15	6%	10%
2.	$86.15	4%	11%
3.	$16,500	5%	$12.20/person; 110 people
4.	$345.96	7%	18.3¢/gal.; 285 gal.

Find the sale price when given the amount of tax and the sales tax rate. **[6.1]**

	Amount of Sales Tax	Sales Tax Rate
5.	$68.04	6%
6.	$14.20	5%
7.	$19.60	7%
8.	$15.75	$4\frac{1}{2}$%

Find the amount of the sale before sales tax was added in each of the following. **[6.1]**

	Total Sale	Sales Tax Rate
9.	$447.32	6%
10.	$133.75	7%
11.	$292.95	5%
12.	$430.56	4%

Complete the following list comparing tax rates. Do not round. [6.2]

	Percent	Per $100	Per $1000	In Mills
13.	______%	$4.06	______	______
14.	______%	______	______	27
15.	1.27%	______	______	______
16.	______%	______	$19.50	______

Find the missing quantity. [6.2]

	Assessed Valuation	Tax Rate	Tax
17.	$426,000	32 mills	________
18.	$98,200	________ per $1000	$1,816.70
19.	________	3.5%	$1,627.50
20.	$140,500	________%	$3,934
21.	________	$3.80 per $100	$3,655.60
22.	$103,600	27 mills	________

Find the taxable income and the tax for each of the following. The letter following the name indicates the marital status. All married people are filing jointly. [6.3]

	Name	Number of Exemptions	Adjusted Gross Income	Total Deductions
23.	G. Eckern, S.	1	$38,415	$4516
24.	The Bridges, M	5	$78,628	$7634
25.	R. McIntosh, M	3	$54,110	$6975
26.	R. Tewell, S	1	$48,752	$3695

Solve each of the following application problems. [6.1–6.2]

27. The Oak Glen Park District budgets on the basis that it will collect $1,978,000. If the total assessed value of the property in the city is $90,550,000, find the tax rate as a percent rounded to the nearest tenth.

28. Total receipts for the day at the Toy Circus are $3442.88. If this includes 6% sales tax, find the amount of the sales tax.

29. A shopping center has a fair market value of $2,608,300. Property in the area is assessed at 28% of fair market value, with a tax rate of $21.50 per $1000. Find the annual property tax.

Find the tax owed in each of the following application problems. [6.3]

30. The Jidobus, married and filing a joint return, have an adjusted gross income of $63,280, six exemptions, and deductions of $4662.

31. Marie Perino had an adjusted gross income of $27,760 last year. She had deductions of $817 for state income tax, $875 for property tax, $1495 in mortgage interest, and $343 in contributions. Perino claims one exemption and files as a single person.

32. Heather and Stephen Hall had total wages and salaries of $69,750, other income of $852, and interest income of $2880.They are allowed an adjustment to income of $2450. Their itemized deductions are $7218 in mortgage interest, $471 in state income taxes, and $1040 in charitable contributions. The Halls are filing a joint return and claim four exemptions.

Find the amount of any refund or tax due for the following people. The letter following the name indicates marital status. Assume a 52-week year and that married people are filing jointly. **[6.3]**

	Name	Taxable Income	Federal Income Tax Withheld from Checks
33.	The Dales, M.	$62,450	$198.50 weekly
34.	Lagera, T., S	$33,825	$533.20 monthly
35.	Rosa, D., S	$28,315	$105.40 weekly
36.	The Taveras, M	$40,180	$537.30 monthly

Chapter 6 Summary Exercise *Financial Planning for Taxes*

Jack Armstrong, owner of All American Truck Stop, is considering two separate locations along the interstate, Anderson and Bentonville. The two locations are about 200 miles apart, and while the sites offer similar business potential, there are differences in land acquisition costs, building costs, and most importantly property taxes. Armstrong feels that he needs 11 acres of land and buildings and improvements that will total 90,000 square feet. The land and building costs and property tax information are as follows.

	Anderson	Bentonville
Land cost (per square foot)	$0.40	$0.35
Building and improvement cost (per square foot)	$32.80	$36.90
Assessment rate	25%	20%
Tax rate	32 mills	$2.95 per $100

Knowing that there are 43,560 square feet in an acre and that the total cost of land improvements will be used as fair market value in both locations, Armstrong needs to answer the following questions to help him in his decision.

(a) What is the cost of the land and improvements in each location?
(b) Find the assessed valuation of the land and improvements in each location.
(c) Find the annual property tax in each location.
(d) What is the total cost including land, building, and property taxes over a ten-year period in each location?
(e) On the basis of cost over a 10-year period, which location should Armstrong select?

Net Assets Business on the Internet

Round Table Pizza

Statistics

- 1959: First store opens in Menlo Park, California
- 1979: Locations total 150
- 1999: Over 530 locations
- 2000: 20 Round Table Pizza locations in Las Vegas

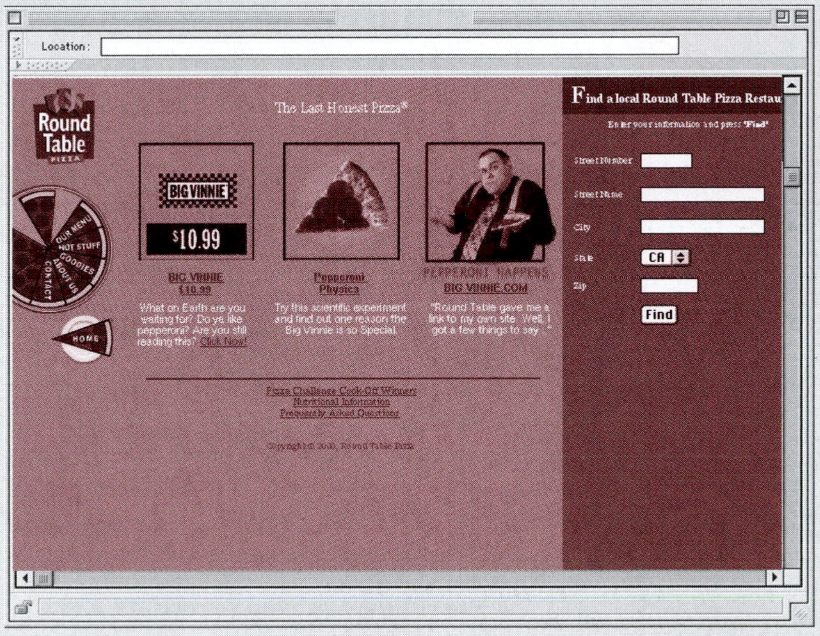

The first Round Table Pizza restaurant was opened in 1959 in Menlo Park, California. Starting with an old English theme and a $2500 loan, the Round Table system was being franchised just three years later. Today, Round Table is one of the nation's largest pizza chains with more than 530 franchised and company-owned restaurants in the western United States. In addition, they have restaurants in Asia and the Middle East. Most restaurants offer delivery/carry-out service as well as dine-in service.

Round Table Pizza continues to stand out in the pizza marketplace by stressing high-quality, innovative products and ingredients and by introducing at least two new products per year. Recent additions to their product line have been the Gourmet Sandwiches and Round Table Chicken Wings; however, the company maintains an unswerving commitment to the high quality of their pizza, and advertises itself as the restaurant serving The Last Honest Pizza.

1. Total sales for one day at the Round Table Pizza beverage bar were $1148.40 including 5% sales tax charged on all purchases. Find the amount that is sales tax.

2. A prime commercial property in the neighboring county has an assessed valuation of $132,800 and an annual property tax of $3691.84. Find the tax rate per $1000.

3. A Round Table Pizza restaurant is on property with a fair market value of $476,000. The property is in an area that is assessed at 35% of market value. The property tax rate is $3.10 per $100. Find the property tax.

4. Brian Katavich, owner of a Round Table Pizza restaurant, has annual income of $68,730 from his business, $1586 in interest income, $862 in miscellaneous income, and $315 in dividend income. If Katavich has adjustments to income of $427, find his adjusted gross income.

7 Risk Management

The amount of money paid toward insurance premiums by businesses is quite small when compared to other business expenses. Likewise, the average household pays only a small portion of its budget on insurance premiums. Figure 7.1 shows the percent of total household spending going to pay for insurance coverage.

What are the risks of not having insurance or of having inadequate insurance coverage?

People buy insurance to protect against risk. In the event that some undesirable event occurs, **peril insurance** provides financial compensation. Perils that are insurable include illness, death, fire, flood, theft, automobile collision, property damage, and personal liability. A person or business buying the insurance, the **insured** or **policyholder**, pays a relatively small amount of money—called the premium—to provide protection against a large loss. If the undesirable event occurs, the insurance company, the **insurer** or **carrier**, pays the insured for the loss up to the face value or stated amount of the policy.

Insurance is based on the idea that many pay into a fund while a few draw out of the fund. For example, a business may pay a fire insurance premium of a few hundred dollars

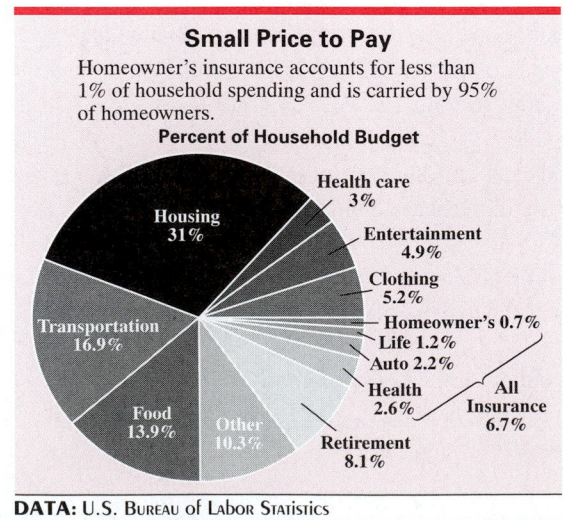

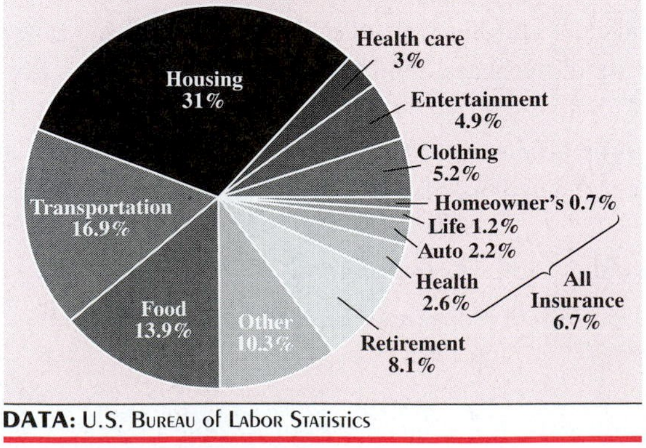

Figure 7.1

for several years without ever having a fire loss. However, should a fire occur, the loss may result in many thousands of dollars being paid to the business.

This chapter looks first at business insurance, including fire and liability coverage; next, motor vehicle insurance is discussed; then, the many types of life insurance policies are examined. Nonforfeiture options and settlement options are covered in the last section of the chapter.

7.1 Business Insurance

Objectives

1. *Define the terms: policy, face value, and premium.*
2. *Identify the factors that determine the premium.*
3. *Find the annual premium for fire insurance given rating and property values.*
4. *Calculate short-term rates and cancellations.*
5. *Calculate prorated insurance premium cancellations.*
6. *Use the coinsurance formula to solve problems.*
7. *Find the insurance liability when there are multiple carriers.*
8. *Find liability of multiple carriers when coinsurance requirement is not met.*
9. *List additional risks against which a business may be insured.*

There is only a slight chance that any particular building will suffer fire damage during a given year. However, if such fire damage were to occur, the financial loss to the owner could be very large. To protect against this small chance of a large loss, people pay an amount equal to a small percent of the value of their property to a fire insurance company. The company collects money from a large number of property owners, then pays for expenses due to fire damage for those few buildings which are damaged.

Objective 1 *Define the Terms: Policy, Face Value, and Premium.* The contract between the owner of a building and a fire insurance company is called a **policy**. The amount of insurance provided by the company is called the **face value** of the policy. The charge for the policy is called the **premium**.

Objective 2 *Identify the Factors That Determine the Premium.* The amount of premium charged by the insurance company depends on several factors, such as the type of construction of the building, the contents and use of the building, the location of the building, and the type of fire protection that is available. Wood-frame buildings are generally more likely to be damaged by fire, and thus require a larger premium than masonry buildings. Categories are assigned to building types by insurance company employees called **underwriters**. These categories are usually named by letters, such as A, B, C, and so on. Underwriters also assign ratings, called territorial ratings, to each area served, which describe the quality of fire protection in the area. While fire insurance rates vary from state to state, the rates in Table 7.1 are typical.

Objective 3 *Find the Annual Premium for Fire Insurance Given Rating and Property Values.* The annual premium rate for fire insurance is expressed as a certain amount for each \$100 in value. The basic percent equation is used to find the annual insurance premium.

The value of the building in hundreds of dollars is the base, the insurance premium per $100 of fire insurance is the rate, and the annual insurance premium is the part.

$$\text{Annual Premium} = \underset{\text{(in \$100s)}}{\text{Building value}} \times \underset{\text{(per \$100s)}}{\text{Insurance premium}}$$

$$P = B \times R$$

Example 1 Finding the Annual Fire Insurance Premium

Flowers For You is in a building with a class B rating. The territory is rated 3. Find the annual premium to insure the building that is worth $378,000 and has contents valued at $92,000.

Solution From Table 7.1, the rates per $100 for a class B building in area 3 are $0.54 for the building and $0.60 for the contents. The premium for the building is found as follows.

$$\text{Value of building} = \$378{,}000 \div 100 = 3780 \text{ hundreds}$$
$$\text{Rate for building (from table)} = \$0.54$$
$$\text{Premium for building} = \text{Value (in hundreds)} \times \text{Rate}$$
$$= 3780 \times \$0.54 = \$2041.20$$

The premium for the contents can be found in the same way.

$$\text{Value of contents} = \$92{,}000 \div 100 = 920 \text{ hundreds}$$
$$\text{Rate for contents} = \$0.60 \text{ (from table)}$$
$$\text{Premium for contents} = \text{Value (in hundreds)} \times \text{Rate}$$
$$= 920 \times \$0.60 = \$552$$
$$\text{Total premium} = \$2041.20 \text{ (building)} + \$552 \text{ (contents)}$$
$$= \$2593.20 \text{ (building and contents)}$$
$$= \$2593 \text{ (rounded)}$$

NOTE Fire insurance premiums are rounded to the nearest dollar.

Objective 4 *Calculate Short-Term Rates and Cancellations.* Insurance is sometimes purchased for part of a year, perhaps even for just a few months. Perhaps only a short period of time re-

Table 7.1 ANNUAL RATES FOR EACH $100 OF FIRE INSURANCE

	Building Rating: A		Building Rating: B		Building Rating: C	
Territorial Rating	Building	Contents	Building	Contents	Building	Contents
1	$0.25	$0.32	$0.36	$0.49	$0.45	$0.60
2	$0.30	$0.44	$0.45	$0.55	$0.54	$0.75
3	$0.37	$0.46	$0.54	$0.60	$0.63	$0.80
4	$0.50	$0.52	$0.75	$0.77	$0.84	$0.90
5	$0.62	$0.58	$0.92	$0.99	$1.14	$1.05

Table 7.2 SHORT-TERM RATE SCHEDULE

Time in Months	Percent of Annual Premium	Time in Months	Percent of Annual Premium
1	18%	7	75%
2	35%	8	80%
3	45%	9	85%
4	55%	10	90%
5	65%	11	95%
6	70%	12	100%

mains on a lease. Also, if a business is sold or the owner wishes to change insurance carriers during the period of a policy, the existing policy must be canceled. In each of these cases, the insurance company will charge a **short-term** or **cancellation rate**. When the short-term rate is used, a penalty results.

As shown in Table 7.2, one month's insurance costs 18% of an annual premium while one month is only $8\frac{1}{3}\%$ of a year ($1 \div 12 = 0.083333$). The premium for a 6-month policy or a policy canceled after 6 months costs 70% of the annual premium.

Example 2 Determining Short-Term Rates

Bob Garrett sold his Irving, Texas, grocery store. Because of the sale he canceled his fire insurance after 4 months. The annual premium was $4680. Use the short-term rate schedule (Table 7.2) to find the amount of refund to the insured.

Solution The short-term rate for 4 months is 55% of the annual premium.

$$\$4680 \text{ (annual premium)} \times 0.55 = \$2574 \text{ (premium for 4 months)}$$

The refund is found by subtracting the 4-month premium from the annual premium.

$$\$4680 \text{ (annual premium)} - \$2574 \text{ (4-month premium)}$$
$$= \$2106 \text{ (refund)}$$

Objective 5 *Calculate Prorated Insurance Premium Cancellations.* Occasionally an insurance company may cancel an insurance policy. This is normally the result of fraud on the part of the insured or any violation of the insurance agreement with the insurance company.

When the insurance company initiates a policy cancellation, the insured is not penalized as with the short-term or cancellation rate. Instead, the insured is charged only for the exact amount of time that the insurance was in force. It is normal for this proration to be to the exact day. Here, we will prorate on a monthly basis, which results in the insured paying only for the number of months that the insurance was provided.

Example 3 Calculating Prorated Insurance Cancellations

Your Creations Art Supplies had a fire insurance policy with an annual premium of $2832. Because the insured was in violation of fire codes, the insurance company canceled the policy after 7 months and prorated the premium. Find (a) the amount of the premium retained by the insurance company and (b) the amount of refund to the insured.

Solution Since the cancellation is after 7 months, the insured is charged for $\frac{7}{12}$ of the year.

(a) The amount of the premium retained by the company is found by multiplying the annual premium by $\frac{7}{12}$.

$$\$2832 \text{ (annual premium)} \times \frac{7}{12} = \$1652 \text{ (premium for seven months)}$$

(b) The refund is found by subtracting the premium for 7 months from the annual premium.

$$\$2832 \text{ (annual premium)} - \$1652 \text{ (7-month premium)} = \$1180 \text{ (refund)}$$

The refund is equal to $\frac{5}{12}$ of the annual premium.

Objective 6 ***Use the Coinsurance Formula to Solve Problems.*** Most fires damage only a portion of a building and its contents. Since complete destruction of a building is rare, many owners save money by buying insurance for only a portion of the value of the building and contents. Realizing this, insurance companies place a **coinsurance clause** in fire insurance policies. With coinsurance, part of the risk of fire, under certain conditions, is assumed by the business firm taking out the insurance. For example, an 80% coinsurance clause provides that for full protection, the amount of insurance taken out must be at least 80% of the replacement cost of the building and contents insured.

If the amount of insurance is less than 80% of the replacement cost, the insurance company pays only a portion of any loss. For example, if a business firm took out insurance with a face value of only 40% of the replacement cost of the building insured and then had a loss, the insurance company would pay only half the loss, since 40% is half of 80%.

Use the following coinsurance formula to find the portion of a loss that will be paid by the insurance company.

$$\text{Amount insurance companies will pay (assuming 80\% coinsurance)}$$
$$= \text{Amount of loss} \times \frac{\text{Amount of policy}}{\text{80\% of replacement cost}}$$

NOTE The company will never pay more than the face value of the policy, nor will the company pay more than the amount of the loss. ■

Example 4 Using the Coinsurance Formula

Helen Dale owns an industrial building valued at $760,000. Her fire insurance policy (with an 80% coinsurance clause) has a face value of $570,000. The building suffers a fire loss of $144,000. Find the amount of the loss that the insurance company will pay and the amount that Dale must pay.

Solution The policy should have been for at least 80% of the value of the building, or

$$0.80 \times \$760{,}000 = \$608{,}000.$$

Since the face value of the policy is less than 80% of the value of the building, the company will pay only a portion of the loss. Use the coinsurance formula.

$$\text{Amount insurance company pays} = \$144{,}000 \times \frac{\$570{,}000}{\$608{,}000} = \$135{,}000$$

The company will pay $135,000 toward the loss, and Dale must pay the additional $9000 ($144,000 − $135,000).

Scientific *Calculator Approach*

The calculator solution to Example 4 uses chain calculations and parentheses to set off the denominator. The result is then subtracted from the fire loss.

144000 [×] 570000 [÷] [(] 80 [%] [×] 760000 [)] [=] 135000

144000 [−] 135000 [=] 9000

Example 5 Finding the Amount of Loss Paid by the Insurance Company

A Swedish investment group owns a warehouse valued at $3,450,000. The company has a fire insurance policy with a face value of $2,950,000. The policy has an 80% coinsurance feature. If the firm has a fire loss of $233,500, find the part of the loss paid by the insurance company.

Solution The value of the warehouse is $3,450,000. Take 80% of this value.

$$0.80 \times \$3{,}450{,}000 = \$2{,}760{,}000$$

The business has a fire insurance policy with a face value of more than 80% of the value of the store. Therefore, the insurance company will pay the entire $233,500 loss.

Objective 7 ***Find the Insurance Liability When There Are Multiple Carriers.*** A business may have fire insurance policies with several companies at the same time. Perhaps additional insurance coverage was purchased over a period of time, as new additions were made to a factory or building complex. Or perhaps the value of the building is so high that no one insurance company wants to take the entire risk by itself, so several companies each agree to take a portion of the insurance coverage and thereby share the risk. In either event, the insurance coverage is divided among **multiple carriers**. When an insurance claim is made against multiple carriers, each insurance company pays its fractional portion of the total claim on the property equal to its prorated share of the total coverage.

Example 6 Understanding Multiple Carrier Insurance

World Recycling Conglomerate (WRC) has an insured loss of $1,800,000 while having insurance coverage greater than its coinsurance requirement. The insurance is divided among Company A with $5,900,000 coverage, Company B with $4,425,000 coverage, and Company C with $1,475,000 coverage. Find the amount of the loss paid by each of the insurance companies.

Solution Start by finding the total face value of all three policies.

$$\$5{,}900{,}000 + \$4{,}425{,}000 + \$1{,}475{,}000 = \$11{,}800{,}000 \text{ total face value}$$

$$\$5{,}900{,}000 \qquad \text{Company A pays } \frac{\$5{,}900{,}000}{\$11{,}800{,}000} = \frac{1}{2} \text{ of the loss}$$

$$\$4{,}425{,}000 \qquad \text{Company B pays } \frac{\$4{,}425{,}000}{\$11{,}800{,}000} = \frac{3}{8} \text{ of the loss}$$

$$+\ \$1{,}475{,}000 \qquad \text{Company C pays } \frac{\$1{,}475{,}000}{\$11{,}800{,}000} = \frac{1}{8} \text{ of the loss}$$

$$\$11{,}800{,}000 \text{ total face value}$$

Since the insured loss is $1,800,000 the amount paid by each of the multiple carriers follows.

$$\text{Company A} \qquad \frac{1}{2} \times \$1{,}800{,}000 = \$900{,}000$$

$$\text{Company B} \qquad \frac{3}{8} \times \$1{,}800{,}000 = \$675{,}000$$

$$\text{Company C} \qquad \frac{1}{8} \times \$1{,}800{,}000 = \underline{\$225{,}000}$$

$$\text{Total loss} = \$1{,}800{,}000$$

Objective 8 ***Find Liability of Multiple Carriers When Coinsurance Requirement Is Not Met.*** If the coinsurance requirement is not met, the total amount of the loss paid by the insurance coverage is found, and then the amount that each of the carriers pays is found by the method shown in Example 7.

Example 7 Finding Liability of Multiple Carriers When Coinsurance Requirements Are Not Met

The Carpet Solution warehouse is valued at $2,000,000 and is insured under an 80% coinsurance clause for $1,200,000. The insurance consists of an $800,000 policy with Company A and a $400,000 policy with Company B. If the warehouse suffers a loss of $200,000, find (a) the part of any loss that is covered, (b) the amount of the loss the insurance companies will pay, (c) each insurance company's portion of the $200,000 loss, and (d) the amount paid by the insured.

Solution

(a) First, find the amount of insurance needed to satisfy the 80% coinsurance clause.

$$0.80 \times \$2{,}000{,}000 = \$1{,}600{,}000$$

Since the face value of the policy ($1,200,000) is less than 80% ($1,600,000), the insurance companies will only pay a portion of the loss.

$$\text{Part insurance companies pay} = \frac{\$1{,}200{,}000}{\$1{,}600{,}000}$$

(b) Use the coinsurance formula to find the amount of the loss that the insurance companies will pay.

$$\text{Amount insurance companies pay} = \$200{,}000 \times \frac{\$1{,}200{,}000}{\$1{,}600{,}000} = \$150{,}000$$

(c) The total face value of the insurance is $1,200,000. Since the amount of the loss that the insurance companies will pay is $150,000, the amount paid by each of the multiple carriers is as follows.

$$\frac{\$800{,}000}{\$1{,}200{,}000} \times \$150{,}000 = \$100{,}000 \quad \text{Company A}$$

$$\frac{\$400{,}000}{\$1{,}200{,}000} \times \$150{,}000 = \underline{\$50{,}000} \quad \text{Company B}$$

$$\$150{,}000 \quad \text{Amount of loss paid}$$

(d) The Carpet Solution must pay $50,000, the difference between the loss and the amount paid by the insurance companies.

Objective 9 ***List Additional Risks Against Which a Business May Be Insured.*** There are many types of insurance coverage that a business might want. One of the most common is liability coverage, which protects against monetary awards from personal-injury lawsuits caused by the business; another common coverage protects property against damage caused by windstorm, hail, or fire. Homeowners usually buy a **homeowner's policy**, which protects against these losses and many others, including all credit cards and automated teller cards, business property brought home, and medical costs for guests who are injured. Other policies are designed for condominium owners, rental property owners, and apartment dwellers. Many types of additional coverage are available to give complete and comprehensive insurance coverage.

A business owner's package policy, known in the insurance industry as a **special multiperils policy** or **SMP**, typically includes coverage of the following:

- Replacement cost for the building and contents
- Contents coverage that provides for a 25% peak-season increase
- Business property that is in transit or temporarily away from the premises
- Money, securities, accounts receivable, and other valuable papers up to $1000
- Loss of income, including coverage for rents and interruption of business for up to 12 months
- Liability and medical coverage resulting from personal injury, advertising injury, and medical malpractice

In addition to coverage of these standard risks, a list of optional coverages is also available. The businessperson may select those which he or she feels are necessary. A few of these are:

- Replacement cost coverage on exterior signs
- Replacement cost coverage for glass
- Minicomputer coverage
- Coverage for loss of refrigeration

- Professional liability coverage for barbers, beauticians, pharmacists, hearing aid sellers, morticians, optometrists, and veterinarians
- Coverage for nonowned and hired automobiles
- Liquor liablity coverage

In addition to the many coverages for business property and personal liability, the employer may be required to provide **worker's compensation insurance** for employees, which provides payments to an employee who is unable to work because of a job-related injury or illness.

An employer may pay the entire premium or part of the premium for employee health insurance, dental insurance, and group life insurance. Most often these **group insurance plans** offer slightly reduced premiums to those participating in the plan. Participation in group insurance plans is sometimes an incentive for remaining with an employer, since changing jobs may eliminate participation in the group insurance plan.

7.1 Exercises

Find the total annual premium for fire insurance for each of the following. Round to the nearest dollar. Use Table 7.1.

	Territorial Rating	Building Classification	Building Value	Contents Value
1.	4	C	$140,000	$75,000
2.	1	C	$285,000	$152,000
3.	3	A	$596,400	$206,700
4.	3	C	$220,500	$105,000
5.	5	B	$782,600	$212,000
6.	2	B	$345,700	$174,500
7.	5	C	$583,200	$221,400
8.	4	A	$850,500	$425,800

Find the amount of refund to the insured using the Short-Term Rate Schedule (Table 7.2).

	Annual Premium	Months in Force
9.	$2162	8
10.	$1382	2
11.	$807	1
12.	$964	2
13.	$1507	3
14.	$1866	10
15.	$4860	11
16.	$3760	6

Find (a) the amount of premium retained by the company and (b) the amount of refund to the insured using proration.

	Annual Premium	Months in Force
17.	$2680	9
18.	$1936	8
19.	$4375	7
20.	$876	11
21.	$5308	6
22.	$3192	3

Find the amount of each of the following losses that will be paid by the insurance company. Assume that each policy includes an 80% coinsurance clause.

	Value of Building	Face Value of Policy	Amount of Loss
23.	$277,000	$223,500	$19,850
24.	$780,000	$585,000	$10,400
25.	$78,500	$47,500	$1,500
26.	$750,000	$500,000	$56,000
27.	$218,500	$195,000	$36,500
28.	$124,800	$80,000	$25,000
29.	$147,850	$100,000	$14,850
30.	$285,000	$150,000	$18,500

Find the amount paid by each insurance company in the following problems involving multiple carriers. Assume that the coinsurance requirement is met. Round all answers to the nearest dollar.

	Insurance Loss	Companies and Coverage	
31.	$80,000	Company A	$750,000
		Company B	$250,000
32.	$360,000	Company 1	$1,200,000
		Company 2	$800,000
33.	$650,000	Company 1	$1,350,000
		Company 2	$1,200,000
		Company 3	$450,000
34.	$1,600,000	Company A	$4,800,000
		Company B	$800,000
		Company C	$2,400,000

Find (a) the amount of the loss paid by the insurance companies, (b) each insurance company's payment, and (c) the amount paid by the insured in each of the following problems involving coinsurance and multiple carriers. Assume an 80% coinsurance clause and round all answers to the nearest dollar.

	Property Value	Insurance Loss	Companies and Coverage	
35.	$90,000	$36,000	Company A	$35,000
			Company B	$25,000
36.	$160,000	$70,000	Company A	$90,000
			Company B	$30,000
37.	$250,000	$20,000	Company 1	$75,000
			Company 2	$50,000
38.	$480,000	$100,000	Company 1	$180,000
			Company 2	$60,000

Find the annual fire insurance premium for each of the following application problems. Round to the nearest dollar. Use Table 7.1.

39. FIRE-INSURANCE PREMIUM Stephanie Wetherbee owns a class B building worth $165,400. Contents are valued at $128,000. The territorial rating is 3.

40. INDUSTRIAL FIRE INSURANCE Valley Crop Dusting owns a class B building worth $107,500. Contents are worth $39,800. The territorial rating is 2.

41. RESTAURANT FIRE INSURANCE The Rocklin Grill owns a building worth $84,000. The contents are worth $18,500. The building is class B, with a territorial rating of 1.

42. INDUSTRIAL BUILDING INSURANCE London's Dredging Equipment is in a C-rated building with a territorial rating of 4. The building is worth $105,000 and the contents are worth $682,000.

Find the amount of refund to the insured using the Short-Term Rate Schedule (Table 7.2) in each of the following.

43. REAL ESTATE OFFICE Re Max Realty pays an annual fire insurance premium of $2350. They transfer insurance companies after 4 months.

44. JOB PRINTING Postal Printers pays an annual fire insurance premium of $1960. The business is sold and insurance canceled after 6 months.

45. RANCH SUPPLY Minkner Ranch Supply cancels their fire insurance after 9 months. Their annual premium is $2750.

46. HORSE STABLES Martinez Horse Stables pays an annual fire insurance premium of $3960. They change insurance companies after 2 months.

Find (a) the amount of premium retained by the company and (b) the amount of refund to the insured using proration.

47. CONSTRUCTION BUSINESS West Construction pays an annual fire insurance premium of $2670. The insurance company cancels the policy after 5 months.

48. SPORTING GOODS The Sports Center has had their fire insurance canceled after 10 months. Their annual premium is $3380.

49. DRUG STORE As the result of a recent claim, Buy-Rite Drug Store has had their fire insurance policy canceled after 7 months. Their annual premium is $1944.

50. COFFEE SHOP Java City Coffee pays an annual fire insurance premium of $4270. The insurance company cancels the policy after 3 months.

51. Describe three factors that determine the premium charged for fire insurance. (See Objective 2.)

52. Explain the coinsurance clause and describe how coinsurance works. (See Objective 6.)

In each of the following application problems, find the amount of the loss paid by (a) the insurance company and (b) the insured. Assume an 80% coinsurance requirement.

53. GIFT SHOP FIRE LOSS Indonesian Wonder Gift Shop has a value of $395,000. The shop is insured for $280,000. Fire loss is $22,500.

54. WELDING FIRE LOSS Flashpoint Welding Supplies owns a building valued at $540,000 and is insured for $308,000. Fire loss is $34,000.

55. SALVATION ARMY LOSS The main office of the Salvation Army suffers a loss from fire of $45,000. The building is valued at $550,000 and is insured for $300,000.

56. APARTMENT FIRE LOSS Kathy Stephenson owns rental units valued at $185,000 and they are insured for $111,000. Fire loss is $28,000.

57. Explain in your own words multiple carrier insurance. Give two reasons for dividing insurance among multiple carriers. (See Objective 7.)

58. Several types of insurance coverage beyond basic fire coverage are included in a homeowner's policy. List and explain three losses that would be covered. (See Objective 9.)

In each of the following, find the amount paid by each of the multiple carriers. Assume that the coinsurance requirement has been met and round to the nearest dollar.

59. COINSURED FIRE LOSS Camp Curry Stable had an insured fire loss of $548,000. It had insurance coverage as follows: Company A, $600,000; Company B, $400,000; and Company C, $200,000.

60. FIRE LOSS The Cycle Center had an insured fire loss of $68,500. They had insurance as follows: Company 1, $60,000; Company 2, $40,000; and Company 3, $30,000.

In each of the following application problems, find (a) the amount that the insured would receive and (b) the amount that each of the insurance companies would pay. Round to the nearest dollar.

61. AUTO DEALERSHIP The John L. Sullivan Chevrolet dealership is valued at $4,000,000, and the fire insurance policies contain an 80% coinsurance clause. The fire policies include $1,800,000 with Company A and $600,000 with Company B. The dealership suffers a $500,000 fire loss.

62. MANUFACTURING COMPANY The fire insurance policies on the Global Manufacturing Company contain an 80% coinsurance clause and the warehouse is valued at $2,400,000. Fire policies on the warehouse are $900,000 with Company A and $300,000 with Company B. Global Manufacturing has a fire loss of $800,000.

63. RESTAURANT FIRE LOSS Jack Pritchard's Steak House is valued at $360,000. The fire policies are $100,000 with Company 1, $50,000 with Company 2, and $30,000 with Company 3, while each contains an 80% coinsurance clause. There is a fire at the steakhouse causing a $120,000 loss.

64. INDUSTRIAL FIRE LOSS The main foundry of Delta Steel is appraised at $5,500,000. Fire insurance policies on the plant are $1,500,000 with Company 1, $1,000,000 with Company 2, and $800,000 with Company 3. The policies contain an 80% coinsurance clause and the foundry suffers a $1,200,000 fire loss.

7.2 Motor Vehicle Insurance

Objectives

1. *Describe the factors that affect the cost of motor vehicle insurance.*
2. *Define liability insurance and determine the premium.*
3. *Define property damage insurance and determine the premium.*
4. *Describe comprehensive and collision insurance and determine the premium.*
5. *Define no-fault and uninsured motorist insurance.*
6. *Apply youthful operator factors.*
7. *Calculate a motor vehicle insurance premium.*
8. *Find the amounts paid by the insurance company and the insured.*

Objective 1 *Describe the Factors That Affect the Cost of Motor Vehicle Insurance.* Automobile accidents increase in number each year, and the average cost of repairing a motor vehicle after an accident continues to rise dramatically. Businesses and individuals buy motor vehicle insurance to protect against the possible large cost of an accident. The cost of this insurance, the **premium**, is determined by people called **actuaries**, who classify accidents according to location, age and sex of the drivers, and other factors. Insurance companies use these results to determine the premiums. For example, there are more accidents in heavily populated cities than in rural areas. Certain makes and models of automobiles are stolen more often than others. Young male drivers (16–25 years of age) are involved in many more accidents than they should be, considering their proportion of the population. The more expensive a vehicle and the newer a vehicle, the more it costs to repair. These are several of the factors that determine the cost of motor-vehicle insurance.

Objective 2 *Define Liability Insurance and Determine the Premium.* **Liability** or **bodily injury insurance** protects the insured in case he or she injures someone with a car. Many states have minimum amounts of liability insurance coverage set by law. The amount of liability insurance is expressed as two numbers with a slash between them, such as 15/30. The numbers 15/30 mean that the insurance company will pay up to $15,000 for injury to one person, and a maximum of $30,000 for all persons injured in the same accident. Table 7.3 shows typical premium rates for various amounts of liability coverage.

Table 7.3 ANNUAL LIABILITY AND MEDICAL INSURANCE PREMIUMS FOR VARIOUS TYPES OF COVERAGE

Territory	15/30 $1000	25/50 $2000	50/100 $3000	100/300 $5000	250/500 $10,000
1	$207	$222	$253	$282	$308
2	148	156	168	196	198
3	310	314	375	398	459
4	216	218	253	284	310

The following graph shows the number of pedestrians who were killed or injured by automobiles in the United States in a 1-year period. Liability insurance protects the insured for these types of accidents.

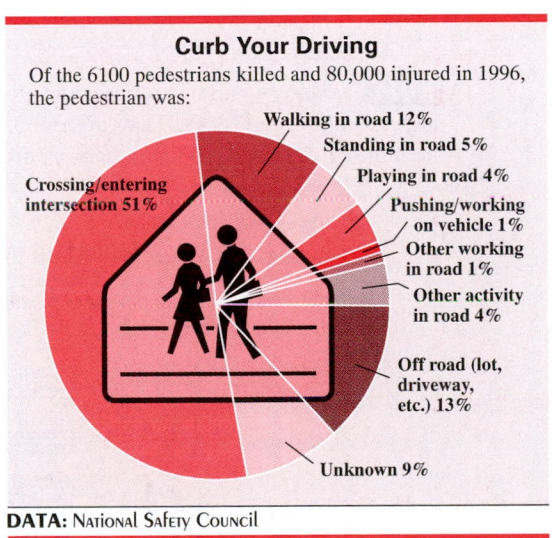

Curb Your Driving

Of the 6100 pedestrians killed and 80,000 injured in 1996, the pedestrian was:

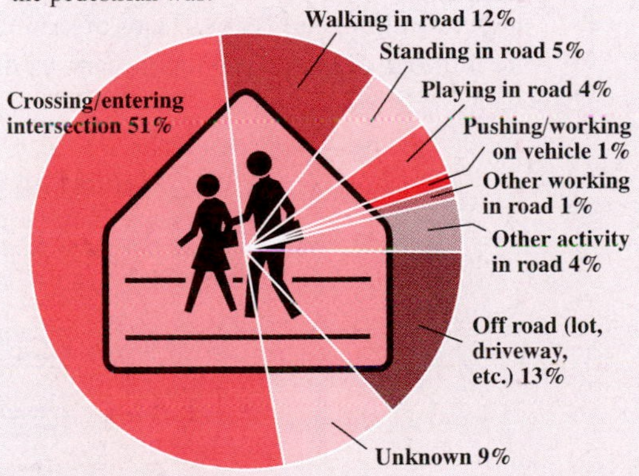

DATA: National Safety Council

Medical insurance is included in the cost of the liability insurance provided to the driver and passengers of a vehicle in case of injury. For example, the column of the table headed "15/30" shows that the insured can also receive reimbursement for up to $1000 of his or her own medical expenses in an accident.

For purposes of setting premiums, insurance companies divide the nation into territories, as many as thirty or more. These territories are established on the basis of population, the number of motor vehicles, and the number of accidents and other claims within the territory. Four territories are shown in Table 7.3. All tables in this section show annual premiums.

Example 1 Finding the Liability and Medical Premium

Zallia Todd, owner of Flowers For You, is in territory 2 and wants 100/300 liability coverage. Find the amount of the premium for this coverage and the amount of medical coverage included.

Solution Look up territory 2 in Table 7.3 and 100/300 coverage to find a premium of $196, which includes $5000 of medical coverage.

Objective 3 *Define Property Damage Insurance and Determine the Premium.* **Property damage insurance** pays for damages caused to another vehicle or other property. Table 7.4 shows the premiums for property damage insurance. The coverage amount is the maximum amount that the insurance company will pay. If a claim for damages exceeds this maximum amount, the insured must pay the excess.

Example 2 Finding the Premium for Property Damage

Zallia Todd, in territory 2, wants property damage coverage of $50,000. Find the premium.

Solution Property damage coverage of $50,000 in territory 2 requires a premium of $76, as Table 7.4 shows.

NOTE Many insurance companies give discounts on insurance premiums. You must be certain that you receive any discounts you are entitled to. The following graph shows the number of states requiring various discounts on insurance premiums.

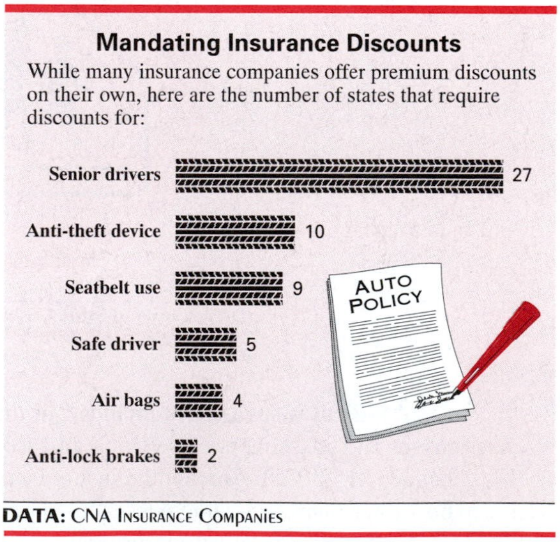

Table 7.4 PROPERTY DAMAGE INSURANCE

Territory	$10,000	$25,000	$50,000	$100,000
1	$88	$93	$97	$103
2	64	69	76	84
3	129	134	145	158
4	86	101	112	124

Table 7.5 COMPREHENSIVE AND COLLISION INSURANCE

		Comprehensive			Collision ($250 Deductible)		
Territory	Age Group	6	7	8	6	7	8
1	1	$58	$64	$90	$153	$165	$184
	2–3	50	56	82	135	147	171
	4–5	44	52	76	116	128	147
	6	34	44	64	92	110	128
2	1	$26	$28	$40	$89	$95	$104
	2–3	22	24	36	80	86	98
	4–5	20	24	34	71	77	86
	6	16	20	28	60	68	77
3	1	$70	$78	$108	$145	$157	$174
	2–3	60	66	90	128	139	162
	4–5	52	64	92	111	122	139
	6	42	52	78	88	105	122
4	1	$42	$46	$66	$97	$104	$124
	2–3	36	40	58	87	94	107
	4–5	32	38	54	77	84	94
	6	26	32	46	64	74	84

Objective 4 *Describe Comprehensive and Collision Insurance and Determine the Premium.* **Comprehensive insurance** pays for damage to the insured's vehicle caused by fire, theft, vandalism, falling trees, and other such events. **Collision insurance** pays for repairs to the insured's vehicle in case of an accident. Collision insurance often includes a **deductible**. The deductible is paid by the insured in the event of a claim, with the insurance company paying all amounts above the deductible. Common deductible amounts are $100, $250, and in some cases $500 or $1000. For example, if the cost of repairing damage caused by an accident is $1045 and the deductible amount is $250, the insured pays $250, and the insurance company pays $795 ($1045 − $250 = $795).

NOTE The higher the deductible amount, the lower the cost of the collision coverage. The insured shares a greater portion of the risk as the deductible amount increases.

Table 7.5 shows some typical rates for comprehensive and collision insurance. Rates are determined not only by territories, but also by age group and cost category of the vehicle. Age group 1 is a vehicle that is 1 year old or less and age group 6 is a vehicle 6 years old or older.

NOTE A Ford Escort might be a category 6 and a Lincoln Continental might be a category 8. The collision coverage in Table 7.5 is for $250 deductible coverage.

Example 3 Finding the Comprehensive and Collision Premium

Zallia Todd, owner of Flowers For You, is in territory 2 and has a 2-year-old category 8 minivan. Use Table 7.5 to find the annual premium for (a) comprehensive coverage and (b) collision coverage.

Solution

(a) The annual premium for comprehensive coverage is $36.
(b) The annual premium for collision coverage is $98.

Top Metro Areas for Vehicle Thefts

1. Jersey City, NJ
2. Fresno, CA
3. Miami, FL
4. Memphis, TN
5. New York City, NY
6. Phoenix-Mesa, AZ
7. New Orleans, LA
8. Tucson, AZ
9. Pine Bluff, AK
10. Los Angeles-Long Beach, CA
11. Sacramento, CA
12. Detroit, MI
13. Stockton-Lodi, CA
14. Albuquerque, NM
15. Jackson, MS

Data: National Insurance Crime Bureau, 1996.

Objective 5 *Define No-Fault and Uninsured Motorist Insurance.* With **no-fault insurance**, the insured is reimbursed for medical expenses and all costs associated with an accident by his or her own insurance company, no matter who caused the accident. Any damages for pain and suffering are eliminated except in cases of permanent injury or death. Insurance companies argue that no-fault insurance removes lawyers and the courts and results in easier and less expensive settlements. On the other hand, trial lawyers and some consumer groups contend that no-fault insurance leaves accident victims unable to recover all of their damages and unprotected from the abuses of some insurance companies.

In the states that do not have no-fault insurance a driver must be concerned about an accident with an uninsured driver. Drivers in these states need **uninsured motorist insurance**, which protects the vehicle owner in a collision with a vehicle that is not insured. Some insurance companies offer **underinsured motorist insurance**, which provides protection in the event that there is a collision with a vehicle that is underinsured. Typical costs for uninsured motorist insurance are shown in Table 7.6.

Table 7.6 UNINSURED MOTORIST INSURANCE

Territory	Basic Limit
1	$66
2	44
3	76
4	70

Example 4 Determining the Premium for Uninsured Motorist Coverage

Zallia Todd, living in territory 2, wants uninsured motorist coverage. Find the premium in Table 7.6.

Solution The annual premium for uninsured motorist coverage in territory 2 is $44.

Objective 6 *Apply Youthful-Operator Factors.* The graph below helps to explain why most insurance companies distinguish between **youthful** and **adult operators**. Although the age at which a driver becomes an adult varies from company to company, drivers of age 25 or less are usually considered youthful drivers and drivers over 25 are considered adults. Due to the higher proportion of accidents in the 25-and-under bracket, insurance companies add an additional amount to the insurance premium. In Table 7.7, there are two categories of youthful drivers, age 20 or less, and age 21–25. Consideration is also given to the youthful operator who has had driver's training. Some companies give discounts to youthful drivers who are "good students" (a B average or better). To use the youthful operator table, first determine the premium for all coverage desired and then multiply this premium by the appropriate youthful-operator factor to find the total premium.

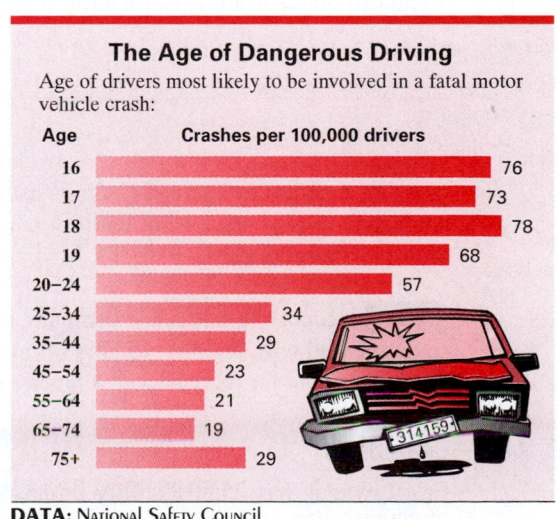

Table 7.7 YOUTHFUL-OPERATOR FACTOR

Age	With Driver's Training	Without Driver's Training
20 or less	1.55	1.75
21–25	1.15	1.40

Objective 7 *Calculate a Motor Vehicle Insurance Premium.* The total annual insurance premium is found by adding the costs of each type of insurance coverage.

Example 5 Using the Youthful-Operator Factor

James Ito lives in territory 4, is 22 years old, has had driver's training, and drives a 5-year-old car in category 7. He wants a 25/50 liability policy, $10,000 property damage coverage, a comprehensive and collision policy, and uninsured motorist coverage. Find his annual insurance premium.

Solution As shown in Table 7.3, his annual premium for 25/50 liability insurance is $218. In Table 7.4, his annual premium for $10,000 property damage coverage is $86. In Table 7.5, comprehensive insurance costs $38 and the premium for collision is $84. Uninsured motorist insurance from Table 7.6 is $70. The youthful-operator factor for a 22 year old with driver's training from Table 7.7 is 1.15. First add the premiums from the various tables.

$$\$218 + \$86 + \$38 + \$84 + \$70 = \$496$$

Then multiply by the youthful-operator factor of 1.15, found in Table 7.7.

$$\text{Total premium} = \$496 \times 1.15 = \$570.40$$

Scientific *Calculator Approach*

The calculator solution to Example 5 uses parentheses and chain calculations.

[( ] 218 [+] 86 [+] 38 [+] 84 [+] 70 [ )] [×] 1.15 [=] 570.4

Objective 8 *Find the Amounts Paid by the Insurance Company and the Insured.* The cost of increasing insurance coverage limits is usually quite small. For example, in Table 7.3 the additional cost of increasing liability coverage in territory 1 from 50/100 to 100/300 is only $29 per year ($282 − $253). Medical coverage would also be increased.

NOTE Since the insurance company pays only to the maximum amount of insurance coverage and with the driver liable for all additional amounts, many people pay an additional premium for increased coverage. ■

Example 6 Finding the Amounts Paid by the Insurance Company and the Insured

Eric Liwanag has 25/50 liability limits, $25,000 property damage limits, and $250 deductible collision insurance. While on vacation he was at fault in an accident that caused $5800 damage to his car and $3380 in damage to another car and resulted in severe injuries to the other driver and her passenger. A subsequent lawsuit for injuries resulted in a judgment of $45,000 and $35,000, respectively, to the other parties. Find the amounts that the insurance company will pay for (a) repairing Liwanag's car, (b) repairing the other car, and (c) paying the court judgment resulting from the lawsuit. (d) How much will Liwanag have to pay to the injured parties?

Solution

(a) The insurance company will pay $5550 ($5800 − $250 deductible) to repair Liwanag's car.
(b) Repairs on the other car will be paid to the property damage limits; here, the total repairs of $3380 are paid.
(c) Since more than one person was injured, the insurance company pays the limit of $50,000 ($25,000 to each of the two injured parties).
(d) Liwanag is liable for $30,000 ($80,000 − $50,000), the amount awarded over the insurance limits ($45,000− $25,000 = $20,000 and $35,000 − $25,000 = $10,000.)

Additional factors may affect the annual premium for motor vehicle insurance, such as whether the vehicle is used for pleasure, as transportation to and from work or for business purposes, how far the vehicle is driven each year, whether the youthful driver is male or female, the marital status of the male youthful driver, the past driving record of the driver, and whether the driver has more than one car insured with the insurance company. Quite often, discounts are given to nonsmokers and good students. There are also discounts for automobiles equipped with air bags and antilock braking systems. Many insurance companies charge an annual policy fee, which covers the cost of processing the policy each year.

Where is the automobile accident death rate the highest? The number of driving deaths per 100,000 drivers is 16.4 in the United States. The following graph shows that there are some European countries that have much higher death rates than the United States.

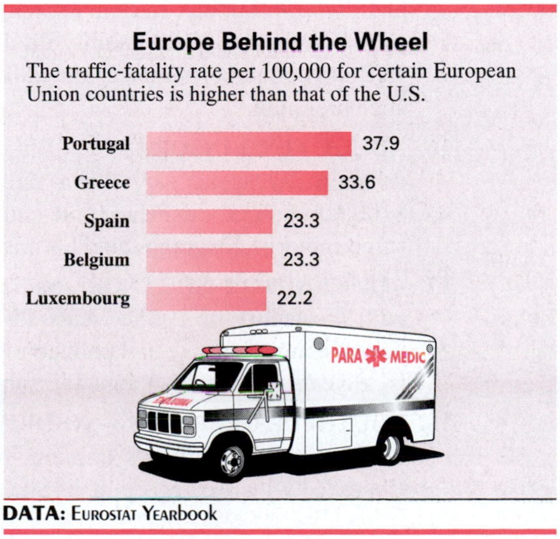
Europe Behind the Wheel
The traffic-fatality rate per 100,000 for certain European Union countries is higher than that of the U.S.
Portugal 37.9
Greece 33.6
Spain 23.3
Belgium 23.3
Luxembourg 22.2
PARA MEDIC
DATA: Eurostat Yearbook

7.2 Exercises

Find the annual premium for each of the following people.

	Name	Territory	Age	Driver Training	Liability	Property Damage	Comprehensive Collision Age Group	Category	Uninsured Motorist
1.	Maxey	4	35	—	100/300	$50,000	2	8	yes
2.	Morrissey	1	20	yes	25/50	$25,000	4	7	no
3.	Shraim	3	52	—	250/500	$50,000	2	8	yes
4.	Waldron	2	67	—	50/100	$100,000	1	6	yes
5.	Carter	1	35	—	100/300	$25,000	5	6	no
6.	Tao	2	24	yes	100/300	$50,000	5	7	yes
7.	Baeta	4	52	—	250/500	$50,000	1	8	yes
8.	Gualco	1	31	—	250/500	$50,000	none	none	no
9.	Harrison	3	44	—	50/100	$25,000	5	6	yes
10.	Ballinger	2	17	yes	15/30	$10,000	4	8	yes
11.	Green	2	43	—	100/300	$100,000	6	6	no
12.	Rodriquez	3	60	—	50/100	$100,000	5	7	yes

13. Describe four factors that determine the premium on an automobile insurance policy. (See Objective 1.)

14. Explain the difference between comprehensive insurance and collision insurance. (See Objective 4.)

Solve each of the following application problems.

15. YOUTHFUL-OPERATOR AUTO INSURANCE Laura Coaty is 21 years old, has had driver's training, lives in territory 1, and drives a 4-year-old car with a category of 6. She wants 50/100 liability limits, $25,000 property damage limits, comprehensive and collision insurance, and uninsured motorist coverage. Find her annual insurance premium.

16. ADULT AUTO INSURANCE Bill Poole is 47 years old, lives in territory 4, and drives a 2-year-old car with a category of 7. He wants 250/500 liability limits, $100,000 property damage limits, comprehensive and collision insurance, and uninsured motorist coverage. Find his annual insurance premium.

17. ADULT AUTO INSURANCE Sadie Simms lives in territory 3 and drives a new car with a category of 7. She wants 250/500 liability limits, $50,000 property damage limits, comprehensive and collision insurance, and uninsured motorist coverage. She is 38 years old. Find her annual insurance premium.

18. YOUTHFUL OPERATOR—NO DRIVER TRAINING Michelle Massa is 17 years old, has not had driver's training, lives in territory 2, and purchased a new Jeep Cherokee with a category of 6. She wants 50/100 liability limits, $25,000 property damage limits, comprehensive and collision insurance, and uninsured motorist coverage. Find her annual insurance premium.

19. **BODILY INJURY INSURANCE** Suppose your bodily injury policy has limits of 25/50, and you injure a person on a bicycle. The judge awards damages of $36,500 to the cyclist. (a) How much will the company pay? (b) How much will you pay?

20. **BODILY INJURY INSURANCE** Your best friend causes injury to three people and they receive damages of $50,000 each. She has a policy with bodily injury limits of 100/300. (a) How much will the company pay to each person? (b) How much must your friend pay?

21. **INSURANCE CLAIM'S PAYMENT** A reckless driver caused Leslie Silva to collide with a car in another lane. Silva had 50/100 liability limits, $25,000 property damage limits, and collision coverage with a $250 deductible. Silva's car had damage of $1878, while the other car suffered $6936 in damage. The resulting lawsuit gave injury awards of $60,000 and $55,000, respectively, to the two people in the other car. Find the amount that the insurance company will pay for (a) repairing Silva's car, (b) repairing the other car, and (c) personal injury damages. (d) How much must Silva pay beyond her insurance coverage, including the collision deductible?

22. **INSURANCE CLAIMS PAYMENT** Driving a dangerous vehicle at excessive speed caused Bob Armstrong to crash into another car. Armstrong had 15/30 liability limits, $10,000 property damage limits, and collision coverage with a $250 deductible. Damage to Armstrong's car was $2980; the other car, with a value of $22,800, was totaled. The results of a lawsuit awarded $75,000 and $45,000, respectively, in damages for personal injury to the two people in the other car. Find the amount that the insurance company will pay for (a) repairing Armstrong's car, (b) repairing the other car, and (c) personal injury damages. (d) How much must Armstrong pay beyond his insurance coverage?

23. **COMPANY/INSURED LIABILITY** An employee for Safeco Security was driving to a job when a ladder fell from the truck into the path of an oncoming car. Damage to the car was $10,250. Both the driver and the passenger suffered injuries and were given court awards of $20,000 and $40,000, respectively. The security company had 25/50 liability insurance limits and $10,000 property damage limits. (a) Find the total amount paid by the insurance company for both property damage and liability and (b) find the total amount beyond the insurance limits for which the business owner was liable.

24. **COMPANY/INSURED LIABILITY** A trailer-mounted concrete mixer being towed by a contractor broke loose on the beltway and caused a serious accident involving a car and three occupants. Damage to the car was $10,807. The driver and two passengers were given court awards for personal injury of $25,000, $35,000, and $38,000, respectively. The contractor had 25/50 liability insurance limits and $10,000 property damage limits. Find (a) the total amount paid by the insurance company for both property damage and liability, and (b) the total amount beyond the insurance limits for which the contractor was liable.

25. Explain why insurance companies charge a higher premium on auto insurance sold to a youthful operator. Do you think that this higher premium is a good idea or not? (See Objective 6.)

26. Propety damage pays for damage caused by you to the property of others. Since the average cost of a new car today is over $20,000, what amount of property damage coverage would you recommend to a friend who owns her own business? Why is this your recommendation? (See Objective 3.)

27. Read Section 3 of the foldout. (a) Name three types of insurance coverage that you should have. (b) List three suggestions to be used when shopping for insurance.

7.3 Life Insurance

Objectives

1. *List reasons for purchasing life insurance.*
2. *Define term, decreasing term, and whole life policies.*
3. *Understand universal life, variable life, limited payment, and endowment policies.*
4. *Find the annual premium for life insurance.*
5. *Use premium factors with different modes of premium payment.*
6. *Describe discounts, conditions, and additional coverage.*
7. *Understand nonforfeiture options.*
8. *Calculate income under various settlement options.*

Objective 1 *List Reasons for Purchasing Life Insurance.* Individuals buy life insurance for a variety of reasons. Most often the insured wants to provide for the needs of others in the event of early death or disability. Parents may want to guarantee that their children will have enough money for college even if the parents die. Also, some types of life insurance provide paybacks upon retirement, paybacks that allow a retired person to live better than he or she might otherwise live. Insurance money can also be used to pay off mortgages. According to the *World Almanac and Book of Facts*, the average amount of life insurance per household in the United States today is $112,400.

Life insurance is perhaps even more important for a person in business, particularly for an owner or partner in a small business. A business often takes a number of years to grow and may be the owner's main asset. The unexpected death of the owner might leave the business without proper guidance and control, and the business may suffer drastically before it can be sold. Life insurance on the partners in a business supplies the surviving partner with the necessary money to buy out a deceased partner's interest in the partnership.

Objective 2 *Define Term, Decreasing Term, and Whole Life Policies.* There are several types of life insurance policies available. The most common types are term, decreasing term, and whole life insurance.

Term insurance provides protection for a fixed length of time, such as 1 year, 5 years, or 10 years. At the end of the fixed period of time, the policy can usually be renewed for an additional period of time at a higher premium. Some term policies provide that on the expiration of the term stated in the policy, the insurance can be converted to one of the following types of insurance. Term insurance, the least expensive of the types of insurance listed, accounts for 20% of all policies and gives the greatest amount of life insurance coverage for the premium dollar. At the expiration of a term insurance policy, however, the insured receives nothing from the insurance company except a request to buy more insurance. The nation's leading causes of death are shown on p. 261.

Table 7.8 A TYPICAL DECREASING TERM INSURANCE POLICY

Age of Insured	Coverage
Under 29	$40,000
30–34	35,000
35–39	30,000
40–44	25,000
45–49	18,000
50–54	11,000
55–59	7,000
60–66	4,000
67 and over	0

Decreasing term insurance is a modification of term insurance where the insured pays a fixed premium until age 60 or 65, with the amount of life insurance decreasing periodically. This policy is designed to fit the ages and stages of life as life insurance needs change. For the person just starting out, it gives more protection for less money. A typical policy, costing $11 per month, is shown in Table 7.8. Decreasing term insurance is commonly available to employees of large companies as a fringe benefit, paid for by either the employee, the employer, or both. Most mortgage insurance policies are this type. The amount of life insurance coverage decreases as the amount of the mortgage is reduced.

Nation's Top Killers

The 10 leading causes of death in the United States, ranked according to the number of lives lost.

1. Heart disease
2. Cancer
3. Stroke
4. Lung disease
5. Accidents
6. Pneumonia and influenza
7. Diabetes
8. AIDS
9. Suicide
10. Liver disease

Data: National Center for Health Statistics, 1999.

Whole life insurance (also called **straight life** or **ordinary life insurance**) combines life insurance protection with savings. The insured pays a constant premium until death or until retirement, whichever occurs sooner. Upon retirement, monthly payments may be made by the company to the insured until his or her death.

Whole-life insurance builds up **cash value**, or money used to pay retirement benefits to the insured. Also, these cash values can be borrowed by the insured at favorable interest rates. Cash value accumulation is guaranteed by the company. The rate of interest used to calculate cash values by the company is very conservative. For this reason many consumer finance experts recommend the purchase of term insurance, with the difference in premiums between term insurance and whole life invested in a good no-load mutual fund or money market fund.

NOTE Many people plan to save the additional money they would have paid for whole-life insurance, but they neglect to do so. The advantage of whole life insurance is that regular payments are required. Remember also that term insurance has no cash value whereas whole life does. ■

The following graph shows the increase in life expectancy in the United States from 1930 to 1996.

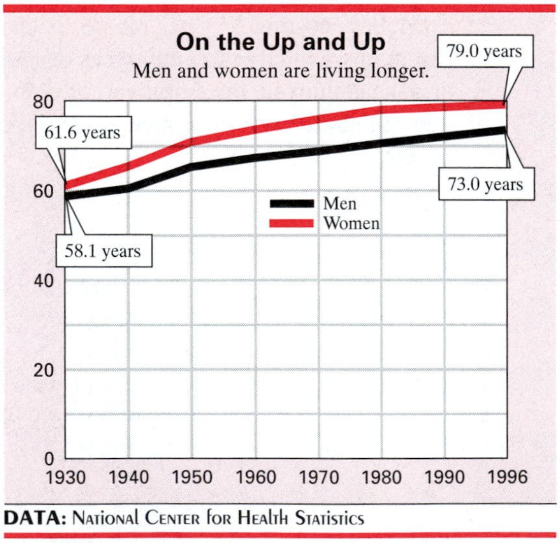

Objective 3 *Understand Universal Life, Variable Life, Limited Payment, and Endowment Policies.*

Universal and **variable life policies** provide the life insurance coverage of term insurance (high coverage per premium dollar), plus a tax-deferred way to accumulate assets and earn interest at money market rates. Unlike traditional whole life insurance policies, universal life insurance allows the insured to vary the amount of premium depending on the changing needs of the insured. A younger insured person with limited funds may want maximum insurance protection for the family. At a later date, the insured may want to begin actively building assets and may increase the premium to build cash value for retirement benefits. Universal life insurance is sensitive to interest rate changes. The portion of the premium going into retirement benefits receives money market interest rates and is usually guaranteed a minimum rate of return, regardless of what happens to market rates. The insured person profits from higher interest rates but is also protected if interest rates drop below the guaranteed rate. The idea is that returns will be greater than those given to ordinary life policyholders.

Variable life is the latest attempt to encourage sales of the insurance industry's main product, whole life insurance. It allows you to allocate your premiums among one or more separate investments that offer varying degrees of risk and reward—stocks, bonds, combinations of both, or accounts that provide for guarantees of interest and principal. Typical policies available today allow the policyholder to switch investments from one fund to another twice each year. These features coupled with some tax benefits have resulted in the variable life policy accounting for 25% and 35%, respectively, of the new policies sold in recent years by two of the largest life insurance companies.

Limited-payment life insurance is similar to ordinary life, except that premiums are paid for only a certain fixed number of years, such as 20 years. For this reason, this insurance is often called "20-pay life," representing payments of 20 years. The premium for limited-payment life is higher than for ordinary life policies. Limited-payment life is commonly used by athletes, actors, and others whose income is likely to be high for several years and then decline.

An **endowment policy** is the most expensive type of policy. These policies guarantee payment of a fixed amount of money to a given individual, whether or not the insured lives. Endowment policies might be taken out by parents to guarantee a sum of money for their children's college education. Because of the high premiums, this is one of the least popular types of policies today.

It is not always easy to decide on the best type of policy. While term insurance gives the greatest amount of insurance for each premium dollar, it pays only when the insured dies. Since term insurance costs increase rapidly with age, most people buying term insurance drop it before they die. Ordinary life insurance provides less coverage for each premium dollar in the event of death; however, it does provide a return to the insured at retirement. The only certain method for determining which policy will give the best return is for the insured to know when he or she will die, and this is not possible.

The Commissioners 1980 Standard Ordinary Table of Mortality (shown in Table 7.9) shows the number of deaths per 1000 and the remaining life expectancy in years for both males and females among the people in the United States having life insurance. Insurance companies use this **mortality table** to evaluate their life insurance reserves, from which benefits are paid. This table is updated periodically. The most recent one is shown here.

NOTE Life expectancy for women is greater than for men, so a woman pays a lower life insurance premium than does a man of the same age. Find the insurance premium for a woman by subtracting 5 years from her age before using the table of premiums. ■

Objective 4 *Find the Annual Premium for Life Insurance.* Calculation of life insurance rates and premiums uses fairly involved mathematics and is done by actuaries. The results of such calculations are published in the tables of premiums. A typical table is shown in Table 7.10.

The premium for a life insurance policy is found with the following formula.

$$\text{Annual premium} = \text{Number of thousands} \times \text{Rate per } \$1000$$

Table 7.9 COMMISSIONERS 1980 STANDARD ORDINARY TABLE OF MORTALITY SHOWING THE LIFE EXPECTANCY OF MEN AND WOMEN AT VARIOUS AGES

	Males		Females			Males		Females	
Age	Deaths per 1000	Expectation of Life (Years)	Deaths per 1000	Expectation of Life (Years)	Age	Deaths per 1000	Expectation of Life (Years)	Deaths per 1000	Expectation of Life (Years)
0	4.18	70.83	2.89	75.83	41	3.29	33.16	2.64	37.46
1	1.07	70.13	0.87	75.04	42	3.46	32.26	2.87	36.55
2	0.99	69.20	0.81	74.11	43	3.87	31.38	3.09	35.66
3	0.98	68.27	0.79	73.17	44	4.19	30.50	3.32	34.77
4	0.95	67.34	0.77	72.23	45	4.55	29.62	3.56	33.88
5	0.90	66.40	0.76	71.28	46	4.92	28.76	3.80	33.00
6	0.86	65.46	0.73	70.34	47	5.32	27.90	4.05	32.12
7	0.80	64.52	0.72	69.39	48	5.74	27.04	4.33	31.25
8	0.76	63.57	0.70	68.44	49	6.21	26.20	4.63	30.39
9	0.74	62.62	0.69	67.48	50	6.71	25.36	4.96	29.53
10	0.73	61.66	0.68	66.53	51	7.30	24.52	5.31	28.67
11	0.77	60.71	0.69	65.58	52	7.96	23.70	5.70	27.82
12	0.85	59.75	0.72	64.62	53	8.71	22.89	6.15	26.98
13	0.99	58.80	0.75	63.67	54	9.56	22.08	6.60	26.14
14	1.15	57.86	0.80	62.71	55	10.47	21.29	7.09	25.31
15	1.33	56.93	0.85	61.76	56	11.46	20.51	7.57	24.49
16	1.51	56.00	0.90	60.82	57	12.49	19.74	8.03	23.67
17	1.67	55.09	0.95	59.87	58	13.59	18.99	8.47	22.86
18	1.78	54.18	0.98	58.93	59	14.77	18.24	8.94	22.05
19	1.86	53.27	1.02	57.98	60	16.08	17.51	9.47	21.25
20	1.90	52.37	1.05	57.04	61	17.54	16.79	10.13	20.44
21	1.91	51.47	1.07	56.10	62	19.19	16.08	10.96	19.65
22	1.89	50.57	1.09	55.16	63	21.06	15.38	12.02	18.86
23	1.86	49.66	1.11	54.22	64	23.14	14.70	13.25	18.08
24	1.82	48.75	1.14	53.28	65	25.42	14.04	14.59	17.32
25	1.77	47.84	1.16	42.34	66	27.85	13.39	16.00	16.57
26	1.73	46.93	1.19	51.40	67	30.44	12.76	17.43	15.83
27	1.71	46.01	1.22	50.46	68	33.19	12.14	18.84	15.10
28	1.70	45.09	1.26	49.52	69	36.17	11.54	20.36	14.38
29	1.71	44.16	1.30	48.59	70	39.51	10.96	22.11	13.67
30	1.73	43.24	1.35	47.65	71	43.30	10.39	24.23	12.97
31	1.78	42.31	1.40	46.71	72	47.65	9.84	26.87	12.28
32	1.83	41.38	1.45	45.78	73	52.64	9.30	30.11	11.60
33	1.91	40.46	1.50	44.84	74	58.19	8.79	33.93	10.95
34	2.00	39.54	1.58	43.91	75	64.19	8.31	38.24	10.32
35	2.11	38.61	1.65	42.98	76	70.53	7.84	42.97	9.71
36	2.24	37.69	1.76	42.05	77	77.12	7.40	48.04	9.12
37	2.40	36.78	1.89	41.12	78	83.90	6.97	53.45	8.55
38	2.58	35.87	2.04	40.20	79	91.05	6.57	59.35	8.01
39	2.79	34.96	2.22	39.28	80	98.84	6.18	65.99	7.48
40	3.02	34.05	2.42	38.36	81	107.48	5.80	73.60	6.98

(continued)

Table 7.9 *(continued)* COMMISSIONERS 1980 STANDARD ORDINARY TABLE OF MORTALITY SHOWING THE LIFE EXPECTANCY OF MEN AND WOMEN AT VARIOUS AGES

	Males		Females			Males		Females	
Age	Deaths per 1000	Expectation of Life (Years)	Deaths per 1000	Expectation of Life (Years)	Age	Deaths per 1000	Expectation of Life (Years)	Deaths per 1000	Expectation of Life (Years)
82	117.25	5.44	82.40	6.49	91	236.98	2.94	208.87	3.15
83	128.26	5.09	92.53	6.03	92	253.45	2.70	228.81	2.85
84	140.25	4.77	103.81	5.59	93	272.11	2.44	251.51	2.55
85	152.95	4.46	116.10	5.18	94	295.90	2.17	279.31	2.24
86	166.09	4.18	129.29	4.80	95	329.96	1.87	317.32	1.91
87	179.55	3.91	143.32	4.43	96	384.55	1.54	375.74	1.56
88	193.27	3.66	158.18	4.09	97	480.20	1.20	474.97	1.21
89	207.29	3.41	173.94	3.77	98	657.98	0.84	655.85	0.84
90	221.77	3.18	190.75	3.45					

Data: American Council of Life Insurance, *Life Insurance Fact Book* (Washington, D.C., 1999).

Table 7.10 ANNUAL PREMIUM RATE* PER $1000 OF LIFE INSURANCE

Age	Renewable Term	Whole Life	Universal Life	20-Pay Life
20	2.28	4.07	3.48	12.30
21	2.33	4.26	3.85	12.95
22	2.39	4.37	4.10	13.72
23	2.43	4.45	4.56	14.28
24	2.52	4.68	4.80	15.95
25	2.58	5.06	5.11	16.60
30	2.97	5.66	6.08	18.78
35	3.41	7.68	7.45	21.60
40	4.15	12.67	10.62	24.26
45	4.92	19.86	15.24	28.16
50		26.23	21.46	32.59
55		31.75	28.38	38.63
60		38.42	36.72	45.74

*For women, subtract 5 years from the actual age. For example, rates for a 30-year-old woman are shown for age 25 in the table.

Example 1 Finding the Life Insurance Premium

Judith Allen, owner of Canadian Book Sales, is 40 years old and wants to buy a life insurance policy with a face value of $50,000. Use Table 7.10 to find her annual premium for (a) a renewable term policy, (b) a whole life policy, (c) a universal life policy, and (d) a 20-pay life plan.

Solution Use Table 7.10 and the life insurance premium formula. Since the table gives rates per $1000 of face value, first find the number of thousands in $50,000.

$$50{,}000 = 50 \text{ thousands}$$

(a) The rate per $1000 for a 40-year-old woman (use 40 − 5 = 35 age) for a renewable term plan is $3.41. The total annual premium is thus

$$50 \times \$3.41 = \$170.50.$$

(b) For a whole life policy, the rate is $7.68 per $1000, for a total annual premium of

$$50 \times \$7.68 = \$384.$$

This premium is higher than for renewable term insurance, since whole life insurance builds up cash values and term insurance does not.

(c) The rate for a universal life policy is $7.45 per $1000 for an annual premium of

$$50 \times \$7.45 = \$372.50.$$

This premium is higher than for renewable term but less than the premium for whole life insurance.

(d) For 20-pay life, the rate per $1000 is $21.60 for an annual premium of

$$50 \times \$21.60 = \$1080.$$

Allen would pay $1080 annually for 20 years, at which time the plan is paid up. She then has insurance protection until retirement, with retirement income thereafter.

NOTE Remember to subtract 5 years from the age of a female before using the table of premiums.

Objective 5 *Use Premium Factors with Different Modes of Premium Payment.* The annual life insurance premium is not always paid in one single payment. Many companies give the insured the option of paying the premium semiannually, quarterly, or monthly. For this convenience, the policyholder pays an additional amount, determined by a **premium factor**. Table 7.11 shows typical premium factors.

Table 7.11 PREMIUM FACTORS

Mode of Payment	Premium Factor
Semiannually	0.51
Quarterly	0.26
Monthly	0.0908

Example 2 Using a Premium Factor

The annual insurance premium for Erica Gheen is $384. Use Table 7.11 to find the amount of the premium and the total annual cost if she pays at the following periods: **(a)** semiannually, **(b)** quarterly, **(c)** monthly.

Solution

(a) The semiannual premium factor is 0.51. So, her semiannual premium is

$$\$384 \times 0.51 = \$195.84$$

The total annual cost is $391.68 ($195.84 × 2).

(b) Since the quarterly premium factor is 0.26, her quarterly premium is

$$\$384 \times 0.26 = \$99.84$$

The total annual cost is \$399.36 (\$99.84 × 4).

(c) The monthly premium factor is 0.0908, making the monthly premium

$$\$384 \times 0.0908 = \$34.87$$

The total annual cost is \$418.44 (\$34.87 × 12).

Objective 6 *Describe Discounts, Conditions, and Additional Coverage.* Many companies today offer a **nonsmokers discount** because they feel that nonsmokers are better insurance risks. Normally, not having smoked for 12 months qualifies one as a nonsmoker. Most policies also contain a **suicide clause**. This clause states that suicide is not covered, usually for the first 2 years of the policy.

Additional coverage is often available for small increases in the premium. The **accidental death benefit** coverage will pay an additional death benefit if the insured dies as the result of an accident. An optional benefit known as **waiver of premium** allows the life insurance policy to remain in force without payment of premium when the insured becomes disabled. A **guaranteed conversion privilege** lets the insured convert term insurance to any type of whole life or universal life insurance without physical examination. **Companion** or **spouse insurance** allows an insured to add a companion or spouse to a policy and results in both being insured on one policy.

Objective 7 *Understand Nonforfeiture Options.* Most insurance, such as fire, motor vehicle, and even term life insurance, protects against a specific hazardous event. The insured or the heirs of the insured collect only if the event takes place. Life insurance is different; the insured often receives benefits while living. These benefits build over the life of the policy and are available even if the insured stops paying premiums and cancels the policy. The benefits available upon cancellation are called **nonforfeiture options**. The life insurance company invests the premium dollars remaining after death benefits and operating expenses are paid. The invested money plus interest become the cash values for the insurance policyholders.

Cash value that has accrued to the insured may be received in one of several forms when the policy is canceled.

1. The policyholder may decide on a cash settlement option when canceling the insurance policy. Often the insured borrows against the cash value, leaving the policy in force as an alternative to canceling the policy.
2. The cash value of the policy may be used to buy paid-up insurance for a smaller policy amount. This paid-up insurance remains in force for the life of the insured.
3. With an extended-term option, the insured purchases insurance of the same face value for a specific period of time. The duration of this extended-term insurance depends on the amount of the cash value of the insured's policy.

Typical nonforfeiture options for a policy issued at age 25 appear in Table 7.12.

Table 7.12 NONFORFEITURE OPTIONS PER $1000/POLICY ISSUED AT AGE 25

	Whole Life				20-Pay Life				Universal Life
			Ext. Term				Ext. Term		
Years in Force	Cash Value	Paid-up Ins.	Years	Days	Cash Value	Paid-up Ins.	Years	Days	CashValue
3*	$5	$16	1	192	$29	$93	11	18	$40
5	28	84	10	190	70	228	20	96	96
10	96	258	18	112	187	554	29	115	310
15	169	415	20	312	339	789	33	215	680
20	283	579	23	315	491	1000	Life	Life	1125
25	394	637	25	130	506	1000			1495
30	491	698	26	210	523	1000			1968

*Normally, there is no cash value accrued in the first 2 years. The cash value of universal life is an estimate of the value, not a guarantee, and will vary depending on the performance of the portfolio of investments chosen by the insured.

Example 3 Determining Nonforfeiture Options

Lois Stevens purchased an $80,000 whole life insurance policy when she was 25 years old and has paid on the policy for 20 years. Determine the following values for her policy: (a) cash value, (b) the amount of paid-up insurance that she could receive, and (c) the time period for which she could have extended term insurance.

Solution

(a) The cash value found in Table 7.12 under whole life for 20 years is $283 per $1000 of insurance. The cash value of her policy is

$$\$283 \times 80 \text{ (number of \$1000s)} = \$22{,}640.$$

(b) Again, from Table 7.12, the amount of paid-up insurance that she could receive is $579 per $1000, or

$$\$579 \times 80 = \$46{,}320 \text{ paid-up insurance}$$

This coverage would remain in force until her death without paying any additional premium.

(c) The time period for which she could have extended-term insurance of the same amount, $80,000, is 23 years and 315 days.

There are two types of insurance companies—**mutual companies** and **stock companies**. The policyholders are the owners in a mutual company, with the policies called **participating policies**. The owners (policyholders) share in any profits of the company; the profits are paid in the form of dividends. If the company prospers, the policyholders receive a dividend or refund of premium. In a stock company, the stockholders are the owners of the company, with the policies called **nonparticipating policies**. If the company prospers, the stockholders, not the policyholders, receive a dividend. The distinction is important in determining the **net cost of insurance policies**.

Table 7.13 MONTHLY PAYMENTS PER $1000 OF FACE VALUE

Options 1 and 2: Fixed Amount or Fixed Number of years		Options 3 and 4: Income for Life				
		Age when Payments Begin				
Years	Amount	Male	Female	Life Annuity	Life with 10 Years Certain	Life with 15 Years Certain
10	$9.78	50	55	$4.63	$4.49	$4.38
12	8.46	55	60	5.36	5.16	4.78
14	7.63	60	65	5.94	5.73	4.91
16	6.91	65	70	6.93	6.51	5.34
18	6.07	70	75	7.86	6.93	6.28
20	5.78					

Objective 8 *Calculate Income Under Various Settlement Options.* At the death of a life insurance policyholder, the **beneficiary**, the individual chosen by the insured to receive benefits upon death, has several **settlement options** when choosing how the death benefits are to be received. In many cases the beneficiary elects to receive a single lump sum payment of the face value of the policy. In other cases, the beneficiary allows the life insurance company to invest the face value and to pay the beneficiary the proceeds and interest over a period of time in the form of an **annuity**. These are the more common options available.

1. A fixed-amount annuity may be paid each month. The monthly payments continue, including interest, until all the proceeds are used up.
2. The beneficiary may prefer the payments of a fixed-period annuity. The insurance company determines the amount that may be paid monthly, for example, for 10 years. The payment continues for exactly that period of time, even if the beneficiary dies.
3. Payments for life is another option. Based upon the age and sex of the beneficiary, the insurance company calculates an amount to be paid to the beneficiary for as long as he or she lives.
4. A last option is payments for life with a guaranteed number of years. Here, if the beneficiary dies before receiving the benefits for the guaranteed time period, the payments continue to the beneficiaries' heirs until the guarantee is satisfied. The guarantees usually range from 5 to 25 years.

Table 7.13 shows the monthly income per $1000 of insurance coverage under various settlement options.

Example 4 Finding Settlement Options

Chris Bowler is the beneficiary of a $40,000 life insurance policy. Find (a) the monthly payment if he decides to receive payments for 18 years and (b) the number of years payments will continue if he selects a monthly payment of $300.

Solution

(a) The monthly payment from Table 7.13 for 18 years is \$6.07 per \$1000 of face value. The monthly payment he receives is

$$\$6.07 \times 40 \text{ (thousands)} = \$242.80$$

(b) A monthly payment of \$300 is equivalent to

$$\frac{\$300}{40 \text{ (thousands)}} = \$7.50 \text{ per } \$1000 \text{ face value}$$

Reading the amount column under Options 1 and 2, find \$7.63 (closest to \$7.50). The \$300 payment will continue a little over 14 years.

Example 5 Finding Settlement Options

Clarence Hanks, 60 years of age, is the beneficiary of a \$20,000 life insurance policy. Find (a) his monthly payment from a life annuity and (b) his monthly payment from a life annuity with 15 years certain.

Solution

(a) Under Options 3 and 4, in Table 7.13, look up male, age 60. Look across to the life annuity column, to find \$5.94. His monthly payment for life is

$$\$5.94 \times 20 \text{ (thousands)} = \$118.80$$

(b) The monthly payment from a life annuity with 15 years certain is

$$\$4.91 \times 20 \text{ (thousands)} = \$98.20$$

7.3 Exercises

Find the annual premium, the semiannual premium, the quarterly premium, and the monthly premium for each of the following. (Note: Subtract 5 years for females.) Use Tables 7.10 and 7.11.

	Face Value of Policy	Age of Insured	Sex of Insured	Type of Policy
1.	\$30,000	60	F	Whole life
2.	\$60,000	30	M	Whole life
3.	\$35,000	40	M	20-pay life
4.	\$80,000	40	F	Universal life
5.	\$85,000	30	M	Universal life
6.	\$150,000	45	F	Renewable term
7.	\$100,000	35	F	Renewable term
8.	\$10,000	22	M	20-pay life
9.	\$50,000	45	F	Universal life

	Face Value of Policy	Age of Insured	Sex of Insured	Type of Policy
10.	$40,000	29	F	20-pay life
11.	$70,000	50	M	Whole life
12.	$60,000	35	M	Renewable term
13.	$100,000	60	F	Whole life
14.	$100,000	27	F	Universal life

15. Explain in your own words the advantages and disadvantages of buying renewable term life insurance. Would you buy renewable term life insurance for yourself? Why or why not? (See Objective 1.)

16. If you were the beneficiary of a $40,000 life insurance policy, what settlement option would you choose? Why? (See Objective 8.)

Find the nonforfeiture values of the following policies. The policies were issued at age 25. Use Table 7.12.

	Years in Force	Type of Policy	Face Value	Nonforfeiture Option
17.	10	Universal life	$50,000	Cash value
18.	15	Universal life	$75,000	Cash value
19.	30	20-pay life	$30,000	Paid-up insurance
20.	15	Whole life	$35,000	Extended term
21.	30	Universal life	$100,000	Cash value
22.	15	20-pay life	$25,000	Paid-up insurance
23.	20	Whole life	$100,000	Extended term
24.	10	Whole life	$40,000	Extended term

Find the monthly payment or period of payment under the following policy settlement options. Use Table 7.13.

	Beneficiary		Face Value	Settlement Option	Monthly Payment	
	Age	Sex			Years	Amount
25.	55	M	$50,000	Fixed amount per month	20	______
26.	65	F	$75,000	Life, 10 years certain	10	______
27.	65	M	$10,000	Fixed number of years	______	$60.70
28.	60	M	$40,000	Fixed number of years	______	$338.40
29.	70	F	$30,000	Life, 15 years certain	15	______
30.	60	F	$100,000	Fixed amount per month	10	______

Solve each of the following application problems.

31. WHOLE LIFE POLICY Linda Davis buys a $50,000 whole life policy at age 40. Her son Matthew is the beneficiary, and will collect the face value of the policy. (a) Find the annual premium. (b) How much will Matthew get if his mother dies after paying premiums for 9 years?

32. 20-PAY LIFE POLICY Luan Lee buys a $100,000, 20-pay life policy at age 45. Her son Bryan is the beneficiary and will collect the face value of the policy. (a) Find the annual premium. (b) How much will Bryan get if his mother dies after making payments for 12 years?

33. EMPLOYEE LIFE INSURANCE Ozark Steel Foundary feels that it would suffer considerable hardship if the firm's head moldmaker died suddenly. Therefore, the firm takes out a $90,000 policy on the moldmaker's life. The moldmaker is a 45-year-old woman, and the company buys a renewable term policy. Find the semiannual premium.

34. MANAGER LIFE INSURANCE City Cellular purchased a $70,000 whole life policy on the finance manager's life. If the finance manager is a 35-year-old man, find the quarterly premium.

35. 20-PAY LIFE INSURANCE Henry Hernandez takes out a 20-pay life policy with a face value of $75,000. He is 50 years old. Find the monthly premium.

36. UNIVERSAL LIFE INSURANCE Richard Gonsalves takes out a universal life policy with a face value of $50,000. He is 40 years old. Find the monthly premium.

37. PREMIUM FACTORS The annual premium for a whole life policy is $872. Using premium factors, find (a) the semiannual premium, (b) the quarterly premium, (c) the monthly premium, and (d) the total annual cost of each of the plans.

38. UNIVERSAL LIFE A universal life policy has an annual premium of $1806. Use premium factors to find (a) the semiannual premium, (b) the quarterly premium, (c) the monthly premium, and (d) the total annual cost for each of the plans.

39. NONFORFEITURE OPTIONS Catherine Konradt purchased a $20,000 whole life policy 20 years ago when she was 25 years old. Use the table of nonforfeiture options (Table 7.12) to determine the (a) cash value, (b) the amount of paid-up insurance which she could receive, and (c) the time period for which she could have paid-up insurance.

40. NONFORFEITURE OPTIONS Lee Hardesty purchased a 20-pay life policy 15 years ago when he was 25 years old. The face value of the policy is $80,000. Find the following values using the nonforfeiture options table (Table 7.12): (a) cash value, (b) the amount of paid-up insurance that he could have, and (c) the time period for which he could have paid-up term insurance.

41. POLICY CASH VALUE The face value of a universal life policy purchased by Patty Gillette is $100,000. She purchased the policy 25 years ago when she was 25 years of age. Use the nonforfeiture options table (Table 7.12) to find the cash value of the policy today.

42. NONFORFEITURE OPTIONS When he was 25 years old, Chuck Manly purchased a $40,000 whole life policy. Now that 30 years have passed, Manly wants to know the options available when canceling his policy. Use the nonforfeiture options table (Table 7.12) to find the (a) cash value, (b) the amount of paid-up insurance he could have, and (c) the time period for which he could have paid-up term insurance.

43. **SETTLEMENT OPTIONS** Ryan Polstra is the beneficiary of a $25,000 life insurance policy. Polstra is 50 years of age and is considering the various settlement options available. Use Table 7.13 to find (a) the monthly payment if he selects payments for 12 years, (b) the number of years he will receive payments of $145 per month, (c) the monthly payment from a life annuity, and (d) the amount he would receive monthly if he chooses a life annuity with 15 years certain.

44. **SETTLEMENT OPTIONS** Jenn Luan is the beneficiary of a $30,000 life insurance policy. Luan is 65 years of age and is considering the various settlement options available. Use Table 7.13 to find (a) the monthly payment she would receive if she selects a fixed amount for 10 years, (b) the number of years she can receive $225 per month, (c) the amount she can receive per month as a life annuity, and (d) the monthly amount she could receive as a life annuity with 10 years certain.

45. **SETTLEMENT OPTIONS** James Marcotte is the beneficiary of a $50,000 life insurance policy. Marcotte is 60 years of age and is considering the various settlement options available. Use Table 7.13 to find (a) the monthly payment he would receive if he selects a fixed amount for 16 years, (b) the number of years he can receive $305 per month, (c) the amount he can receive per month as a life annuity, and (d) the monthly amount he could receive as a life annuity with 10 years certain.

46. **SETTLEMENT OPTIONS** Meghan Anderson is the beneficiary of a $70,000 life insurance policy. Anderson is 60 years of age and is considering the various settlement options available. Use Table 7.13 to find (a) the monthly payment she would receive if she selects a fixed amount for 18 years, (b) the number of years she can receive $405 per month, (c) the amount she can receive as a life annuity, and (d) the monthly amount she could receive as a life annuity with 10 years certain.

47. The greatest advantage of a renewable term policy is that you get the most coverage for your premium dollar. However, renewable term provides no cash value for retirement. Explain how you would provide for retirement to offset this disadvantage of a renewable term life policy. (See Objective 2.)

48. The additional charge for paying an insurance premium semiannually, quarterly, or monthly is determined by a premium factor. Would you select other than the single premium payment? Why or why not? How could you justify the additional charge (premium factor)? (See Objective 5.)

49. Read Section 8 of the foldout. List three possible consequences to surviving family members when a person dies without a will.

Chapter 7 Quick Review

Chapter Terms

Review the following terms to test your understanding of the chapter. For each term you do not know, refer to the page number found next to that term.

Concepts / Examples

7.1 Annual premium for fire insurance

Use the building and territorial rating in Table 7.1 to find the premiums per $100 for the building and for the contents. Add the two premiums.

Building value, $80,000, contents, $35,000; territorial rating, 4; building rating, B. Find the annual premium.

Building: $800 \times \$0.75 = \$600$

Contents: $350 \times \$0.77 = \$269.50$

Total premium: $\$600 + \$269.50 = \$869.50$

7.1 Short-term rates and cancellations

Annual premium is multiplied by the short-term rate. Use Table 7.2.

Annual premium is $2320. Short-term rate for 9 months is 85%. Premium for 9 months is

$\$2320 \times 0.85 = \$1972$ premium

$\$2320 - \$1972 = \$348$ refund

7.1 Calculate prorated insurance premium cancellations

Multiply annual premium by a fraction with months of insurance in force as the numerator and 12 as the denominator.

Annual premium, $1620; policy canceled after 4 months. Find refund. Premium for 4 months is

$$\$1620 \times \frac{4}{12} = \$540$$

$\$1620 - \$540 = \$1080$ refund

7.1 Coinsurance formula

Part of the risk is taken by the insured. An 80% coinsurance clause is common.

$$\text{Loss paid by insurance company} = \text{Amount of loss} \times \frac{\text{Policy amount}}{80\% \text{ of replacement cost}}$$

Building value, \$125,000; policy amount, \$75,000; loss, \$40,000; 80% coinsurance clause. Find the amount of loss paid by insurance company.

$$\$40{,}000 \times \frac{\$75{,}000}{\$100{,}000} = \$30{,}000$$

Insurance company pays \$30,000.

7.1 Multiple carriers

Several companies insure the same property to limit their risk; each company pays its fractional portion of any claim.

Insured loss, \$500,000; insurance is Company A with \$1,000,000; Company B with \$750,000; Company C with \$250,000. Find the amount of loss paid by each company.

Total insurance is

	\$1,000,000
	750,000
+	250,000
	\$2,000,000

Company A pays

$$\frac{\$1{,}000{,}000}{\$2{,}000{,}000} \times \$500{,}000 = \$250{,}000$$

Company B pays

$$\frac{\$750{,}000}{\$2{,}000{,}000} \times \$500{,}000 = \$187{,}500$$

Company C pays

$$\frac{\$250{,}000}{\$2{,}000{,}000} \times \$500{,}000 = \$62{,}500$$

7.2 Annual auto insurance premium

Most drivers are legally required to purchase automobile insurance. The premium is determined by the types of coverage selected, the type of car, geographic territory, past driving record, and other factors. See Tables 7.3 to 7.6.

Determine the premium: territory, 2; liability, 50/100; property damage, \$50,000; comprehensive and collision, 3-year-old car with a category of 8; uninsured motorist coverage; driver, age 23 with driver's training.

	\$168	Liability
	76	Property damage
	36	Comprehensive
	98	Collision
+	44	Uninsured motorist
	\$422	× 1.15 youthful-operator factor
		= \$485.30

7.2 Amount paid by insurance company and by insured

Company pays up to maximum amount of insurance coverage; insured pays balance.

Policy terms: liability, 15/30; property damage, \$10,000; collision, \$250 deductible. Accident caused \$2850 damage to insured's car; \$3850 to other car; injury liability of \$20,000 and \$25,000, respectively.

Company pays \$2600 (\$2850 − \$250) to repair insured's car.

Company pays \$3850 to repair other car (\$10,000 limit).

Company pays \$30,000 for two injured people (\$15,000 each).

Insured pays \$15,000 (\$45,000 − \$30,000), amount over limit.

7.3 Annual life insurance premium

There are several types of life policies. Use Table 7.10 and multiply by the number of \$1000s of coverage. Subtract 5 years from the age of females. Premium = number of thousands × rate per \$1000.

Find the premiums on a \$50,000 policy for a 30-year-old male.

Renewable term: 50 × \$2.97 = \$148.50

Whole life: 50 × \$5.66 = \$283

Universal life: 50 × \$6.08 = \$304

20-Pay life: 50 × \$18.78 = \$939

7.3 Premium factors

If not paid annually, life insurance premiums may be paid semiannually, quarterly, or monthly. The annual premium is multiplied by the premium factor to determine the premium amount. Use Table 7.11.

The annual life insurance premium is \$740. Find the semiannual, quarterly, and monthly premium.

Semiannual: \$740 × 0.51 = \$377.40

Quarterly: \$740 × 0.26 = \$192.40

Monthly: \$740 × 0.0908 = \$67.19

7.3 Nonforfeiture options

Upon cancellation of a policy the insured may receive a cash settlement, paid-up insurance, or extended term insurance as a nonforfeiture option. Use Table 7.12.

Insurance is a \$40,000, 20-pay life policy; in force for 10 years; issued at age 25.

Cash value: \$187 × 40 = \$7480

Paid-up insurance: \$554 × 40 = \$22,160

Extended term insurance period is 29 years, 115 days.

7.3 Settlement options

Upon the death of the insured, the beneficiary often has choices as to how the money may be received. These range from all cash to various types of payment arrangements. Use Table 7.13.

Insurance is a \$30,000 policy; female beneficiary, age 60. Find monthly payments for:

16 years: 30 × \$6.91 = \$207.30

Life annuity: 30 × \$5.36 = \$160.80

Life with 15 years certain: 30 × \$4.78 = \$143.40

For how many years would \$200 a month be paid?

$$\frac{\$200}{30} = \$6.67 \text{ per } \$1000$$

A little more than 16 years: \$6.67 is closest to \$6.91.

Chapter 7 Review Exercises

Find the total annual premium for fire insurance for each of the following. Use Table 7.1. **[7.1]**

	Territorial Rating	Building Classification	Building Value	Contents Value
1.	4	A	$640,000	$275,000
2.	2	C	$375,000	$198,000
3.	3	A	$80,000	$30,000
4.	1	B	$193,000	$68,000

Find the amount of refund to the insured using the Short-Term Rate Schedule (Table 7.2). **[7.1]**

	Annual Premium	Months in Force
5.	$1773	2
6.	$1078	6
7.	$1486	9
8.	$2878	5

Find (a) the amount of premium retained by the company and (b) the amount of refund to the insured using proration. **[7.1]**

	Annual Premium	Months in Force
9.	$3150	5
10.	$1975	9
11.	$1476	10
12.	$2784	2

Find the amount of each of the following losses that will be paid by the insurance company. Assume that each policy includes an 80% coinsurance clause. **[7.1]**

	Value of Building	Face Value of Policy	Amount of Loss
13.	$456,000	$320,000	$45,000
14.	$277,500	$165,000	$97,800
15.	$186,700	$120,000	$3,400
16.	$325,000	$220,000	$42,200

Find the annual motor vehicle insurance premium for the following people. [7.2]

Name	Territory	Age	Driver Training	Liability	Property Damage	Comprehensive Collision Age Group	Category	Uninsured Motorist
17. Larik	1	42	—	50/100	$100,000	1	8	yes
18. Ramos	3	18	yes	15/30	$10,000	5	7	yes
19. Verano	1	24	no	25/50	$25,000	2	8	yes
20. Wilson	4	29	—	250/500	$100,000	1	6	yes

Find the annual premium for each of the following life insurance policies. [7.3]

21. Carolyn Phelps: 20-pay life; $70,000 face value; age 55

22. Ralph Todd: renewable term; $50,000 face value; age 45

23. Gilbert Eckern: whole life: $30,000 face value; age 23

24. Irene Chang: universal life; $40,000 face value; age 29

Solve the following application problems.

25. Dave's Body and Paint has an insurable loss of $72,000, while having insurance coverage beyond coinsurance requirement. The insurance is divided between Company A with $250,000 coverage, Company B with $150,000 coverage, and Company C with $100,000 coverage. Find the amount of loss paid by each of the insurance companies. [7.1]

26. The headquarters building of Western States Life is valued at $820,000, and the fire insurance policies contain an 80% coinsurance clause. The policies include $350,000 with company 1 and $200,000 with company 2. The Western building suffered a $150,000 fire loss. Find (a) the amount that Western would receive after the loss and (b) the amount of loss paid by each insurance company. Round to the nearest dollar. [7.1]

27. Your cousin who has a bodily injury policy with limits of 25/50 injures a bicycle rider. The judge awards damages of $34,000 to the injured cyclist. (a) How much will the company pay and (b) how much will your cousin pay? [7.2]

28. Three people are injured in an automobile accident and receive $15,000 each in damages. The driver has bodily injury insurance of 50/100. (a) How much will the company pay to each person and (b) how much must the driver pay? [7.2]

29. Some scaffolding falls off a truck and into the path of a car, resulting in serious damage and injury. The car had damage of $16,800, while the driver and passenger of the car were given court awards of $25,000 and $35,000, respectively. The driver of the truck had 15/30 liability insurance limits and $10,000 property damage limits. Find (a) the total amount paid by the insurance company for both property damage and liability and (b) the total amount beyond the insurance limits for which the driver was liable. [7.2]

30. The annual premium for a whole life policy is $970. Use premium factors to find (a) the semiannual premium, (b) the quarterly premium, (c) the monthly premium, and (d) the total annual cost for each of the plans. [7.3]

31. Lori Johnson purchased a universal life policy 20 years ago when she was 25 years old. The face value of the policy is $60,000. Find the cash value of the policy using the nonforfeiture options table (Table 7.12). [7.3]

32. Glenn Lewis purchased a $40,000 whole life policy 20 years ago when he was 25 years old. Use Table 7.12 to determine (a) the cash value, (b) the amount of paid-up insurance he could receive, and (c) the time period for which he could have paid-up term insurance. [7.3]

33. Jim Jordan is the beneficiary of an $80,000 life insurance policy. Jordan is 60 years old and is considering the various settlement options available. Use Table 7.13 to find (a) the monthly payment he would receive if he selects a fixed payment for 16 years, (b) the number of years he can receive $675 per month, (c) the amount he can receive as a life annuity, and (d) the monthly amount he could receive as a life annuity with 15 years certain. **[7.3]**

34. Ann-Marie Sargent is the beneficiary of a $40,000 life insurance policy. Sargent is 55 years old and is considering the various settlement options available. Use Table 7.13 to find (a) the monthly payment she would receive if she selects a fixed payment for 20 years, (b) the number of years she can receive $245 per month, (c) the amount she can receive as a life annuity, and (d) the monthly amount she could receive as a life annuity with 10 years certain. **[7.3]**

Chapter 7 Summary Exercise
Financial Planning for Insurance

Childcare Playground Toys imports parts from Thailand and Malaysia and assembles quality playground equipment and riding toys. Planning ahead, the company set aside $41,700 to pay fire insurance premiums on the company property and a semiannual life insurance premium for the president. Both were due in the same month. Find each of the following.

(a) The building occupied by the company is a class B building worth $1,730,000. The contents are worth $3,502,000 and the territorial rating is 4. Find the annual insurance premium.

(b) The president of Childcare Playground Toys is a 45-year-old woman and the company is buying a $175,000, renewable term life insurance policy on the president's life. Find the semiannual premium.

(c) Find the total amount needed to pay the fire insurance premium and the semiannual life insurance premium.

(d) How much more than the amount needed had the company set aside to pay these expenses?

Net Assets Business on the Internet

Quicken InsureMarket

Statistics

- Number one in insurance quote comparisons
- Assesses insurance needs
- Comparison shop for policies
- Offers life, auto, health, and more

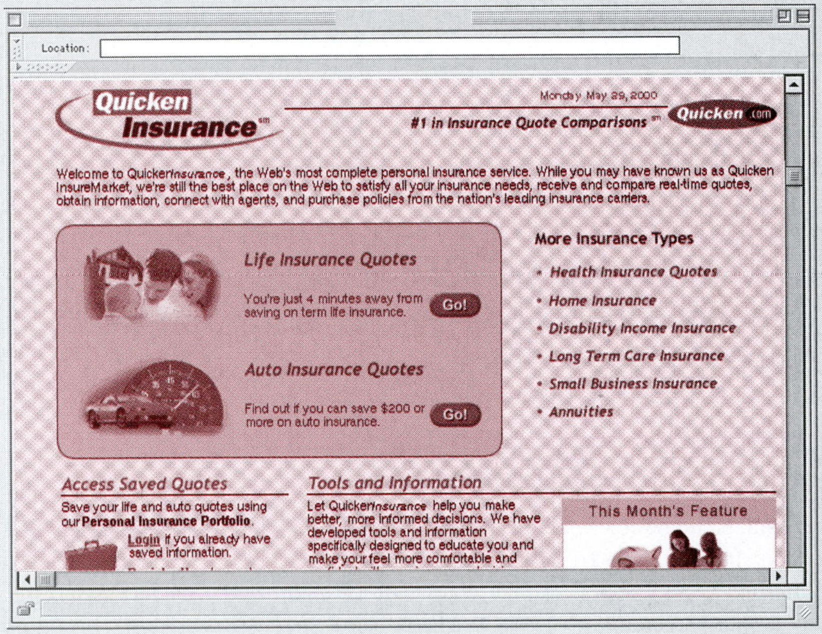

As part of today's growing electronic commerce, Quicken InsureMarket offers insurance shopping over the Internet. The company educates the customer regarding types of insurance, assesses insurance needs, and gives a price quote for the type of insurance coverage desired.

Insurance products are offered by over seventeen insurance companies. After a series of questions are answered by the customer regarding a specific insurance policy, the best price quote is given by the participating companies. The company offers life, auto, home, health, business, and other types of insurance.

1. A building has a value of $180,000, and the contents are valued at $78,000. If the building has a class rating of C with a territorial rating of 2, find the annual insurance premium. Use Table 7.1.

2. The office building owned by antiques.com was sold six months after renewing the fire insurance policy. If the annual premium was $7330, use the short-term rate schedule (Table 7.2) to find the amount of refund to the policy holder.

3. Suppose the bodily injury coverage on your car insurance is 50/100 and you injure a motorcyclist in an accident. If the court awards $63,000 in damages to the motorcyclist, find **(a)** the amount that the insurance company will pay and **(b)** the amount that you will pay.

4. Gloria Nobles, a 45-year-old woman, purchased a whole-life policy with a face value of $200,000. Use the premium factors to find **(a)** the semiannual premium, **(b)** the quarterly premium, **(c)** the monthly premium, and **(d)** the total annual cost of each of the plans.

Part 2 Cumulative Review Chapters 4–7

The following credit card transactions were made at Gifts and Such. Answer questions 1 to 5 using this information. [4.2]

	Sales		Credits
$93.50	$315.26	$22.51	$99.84
$117.75	$38.00	$162.15	$72.68
$173.05	$92.18		$35.63

1. Find the total amount of the sales slips.
2. What is the total amount of the credit slips?
3. Find the total amount of the deposit.
4. Assuming that the bank charges the retailer a $2\frac{1}{4}\%$ discount charge, find the amount of the discount charge at the statement date.
5. Find the amount of the credit given to the retailer after the fee is subtracted.

Solve the following application problems.

6. The bank statement of Adam Hamel Productions shows a balance or $16,298.60, a returned check amounting to $673.60, a service charge of $28.32, and an interest credit of $40.72. There were unrecorded deposits of $3631.28 and $7136.60, and checks outstanding are $2570.20, $4331.92, $293, and $1887.32. The checkbook shows a balance of $18,645.24. Use the T-account form, Figure 4.13, to find the adjusted balance of this checking account. [4.3]
7. Angela Perez worked 7 hours on Monday, 10 hours on Tuesday, 8 hours on Wednesday, 9 hours on Thursday, and 10 hours on Friday. Her regular hourly pay is $12.80. Find her gross earnings for the week if Perez is paid overtime for all hours over 8 worked in a day. [5.1]
8. Laura Rogers is a commission salesperson for Education Specialties, which allows her a draw of $1200 per month. Her commission is 5% of the first $5000 in sales, 8% of the next $15,000, and 15% of all sales over $20,000. Her sales for the month were $28,400. Find (a) her total commission and (b) the earnings due at the end of the month after repaying the drawing account balance of $1200. [5.2]
9. Eric Means is paid $2.18 for each computer keyboard assembled, is charged $1.05 for each rejection, and is paid time and a half for overtime production. Find his gross earnings for the week when he assembles 268 keyboards at the regular rate and 32 keyboards at the overtime rate and has 9 chargebacks. [5.3]
10. Scott Samuel is a salesperson for Novartis Pharmaceuticals (a Swedish Company) and is paid $650 per week plus a commission of 3% of all sales. His sales last week were $33,482. Find the amount of (a) his Social Security tax (6.2%), (b) his Medicare tax (1.45%), and (c) his state disability (1%) for the pay period. [5.4]
11. Jenna DeMarco has earnings of $722 in one week of April. She is single and claims three withholding allowances. Her deductions include FICA (6.2%), Medicare (1.45%), federal withholding, state disability insurance (1%), state withholding of 2.8%, a United Way contribution of $10, and a savings bond of $50. Use the percentage method of withholding to find her net pay for the week. [5.5]
12. The employees of Highland Farms paid a total of $968.50 in Social Security tax last month, $223.50 in Medicare tax, and $1975.38 in federal withholding tax. Find the total amount that the employer must send to the Internal Revenue Service. [5.6]
13. The Custom Fireside Shop charges $286.96, including 5.5% sales tax for a custom fireplace screen. Find the price of the fireplace screen without the tax. [6.1]

Complete the following list comparing tax rates. [6.2]

Percent	Per $100	Per $1000	In Mills
14. (a)______	$2.68	(b)______	(c)______
15. 4.62%	(a)______	(b)______	(c)______

16. Phyllis Beaton earned $38,514.75 last year from Bel Air Supermarket as assistant manager and $675.18 in interest from her credit union. She had $1800 in IRA contributions. Find her adjusted gross income. [6.3]

17. Feather's Custom Cabinets owns a class A building worth $179,480. The contents are valued at $83,300 and the territorial rating is 4. Find the annual fire insurance premium. [7.1]

18. The Outback Restaurant is valued at $720,000. The fire policies are $200,000 with Company A, $100,000 with Company B, and $60,000 with Company C, while each contains an 80% coinsurance clause. There is a fire at the restaurant causing a $240,000 loss. Find (a) the amount that the insured would receive and (b) the amount that each of the insurance companies would pay. Round to the nearest dollar. [7.1]

19. Del Nelson was distracted by his cell phone as he drove along the interstate. He didn't notice that cars were braking and he crashed into the car in front of him. Nelson had 25/50 liability limits, $25,000 property damage limits, and collision coverage with a $250 deductible. Damage to Nelson's car was $6340, while the car in front of him, with a value of $34,800, was totaled. The results of a lawsuit awarded $100,000 and $65,000, respectively, in damages for personal injury to the two people in the other car. Find the amount that the insurance company will pay for (a) repairing Nelson's car, (b) repairing the other car, and (c) personal injury damages. (d) How much must Nelson pay beyond his insurance coverage? [7.2]

20. Don Carlson purchased a 20-pay life policy with a face value of $100,000. He is 50 years old. Use the premium factors to find (a) the semiannual premium, (b) the quarterly premium, (c) the monthly premium, and (d) the total annual cost of each of the plans. [7.3]

Part 3 Mathematics of Retailing

8 Mathematics of Buying

Retail businesses make a profit by purchasing items and then selling them for more than they cost. There are several steps in this process: **manufacturers** buy raw materials and component parts and assemble them into products that can be sold to other manufacturers or wholesalers. The **wholesaler**, often called a "middleman," buys from manufacturers or other wholesalers and sells to the retailer. **Retailers** sell directly to the ultimate user, the **consumer**.

The retail advertisement shown in Figure 8.1 indicates that the retail store received some very large trade discounts, so they are able to offer customers as much as 50% to 70% off the list price. These high discounts may have resulted from very large quantities of merchandise purchased or perhaps it was the end of the season or the last of a production cycle for the manufacturer.

What types of products do you normally see at these drastically reduced prices?

Documents called invoices help businesses keep track of sales and various types of discounts help businesses to buy products at lower costs so that they can increase profits. Recent technology has enabled businesses to replace much of their paper-based business

save
50-70%
off original prices
when you take an
extra 33% off
Here's an example
of how you save:
20.00 Reg. price
14.99 Sale price
-5.00 33% off
9.99 Final price

Figure 8.1

processes with electronic solutions, known collectively as **electronic commerce (EC)**. Expect to see further changes in how business is conducted in the future. This chapter covers the mathematics needed for working with invoices and discounts—the mathematics of buying.

8.1 Invoices and Trade Discounts

Objectives

1. *Complete an invoice.*
2. *Understand common shipping terms.*
3. *Identify invoice abbreviations.*
4. *Calculate net cost and trade discounts.*
5. *Differentiate between single and series discounts.*
6. *Calculate each series discount separately.*
7. *Use complements to calculate series discounts.*
8. *Use a table to find the net cost equivalent of series discounts.*

An **invoice** is a printed record of a purchase and sale. For the seller it is a **sales invoice** and records a sale; for the buyer it is a **purchase invoice** and records a purchase. The invoice identifies the seller and the buyer, includes a description of the items purchased, the quantity purchased, the unit price of each item, the **extension total** (the number of items purchased times the price per unit), any discounts, the shipping and insurance charges, and the **invoice total** (the sum of all the extension totals).

Objective 1 *Complete an Invoice.* The document on page 286 serves as a sales invoice for J. B. Sherr Company and as a purchase invoice for Oaks Hardware. The numbers of items shipped column multiplied by the unit price give the amount or totals for each item. The **total invoice amount** is the sum of the amount totals.

Trade and cash discounts, discussed later in this section, are never applied to shipping and insurance charges. For this reason shipping and insurance charges are often not included in the invoice total, and the purchaser must add them to the invoice total to find the total amount due.

Objective 2 *Understand Common Shipping Terms.* A common shipping term appearing on invoices is **free on board (FOB)**, followed by the words "shipping point" or "destination." The term "FOB shipping point" means that the *buyer* pays for shipping and that ownership of the merchandise passes to the purchaser when the merchandise is given to the shipper. The term "FOB destination" means that the *seller* pays the shipping charges and retains ownership until the goods reach the destination. This distinction is important in the event that the merchandise is lost or damaged during shipment.

The shipping term **COD** means **cash on delivery**. Here, the shipper makes delivery to the purchaser on the receipt of enough cash to pay for the goods. A shipping term used when goods are moved by ship is **FAS**, which means **free alongside ship**. Here, the goods are delivered to the dock with all freight charges to that point paid by the shipper.

Objective 3 *Identify Invoice Abbreviations.* A number of abbreviations are used on invoices to identify measurements, quantities of merchandise, shipping terms, and additional discounts. Those most commonly used are shown in Table 8.1. (Some of these measure-

J. B. SHERR Co.

SHOWROOM AND WAREHOUSE
1704 ROLLINS ROAD
BURLINGAME, CA 94010
TELEPHONE: (650) 697-3430
TO ORDER 1-800-660-1422

INVOICE

SOLD TO:
OAKS HARDWARE OAK007
10136 FAIR OAKS BLVD.
FAIR OAKS CA 95628

PAGE NO. 1 OF

INVOICE DATE	INVOICE NO.
03/17	0002271-IN

SHIP TO:
OAKS HARDWARE
10136 FAIR OAKS BLVD.
FAIR OAKS CA 95628

TERMS 1% 15 DAYS, NET 30

SALESMAN	ENTRY NUMBER	ENTRY DATE	SHIPPING DATE	SHIPPED VIA	CUSTOMER ORDER NO./DEPT.
0008	0002280	03/17		UPS	3-13

TAG #	QTY. ORD.	SHIPPED	UNIT	STOCK NUMBER	DESCRIPTION	SUGG. RETAIL	UNIT PRICE	AMOUNT
1	12	12	EACH	736-080	IMPT NATURAL SEA SPONGE	.00	2.400	28.80
2	1	0	EA	267-6682	ACRYLIC BUTTER DISH	6.39	3.760	.00
3	1	1	EA	267-6683	ACRYLIC CREAM & SUGAR S	6.39	3.760	3.76
4	1	1	EA	267-6684	ACRYLIC NAPKIN HOLDER	5.29	3.140	3.14
5	6	6	EACH	694-322	IMPT WENOL METAL POLISH	7.79	4.650	27.90
6	6	6	EACH	694-353	IMPTRED BEAR POLISH	7.39	4.370	26.22
7	1	1	EA	274-10012	FRIENDSHIP MIXING BOWL	31.50	18.750	18.75
8	1	0	EACH	274-10014	FRIENDSHIP MIXING BOWL	44.90	26.500	.00
9	2	2	EA	589-31008	FLEXIBLE CHOPPING MATS	3.93	2.360	4.72
10	1	1	EA	589-22153	CORNER SINK SHELF W/SUC	2.79	1.600	1.60
11	6	0	EACH	281-7950	GEMCO JUICER W/GLASS JA	4.59	2.730	.00
12	2	2	EACH	54-611	ARDEN WAFFLE TOWEL-BLUE	3.19	1.900	3.80
13	2	2	EACH	54-612	ARDEN WAFFLE TOWEL-GREE	3.19	1.900	3.80
14	2	2	EACH	60-6	ASHLAND TIRE MAT 18.5 X	18.75	11.250	22.50
15	2	2	EACH	998-713	3 HALF/RD DRAGON 18 X 30	.00	5.200	10.40
16	2	2	EACH	998-143	3 WELCOME MAT 18 X 30	.00	6.910	13.82
17	3	3	EACH	998-303	3 MB PLAIN MAT 18 X 30	.00	8.640	25.92
18	2	2	EACH	998-504	4 HALF/RD MB PLAIN 20 X 3	.00	10.550	21.10
19								
20								

NON-TAX TOTAL	TAXABLE TOTAL	SALES TAX	FREIGHT	MISC.	INVOICE TOTAL
216.23	.00	.00	19.45		235.68

ALL ITEMS NOT SHIPPED ARE CANCELLED. PLEASE REORDER

PLEASE PAY FROM THIS INVOICE

ALL ORDERS SUBJECT TO CREDIT ACCEPTANCE. PRICES SUBJECT TO CHANGE WITHOUT NOTICE. SHORTAGES MUST BE REPORTED WITHIN 10 DAYS. NO RETURNS ACCEPTED WITHOUT PRIOR AUTHORIZATION. PAST DUE ACCOUNTS SUBJECT TO INTEREST (1½% PER MO.) PLUS COLLECTION CHARGES.

CUSTOMER'S COPY

ments are from the metric system. These measurements often appear on invoices for imported goods.)

Objective 4 *Calculate Net Cost and Trade Discounts.* **Trade discounts** are offered to businesses or individuals who buy an item that is to be sold or used to produce an item that will then be sold. Normally, the seller prices an item at its **list price** (the suggested price at which the item is to be sold to the public). Then the seller gives a trade discount that is subtracted from the list price to get the net cost (the amount to be paid by the buyer). Find the net cost by using the following formula.

$$\text{Net cost} = \text{List price} - \text{Trade discount} \quad \text{or} \quad \begin{array}{r} \text{List price} \\ \underline{-\ \text{Trade discount}} \\ \text{Net cost} \end{array}$$

Table 8.1 COMMON INVOICE ABBREVIATIONS

ea.	= each	drm.	= drum
doz.	= dozen	cs.	= case
gro.	= gross (144 items)	bx.	= box
gr. gro.	= great gross (12 gross)	sk.	= sack
qt.	= quart	pr.	= pair
gal.	= gallon (4 quarts)	C	= Roman numeral for 100
bbl.	= barrel ($31\frac{1}{2}$ gallons)	M	= Roman numeral for 1000
ml	= milliliter	cwt.	= per hundred weight
cl	= centiliter	cpm.	= cost per thousand
L	= liter	@	= at
in.	= inch	lb.	= pound
ft.	= foot	oz.	= ounce
yd.	= yard	g	= gram
mm	= millimeter	kg	= kilogram
cm	= centimeter	FOB	= free on board
m	= meter	ROG	= receipt-of-goods
km	= kilometer	EOM	= end-of-month
ct.	= crate	ex. or x	= extra dating
cart.	= carton	COD	= cash on delivery
ctn.	= carton	FAS	= free alongside ship

NOTE The terms *net cost* and *net price* both refer to the amount paid by the buyer. However, net cost is the preferred term since this is the cost of an item to the business. ■

Example 1 Calculating a Single Trade Discount

The list price of a Shredmaster Home/Office Shredder is \$99.80, and the trade discount is 25%. Find the net cost.

Solution First find the amount of the trade discount by taking 25% of \$99.80.

$$R \times B = P$$

$$25\% \times \$99.80 = 0.25 \times \$99.80 = \$24.95$$

Find the net cost by subtracting \$24.95 from the list price of \$99.80.

$$\$99.80 \text{ (list price)} - \$24.95 \text{ (trade discount)} = \$74.85 \text{ (net cost)}$$

The net cost of the shredder is \$74.85.

Objective 5 ***Differentiate Between Single and Series Discounts.*** In Example 1 a **single discount** of 25% was offered. Another type of discount combines two or more discounts into a **series** or **chain discount**. A series discount written as 20/10 means that a 20% discount is subtracted from the list price, and *from this difference* another 10% discount is subtracted.

Discounts are sometimes added to or subtracted from a series discount. For example, another discount of 5% could be added to the series discount 20/10, for a new series discount of 20/10/5.

Why Trade Discounts May Change

Price changes may cause trade discounts to be raised or lowered.

As the *quantity purchased* increases, the discount may increase.

The buyer's position in *marketing channels* may determine the amount of discount offered (a wholesaler would receive a larger discount than the succeeding retailer).

Geographic location may influence the trade discount. An additional discount may be offered to increase sales in one particular area.

Seasonal fluctuations in sales may influence the trade discounts offered.

Competition from other companies may cause trade discounts to be raised or lowered.

Objective 6 *Calculate Each Series Discount Separately.* Three methods can be used to calculate a series discount and the net cost. The first of these is to *calculate each series discount separately*.

Example 2 Calculating Series Trade Discounts

Oaks Hardware is offered a series discount of 20/10 on a Porter-Cable cordless drill with a list price of \$150. Find the net cost after the series discount.

Solution First multiply the decimal equivalent of 20% (0.2) by \$150. Then subtract the product (\$30) from \$150, getting \$120. Then multiply the decimal equivalent of the second discount, 10% (0.1), by \$120. Subtract the product (\$12) from \$120, getting \$108, the net cost. Write this calculation as follows.

\$150 list price		Discount: 20/10
− 30 (0.2 × \$150)	←	20
\$120		
− 12 (0.1 × \$120)	←	10
\$108 net cost		

After the first discount, each discount is applied to the balance remaining after the preceding discount or discounts have been subtracted. This method demonstrates how trade discounts are applied but is usually *not* the preferred method for finding the invoice amount because it involves too many steps.

NOTE Single discounts in a series are *never* added together; for example, a series discount of 20/10 is *not the same* as a discount of 30%.

Objective 7 *Use Complements to Calculate Series Discounts.* The second method is to *use complements*. First, find the **complement** (with respect to 1, or 100%) of each single discount. The complement is the number that must be added to a given discount to get 1 or 100%. The complement of 20% is 80% since 20% + 80% = 100%. The complement of 40% is 60%.

Discounts with fractions are occasionally used. The complement (with respect to 1) of $22\frac{1}{2}\%$ is $77\frac{1}{2}\%$ or 0.775 ($22\frac{1}{2}\% = 22.5\% = 0.225$). When a fractional discount such as $16\frac{2}{3}\%$ is used, the complement (with respect to 1) of 0.1666 is not exact, but a repeating decimal, and if used can cause errors. In this case the fraction $\frac{5}{6}$ is used as the comple-

Table 8.2 TYPICAL COMPLEMENTS WITH RESPECT TO 1

Discount	Decimal Equivalent	Complement with Respect to 1
10%	0.1	0.9
$12\frac{1}{2}\%$	0.125	0.875
15%	0.15	0.85
25%	0.25	0.75
30%	0.3	0.7
$33\frac{1}{3}\%$	$0.333\overline{3}$ $(\frac{1}{3})$	$0.666\overline{6}$ $(\frac{2}{3})$
35%	0.35	0.65
50%	0.5	0.5

ment. Since $16\frac{2}{3}\%$ equals $\frac{1}{6}$ of 100%, the complement (with respect to 1) is $\frac{5}{6}$. Use of the fraction equivalents of these repeating decimals will result in fewer errors. Other typical complements (with respect to 1) are shown in Table 8.2.

The complement of the discount is the portion actually paid. For example,

- 10% discount means 90% paid
- 25% discount means 75% paid
- $33\frac{1}{3}\%$ discount means $66\frac{2}{3}\%$ $(\frac{2}{3})$ paid
- 50% discount means 50% paid

Multiply the complements of the single discounts together to get the **net cost equivalent** or **percent paid**. Then multiply the net cost equivalent (percent paid) by the list price to obtain the net cost, as shown in the formula.

$$\text{Percent paid} \times \text{List price} = \text{Net cost}$$
$$R \times B = P$$

Example 3 Using Complements to Solve Series Discounts

Oaks Hardware is offered a series discount of 20/10 on a Porter-Cable cordless drill with a list price of $150. Find the net cost after the series discount.

Solution For a series discount of 20/10, the complements (with respect to 1) of 20% and 10% are 0.8 and 0.9. Multiplying the complements together gives $0.8 \times 0.9 = 0.72$, the net cost equivalent. (In other words, a series discount of 20/10 is the same as paying 72% of the list price.) To find the net cost, multiply 0.72 by the list price of $150, which gives $108 as the net cost. Write this calculation as follows.

20/10 ← Series discount

$0.8 \times 0.9 = 0.72$ Net cost equivalent (percent paid)

Complements with respect to 1

Use the formula Percent paid × List price = Net cost.

$$R \times B = P$$

$$0.72 \times \$150 = \$108 \quad \text{Net cost}$$

Find the amount of the discount by subtracting the net cost from the list price.

$150 (list price) − $108 (net cost) = $42 (amount of discount)

Scientific *Calculator Approach*

On many calculators you can subtract the discount percents from the list price in a series calculation. For Example 3 this would be calculated as:

150 [−] 20 [%] [−] 10 [%] [=] 108

Example 4 Using Complements to Solve Series Discounts

The list price of an outdoor patio furniture set is $210. Find the net cost if a series discount of 20/10/10 is offered.

Solution Start by finding the complements with respect to 1 of each discount.

20/10/10 — Series discount

$$0.8 \times 0.9 \times 0.9 = 0.648$$

Net cost equivalent (percent paid)

Complements with respect to 1

Using the formula Percent paid × List price = Net cost.

$$R \times B = P$$

$$0.648 \times \$210 = \$136.08 \quad \text{Net cost}$$

Problem-Solving Hint *Never round the net cost equivalent.* If a repeating decimal results, use the fraction equivalent. In Example 4, if the net cost equivalent had been rounded to 0.65, the resulting net cost would have been $136.50 (0.65 × $210). This error of $0.42 demonstrates the importance of not rounding the net cost equivalent.

Objective 8 ***Use a Table to Find the Net Cost Equivalent of Series Discounts.*** The third method is to *use net cost equivalents.* For example, to use Table 8.3 for a series discount of 20/10/10, find the number located to the right of 10/10 and below 20%. The number is 0.648, the net cost equivalent for a discount of 20/10/10. Multiply this number by the list price to get the net cost.

The order of the discounts in the series makes no difference. A 10/20 series is the same as a 20/10 series, and a 15/10/20 series is identical to a 20/15/10 series. This is true because changing the order in which numbers are multiplied does not change the answer.

Table 8.3 NET COST EQUIVALENTS OF SERIES DISCOUNTS

	5%	10%	15%	20%	25%	30%	35%	40%
5	0.9025	0.855	0.8075	0.76	0.7125	0.665	0.6175	0.57
10	0.855	0.81	0.765	0.72	0.675	0.63	0.585	0.54
10/5	0.81225	0.7695	0.72675	0.684	0.64125	0.5985	0.55575	0.513
10/10	0.7695	0.729	0.6885	0.648	0.6075	0.567	0.5265	0.486
15	0.8075	0.765	0.7225	0.68	0.6375	0.595	0.5525	0.51
15/10	0.72675	0.6885	0.65025	0.612	0.57375	0.5355	0.49725	0.459
20	0.76	0.72	0.68	0.64	0.6	0.56	0.52	0.48
20/15	0.646	0.612	0.578	0.544	0.51	0.476	0.442	0.408
25	0.7125	0.675	0.6375	0.6	0.5625	0.525	0.4875	0.45
25/20	0.57	0.54	0.51	0.48	0.45	0.42	0.39	0.36
25/25	0.534375	0.50625	0.478125	0.45	0.421875	0.39375	0.365625	0.3375
30	0.665	0.63	0.595	0.56	0.525	0.49	0.455	0.42
40	0.57	0.54	0.51	0.48	0.45	0.42	0.39	0.36

The net cost equivalent (percent paid) found using Table 8.3 and the list price are multiplied to find the net cost.

Example 5 Using a Table of Net Cost Equivalents

Use Table 8.3 to find the net cost equivalent for the following series discounts.

(a) 10/20 **(b)** 10/10/40 **(c)** 25/20/10 **(d)** 35/20/15

Solution

(a) 0.72 **(b)** 0.486 **(c)** 0.54 **(d)** 0.442

NOTE *Do not round any of the net cost equivalents.* Doing so will cause an error in the net cost. ■

8.1 Exercises

ABBREVIATIONS ON INVOICES *What do each of the following abbreviations represent?*

1. ft.
2. mm
3. sk.
4. qt.
5. gr. gro.
6. kg
7. cs.
8. gro.
9. drm.
10. yd.
11. cpm.
12. bbl.
13. gal.
14. cwt.
15. COD
16. FOB

17. USING INVOICES Compute each of the following extension totals, find the invoice total, and the total amount due.

J & K'S MUSTANG PARTS
New and Used

Sold to: Dave's Auto Body & Paint
4443-B Auburn Blvd.
Sacramento, CA 95841

Date: July17
Order. No.: 100603
Shipped by: Emery
Terms: Net

Quantity	Order No./Description	Unit Price	Extension Total
24	filler tube gaskets	$2.25 ea.	
12 pr.	taillight lens gaskets	$4.75 pr.	
6 pr.	taillight bezels to body	$10.80 pr.	
2 gr.	door panel fasteners	$14.20 gr.	
18	bumper bolt kits	$16.50 ea.	
		Invoice Total	
		Shipping and Insurance	$23.75
		Total Amount Due	

18. USING INVOICES Compute each of the following extension totals, find the invoice total, and the total amount due.

HOME DECORATOR'S SUPPLY

Sold to: Wallpaper Plus
232 Main Street
Portland, ME 04103

Date: June 10
Order. No.: 796152
Shipped by: UPS
Terms: Net

Quantity	Order No./Description	Unit Price	Extension Total
6 doz.	bristle brush, wide	$37.80 doz.	
3 gro.	blades, single edge	$12.60 gro.	
9 doz.	wallpaper paste	$14.04 doz.	
8	easyoff steamer	$106.12 ea.	
53 pr.	step horse, low top	$68.12 pr.	
		Invoice Total	
		Shipping and Insurance	$37.45
		Total Amount Due	

19. Explain the difference between "FOB shipping point" and "FOB destination." (See Objective 2.)

20. Name six items that appear on an invoice. Try to do this without looking at an invoice. (See Objective 1.)

Using complements (with respect to 1) of the single discounts, find the net cost equivalent for each of the following discounts. Use Table 8.3 for the first four problems.

21. 10/20 **22.** 10/10 **23.** 20/20/20

24. 10/15/20 **25.** 10/20/25 **26.** 40/20/10

27. 30/42$\frac{1}{2}$ **28.** 10/16$\frac{2}{3}$ **29.** 20/30/5

30. 20/20/10 **31.** 50/10/20/5 **32.** 25/10/20/10

Find the net cost for each of the following. Round to the nearest cent.

33. \$418 less 20/20

34. \$148 less 25/10

35. \$8.20 less 5/10

36. \$860 less 20/40

37. \$9.80 less 10/10/10

38. \$8.80 less 40/10/20

39. \$1630 less 10/5/10

40. \$15.70 less 5/10/20

41. \$25 less 30/32$\frac{1}{2}$

42. \$590 less 10/12$\frac{1}{2}$/10

43. \$1250 less 20/20/20

44. \$1410 less 10/20/5

45. Explain the difference between a single trade discount and a series or chain trade discount. (See Objectives 4 and 5.)

46. Identify and explain four reasons that might cause series trade discounts to change. (See Objective 5.)

47. Explain what a complement (with respect to 1 or 100%) is. Give an example. (See Objective 7.)

48. Using complements, explain how to find the net cost equivalent of a 25/20 series discount. Explain why a 25/10/10 series discount is not the same as a 25/20 series discount. (See Objective 7.)

Solve each of the following application problems in trade discount. Round to the nearest cent.

49. PURCHASING GOLF CLUBS The list price of a Pro Select Tour Classic II golf set is \$399.99. If the series discount offered is 10/10/25, what is the net cost after trade discounts?

50. ADJUSTABLE WALKERS Roger Wheatley, a restorative nurse assistant (RNA), finds that the list price of one dozen adjustable walkers is \$1680. Find the cost per walker if a series discount of 40/25 is offered.

51. HARDWARE PURCHASE Oaks Hardware purchases an extension ladder list priced at \$120. It is available at either a 10/15/10 discount or a 20/15 discount. (a) Which discount gives the lower price? (b) Find the difference.

52. TRIPOD PURCHASE The list price of an aluminum tripod is \$65. It is available at either a 15/10/10 discount or a 15/20 discount. (a) Which discount gives the lower price? (b) Find the difference.

53. BULK CHEMICALS Brazilian Chemical Supply offers a series discount of 20/20/10 on all bulk purchases. A tank (bulk) of industrial solvent is list-priced at \$28,500. What is the net cost after trade discounts?

54. DANCE SHOES How much will Giselle Papalewis, a dance instructor, pay for three dozen pairs of dance shoes if the list price is \$144 per dozen and a series discount of 10/25/30 is offered?

55. AUTOMOTIVE SUPPLIES Kaci Salmon, an automotive mechanics student, is offered mechanic's net prices on all purchases at Gilbert Tool Supply. If mechanic's net prices mean a 20/10 discount, how much will Salmon spend on a Black and Decker drill that is list priced at \$47?

56. BILLING ERROR The Office Depot offers a series trade discount of 30/20 to its regular customers. Chris Hutchinson, a new man in the billing department, understood the 30/20 terms to mean 50% and computed this trade discount on a list price of \$5440. How much difference did this error make in the amount of the invoice?

57. STICKER PURCHASES The AAA Foto and Copy Shop purchases stickers at a list price of \$135 per 1000. If they receive a trade discount of $40/33\frac{1}{3}$, find the net cost of 3500 stickers.

58. STATIONERY SALES One brand of file folders is list-priced at \$6.60 per dozen. A wholesale stationer offers a trade discount of 10/5/15 on the folders. Find the net cost of $5\frac{1}{2}$ dozen folders.

59. BRASS DEADBOLTS The list price of brass deadbolts is \$9.95. A retailer has a choice of two suppliers, one offering a discount of 10/25/15 and the other offering a discount of 20/15/10. If the retailer purchases four dozen deadbolts, find (a) the total cost of the less expensive supplier and (b) the amount saved by selecting the lower price.

60. FIBER OPTICS Cindy Herring has a choice of two suppliers of fiber optics for her business. Tyler Supplies offers a 20/10/25 discount on a list price of \$5.70 per unit. Irving Optics offers a 30/20 series discount on a list price of \$5.40 per unit. (a) Which supplier gives her the lower price? (b) How much does she save if she buys 12,500 units from the lower priced supplier? (*Hint:* Do not round.)

8.2 Single Discount Equivalents

Objectives

1. *Express a series discount as an equivalent single discount rate.*
2. *Find the net cost by multiplying list price by the complements of single discounts.*
3. *Find the list price if given the series discount and the net cost.*
4. *Determine a single trade discount rate.*
5. *Find the trade discount that must be added to match a competitor's price.*

Objective 1 *Express a Series Discount as an Equivalent Single Discount Rate.* Series or chain discount rates must often be expressed as a single discount rate. Find a **single discount equivalent** to a series discount by multiplying the complements (with respect to 1 or 100%) of the individual discounts. As in the previous section, the result is the net cost equivalent. Then, subtract the net cost equivalent from 1. The result is the single discount that is equivalent to the series discount. *The single discount is expressed as a percent.*

$$\text{Single discount equivalent} = 1 - \text{Net cost equivalent}$$

Example 1 Finding a Single Discount Equivalent

If Air Clean Manufacturing offered a 20/10 discount to wholesale accounts on all heater filters, what would the single discount equivalent be?

Solution Find the net cost equivalent (percent paid).

20/10

$$0.8 \times 0.9 = 0.72 \quad \text{Net cost equivalent}$$

Subtract the net cost equivalent from 1.

$$1.00 \text{ (base } - 0.72 \text{ remains)} = 0.28$$

The single discount equivalent of a 20/10 series discount is 28%.

This method may also be used with the table of net cost equivalents of series discounts (Table 8.3). For example, Table 8.3 shows 0.72 as the net cost equivalent for the series discount 20/10. The single discount is therefore 28% (1.00 − 0.72 = 0.28 = 28%).

Objective 2 ***Find the Net Cost by Multiplying List Price by the Complements of Single Discounts.*** The net cost can also be found by multiplying the list price by the complements of each of the single discounts in a series, as shown in Example 2.

Example 2 Finding the Net Cost Using Complements

The list price of an oak entertainment center is \$970. Find the net price if trade discounts of $20/15/27\frac{1}{2}$ are offered.

Solution

Net cost = List price × Complements of individual discounts

Net cost = \$970 × 0.8 × 0.85 × 0.725

(complements of $20/15/27\frac{1}{2}$)

= \$478.21

Scientific *Calculator Approach*

For the calculator solution to Example 2, enter the list price and multiply by the complements to find the net price.

970 [×] .8 [×] .85 [×] .725 [=] 478.21

Objective 3 ***Find the List Price If Given the Series Discount and the Net Cost.*** Sometimes the net cost after trade discounts is known along with the series discount, and the list price must be found.

Example 3 Solving for List Price

Find the list price of a handmade rug from Pakistan that has a net cost of \$544 after trade discounts of 20/20.

Solution Use a net cost equivalent, along with knowledge of percent. Start by finding the percent paid, using complements.

20/20 — Series discount

0.8 × 0.8 = 0.64 — Net cost equivalent (percent paid)

(0.8 and 0.8: Complements with respect to 1)

As the work shows, 0.64, or 64% of the list price, or $544, was paid. Use the basic percent equation to find the list price.

$$R \times B = P$$

$$0.64 \times \text{List price} = \$544$$

$$0.64 \times B = \$544$$

$$0.64B = \$544$$

$$\frac{0.64B}{0.64} = \frac{\$544}{0.64} = \$850 \quad \text{List price}$$

Check the answer.

$850	List price
− 170	(0.2 × $850)
$680	
− 136	(0.2 × $680)
$544	Net cost

The list price of the rug is $850.

Example 4 Solving for List Price

Find the list price of a 6-foot fiberglass stepladder having a series discount of 10/30/20 and a net cost of $45.36.

Solution Find the percent paid.

10/30/20 Series discount

$$0.9 \times 0.7 \times 0.8 = 0.504 \quad \text{Net cost equivalent (percent paid)}$$

Complements with respect to 1

Therefore 0.504 of the list price is $45.36. Now use the basic percent equation.

$$R \times B = P$$

$$0.504 \times \text{List price} = \$45.36$$

$$0.504 \times B = \$45.36$$

$$0.504B = \$45.36$$

$$\frac{0.504B}{0.504} = \frac{\$45.36}{0.504} = \$90 \quad \text{List price}$$

The list price of the stepladder is $90. Check this answer as in the previous example.

Problem-Solving Hint Notice that Examples 3 and 4 are decrease problems similar to those shown in Chapter 3, Section 5. They are still base problems but may look different because the discount is now shown as a series of two or more discounts rather than a single percent decrease as in Chapter 3. If you need help, refer to Section 3.5.

Objective 4 | *Determine a Single Trade Discount Rate.*

Example 5 Finding the Single Trade Discount Rate

The list price of a compact disc player is \$550. If the wholesaler offers the system at a net cost of \$341, find the single trade discount rate being offered.

Solution Use the following formula.

$$\text{Percent paid} \times \text{List price} = \text{Net cost}$$
$$P \times L = N$$
$$P \times \$550 = \$341$$
$$\$550P = \$341$$
$$\frac{\cancel{550}P}{\cancel{550}} = \frac{341}{550} = 0.62 \text{ or } 62\% \qquad \text{Percent paid}$$

Since 62% is paid, the discount offered is 38% (100% − 62% = 38%).

For an alternative approach, first find the amount of discount, or \$550 − \$341 = \$209. Next, find the rate of discount by using the basic percent equation

$$R \times B = P$$
$$\underline{\qquad\qquad}\% \text{ of List price is Discount}$$
$$\underline{\qquad\qquad}\% \text{ of } \$550 = \$209$$
$$R \times \$550 = \$209$$
$$\$550R = \$209$$
$$\frac{\cancel{550}R}{\cancel{550}} = \frac{209}{550}$$
$$R = \frac{209}{550} = 0.38 = 38\%$$

By either method, the discount is 38%.

Objective 5 | *Find the Trade Discount That Must Be Added to Match a Competitor's Price.*

Example 6 Adding a Discount to Match a Competitor's Price

S and B Distributors offered a 20% trade discount on small compressors list-priced at \$450. Find the trade discount that must be added to match a competitor's price of \$342.

Solution First use the formula

$$P \times L = N$$

to find the single discount needed. The percent paid is found by multiplying together the complements of the 20% discount already given and the new, unknown discount, or

$$0.8 \times \text{Complement of discount} \times L = N$$

$$0.8 \times d \times \$450 = \$342$$

$$\$360d = \$342$$

$$\frac{\cancel{360}d}{\cancel{360}} = \frac{342}{360}$$

$$d = 0.95 \text{ or } 95\%$$

Therefore, 95% is the complement of the trade discount that must be added. The additional discount needed is 5% (100% − 95%). To match the competition, S and B Distributors must give a 20/5 series discount.

8.2 Exercises

Find the net cost equivalent and the percent form of the single discount equivalent for each of the following series discounts.

1. 10/20 **2.** 10/10 **3.** 15/35

4. 10/50 **5.** 20/20 **6.** 20/20/20

7. 20/20/10 **8.** 15/5/10 **9.** 25/10

10. $30/37\frac{1}{2}$ **11.** $16\frac{2}{3}/10$ **12.** 30/25

13. 10/10/20 **14.** 20/20/10 **15.** 55/40/10

16. 10/30/10 **17.** 40/25 **18.** 5/5/5

19. $20/12\frac{1}{2}$ **20.** $10/33\frac{1}{3}$ **21.** 20/10/20/10

22. 10/20/25/10 **23.** 5/20/30/5 **24.** 10/5/30/20

25. Using complements, show that the single discount equivalent of a 20/25/10 series discount is 46%. (See Objective 1.)

26. Suppose that you own a business and are offered a choice of a 10/20 trade discount or a 20/10 trade discount. Which do you prefer? Why? (See Objective 1.)

Find the list price given the net cost and the series discount.

27. Net cost, $518.40; trade discount, 20/10

28. Net cost, $343.35; trade discount, 10/30

29. Net cost, $279.30; trade discount, 40/5/30

30. Net cost, $5250; trade discount, $25/33\frac{1}{3}$

31. Net cost, $1313.28; trade discount, 5/10/20

32. Net cost, $2697.30; trade discount, 10/10/10

Solve each of the following application problems in trade discount.

33. COMPARING DISCOUNTS A Uniden 40-channel cordless phone with a list price of $89.95 is offered to wholesalers with a series discount of 20/10/10. The same cordless phone is offered to retailers with a series discount of 20/10. (a) Find the wholesaler's price. (b)Find the retailer's price. (c) Find the difference between the two prices.

34. COMPARING DISCOUNTS Kathy Miller is offered oak stair railing by The Turning Point for \$1370 less 30/10. Sierra Stair Company offers the same railing for \$1220 less 10/10. (a) Which offer is better? (b) How much does Miller save by taking the better offer?

35. PRICING POTTED PLANTS Irene's Plant Place paid a net price of \$414.40 for a shipment of potted plants after a trade discount of 30/20 from the list price. Find the list price.

36. VITAMINS, MINERALS, AND SUPPLEMENTS SJ's Nutrition Center received a shipment of vitamins, minerals, and diet supplements at a net cost of \$1125. This cost was the result of a trade discount of 25/20 from list price. Find the list price of this shipment.

37. TIRE PRICES All season radial tires, size P205/75R15, with a list price of \$61.90 are offered to wholesalers with a series discount of 10/20/10. The same tire is offered to retailers with a series discount of 10/20. (a) Find the wholesalers price. (b) Find the retailers price. (c) Find the difference between the two prices.

38. SATELLITE DISH PRICING A satellite dish is list priced at \$1995. The manufacturer offers a series discount of 25/20/10 to wholesalers and a 25/20 series discount to retailers. (a) What is the wholesaler's price? (b)What is the retailer's price? (c) What is the difference between the prices?

39. PORTABLE GENERATORS A portable generator with a list price of \$295.95 is sold by a wholesaler at a net cost of \$221.95. Find the single trade discount rate being offered. Round to the nearest tenth of a percent.

40. TRUCK-BED LINERS Truck Stuff offers a fiberglass truck-bed liner at a net cost of \$180. If the list price of the bed liner is \$281.25, find the single trade discount rate.

41. SECURITY ALARM SYSTEMS Capitol Alarm purchased a security alarm system at a net cost of \$2733.75 and a series discount of $10/10/12\frac{1}{2}$. Find the list price.

42. FINDING LIST PRICE Find the list price of a 32-inch color television having a net cost of \$447.12 and a series discount of 10/20/10.

43. MOTOR OIL PRICING An auto wholesaler offers a 10% trade discount on a case of oil priced at \$27.60. A competitor offers the same oil at \$23.60. What additional trade discount must be given to meet the competitor's price? Round to the nearest tenth of a percent.

44. PERSONAL COMPUTERS A personal computer distributor has offered a computer system at a \$1450 list price less a 30% trade discount. Find the additional trade discount needed to meet a competitor's net price of \$933.80.

8.3 Cash Discounts: Ordinary Dating Method

Objectives

1. *Calculate net cost after discounts.*
2. *Use the ordinary dating method.*
3. *Determine whether cash discounts are earned.*
4. *Use postdating when calculating cash discounts.*
5. *Determine the amount due when goods are returned.*

Objective 1 *Calculate Net Cost After Discounts.* **Cash discounts** are offered by sellers to encourage prompt payment by customers. Since businesses must often borrow money for their operation, the prompt receipt of cash payment from customers increases the efficiency of the business and decreases the need for borrowed money. Saving interest on borrowed funds is a main reason that a cash incentive is often given to customers. In effect, the seller is saying, "Pay me quickly and receive a discount."

To find the net cost when a cash discount is offered, begin with the list price and subtract any trade discounts. From this amount subtract the cash discount. Use the following formula to find the net cost.

$$\text{Net cost} = (\text{List price} - \text{Trade discount}) - \text{Cash discount}$$

NOTE If an invoice amount includes shipping and insurance charges, subtract these charges first, before a cash discount is taken. Then add them back to find net cost after the cash discount is subtracted. ▬

The type of cash discount appears on the invoice under "Terms," which can be found in the bottom right-hand corner of the Hershey Chocolate U.S.A. invoice in Figure 8.2.

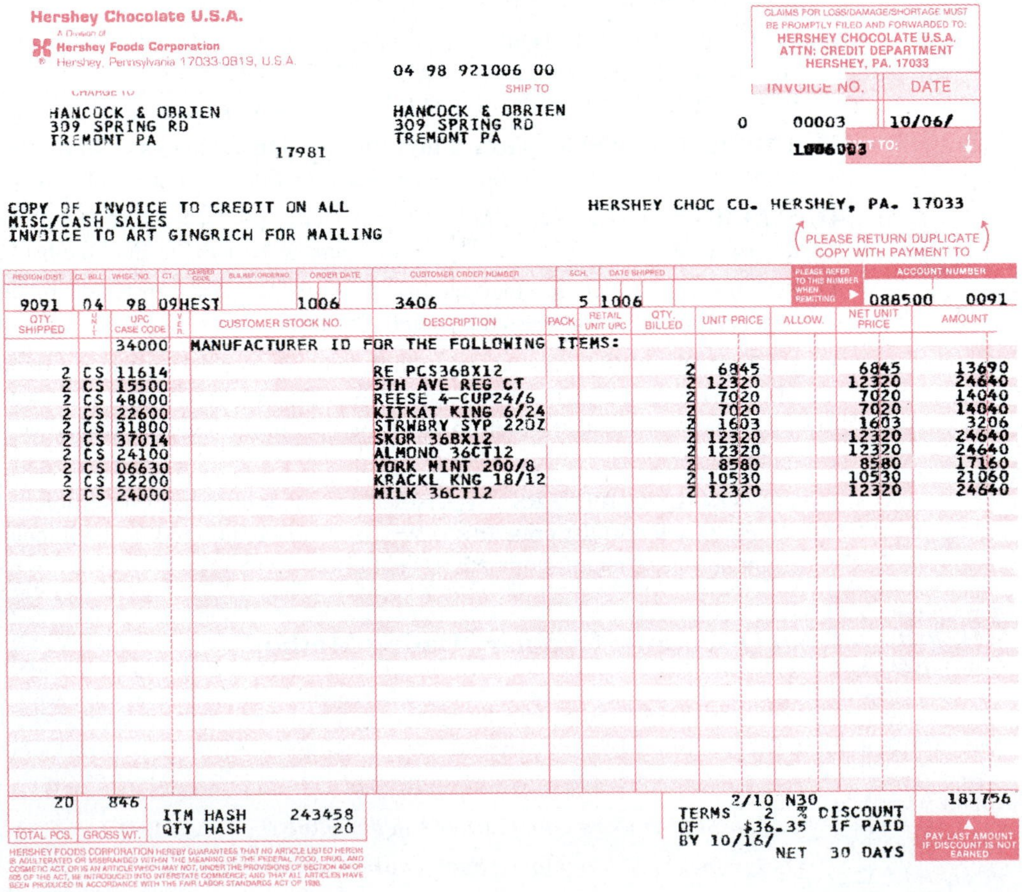
Hershey Chocolate U.S.A.
A Division of
Hershey Foods Corporation
Hershey, Pennsylvania 17033-0819, U.S.A.

04 98 921006 00

CLAIMS FOR LOSS/DAMAGE/SHORTAGE MUST BE PROMPTLY FILED AND FORWARDED TO:
HERSHEY CHOCOLATE U.S.A.
ATTN: CREDIT DEPARTMENT
HERSHEY, PA. 17033

CHARGE TO
HANCOCK & OBRIEN
309 SPRING RD
TREMONT PA
17981

SHIP TO
HANCOCK & OBRIEN
309 SPRING RD
TREMONT PA

INVOICE NO.	DATE
0 00003	10/06/

1006003

COPY OF INVOICE TO CREDIT ON ALL
MISC/CASH SALES
INVOICE TO ART GINGRICH FOR MAILING

HERSHEY CHOC CO. HERSHEY, PA. 17033

PLEASE RETURN DUPLICATE COPY WITH PAYMENT TO

REGION/DIST.	CL. BILL	WHSE. NO.	CT.	CARRIER CODE	B/L & REF. ORDER NO.	ORDER DATE	CUSTOMER ORDER NUMBER	SCH.	DATE SHIPPED	PLEASE REFER TO THIS NUMBER WHEN REMITTING	ACCOUNT NUMBER
9091	04	98	09	HEST		1006	3406	5	1006		088500 0091

QTY. SHIPPED	UNIT	UPC CASE CODE	VER.	CUSTOMER STOCK NO.	DESCRIPTION	PACK	RETAIL UNIT UPC	QTY. BILLED	UNIT PRICE	ALLOW.	NET UNIT PRICE	AMOUNT
		34000		MANUFACTURER ID FOR THE FOLLOWING ITEMS:								
2	CS	11614			RE PCS36BX12			2	6845		6845	13690
2	CS	15500			5TH AVE REG CT			2	12320		12320	24640
2	CS	48000			REESE 4-CUP24/6			2	7020		7020	14040
2	CS	22600			KITKAT KING 6/24			2	7020		7020	14040
2	CS	31800			STRWBRY SYP 22OZ			2	1603		1603	3206
2	CS	37014			SKOR 36BX12			2	12320		12320	24640
2	CS	24100			ALMOND 36CT12			2	12320		12320	24640
2	CS	06630			YORK MINT 200/8			2	8580		8580	17160
2	CS	22200			KRACKL KNG 18/12			2	10530		10530	21060
2	CS	24000			MILK 36CT12			2	12320		12320	24640

TOTAL PCS.	GROSS WT.		
20	846	ITM HASH	243458
		QTY HASH	20

HERSHEY FOODS CORPORATION HEREBY GUARANTEES THAT NO ARTICLE LISTED HEREIN IS ADULTERATED OR MISBRANDED WITHIN THE MEANING OF THE FEDERAL FOOD, DRUG, AND COSMETIC ACT, OR IS AN ARTICLE WHICH MAY NOT, UNDER THE PROVISIONS OF SECTION 404 OR 505 OF THE ACT, BE INTRODUCED INTO INTERSTATE COMMERCE; AND THAT ALL ARTICLES HAVE BEEN PRODUCED IN ACCORDANCE WITH THE FAIR LABOR STANDARDS ACT OF 1938.

2/10 N30
TERMS 2 % DISCOUNT
OF $36.35 IF PAID
BY 10/16/
NET 30 DAYS

181756

PAY LAST AMOUNT IF DISCOUNT IS NOT EARNED

Figure 8.2

Many companies using automated billing systems state the exact amount of the cash discount at the bottom of the invoice, to eliminate all calculations on the part of the buyer. This Hershey invoice is an example of an invoice stating the exact amount of the cash discount. Not all businesses do this, however, so it is important to know how to determine cash discounts.

There are many methods for determining cash discounts, but nearly all of these are based on the "ordinary dating method." The methods discussed in this section and the methods discussed in the next section, are the ones most commonly used today.

Objective 2 *Use the Ordinary Dating Method.* The **ordinary dating method** of cash discount, for example, is expressed on an invoice as

2/10, n/30 or sometimes 2/10, net/30

and is read "two ten, net thirty." The first digit is the rate of discount (2%), the second digit is the number of days allowed to take the discount (10 days), and n/30 or net/30 is the total number of days given to pay the invoice in full, if the buyer does not use the cash discount. The 2% discount may be subtracted from the amount owed if the invoice is paid within 10 days from the date of the invoice. If payment is made between the 11th and 30th days from the invoice date, the entire amount of the invoice is due. After 30 days from the date of the invoice, the invoice is considered overdue and may be subject to a late charge.

To find the due date of an invoice, use the number of days in each month, given in Table 8.4.

NOTE Leap years occur every 4 years. They are the same as Summer Olympic years and presidential-election years in the United States. If a year is evenly divisible by the number 4, it is a leap year. The years 2004 and 2008 are both leap years because they are evenly divisible by 4. ■

Objective 3 *Determine Whether Cash Discounts Are Earned.* Find the date that an invoice is due by counting from the next day after the date of the invoice. *The date of the invoice is never counted.* Another way to determine due dates is to add the given number of days to the starting date. For example, to determine 10 days from April 7, add the number of days to the date (7 + 10 = 17). The due date, or 10 days from April 7, is April 17.

When the discount date or net payment date falls in the next month, find the number of days remaining in the current month by subtracting the invoice date from the number of days in the month. Then find the number of days in the next month needed to equal the

Table 8.4 THE NUMBER OF DAYS IN EACH MONTH

30-Day Months	31-Day Months		Exception
April	January	August	February
June	March	October	(28 days normally; 29 days in leap years)
September	May	December	
November	July		

discount period or net payment period. For example, find 15 days from October 20 as follows.

31	Days in October
− 20	The beginning date, October 20
11	Days remaining in October
+ 4	Additional days needed in November to equal 15 days
15	Days

Finally, November 4 is 15 days from October 20.

Example 1 Finding Cash Discount Dates

A Hershey Chocolate invoice is dated January 2 and offers terms of 2/10, net 30. Find (a) the last date on which the 2% discount may be taken and (b) the net payment date.

Solution

(a) Beginning with the invoice date, January 2, the last date for taking the discount is January 12 (2 + 10).

(b) The net payment date is February 1 (29 days remaining in January plus 1 day in February).

Example 2 Finding the Amount Due on an Invoice

An invoice received by Oaks Hardware for $840 is dated July 1 and offers terms of 2/10, n/30. If the invoice is paid on July 8 and the shipping and insurance charges, which were "FOB shipping point," are $18.70, find the total amount due.

Solution The invoice was paid 7 days after its date (8 − 1 = 7); therefore, the 2% cash discount may be taken. The discount is $840 × 0.02 = $16.80. The cash discount is subtracted from the invoice amount to determine the amount due.

$840 (invoice amount) − $16.80 (cash discount of 2%) = $823.20 (amount due)

The shipping and insurance charges are added to find the total amount due.

$823.20 (amount due) + $18.70 (shipping and insurance) = $841.90 (total amount due)

NOTE When the terms of an invoice are 2/10, only 98% (100% − 2%) of the invoice must be paid during the first 10 days. In Example 2, the amount due may be found as follows.

$$\underbrace{\$840}_{\text{Invoice amount}} \times \underbrace{0.98}_{\text{Complement of 2\%}} = \underbrace{\$823.20}_{\text{Amount due}} + \underbrace{\$18.70}_{\text{Shipping and insurance}} = \$841.90$$

Problem-Solving Hint A cash discount is never taken on shipping and insurance charges. Be certain that shipping and insurance charges are excluded from the invoice amount before calculating the cash discount. Shipping and insurance charges must then be added back to find the total amount due.

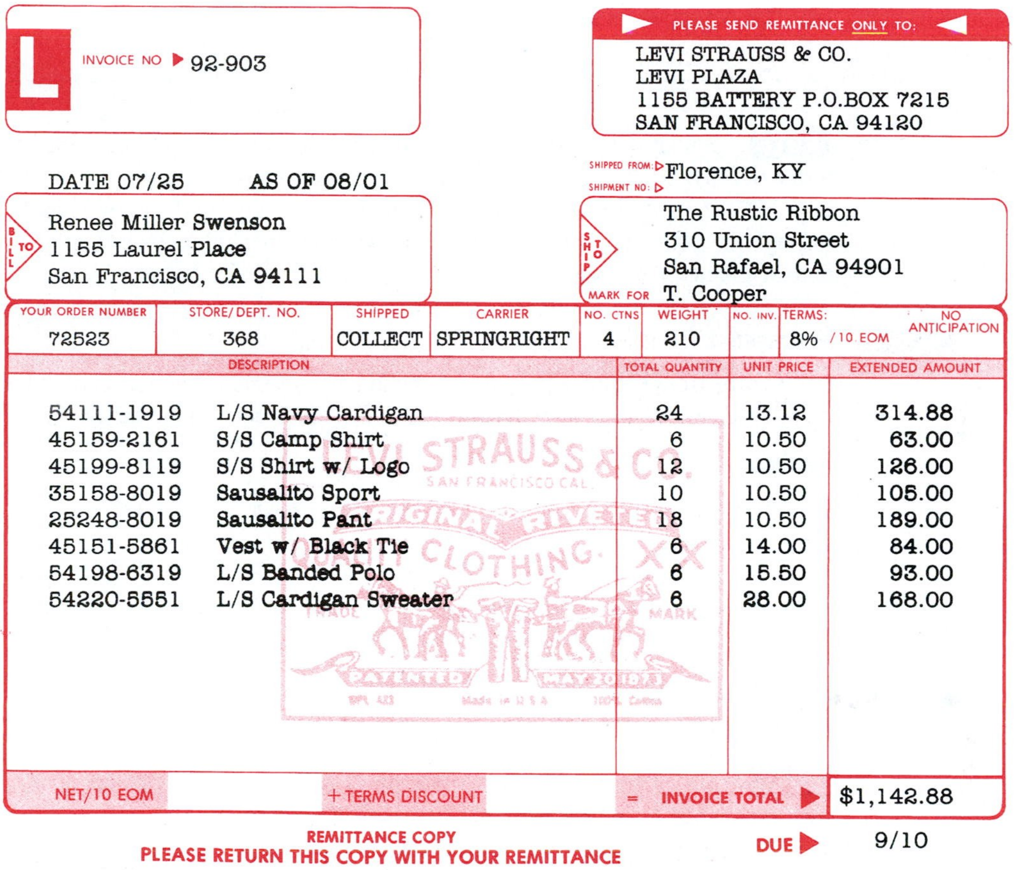

INVOICE NO ▶ 92-903

PLEASE SEND REMITTANCE ONLY TO:
LEVI STRAUSS & CO.
LEVI PLAZA
1155 BATTERY P.O.BOX 7215
SAN FRANCISCO, CA 94120

DATE 07/25 AS OF 08/01

SHIPPED FROM: Florence, KY
SHIPMENT NO:

BILL TO: Renee Miller Swenson
1155 Laurel Place
San Francisco, CA 94111

SHIP TO: The Rustic Ribbon
310 Union Street
San Rafael, CA 94901
MARK FOR T. Cooper

YOUR ORDER NUMBER	STORE/DEPT. NO.	SHIPPED	CARRIER	NO. CTNS	WEIGHT	NO. INV.	TERMS:	NO ANTICIPATION
72523	368	COLLECT	SPRINGRIGHT	4	210		8% /10 EOM	

	DESCRIPTION	TOTAL QUANTITY	UNIT PRICE	EXTENDED AMOUNT
54111-1919	L/S Navy Cardigan	24	13.12	314.88
45159-2161	S/S Camp Shirt	6	10.50	63.00
45199-8119	S/S Shirt w/ Logo	12	10.50	126.00
35158-8019	Sausalito Sport	10	10.50	105.00
25248-8019	Sausalito Pant	18	10.50	189.00
45151-5861	Vest w/ Black Tie	6	14.00	84.00
54198-6319	L/S Banded Polo	6	15.50	93.00
54220-5551	L/S Cardigan Sweater	6	28.00	168.00

NET/10 EOM + TERMS DISCOUNT = INVOICE TOTAL ▶ $1,142.88

REMITTANCE COPY
PLEASE RETURN THIS COPY WITH YOUR REMITTANCE

DUE ▶ 9/10

Figure 8.3

Objective 4 *Use Postdating When Calculating Cash Discounts.* In the ordinary dating method, the cash discount date and net payment date are both counted from the date of the invoice. Occasionally, the seller places a date later than the actual invoice date, sometimes labeling it "**as of**." This is called **postdating**. Notice that the Levi Strauss invoice in Figure 8.3 is dated "07/25 as of 08/01." The cash discount period and the net payment date are counted from 08/01 (August 1). This results in giving additional time for the purchaser to pay the invoice and receive the discount. The date due (9/10) is the due date on the invoice. The "terms:" of 8%/10 EOM are explained in Section 8.4.

Example 3 Using Postdating "as of" with Invoices

An invoice for some Australian glassware is dated October 21 as of November 1 with terms of 3/15, n/30. Find (a) the last date on which the cash discount may be taken and (b) the net payment date.

Solution

(a) Beginning with the postdate (as of) of November 1, the last date for taking the discount is November 16 (1 + 15).

(b) The net payment date is December 1 (29 days remaining in November and 1 day in December).

NOTE Sometimes a sliding scale of cash discounts is offered. For example, with terms of 3/10, 2/20, 1/30, n/60, a discount of 3% is given if payment is made within 10 days, 2% if paid from the 11th through the 20th day, and 1% if paid from the 21st through the 30th day. The entire amount (net) must be paid no later than 60 days from the date of the invoice. ■

Example 4 Determining Cash Discount Due Dates

An invoice from Cellular Products is dated May 18 and offers terms of 4/10, 3/25, 1/40, n/60. Find (a) the three final dates for cash discounts and (b) the net payment date.

Solution

(a) The three final cash discount dates are as follows.

4% if paid by May 28 (10 days from May 18)
3% if paid by June 12 (25 days from May 18)
1% if paid by June 27 (40 days from May 18)

(b) The net payment date is July 17.

NOTE Never take more than one of the cash discounts. With all methods of giving cash discounts, if the net payment period is not given, the net payment due date is assumed to be 20 days beyond the cash discount period. After that date, the invoice is considered overdue. If either the final discount date or the net payment date is on a Sunday or holiday, the next business day is used. Many companies insist that payment is made when it is received. It is general practice, however, to consider payment made when it is mailed. ■

Objective 5 *Determine the Amount Due When Goods Are Returned.* A buyer receiving incorrect or damaged merchandise may return the goods to the seller. The value of the returned goods must be subtracted from the amount of the invoice before calculating the cash discount.

Example 5 Finding the Amount Due When Goods Are Returned

An invoice from Homeproducts.com amounts to $380, is dated March 9, and offers terms of 4/10, net 30. If $75 of goods are returned and the invoice is paid on March 17, what amount is due?

Solution The invoice was paid 8 days after its date ($17 - 9 = 8$), so the 4% cash discount is taken. The discount is taken on $305.

$$\$380 \text{ (invoice amount)} - \$75 \text{ (goods returned)} = \$305 \text{ (goods retained)}$$

The cash discount is subtracted from the $305. Since $\$305 \times 0.04 = \$12.20$, the amount due is

$$\$305 \text{ (goods retained)} - \$12.20 \text{ (cash discount)} = \$292.80 \text{ (amount due)}$$

8.3 Exercises

Find the final discount date and the net payment date for each of the following problems.

	Invoice Date	As of	Terms
1.	Oct. 8		3/10, n/30
2.	April 12		2/10, n/30
3.	Feb. 25	Mar. 10	3/15, n/20
4.	Nov. 7	Nov. 18	3/10, n/30
5.	Sept. 11		4/10, n/60
6.	Mar. 23		2/10, n/30
7.	Jan. 14		2/10, n/60
8.	Oct. 3		3/10, net 15
9.	Dec. 7	Jan. 5	5/15, n/60
10.	July 31	Aug. 15	3/10, n/30

Solve for the amount of discount and the amount due on each of the following invoices.

	Invoice Amount	Invoice Date	Terms	Shipping and Insurance	Goods Returned	Date Invoice Paid
11.	$151.35	June 6	2/10 net 30	$12.58		June 15
12.	$66.10	Mar. 8	6/10, n/30	$4.39		Mar. 14
13.	$96.06	Nov. 30	Net 30	$5.22		Dec. 20
14.	$148	July 19	3/15, 1/25, n/60	$7.45		Aug. 16
15.	$724	Jan. 20	5/10, 2/20, n/30	$38.14		Feb. 5
16.	$1282	July 1	4/15, n/40	$21.40		July 7
17.	$780.70	May 5	net 30	$3.80	$125	June 1
18.	$162	Jan. 15	2/15, net 30	$8.18	$12	Jan. 23
19.	$635	Oct. 10	5/10, n/30	$53.18	$52	Oct. 18
20.	$1623.08	Nov. 12	2/10, n/30	$122.14	$187	Nov. 25

21. Describe the difference between a trade discount and a cash discount. Why are cash discounts offered? (See Objective 1.)

22. Using 2/10, n/30 as an example, explain what an ordinary cash discount means. (See Objective 2.)

Solve each of the following application problems.

23. HOME BAKING Grainworks Suppliers offer cash discounts of 2/10, 1/15, net 30 to all customers. An invoice dated May 18 amounting to $2010.70 is paid on June 1. Find the amount needed to pay the invoice.

24. **RUSSIAN ELECTRICAL SUPPLIES** A shipment of electrical supplies is received from the Lyskovo Electrotechnical Works. The invoice is dated March 8, amounts to $6824.58, and has terms of 2/15, 1/20 as of March 20. Find the amount needed to pay the invoice on April 2.

25. **AGRICULTURAL PRODUCTS** Agricultural Wholesale Products offers customers a trade discount of 10/20/5 on all products, with terms of net 30. Find the customers price for products with a total list price of $986 if the invoice was paid within 30 days.

26. **FISHING EQUIPMENT** Joe Nejad received an invoice for $586.12 for fishing tackle. The invoice was dated February 21, as of March 4, with terms of 4/10, 3/30, n/60. Find the total amount necessary to pay the invoice in full on March 22.

27. **SMALL MOTORS** A $1\frac{1}{2}$-horsepower motor is list priced at $215.80 with trade discounts of 20/20 and terms of 4/15, n/30. If a retailer takes both of these discounts, find the net cost of the motor.

28. **MARINE PARTS** An invoice for $12,472 from SeaRay Products is dated September 12 and has cash terms of 6/20, 4/35, 2/60, n/90. Find the amount necessary to pay in full on November 7.

29. **RECREATION EQUIPMENT** The list price of a popular brand of snowmobile is $5190. If a dealer can obtain trade discounts of 10/20/30 and cash terms of 4/10, n/30, find the lowest possible net cost.

30. **AUTOSOUND SYSTEMS** The list price of Road Thunder Speakers is $120. If the manufacturer offers trade discounts of 30/10 and terms of 3/15, n/30, find the net cost of the speakers assuming both discounts are taken.

31. **NUTRITIONAL PRODUCTS** James Nutrition offers retailers a trade discount of 10/20/10 on all purchases, with terms of 3/10, n/30. If the total list price of an order is $3215.80, find the retailers net cost if both discounts are earned.

32. **AUTO-SEAT COVERS** Sheepskin seat covers are list priced at $79.90 with a trade discount of 10/10/25 and terms of 3/10, n/30. Find the net cost assuming that both discounts are earned.

33. **RECREATION PRODUCTS** An invoice from Tower Recreation is dated April 14 and offers terms of 6/10, 4/20, 1/30, n/50. Find (a) the three final discount dates and (b) the net payment date.

34. **INVOICE TERMS** An invoice with terms of 4/15, 3/20, 1/30, n/60 is dated September 4. Find (a) the three final discount dates and (b) the net payment date.

35. **INVOICE TERMS** Truck Stuff receives an invoice dated March 28 AS OF April 5 with terms of 4/20, n/30. Find (a) the final discount date and (b) the net payment date.

36. **AS OF DATING** An invoice is dated May 20 AS OF June 5 with terms of 2/10, n/30. Find (a) the final discount date and (b) the net payment date.

37. **NURSERY SUPPLIES** Valley Nursery received an invoice for supplies amounting to $3724.40. The invoice is dated October 19 AS OF November 10 and offers terms of 2/20, n/40. Find the amount necessary to pay in full on November 28 if $104.50 worth of supplies are returned.

38. **FIREPLACE ACCESSORIES** The Fireside Shop received an invoice for hardware amounting to $218.80. The invoice is dated July 12 AS OF July 20 and offers terms

of 3/20, n/60. Find the amount necessary to pay in full on August 3 if $24.30 worth of goods are returned.

39. APPLIANCE REPAIRS An invoice received by Capital Appliance for repair parts amounts to $3322.80. The invoice is dated August 22 AS OF September 10 and offers terms of 4/10, 2/20, n/30. Find the amount necessary to pay in full on September 26 if $152.80 worth of goods are returned.

40. CHILDRENS RETAILER An invoice received by Sydney's Baby Barn amounts to $380.50. The invoice is dated October 25 AS OF November 2 and offers terms of 3/15, n/30. Find the amount necessary to pay in full on November 18 if $56.50 worth of goods are returned.

41. How do you remember the number of days in each month of the year? List the months and the number of days in each.

42. Explain how "AS OF" dating (postdating) works. Why is it used? (See Objective 4.)

8.4 Cash Discounts: Other Dating Methods

Objectives

1. *Solve cash discount problems with end-of-month dating.*
2. *Use receipt-of-goods dating to solve cash discount problems.*
3. *Use extra dating to solve cash discount problems.*
4. *Determine credit given for partial payment of an invoice.*

In addition to the ordinary dating method of cash discounts, there are several other cash discount methods that are in common use.

Objective 1 *Solve Cash Discount Problems with End-of-Month Dating.* **End-of-month dating** and **proximo dating**, abbreviated **EOM** and **prox.**, are both treated the same way. For example, both "3/10 EOM" and "3/10 prox." mean that a 3% cash discount may be taken if payment is made within 10 days. However, the 10 days are counted from the end of the month in which the invoice is dated. For example, an invoice dated July 14 with terms of 3/10 EOM would have a discount date 10 days from the end of the month, or August 10.

Since this is a method of increasing the length of time during which a discount may be taken, it has become common business practice with EOM and prox. dating to add a month *when the date of an invoice is the twenty-sixth of the month or later*. For example, if an invoice is dated March 25 and the discount offered is 3/10 prox., the last date on which the discount may be taken is April 10. However, if the invoice is dated March 26 (or any later day in March) and the cash discount offered is 3/10 prox., then the last date on which the discount may be taken is May 10.

Problem-Solving Hint The practice of adding an extra month when the invoice is dated the 26th of a month or after is used *only* with the end-of-month (proximo) dating cash discount. It does *not* apply to any of the other cash discount methods.

Example 1 Using End-of-Month Dating

An invoice from Harley-Davidson is dated June 10, with terms of 3/20 EOM. Find (a) the final date on which the cash discount may be taken and (b) the net payment date.

Solution

(a) The discount date is July 20 (20 days after the end of June).

(b) The net payment date is August 9 (20 days after the last discount date, July 20, since the net payment date is not otherwise given).

NOTE With all methods of cash discounts, if the net payment period is not given, the net payment due date is assumed to be 20 days beyond the cash discount date.

Example 2 Understanding Proximo Dating

Find the amount due on an invoice of $782 for some threaded fasteners that is dated August 3 if terms are 1/10 prox., and the invoice is paid on September 4.

Solution The last date on which the discount may be taken is September 10 (10 days after the end of August). September 4 is within the discount period, so the discount is earned. The 1% cash discount is computed on $782, the amount of the invoice. Subtract the discount, $7.82 ($782 × 0.01 = $7.82), from the invoice amount to find the amount due.

$$\$782 \text{ (invoice amount)} - \$7.82 \text{ (cash discount)} = \$774.18 \text{ (amount due)}$$

Objective 2 *Use Receipt-of-Goods Dating to Solve Cash Discount Problems.* **Receipt-of-goods dating**, abbreviated **ROG**, offers discounts determined from the date the goods are actually received. This method is often used when shipping time is long. The invoice might arrive overnight by mail, but the goods may take several weeks. Under the ROG method of cash discount, the buyer is given time to receive and inspect the merchandise and then benefit from a cash discount. For example, the discount "3/15 ROG" allows a 3% cash discount if paid within 15 days from receipt of goods. If the invoice was dated March 5 and goods were received on April 7, the last date to take the 3% cash discount is April 22 (April 7 plus 15 days). The net payment date, since it is not stated, is 20 days after the last discount date, or May 12 (April 22 plus 20 days).

Example 3 Using Receipt-of-Goods Dating

Java City received an invoice dated December 12, with terms of 2/10 ROG. The goods were received on January 2. Find (a) the final date on which the cash discount may be taken and (b) the net payment date.

Solution

(a) The discount date is January 12 (10 days after receipt of goods, January 2 + 10 days).

(b) The net payment date is February 1 (20 days after the last discount date).

Example 4 Working with ROG Dating

Find the amount due on an invoice of $285 for some printing services, with terms of 3/10 ROG, if the invoice is dated June 8, the goods are received June 18, and the invoice is paid June 30.

Solution The last date to take the 3% cash discount is June 28, 10 days after June 18. Since the invoice is paid on June 30, 2 days after the last discount date, no cash discount may be taken. The entire amount of the invoice must be paid.

$285 (invoice amount) − $0 (no discount) = $285 (amount due)

Objective 3 ***Use Extra Dating to Solve Cash Discount Problems.*** **Extra dating**, abbreviated **extra**, **ex.**, or **x**, gives additional time to the buyer to take a cash discount. For example, the discount "2/10-50 extra" or "2/10-50 ex." or "2/10-50 x" allows a 2% cash discount if paid within (10 + 50) or 60 days from the date of the invoice. The discount is written in this form rather than combining the 10 + 50 and writing 2/60 to show that the 50 days are *extra*, or in addition to the normal 10 days offered.

There are several reasons for using extra dating. A supplier might extend the discount period during a slack season to generate more sales or perhaps to gain a competitive advantage. For example, Christmas merchandise might be offered with extra dating, allowing the buyer to take the cash discount after the holiday selling period.

Example 5 Using Extra Dating

An invoice for paint accessories is dated November 23 with terms 2/10–50 ex. Find (a) the final date on which the cash discount may be taken and (b) the net payment date.

Solution

(a) The discount date is January 22 (7 days remaining in November + 31 days in December = 38; thus, 22 more days are needed in January to total 60).

(b) The net payment date is February 11 (20 days after the last discount date).

Example 6 Understanding Extra Dating

An invoice for some plumbing parts is dated August 5, amounts to $2250, offers terms of 3/10–30 x, and is paid on September 12. Find the net payment.

Solution The last day to take the 3% cash discount is September 14 (August 5 + 40 days = September 14). Since the invoice is paid on September 12, the 3% discount may be taken. The 3% cash discount is computed on $2250, the amount of the invoice.

$$3\% \times \$2250 = 0.03 \times \$2250 = \$67.50$$

The discount to be taken is $67.50. The cash discount is subtracted from the invoice amount to determine the amount of payment.

$2250 (invoice amount) − $67.50 (cash discount) = $2182.50 (amount of payment)

When customers pay invoices quickly, there is less need for a business to borrow money. In certain industries it is common to deduct interest that would have to be paid on borrowed money from the invoice amount. To do this, the company uses the current rate of interest and calculates the amount of interest over the remaining days on which the cash discount is allowable. This deduction, known as anticipation, is taken in addition to the cash discount earned. Anticipation involves the use of simple interest, which is discussed in Chapter 11.

Objective 4 *Determine Credit Given for Partial Payment of an Invoice.* Occasionally, a customer may pay only a portion of the total amount due on an invoice. If this **partial payment** is made within a discount period, the customer is entitled to a discount on the portion of the invoice that is paid.

If the terms of an invoice are 3%, 10 days, then only 97% (100% − 3%) of the invoice amount must be paid during the first 10 days. So, for each $0.97 paid, the customer is entitled to $1.00 of credit. When a partial payment is made, the credit given for the partial payment is found by dividing the partial payment by the complement of the cash discount percent. Then, to find the balance due, subtract the credit given from the invoice amount. The cash discount is found by subtracting the partial payment from the credit given.

Example 7 Finding Credit for Partial Payment

Daves Body and Paint receives an invoice for $1140, dated March 8, that offers terms of 2/10 proximo. A partial payment of $450 is made on April 5. Find (a) the amount credited for the partial payment, (b) the balance due on the invoice, and (c) the cash discount earned.

Solution

(a) The cash discount is earned on the $450 partial payment made on April 5 (April 10 was the last discount date).

$$100\% - 2\% = 98\% = 0.98$$

$$\text{Amount paid} = 0.98 \times \text{Credit given}$$

$$\$450 = 0.98 \times C$$

$$\$450 = 0.98C$$

$$\frac{\$450}{0.98} = \frac{\cancel{0.98}C}{\cancel{0.98}}$$

$$\frac{\$450}{0.98} = C$$

$$= \$459.18 \quad \text{Credit given (rounded)}$$

(b) Balance due = Invoice amount − Credit given

$$\text{Balance due} = \$1140 - \$459.18$$

$$= \$680.82$$

(c) Cash discount = Credit given − Partial payment

$$\text{Cash discount} = \$459.18 - \$450$$

$$= \$9.18$$

Scientific
Calculator Approach

A calculator solution to Example 7 will include these three steps.

(a) First, find the amount of credit given.

450 [÷] .98 [=] 459.18 (rounded)

(b) Next, store the amount of credit and subtract this amount from the invoice amount to find the balance due.

STO 1140 − RCL = 680.82 (rounded)

(c) Finally, subtract the partial payment from the amount of credit given to find the cash discount.

RCL − 450 = 9.18 (rounded)

NOTE Cash discounts are important, and a business should make the effort to pay invoices early to earn the cash discount. In many cases the money saved through cash discounts has a great effect on the profitability of a business. Often companies will borrow money to enable them to take advantage of cash discounts. The mathematics of this type of loan is discussed in Section 11.1, "Basics of Simple Interest." ■

The cash discounts discussed here are normally not used when selling to foreign customers or purchasing from foreign suppliers. Instead, other types of discounts may be offered to reduce the price of goods sold to foreign buyers. These discounts may be given as allowances for tariffs paid (import duties) by the customer, reimbursement for shipping, and insurance paid by the customer or in the form of an advertising allowance.

8.4 Exercises

Find the discount date and net payment date for each of the following (the net payment date is 20 days after the final discount date).

	Invoice Date	Terms	Date Goods Received
1.	Feb. 8.	3/10 EOM	
2.	July 14	2/15 ROG	Sept. 3
3.	Nov. 22	1/10–20 x	
4.	May 17	6/30 EOM	
5.	April 12	3/15–50 x	
6.	Oct. 30	1/10 ROG	Dec. 12
7.	June 26	2/10 EOM	
8.	Sept. 27	3/15 prox.	
9.	July 6	2/20 ROG	Aug. 4
10.	Jan. 15	3/15 ROG	Feb. 5

Solve for the amount of discount and the amount due on each of the following.

	Invoice Amount	Invoice Date	Terms	Date Goods Received	Date Invoice Paid
11.	$682.28	June 4	3/20 ROG	July 25	Aug. 10
12.	$356.20	May 17	3/15 prox.		June 12
13.	$194.04	Aug. 22	5/10–60 ex.		Nov. 5
14.	$9,240.40	Jan. 8	1/20 ROG	Mar. 10	Mar. 23
15.	$2,960	Oct. 31	2/10 EOM		Dec. 5
16.	$127.50	Feb. 17	3/20 ROG	Mar. 19	Apr. 10
17.	$4,220	Oct. 4	4/15 prox.		Nov. 10
18.	$256.50	July 17	3/10–40 extra		Sept. 2
19.	$12.38	Mar. 29	2/15 ROG	April 15	April 30
20.	$11,480	April 6	2/15 prox.		April 30
21.	$3,250.60	Oct. 17	3/15–20 x		Oct. 20
22.	$8,318	June 9	3/20 EOM		July 18
23.	$1,708.18	Nov. 13	4/10 prox.		Dec. 10
24.	$13,728.34	April 6	2/10 ROG	April 28	May 6

25. Quite often there is no mention of a net payment date on an invoice. Explain the common business practice when no net payment date is given. (See Objective 1.)

26. Describe why receipt-of-goods dating (ROG) is offered to customers. Use an example in your description. (See Objective 2.)

Find the credit given and the balance due on the invoice after making the following partial payments.

	Invoice Amount	Invoice Date	Terms	Date Invoice Paid	Partial Payment
27.	$3150	Jan. 9	2/10 EOM	Feb. 8	$1862
28.	$484	June 6	5/10–30 x	July 10	$209
29.	$1750	Aug. 12	5/10–30 x	Sept. 15	$684
30.	$920	Jan. 11	3/10, 2/15, n/30	Jan. 23	$450.80
31.	$160	Dec. 8	3/10, n/30	Dec. 15	$97
32.	$8120	Oct. 4	4/20 prox.	Nov. 15	$2016

33. Write a short explanation of partial payment. Why would a company accept a partial payment? Why would a customer make a partial payment? (See Objective 4.)

34. Of all the different types of cash discounts presented in this section, which type seemed most interesting to you? Explain your reasons.

Solve each of the following application problems in cash and trade discounts.

35. BATHROOM DECOR An invoice received by Bath and Home is dated August 18 with terms of 2/15 EOM. Find (a) the final date on which the discount may be taken and (b) the net payment date.

36. RADIATOR SUPPLIES An invoice from Steamers Radiator Supply is dated February 23, with terms of 3/20 ROG, and the goods are received on March 2. Find (a) the final date on which the cash discount may be taken and (b) the net payment date.

37. GLASS REPAIR SUPPLIES An invoice from Carmichael Glass is dated November 11 with terms of 3/20 ROG, and the goods are received on December 3. Find (a) the final date on which the cash discount may be taken and (b) the net payment date.

38. STAINED GLASS SUPPLIES An invoice received from the Stained Glass Exchange is dated June 16, with terms of 2/10 prox. Find (a) the final date on which the cash discount may be taken and (b) the net payment date.

39. RECEIPT OF GOODS DATING Find the amount due on an invoice of $1525 with terms of 1/20 ROG. The invoice is dated October 20, goods are received December 1, and the invoice is paid on December 20.

40. CANADIAN FOOD PRODUCTS Abigal Shellist, a wholesaler of Canadian food products, offers terms of 4/15–40 ex. to encourage the sales of her products. In a recent order, a retailer purchased $9864.18 worth of Canadian foods and was offered the above terms. If the invoice was dated March 10, find (a) the final date on which the cash discount may be taken and (b) the amount paid if the discount was earned.

41. POWER SAW BLADES Big Tooth Blades, a power saw blade distributor, offers terms of 2/10–30 x to stimulate slow sales in the winter months. Oaks Hardware purchased $970.68 worth of saw blades and was offered the above terms. If the invoice was dated November 3, find (a) the final date on which the cash discount may be taken and (b) the amount paid if the discount was earned.

42. EXTRA DATING INVOICE Find the payment that should be made on an invoice dated September 28, amounting to $4680, offering terms of 2/10–50 x and paid on November 25.

43. PLAYGROUND EQUIPMENT A recent invoice for some playground equipment amounted to $4358.50, was dated February 20, and offered terms of 2/20 ROG. If the equipment was received on March 20 and the invoice was paid on April 8, find the amount due.

44. OFFICE SUPPLIES Claudia Aldea purchased some copy paper and other supplies for her insurance office and was offered a cash discount of 2/10 EOM. The invoice amounted to $178.72 and was dated June 2. The forms were received 7 days later, and the invoice was paid on July 7. Find the amount necessary to pay the invoice in full.

45. INVOICE PARTIAL PAYMENT An invoice amounting to $1920 is dated July 23 by Lexington Foot Locker and offers cash terms of 6/30–120 x. If a partial payment of $940 is made on November 28, find (a) the credit given for the partial payment and (b) the balance due on the invoice.

46. LIGHTING SALES Lamps For Less receives an invoice amounting to $5832, with cash terms of 3/10 prox. and dated August 9. If a partial payment of $3350 is made

on August 15, find (a) the credit given for the partial payment and (b) the balance due on the invoice.

47. HOCKEY EQUIPMENT An invoice dated December 8 is received with a shipment of hockey equipment from Canada on April 18 of the following year. The list price of the equipment is $2538, with allowed series discounts of 25/10/10. If cash terms of sale are 3/15 ROG, find the amount necessary to pay in full on April 21.

48. WATERFORD CRYSTAL William Glen receives an invoice for some Waterford Crystal from Ireland amounting to $3628.10 and dated May 17. The terms of the invoice are 5/20–90 x and the invoice is paid on September 2. Find the amount necessary to pay the invoice in full.

49. BEAUTY SUPPLIES Michael Anderson Beauty Supplies offers series discounts of 15/10 with terms of 5/15–30 x. On an invoice dated June 4 for items list priced at $128, find the amount necessary to pay the invoice in full on July 15.

50. COPY SUPPLIES The Copy Corner receives an invoice amounting to $388.20, with terms of 8/10, net/30, and dated August 20 AS OF September 1. If a partial payment of $225 is made on September 8, find (a) the credit given for the partial payment and (b) the balance due on the invoice.

51. FROZEN DESSERT SHOP The Frozen Dessert Shop receives an invoice amounting to $526.80, with terms of 2/20 prox. and dated October 5. If a partial payment of $300 is made on November 12, find (a) the credit given for the partial payment and (b) the balance due on the invoice.

52. CONSTRUCTION COMPANY An invoice of $7819.20 with terms of 4/10 prox. is received by Penny Carter Construction and is dated May 2. If a partial payment of $6000 is made on May 8, find (a) the credit given for the partial payment and (b) the balance due on the invoice.

53. PARTIAL INVOICE PAYMENT An invoice received for some precision hand tools from Germany has terms of 3/15–30 x and is dated May 20. The amount of the invoice is $1120.15, and a partial payment of $580 is made on July 1. Find (a) the credit given for the partial payment and (b) the balance due on the invoice.

54. JEEP SUPPLIES Jeepers Supply makes a partial payment of $660 on an invoice of $1491.54. If the invoice is dated April 14 with terms of 4/20 prox. and the partial payment is made on May 13, find (a) the credit given for the partial payment and (b) the balance due on the invoice.

Chapter 8 Quick Review

Chapter Terms

Review the following terms to test your understanding of the chapter. For each term you do not know, refer to the page number found next to that term:

amount [p. 285]
cash discount [p. 287, 299]
cash on delivery (COD) [p. 285]
chain discount [p. 287]
complement [p. 288]
consumer [p. 284]
electronic commerce (EC) [p. 285]
EOM (end of month) [p. 307]
extension total [p. 285]
extra dating (extra, ex., x) [p. 305]
free alongside ship (FAS) [p. 285]
free on board (FOB) [p. 285]
invoice [p. 285]
invoice total [p. 285]
list price [p. 286]
manufacturer [p. 284]
net cost [p. 286]
net cost equivalent [p. 289]
net price [p. 287]
ordinary dating method [p. 301]
partial payment [p. 310]
percent paid [p. 289]
postdated "as of" [p. 303]
proximo (prox.) dating [p. 307]
purchase invoice [p. 285]
receipt-of-goods dating (ROG) [p. 308]
retailer [p. 284]
sales invoice [p. 285]
series discount [p. 287]
single discount [p. 287]
single discount equilvalent [p. 284]
total invoice amount [p. 285]
trade discounts [p. 286]
unit price [p. 285]
units shipped [p. 285]
wholesaler [p. 284]

Concepts	Examples
8.1 Trade discount and net cost First find the amount of the trade discount. Then use the formula: Net cost = List price − Trade discount	List price, \$28; trade discount, 25%. Find the net cost. $\$28 \times 0.25 = \$7$ Net cost = \$28 − \$7 = \$21
8.1 Complements with respect to 1 The complement is the number that must be added to a given discount to get 1 or 100%.	Find the complement with respect to 1 for each of the following. **(a)** $10\% + x = 100\%$ $x = 100\% - 10\%$ $x = 90\%$ **(b)** $12\frac{1}{2}\%$; complement = 87.5% **(c)** 50%; complement = 50%
8.1 Complement and series discounts The complement of a discount is the percent paid. Multiply the complements of the discounts in the series to get the net cost equivalent.	Series discount, 10/20/10. Find the net cost equivalent. 10 / 20 / 10 ↓ ↓ ↓ $0.9 \times 0.8 \times 0.9 = 0.648$
8.1 Net cost equivalent (percent paid) and the net cost Multiply the net cost equivalent (percent paid) by the list price to get the net cost. Percent paid × List price = Net cost	List price, \$280; series discount 10/30/20. Find the net cost. 10 / 30 / 20 ↓ ↓ ↓ $0.9 \times 0.7 \times 0.8 = 0.504$ percent paid $0.504 \times \$280 = \$141.12$

8.2 Single discount equivalent to a series discount

Often needed to compare one series discount to another, the single discount equivalent is found by subtracting the net cost equivalent from 1.

1 − Net cost equivalent = Single discount equivalent

What single discount is equivalent to a 10/20/20 series discount?

10 / 20 / 20
↓ ↓ ↓
$$0.9 \times 0.8 \times 0.8 = 0.576$$
$$1 - 0.576 = 0.424 = 42.4\%$$

8.2 Finding net cost using complements of individual discounts

To find the net cost, multiply the list price by the product of the complements of the individual discounts.

List price, $510; series discount, 30/10/5. Find the net cost.

30 / 10 / 5
↓ ↓ ↓
$$\$510 \times 0.7 \times 0.9 \times 0.95 = \$305.24 \text{ (rounded)}$$

8.2 Finding list price if given the series discount and the net cost

First, find the net cost equivalent (percent paid), then use the formula to find the list price.

$$P \times L = N$$

Percent paid × List price = Net cost

Net cost, $224; series discount, 20/20. Find list price.

20 / 20
↓ ↓
$$0.8 \times 0.8 = 0.64$$
$$0.64 \times \text{List price} = \$224$$
$$0.64L = \$224$$
$$L = \$350$$

8.2 Determining the trade discount that must be added to meet a competitor's price

First use the formula $P \times L = N$ to find the single discount needed. The answer is the complement of the discount that must be added; subtract it from 100% to get the discount.

List price $640; trade discount 25%. Find the trade discount that must be added to match the competitor's price of $432.

$$P \times L = N$$
$$0.75 \times \text{Complement of discount} \times L = N$$
$$0.75 \times d \times \$640 = \$432$$
$$\$480d = \$432$$
$$d = 0.9 = 90\%$$
$$100\% - 90\% = 10\% \text{ additional discount}$$

8.3 Determining number of days and dates

30-day months	31-day months
April	All the rest except
June	February with 28 days
September	(29 days in leap year)
November	

Date, July 24. Find 10 days from date.

July 31 − 24 = 7 remaining in July

10	Total number of days
− 7	Days remaining in July
3	August—future date

8.3 Ordinary dating and cash discounts With ordinary dating, count days from the date of the invoice. Remember: 2 / 10, *n* / 30 ↓ ↓ ↓ ↓ % days net days	Invoice amount $182; terms 2/10, n/30. Find cash discount and amount due. Cash discount: $\$182 \times 0.02 = \$3.64$ Amount due: $\$182 - \$3.64 = \$178.36$
8.3 Returned goods Subtract returned goods amount from invoice before calculating the cash discount.	Invoice amount, $95; returned goods, $15; terms, 3/15, n/30. Find amount due if discount is earned. Cash discount: $(\$95 - \$15) \times 0.03 = \$2.40$ Amount due: $\$95 - \$15 - \$2.40 = \$77.60$
8.4 Cash discounts with end-of-month dating (EOM or proximo) Count the final discount date and the net date from the end of the month. If the invoice is dated the 26th or after, add the entire following month when determining the dates. If not stated, the net date is 20 days beyond the discount date.	Terms, 2/10 EOM; invoice date, Oct. 18. Find the final discount date and the net payment date. Final discount date: November 10, which is 10 days from the end of October Net payment date: November 30, which is 20 days beyond the discount date
8.4 Receipt-of-goods dating and cash discounts (ROG) Time is counted from the date goods are received to determine the final cash discount date and the net payment date.	Terms, 3/10 ROG; invoice date, March 8; goods received, May 10. Find the final discount date and the net payment date. Final discount date: May 20 (May 10 + 10 days) Net payment date: June 9 (May 20 + 20 days)
8.4 Extra dating and cash discounts Extra dating adds extra days to the usual cash discount period; for example, 3/10–20 x means 3/30.	Terms, 3/10–20 x; invoice date, January 8. Find the final discount date and the net payment date. Final discount date: February 7 (23 days in January + 7 days in February = 30) Net payment date: February 27 (February 7 + 20 days)
8.4 Partial payment credit When only a portion of an invoice amount is paid within the cash discount period, credit is given for the partial payment. Use the formula Amount paid = (1 − Discount rate) × Credit given	Invoice, $400; terms, 2/10, n/30; invoice date, Oct. 10; partial payment of $200 on Oct. 15. Find credit given for partial payment and the balance due on the invoice. $\$200 = (1 - 0.02) \times \text{Credit given}$ $\$200 = 0.98C$ $C = \frac{200}{0.98} = \$204.08$ (rounded) Credit $= \$400 - \$204.08 = \$195.92$ Balance due

Chapter 8 Review Exercises

Find the net cost (invoice amount) for each of the following. Round to the nearest cent. [8.1]

1. List price: $480 less 20/10
2. List price: $276 less $10/12\frac{1}{2}/10$
3. List price: $2830 less 5/15/20
4. List price: $1620 less 20/25/15

Find the net cost equivalent and the percent form of the single discount equivalent for each of the following series discounts. [8.2]

5. 25/15
6. 20/10/20
7. $20/32\frac{1}{2}$
8. 10/20/10/30

Find the list price, given the net cost and the series discount. [8.2]

9. Net cost, $361.50; trade discount, 10/20
10. Net cost, $1050.74; trade discount, 15/20
11. Net cost, $328.70; trade discount, 10/20/15
12. Net cost, $1289.40; trade discount, 5/20/25

Find the final discount date and net payment date for each of the following. (The net payment date is 20 days after the final discount date.) [8.3 and 8.4]

	Invoice Date	Terms	Date Goods Received
13.	Feb. 10	4/15 EOM	Feb. 16
14.	May 8	2/10 ROG	May 20
15.	Dec. 4	3/10 prox	Dec. 8
16.	Oct. 20	2/20–40 extra	Oct. 31

Solve for the amount of discount and the amount due on each of the following. [8.4]

	Invoice Amount	Invoice Date	Terms	Shipping and Insurance	Date Goods Received	Date Invoice Paid
17.	$1280.40	March 16	3/15 ROG	$76.18	April 7	April 20
18.	$945.60	May 9	3/15 proximo		May 20	June 12
19.	$875.50	Feb. 20	4/15 EOM	$67.18	Mar. 1	Mar. 12
20.	$2210.60	Aug. 5	2/10–60 x		Sept. 10	Oct. 13

Find the credit given and the balance due on the invoice after making the following partial payments. Round to the nearest cent. [8.4]

	Invoice Amount	Invoice Date	Terms	Date Invoice Paid	Partial Payment
21.	$660	February 2	2/10, n/30	February 10	$300
22.	$5310	April 22	3/15 EOM	May 14	$2520
23.	$860	July 23	1/10 prox.	August 5	$500
24.	$3850	September 17	3/10–40 x	November 2	$2050

Solve each of the following application problems in cash and trade discounts.

25. The following invoice was paid on November 15. Find (a) the invoice total, (b) the amount that should be paid after the cash discount, and (c) the total amount due, including shipping and insurance. [8.1 and 8.3]

HOLIDAY APPLIANCE REPAIR PARTS

Terms: 2/10, 1/15, n/60 November 6

Quantity	Description	Unit Price	Extension Total
16	M-2 mixers	@ 17.50 ea.	______
8	shelf brackets	@ 3.25 ea.	______
4	blender, model L	@ 12.65 ea.	______
12	bowls, 1 qt. stainless	@ 3.15 ea.	______
		Invoice Total	______
		Cash Discount	______
		Due after Cash Discount	______
		Shipping and Insurance	$ 11.55
		Total Amount Due	______

26. Fireside Shop offers chimney caps for $120 less 25/10. The same chimney cap is offered by Builders Supply for $111 less 25/5. Find (a) the firm that offers the lower price and (b) the difference in price. [8.1]

27. Kelly Melcher Furnishings made purchases at a net cost of $36,458 after a series discount of 20/20/10. Find the list price. [8.2]

28. Restaurant Distributing offers a commercial pasta maker for $980 with a trade discount of 25%. Find the trade discount that must be added to match a competitor's price of $661.50. [8.2]

29. An invoice amounting to $2018 is dated September 18 and offers terms of 3/20 EOM. If $183 of goods are returned and the invoice is paid on October 18, what amount is due? [8.4]

30. Freitas Pneumatic Service receives an invoice dated April 22 for $1854 with terms of 3/15 EOM. If the invoice is paid on May 12, find the amount necessary to pay the invoice in full. [8.4]

31. An invoice of $838 from Kara-Dolls has cash terms of 2/15 EOM and is dated March 8. Find (a) the final date on which the cash discount may be taken and (b) the amount necessary to pay the invoice in full if the cash discount is earned. [8.4]

32. An invoice from Round Table Pizza Products amounts to $5280, was dated November 1, and offers terms of 4/15 proximo. A partial payment of $1800 is made on December 12. Find (a) the amount credited for the partial payment, (b) the balance due on the invoice, and (c) the cash discount earned. **[8.4]**

Chapter 8 Summary Exercise
The Retailer: Invoices, Trade Discounts, and Cash Discounts

Dick and Jeanne Hill of Oaks Hardware order most of their merchandise from Ace Hardware Wholesalers. In early May they order patio and garden items having a total list price of $2893 and general hardware merchandise having a total list price of $3138. Ace Hardware Wholesalers offers trade discounts of 25/15 on these items and charges for shipping.

The invoice for this order arrives a few days later, is dated May 12, has terms of 3/15 EOM, and shows a shipping charge of $175.14. Oaks Hardware will need to know all of the following.

(a) The total amount of the invoice excluding shipping.
(b) The final discount date.
(c) The net payment date.
(d) The amount necessary to pay the invoice in full on June 11 including the shipping.
(e) Suppose that on June 11 the invoice is not paid in full, but a partial payment of $2500 is made instead. Find the credit given for the partial payment and the balance due on the invoice including shipping.

Net Assets Business on the Internet

Ace Hardware

Statistics

- 1924: Established
- 1999: 5100 retail stores in 61 countries
- 2000: $3.0 billion in annual sales

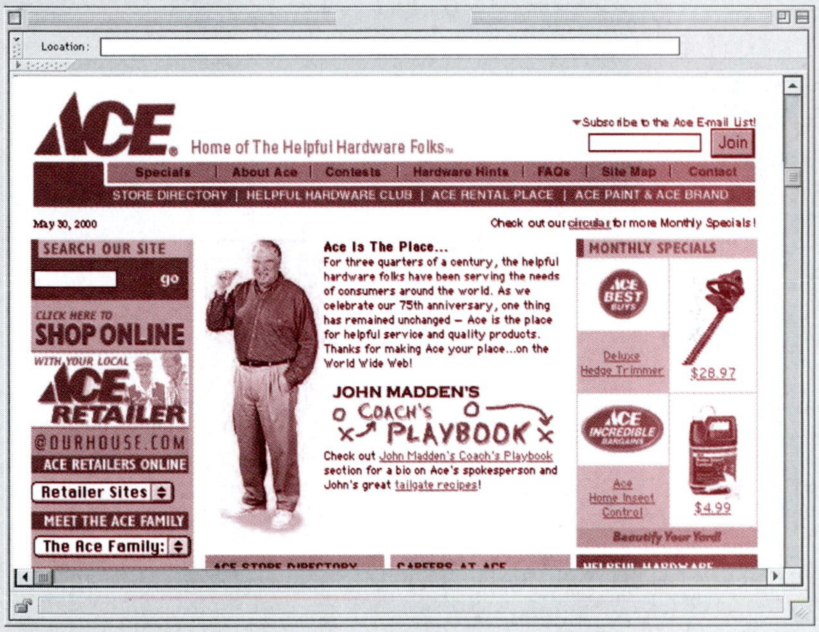

In the entrepreneurial spirit of the 1920s, four Chicago men formed Ace Hardware Stores. Each of the men had operated his own independent hardware business, but in 1924 they joined together to improve their buying power, increase their profits, and share common costs. This marked the beginning of Ace Hardware Stores, Inc., a name selected for the determination and outstanding qualities shown by the ace fighter pilots of World War I.

Today, Ace Hardware faces a new breed of competition—large warehouse-type chains. To meet and beat this new challenge head-on, the 5100 independent Ace Hardware Store owners have launched The New Retail Age of Ace, a strategic plan to provide consumers with the best products and service in the marketplace.

1. An Ace Hardware Store owner purchased adjustable shower-curtain rods that were list priced at $8.50. If the supplier offered a trade discount of 20/20, find the cost of one dozen shower-curtain rods.

2. The list price of a high-pressure power washer is $398. The manufacturer gives a 25/10 trade discount and offers a cash discount of 3/15, net/30. Find the cost to the hardware store if both discounts are earned and taken.

3. An invoice from a supplier is dated July 15 and offers cash discount terms of 2/20 EOM. Find **(a)** the final discount date and **(b)** the net payment date.

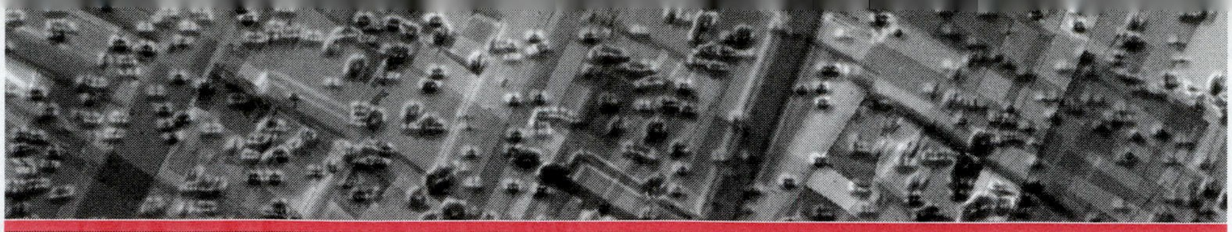

9 Markup

The success of a business depends on many things. Certainly two of the most important are the products that it sells and the price that the business charges for its goods and services. Prices must be low enough to attract customers, yet high enough to cover all operating expenses and provide a profit. Likewise, the product selection must match the needs of the customer. For example, Figure 9.1 shows several favorite gift choices for fathers. This is valuable information to manufacturers, retailers, and all mass merchandisers.

What types of decisions might a business owner make as a result of having this information?

The difference between the price a business pays for an item and the price at which the item is sold is called markup. For example, if a store buys a package of blank videotapes for $11 from a wholesaler and sells it for $15, the markup is $4. There are two standard methods of calculating markup—as a percent of cost and as a percent of selling price. This chapter discusses these two methods, along with how to convert markups from one method to the other and how to use markup to allow for spoilage.

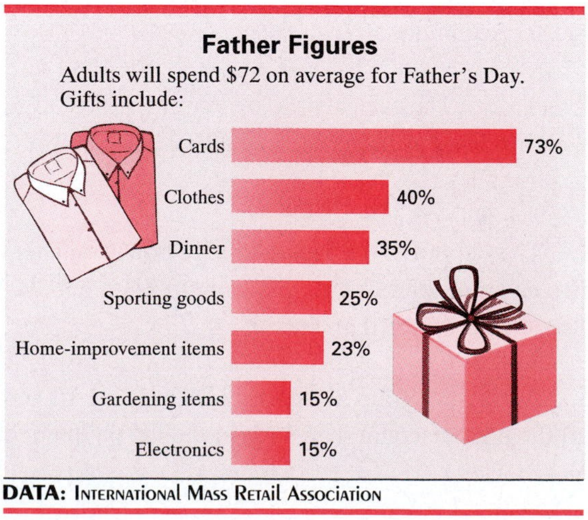

Figure 9.1

9.1 Markup on Cost

Objectives

1. *Recognize the terms used in selling.*
2. *Know the basic markup formula.*
3. *Calculate markup based on cost.*
4. *Apply percent to markup problems.*

Objective 1 *Recognize the Terms Used in Selling.* The following terms are used in markup.

Cost is the price paid to the manufacturer or supplier after trade and cash discounts have been taken. Shipping and insurance charges are included in the cost. This is often called the **wholesale price**.

Selling price, or **retail price**, is the price at which merchandise is offered for sale to the public.

Markup, also called **margin** or **gross profit**, is the difference between the cost and the selling price. These three terms are often used interchangeably.

Operating expenses, (or **overhead**), include the many expenses of business operation, such as wages and salaries of employees, rent for buildings and equipment, utilities, insurance, and advertising. Even an expense item like postage can add up. Mailing costs average from 6.2% of operating expense for small companies to as high as 9.2% for the largest companies. Postal rates have gone up considerably during the last three decades. These postal rates, along with the most current postal rates, are shown below.

Net profit (net earnings) is the amount, if any, remaining for the business after the cost of goods and operating expenses have been paid. (Income tax for the business is computed on net profit.)

U.S. Postal Rates Since 1971

May 16, 1971	8 cents
March 2, 1974	10 cents
Dec. 31, 1975	13 cents
May 29, 1978	15 cents
March 22, 1981	18 cents
Nov. 1, 1981	20 cents
Feb. 17, 1985	22 cents
April 3, 1988	25 cents
Feb. 3, 1991	29 cents
Jan. 1, 1995	32 cents
Jan. 1, 1999	33 cents

Cancel That

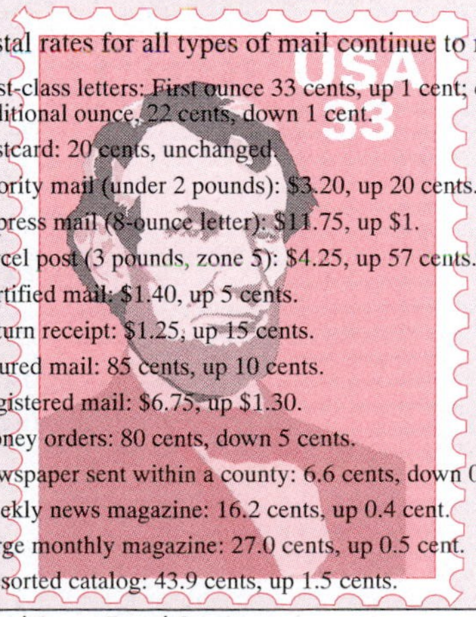

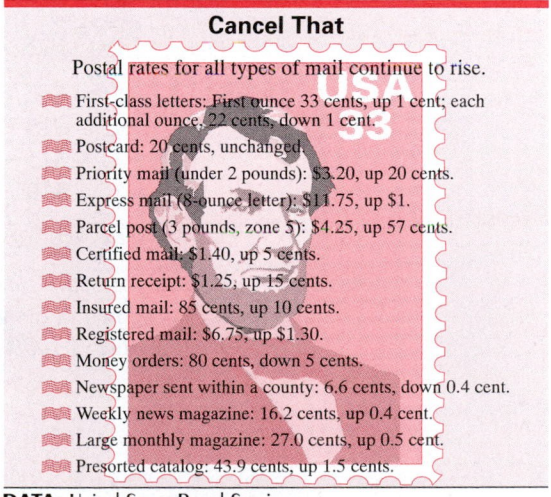

Postal rates for all types of mail continue to rise.

- First-class letters: First ounce 33 cents, up 1 cent; each additional ounce, 22 cents, down 1 cent.
- Postcard: 20 cents, unchanged.
- Priority mail (under 2 pounds): $3.20, up 20 cents.
- Express mail (8-ounce letter): $11.75, up $1.
- Parcel post (3 pounds, zone 5): $4.25, up 57 cents.
- Certified mail: $1.40, up 5 cents.
- Return receipt: $1.25, up 15 cents.
- Insured mail: 85 cents, up 10 cents.
- Registered mail: $6.75, up $1.30.
- Money orders: 80 cents, down 5 cents.
- Newspaper sent within a county: 6.6 cents, down 0.4 cent.
- Weekly news magazine: 16.2 cents, up 0.4 cent.
- Large monthly magazine: 27.0 cents, up 0.5 cent.
- Presorted catalog: 43.9 cents, up 1.5 cents.

DATA: United States Postal Service

Most manufacturers, many wholesalers, and some retailers calculate markup as a percent of cost, called **markup on cost**. Manufacturers, who usually evaluate their inventories on the basis of cost, find this method to be most consistent with their business operations. Retailers, on the other hand, usually compute **markup on selling price** since retailers compare most areas of their business operations to sales revenue. Such items as sales commissions, sales taxes, advertising, and other items of expense are expressed as a percent of sales. It is reasonable, then, for the retailer to express markup as percent of sales. Wholesalers, however, use either cost or selling price, so be sure to find out which is being used.

Objective 2 *Know the Basic Markup Formula.* Whether markup is based on cost or on selling price, the same basic **markup formula** is always used. This formula is as follows.

$$\text{Cost} + \text{Markup} = \text{Selling price}$$
$$C + M = S$$

The markup formula is illustrated in Figure 9.2. Most problems in markup give two of the items in the formula and ask for the third.

Example 1 Using the Basic Markup Formula

The Red Balloon Bookstore received some new children's books. Use the markup formula to determine the selling price, markup, and cost of the books in the following problems.

(a) Cost = \$10
Markup = \$5
Selling price = ________

(b) Cost = \$10
Selling price = \$15
Markup = ________

(c) Markup = \$5
Selling price = \$15
Cost = ________

Solution

(a)
$$C + M = S$$
$$\$10 + \$5 = S$$
$$\$15 = S$$

(b)
$$C + M = S$$
$$\$10 + M = \$15$$
$$M = \$15 - \$10$$
$$M = \$5$$

(c)
$$C + M = S$$
$$C + \$5 = \$15$$
$$C = \$15 - \$5$$
$$C = \$10$$

Objective 3 *Calculate Markup Based on Cost.* *Markup based on cost* is expressed as a percent of cost. As shown in the work with percent, *the base (that to which a number is being compared) is always 100%, so cost will have a value of 100%.* Markup and selling price will

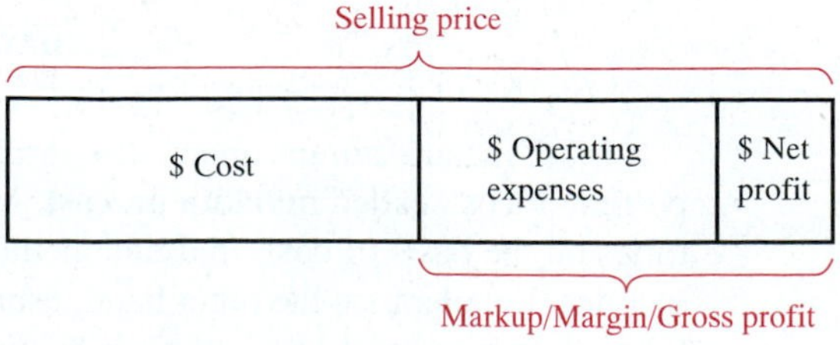

Figure 9.2

also have percent values found by comparing their dollar values to the dollar value of cost. Solve markup problems with the basic markup formula $C + M = S$.

Objective 4 *Apply Percent to Markup Problems.*

Example 2 Calculating Markup on Cost

The manager of Roseville Appliance bought a coffee maker manufactured in Spain for \$15 and will sell it for \$18.75. Find the percent of markup based on cost.

Solution Use the markup formula

$$C + M = S$$

with $C = \$15$ and $S = \$18.75$.

$$\begin{aligned} \$15 + M &= \$18.75 \\ M &= \$18.75 - \$15 \\ M &= \$3.75 \end{aligned}$$

The markup is \$3.75. Now, find the percent of markup based on cost with the basic percent equation, $R \times B = P$. The base is the cost, or \$15, and the markup is the part, or \$3.75. Substitute these values into $R \times B = P$.

$$\begin{aligned} R \times B &= P \\ R \times \$15 &= \$3.75 \\ \$15R &= \$3.75 \\ \frac{\cancel{15}R}{\cancel{15}} &= \frac{3.75}{15} \\ R &= 0.25 = 25\% \end{aligned}$$

The coffee maker costing \$15 and selling for \$18.75 has a 25% markup based on cost.

Scientific *Calculator Approach*

The calculator solution for Example 2 uses the parentheses to find the markup and then divides by the cost.

[( ] 18.75 [−] 15 [ )] [÷] 15 [=] 0.25

NOTE The markup formula and the basic percent equation can be used for solving various types of problems involving markup, as shown in the next examples.

Example 3 Finding Cost When Cost Is Base

Olympic Sports and Leisure has a markup on a basketball of \$14, which is 50% based on cost. Find the cost and the selling price.

Solution The markup is 50% of the cost.

$$\begin{aligned} P &= R \times B \\ M &= 50\% \times C \quad \text{Markup is part, cost is base.} \\ M &= 0.50C \end{aligned}$$

Since the markup is \$14, substitute 14 for M.

$$\$14 = 0.50C$$
$$\frac{14}{0.50} = \frac{\cancel{0.50}C}{\cancel{0.50}}$$
$$28 = C$$

The cost of the basketball is \$28.

Now use the basic markup formula to find the selling price.

$$C + M = S$$
$$\$28 + \$14 = S$$
$$\$42 = S$$

The selling price of the basketball is \$42.

Example 4 Finding the Markup and the Selling Price

Find the markup and selling price for a Texas Instruments financial calculator (assembled in Italy). The cost is \$23.60 and the markup is 25% of cost.

Solution Since the markup is 25% of cost,

$$M = 0.25(\$23.60)$$
$$M = \$5.90$$

The markup is \$5.90. Now use the markup formula, with $C = \$23.60$ and $M = \$5.90$.

$$C + M = S$$
$$\$23.60 + \$5.90 = S$$
$$\$29.50 = S$$

The selling price of the calculator is \$29.50.

Scientific
Calculator Approach

The calculator solution for Example 4 uses the percent add-on feature found on many calculators.

23.6 [+] 25 [%] [=] 29.5

Example 5 Finding Cost When Cost Is Base

Olympic Sports and Leisure is selling a Wilson baseball glove for \$42, which is 140% of the cost. How much did Olympic Sports and Leisure pay for the baseball glove?

Solution If the selling price is 140% of cost, then the markup must be 40% of cost. (The cost is always 100% when markup is based on cost.)

$$M = 0.4C$$

Now use the basic formula.

$$\begin{aligned} C + M &= S \\ C + 0.4C &= \$42 \\ 1.4C &= \$42 \\ \frac{1.4C}{1.4} &= \frac{\$42}{1.4} \\ C &= \$30 \end{aligned}$$

The cost of the glove is \$30. Check: 0.40 (\$30) = \$12 markup, and \$30 + \$12 = \$42.

Example 6 Finding the Cost and the Markup

The retail selling price of a Dell Computer is \$978.75. If the markup is 35% of cost, find the cost and the markup.

Solution Use the formula, with $M = 0.35C$.

$$\begin{aligned} C + M &= S \\ C + 0.35C &= \$978.75 \\ 1C + 0.35C &= \$978.75 \\ 1.35C &= \$978.75 \\ \frac{1.35C}{1.35} &= \frac{\$978.75}{1.35} \\ C &= \$725 \end{aligned}$$

The cost of the computer is \$725.

Now find markup.

$$\begin{aligned} C + M &= S \\ \$725 + M &= \$978.75 \\ M &= \$978.75 - \$725 \\ M &= \$253.75 \end{aligned}$$

The markup is \$253.75.

NOTE Remember, when calculating markup on cost, cost is always the base, 100%.

9.1 Exercises

Find the missing quantities. Round rates to the nearest tenth of a percent and money to the nearest cent.

	Cost Price	Markup	% Markup on Cost	Selling Price
1.	$12.40	______	40%	______
2.	______	$7.20	______	$43.20
3.	$23.50	$11.75	______	______
4.	______	______	100%	$68.98
5.	$158.70	______	______	$198.50
6.	______	$14.40	60%	______
7.	______	$13.50	______	$81
8.	$33.80	______	25%	______
9.	$210	______	______	$328
10.	______	$25.25	______	$73.80
11.	$495	______	27%	______
12.	______	______	16%	$90.83

13. Markup may be calculated on cost or on selling price. Explain why most manufacturers prefer to use cost as base when calculating markup. (See Objective 1.)
14. Write the basic markup formula. Define each term. (See Objective 2.)

Solve each of the following application problems using cost as base. Round rates to the nearest tenth of a percent and money to the nearest cent.

15. J. PETERMAN MERCHANDISE The J. Peterman Company offers a Four-Toggle Jacket (No. 61A6433) in women's sizes S, M, L for $138. If the markup is 35% of cost, find the cost.
16. DENTAL TOOLS The cost of a dentist's handpiece is $560 and the markup is 25% of cost. Find the markup.
17. GARDEN TOOLS Orchard Supply sells garden sprayers at a price of $18.95. If markup is 38% of cost, find the cost.
18. LAWN FERTILIZER Orchard Supply sells lawn fertilizer at a price of $12.50 per bag. If the markup is 25% of cost, find the cost.
19. SOUTH KOREAN PRODUCTS The cost of some hand tools imported from South Korea is $11.96 per set. The Tool Shed decides to use a markup of 25% on cost. Find the selling price of the tool set.
20. RECREATION EQUIPMENT Water Sports purchases jet skis at a cost of $2880 each. If its operating expenses are 25% of its cost, and it wishes to make a net profit of 15% of its cost, find the selling price.
21. NOTEBOOK COMPUTERS What percent of markup on cost must be used if a Hitachi 5290 PC Notebook costing $1740 is sold for $2049.20.

22. MARKUP ON ASPIRIN The Tower Market sells aspirin (100 tablet size) for $3.38 per bottle. If they pay $2.60 per bottle, find the markup percent on cost.

23. OUTDOOR LIGHTING Patios Plus sold an outdoor lighting set for $119.95. The markup on the set was $23.99. Find (a) the cost, (b) the markup percent on cost, and (c) the selling price as a percent of cost.

24. ATHLETIC SHOES Fleet Feet had a markup of $11.66 on some shoes that they sold for $55.66. Find (a) the cost, (b) the markup percent on cost, and (c) the selling price as a percent of cost.

25. FARM EQUIPMENT Fairfield Tractor put a markup of 32% on cost on some parts for which they paid $73.50. Find (a) the selling price as a percent of cost, (b) the selling price, and (c) the markup.

26. BRASS LAMPS A lighting manufacturer offers brass lamps at a selling price that is 175% of the cost. The markup is $61.50. Find (a) the markup pecent on cost, (b) the cost, and (c) the selling price.

27. HIKING BOOTS Nature Trails, a manufacturer of hiking equipment, prices lightweight hiking boots at $44.52, which is 127.2% of their cost. Find (a) the cost, (b) the markup as a percent of cost, and (c) the markup.

28. COIN COLLECTING North Area Coins priced a proof coin at $868, which was 112% of cost. Find (a) the cost, (b) the markup as a percent of cost, and (c) the markup.

29. WELDING SUPPLIES Welder's Supply purchases arc welding units for $3860 each. The company has operating expenses of 22% of cost, and a net profit of 12% of cost. Find the selling price of each arc welding unit.

30. PRICING MOUNTAIN BIKES Olympic Sports and Leisure purchases mountain bikes at a cost of $280 each. The company's operating expenses are 16% of cost, and a net profit of 7% of cost is desired. Find the selling price of one mountain bike.

31. RETAIL HARDWARE Bell Hardware has operating expenses of 18% of cost and desires a 17% net profit on cost. The selling price of a tube of 20-year silicone caulk is $8.95. Find the cost.

32. SLIDING PATIO DOORS American Glass Company sells 8-foot sliding patio doors for $299.90. Their operating expenses are 15% of cost and their net profit is 15% of cost. Find the cost.

9.2 Markup on Selling Price

Objectives

1. *Calculate markup based on selling price.*
2. *Solve markup problems when selling price is base.*
3. *Use the markup formula to solve variations of markup problems.*
4. *Determine percent markup on cost and the equivalent percent markup on selling price.*
5. *Convert markup percent on cost to markup percent on selling price.*
6. *Convert markup percent on selling price to markup percent on cost.*

As mentioned in the previous section, wholesalers sometimes calculate markup based on cost and other times calculate markup based on selling price. Retailers use sales figures in almost all aspects of their business. Almost all expense and income amounts are calculated as a percent of sales. Therefore it is common for retailers to calculate markup based on selling price. In each case, markup will be given as "on cost" or "on selling price." Remember that if markup is based on selling price, then selling price is the base. Since the base is 100%, selling price will have a value of 100%. This section discusses markup on selling price.

Objective 1 *Calculate Markup Based on Selling Price.* The same basic markup formula is used with markup on selling price.

$$\text{Cost} + \text{Markup} = \text{Selling price}$$
$$C + M = S$$

Objective 2 *Solve Markup Problems When Selling Price Is Base.*

Example 1 Solving for Markup on Selling Price

To remain competitive, Olympic Sports and Leisure must sell a 12-pack of golf balls for \$15. They pay \$10 for the golf balls and calculate markup on selling price. Find the amount of markup and the percent of markup on selling price.

Solution First, solve for markup.

$$C + M = S$$
$$\$10 + M = \$15$$
$$M = \$15 - \$10$$
$$M = \$5$$

Now solve for percent of markup on selling price. Use the basic percent equation, $R \times B = P$. In this example, P is the markup, or \$5, and the base B is the selling price, or \$15. Substitute these values into $R \times B = P$, and solve the equation.

$$R \times \$15 = \$5$$
$$R = \frac{\$5}{\$15}$$
$$R = \frac{1}{3} = 0.333\ldots$$
$$= 33\frac{1}{3}\%$$

The percent of markup on selling price is $33\frac{1}{3}\%$.

Objective 3 *Use the Markup Formula to Solve Variations of Markup Problems.* As with problems where markup is based on cost, the basic formula $C + M = S$ may be used for all variations of markup problems when selling price is the base.

Example 2 Finding Markup When Selling Price Is Given

Find the markup on a bottle of vitamin E capsules if the selling price is \$10.29 and the markup is 30% of selling price.

Solution Since the markup is 30% of the selling price,

$$M = 0.3S$$

or

$$M = 0.3(\$10.29) = \$3.09 \quad \text{Rounded}$$

The markup is $3.09.

Example 3 Finding Cost When Selling Price Is Base

A bookstore employee knows that the three-hole binders have a markup of $2.38, which is 40% based on selling price. Find the cost of the binders.

Solution Start by finding selling price and then subtract markup to find cost. Here $R = 40\%$ (the rate of markup) and $P = \$2.38$ (the markup). Use $R \times B = P$ as follows.

$$R \times B = P$$
$$40\% \times S = \$2.38$$
$$0.4S = \$2.38$$
$$S = \frac{\$2.38}{0.4}$$
$$S = \$5.95$$

The selling price is $5.95. Now solve for cost.

$$C + M = S$$
$$C + \$2.38 = \$5.95$$
$$C + \$2.38 - \$2.38 = \$5.95 - \$2.38$$
$$C = \$5.95 - \$2.38$$
$$C = \$3.57$$

The cost is $3.57.

Example 4 Finding the Selling Price and the Markup When Cost Is Given

An employee at Olympic Sports and Leisure is told to calculate the selling price and the markup on a pair of athletic socks if the cost is $3.16 and the markup is 20% of selling price.

Solution Use the formula $C + M = S$. Since the markup is 20% of the selling price, $M = 0.2S$.

$$\$3.16 + 0.2S = S$$
$$\$3.16 + 0.2S - 0.2S = 1S - 0.2S$$
$$\$3.16 = 1S - 0.2S$$
$$\$3.16 = 0.8S$$
$$\frac{3.16}{0.8} = \frac{\cancel{0.8}S}{\cancel{0.8}}$$
$$\$3.95 = S$$

The selling price is \$3.95.

Now find the markup using the markup formula.

$$C + M = S$$
$$\$3.16 + M = \$3.95$$
$$M = \$3.95 - \$3.16$$
$$M = \$0.79$$

The markup is \$0.79.

Markups vary widely from industry to industry and from business to business. This variation is a result of different costs of merchandise, operating costs, level of profit margin, and local competition. Table 9.1 shows average markups for different types of retail stores.

Objective 4 *Determine Percent Markup on Cost and the Equivalent Percent Markup on Selling Price.* Sometimes a markup based on cost must be compared to a markup based on selling price. Such a conversion might be necessary for a manufacturer who thinks in terms of cost and wants to understand a wholesaler or retail customer. Or perhaps a retailer or wholesaler might convert markup on selling price to markup on cost to better understand the manufacturer. Make these comparisons by first computing the markup on cost, then computing the markup on selling price.

Example 5 Determining Equivalent Markups

Claire Magersky sells fishing lures to both fishing-equipment wholesalers and sporting-goods stores. If the lure costs her \$4.20 and she sells it for \$5.25, what is the percent of markup on cost? What is the percent of markup on selling price?

Solution To solve for markup, use the formula $C + M = S$, with $C = \$4.20$ and $S = \$5.25$.

$$C + M = S$$
$$\$4.20 + M = \$5.25$$
$$M = \$5.25 - \$4.20$$
$$M = \$1.05$$

Table 9.1 AVERAGE MARKUPS FOR RETAIL STORES (MARKUP ON SELLING PRICE)

Type of Store	Markup	Type of Store	Markup
General merchandise stores	29.97%	Furniture and home furnishings	35.75%
Grocery stores	22.05%	Drinking places	52.49%
Other food stores	27.31%	Eating places	56.35%
Motor vehicle dealers (new)	12.83%	Drug and proprietary stores	30.81%
Gasoline service stations	14.47%	Liquor stores	20.19%
Other automotive dealers	29.57%	Sporting goods and bicycle shops	29.72%
Apparel and accessories	37.64%	Gift, novelty, and souvenir shops	41.86%

Data: Sole Proprietorship Income Tax Returns, U.S. Treasury Dept., Internal Revenue Service, Statistics Division.

The markup is $1.05.

Next, to solve for the percent of markup on cost, use the percent equation, $R \times B = P$, with $B = \$4.20$ (the cost, since markup is on cost) and $P = \$1.05$ (the markup).

$$R \times B = P$$
$$R \times \$4.20 = \$1.05$$
$$\$4.20R = \$1.05$$
$$R = \frac{1.05}{4.2}$$
$$R = 0.25 = 25\%$$

The markup on cost is 25%.

To find the percent of markup on selling price, use the percent equation again. While P is still $1.05, B changes to $5.25 since the markup on selling price must be found. Substitute into the equation as follows.

$$R \times B = P$$
$$R \times \$5.25 = \$1.05$$
$$\$5.25R = \$1.05$$
$$R = \frac{1.05}{5.25}$$
$$R = 0.2 = 20\%$$

The markup on selling price is 20%.

NOTE In Example 5 the markup on cost was determined first (25%). The problem was then reworked with the same dollar amounts but with the selling price as base. The result was 20%. This shows that a markup of 25% on cost is equivalent to a markup of 20% on selling price. ▬

Objective 5 *Convert Markup Percent on Cost to Markup Percent on Selling Price.* Another method for making markup comparisons is to use **conversion formulas**. *No dollar amounts are needed to use these formulas.* Only the percent of markup is needed. If you have markup percent on cost, you can convert the markup percent on cost to markup percent on selling price with the following formula.

$$\text{\% Markup on selling price} = \frac{\text{\% Markup on cost}}{100\% + \text{\% Markup on cost}}$$

Or, if M_c represents markup on cost and M_s represents markup on selling price,

$$M_s = \frac{M_c}{100\% + M_c}.$$

Example 6 Converting Markup on Cost to Markup on Selling Price

Convert a markup of 25% on cost to its equivalent percent markup on selling price.

Solution Use the formula for converting markup on cost to markup on selling price:

$$M_s = \frac{M_c}{100\% + M_c}$$

with $M_c = 25\%$.

$$\frac{25\%}{100\% + 25\%} = \frac{25\%}{125\%} = \frac{0.25}{1.25} = \frac{1}{5} = 20\% \quad \text{Markup on selling price}$$

This formula shows that a markup of 25% on cost is equivalent to a markup of 20% on selling price.

Scientific *Calculator Approach*

The markup on cost (25%) is divided by 100% plus the markup on cost. The parentheses keys are used here.

25 [%] [÷] [(] 100 [%] [+] 25 [%] [)] [=] 0.2

Objective 6 ***Convert Markup Percent on Selling Price to Markup Percent on Cost.*** Convert markup percent on selling price to markup percent on cost with the following formula.

$$\%\text{ Markup on cost} = \frac{\%\text{ Markup on selling price}}{100\% - \%\text{ Markup on selling price}}$$

Or, if M_c represents markup on cost and M_s represents markup on selling price,

$$M_c = \frac{M_s}{100\% - M_s}.$$

Example 7 Converting Markup on Selling Price to Markup on Cost

Convert a markup of 20% on selling price to its **equivalent markup** on cost.

Solution Use the formula for converting markup on selling price to markup on cost:

$$M_c = \frac{M_s}{100\% - M_s}$$

with $M_s = 20\%$.

$$\frac{20\%}{100\% - 20\%} = \frac{20\%}{80\%} = \frac{0.2}{0.8} = \frac{1}{4} = 25\% \quad \text{Markup on cost}$$

A markup of 20% on selling price is equivalent to a markup of 25% on cost.

Scientific *Calculator Approach*

For Example 7, the markup on selling price (20%) is divided by 100% minus the markup on selling price. The parentheses keys are used here.

20 [%] [÷] [(] 100 [%] [−] 20 [%] [)] [=] 0.25

Table 9.2 shows common markup equivalents expressed as percents on cost and also on selling price.

Table 9.2 MARKUP EQUIVALENTS

Markup on Cost	Markup on Selling Price
20%	$16\frac{2}{3}$%
25%	20%
$33\frac{1}{3}$%	25%
50%	$33\frac{1}{3}$%
$66\frac{2}{3}$%	40%
75%	$42\frac{6}{7}$%
100%	50%

9.2 Exercises

Find the missing quantities. Round rates to the nearest tenth of a percent and money to the nearest cent.

	Cost Price	Markup	% Markup on Selling Price	Selling Price
1.	$21	______	25%	______
2.	$145	______	______	$250
3.	______	$112	46%	______
4.	______	$72	______	$189.50
5.	$18.60	______	$66\frac{2}{3}$%	______
6.	$17.28	______	______	$29.95
7.	______	______	35%	$71.32
8.	$178	______	$33\frac{1}{3}$%	______
9.	______	$42.18	______	$120
10.	$193.15	______	42.5%	______

Find the missing quantities by first computing the markup on one base and then computing the markup on the other. Round rates to the nearest tenth of a percent and money to the nearest cent.

	Cost	Markup	Selling Price	% Markup on Cost	% Markup on Selling Price
11.	______	$57.50	______	25%	20%
12.	______	$0.23	$0.73	______	______
13.	$13.80	______	______	______	38%
14.	$33.75	______	$67.50	______	______
15.	______	$300	______	40%	______
16.	$5.15	______	$15.45	______	______

	Cost	Markup	Selling Price	% Markup on Cost	% Markup on Selling Price
17.	_______	\$78.48	\$436	_______	18%
18.	_______	\$480	_______	25%	_______

Find the equivalent markups on either cost or selling price using the appropriate formula. Round to the nearest tenth of a percent.

	Markup on Cost	Markup on Selling Price
19.	______	20%
20.	______	50%
21.	______	26%
22.	______	15.3%
23.	50%	______
24.	$33\frac{1}{3}\%$	______
25.	______	40%
26.	______	$16\frac{2}{3}\%$

27. Why do you suppose that grocery stores have an average markup of 22.05%, while eating places have an average markup of 56.35%?

28. To have a markup of 100% or greater, the markup must be calculated on cost. Show why this is always true. (See Objectives 5 and 6.)

Solve each of the following application problems. Round rates to the nearest tenth of a percent and money to the nearest cent.

29. EXTERIOR HOUSE PAINT The cost of a 5-gallon bucket of exterior house paint is \$41.40. The markup is 28% on selling price. Find the selling price.

30. AUTO PARTS DEALER An auto parts dealer pays \$7.14 per dozen gallons of windshield washer fluid and the markup is 50% on selling price. Find the selling price per gallon.

31. RUG SALES Old World Rugs sells an area rug for \$595 and maintains a markup of 35% on selling price. Find the cost.

32. HOME GYMNASIUM EQUIPMENT Olympic Sports and Leisure sells a home gymnasium package for \$3522 and maintains a markup of 35% on selling price. Find the cost.

33. CEILING FANS The cost of a ceiling fan is \$92.82 and the markup is 22% on selling price. Find the selling price.

34. SPORTING GOODS SALES Field and Stream Sports pays \$20.80 for a fly rod and sells it for \$32. Find the percent of markup on selling price.

35. BELGIAN FLOOR TILE Handmade floor tile from Belgium has a markup of 36% on the selling price. The tile has a markup of \$1.62 per tile. Find (a) the selling price, (b) the cost, and (c) the cost as a percent of the selling price per tile.

36. SONY DISKMAN Fry's Electronics pays $98.39 for a Sony Diskman. The markup is 18% on selling price. Find (a) the cost as a percent of selling price, (b) the selling price, and (c) the markup.

37. CLOCK RADIOS Best Products buys clock radios for $258 per dozen and has a gross profit of $7.74 per clock radio. Find the percent of gross profit based on selling price.

38. BASEBALL HATS A retailer pays $87.36 per dozen for baseball hats and has a gross profit of $1.68 per hat. Find the percent of gross profit based on selling price.

39. BICYCLE HELMETS The Cycle Center placed a selling price of $38.50 on a bicycle helmet that cost 72% of the selling price. Find (a) the cost, (b) the markup, and (c) the markup as a percent of selling price.

40. LEVI'S DOCKER SHIRTS Mervyns Department Store priced some Levi Docker shirts at $29.95. The cost of the shirts was 58% of the selling price. Find (a) the cost, (b) the markup, and (c) the markup as a percent of selling price.

41. RECYCLING ALUMINUM Recyclable aluminum can be sold for $2880 per ton (1 ton = 2000 pounds). If Alcan Recycling Plant wants a 50% markup on selling price, (a) how much per pound can it pay local residents for their recycled aluminum? (b) What is the equivalent markup percent on cost?

42. RETAIL SILK FLOWERS A retailer purchases silk flowers for $31.56 per dozen and sells them for $4.78 each. (a) Find his percent of markup on selling price. (b) What is the equivalent markup on cost?

43. RIVER RAFT SALES White Water Supply purchased a job lot of 380 river rafts for $7600. If they sold 158 of the rafts at $45 each, 74 at $35 each, 56 at $30 each, and the remainder at $25 each, what is (a) the total amount received for the rafts, (b) the total markup, (c) the markup percent on selling price, and (d) the equivalent markup percent on cost?

44. ITALIAN SILK TIES Dress for Success purchased 240 Italian silk ties for $2280. They sold 162 ties at $25 each, 45 ties at $15 each, 20 ties at $10 each, and the remainder at $5 each. Find (a) the total amount received for the ties, (b) the total markup, (c) the markup percent on selling price, and (d) the equivalent markup percent on cost.

45. REFRIGERATOR REPAIR General Electric Parts Department sells ice cube maker repair kits for $27.90 each, which reflects a markup of 50% on selling price. Find (a) the cost and (b) the percent of markup on cost.

46. SMOKING SUPPLIES The Tinder Box Smoke Shop buys a special blend of pipe tobacco in 10-pound tins at a cost of $24 per tin. The shop sells the tobacco for $1.20 per ounce. Find (a) the markup on selling price, and (b) the equivalent markup on cost. (*Hint:* 1 pound = 16 ounces.)

47. RESTAURANT SUPPLIES A restaurant supplier sells coffee filters for $6.90 per box. If the cost of the filters is $4.80, find (a) the percent of markup on cost and (b) the equivalent percent of markup on selling price.

48. ALL-PURPOSE FERTILIZER Home Base sells a 40-pound bag of Pax All-Purpose Fertilizer for $10.98. The cost of the fertilizer is $7.32 per bag. Find (a) the percent of markup on cost and (b) the equivalent percent of markup on selling price.

49. COMMUNICATION EQUIPMENT A discount store purchased touch-tone wall phones at a cost of $288 per dozen. If they need 20% of cost to cover operating expenses and 15% of cost for net profit, what is (a) the selling price per phone and (b) the percent of markup on selling price?

50. BOWLING EQUIPMENT The Bowlers Pro-Shop determines that operating expenses are 23% of selling price and desires a net profit of 12% of selling price. If the cost of a team shirt is $29.25, what is (a) the selling price and (b) the percent of markup on cost?

51. MOUNTAIN BIKE SALES Cycle City advertises mountain bikes for $199.90. If their cost is $2100 per dozen, what is (a) the markup per bicycle, (b) the percent of markup on selling price, and (c) the percent of markup on cost?

52. HAND TOOL SALES The Tool Shed advertises standard/metric socket sets (manufactured in the U.S.A.) for $39. Their cost is $351 per dozen sets. Find (a) the markup per set, (b) the percent of markup on selling price, and (c) the percent of markup on cost.

9.3 Markup with Spoilage

Objectives

1 *Solve markup problems when items are unfit for sale.*

2 *Solve markup problems when a certain percent of items are unsaleable.*

3 *Calculate markup when a percent of the merchandise must be sold at a reduced price.*

Objective 1 *Solve Markup Problems When Items Are Unfit for Sale.* Merchandise that is perishable, becomes damaged or soiled, or is manufactured with a blemish causes problems for many businesses. To a nursery or garden center, produce buyer, food processor, and clothing manufacturer, such problems are a common occurrence. These items, called **irregulars**, cannot be sold at the regular price. They are often sold at a reduced price. If such items are unsaleable, they represent a total loss. In either case, the markup applied to the items sold at regular price must allow for **spoilage** and damaged items. The result is that perfect items are sold at a higher price to make up for the unsold items in a process called **markup with spoilage**.

Example 1 Finding Selling Price with Spoilage

Village Nursery purchases 105 5-gallon-size juniper shrubs for $351. If a markup of 40% on selling price is necessary and 15 of the plants will be unfit for sale, what is the selling price per plant?

Solution First, find the total selling price of all the plants using a 40% markup on selling price and a cost of $351.

$$C + M = S$$
$$C + 40\%S = S$$
$$\$351 + 0.4S = S$$
$$\$351 = 1.00S - 0.4S$$
$$\$351 = 0.6S$$
$$\frac{\$351}{0.6} = S$$
$$\$585 = S$$

The total selling price is \$585. Now, divide the total selling price by the number of saleable plants to find the selling price per plant.

$$\frac{\$585}{105 \text{ purchased} - 15 \text{ unsaleable}} = \frac{\$585}{90} = \$6.50 \quad \text{Selling price per plant}$$

The junipers must be priced at \$6.50 each to realize a markup of 40% on selling price and to allow for 15 plants that cannot be sold.

NOTE Controlling loss due to spoilage is critical in business. As spoilage increases, profits will fall. These additional costs due to spoilage may be added to the price of the products sold, but the resulting higher price may be too high and no longer competitive. ▬

Objective 2 | ***Solve Markup Problems When a Certain Percent of Items Are Unsaleable.***

Example 2 Finding Selling Price When a Percent of Items Is Unsaleable

The Bagel Boys bakes 60 dozen bagels at a cost of \$2.16 per dozen. If a markup of 50% on selling price is needed and 5% of the bagels will not be sold and must be thrown away, what is the selling price per dozen bagels?

Solution To begin, find the total cost of the bagels.

$$\text{Cost} = 60 \text{ dozen} \times \$2.16 = \$129.60$$

Now find the selling price, using a markup of 50% of selling price.

$$C + M = S$$
$$C + 50\%S = S$$
$$\$129.60 + 0.5S = S$$
$$\$129.60 = 1.00S - 0.5S$$
$$\$129.60 = 0.5S$$
$$\frac{\$129.60}{0.5} = S$$
$$\$259.20 = S$$

The total selling price is \$259.20.

Next, find the number of dozen bagels that will be sold. Since 5% will not be sold, 95% (100% − 5%) will be sold.

$$95\% \times 60 \text{ dozen} = 57 \quad \text{Number of dozens sold}$$

The total selling price of \$259.20 must be received from 57 dozen bagels.

Find the selling price per dozen bagels by dividing the total selling price by the number of bagels to be sold.

$$\frac{\text{Total selling price}}{\text{Number saleable}} = \frac{\$259.20}{57} = \$4.55 \quad \text{Selling price per dozen (rounded)}$$

A selling price of \$4.55 per dozen gives the desired markup of 50% on selling price while allowing for 5% of the bagels to be unsold.

Objective 3 | ***Calculate Markup When a Percent of the Merchandise Must be Sold at a Reduced Price.***

Example 3 Finding Selling Price When Some Items Are Sold at a Reduced Price

An athletic shoe manufacturer makes a walking shoe at a cost of $16.80 per pair. Based on past experience, 10% of the shoes will be defective and must be sold as irregulars for $24 per pair. If the manufacturer produces 1000 pairs of shoes and desires a markup of 100% on cost, what is the selling price per pair?

Solution First, find the cost of the total production.

$$1000 \text{ pairs} \times \$16.80 = \$16{,}800 \qquad \text{Total production cost}$$

The total selling price is

$$C + M = S$$
$$C + 100\%C = S$$
$$2C = S$$

Since $C = \$16{,}800$ and $2C = S$, $2 \times \$16{,}800 = \$33{,}600$ is the total selling price of all shoes.

If 10% of the shoes will sell for $24 per pair, then the sales of irregulars will be

$$(0.10 \times 1000) \times \$24 = 100 \times \$24$$
$$= \$2400$$

Calculate the selling price per pair of the regular priced shoe as shown.

$$\$33{,}600 - \$2400 = \$31{,}200 \qquad \text{Total sales of regulars}$$

$$\frac{\$31{,}200}{1000 - 100} = \frac{31{,}200}{900} = \$34.67 \qquad \text{Rounded}$$

A regular selling price of $34.67 will give the manufacturer a 100% markup on cost while allowing for 10% of the production to sell at $24 per pair.

Scientific *Calculator Approach*

The calculator solution to Example 3 requires several steps. First, find the total selling price and place it in memory.

1000 [×] 16.8 [×] 2 [=] [STO]

Next, find the total amount received from the sale of irregulars.

.1 [×] 1000 [×] 24 [=]

Now, subtract the sales of irregulars from the total selling price.

[+/−] [+] [RCL] [=]

Finally, divide by the number of regular pairs.

[÷] 900 [=] 34.67

9.3 Exercises

Find the selling price per item.

	Total Cost	Quantity Purchased	Number Unsaleable	% Markup on Selling Price	Selling Price per Item
1.	$81	90	9	10%	______
2.	$540	36	6	25%	______
3.	$340	1 gr.	8	20%	______
4.	$189	4 doz.	6	40%	______
5.	$120	25	5	$33\frac{1}{3}\%$	______
6.	$2750	120	10	50%	______
7.	$126	8 doz.	6	15%	______
8.	$2025	250	25	25%	______

Find the missing quantities.

	Total Cost	Quantity Purchased	Percent Unsaleable	% Markup on Selling Price	Number to Sell	Selling Price per Item
9.	$161	25	4%	25%	______	______
10.	$342	120	5%	20%	______	______
11.	$198	2 doz.	25%	50%	______	______
12.	$162	20 cs.	10%	30%	______	______
13.	$190	80 pr.	5%	$33\frac{1}{3}\%$	______	______
14.	$8750	100	30%	15%	______	______
15.	$25,200	2000 gal.	5%	20%	______	______
16.	$7200	100 bbl.	10%	80%	______	______

Find the missing quantities. Markup is based on total cost.

	Total Cost	Quantity Purchased	Percent Sold at Reduced Price	% Markup on Total Cost	Number at Regular Price	Number at Reduced Price	Reduced Price	Regular Selling Price
17.	$500	200	20%	25%	______	______	$2.20	______
18.	$360	120	10%	20%	______	______	$2.00	______
19.	$2200	40	15%	35%	______	______	$50.00	______
20.	$270	90	20%	40%	______	______	$4.00	______
21.	$432	1 gr.	25%	25%	______	______	$2.50	______
22.	$3000	600	10%	$33\frac{1}{3}\%$	______	______	$5.00	______
23.	$2800	1000 pr.	30%	50%	______	______	$3.00	______
24.	$7500	1000	50%	25%	______	______	$5.00	______

25. When merchandise is unsaleable it is as if the items were not received. Explain how the price of the saleable items must be adjusted to offset this loss. (See Objective 1.)

26. Explain how we all pay for the spoiled fruit and vegetables in the produce department of the grocery store. (See Objective 1.)

Solve each of the following application problems.

27. POTTERY SHOP SALES The Aztec Pottery Shop finds that 15% of their production cannot be sold. They produce 100 items at a cost of \$2.15 each and desire a markup of 40% on selling price. Find the selling price per item.

28. PRODUCE SALES Country Produce knows that 20% of the strawberries purchased will spoil and must be thrown out. If they buy 200 baskets of strawberries for \$0.24 per basket and want a markup of 50% on selling price, find the selling price per basket of strawberries.

29. NURSERY SALES Fowler Nursery purchases 75 mature palm trees for \$511. If two of the palms are judged unfit for resale and the nursery desires to maintain a $33\frac{1}{3}\%$ markup on selling price, what is the selling price for each palm?

30. IMPORTED CANDLES Cost Plus Imports purchased 30 crates of candles from Mexico for a total cost of \$237.60. Each crate contains 3 dozen candles. If four crates of the candles are sold at the reduced price of \$0.25 per candle and Cost Plus Imports wants a markup of 100% on cost, find the regular price per candle.

31. LONG-STEMMED ROSES Farmers Flowers purchased 12 gross of long-stemmed roses at a cost of \$945. If 25% of the roses must be sold at \$7.50 per dozen and a markup of 100% on cost is needed, find the regular selling price per dozen roses.

32. CUSTOM BASEBALL-CAPS Custom Caps buys 2000 baseball hats at \$2.50 per hat. If a markup of 50% on selling price is needed and 5% of the hats are unsaleable, what is the selling price of each hat?

33. STOVESHELL CASTINGS Bethel Metals finds that 20% of its stove shell castings are unsaleable. If the cost of manufacturing 55 stoves is \$10,450 and they need a markup of 30% on cost, what is the selling price per stove?

34. TRANSMISSION PAN GASKETS Transco Products knows that 4% of their manufactured transmission pan gaskets are rejects and cannot be sold. If the cost of manufacturing 5000 gaskets is \$3168 and they need a markup of 200% on cost, what is the selling price of each gasket?

35. RECAPPED TRUCK TIRES Wheelco Tire recaps truck tires at a cost of \$80.50 per tire. Past experience shows that 12% of the recaps must be sold as blemishes for \$105. If they recap 500 tires and a markup of 110% on cost is desired, what is the regular selling price per tire?

36. ROOFING TILE Solano Tile manufactures roof tile at a cost of \$42 per square. Past experience shows that 8% of a production run are irregulars and must be sold for \$45 per square. If a production run of 10,000 squares is completed and they desire a markup of 80% on cost, what is the selling price per square?

37. BOOK PUBLISHING U.S. Publishing Company prints a book on sports memorabilia at a production cost of $10.80 per copy. They know that 20% of the production will be sold for $11.50 per copy. They print 50,000 books and they want a markup of 50% on selling price. Find the selling price per book. Round to the nearest cent.

38. IMPORTED WALLPAPER Wallpaper Specialty Imports purchased 7000 rolls of wallpaper manufactured in France at a cost of $6.30 per roll. Past experience shows that 25% of the wallpaper will have to be sold for $6.75 per roll. Find the selling price per roll if the store wants a markup of 50% on selling price.

Chapter 9 Quick Review

Chapter Terms

Review the following terms to test your understanding of the chapter. For each term you do not know, refer to the page number found next to that term.

Concepts	Examples
9.1 Markup on cost Use rate × Base = Part ($R \times B = P$), with cost as base (100%), markup % as rate, and markup as part.	Cost, \$160; markup, 25% on cost. Find the markup. $R \times B = P$ $0.25 \times \$160 = \$40$ Markup.
9.1 Calculating the percent of markup Use Cost + Markup = Selling price ($C + M = S$) and the basic percent equation $R \times B = P$.	Cost, \$420; selling price, \$546. Find the percent of markup based on cost. $C + M = S$ $\$420 + M = \$546$ $M = \$126$ $R \times B = P$ $R \times \$420 = \$126$ $R = \frac{126}{420} = 0.3 = 30\%$ Markup on cost.
9.1 Finding the cost and selling price Use $P = R \times B$ to solve for cost; then use $C + M = S$ to find selling price.	Markup, \$56; markup on cost, 50%. Find cost and selling price. $P = R \times B$ $\$56 = 0.5C$ $C = \frac{56}{0.5} = \$112$ Cost. $C + M = S$ $\$112 + \$56 = \$168$ Selling price.
9.2 Markup on selling price Use Rate × Base = part ($R \times B = P$), with selling price as base (100%), markup % as rate, and markup as part.	Selling price, \$6; markup, 25% on selling price. Find the markup. $P = R \times B$ Markup = 25% × Selling price $M = 0.25(\$6) = \$1.50$ Markup

9.2 Finding the cost

Use the formulas $R \times B = P$ and $C + M = S$.

Markup, \$87.50; markup on selling price, 35%. Find the cost.

First, use $R \times B = P$ to find selling price.

$$35\% \times \text{Selling price} = \$87.50$$
$$0.35S = \$87.50$$
$$S = \$250$$

Now find cost.

$$C + M = S$$
$$C + \$87.50 = \$250 \quad \text{Selling price.}$$
$$C = \$250 - \$87.50 = \$162.50 \quad \text{Cost.}$$

9.2 Calculating the selling price and the markup

Use the formula $C + M = S$.

Cost, \$150; markup, 25% of selling price. Find the selling price and the markup.

$$C + M = S$$
$$\$150 + 0.25S = S$$
$$\$150 = 0.75S$$
$$\frac{\$150}{0.75} = \frac{0.75S}{0.75}$$
$$\$200 = S \quad \text{Selling price.}$$
$$C + M = S$$
$$\$150 + M = \$200 \quad \text{Selling price.}$$
$$M = \$200 - \$150$$
$$= \$50 \quad \text{Markup.}$$

9.2 Converting markup on cost to markup on selling price

Use the formula

$$M_s = \frac{M_c}{100\% + M_c}.$$

Markup on cost, 25%. Convert to markup on selling price.

$$M_s = \frac{25\%}{100\% + 25\%}$$
$$= \frac{0.25}{1.25} = 0.2 = 20\%$$

9.2 Converting markup on selling price to markup on cost

Use the formula

$$M_c = \frac{M_s}{100\% - M_s}.$$

Markup on selling price, 20%. Convert to markup on cost.

$$M_c = \frac{20\%}{100\% - 20\%}$$
$$= \frac{0.2}{0.8} = 0.25 = 25\%$$

9.3 Solving markup with spoilage or unsaleable items

1. Find total cost and selling price.
2. Subtract total sales at reduced prices from total sales.
3. Divide the remaining sales amount by the number of saleable units to get selling price per unit.

60 doughnuts cost \$0.15 each; 10 are not sold; 50% markup on selling price. Find selling price per doughnut.

$$\text{Cost} = 60 \times \$0.15 = \$9$$
$$C + M = S$$
$$\$9 + 0.5S = S$$
$$\$9 = 0.5S$$
$$\$18 = S$$
$$\$18 \div 50 = \$0.36 \quad \text{Sale price per doughnut}$$

Chapter 9 Review Exercises

Find the missing quantities. Round rates to the nearest tenth of a percent and money to the nearest cent. [9.1]

	Cost Price	Markup	% Markup on Cost	Selling Price
1.	\$32	______	20%	______
2.	______	\$6.15	25%	______
3.	______	\$73.50	______	\$220.50
4.	\$108	______	______	\$153.90

Find the missing quantities. Round rates to the nearest tenth of a percent and money to the nearest cent. [9.2]

	Cost Price	Markup	% Markup on Selling Price	Selling Price
5.	\$72.32	______	20%	______
6.	______	\$35	______	\$140
7.	______	\$17.35	$33\frac{1}{3}\%$	______
8.	\$283.02	\$177.18	______	______

Find the missing quantities by first computing the markup on one base and then computing the markup on the other. Round rates to the nearest tenth of a percent and money to the nearest cent. [9.2]

	Cost Price	Markup	Selling Price	% Markup on Cost	% Markup on Selling Price
9.	______	\$480	______	25%	20%
10.	\$64.50	______	\$129	______	______
11.	______	\$3.68	\$11.68	______	______
12.	______	\$474.28	______	100%	______

Find the equivalent markups on either cost or selling price using the appropriate formula. Round to the nearest tenth of a percent. [9.2]

	% Markup on Cost	% Markup on Selling Price
13.	______	20%
14.	100%	______
15.	______	15.3%
16.	20%	______

Find the selling price per item. [9.3]

	Total Cost	Quantity Purchased	Number Unsaleable	% Markup on Selling Price	Selling Price per Item
17.	$324	360	36	20%	______
18.	$780	52	12	40%	______
19.	$970	9 doz.	6	30%	______
20.	$12,650	1500 pr.	150 pr.	45%	______

Find the missing quantities. Markup is based on total cost. [9.3]

	Total Cost	Quantity Purchased	Percent Sold at Reduced Price	% Markup on Total Cost	Number at Regular Price	Number at Reduced Price	Reduced Price	Regular Selling Price Each
21.	$750	150	10%	25%	______	______	$2.50	______
22.	$135	90 pr.	20%	40%	______	______	$1.00	______
23.	$1728	2 gr.	25%	20%	______	______	$4.00	______
24.	$2800	1000 pr.	30%	50%	______	______	$3.00	______

Solve each of the following application problems on markup.

25. Olympic Sports and Leisure buys jogging shorts manufactured in Taiwan for $97.50 per dozen pair. Find the selling price per pair if the retailer maintains a markup of 35% on selling price. [9.2]

26. Circuit City sells a dishwasher for $395. The cost of the dishwasher is $334.75. Find the markup as a percent of cost. [9.1]

27. The Computer Service Center sells a DeskJet print cartridge for $18.75. The print cartridge costs the store $11.25. Find the markup as a percent of selling price. [9.2]

28. Wild Rivers offers an inflatable boat for $199.95. If the boats cost $1943.52 per dozen, what is (a) the markup, (b) the percent of markup on selling price, and (c) the percent of markup on cost? Round to the nearest tenth of a percent. [9.1 and 9.2]

29. Raleys Superstore bought 1820 swimming pool blow-up toys for a total cost of $10,010. The toys were sold as follows: 580 at $13.95 each, 635 at $9.95 each, 318 at $8.95 each, 122 at $7.95 each, and the balance at $5.00 each. Find (a) the total selling price of all the toys and (b) the markup as a percent of selling price (to the nearest whole percent). [9.2]

30. Fan Fever purchased 200 posters at a cost of \$360. If 20% of the posters must be sold at \$2 each and a markup of 100% on cost is needed, find the regular selling price of each poster. **[9.3]**

31. Office Depot pays \$3.96 for a package of 20-pound paper and sells it for \$4.95. Find (a) the percent of markup on selling price and (b) the percent of markup on cost. **[9.2]**

32. Fosters' Doughnuts bakes lemon squares at a cost of \$1.93 per dozen. Markup on cost is 100% and 15% of the lemon squares will have to be sold at \$2.40 per dozen. If Fosters bakes 180 dozen lemon squares, what is the regular selling price per dozen? Round to the nearest cent. **[9.3]**

Chapter 9 Summary Exercise
The Retailer: Using Markup to Maintain Net Profit

The Hallmark Shop buys 3400 boxes of holiday greeting cards directly from the manufacturer. The list price of the cards is \$4.95 per box, there is a trade discount of 30/10/20, and a cash discount of 5/10–40 x. The Hallmark Shop earns and receives both discounts.

The cards were sold as follows: 1080 boxes at \$3.95, 1250 boxes at \$2.95, 660 boxes at \$2.50, 230 boxes at \$2.00, and the remaining boxes were unsaleable.

To better manage greeting card sales next year, determine each of the following.

(a) The net cost of the greeting cards after trade and cash discounts

(b) The total sales amount received from all the holiday greeting cards

(c) The amount of net profit from the sales of the cards

(d) The markup as a percent of selling price to the nearest whole percent

(e) The equivalent percent of markup on cost to the nearest whole percent

Net Assets Business on the Internet

REI

Statistics

- 1938: Established by a group of 24 mountaineers
- 1998: $567 million in sales
- 1999: 49 stores located in 21 states
- Named in 100 Best Companies to Work For by *Fortune* magazine

REI was formed in 1938 by a group of 24 mountain climbers from Seattle, Washington. They wanted the finest-quality climbing equipment and formed a buying cooperative (membership group) in order to find the best prices for their equipment. Today, anyone may shop at REI, but members—those who pay a one-time $15 fee to join—share in the company's profits through an annual patronage refund. In 1998, REI declared a total patronage refund of 10.4 percent to 1.6 million active members for a total of $31.3 million. REI's easy-to-navigate Internet store provides access to more than 10,000 outdoor products and offers a variety of interactive education opportunities for outdoor enthusiasts. In addition to selecting from thousands of products and placing secure on-line orders, customers can use gear checklists, interact with gear experts, and learn basic outdoor skills by accessing educational clinics.

1. REI purchased one dozen Jansport backpacks at a cost of $504. If the company uses a markup of 25% on selling price, find the selling price of each backpack.

2. A two-person dome tent with a full rain fly has a wholesale price of $78. If the store has operating expenses of 24.5% of cost and a net profit of 10.5% of cost, find the selling price.

3. A freeze-dried dinner packet is sold for $12.50. If the packet has a cost of $8.75, find **(a)** the markup, **(b)** the percent of markup on selling price, and **(c)** the percent of markup on cost. Round to the nearest tenth of a percent.

4. REI purchased 4 dozen pairs of hiking boots at a cost of $1958.40. The markup on selling price is 40% and it is known that 25% of the hiking boots will have to be sold at $50 per pair. Find the regular selling price of each pair of hiking boots. Round to the nearest cent.

10 Markdown and Inventory Control

Markdowns and sales are the cornerstone of many businesses today. The word "free" in advertising gets the greatest response from consumers, and the word "sale" is not far behind. While there can be many reasons for offering markdowns, the resulting increase in sales volume helps move out existing inventory to make room for new merchandise. The newspaper advertisement shown in Figure 10.1 is designed to generate sales over a 3-day holiday weekend.

How can this retailer offer such high discounts?

Management spends a great deal of time controlling inventory. While on one hand the inventory must include the correct kinds and quantities of products to satisfy customers, it is equally important to control inventory to minimize any unsaleable or surplus

Figure 10.1

merchandise. If inventory is too low, product choices may be limited and customers will go to other stores to do their shopping. However, if inventory is too high, merchandise may become dated while it sits on the shelves, tying up needed business capital. This chapter discusses some of the main ideas of inventory control.

10.1 Markdown

Objectives

1. ***Calculate markdown and reduced price.***
2. ***Calculate percent of markdown.***
3. ***Find original selling price.***
4. ***Identify the terms associated with loss.***
5. ***Determine the break-even point and operating loss.***
6. ***Determine the amount of absolute loss and the percent of absolute loss.***
7. ***Find the maximum percent of markdown to be given.***

Merchandise often does not sell at its marked price, for any of several reasons. The retailer may have ordered too much, or the merchandise may have become soiled or damaged, or perhaps only odd sizes and colors are left. Also, merchandise may not sell because of lower prices at other stores, seasonal changes, economic fluctuations, or changes in fashion.

Objective 1 *Calculate Markdown and Reduced Price.* When merchandise does not sell at its marked price, its price is often reduced. The difference between the original selling price and the reduced selling price is called the **markdown**; the selling price after the markdown is called the **reduced price**, **sale price**, or **actual selling price**. The basic formula for markdown is as follows.

Reduced price = Original price − Markdown

Example 1 Finding the Reduced Price

Lazy Boy has marked down a leather recliner. What is the reduced price if the original price was $960 and the markdown is 25%?

Solution The markdown is 25% of $960, or 0.25 × $960 = $240. The reduced price is

$960 (original price) − $240 (markdown) = $720 (reduced price).

Scientific *Calculator Approach*

The calculator solution to Example 1 uses the complement, with respect to one, of the 25% discount.

960 [×] [(] 1 [−] .25 [)] [=] 720

Objective 2 *Calculate Percent of Markdown.* The next example shows how to find a **percent**, or **rate, of markdown**.

NOTE The original selling price is always the base or 100% and the percent of markdown is always calculated on the original selling price. ■

Example 2 Calculating the Percent of Markdown

The total inventory of Mother's Day cards at a large gift shop has a retail value of \$785. If the cards were sold at reduced prices that totaled \$530, what is the percent of markdown on the original price?

Solution First find the amount of the markdown.

\$785 (Original price) − \$530 (Reduced price) = \$255 (Markdown)

To solve for percent of markdown on original price, use the percent equation: rate × base = part. The base is the original price of \$785, the rate is unknown, and the part is the amount of markdown, or \$255.

$$R \times B = P$$
$$\_\_\_\_\% \times \$785 = \$255$$
$$R \times \$785 = \$255$$
$$\$785R = \$255$$
$$R = \frac{255}{785}$$
$$R = 0.3248 = 32\% \quad \text{Rounded to the nearest whole percent}$$

The cards were sold at a markdown of 32%.

Scientific *Calculator Approach*

In the calculator solution to Example 2, the reduced price is subtracted from the original price using parentheses. This results in the markdown, which is then divided by the original price.

[( ] 785 [−] 530 [ )] [÷] 785 [=] 0.3248

Objective 3 | ***Find Original Selling Price.***

Example 3 Finding the Original Price

Find the original price if a child's raincoat is offered at the reduced price of \$18 after a 40% markdown from the original price.

Solution After the 40% markdown, the reduced price represents 60% of the original price. Find the original price, which is the base.

$$R \times B = P$$
$$60\% \times B = \$18$$
$$0.6B = \$18$$
$$B = \frac{18}{0.6}$$
$$B = \$30$$

The original price of the raincoat was \$30.

Problem-Solving Hint In Example 3, notice that 60%, not 40%, is used in the formula. The reduced price, \$18, is represented by 60%.

Objective 4 | *Identify the Terms Associated with Loss.* The amount of markdown must be large enough to sell the merchandise while providing as much profit as possible. Merchandise that is marked down will result in either a reduced net profit, a break-even point, an operating loss, or an absolute loss. Figure 10.2 illustrates the meanings of these terms.

Reduced net profit results when the reduced price is still within the net profit range; that is, when it is greater than the total cost plus operating expenses.

The **break-even point** is the point at which the reduced price just covers cost plus overhead (operating expenses). At this point the business neither makes money nor loses money.

An **operating loss** occurs when the reduced price is less than the break-even point. The operating loss is the difference between the break-even point and the reduced selling price.

An **absolute loss** is the result of a reduced price that is below the cost of the merchandise alone. The absolute loss is the difference between the cost and reduced selling price.

The following formulas are helpful when working with markdowns.

$$\text{Break-even point} = \text{Cost} + \text{Operating expenses}$$

$$\text{Operating loss} = \text{Break-even point} - \text{Reduced selling price}$$

$$\text{Absolute loss} = \text{Cost} - \text{Reduced selling price}$$

Objective 5 | *Determine the Break-Even Point and Operating Loss.*

Example 4 Determining a Profit or a Loss

Cordova Appliance paid \$40 for a garbage disposal. If operating expenses are 30% of cost and the garbage disposal is sold for \$50, find the break-even point and the amount of loss.

Solution Use the following formula to find the break-even point.

$$\text{Cost} + \text{Operating expenses} = \text{Break-even point}$$

$$\$40 + (0.3 \times \$40) = \$40 + \$12 = \$52 \quad \text{Break-even point.}$$

Since the break-even point is \$52 and the selling price is \$50, there is a loss of

$$\$52 - \$50 = \$2$$

This \$2 loss is an operating loss since the selling price is less than the break-even point but greater than the cost. See Figure 10.3.

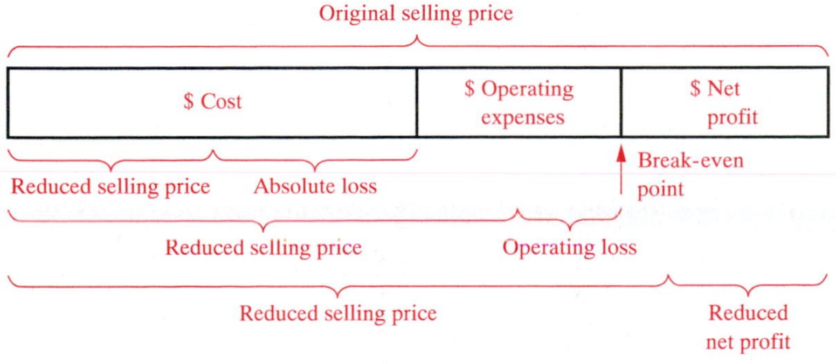

Figure 10.2

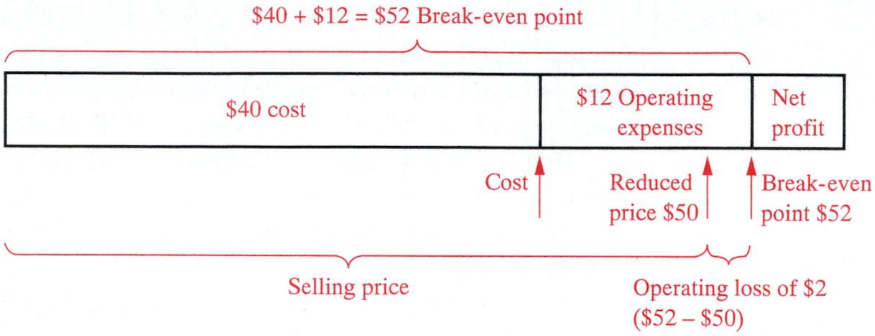

Figure 10.3

Scientific *Calculator Approach*

The calculator solution to Example 4 is as follows.

40 [+] [(] .3 [×] 40 [)] [−] 50 [=] 2

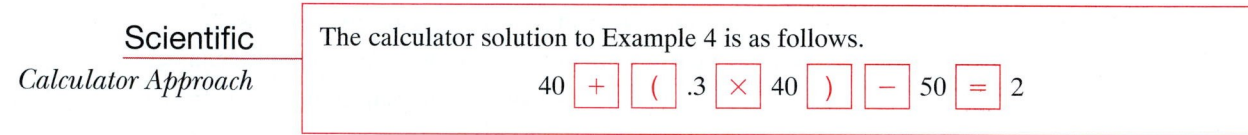

Objective 6 *Determine the Amount of Absolute Loss and the Percent of Absolute Loss.*

Example 5 Determining the Operating Loss and the Absolute Loss

A set of graphite golf clubs normally selling for $360 at the Oakridge Pro-Shop is marked down 30%. The cost of the golf clubs is $260 and the operating expenses are 20% of the cost. Find (a) the operating loss, (b) the absolute loss, and (c) the percent of absolute loss based on cost (round to the nearest percent).

Solution

(a) The break-even point (cost + operating expenses) is $312 ($260 + 0.2 × $260 = $260 + $52). The reduced price is

$$\$360 - (0.3 \times \$360) = \$360 - \$108 = \$252$$

The operating loss is

$$\$312 \text{ (Break-even point)} - \$252 \text{ (Reduced price)} = \$60 \text{ (Operating loss)}$$

(b) The absolute or gross loss is the difference between the cost and the reduced price.

$$\$260 \text{ (Cost)} - \$252 \text{ (Reduced price)} = \$8 \text{ (Absolute loss)}$$

(c) The percent of absolute loss is always expressed as a percent of cost, so find the percent of cost that is the absolute loss using the percent equation.

$$R \times B = P$$
$$\_\_\_\_\_\_\% \times \$260 = \$8$$
$$\$260R = \$8$$
$$R = \frac{8}{260}$$
$$R = 0.0308 = 3\% \quad \text{Rounded}$$

The rate of absolute loss is approximately 3%. See Figure 10.4.

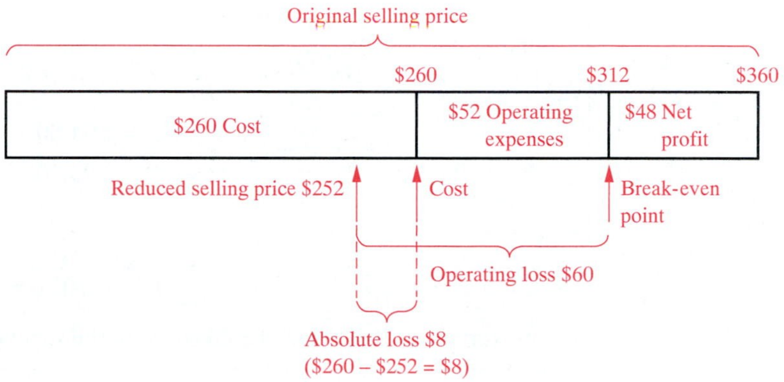

Figure 10.4

Objective 7 | *Find the Maximum Percent of Markdown to Be Given.*

Example 6 Finding the Original Price and Maximum Percent of Markdown to Be Given

A wood-burning pellet stove cost a retailer $630. If the store's operating expenses are 25% of cost, and net profit is 15% of cost, what is (a) the original selling price of the stove and (b) the maximum percent of markdown that may be given without taking an operating loss (round to the nearest tenth of a percent)?

Solution

(a) The markup is 40% (25% + 15%) on cost. The selling price is found from the basic markup formula.

$$C + M = S$$
$$\$630 + (0.4 \times \$630) = S$$
$$\$630 + \$252 = S$$
$$\$882 = S$$

The original selling price is $882.

(b) The break-even point is the sum of cost and operating expenses.

$$\$630 + (0.25 \times \$630) = \$630 + \$157.50$$
$$= \$787.50$$

The break-even point is $787.50.

The maximum amount of markdown that may be taken without an operating loss is as follows.

$882 (Selling price) − $787.50 (Break-even point) = $94.50 (Markdown)

Since the maximum markdown is $94.50, the maximum percent of markdown can be found by asking this question: What % of original selling price is markdown?

$$R \times B = P$$

$$\underline{\qquad\qquad}\% \times \$882 = \$94.50$$

$$\$882R = \$94.50$$

$$R = \frac{94.50}{882}$$

$$R = 0.1071 = 10.7\% \quad \text{Rounded.}$$

A maximum of 10.7% may be taken as markdown without an operating loss.

Stores can't always tell what customers will and won't buy. Sometimes the store's goal in offering huge markdowns is to minimize loss.

Ask Marilyn

It drives me nuts when I see the postholiday season sale prices, especially on clothes. It amazes me that they have that much of a markup the rest of the time, and it makes me feel like such a fool for ever paying the full price! Am I missing something?

J. W. San Diego, Calif.

Yes! Those sale prices are great bargains. Many people believe that if a tie first sells for $50, then gets reduced to $35 and finally goes down to $20 if it still hasn't sold, the store owners must have paid far less than $20 for it in the first place (or they wouldn't sell it for $20). But what would be their alternative? Even if the store paid $40 for the tie, it would be better to sell it for $20 than to discard it, which would add $0 to their bank account.

Let's say that a store's tie-buyer underestimates the customers' taste and pays $40 each for 50 ties with little smiley faces painted on them, pricing them at $50 each. By the end of the holiday season, 49 ties are left unsold. (One was sold to a woman who couldn't stand her husband.) Dismayed, the store reduces the price to $25 each to get rid of the darned things. (So far, the store has spent $2000 and taken in $50.) At the end of the sale season, 48 ties are left unsold. (Another one was sold to a woman who couldn't stand her son-in-law.)

After transferring the buyer to the children's department, the store reduces the price to $10 each. (It has now spent $2000 and taken in $75.) At this point, someone like you walks in, picks up one of the ties and says, "The *nerve* of these people! They must be making a *fortune* in this place. No *wonder* they've got Newt Gingrich's face painted on all these ties."

10.1 Exercises

Find the missing quantities. Round rates to the nearest whole percent and money to the nearest cent.

	Original Price	% Markdown	$ Markdown	Reduced Price
1.	$860	______	$215	______
2.	$240	______	$96	______
3.	$30.80	______	______	$16.94
4.	$1450	______	______	$725
5.	______	20%	______	$5.20
6.	______	$66\frac{2}{3}\%$	______	$3.10
7.	______	40%	$1.08	______
8.	______	20%	$0.77	______
9.	$43.50	50%	______	______
10.	$2327.50	44%	______	______
11.	______	______	$175	$682
12.	______	______	$276.93	$1261.57

Complete the following. If there is no operating loss or absolute loss, write none. Use Figure 10.2 as a guide.

	Cost	Operating Expense	Break-even Point	Reduced Price	Operating Loss	Absolute Loss
13.	$48	$12	______	$50	______	______
14.	$25	$8	______	$22	______	______
15.	$50	______	$66	$44	______	______
16.	$12.50	______	$16.50	$11	______	______
17.	$310	$75	______	______	$135	______
18.	$78	$22	______	______	$30	______
19.	______	______	______	$25	$14	$4
20.	______	______	______	$100	$56	$16

21. Describe five reasons why a store will reduce the price of merchandise (markdown) to get it sold.

22. As a result of a markdown, there are four possible results: reduced net profit, breaking even, operating loss, and absolute loss. As a business owner, which would concern you the most? Explain. (See Objective 4.)

Solve each of the following application problems. Round rates to the nearest whole percent and money to the nearest cent.

23. GAS BARBECUE MARKDOWN A Sunbeam Grill Master gas barbecue is marked down $76.48, a reduction of 32%. Find the original price.

24. VIDEO CAMCORDER An RCA camcorder is marked down 20% to $409.60. Find the original price.

25. OAK FURNITURE An oak bedroom set originally priced at $1675 is reduced to $1425. Find the percent of markdown on the original price.

26. VIDEOCASSETTE RECORDERS World Electronics prices their entire inventory of last year's model videocassette recorders (VCRs) at $44,503. If the original price of these VCRs was $75,428, find the percent of markdown on the original price.

27. IMPORTED RUGS The Persian and Oriental Rug Gallery paid $2211 for an imported rug from China. Their operating expenses are $33\frac{1}{3}\%$ of cost. If they sell the rug at a clearance price of $2650, find the amount of profit or loss.

28. EXERCISE BICYCLES Ship Shape Shop has an end-of-season sale during which it sells a stationary bicycle for $265. If the cost was $198 and the operating expenses were 25% of cost, find the amount of profit or loss.

29. ANTIQUES American Antique paid $153.49 for a fern stand. Their original selling price was $208.78, but this was marked down 46% in order to make room for incoming merchandise. Operating expenses are 14.9% of cost. Find (a) the operating loss, (b) the absolute loss, and (c) the percent of absolute loss based on cost.

30. TRUCK BED LINERS Pep Boys Automotive paid $208.50 for a pick-up truck bed liner. The original selling price was $291.90, but this was marked down 35%. Operating expenses are 28% of cost. Find (a) the operating loss, (b) the absolute loss, and (c) the percent of absolute loss based on cost.

31. PHOTOGRAPHIC ENLARGERS Photo Supply, a retailer, pays $190 for an enlarger. The store's operating expenses are 20% of cost and net profit is 15% of cost. Find (a) the selling price of the enlarger and (b) the maximum percent of markdown that may be given without taking an operating loss (round to the nearest whole percent).

32. AIR CONDITIONERS A room air conditioner cost a retailer $278. The store's operating expenses are 30% of cost and net profit is 10% of cost. Find (a) the selling price of the air conditioner and (b) the maximum percent of markdown that may be given without taking an operating loss.

10.2 Average Inventory and Inventory Turnover

Objectives

1. *Determine average inventory.*
2. *Calculate stock turnover.*
3. *Identify considerations in stock turnover.*

The average time needed for merchandise to sell is a common measure of the efficiency of a business. The number of times that the merchandise sells and is replaced during a certain period of time is called the **inventory turnover** or the **stock turnover**. Businesses such as florist shops or produce departments have a very fast turnover of merchandise, perhaps just a few days. On the other hand, a furniture store will normally have a much slower turnover, perhaps several months.

If inventory is kept at a minimum level, less capital is needed to operate the business and the speed of selling the inventory is increased. In addition, the risk of having old and unsaleable merchandise is decreased. On the other hand, if the inventory is too low, sales may be lost due to poor selection and inadequate quantities of merchandise offered to the customer.

Objective 1 *Determine Average Inventory.* Find stock turnover by first calculating the **average inventory**. The average inventory for a certain time period is found by adding the inventories taken during the time period and then dividing the total by the number of times that the inventory was taken.

Example 1 Determining Average Inventory

The inventory value at Sports and More was $285,672 on April 1 and $198,560 on April 30. What was the average inventory?

Solution First, add the inventory values.

$285,672	(April 1)
+ $198,560	(April 30)
$484,232	

Then divide by the number of times inventory was taken.

$$\frac{\$484{,}232}{2} = \$242{,}116$$

The average inventory was $242,116.

Example 2 Finding Average Inventory for the Quarter

The retail value of inventory at Federal Drug was $22,615 on January 1, $18,321 on February 1, $26,718 on March 1, and $16,228 on March 31. Find the average inventory for the first quarter of the year.

Solution First add the inventories and then divide by the number of inventories taken.

$$\frac{\$22{,}615 + \$18{,}321 + \$26{,}718 + \$16{,}228}{4} = \$20{,}970.50$$

The average inventory for the first quarter of the year is $20,970.50.

NOTE In Example 2, inventory was taken four times to find the average inventory for the first quarter of the year. To find the average inventory for a period of time, an inventory must be taken at the beginning of the period and one final time at the end of the period. To find average inventory for a full year, it is common to find inventory on the first day of each month and on the last day of the last month. Average inventory is found by adding all these inventory amounts (13 of them) and then dividing by 13, the number of inventories taken. Methods of taking inventory and inventory valuation are discussed in the next section. ▬

Keeping a close watch on inventory is an ongoing concern of management. Year-end sales promotions are designed to bring about a major reduction in a store's inventory.

MONDAY–FRIDAY 10AM–9PM SATURDAY 10AM–8PM & SUNDAY 11AM–6PM. PRICES GOOD 'TIL TUESDAY!
GET HUGE STOREWIDE SAVINGS!
YEAR-END SALE!
No Money Down, No Interest & No Payment 'til June
...On Every Item ...On Every Room!
Same As Cash Option. On Approved Credit With No Down Payment, Interest Accrues From Delivery Date if not Paid in Full by June

Objective 2 *Calculate Stock Turnover.* Most inventories are taken at retail because products or shelf locations are marked with the retail price and not the cost. Also, statistical averages are usually shown as average turnover at retail. Many businesses however, value inventory at cost. For this reason, stock turnover is found by either of these formulas.

$$\text{Turnover at retail} = \frac{\text{Retail sales}}{\text{Average inventory at retail}}$$

$$\text{Turnover at cost} = \frac{\text{Cost of goods sold}}{\text{Average inventory at cost}}$$

The turnover ratio may be identical by either method. The variation that often exists is caused by stolen merchandise (called *inventory shrinkage*) or merchandise that has been marked down or become unsaleable. Normally, **turnover at retail** is slightly lower than **turnover at cost**. For this reason, many businesses prefer this more conservative figure.

Example 3 Finding Stock Turnover at Retail

During May, the Skater's World has sales of \$32,032 and an average retail inventory of \$9856. Find the stock turnover at retail.

Solution

$$\text{Turnover at retail} = \frac{\text{Retail sales}}{\text{Average inventory at retail}} = \frac{\$32{,}032}{\$9856} = 3.25 \text{ at retail}$$

On average, the store turned over its entire inventory 3.25 times during the month.

Example 4 Finding Stock Turnover at Cost

If the Skater's World in Example 3 used a markup of 40% on selling price and if the cost of goods sold was \$19,396, what is the stock turnover on cost?

Solution Inventory at retail in the store was \$9856. Since markup is 40% on selling price, the cost is 60% of the inventory.

$$C = 0.6 \times \$9856$$
$$= \$5913.60$$

Inventory value at cost is \$5913.60.

$$\text{Turnover at cost} = \frac{\text{Cost of goods sold}}{\text{Average inventory at cost}}$$
$$= \frac{\$19{,}396}{\$5913.60} = 3.28 \text{ at cost} \qquad \text{Rounded}$$

The turnover on cost is 3.28.

NOTE In Example 4, the average inventory at cost needed to be calculated. To do this the average markup was used along with the cost of goods sold. Since the markup was 40% on selling price, the cost of goods sold was 60% of the selling price.

Objective 3 ***Identify Considerations in Stock Turnover.*** Stock turnover is useful for comparison purposes only. Many trade organizations publish such operating statistics to permit businesses to compare their operation with the industry as a whole. In addition to this, management will compare turnover from period to period and from department to department.

A rapid stock turnover is usually given high priority by management. Here are some of the benefits.

- Capital invested in inventory is kept at a minimum. This allows additional funds to be used for special purchases and discounts.
- Items in inventory are up-to-date, fresher, and are therefore less likely to be sold at a reduced price or loss.
- Costly storage space and the other expenses of handling inventory can be minimized.

On the other hand, a high inventory turnover may cause problems that result in reduced profits.

- Orders for smaller quantities of goods from wholesalers may result in losses of quantity discounts and thus incur additional processing, handling, and bookkeeping expenses.
- Items can be sold out and customers may be dissatisfied.

Both inventory selection and inventory turnover are very important management decisions, and much attention is typically given to this part of the business.

10.2 Exercises

Find the average inventory in each of the following.

	Date	Inventory Amount at Retail
1.	April 1	$10,603
	April 30	$12,757
3.	July 1	$18,300
	October 1	$26,580
	December 31	$23,139
5.	January 1	$16,250
	March 1	$20,780
	May 1	$28,720
	July 1	$24,630
	September 1	$23,550
	November 1	$34,800
	December 31	$22,770

	Date	Inventory Amount at Retail
2.	January 1	$42,312
	July 1	$38,514
	December 31	$30,219
4.	January 31	$69,480
	April 30	$55,860
	July 31	$80,715
	October 31	$88,050
	January 31	$63,975
6.	January 1	$65,430
	April 1	$58,710
	July 1	$53,410
	October 1	$78,950
	December 31	$46,340

Find the stock turnover at retail and turnover at cost for each of the following. Round to the nearest hundredth.

	Average Inventory at Cost	Average Inventory at Retail	Cost of Goods Sold	Retail Sales
7.	$26,745	$53,085	$75,591	$149,175
8.	$15,140	$24,080	$67,408	$106,193
9.	$22,390	$32,730	$178,687	$259,876
10.	$30,280	$48,160	$134,816	$212,386
11.	$26,400	$42,660	$270,600	$437,260
12.	$72,120	$138,460	$259,123	$487,379
13.	$180,600	$256,700	$846,336	$1,196,222
14.	$411,580	$780,600	$1,905,668	$3,559,536

15. Identify three types of businesses that you think would have a high turnover. Identify three types of businesses that you think would have a low turnover.

16. Which departments in a grocery store do you think have the highest turnover? Which ones have the lowest turnover?

Solve each of the following application problems. Round stock turnover to the nearest hundredth.

17. **MUSIC STORE INVENTORY** Music Central took inventory at the first of each month for the full year. The sum of these inventories was $655,974. On December

31, inventory was again taken and amounted to $52,476. Find the average inventory for the year.

18. PHARMACY INVENTORY Inventory at retail at Sid's Pharmacy was $38,864 on April 1, $47,536 on May 1, and $26,128 on June 1. Find the average inventory.

19. GLASS SHOP The Glass Works has an average inventory at cost of $15,730 and cost of goods sold for the same period is $85,412. Find the stock turnover at cost.

20. RETAIL GROCERIES The Jumbo Market has an average canned fruit inventory of $2320 at retail. Sales of canned fruit for the year were $98,669. Find the stock turnover at retail.

21. METAL PLATING Capital Plating had an average inventory of $27,250 at retail. The cost of goods sold for the year was $103,400 and a markup of 40% on selling price was used. What was the turnover at cost?

22. UNDERWATER DIVING The Associated Divers uses a markup of 30% on selling price. If their average inventory is $15,650 at retail and the cost of goods sold is $53,023, what is the stock turnover at cost?

23. ELECTRICAL SUPPLIES The cost of goods sold at Capitol Electric was $2,108,410. The following inventories were taken at cost: $208,180, $247,660, and $114,438. Find the stock turnover at cost.

24. HOBBY AND SPORTS Inventory at Harbortown Hobby and Sports was taken at retail value four times and found to be at $53,820, $49,510, $60,820, and $56,380. Sales during the same period were $252,077. Find the stock turnover at retail.

25. POSTERS AND CARDS Posters and Cards had the following inventories at retail: $33,820, $46,240, $39,830, $52,040, and $48,700. This business uses a 35% markup on cost and the cost of goods sold during this period was $136,450. Find the turnover at cost.

26. FLOOR COVERING Inventory at Clinton Floor Covering was taken at retail four different times and found to be $98,500, $135,820, $107,420, and $124,300. The company uses a markup of 25% on cost and the cost of goods sold for this period was $305,920. Find the inventory turnover at cost.

10.3 Valuation of Inventory

Objectives

1. *Define perpetual inventory.*
2. *Understand uniform product codes (UPC).*
3. *Use specific identification to value inventory.*
4. *Use the weighted average method to value inventory.*
5. *Use the FIFO method to value inventory.*
6. *Use the LIFO method to value inventory.*
7. *Use the gross profit method to estimate inventory.*
8. *Use the retail method to estimate inventory.*

Objective 1 *Define Perpetual Inventory.* Placing a value on the merchandise that a firm has in stock is called **inventory valuation**. It is not always easy to place this value on each of the items in inventory. Many large companies keep a **perpetual inventory** by using a computer. As new items are received, the quantity, size, and cost of each are placed in the computer. Sales clerks enter product codes into the cash register (or uniform product codes are entered automatically with an optical scanner) and the computer processes the information. The result is an up-to-the-moment, or perpetual, inventory value.

Objective 2 *Understand Uniform Product Codes (UPC).* **Uniform product codes (UPC)** are the stripes known as bar codes that appear on many items sold in stores. Each product and product size is assigned its own code number. These UPCs are a great help in keeping track of inventory.

ISBN 0-87720-071 - 8

9 780877 200710

A Cracker Jack box has a UPC number (also called the *bar code*) of 0-87720-071-8. The checkout clerk in a retail store passes the coded lines over an optical scanner. The numbers are picked up by a computer, which recognizes the product by its code. The computer then forwards the price of the item to the cash register. At the same time the price is being recorded, the computer is subtracting the item automatically from inventory. After all the items being purchased have passed over the scanner, the customer receives a detailed cash-register receipt that gives a description of each item, the price of each item, and a total purchase price. Since the computer keeps track of stock on hand and is programmed to respond when inventory gets low, it provides more accurate inventory control and lower labor costs for the store.

Most businesses take a **physical inventory** (an actual count of each item in stock at a given time) at regular intervals. For example, inventory may be taken monthly, quarterly, semiannually, or just once a year. An inventory taken at regular time intervals is called a **periodic inventory**.

There are four major methods used for inventory valuation. They are: the specific identification method, the weighted-average method, the first-in, first-out method, and the last-in, first-out method.

Objective 3 *Use Specific Identification to Value Inventory.* The **specific-identification method** of inventory valuation is useful where items are easily identified and costs do not fluctuate. If the number of items is large and the exact cost of each unit is not known, then it may be difficult or impossible to use this method of inventory. With specific identification, each item is cost-coded, either with numerals or a letter code. The cost may be included in a group of numbers written on the item or in a 10-letter code (where each letter is different).

For example, if a store uses the code SMALPROFIT, or

1	2	3	4	5	6	7	8	9	0
S	M	A	L	P	R	O	F	I	T

then an item bearing the cost code ARF would have a cost of

$$\begin{array}{ccc} A & R & F \\ \downarrow & \downarrow & \downarrow \\ 3 & 6 & 8 \end{array}$$

or \$3.68. An item coded PMITT would have a cost of \$529.

Example 1 Finding the Value of Inventory

Hoig's Marine has four fishing boats in stock. Cost codes indicate the following costs to the store.

Model A	$2718
Model B	$2571
Model C	$3498
Model D	$3974

Find the value of the fishing boat inventory using the specific-identification method.

Solution The value of the inventory is found by adding the costs of the four fishing boats.

Model A	$ 2,718	
Model B	2,571	
Model C	3,498	
Model D	+ 3,974	
	$12,761	Total value of inventory

Objective 4 *Use the Weighted-Average Method to Value Inventory.* Since the cost of many items changes over time, there may be several of the same items that were purchased at different costs. Because of this variation, there are several common methods used to value an inventory. One method, the **weighted-average (average-cost) method**, values the items in an inventory at the average cost of buying them.

Example 2 Using Weighted-Average (Average-Cost) Inventory Valuation

Olympic Sports and Leisure made the following purchases of the Explorer External Frame backpack during the year.

Beginning inventory	20 backpacks at $70
January	50 backpacks at $80
March	100 backpacks at $90
July	60 backpacks at $85
October	40 backpacks at $75

At the end of the year there are 75 backpacks in inventory. Use the weighted-average method to find the inventory value.

Solution Find the total cost of all the backpacks.

Beginning inventory	20 × $70 =	$1,400
January	50 × $80 =	$4,000
March	100 × $90 =	$9,000
July	60 × $85 =	$5,100
October	40 × $75 =	$3,000
Total	270	$22,500

Find the average cost per backpack by dividing this total cost by the number purchased. The average cost per backpack is

$$\frac{\$22{,}500}{270} = \$83.33 \quad \text{Rounded}$$

Since the average cost is \$83.33 and 75 backpacks remain in inventory, the weighted-average method gives the inventory value of the remaining backpacks as \$83.33 × 75 = \$6249.75.

Scientific
Calculator Approach

The calculator solution to Example 2 has several steps. First, find the total number of backpacks purchased and place the total in memory.

20 [+] 50 [+] 100 [+] 60 [+] 40 [=] 270 [STO]

Next, find the total cost of all the backpacks purchased and divide by the number stored in memory. This gives the average cost per backpack.

20 [×] 70 [+] 50 [×] 80 [+] 100 [×] 90 [+] 60 [×] 85

[+] 40 [×] 75 [=] [÷] [RCL] [=] 83.3333

Finally, round the average cost to the nearest cent and multiply by the number of backpacks in inventory to get the weighted average inventory value.

83.33 [×] 75 [=] 6249.75

Objective 5 *Use the FIFO Method to Value Inventory.* The **first-in, first-out (FIFO) method** of inventory valuation assumes a natural flow of goods through the inventory: the first goods to arrive are the first goods to be sold, so, the most recent items purchased by the company are the items remaining in inventory.

Example 3 Using FIFO Inventory Valuation

Use the FIFO method to find the inventory value of the 75 backpacks from Olympic Sports and Leisure in Example 2.

Solution With the FIFO method, the 75 remaining backpacks are assumed to consist of the last 75 backpacks purchased, or 40 backbacks bought in October and 35 (75 − 40 = 35) backpacks from the previous purchase in July.

The value of the inventory is as follows

October	40 backpacks at \$75 =	\$3000	Value of last 40
July	35 backpacks at \$85 =	\$2975	Value of previous 35
	75 valued at	\$5975	

The value of the backpack inventory is \$5975 using the FIFO method.

Objective 6 *Use the LIFO Method to Value Inventory.* The **last-in, first-out (LIFO) method** of inventory valuation assumes a flow of goods through the inventory that is just the opposite of the FIFO method. With LIFO, the goods remaining in inventory are the first goods purchased.

Example 4 Using LIFO Inventory Valuation

Use the LIFO method to value the 75 backpacks in inventory at Olympic Sports and Leisure (see Example 2).

Solution The calculation starts with the beginning inventory and moves through the year's purchases, resulting in 75 backpacks. The beginning inventory and January purchases come to 70 backpacks, so the cost of 5 more (75 − 70 = 5) backpacks from the March purchase is needed.

Beginning inventory	20 backpacks at \$70 =	\$1400	Value of first 20
January	50 backpacks at \$80 =	\$4000	Value of next 50
March	5 backpacks at \$90 =	\$450	Value of last 5
Total	75 valued at	\$5850	

The value of the backpack inventory is \$5850 using the LIFO method.

Depending on the method of valuing inventories that is used, Olympic Sports and Leisure may show the inventory value of the 75 backpacks as follows.

Average-cost method	\$6249.75
FIFO	\$5975.00
LIFO	\$5850.00

The preferred inventory valuation method would be determined by Olympic Sports and Leisure, perhaps on the advice of an accountant. The comparison of inventory evaluation methods below would help them in selecting the method.

Comparing Inventory Evaluation Methods

- When the market is stable, prices do not fluctuate. Since prices are not changing, all methods give the *same* inventory value.
- In a market where prices are rising and falling, each method gives a *different* inventory value.
- Weighted-average costs tend to *smooth out* inventory values that might result from price fluctuations.
- The FIFO method results in an inventory value that most closely reflects the *current* replacement cost of inventory.
- The LIFO method results in the last cost of inventory being assigned to the cost of merchandise sold. This results in a *closer matching* of current costs with sales revenues.

NOTE Accepted accounting practice insists that the method used to evaluate inventory be stated on the company financial statements. ■

The large quantities and varied types of items that are often in inventory make it time-consuming and expensive to take an actual physical inventory. Where this is the case, a physical inventory may be taken only once or twice a year. However, the need to monitor the performance of the business throughout the year is extremely important. To do this, methods of approximating inventory value have been developed. Two common methods for doing this are the gross-profit method and the retail method.

Objective 7 *Use the Gross Profit Method to Estimate Inventory.* An estimate of inventory using the **gross-profit method** is found as follows.

	Beginning inventory (at cost)
+	Purchase (at cost)
	Merchandise available for sale (at cost)
−	Cost of goods sold
	Ending inventory (at cost)

The beginning inventory and amount of purchases are taken from company records and the cost of goods sold is normally determined by applying the rate of markup to the amount of net sales.

Example 5 Estimating Inventory Value Using the Gross-Profit Method

Inventory on June 30 was \$242,000. During the next three months, the company had purchases of \$425,000 and net sales of \$528,000. Use the gross-profit method to estimate the value of the inventory on September 30 if the company uses a markup of 25% on selling price.

Solution First, find the cost of goods sold. Since markup is 25% of selling price, find cost.

$$C + M = S$$
$$C + 0.25S = S$$
$$C + 0.25S - 0.25S = S - 0.25S$$
$$C = S - 0.25S$$
$$C = 0.75S$$
$$C = 0.75 \times \$528{,}000$$
$$C = \$396{,}000$$

The cost of goods sold is \$396,000. Now use the gross profit method.

	\$242,000	Beginning inventory (June 30)
+	425,000	Purchases
	\$667,000	Merchandise available for sale
−	396,000	Cost of goods sold
	\$271,000	Ending inventory (September 30)

The estimated value of the inventory on September 30 is \$271,000.

Objective 8 *Use the Retail Method to Estimate Inventory.* The **retail method** of estimating inventory requires that a business keep records of all purchases at both cost and retail prices.

The format for estimating inventory value at retail is the same as that used in estimating inventory using the gross profit method. The difference however, is that all amounts used are at retail. Inventory at retail is estimated as follows.

	Beginning inventory (at retail)
+	Purchases (at retail)
	Merchandise available for sale (at retail)
−	Net sales
	Ending inventory (at retail)

Notice that net sales are used in this method instead of cost of goods sold. Also, ending inventory value at retail must now be changed to an estimated value at cost. This is done by multiplying the ending inventory value at retail by a ratio determined by comparing the merchandise available for sale at cost to the merchandise available for sale at retail.

Example 6 Estimating Inventory Value Using the Retail Method

At one store, inventory on March 31 was valued at $9000 at cost and $15,000 at retail. During the next 3 months, the company made purchases of $36,000 at cost, or $60,000 at retail, and had total sales of $54,000. Use the retail method to estimate the value of inventory at cost on June 30.

Solution

At Cost	*At Retail*	
	$15,000	Beginning inventory
$9,000	+ 60,000	Purchases
+ 36,000	$75,000	Merchandise available for sale
$45,000	− 54,000	Net sales
	$21,000	June 30 inventory (at retail)

Now find the ratio of merchandise available for sale at cost to merchandise available for sale at retail.

$$\frac{\$45,000}{\$75,000}$$ Merchandise available for sale at cost / Merchandise available for sale at retail

Finally, the estimated inventory value at cost on June 30 is found by multiplying inventory at retail on June 30 by this ratio.

$$\$21,000 \times \frac{45,000}{75,000} = \$12,600$$ June 30 inventory (at cost)

The retail method of estimating inventory value assumes that the ratio of the merchandise available for sale at cost to the merchandise available for sale at retail is the same as the ratio of the ending inventory at cost to ending inventory at retail. Both the gross-profit method and the retail method of estimating inventories must be updated from time to time with physical inventories to assure accurate record keeping.

10.3 Exercises

Find the inventory value in each of the following using the specific-identification method.

	Description	Cost
1.	Fair	$208
	Good	$274
	Excellent	$345
3.	Good	$79
	Excellent	$186
	Mint	$295

	Description	Cost
2.	Economy	$1215
	Standard	$1509
	Luxury	$1873
	Deluxe	$2116
4.	Incomplete	$835
	Cheap	$972
	Improved	$1170
	Tolerable	$1360

Find the inventory values using (a) the weighted-average method, (b) the FIFO method, and (c) the LIFO method for each of the following. Round to the nearest dollar.

	Purchases		Now in Inventory
5.	Beginning inventory:	10 units at $8	
	June:	25 units at $9	
	August:	15 units at $10	20 units
6.	Beginning inventory:	80 units at $14.50	
	July:	50 units at $15.80	
	October:	70 units at $13.90	90 units
7.	Beginning inventory:	50 units at $30.50	
	March:	70 units at $31.50	
	June:	30 units at $33.25	
	August:	40 units at $30.75	75 units
8.	Beginning inventory:	700 units at $1.25	
	May:	400 units at $1.75	
	August:	500 units at $2.25	
	October:	600 units at $3.00	720 units

Solve each of the following application problems. Round to the nearest dollar.

9. SPECIFIC-IDENTIFICATION INVENTORY VALUE Towne Furniture had the following coffee tables in stock.

Model Number	Cost
P6251	$182
C3852	$210
PC623	$132
RW778	$921
WO335	$325

Find the inventory value using the specific-identification method.

10. **INVENTORY VALUE** Foothill Heat and Cool had the following refrigeration compressors in stock. Use the specific identification method to find the inventory value.

Model Number	Cost
AC129	$428
AC428	$715
AC2207	$526
AC78C	$1718
AC3107	$635

11. **BANDAGE INVENTORY** Rite-Aid Drug Store made the following purchases of elastic bandages made in Brazil.

Beginning inventory:	24 units at $1.50
June:	40 units at $1.35
August:	48 units at $1.25
November:	18 units at $1.60

Inventory at the end of the year shows that 35 units remain. Find the inventory value using (a) the weighted-average method, (b) the FIFO method, and (c) the LIFO method.

12. **PAINT INVENTORY** The Graphic Hobby House made purchases of assorted colors of spray paint during the year as follows.

Beginning inventory:	200 cans at $1.10
March:	400 cans at $1.20
May:	700 cans at $1.00
August:	500 cans at $1.15
November:	300 cans at $1.30

At the end of the year, they had 450 cans of spray paint in stock. Find the inventory value using (a) the weighted-average method, (b) the FIFO method, and (c) the LIFO method.

13. **APPLIANCE PARTS INVENTORY** B&B Appliance Repair made the following purchases of dishwasher pumps.

Beginning inventory:	200 units at $3.10
July:	250 units at $3.50
August:	300 units at $4.25
October:	280 units at $4.50

Inventory at the end of October shows that 320 units remain. Find the inventory value using (a) the weighted-average method, (b) the FIFO method, and (c) the LIFO method.

14. **AUTOMOBILE MUFFLERS** Marco Muffler Wholesalers made purchases of automobile mufflers throughout the year as follows:

Beginning inventory:	300 units at $21.60
March:	400 units at $24.00
August:	450 units at $24.30
November:	350 units at $22.50

An inventory at the end of December shows that 530 mufflers remain. Find the inventory value using (a) the weighted-average method, (b) the FIFO method, and (c) the LIFO method.

15. **AUTO SUPPLIES INVENTORY** Central States Auto Wholesalers made the following purchases of antifreeze.

Beginning inventory:	350 cases at $8.25
October:	300 cases at $9.50
November:	360 cases at $11.45
December:	240 cases at $10.10

The January inventory found that there were 625 cases remaining. Find the inventory value using (a) the weighted-average method, (b) the FIFO method, and (c) the LIFO method.

16. **LINEN SUPPLY** Industrial Linen Supply made the following purchases of shop towels.

Beginning inventory:	650 towels at $3.80
June:	500 towels at $4.20
September:	450 towels at $3.95
December:	600 towels at $4.05

In January an inventory found that there were 775 towels remaining. Find the inventory value using (a) the weighted-average method, (b) the FIFO method, and (c) the LIFO method.

17. **SECURITY DOORS** The inventory on December 31 at Safe Security Doors was $136,000 at cost. During the next 3 months, the company made purchases of $148,000 (cost) and had net sales of $236,000. Use the gross-profit method to estimate the value of the inventory at cost on March 31 if the company uses a markup of 35% on selling price.

18. **TOY STORE INVENTORY** Happy Toys has an inventory of $52,000 at cost on June 30. Purchases during the next 3 months were $68,000 (cost) and net sales were $126,000. If the company uses a 40% markup on selling price, use the gross-profit method to estimate the value of the inventory at cost on September 30.

19. **PIANO REPAIR** The September 30 inventory at Liverpool Piano Repair was $43,750 at cost and $62,500 at retail. Purchases during the next 3 months were $51,600 at cost, $73,800 at retail, and net sales were $92,500. Use the retail method to estimate the value of inventory at cost on December 31.

20. **EVALUATING INVENTORY** Uptown Sound had an inventory of $27,000 at cost and $43,000 at retail on March 31. During the next 3 months, they made purchases of $108,000 at cost and $180,000 at retail and had net sales of $162,000. Use the retail method to estimate the value of inventory at cost on June 30.

21. In your opinion, what are the benefits to a merchant who is using uniform product codes (UPC)? (See Objective 2.)

22. List the four inventory valuation methods discussed in this section. Explain how one of these methods is used to determine inventory value.

Chapter 10 Quick Review

Chapter Terms

Review the following terms to test your understanding of the chapter. For each term you do not know, refer to the page number found next to that term.

absolute loss [p. 353]
actual selling price [p. 351]
average inventory [p. 359]
break-even point [p. 353]
first-in, first-out (FIFO) method [p. 366]
formula for markdown [p. 351]
gross-profit method [p. 368]
inventory turnover [p. 358]
inventory valuation [p. 363]
last-in, first-out (LIFO) method [p. 366]
markdown [p. 351]
operating loss [p. 353]
percent of markdown [p. 351]
periodic inventory [p. 364]
perpetual inventory [p. 363]
physical inventory [p. 364]
rate of markdown [p. 351]
reduced net profit [p. 353]
reduced price [p. 351]
retail method [p. 368]
sale price [p. 351]
specific-identification method [p. 364]
stock turnover [p. 358]
turnover at cost [p. 360]
turnover at retail [p. 360]
uniform product code (UPC) [p. 364]
weighted-average (average-cost) method [p. 365]

Concepts	Examples
10.1 Percent of markdown Markdown is always a percent of the original price. Use the following formula $\text{Markdown percent} = \frac{\text{Markdown amount}}{\text{Original price}}$	Original price, \$76; markdown, \$19. Find the percent of markdown. $\text{Markdown percent} = \frac{19}{76} = 0.25 = 25\%$
10.1 Break-even point Add cost to operating expenses to find the break-even point.	Cost, \$54; operating expenses, \$16. Find the break-even point. \$54 + \$16 = \$70 Break-even point
10.1 Operating loss When the reduced price is below the break-even point, subtract the reduced price from the break-even point to find the operating loss.	Break-even point, \$70; reduced price, \$58. Find the operating loss. \$70 − \$58 = \$12 Operating loss
10.1 Absolute loss When the reduced price is below cost, subtract the reduced price from the cost to find the absolute loss.	Cost, \$54; reduced price, \$48. Find the absolute loss. \$54 − \$48 = \$6 Absolute loss
10.2 Average inventory Take inventory two or more times; add the totals and divide by the number of inventories to get the average.	Inventories: \$22,635, \$24,692, and \$18,796. Find the average inventory. $\frac{\$22{,}635 + \$24{,}692 + \$18{,}796}{3} = \frac{\$66{,}123}{3} = \$22{,}041$ Average inventory
10.2 Turnover at retail Use the following formula $\text{Turnover} = \frac{\text{Retail sales}}{\text{Average inventory at retail}}$	Sales, \$78,496; average inventory at retail, \$18,076. Find turnover at retail. $\frac{\$78{,}496}{\$18{,}076} = 4.34$ at retail Rounded

10.2 Turnover at cost Use the following formula. $$\text{Turnover} = \frac{\text{Cost of goods sold}}{\text{Average inventory at cost}}$$	Cost of goods sold, \$26,542; average inventory at cost, \$6592. Find the turnover at cost. $$\frac{\$26{,}542}{\$6592} = 4.03 \text{ at cost}$$ Rounded
10.3 Specific identification to value inventory Each item is cost-coded and the cost of each of the items is added to find total inventory.	Costs: item 1, \$593; item 2, \$614; item 3, \$498. Find total value of inventory. \$593 + \$614 + \$498 = \$1705. Total value of inventory
10.3 Weighted-average method of inventory valuation This method values items in an inventory at the average cost of buying them.	Beginning inventory of 20 at \$75; purchases of 15 at \$80; 25 at \$65; 18 at \$70; 22 remain in inventory. Find the inventory value. $20 \times \$75 = \$1500$ $15 \times \$80 = \$1200$ $25 \times \$65 = \$1625$ $18 \times \$70 = \$1260$ Totals 78 \$5585 $$\frac{\$5585}{78} \times 22 = \$1575.26$$
10.3 First-in, first-out (FIFO) method of inventory valuation First items in are first sold. Inventory is based on cost of most recent items purchased.	Beginning inventory of 25 items at \$40; purchase on Aug. 7, 30 items at \$35; 35 remain in inventory. Find the inventory value. $30 \times \$35 = \$1050$ Value of last 30 $5 \times \$40 = \$200$ Value of previous 5 Totals 35 \$1250 Value of inventory
10.3 Last-in, first-out (LIFO) method of inventory valuation The goods remaining in inventory are the first goods purchased.	Beginning inventory of 48 items at \$20 each; purchase on May 9, 40 items at \$25 each; 55 remain in inventory. Find the inventory value. $48 \times \$20 = \$960$ Value of first 48 $7 \times \$25 = \$175$ Value of last 7 Totals 55 \$1135 Value of inventory

Chapter 10 Review Exercises

Find the missing quantities. Round rates to the nearest whole percent and money to the nearest cent. **[10.1]**

	Original Price	% Markdown	$ Markdown	Reduced Price
1.	$96	30%	______	______
2.	______	$33\frac{1}{3}\%$	______	$10
3.	______	50%	$2.70	______
4.	$2340	______	______	$1755

Complete the following. If there is no operating loss or absolute loss, write "none." Use Figure 10.2 as a guide. **[10.1]**

	Cost	Operating Expense	Breakeven Point	Reduced Price	Operating Loss	Absolute Loss
5.	$150	______	$198	$132	______	$18
6.	$80	$20	______	$93	______	______
7.	$78	$22	______	______	$30	______
8.	$5	$1.25	______	$5.50	______	______

Find the average inventory in each of the following. **[10.2]**

	Date	Inventory Amount at Retail
9.	Beginning inventory	$44,398
	March 31	$37,704
11.	Beginning inventory	$77,159
	April 1	$67,305
	July 1	$80,664
	October 1	$95,229
	December 31	$61,702

	Date	Inventory Amount at Retail
10.	Beginning inventory	$316,481
	July 1	$432,185
	December 31	$296,738
12.	Beginning inventory	$36,502
	April 1	$27,331
	July 1	$28,709
	October 1	$32,153
	December 31	$39,604

Find the stock turnover at retail and at cost in each of the following. Round to the nearest hundredth. **[10.2]**

	Average Inventory at Cost	Average Inventory at Retail	Cost of Goods Sold	Retail Sales
13.	$14,120	$25,572	$81,312	$146,528
14.	$11,195	$16,365	$89,343	$129,938
15.	$90,300	$128,350	$423,168	$598,111
16.	$102,895	$195,150	$476,417	$889,884

Find the inventory value in each of the following using the specific-identification method. [10.3]

	Description	Cost		Description	Cost
17.	Small	$122	**18.**	Poor	$314
	Medium	$199		Fair	$422
	Large	$235		Good	$506
19.	Economy	$795	**20.**	Good	$1283
	Standard	$850		Better	$1398
	Luxury	$915		Best	$1564
	Deluxe	$1080		Designer	$1772

Solving the following application problems.

21. An industrial floor buffer originally priced at $1850 is marked down to $1332. Find the percent of markdown on the original price. [10.1]

22. A cordless telephone costs a retailer $56. The store's operating expenses are 30% of cost and net profit is 10% of cost. Find (a) the selling price of the cordless telephone and (b) the maximum percent of markdown that may be given without taking an operating loss. (Round to the nearest whole percent.) [10.1]

23. Party Supplies had an inventory of $29,332 on January 1, $36,908 on July 1, and $31,464 on December 31. Find the average inventory. [10.2]

24. The Natural Grocer has an average inventory of $8460 at cost. If the cost of goods sold for the year was $91,125, find the stock turnover at cost. Round to the nearest hundred. [10.2]

25. Inventory at a local store was taken at retail value four times and was found to be $53,820, $49,510, $60,820, and $56,380. Sales during the same period were $252,077. Find the stock turnover at retail. Round to the nearest hundredth. [10.3]

26. Thunder Manufacturing made the following purchases of rivet drums during the year: 25 at $135 each, 40 at $165 each, 15 at $108.50 each, and 30 at $142 each. An inventory shows that 45 rivet drums remain. Find the inventory value using (a) the weighted-average method, (b) the FIFO method, and (c) the LIFO method. [10.3]

27. The inventory on December 31 at Modern Clothiers was $118,000 at cost. During the next 3 months, the company made purchases of $186,000 (cost) and had net sales of $378,000. Use the gross profit method to estimate the value of the inventory at cost on March 31 if the company uses a markup of 50% on selling price. [10.3]

28. Tap Plastics had an inventory of $54,000 at cost and $90,000 at retail on June 30. During the next 3 months, they made purchases of $216,000 at cost, $360,000 at retail, and had net sales of $324,000. Use the retail method to estimate the value of the inventory at cost on September 30. [10.3]

Chapter 10 Summary Exercise *Markdown: Reducing Prices to Move Merchandise*

Olympic Sports and Leisure purchased two dozen pairs of Roller Derby Baja adult in-line skates at a cost of \$1950. Operating expenses for the store are 25% of cost while total markup on this type of product is 35% of selling price. Only six pairs of the skates sell at the original price and the manager decides to mark down the remaining skates. The price is reduced 25% and six more pairs sell. The remaining 12 pairs of skates are marked down 50% of the original selling price and are finally sold.

(a) Find the original selling price of each pair of skates.

(b) Find the total sales of all the skates.

(c) Find the operating loss.

(d) Find the absolute loss.

Net Assets Business on the Internet

CDW Computer Centers

Statistics

- 1984: Established
- 1995: launched e-commerce site
- 1998: 28 inventory turns in fourth quarter
- 2000: 48,000 computer products
- 10,000 orders shipped daily

CDW Computer Centers is the country's leading direct solutions provider, offering complete, customized computing solutions for business and consumers nationwide. The company began as a one-man, home-based business in 1984 and in 1998 was selected by *Fortune* magazine as one of the "100 Best Companies to Work For."

CDW offers more than 45,000 brand-name computer products and is the number one direct source of Compaq, IBM, Microsoft, Toshiba, Computer Associates, and other top name brands. At CDW, "it's all about people." From the beginning, CDW's strength has always been building win/win/win relationships with customers, vendors, coworkers, and the community.

1. A Compaq Armada 1750 computer originally priced at $2097.95 is marked down to $1846.20. Find the percent of markdown.

2. The cost of a laptop computer is $1225. The store's operating expenses are 15% of cost and the net profit is 10% of cost. Find **(a)** the selling price of the laptop computer and **(b)** the maximum percent of markdown that may be given without taking an operating loss (round to the nearest whole percent).

3. The sales of CDW Computer Centers were $519.9 million in the fourth quarter of last year. If the average inventory at retail was $18.8 million during the same quarter, find the turnover at retail. Round to the nearest hundredth.

4. CDW Computer Centers made the following purchases of a certain model of computer printer: 40 at $135 each, 25 at $128 each, 50 at $140.50 each, and 35 at $131 each. An inventory shows that 45 printers remain. **(a)** Find the inventory value using the weighted average method. **(b)** Find the inventory value using the FIFO method. **(c)** Find the value of the inventory using the LIFO method.

Part 3 Cumulative Review Chapters 8–10

Find the net cost (invoice amount) for each of the following. [8.1]

1. List price \$280, less 10/20

2. List price \$375, less 25/10/5

Find the net cost equivalent and the percent form of the single discount equivalent for each of the following series discounts. [8.2]

3. 10/20 **4.** 20/20 **5.** 20/30/5 **6.** 50/40/10

Solve for the amount due on each of the following. [8.4]

	Invoice Amount	Invoice Date	Terms	Shipping and Insurance	Date Goods Received	Date Invoice Paid
7.	\$740.58	Mar. 20	2/10/ROG	\$36.80	April 6	Apr. 14
8.	\$874.22	Feb. 12	4/15 proximo		Feb. 22	Mar. 12
9.	\$3788.20	Jul. 19	3/15 EOM	\$71.18	Aug. 7	Aug. 9
10.	\$4692.50	Nov. 5	2/15-50x		Nov. 10	Jan. 6

Find the missing quantities by first computing the markup on one base and then computing the markup on the other. Round rates to the nearest tenth of a percent and money to the nearest cent. [9.2]

	Cost Price	Markup	Selling Price	% Markup on Cost	% Markup on Selling Price
11.	______	\$288.14	\$576.28	______	______
12.	______	\$38.22	______	$33\frac{1}{3}\%$	______

Find the missing quantities. Markup is based on total cost. [9.3]

	Total Cost	Quantity Purchased	Percent Sold at Reduced Price	% Markup on Total Cost	Number at Regular Price	Number at Reduced Price	Reduced Price	Regular Selling Price Each
13.	\$1400	1000 pr.	30%	50%	______	______	\$1.50	______
14.	\$2250	150	10%	25%	______	______	\$7.50	______

Complete the following. If there is no operating loss or absolute loss, write "none." Use Figure 10.2 as a guide. [10.1]

	Cost	Operating Expense	Break-even Point	Reduced Price	Operating Loss	Absolute Loss
15.	\$150	______	\$198	\$132	______	______
16.	\$39	\$11	______	______	\$15	______

Find the average inventory in each of the following. [10.2]

	Date	Inventory Amount at Retail
17.	Beginning inventory	$74,422
	July 1	$58,320
	December 31	$61,889
18.	Beginning inventory	$218,143
	April 1	$186,326
	July 1	$275,637
	October 1	$207,448
	December 31	$172,351

Find the stock turnover at retail and at cost in each of the following. Round to the nearest hundredth. [10.2]

	Average Inventory at Cost	Average Inventory at Retail	Cost of Goods Sold	Retail Sales
19.	$14,120	$25,572	$81,312	$146,528
20.	$45,150	$64,175	$211,584	$299,056

21. A home-security system is list-priced at $2995. The manufacturer offers a series discount of 20/20/20 to wholesalers and a 20/20 series discount to retailers. (a) What is the wholesaler's price? (b) What is the retailer's price? (c) What is the difference between the prices? [8.2]

22. Pet Supply Wholesalers receives an invoice dated June 22 for $3578 with terms of 2/10 EOM. The invoice is paid on July 8. Find the amount necessary to pay the invoice in full. [8.4]

23. University Equipment Company sells an Infocus Projector for $5250 while using a 25% markup on cost. Find the cost. [9.1]

24. Office Max sells a Texas Instruments financial calculator for $19.95. The calculator costs the store $15.96. Find the markup as a percent of selling price. [9.2]

25. Area Rugs Galore offers an 8- by 10-foot Persian rug for $499.95. If the rugs cost $4499.55 per dozen, find (a) the markup, (b) the percent of markup on selling price, and (c) the percent of markup on cost. Round to the nearest tenth of a percent. [9.1 and 9.2]

26. A commercial riding lawn mower, originally priced at $9250, is marked down to $6660. Find the percent of markdown on the original price. [10.1]

27. John Cross Pool Supply, a retailer, pays $285 for a diving board. The original selling price was $399, but was marked down 40%. If operating expenses are 30% of cost, find (a) the operating loss and (b) the absolute loss. [10.1]

28. Red Cross Medical Supplies had an inventory of $58,664 on January 1, $73,815 on July 1, and $62,938 on December 31. Find the average inventory. [10.2]

Round to the nearest dollar amount.

29. Clutch Masters made the following purchases of universal joints during the year: Beginning inventory, 30 at $18.50 each; June, 25 at $21.80 each; September, 20 at $20.50 each; and November, 30 at $21.25 each. An inventory shows that 55 universal joints remain. Find the inventory value using the weighted-average method. [10.3]

30. Find the inventory value listed in Exercise 29 using (a) the FIFO method and (b) the LIFO method. [10.3]

Part 4 Mathematics of Finance

11 Simple Interest

Some of the oldest documents in existence—clay tablets dating back almost 5000 years—show the calculation of interest charges. Interest, a fee for borrowing money, is about as old as civilization itself.

The largest and financially most secure companies, such as IBM and AT&T, borrow at or near the most favorable interest rate for short-term loans, known as the **prime rate**. The prime rate is an important factor in determining the rates of interest paid to depositors on savings and the rates of interest charged to borrowers on loans. As you can see in Figure 11.1, the prime rate fluctuates widely, although it typically remains between 6% and 15%. Most short-term interest rates move up and down with the prime rate. For example, when the prime rate moves up, it will likely cost you more to finance a car. A good understanding of interest is important since interest charges can represent a significant cost for both individuals and firms.

Two basic types of interest are in common use today: **simple interest** and **compound interest**. Simple interest is interest paid on only the principal. Compound interest is interest paid on *both principal and past interest*. This chapter discusses simple interest, and Chapter 13 covers compound interest.

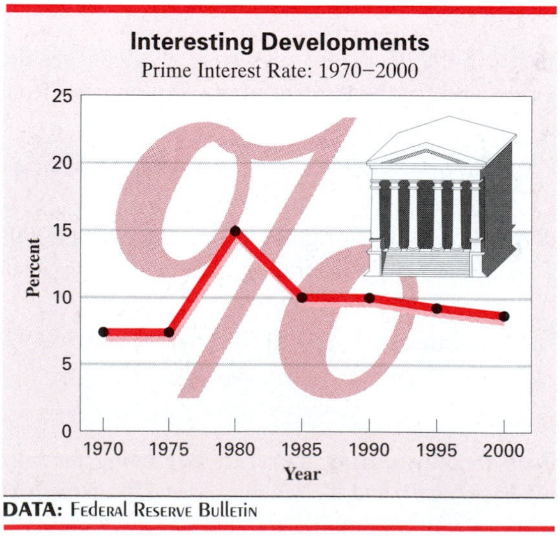

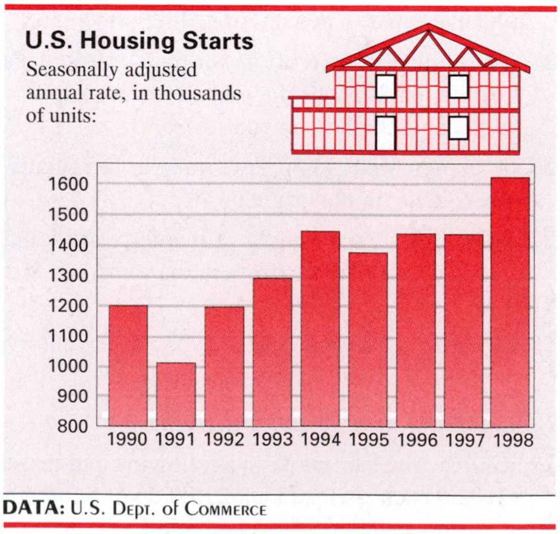

Figure 11.1

Why do you think that the interest rate has an effect on the number of new homes that are constructed in any particular year?

11.1 Basics of Simple Interest

Objectives

1. *Find simple interest.*
2. *Find interest for less than a year.*
3. *Find principal if given rate and time.*
4. *Find rate if given principal and time.*
5. *Find time if given principal and rate.*

Interest is the price paid for borrowing money. Interest rates are usually expressed as a percent of the amount borrowed. For example, in some states, retail companies such as Sears, Wards, and Penney's charge up to 21% interest per year on money owed to them. The amount borrowed is called the **principal**. The percent of interest charged is called the **rate of interest**. The number of years or fraction of a year for which the loan is made is called the **time**.

Objective 1 *Find Simple Interest.* **Simple interest,** interest charged on the entire principal for the entire length of the loan, is found by the following formula that is simply a modification of the basic percent formula.

The simple interest I, on a principal of P dollars at a rate of interest R percent per year for T years is given as follows.

$$I = P \cdot R \cdot T = PRT$$

The rate R, is expressed as a decimal or fraction, and time T, is expressed as the number of years, or the fraction of a year.

NOTE Simple interest is usually used for short-term loans that last less than a year. It is important to remember that time is in years. This means that a time period given in months or days must be converted to a fraction of a year before being substituted into the formula for T. ■

Example 1 Finding Simple Interest

Gilbert Construction Company must borrow \$60,000 to build an 1800-square-foot home. The owner, Susan Gilbert, is considering whether she should borrow the funds at (a) 8% per year for 1 year or (b) $8\frac{1}{2}\%$ per year for $1\frac{1}{2}$ years. Find the simple interest on both loans.

Solution

(a) Use the formula $I = PRT$. Substitute \$60,000 for P, 0.08 (the decimal form of 8%) for R, and 1 for T.

$$\begin{aligned} I &= PRT \\ &= \$60{,}000 \times 0.08 \times 1 \\ &= \$4800 \text{ simple interest} \end{aligned}$$

Check the answer by dividing \$4800 by \$60,000 to find 8% interest.

(b) Again use the simple interest formula; however, now use $R = 0.085$ ($8\frac{1}{2}\%$) and $T = 1.5$ ($1\frac{1}{2}$ years).

$$\begin{aligned} I &= PRT \\ &= \$60{,}000 \times 0.085 \times 1.5 \\ &= \$7650 \text{ simple interest} \end{aligned}$$

Gilbert chooses the 1-year loan since she believes that she can build and sell the home within 1 year.

Objective 2 ***Find Interest for Less Than a Year.*** Notice in part (a) of the next example how 9 months is written as $\frac{9}{12}$ of a year and in part (b) that 13 months (obviously more than 1 year) is written as $\frac{13}{12}$ or $1\frac{1}{12}$ of a year.

Example 2 Finding Simple Interest Using Months

Find the interest on a loan of $2800 at 8% simple interest for a period of (a) 9 months and a period of (b) 13 months.

Solution

(a) Since there are 12 months in a year, 9 months is $\frac{9}{12}$ or 0.75 of a year. Find the interest as follows.

$$\begin{aligned} I &= PRT \\ &= \$2800 \times 0.08 \times \frac{9}{12} \\ &= \$168 \end{aligned}$$

(b) Use $\frac{13}{12}$ to represent 13 months in terms of number of years.

$$\begin{aligned} I &= PRT \\ &= \$2800 \times 0.08 \times \frac{13}{12} \\ &= \$242.67 \quad \text{Rounded} \end{aligned}$$

Scientific *Calculator Approach*

The calculator solution to part (a) of Example 2 follows.

2800 [×] .08 [×] 9 [÷] 12 [=] 168

Objective 3 ***Find Principal If Given Rate and Time.*** Sometimes the amount of interest is known, but the principal, rate, or time must be found. Do this with the following modifications of the formula for simple interest.

$$I = PRT$$

$$\frac{I}{RT} = \frac{PRT}{RT} \quad \text{Divide both sides by } RT.$$

$$\frac{I}{RT} = P \quad \text{or} \quad P = \frac{I}{RT}$$

Similarly, dividing both sides of $I = PRT$ by PT gives the formula for R.

$$R = \frac{I}{PT}$$

And dividing both sides of $I = PRT$ by PR gives the formula for T.

$$T = \frac{I}{PR}$$

Problem-Solving Hint Don't forget, for an annual interest rate, the time is ***always in years.***

NOTE You do not have to remember all of these formulas. Just remember $I = PRT$ and use algebra to solve for the unknown.

Example 3 Finding the Principal

Gilbert Construction Company borrows funds at 12% for 10 months to build a home. Find the principal that results in interest of $8000.

Solution Find the principal by dividing both sides of $I = PRT$ by RT.

$$P = \frac{I}{RT}$$

$$= \frac{\$8000}{0.12 \times \frac{10}{12}} \quad \text{Substitute values for variables.}$$

$$= \$80{,}000$$

The principal or loan amount is $80,000.

Scientific *Calculator Approach*

For the calculator solution to Example 3, divide using the chain calculation.

8000 ÷ (.12 × 10 ÷ 12) = 80000

Objective 4 ***Find Rate If Given Principal and Time.*** The next example shows how to find the interest rate if the principal and time are given.

Example 4 Finding the Rate

Jeff Guerrant took out a loan of $9000 for 9 months for a used red Grand Am and had an interest charge of $540. What was the interest rate?

Solution Divide both sides of $I = PRT$ by PT to get the following.

$$R = \frac{I}{PT}$$

$$= \frac{\$540}{\$9000 \times \frac{9}{12}} \quad \text{Substitute values for variables.}$$

$$= 0.08$$

The rate was 8%.

NOTE In order to avoid rounding errors, it is important not to round off any calculations until the very end when solving these types of problems.

Objective 5 *Find Time If Given Principal and Rate.* The simple interest formula can also be used to find the time of a loan.

Example 5 Finding the Time

Eric Thomas, a loan officer at Midwest Bank, made a loan of $4800 at 10%. How many months will it take to produce $280 in interest?

Solution Use $I = PRT$ and divide both sides by PR to get

$$T \text{ (in years)} = \frac{I}{PR}$$

We included "(in years)" to emphasize this important point—time of the loan is measured in years. Now substitute $280 for I, $4800 for P, and 0.10 for R.

$$T \text{ (in years)} = \frac{\$280}{\$4800 \times 0.10} \quad \text{Substitute values for variables.}$$

$$T \text{ (in years)} = \frac{\$280}{\$480}$$

Reduced to lowest terms

$$T \text{ (in years)} = \frac{\$280}{\$480} = \frac{7}{12} \text{ year} = 7 \text{ months}$$

NOTE In all the examples of this section the time has been expressed in years or months. Section 11.2 will deal with interest problems in which the time is expressed in days.

11.1 Exercises

Find the simple interest for each of the following. Round to the nearest cent.

1. $6800 at 10% for $1\frac{1}{4}$ years
2. $9500 at 12% for $1\frac{1}{2}$ years
3. $12,400 at $9\frac{1}{2}$% for 8 months
4. $1200 at 8% for 8 months
5. $8250 at 13% for 15 months
6. $4270 at 18% for 20 months
7. $9874 at $7\frac{1}{8}$% for 11 months
8. $10,745 at $4\frac{5}{8}$% for 9 months
9. $74,986.15 at 12.23% for 5 months
10. $39,072.76 at 11.23% for 7 months

Complete this chart. Round money to the nearest cent, rate to the nearest tenth of a percent, and time to the nearest month.

	Principal	Interest	Rate	Time in Months
11.	$10,000	______	10%	6
12.	$15,000	______	$12\frac{1}{2}$%	10

	Principal	Interest	Rate	Time in Months
13.	______	\$162.50	8%	6
14.	______	\$84	10%	7
15.	\$3800	\$199.50	______	9
16.	\$2600	\$144.08	______	7
17.	\$5350	\$749	12%	______
18.	\$8200	\$1722	14%	______

Solve each application problem. Round money to the nearest cent, rate to the nearest tenth of a percent, and time to the nearest month.

19. Find the time, in years, necessary for an \$8200 loan to produce \$1353 in interest at 11%.

20. A \$10,400 loan resulted in interest charges of \$1248 at 12% interest. Find the time in years.

21. Find the principal if \$245 in interest is due after 10 months at 7%.

22. What amount of principal produces \$55 in interest after 1 month at 11%?

23. What time (in months) is necessary for a principal of \$840 to produce \$77 in interest at 10%?

24. How many months are necessary for \$4800 to produce \$44 in interest at 11%?

25. Find the rate if a principal of \$1890 produces \$138.60 in interest after 11 months.

26. A loan of \$8500 produces interest of \$595 in 7 months. Find the rate.

27. LATE-PAYMENT PENALTY Casey Watkins is 3 months late on a car payment to her uncle's finance company. She has agreed to a penalty of 10% of the balance of \$5850. Find the penalty.

28. CONSTRUCTION Denver Construction, Inc. is 2 months late on a payment. The penalty is based on a loan amount of \$682,500 and a rate of 14%. Find the penalty.

29. LOAN TO BUY TIMBER Georgia Plantations borrows funds to buy land and agrees to pay interest of \$114,375 in 9 months at 10% interest. Find the principal.

30. AUTOMOBILE LOAN Michael George borrows money to replace his Chevrolet. He agrees to pay \$1170 in interest at 9% in 10 months. Find the principal (loan amount).

31. LAWN MOWERS The Green Care Company needs to borrow \$7000 to buy new lawn mowers. The firm wants to pay no more than \$560 in interest. If the interest rate is 12%, what is the longest time for which the money may be borrowed?

32. REMODELING A STOREFRONT The Mall Frame Shop wants to borrow \$9300 to remodel the front of the store. Given that the owner has bad credit, the least expensive interest rate he can find is 14%. Find the longest time for which the money may be borrowed if the firm can pay only \$1193.50 in interest.

33. SAVINGS Sam Casio deposits \$5400 in a savings account at the Friendly Savings and Loan Bank. When he withdraws his money 8 months later, he receives a check for \$5544. What rate of interest does the bank pay?

34. DIVIDENDS Dorothy Duerr invested \$7800 in an account that pays dividends monthly, based on simple interest. If Duerr has received a total of \$276.25 over the past 10 months, what rate of interest does the account pay?

35. A \$2000 loan was made for 3 months at 7% simple interest.
A student calculates the interest due on the loan as follows.

$$I = \$2000 \times 0.07 \times 3 = \$420$$

Is this correct? If not, explain why not and state the correct answer. (See Objective 1.)

36. Are you willing to invest your own funds in an account paying 1% per year for several years? Why or why not? What interest rate should a bank charge you when you borrow? Why?

37. SURGERY Lupe Garcia needs to borrow \$3200 for 8 months for an operation that she needs this month. She can choose between a credit union charging 10% or a loan company charging 18%. How much will she save by using the credit union?

38. DIFFERENT LOAN RATES You need to borrow \$800 for 4 months and can go to a pawn shop charging 20% interest or to your uncle who will charge 12% interest. Find the difference in the interest charges.

39. COMPUTER UPGRADES An accounting firm wishes to borrow \$6800 for computer upgrades and they believe that they will only be able to repay a total of \$7276 in 6 months. What is the highest interest rate at which they can afford to borrow?

40. STUDENT LOAN Your roommate wishes to borrow \$7875. She believes that she will be able to repay as much as \$10,000 in 18 months after she graduates and has several months to save. Find the maximum interest rate she can afford. (Round to the nearest percent.)

41. DRESS SHOP Gladys' Dress Shop received an invoice for \$2543, with terms of 2/15, n/60. The store can borrow money at the bank at 12%. How much would be saved by borrowing money for 60 − 15 = 45 days to take advantage of the cash discount?

42. PET SUPPLIES Westside Pet Supplies received an invoice of \$1796, with terms 1/10, n/30. Due to a poor credit history, the store must pay 14.5% interest. How much would the store save by borrowing money for 30 − 10 = 20 days to take advantage of the cash discount?

43. HOME ADDITION Ricky and Gina Hardin borrow \$24,900 to add a room to their home. They finance the loan at 10% simple interest for 120 days. They hope to pay the loan off then using proceeds from the sale of a house. Find the interest and the maturity value.

11.2 Simple Interest for a Given Number of Days

Objectives

1. *Find the number of days from one date to another using a table.*
2. *Find the number of days from one date to another using the actual number of days.*
3. *Find exact and ordinary interest.*

The previous section showed how to find simple interest for loans of a given number of months or years. In this section, *loans for a given number of days* are discussed. In business, it is common for loans to be for a given number of days, such as "due in 90 days," or else to be due at some fixed date in the future, such as "due on April 17." You will find this topic useful if you own your own business or if you are involved in the finances of a business.

Objective 1 *Find the Number of Days from One Date to Another Using a Table.* There are two ways to find the number of days from one date to another. One way is by the use of Table 11.1. This table assigns a different number to each day of the year. For example, the number for June 11 is found by locating 11 at the left, and June across the top. You should find that June 11 is day 162. Also, December 29 is day 363. The number of days from June 11 to December 29 is found by subtracting.

December 29 is day	363	
June 11 is day	− 162	Subtract.
	201	

There are 201 days from June 11 to December 29. (Throughout this book, ignore leap years unless otherwise stated.)

Table 11.1 THE NUMBER OF EACH OF THE DAYS OF THE YEAR (ADD 1 TO EACH DATE AFTER FEBRUARY 29 FOR A LEAP YEAR)

Day of Month	Jan.	Feb.	March	April	May	June	July	Aug.	Sept.	Oct.	Nov.	Dec.	Day of Month
1	1	32	60	91	121	152	182	213	244	274	305	335	1
2	2	33	61	92	122	153	183	214	245	275	306	336	2
3	3	34	62	93	123	154	184	215	246	276	307	337	3
4	4	35	63	94	124	155	185	216	247	277	308	338	4
5	5	36	64	95	125	156	186	217	248	278	309	339	5
6	6	37	65	96	126	157	187	218	249	279	310	340	6
7	7	38	66	97	127	158	188	219	250	280	311	341	7
8	8	39	67	98	128	159	189	220	251	281	312	342	8
9	9	40	68	99	129	160	190	221	252	282	313	343	9
10	10	41	69	100	130	161	191	222	253	283	314	344	10
11	11	42	70	101	131	162	192	223	254	284	315	345	11
12	12	43	71	102	132	163	193	224	255	285	316	346	12
13	13	44	72	103	133	164	194	225	256	286	317	347	13
14	14	45	73	104	134	165	195	226	257	287	318	348	14
15	15	46	74	105	135	166	196	227	258	288	319	349	15
16	16	47	75	106	136	167	197	228	259	289	320	350	16
17	17	48	76	107	137	168	198	229	260	290	321	351	17
18	18	49	77	108	138	169	199	230	261	291	322	352	18
19	19	50	78	109	139	170	200	231	262	292	323	353	19
20	20	51	79	110	140	171	201	232	263	293	324	354	20
21	21	52	80	111	141	172	202	233	264	294	325	355	21
22	22	53	81	112	142	173	203	234	265	295	326	356	22
23	23	54	82	113	143	174	204	235	266	296	327	357	23
24	24	55	83	114	144	175	205	236	267	297	328	358	24
25	25	56	84	115	145	176	206	237	268	298	329	359	25
26	26	57	85	116	146	177	207	238	269	299	330	360	26
27	27	58	86	117	147	178	208	239	270	300	331	361	27
28	28	59	87	118	148	179	209	240	271	301	332	362	28
29	29		88	119	149	180	210	241	272	302	333	363	29
30	30		89	120	150	181	211	242	273	303	334	364	30
31	31		90		151		212	243		304		365	31

Problem-Solving Hint When counting the number of days of a loan, do not count the day the loan is made, but do count its due date.

Example 1 Finding the Number of Days

Find the number of days from (a) March 24 to July 22 and (b) November 8 to February 17 of the next year.

Solution

(a) The number of days can be estimated. There are about 4 months from March 24 to July 22. Assume 30 days per month and multiply to find about 120 days. Now to find the exact number of days, note that March 24 is day 83 and July 22 is day 203.

July 22 is day	203
March 24 is day	− 83
	120

There are 120 days from March 24 to July 22. It turns out that our estimate was the exact number of days. Estimates rarely provide the exact answer, but they can help minimize errors by providing approximate values.

(b) Since November 8 is in one year and February 17 is in the next year, first find the number of days from November 8 to the end of the year.

Last day of the year is number	365
November 8 is day	− 312
	53

There are 53 days from November 8 to the end of the year.

Then find the number of days from the beginning of the next year to February 17. According to the chart, February 17 is the 48th day of the year. The total number of days is found as follows.

November 8 to end of year	53
January 1 to February 17	+ 48
	101

There are 101 days from November 8 to February 17 of the next year.

Objective 2 ***Find the Number of Days from One Date to Another Using the Actual Number of Days.*** An alternate way of finding the number of days, useful when Table 11.1 is not available, is to use the actual number of days in each month, as shown in Table 11.2.

Example 2 Finding the Number of Days from One Date to Another Using Actual Days

Find the number of days from (a) March 12 to June 7 and (b) November 4 to February 21.

Solution

(a) Since March has 31 days, there are $31 - 12 = 19$ days left in March, then 30 days in April, 31 in May, and an additional 7 days in June for a total as follows.

19	Remaining in March
30	April
31	May
+ 7	June
87	Days from March 12 to June 7

(b) Add.

26	Remaining in November
31	December
31	January
+ 21	February
109	

There are 109 days from November 4 to February 21.

Table 11.2 THE NUMBER OF DAYS IN EACH MONTH

31 Days		30 Days	28 Days
January	August	April	February
March	October	June	(29 days in leap year)
May	December	September	
July		November	

Example 3 Finding Specific Dates

Find the date that is 90 days from (a) March 25 and (b) November 7.

Solution

(a) From Table 11.1, March 25 is day 84. Add 90.

March 25 is day	84
	+ 90 days
	174

As shown in Table 11.1, day 174 is June 23, so 90 days from March 25 is June 23. Alternatively, work as follows.

March 25 to end of month	6
April	30
May	+ 31
	67

Since $90 - 67 = 23$, an additional 23 days in June are needed, giving June 23.

(b) November 7 is day 311. Add 90 days to get the following.

$$\begin{array}{r} 311 \\ +\ 90 \\ \hline 401. \end{array}$$

Since there are only 365 days in a year, subtract 365.

$$\begin{array}{r} 401 \\ -\ 365 \\ \hline 36 \end{array} \quad \text{Day of the following year.}$$

Day 36 of the following year is February 5, so that 90 days from November 7 is February 5 of the following year.

In the formula for simple interest, time is always measured in years or parts of years. In the examples of the previous section, time was in months, with T in the formula $I = PRT$ written as

$$T = \frac{\text{Given number of months}}{12}.$$

Objective 3 ***Find Exact and Ordinary Interest.*** Things are not so simple when the loan is given in days. There are two common methods for calculating simple interest for a given number of days: **exact interest**, and **ordinary** or **banker's interest**.

In the formula $I = PRT$, the fraction for time is found as follows.

$$\text{Exact interest} \qquad T = \frac{\text{Exact number of days in the loan}}{365}$$

$$\text{Ordinary or banker's interest} \qquad T = \frac{\text{Exact number of days in the loan}}{360}$$

Government agencies and the Federal Reserve Bank use exact interest, as do many credit unions, while most banks and other financial institutions use ordinary interest. Ordinary interest may have been used originally because it was easier to calculate than exact interest. With the modern use of calculators and computers, however, ordinary interest is probably used today out of tradition and because it produces a *greater dollar amount* of interest than does exact interest, as shown in the next example.

Example 4 Finding Exact and Ordinary Interest

Tyler Radio Shop borrowed $17,650 on May 12. The loan, at an interest rate of 12%, is due on August 27. Find the interest on the loan using (a) exact interest and (b) ordinary interest.

Solution Using the table or calculating the number of days in each month, there are 107 days from May 12 to August 27.

(a) The exact interest is found from $I = PRT$ with $P = \$17{,}650$, $R = 0.12$, and $T = \frac{107}{365}$. (Remember to use 365 as the denominator with exact interest.)

$$I = PRT$$
$$= \$17{,}650 \times 0.12 \times \frac{107}{365}$$
$$= \$620.89 \quad \text{Rounded}$$

(b) Find ordinary interest with the same formula and values, except that $T = \frac{107}{360}$.

$$I = PRT$$
$$I = \$17{,}650 \times 0.12 \times \frac{107}{360}$$
$$I = \$629.52 \quad \text{Rounded}$$

In this example, the ordinary interest is $\$629.52 - \$620.89 = \$8.63$ more than the exact interest.

NOTE If P and R are the same, more interest is generated using ordinary interest than using exact interest. ■

The formulas from Section 11.1 and 11.2 are repeated here for your convenience.

Interest	$I = PRT$	T is in years.
Principal	$P = \frac{I}{RT}$	T is in years.
Rate	$R = \frac{I}{PT}$	T is in years.
Time T (in years)	$T = \frac{I}{PR}$	
Time T (in months)	$T = \frac{I}{PR} \times 12$	T is a fraction of a year.
Time T (in days)	$T = \frac{I}{PR} \times 360$	Use 360 days for banker's or ordinary interest.
Time T (in days)	$T = \frac{I}{PR} \times 365$	Use 365 days for exact interest.

NOTE Throughout the balance of this book, assume ordinary, or banker's, interest unless stated otherwise. ■

11.2 Exercises

Find the exact number of days from the first date to the second. (In Exercises 5–6, assume that the second month given is in the following year.)

1. May 7 to Dec 2

2. April 24 to July 7

3. October 27 to December 2

4. July 12 to October 29

5. September 2 to March 17

6. July 24 to March 30

Find the date that is the indicated number of days from the given date.

7. 120 days from June 14

8. 45 days from July 28

9. 30 days from February 8

10. 90 days from November 18

11. 120 days from December 12

12. 150 days from November 1

How much ordinary interest would you pay for each of the following? Round to the nearest cent.

13. \$1800 at 10% for 90 days

14. \$2600 at $10\frac{1}{2}\%$ for 180 days

15. \$3250 at 12% for 150 days

16. \$8620 at 11.25% for 120 days

17. A loan of \$1250 at 11% made on May 9 and due August 25

18. A loan of \$680.40 at 12% made on June 20 and due December 26

19. A loan of \$1520 at 10% made on February 27 and due August 5

20. A loan of \$3600 at 11% made on December 2 and due February 20

Find the exact interest for each of the following. Round to the nearest cent.

21. \$3800 at 10% for 100 days

22. \$6900 at 8% for 200 days

23. \$4600 at 13% for 60 days

24. \$3150 at 12.5% for 200 days

25. A loan of \$6500 at 7% made on July 12 and due on October 12

26. A loan of \$8120 at 9% made on January 30 and due May 20

27. A loan of \$2050 at 12% made on June 24 and due February 12 of the next year

28. A loan of \$14,000 at 8% made on August 12 and due March 19 of the next year

29. Explain the difference between exact and ordinary interest. (See Objective 3.)

30. As a borrower, would you prefer exact or ordinary interest? Why? (See Objective 3.)

Solve each of the following application problems using exact interest. Round to the nearest cent.

31. IRS PENALTY Fred Thomas accidentally paid \$1800 in employee taxes 40 days late. Unbelievably, the Internal Revenue Service charged a penalty at a rate of 60% per year. Find the penalty.

32. MISSED TAX PAYMENT Bella Steinem missed an income tax payment. The payment was due June 15 and was paid September 7. The penalty was 14% simple interest on the unpaid tax of \$4600. Find the penalty.

33. CONSTRUCTION LOAN Burns Construction borrows \$56,000 on October 23 at 12% simple interest. Find the interest that must be paid on March 15.

34. COMPUTER PURCHASE Future Tech Inc. borrows \$28,000 on November 9 for computer purchases and agrees to repay the 12% interest loan on March 30 of the following year. Find the interest.

Solve each application problem using ordinary interest. Round to the nearest cent.

35. PHOTOGRAPHY STUDIO Nolan Brinkman borrowed \$48,000 on June 10 to open a photography studio. Given a rate of $12\frac{1}{4}\%$, find the interest due on Christmas (December 25).

36. COWBOY OUTFITS On May 18, the Wilson Dude Ranch bought a supply of cowboy outfits for \$11,270. The ranch agreed to pay for them on August 30, with $10\frac{1}{2}\%$ interest. Find the interest owed.

37. CONSTRUCTION Gilbert Construction Company can borrow $80,000 for 120 days at 11% interest. In 1980, the same note would have been at a rate of 20%. Find the difference in the interest charges based on the two rates.

38. CONSTRUCTION IN MEXICO A construction company in Mexico City borrows 300,000 pesos for 90 days at an interest rate of 18%. The same loan would have been at a rate of 35% several years ago. Find the difference in the interest charges based on the two rates.

In Exercises 39–42 find the (a) exact interest and (b) ordinary interest for each of the following, rounded to the nearest cent. Then (c) find the amount by which the ordinary interest is larger.

39. $18,000 at 9% for 120 days

40. $75,000 at 11% for 120 days

41. $145,000 at $9\frac{1}{2}$% for 240 days

42. $250,000 at $12\frac{1}{8}$% for 180 days

Solve each of the following application problems. Round to the nearest cent.

43. JEWELRY INVENTORY Classic Jewelry needs to borrow $160,000 to finance an inventory of jewelry for Christmas sales. The loan will be for 95 days at $9\frac{5}{8}$%. How much more would the interest be at a bank using ordinary interest than at one using exact interest?

44. SKI EQUIPMENT Ski Mountain Ltd. borrows $120,000 for extra inventory for the coming ski season. The owners expect to repay the loan after the ski season is over, in 140 days, at a rate of $10\frac{1}{2}$%. How much more would the firm pay using ordinary interest compared to exact interest?

45. NEW TRUCKS Blaine Trucking wishes to borrow $880,000 for new trucks on September 20 and plans to pay off the loan on May 1 of the following year. State Bank will lend them the funds at 9% based on exact interest calculations. First Bank will lend them the funds at $8\frac{7}{8}$% interest based on ordinary interest calculations. Which bank is asking for less interest? How much less?

46. INTERNATIONAL CONSTRUCTION A Canadian construction firm needs to borrow an additional $3,500,000 for 150 days to build a bridge linking Mexico to the United States. National Bank will lend them the funds at $10\frac{1}{2}$% simple interest based on exact interest calculations, and Laredo Bank will lend them the funds at $10\frac{5}{8}$% simple interest based on ordinary interest calculations. Which bank is asking for less interest? How much less?

11.3 Maturity Value

Objectives

1 *Find maturity value.*
2 *Find principal if given maturity value, time, and rate.*
3 *Find rate if given principal, maturity value, and time.*
4 *Find time if given maturity value, principal, and rate.*

Suppose you borrow \$8000 at 11% interest for 9 months. The interest you owe on the loan is calculated as follows.

$$I = PRT$$
$$= \$8000 \times 0.11 \times \frac{9}{12}$$
$$= \$660$$

Objective 1 *Find Maturity Value.* The total amount that must be repaid in 9 months is the sum of the principal and the interest.

$$\text{Principal} + \text{Interest} = \$8000 + \$660 = \$8660$$

This amount, \$8660, is called the **maturity value**, or **future value**, of the loan. The date the loan is paid off is the **maturity date**. The principal or **present value** is the amount received by the borrower today. The formula for maturity value is as follows. A loan with a principal of P dollars and interest of I dollars has a maturity value M given by

$$M = P + I$$

NOTE The maturity value of a loan always exceeds the principal (the original loan amount) since maturity value is principal *plus* interest. ▬

Example 1 Finding Interest and Maturity Value

Jim Wilcox would like to remodel his small bookstore so that he can serve customers coffee and allow them to sit and browse. To remodel the store, he borrows \$7200 for 21 months at 9.25% interest. Find the interest due on the loan and the maturity value.

Solution Interest due is found using $I = PRT$, where T is in years (21 months $= \frac{21}{12}$ years).

$$I = PRT$$
$$= \$7200 \times 0.0925 \times \frac{21}{12} = \$1165.50$$

Find the maturity value using $M = P + I$, where $P = \$7200$ and $I = \$1165.50$.

$$M = P + I$$
$$= \$7200 + \$1165.50 = \$8365.50$$

The formula for maturity value, $M = P + I$, can be written in a different way if I is replaced with PRT (since $I = PRT$).

$$M = P + I$$
$$= P + PRT \quad \text{Substitute } PRT \text{ for } I.$$
$$= P(1 + RT) \quad \text{Use the distributive property.}$$

Notice that $P(1 + RT) = P + PRT$ by the distributive property of algebra discussed in Chapter 2. Therefore, there are two formulas for maturity value.

The maturity value M, of a principal of P dollars at a rate of interest R for T years is either

$$M = P + I$$

or, since $I = PRT$,

$$M = P(1 + RT)$$

Problem-Solving Hint

Do not round off values too soon when doing interest rate problems. Round *after* finding the final value.

Example 2 Finding Maturity Value

Use the formula $M = P(1 + RT)$ to find the maturity value for a loan of \$6000 for 120 days at 9% interest.

Solution Substitute \$6000 for P, 0.09 for R, and $\frac{120}{360}$ for T (since 120 days is $\frac{120}{360}$ of a year). The maturity value is as follows.

$$\begin{aligned} M &= P(1 + RT) \\ &= \$6000 \times \left[1 + \left(0.09 \times \frac{120}{360}\right)\right] \end{aligned}$$

NOTE Parentheses were placed around $0.09 \times \frac{120}{360}$ to emphasize that these numbers are multiplied as a first step. ■

After multiplying 0.09 and $\frac{120}{360}$, add 1.

$$\begin{aligned} M &= \$6000 \times (1 + 0.03) \\ &= \$6000 \times 1.03 \\ &= \$6180 \end{aligned}$$

The interest can be found by subtracting the principal from the maturity value.

$$I = \$6180 - \$6000 = \$180$$

Scientific *Calculator Approach*

The calculator solution to this example uses parentheses on the calculator in place of the brackets.

6000 [×] [(] 1 [+] .09 [×] 120 [÷] 360 [)] [=] 6180

Subtract the principal from \$6180 to obtain the interest.

6180 [−] 6000 [=] 180

Objective 2 ***Find Principal If Given Maturity Value, Time, and Rate.*** Sometimes the maturity value is given, and either the principal, rate, or time must be found. For example, given the maturity value, rate, and time, find principal as follows.

$$M = P(1 + RT)$$

$$\frac{M}{(1 + RT)} = \frac{P(1 + RT)}{(1 + RT)} \qquad \text{Divide both sides by } (1 + RT).$$

$$\frac{M}{(1 + RT)} = \frac{P\cancel{(1 + RT)}}{\cancel{(1 + RT)}} \qquad \text{Divide out common factors.}$$

$$\frac{M}{1 + RT} = P \quad \text{or} \quad P = \frac{M}{1 + RT}$$

Principal is also called the *present value* of the loan.

Example 3 Finding Principal Given Time, Rate, and Maturity Value

Find the principal that would produce a maturity value of \$1530 in 4 months at 6% interest.

Solution Use the formula above and substitute \$1530 for M, 0.06 for R, and $\frac{4}{12}$ for T.

$$P = \frac{M}{1 + RT}$$

$$= \frac{\$1530}{1 + (0.06 \times \frac{4}{12})}$$

As shown by the parentheses, first multiply 0.06 and $\frac{4}{12}$, and then add 1.

$$P = \frac{\$1530}{1 + 0.02}$$

$$= \frac{\$1530}{1.02}$$

$$= \$1500$$

The principal is \$1500; the interest is \$1530 − \$1500 = \$30.

Scientific *Calculator Approach*

The calculator solution to Example 3 uses parentheses for the entire denominator.

1530 [÷] [(] 1 [+] .06 [×] 4 [÷] 12 [)] [=] 1500

Objective 3 ***Find Rate If Given Principal, Maturity Value, and Time.*** If principal, maturity value, and time are given, rate can be found as follows.

$$M = P(1 + RT)$$

$$M = P + PRT \qquad \text{Use the distributive property.}$$

$$M - P = PRT \qquad \text{Subtract } P \text{ from both sides.}$$

$$\frac{M - P}{PT} = \frac{\cancel{P}R\cancel{T}}{\cancel{PT}} \qquad \text{Divide both sides by } PT.$$

$$\frac{M - P}{PT} = R$$

Example 4 Finding Rate Given Principal, Maturity Value, and Time

Lin Pao invests a principal of \$7200 and receives a maturity value of \$7540 in 200 days. Find the interest rate.

Solution Use the formula above and substitute values.

$$\begin{aligned} R &= \frac{M - P}{PT} \\ &= \frac{\$7540 - \$7200}{\$7200 \times \frac{200}{360}} \\ &= 0.085, \qquad \text{or} \qquad 8.5\% \end{aligned}$$

Scientific *Calculator Approach*

The calculator solution to Example 4 uses parentheses for the numerator (\$7540 − \$7200) and for the denominator (\$7200 × $\frac{200}{360}$).

[( ] 7540 [−] 7200 [ )] [÷] [( ] 7200 [×] 200 [÷] 360 [ )] [=] 0.085

Notice that maturity value minus principal ($M - P$) is interest I. Therefore,

$$R = \frac{M - P}{PT} = \frac{I}{PT}$$

and Example 4 can be solved using this formula.

$$\text{Interest} = \$7540 - \$7200 = \$340$$

$$\begin{aligned} R &= \frac{I}{PT} \\ &= \frac{\$340}{\$7200 \times \frac{200}{360}} \\ &= 0.085, \qquad \text{or} \qquad 8.5\% \end{aligned}$$

Objective 4 ***Find Time If Given Maturity Value, Principal, and Rate.*** Given maturity value, principal, and rate, time can be found as follows.

$$M = P(1 + RT)$$

$$M = P + PRT \qquad \text{Use the distributive property.}$$

$$M - P = PRT \qquad \text{Subtract } P \text{ from both sides.}$$

$$\frac{M - P}{PR} = \frac{PRT}{PR} \qquad \text{Divide both sides by } PR.$$

$$\frac{M - P}{PR} = T \text{ (in years)}$$

This gives a value for time T in *years*. To convert to *days* multiply by 360; for example, $\frac{1}{2}$ of a year is $\frac{1}{2} \times 360 = 180$ days.

$$\text{T (in days)} = \frac{M - P}{PR} \times 360$$

Since $I = M - P$, this is the same as

$$T \text{ (in days)} = \frac{I}{PR} \times 360$$

Example 5 Finding the Time in Days

Neon Lights, Inc. borrowed \$18,250 at $10\frac{1}{8}\%$ interest for the construction of new signs and agreed to repay \$19,687.19. Find the time in days.

Solution Use the formula above and substitute values.

$$\begin{aligned} T &= \frac{M - P}{PR} \times 360 \\ &= \frac{\$19{,}687.19 - \$18{,}250}{\$18{,}250 \times 0.10125} \times 360 \\ &= \frac{1437.19}{1847.81} \times 360 \\ &= 280 \text{ days} \quad \text{Rounded.} \end{aligned}$$

Scientific *Calculator Approach*

The calculator solution to Example 5 uses parentheses to set off both the numerator and the denominator.

(19687.19 − 18250) ÷ (18250 × .10125) × 360 = 280

The formulas from Section 11.3 are repeated here for your convenience.

Interest	$I = PRT$
Maturity Value	$M = P(1 + RT)$
Principal	$P = \frac{I}{RT} = \frac{M - P}{RT} = \frac{M}{1 + RT}$
Rate	$R = \frac{I}{PT} = \frac{M - P}{PT}$
Time (in years)	$\frac{I}{PR} = \frac{M - P}{PR}$
Time (in days)	$\frac{1}{PR} \times 360 = \frac{M - P}{PR} \times 360$

11.3 Exercises

Find the interest and the maturity value for each of the following loans. Round to the nearest cent.

1. \$8500 at 10% for 9 months
2. \$12,200 at $9\frac{1}{2}\%$ for 10 months
3. \$5800 at 8.5% for 140 days
4. \$10,800 at $7\frac{3}{4}\%$ for 220 days

5. \$8640 at 10% for $1\frac{1}{4}$ years

6. \$9500 at 15% for $1\frac{1}{2}$ years

Complete this chart. Round money to the nearest cent, rate to the nearest tenth of a percent, and time to the nearest day.

	Principal	Interest	Rate	Time in Days	Maturity Value
7.	\$4,800	______	10%	100	______
8.	\$7,500	______	12%	120	______
9.	\$8,600	\$133.78	8%	______	______
10.	\$7,400	\$292.92	$9\frac{1}{2}$%	______	______
11.	\$14,000	\$490.00	______	120	______
12.	\$12,800	\$544.00	______	180	______
13.	______	______	$7\frac{7}{8}$%	200	\$17,117.50
14.	______	______	$10\frac{1}{2}$%	240	\$30,602
15.	______	\$666.00	9%	______	\$15,466
16.	______	\$8086.05	9.1%	______	\$89,086.05
17.	\$1,800	______	14%	______	\$1926
18.	\$11,250	______	12.8%	______	\$11,710
19.	______	\$3272.92	______	185	\$45,732.36
20.	______	\$1963.33	______	76	\$76,963.33

Solve the following application problems. Round to the nearest cent.

21. HOT-TUB ENCLOSURE José Garcia borrowed \$12,000 for 140 days at $10\frac{1}{2}$% to add a hot tub and enclosure to his home. Find the interest and the maturity value.

22. FUNERAL AND PROBATE EXPENSES Ben Thompson's invalid grandfather died and Ben had to borrow \$12,000 for 120 days at 11% for funeral and attorney expenses. Find the interest and maturity value.

23. COST OF MARRIAGE Joey Patrick and Lori Hooten borrow \$12,400 at $10\frac{5}{8}$% interest for wedding expenses and a honeymoon to Hawaii. Given that they must repay \$12,912.36, find the time in days and the amount of interest. Round to the nearest day.

24. CAR DEALERSHIP EXPANSION Walt Wilson borrows \$60,000 at 9% interest to expand his car dealership. He agrees to repay \$61,800. Find the time in days and the amount of interest paid.

25. LOAN FROM UNCLE Martha Wheat borrowed \$930 from her uncle and agreed to repay him \$1000 after 150 days. Find the interest rate to the nearest percent and the amount of interest.

26. HEALTH-FOOD STORE Jenny Cronin borrows \$46,000 to open a small health-food store. She agrees to repay \$49,105 in 270 days. Find the interest rate and the amount of interest.

27. MOTOR COMPANY Bill Melton of Melton Motor Company agreed to repay \$854,166.67 at 10% interest in 150 days. Find the principal and the interest.

28. SETTLING A LAWSUIT Jill Phan borrowed some money to settle a lawsuit against her restaurant—a customer slipped on a puddle of water from the icemaker and fell. The 180-day loan was at 9% interest and Phan must repay $44,412.50. Find the principal and the interest.

29. List the two basic equations from which all of the other equations in this section can be derived. Use the two equations to derive the equation for time given maturity value, principal and rate. (See Objectives 1 and 3.)

30. Using $I = PRT$ and $M = P + I$, explain the steps to derive the equation for principal given maturity value, rate, and time. (See Objective 2.)

11.4 Inflation and the Time Value of Money

Objectives

1 *Define inflation and the consumer price index.*
2 *Understand the time value of money.*
3 *Define present value and future value.*
4 *Calculate present and future values using simple interest.*
5 *Find present value for a given maturity value.*
6 *Find present value after a loan is made.*

A family of four would have done *well* with an income of only $500 per month in 1950. Today, this amount will not go far in terms of taking care of a family, i.e., five hundred dollars today will not purchase nearly as much as it did in 1950. Why?

This section gives a brief introduction to inflation and some associated terms and concepts. These terms are then applied to simple interest loans. The concepts presented here are more fully discussed in Chapter 13.

Objective 1 *Define Inflation and the Consumer Price Index.* **Inflation** is the culprit. It results in a continuing rise in the general price level of goods and services. The effect of inflation can be seen by the increasing costs in a grocery store.

Item	1970 Price	2000 Price
Loaf of bread	$0.24	$1.75
$\frac{1}{2}$ gal. of milk	$0.66	$2.25
1 lb. of bacon	$0.95	$2.50

Effectively, a dollar bought less in 2000 than it did in 1970; every year the dollar buys less.

The **consumer price index (CPI)** is calculated by the government annually in the United States and is often referred to as the "cost-of-living index." Other countries have similar indexes. The CPI can be used to track inflation: it measures the average change in prices from one year to the next for a common bundle of goods and services bought by the average consumer on a regular basis. Figure 11.2 shows that the yearly inflation, as measured by the CPI, differs substantially from year to year. Your local librarian can help you find recent CPI data. The CPI is discussed in more detail in Chapter 19.

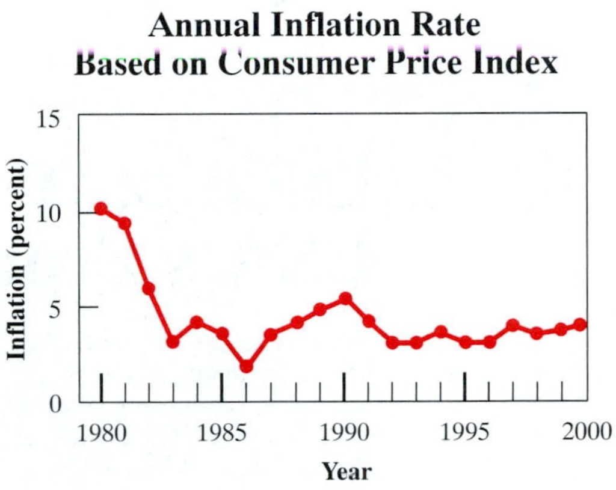

Figure 11.2

Data: U.S. Bureau of Census, 1999.

Example 1 Estimating the Effect of Inflation

Inflation from one year to the next was 4.8% as measured by the CPI. (a) Find the effect of the increase on a family with an annual income and budget of \$19,800 (after taxes). (b) What is the overall effect if the family members only receive a 2% after-tax increase in pay for the year?

Solution

(a) This is a percent problem. The cost of the goods and services that this family buys went up by 4.8% as measured by the CPI.

$$0.048 \times \$19{,}800 = \$950.40$$

Next year these same goods and services will cost the family

$$\$19{,}800 + \$950.40 = \$20{,}750.40$$

(b) The family's income only went up 2% after taxes.

$$0.02 \times \$19{,}800 = \$396$$

Thus, their new income is \$19,800 + \$396 = \$20,196. However, the cost of buying the same goods and services as they bought the previous year is \$20,750.40. In effect, the family has lost \$20,750.40 − \$20,196 = \$554.40 in purchasing power.

As shown in Example 1, inflation slowly erodes fixed incomes or incomes with a small annual increase built into them. Imagine the effect of losing purchasing power every year for 10 years in a row. Inflation can erode people's purchasing power *even as their annual salaries are increasing*. Retired people are particularly concerned with inflation since they must live off Social Security and the assets they have accumulated during their lifetime. Some retired people do not have ways of increasing their income to keep pace with inflation and must lower their standards of living substantially during their 10 to 30 or more years of retirement. Other retired people have investments such as stocks that help them stay ahead of inflation. Newspaper headlines such as those below cause concern for elderly people on fixed incomes.

> **Inflation Threatens USA's Economic Parade**
>
> Even the boy who cried wolf was right once.
>
> That's something that Federal Reserve Chairman Alan Greenspan and his fellow policymakers at the inflation stingy central bank want to . . . prices have been increasing at a 6% annual clip. That compares to a 1.4% rise for consumer prices.
>
> "Reports of inflation's demise were clearly exaggerated, . . ."

Source: August 18, 1998, *USA Today*. Reprinted by permission.

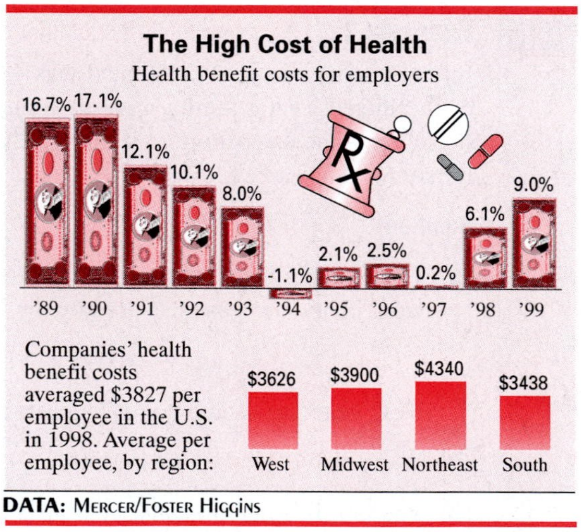

The federal government generally tries to keep inflation at moderate levels because inflation can be so harmful. When the economy becomes overheated and inflationary pressures increase, the Federal Reserve nudges interest rates upward. This reduces borrowing slightly, slows the economy, and reduces inflationary pressures. Conversely, if inflation is low and the economy is growing very slowly or not at all, the Federal Reserve nudges interest rates downward, thereby stimulating the economy and creating more jobs. The Federal Reserve has been assigned the very difficult task of maintaining a growing and healthy economy with low levels of inflation.

Objective 2 *Understand the Time Value of Money.* Would you leave $1000 on deposit with your bank for 2 years (in a savings account) and expect only $1000 back when the money is returned? Probably not! You would expect interest in addition to the return of your principal. The **time value of money** is the idea that the loaning of money to someone or to a firm has value, and that value is typically repaid by returning interest in addition to principal. Interestingly, the Islamic religion does not approve of interest. Banks in some Islamic countries take partial ownership of a company that they lend money to and share in profits until they have been repaid. The bank is not repaid in the event that there are no profits.

The time value of money is a very important concept in modern society. Regular investments of relatively small sums of money can result in large sums at later dates. For example, how would you like to turn \$54,000 into \$339,000? Invest \$150 per month for 30 years, say from age 30 to age 60, at 10% interest compounded monthly. This topic is discussed in Chapters 13 and 14.

Application of the time value of money will help you send your children to college, purchase a home, and prepare for your own retirement. The time value of money is equally important to firms that invest to build factories, develop new technologies, or need capital for many other reasons.

Objective 3 *Define Present Value and Future Value.* The principal amount that must be invested today to produce a given future amount is called the **present value**. The amount that this sum grows to at some future date is called the **future value**. We have been calculating the future value in the earlier sections of this chapter when we were calculating the maturity amount.

$$\text{Future value: } M = P(1 + RT)$$

We have also been calculating the present value in the earlier sections of this chapter when we were calculating the principal with the following form of the above equation.

$$P = \frac{M}{(1 + RT)}$$

Objective 4 *Calculate Present and Future Values Using Simple Interest.*

Example 2 Finding Future Value

Joan Waters loans her inheritance of \$28,500 to a friend at $9\frac{1}{4}\%$ simple interest on May 30. Find the future value of this amount on February 4 of the following year to the nearest cent.

Solution First find the number of days from May 30 to February 4 by finding the number of days from May 30 to the end of the year. Then find the number of days from the beginning of the year to February 4.

The end of the year is day	365
May 30 is day	− 150
May 30 to end of year	215

February 4 is day 35 of the next year. The total number of days is 215 + 35 = 250 days. Now use the formula for the maturity value (future value) using simple interest.

$$\begin{aligned} M &= P(1 + RT) \\ &= \$28{,}500\left(1 + 0.0925 \times \frac{250}{360}\right) \\ &= \$30{,}330.73 \quad \text{Rounded} \end{aligned}$$

The future value on February 4 is \$30,330.73.

Example 3 Finding Present Value

KLTV Television plans to spend $280,000 for new satellite dishes and associated electronic equipment in 18 months. What present value must be invested today at 6% simple interest so that the firm will have the needed future value?

Solution Use the following form of the simple interest formula.

$$P = \frac{M}{(1 + RT)}$$
$$= \frac{\$280{,}000}{(1 + 0.06 \times \frac{18}{12})}$$
$$= \frac{\$280{,}000}{(1 + 0.09)}$$
$$= \frac{\$280{,}000}{1.09}$$
$$= \$256{,}880.73$$

KLTV Television must invest a present value of $256,880.73 today at 6% simple interest in order to have a future value of $280,000 in 18 months.

Scientific *Calculator Approach*

For the calculator solution to Example 3, the denominator $(1 + 0.06 \times \frac{18}{12})$ should be considered a single number. Set it apart using parentheses.

280000 [÷] [(] 1 [+] .06 [×] 18 [÷] 12 [)] [=] 256880.73

Objective 5 ***Find Present Value for a Given Maturity Value.*** Example 3 found the present value of a specific amount needed at a specific future date. Sometimes a firm or a bank needs to sell a loan it has made to a third party. Notes are sold for their present value on the date sold, based on the market interest rate at the time. The market interest rate frequently differs from the interest rate on the original note. To find the present value, first calculate the future value of the loan, using the terms of the loan. Then find the present value of this amount using the market interest rate for the time value of money.

Example 4 Finding Present Value on Date Loan Is Made

Gencor Refinery currently owes $365,000 at 9% simple interest to Bank USA. The note is due in 10 months. Bank USA needs cash and sells the note to an investment company, at the market rate of 12%. Find the present value received by Bank USA.

Solution First, find the maturity value of the loan to Gencor Refinery using the terms of the loan.

$$M = P(1 + RT)$$
$$= \$365{,}000\left(1 + 0.09 \times \frac{10}{12}\right)$$
$$= \$392{,}375$$

The investment company is buying the right to receive \$392,375 from Gencor in 10 months. Find the present value using the market rate of 12%.

$$P = \frac{M}{1 + RT}$$
$$= \frac{\$392{,}375}{1 + 0.12 \times \frac{10}{12}}$$
$$= \frac{\$392{,}375}{1 + 0.1}$$
$$= \frac{\$392{,}375}{1.1}$$
$$= \$356{,}704.55 \quad \text{Rounded}$$

NOTE In Example 4, an agreement was made to loan money at 9%. However, a change in interest rates caused the time value of money to change to 12%. If the time value of money was also 9%, then the present value and the principal would have been the same, or \$365,000. ▬

Scientific *Calculator Approach*

The calculator solution to Example 4 follows.

365000 [×] [(] 1 [+] .09 [×] 10 [÷] 12 [)] [=] 392375

392375 [÷] [(] 1 [+] .12 [×] 10 [÷] 12 [)] [=] 356704.55

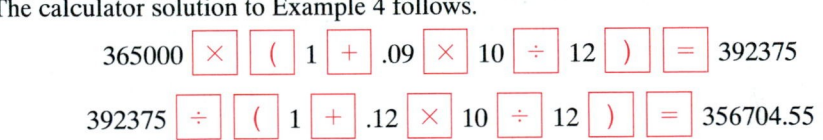

Objective 6 *Find Present Value after a Loan Is Made.* Sometimes you may need to find the present value, the value of the loan on some date after a loan is made, but before it is due. Such a step might be necessary, for example, in preparing a balance sheet. This would give the exact value of the loan on the date the balance sheet was prepared. (Balance sheets are explained in Chapter 17.) Also, a present value might be found before deciding to convert the loan to cash by discounting it at the bank (see Section 12.3).

Example 5 Finding Present Value on a Date after Loan Is Made

An Argentinean businessman makes a loan of \$6500 to a business partner on May 13 for 90 days. The interest rate on the loan is 8%. After the loan is made, the general level of interest rates in Argentina rises quickly so that the time value of money is 12%. Find the present value of the loan on June 26.

Solution First find the maturity value of the loan. The principal is \$6500, the rate is 8%, and the time is 90 days, or $\frac{90}{360}$ year.

$$M = P(1 + RT)$$
$$= \$6500\left[1 + \left(0.08 \times \frac{90}{360}\right)\right]$$
$$= \$6500(1 + 0.02)$$
$$= \$6630$$

The loan was made on May 13, for 90 days. The loan is due on August 11. As shown in Figure 11.3, the number of days from June 26 (the day for which the present value is desired) to August 11 is as follows.

4	Days in June
31	Days in July
+ 11	Days in August
46	Total days.

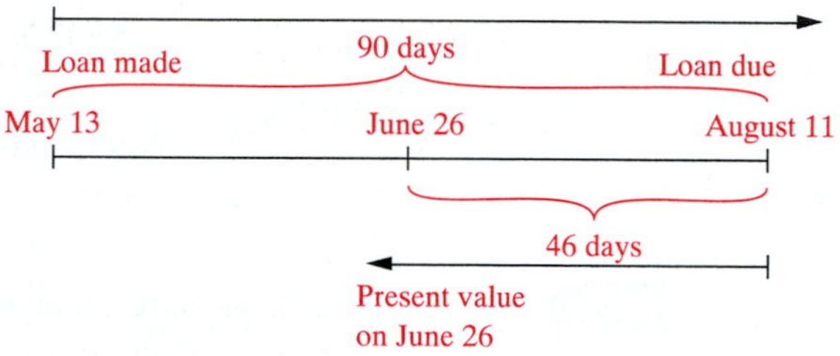

Φιγθρε ++>×

The total number of days could also be found by using Table 11.1.

To find the present value 46 days before the loan is paid off, use the formula for present value, with $M = \$6630$, $R = 12\%$, and $T = \frac{46}{360}$ year.

$$P = \frac{M}{1 + RT}$$

$$= \frac{\$6630}{1 + (0.12 \times \frac{46}{360})}$$

$$= \$6529.88 \quad \text{Rounded}$$

The businessman should be willing to accept \$6529.88 on June 26 in full payment of the loan, both principal and interest since he could invest this sum at 12% and it would grow to \$6630 by August 11.

Problem-Solving Hint Since $\frac{46}{360}$ does not produce a terminating decimal, it is important that it not be rounded at an intermediate step. The calculation should be done in one step and the final answer rounded.

11.4 Exercises

In the following problems, find the present value or future value, as indicated, to the nearest cent.

	Present Value	Interest Rate	Time	Future Value
1.	\$4,800	9%	110 days	______
2.	\$8,000	$8\frac{1}{2}\%$	140 days	______
3.	\$10,500	11%	10 months	______

	Present Value	Interest Rate	Time	Future Value
4.	\$6,800	$12\frac{1}{2}\%$	9 months	______
5.	\$4,100	$8\frac{3}{8}\%$	$1\frac{1}{4}$ years	______
6.	\$7,400	$10\frac{1}{4}\%$	$1\frac{1}{2}$ years	______
7.	______	6%	100 days	\$2,440.00
8.	______	9%	180 days	\$1,985.50
9.	______	$11\frac{1}{8}\%$	6 months	\$8,867.25
10.	______	12%	11 months	\$10,323.00
11.	______	$5\frac{7}{8}\%$	$1\frac{1}{8}$ years	\$9,275.02
12.	______	$7\frac{1}{4}\%$	$1\frac{1}{2}$ years	\$7,096.00

13. Explain the difference between the interest rate on a loan and the time value of money. (See Objective 4.)

14. Explain the meaning of inflation and the CPI. (See Objective 1.)

Solve the following application problems. Round to the nearest cent.

15. AUTO DEALERSHIPS General Motors loaned one of their auto dealers \$6,500,000 on August 4 at $9\frac{1}{2}\%$ simple interest. Find the future value as of December 31.

16. LOAN TO A SUPPLIER Karl's Kameras loaned a supplier \$28,300 on July 14 at 8% simple interest. Find the future value of the loan on March 15 of the following year.

17. BANK OPERATIONS Chase Bank loaned a large Canadian firm \$8.5 million (U.S.) at $9\frac{1}{2}\%$ simple interest for 90 days and simultaneously paid $6\frac{3}{4}\%$ interest for the same funds to their depositors. Find the difference between the interest earned by Chase Bank and the interest paid to their depositors.

18. OUTSTANDING LOAN A finance company carries an outstanding loan, to a bank on the east coast, of \$875,000 at 10% simple interest. They simultaneously lend all \$875,000 to individual customers at a rate of 18% simple interest. (a) Find the difference in interest earned and interest paid out on a daily basis, assuming 365 days in a year. (b) Without rounding the figure in part (a), find the difference between interest earned and interest paid out for a 365-day year.

19. PRESENT VALUE A loan of \$1200 was made for 120 days at 9%. Find the present value of the loan on the day it was made if money is worth 6%.

20. PRESENT VALUE Find the present value, on the date it was made, of a loan of \$14,700 at 12% for 210 days if money is worth 10%.

21. PRESENT VALUE A loan of \$6980 was made on May 24, for 214 days, at 11%. Find the present value of the loan on July 9, if the value of money is 10%.

22. PRESENT VALUE On February 24, an 89-day loan of \$980, at 12% is made. Find the present value of the loan on March 10, if the value of money is 10%.

23. COLLEGE EXPENSES Worried about college expenses, Lupe Martinez sold a diamond ring she had inherited for \$3200 and invested the funds at $7\frac{3}{4}\%$ simple interest for 14 months. Find the future value.

24. **LAWSUIT AWARD** Centron Engineering won a lawsuit and received $255,000, which they invested at $6\frac{3}{4}\%$ simple interest on July 17. Find the future value as of November 30.

25. **INFLATION AND PURCHASING POWER** A family with a spending budget of $26,500 receives an increase in wages of 3% in a year in which inflation was 4.5%. Find the net gain or loss in their purchasing power, ignoring taxes.

26. **GAIN IN PURCHASING POWER** Ben Thomas spends $34,300 and receives a 6% raise in a year in which inflation was 2.5%. Ignoring taxes, find the net gain or loss in his purchasing power.

In the following exercises, three different bids for a single item are given. Find the lowest bid by calculating the present value of each. Assume money has a value of 12%.

27. $5600 today, $5800 in 150 days, $6000 in 210 days.

28. $41,250 today, $43,500 in 210 days, $45,000 in 300 days

29. Do you think inflation will have a significant impact on your finances during your lifetime? Explain.

30. What effects can you foresee of a rapid increase in inflation from 6% to 10%?

Chapter 11 Quick Review

Chapter Terms

Review the following terms to test your understanding of the chapter. For each term you do not know, refer to the page number found next to that term.

banker's interest [p. 392]
compound interest [p. 382]
consumer price index [p. 402]
CPI [p. 402]
exact interest [p. 392]
future value [p. 396]
inflation [p. 402]
interest [p. 383]
maturity date [p. 396]
maturity value [p. 396]
ordinary interest [p. 392]
present value [p. 396]
prime rate [p. 382]
principal [p. 383]
rate of interest [p. 383]
simple interest [p. 383]
time [p. 383]
time value of money [p. 404]

Concept	Example
11.1 Finding simple interest given time expressed in years Use the formula $I = PRT$ with R in decimal form and time in years.	Find the simple interest on \$4000 for $2\frac{1}{2}$ years at 8% per year. $I = PRT$ $= \$4000 \times 0.08 \times 2.5$ $= \$800$
11.1 Finding simple interest given time expressed in months Use $I = PRT$. Express time in years by dividing time in months by 12.	Find the simple interest on \$8600 for 14 months at $6\frac{1}{2}\%$ per year. $I = PRT$ $= \$8600 \times 0.065 \times \frac{14}{12}$ $= \$652.17$ Rounded
11.1 Determining principal given interest, rate, and time Use the formula $P = \frac{I}{RT}$ with R in decimal form and time in years.	Find the principal that would produce an interest of \$150 in 8 months at 6%. $P = \frac{I}{RT}$ $= \frac{\$150}{0.06 \times \frac{8}{12}}$ $= \$3750$
11.1 Finding rate given interest, principal, and time Use the formula $R = \frac{I}{PT}$ with T in years.	Find the rate on a \$15,000 loan for 60 months if the interest was \$6750. $R = \frac{I}{PT}$ $= \frac{\$6750}{\$15{,}000 \times \frac{60}{12}}$ $= 0.09$, or 9%

11.1 Determining time given interest, principal, and rate

Use the formula

$$T = \frac{I}{PR}$$

with R in decimal form.

Find the time for a loan of \$12,000 at 10% to produce \$2400 in interest.

$$T = \frac{I}{PR} = \frac{\$2400}{\$12{,}000 \times 0.10} = 2 \text{ years}$$

11.2 Finding the number of days from one date to another using a table

Find the day corresponding to each date and subtract.

Find the number of days from February 15 to July 28.

July 28 is day	209
Feb. 15 is day	− 46
	163

11.2 Determining the number of days from one date to another using the actual number of days in a month

Add the actual number of days in each month or partial month from one date to the next.

Find the number of days from January 17 to April 15.

14	Remaining in January
28	February
31	March
+ 15	April
88	

11.2 Finding exact interest

Use the formula $I = PRT$ with

$$T = \frac{\text{Time in days}}{365}$$

Find the exact interest on \$7300 at 7% for 100 days.

$$I = PRT = \$7300 \times 0.07 \times \frac{100}{365} = \$140$$

11.2 Finding ordinary interest

Use the formula $I = PRT$ with

$$T = \frac{\text{Time in days}}{360}$$

Find the ordinary interest on \$1100 at 7% for 90 days.

$$I = PRT = \$1100 \times 0.07 \times \frac{90}{360} = \$19.25$$

11.3 Determining the maturity value of a loan, when principal, rate, and time are known

Use the formula

$$M = P(1 + RT)$$

with R in decimal form and T in years.

Find the maturity value of a \$2500 loan that is made for 2 years at 6%.

$$M = P(1 + RT) = \$2500[1 + (0.06 \times 2)] = \$2800$$

11.3 Finding the principal when maturity value, rate, and time are known

Use the formula

$$P = \frac{M}{1 + RT}$$

with R in decimal form and T in years.

Find the principal that would produce a maturity value of \$1500 in 9 months at 4%.

$$P = \frac{M}{1 + RT} = \frac{\$1500}{1 + (0.04 \times \frac{9}{12})} = \$1456.31 \quad \text{Rounded}$$

11.3 Determining the time in days when maturity value, rate, and principal are known

Use the formula

$$T = \frac{M - P}{PR} \times 360$$

to find time in days.

A principal of \$12,000 produces a maturity value of \$12,540 at a 6% rate. Find the time of the loan in days.

$$\begin{aligned} T &= \frac{M - P}{PR} \times 360 \\ &= \frac{\$12{,}540 - \$12{,}000}{\$12{,}000 \times 0.06} \times 360 \\ &= 0.75 \times 360 \\ &= 270 \text{ days} \end{aligned}$$

11.3 Finding rate given principal, maturity value, and time

Use the formula

$$R = \frac{M - P}{PT}$$

with time expressed in years.

A principal of \$8400 produces a maturity value of \$8596 in 210 days. Find the rate of interest.

$$\begin{aligned} R &= \frac{M - P}{PT} \\ &= \frac{\$8596 - \$8400}{\$8400 \times \frac{210}{360}} \\ &= 0.04, \text{ or } 4\% \end{aligned}$$

11.3 Finding time in days when principal, rate, and interest are known

Use the formula

$$T = \frac{I}{PR} \times 360$$

with R in decimal form.

Find the time in days that \$1500 was invested at 8% if \$25 interest was earned.

$$\begin{aligned} T &= \frac{I}{PR} \times 360 \\ &= \frac{\$25}{\$1500 \times 0.08} \times 360 \\ &= 75 \text{ days} \end{aligned}$$

11.4 Finding the effect of inflation on the cost of living.

Use the percent formula.

Find the increase in the cost of living for a family with a budget of \$24,600 in a year with $3\frac{1}{2}\%$ inflation.

$$0.035 \times \$24{,}600 = \$861$$

They experience an \$861 increase in the cost of living.

11.4 Determining present value given maturity or future value, rate, and time

Use the formula

$$P = \frac{M}{1 + RT}$$

with R in decimal form and T in years.

A debt of \$8000 must be paid in 9 months. Find the amount that could be deposited today at 9% interest so that enough money will be available.

$$\begin{aligned} P &= \frac{M}{1 + RT} \\ &= \frac{\$8000}{1 + (0.09 \times \frac{9}{12})} \\ &= \$7494.15 \end{aligned}$$

11.4 Finding present value given rate and time value of money

Use the formula

$$M = P(1 + RT)$$

with R in decimal form and time in years. Then use the formula

$$P = \frac{M}{1 + RT}$$

with R = time value of money and T in years.

A loan of \$750 at 12% is due in 15 months. Find the present value of the loan if the time value of money is 9%.

$$\begin{aligned} M &= P(1 + RT) \\ &= \$750 \times \left[1 + \left(0.12 \times \frac{15}{12}\right)\right] \\ &= \$862.50 \end{aligned}$$

$$\begin{aligned} P &= \frac{M}{1 + RT} \\ &= \frac{\$862.50}{1 + (0.09 \times \frac{15}{12})} \\ &= \$775.28 \end{aligned}$$

11.4 Finding present value after a loan is made

Use the formula $M = P(1 + RT)$ with R in decimal form and T in years. Next, use the formula

$$P = \frac{M}{1 + RT}$$

with R = time value of money, and T = fraction of a year using days from present value date to due date.

An \$8000 loan is made on June 5 for 90 days at 9%. The time value of money is 11%. Find the present value of the loan on August 10.

$$\begin{aligned} M &= P(1 + RT) \\ &= \$8000 \times \left[1 + \left(0.09 \times \frac{90}{360}\right)\right] \\ &= \$8180 \end{aligned}$$

Now, find time in days.

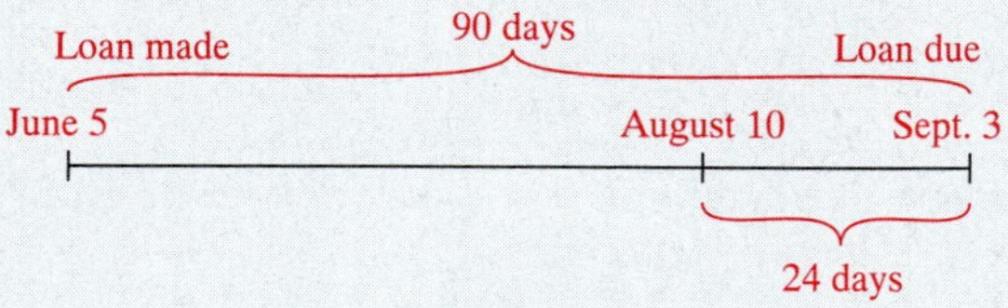

$$\begin{aligned} P &= \frac{M}{1 + RT} \\ &= \frac{\$8180}{1 + (0.11 \times \frac{24}{360})} \\ &= \$8120.45 \end{aligned}$$

Chapter 11 Review Exercises

Complete this table. Round money to the nearest cent, rate to the nearest tenth of a percent, and time to the nearest month. **[11.1]**

	Interest	Principal	Rate	Time
1.	______	$8,400	9%	2 years
2.	______	$9,600	$10\frac{1}{2}\%$	10 months
3.	$696.80	______	12%	8 months
4.	$144	______	8%	9 months
5.	$810	$12,000	______	9 months
6.	$750	$8,000	______	15 months
7.	$540	$12,000	6%	______
8.	$1600	$8,000	8%	______

Find the number of days from the first date to the second. **[11.2]**

9. April 15 to August 7

10. July 12 to November 4

Find the exact interest for each of the following. Round to the nearest cent. **[11.2]**

	Interest	Principal	Rate	Time
11.	______	$10,500	$7\frac{1}{2}\%$	120 days
12.	______	$8,400	$9\frac{3}{4}\%$	150 days
13.	______	$7,200	8%	From July 12 to November 30
14.	______	$6,800	$7\frac{1}{2}\%$	From February 4 to May 9

Find the ordinary interest for each of the following. Round to the nearest cent. **[11.2]**

	Interest	Principal	Rate	Time
15.	________	$7,400	7%	30 days
16.	________	$52,000	10.2%	220 days

Complete the following table. Round money to the nearest cent, rate to the nearest tenth of a percent, and time to the nearest day. **[11.3]**

	Principal	Rate	Time	Interest
17.	________	$7\frac{1}{2}\%$	80 days	$203.33
18.	________	11%	180 days	$340
19.	$6,000	______	60 days	$70
20.	$8,400	______	120 days	$231
21.	$7,800	9%	______	$78
22.	$4,900	$10\frac{1}{8}\%$	______	$124.03

Complete the following table. Round money to the nearest cent and rate to the nearest tenth of a percent. **[11.3]**

	Maturity Value	Principal	Rate	Time
23.	________	\$5,500	8%	$1\frac{3}{4}$ years
24.	________	\$6,900	7.2%	9 months
25.	\$12,180.00	________	9%	60 days
26.	\$6,752.78	________	10%	140 days
27.	\$8,120.00	\$8,000	________	60 days
28.	\$17,537.50	\$15,250	________	15 months

Find the present value of each of the following. Round to the nearest cent. **[11.4]**

	Present Value	Maturity Value	Rate	Time
29.	________	\$6,600	$8\frac{1}{2}$%	120 days
30.	________	\$12,000	12%	100 days

Find the present value of the following loans on the day they are made. Round to the nearest cent. **[11.4]**

	Present Value	Principal Amount	Interest Rate of Loan	Length of Loan	Value of Money
31.	________	\$8,000	$9\frac{1}{2}$%	10 months	8%
32.	________	\$20,000	7%	15 months	9%

Find the present value for the following loans on the indicated dates. Assume the value of money is 9%. Round to the nearest cent. **[11.4]**

	Present Value	Amount	Interest Rate of Loan	Length of Loan	Loan Made	Find Present Value On
33.	________	\$800	10%	60 days	Feb. 1	March 15
34.	________	\$15,000	8%	300 days	Apr. 1	October 15

Solve the following application problems. Round money to the nearest cent, rate to the nearest tenth of a percent, and time to the nearest day.

35. Christina Barrett borrowed \$4500 at $7\frac{1}{2}$% for 90 days to go on an exchange trip to Seoul, Korea. Find the simple interest. **[11.1]**

36. CenterStage Music borrowed \$18,600 at 9% simple interest for 150 days to obtain a trade discount on guitars purchased from Martin. Find the maturity value. **[11.3]**

37. A \$14,700 loan at $8\frac{1}{2}$% has a maturity value of \$15,567.71. Find the term in days. **[11.3]**

38. An 8-month, 9% loan had a maturity value of \$3816. Find the amount originally borrowed. **[11.3]**

39. Collins Dairy has just bought new milking machines. The machines must be paid for with a single payment of \$12,000 in 270 days. The firm has \$11,200 that it can invest today. What rate of interest must it earn on this deposit to have the necessary \$12,000? **[11.2]**

40. Quik Print is ordering a new $27,500 printing press. The seller of the press wants payment in 120 days. The print shop has $26,000 available for investment today. What rate of interest must it earn on this deposit to have the needed $27,500? **[11.2]**

41. How many days will it take for $17,600 to become $18,348 at 17% interest? **[11.3]**

42. Find the present value of a loan on the day it was made if the loan is for $9100 at 11% for 85 days. Assume that money has a time value of 9.7% **[11.4]**

43. A loan of $19,250 is made on October 15, at 12%. The loan is for 75 days. Find the present value of the loan on November 27 if money has a time value of 11.7%. **[11.4]**

44. Suppose a loan of $11,800 is made on July 12, for 153 days, at 9%. Find the present value of the loan on September 20, if money has a time value of 11%. **[11.4]**

45. Sherri Woods makes and spends $28,400 per year. She receives an increase in her annual salary of 6% during a year in which inflation is 3%. Ignoring taxes, find her net gain or loss in purchasing power. **[11.4]**

46. Bob and Jane Shaw have been spending their entire after-tax income of $46,850. Assume they receive no increase in their after-tax salary for a particular year during which inflation was 6%. Find the net gain or loss in their purchasing power. **[11.4]**

47. Explain the possible effects of inflation on individuals that are retired. **[11.4]**

48. Which of the following would you prefer in a period with high inflation? (a) A salary of $25,000 per year with no raises for 10 years, or (b) a salary of $22,000 per year with annual raises, in each of the 10 years, that slightly exceeds the rate of inflation. Explain. **[11.4]**

Chapter 11 Summary Exercise *Multiple Loans*

Wes Whitmeyer started up an appraisal company to appraise (estimate) the value of homes and commercial real estate. During the first few months of operation of his new business he received the following simple-interest loans.

Loan Number	Date Loan Made	Principal	Length of Loan	Interest Rate	Purpose of Loan
1	March 3	\$12,000	6 months	10%	Used Truck
2	April 17	\$6,200	90 days	12%	Computer Equipment
3	July 9	\$18,400	200 days	$9\frac{1}{2}\%$	Remodel the Office

Complete the following table. Round to the nearest cent if needed.

Loan Number	Maturity Date	Simple Interest	Maturity Value
1	________	________	________
2	________	________	________
3	________	________	________
	Totals:	________	________

Net Assets Business on the Internet

U.S. Home

Statistics

- 1954: Established
- 1998: Revenues of $1.5 billion
- Average sale price of homes: $182,600
- Named National Builder of the Year by *Professional Builder* magazine

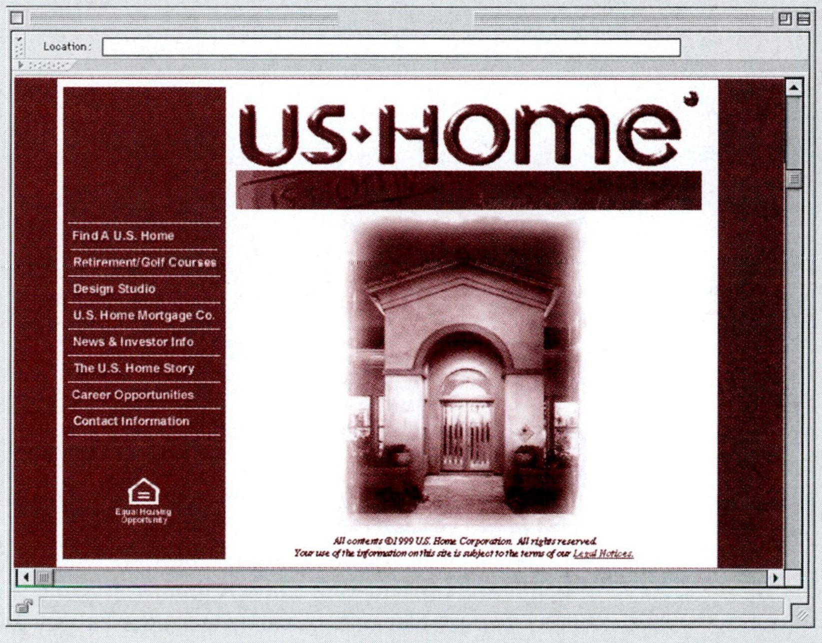

U.S. Home was established in 1954 with the intention of becoming a nationwide home-building company that provides unique home design in a variety of locations across the United States. Having achieved this goal, the firm has built nearly 300,000 homes in 12 states. The company builds homes in more than 190 communities and has become one of the nation's largest retirement and active-adult home builders.

With a mortgage company as part of the organization, U.S. Home is also able to provide funding to thousands of home buyers each year. U.S. Home Mortgage Corporation, finances over 80% of the homes that the company builds. The mortgage group helps individuals finance houses using FHA, VA, or conventional loans. The company borrows money to purchase and develop large tracts of acreage. At the end of 1998, long-term debt was $425 million.

1. U.S. Home plans to build 10,000 homes in 2001. If U.S. Home Mortgage Corporation finances 83% of these, how many homes should they prepare to finance in 2001? Assume an average sales price of $185,000 and find the total value of the homes U.S. Home Mortgage Corporation should expect to finance in 2001.

2. Assume U.S. Home borrows $28,000,000 at 8% on a simple interest note from Bank of America for 9 months. Find the interest and maturity value.

3. In question 2, assume the funds are borrowed for 9 months at $9\frac{1}{2}$ rather than at 8%. Find the increase in interest.

12 Notes and Bank Discount

Almost every organization borrows money. Figure 12.1 shows nonfinancial corporate borrowing in *trillions of dollars*. As you can see, corporations borrow a lot of money.

Why do you think a large and successful corporation such as McDonald's would need to borrow money?

When an individual or business borrows money, written proof of the loan often takes the form of a **promissory note**. A promissory note is a legal and frequently transferable document in which one person or firm agrees to pay a stated amount of money, at a stated time and interest rate, to another. Banks typically require business owners to sign promissory notes when lending them money. Sometimes simple interest notes are used, while other times simple discount notes are used. Both types of notes are discussed in this chapter.

NOTE A year is assumed to have 360 days for all calculations in this chapter.

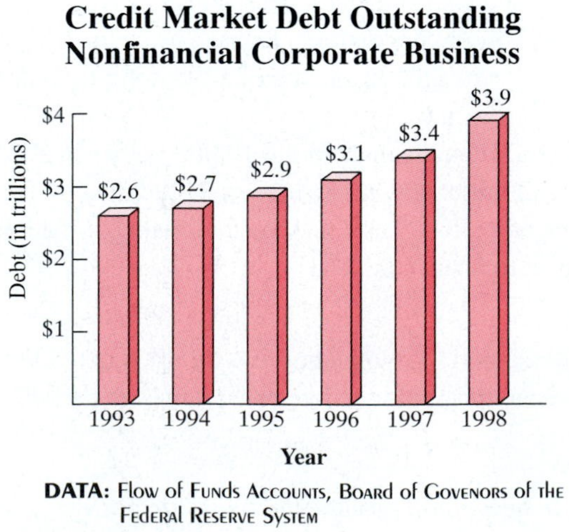

Figure 12.1

12.1 Simple Interest Notes

Objectives

1. *Identify the parts of a simple interest promissory note.*
2. *Find the due date of a note.*
3. *Find the face value, time, and rate of a note.*

An example of a promissory note is shown in Figure 12.2.

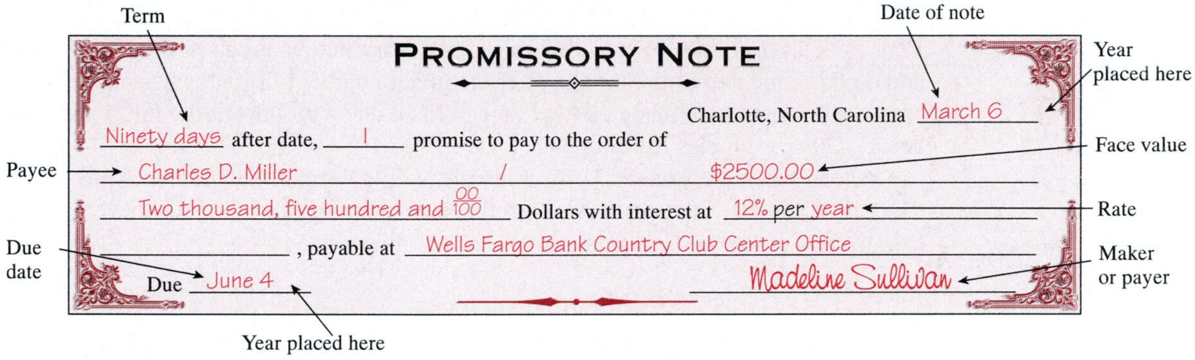
Term
Date of note
PROMISSORY NOTE
Year placed here
Charlotte, North Carolina March 6
Ninety days after date, I promise to pay to the order of
Face value
Payee
Charles D. Miller
$2500.00
Two thousand, five hundred and 00/100 Dollars with interest at 12% per year
Rate
Due date
, payable at Wells Fargo Bank Country Club Center Office
Maker or payer
Due June 4
Madeline Sullivan
Year placed here

Figure 12.2

Objective 1 *Identify the Parts of a Simple Interest Promissory Note.* The person borrowing the money is called the **maker** or **payer** of the note (Madeline Sullivan for this note). The person who loaned the money, and who will receive the payment, is called the **payee** (Charles D. Miller for this note). The length of time until the note is due is called the **term of the note** (90 days for this note). The **face value**, or principal, of the note ($2500 for this note) is the amount written on the line in front of *dollars*. The interest rate on the note is 12% per year.

The **maturity value** of the loan is the face value plus any interest that is due. Since the interest for this note is found by using formulas for simple interest, this note is a **simple interest note**. When using the formulas for simple interest, the face value of the note is used as the value for the principal, *P*. Find the interest on the note shown in Figure 12.2 as follows.

$$\text{Interest} = \text{Face value} \times \text{Rate} \times \text{Time}$$

$$= \$2500 \times 0.12 \times \frac{90}{360} = \$75$$

The maturity value of the loan is

$$\text{Maturity value} = \text{Face value} + \text{Interest}$$

$$= \$2500 + \$75 = \$2575$$

Madeline Sullivan must pay $2575 to Charles D. Miller at the note's maturity, or June 4, which is 90 days after March 6.

The promissory note shown in Figure 12.2 contains all the information needed to calculate interest owed and maturity value. However, banks and other financial institutions use a more comprehensive note containing very detailed listings of necessary payment dates and amounts, as well as paragraphs describing the bank's rights in case of nonpayment of the note. A typical promissory note from a large bank is shown in Figure 12.3.

Almost all notes written by banks are secured by **collateral**. That is, the person borrowing the money *must pledge assets* such as cars, stock, or real estate that are of equal to or of greater value than the amount of the loan. The collateral for the note in Figure 12.3 is Lot 167, Bill Morris Survey. In the event of nonpayment, the bank will take the collateral and sell or liquidate it. The bank then uses the proceeds to pay off the note—any excess is returned to the maker of the note.

Objective 2 *Find the Due Date of a Note.* When a promissory note is given in months, the loan is due on the same day of the month, after the given number of months has passed. For example, a 4-month note made on May 25 would be due 4 months in the future on September 25. Other examples follow.

A loan made on January 31 for 3 months would normally be due on April 31. However, there are only 30 days in April, so the loan is due on April 30. Whenever a due date

FIXED RATE CONSUMER NOTE, DISCLOSURE AND SECURITY AGREEMENT

BORROWER
Joan Waters

ADDRESS
2705 4th St., Kansas City, MO

TELEPHONE NO.

IDENTIFICATION NO.
HA69724

OFFICER INITIALS	INTEREST RATE	PRINCIPAL AMOUNT	FUNDING DATE	MATURITY DATE	CUSTOMER NUMBER	LOAN NUMBER
GWS	12%	$ 10,000	4/1	8/1	3627014	HA69724

** ANNUAL PERCENTAGE RATE **	** FINANCE CHARGE **	AMOUNT FINANCED	TOTAL OF PAYMENTS
THE COST OF THE CREDIT AS A YEARLY RATE. 12%	THE DOLLAR AMOUNT THE CREDIT WILL COST. $ 1957.04	THE AMOUNT OF CREDIT PROVIDED TO THE BORROWER OR ON BORROWER'S BEHALF. $ 10,000	THE AMOUNT BORROWER WILL HAVE PAID AFTER PAYMENTS HAVE BEEN MADE AS SCHEDULED. $ 11,957.04

NUMBER OF PAYMENTS	AMOUNT OF PAYMENTS	WHEN PAYMENTS ARE DUE
36	$ 332.14 monthly	beginning April 11

DEMAND FEATURE: ☐ This Note has a demand feature.

REQUIRED DEPOSIT ACCOUNT: ☐ If checked, the Lender requires that Borrower maintain a deposit balance with Lender. The Annual Percentage Rate does not take into account required deposits.

SECURITY: A security interest has been granted in: ☐ Collateral securing other loans with Lender may also secure this loan; ☐ Any deposit accounts of Borrower with Lender; ☒ The goods or property being purchased; ☐ Other (describe): LOT 167, BILL MORRIS SURVEY

FILING FEES: $ NONE in fees are being paid to public officials in order to research, perfect or release a security interest in the Collateral.

PREPAYMENT: If Borrower pays off early, Borrower ☐ may ☐ will not have to pay a penalty; ☐ may ☐ will not be charged a minimum finance charge; ☐ may ☐ will not be entitled to a refund of part of the finance charge.

ITEMIZATION: Borrower has the right to receive at this time an itemization of the Amount Financed. Borrower ☐ does ☐ does not want an itemization.

LATE CHARGES: If an installment is received more than 10 days late, Borrower will be charged a late payment charge of: ☐ ______ % of the unpaid late installment; ☒ 5 % of the unpaid late installment or $ 15.00, whichever is ☒ greater ☐ less; as permitted by law.

ASSUMPTION: ☐ This loan may not be assumed on its original terms. ☐ This loan may be assumed on its original terms, subject to certain conditions.

See your contract documents for additional information about nonpayment, default, prepayment penalties and refunds and acceleration. e means an estimate

Figure 12.3

does not exist, such as February 30 or November 31, use the last day of the month (February 28 or November 30 in these examples). February 29 must be used if it is a leap year.

Date Made	Length of Loan	Date Due
March 12	5 months	August 12
April 24	7 months	November 24
October 7	9 months	July 7
January 31	3 months	April 30

Example 1 Finding Due Date, Interest, and Maturity Value

Find the due date, interest, and maturity value for a loan made September 30 for 5 months at 9.5% with a face value of \$2380.

Solution Counting 5 months from September 30 produces February 30. Since February has only 28 days (if not in a leap year), the note is due on the last day of February, or February 28. Interest for 5 months is

$$I = \$2380 \times 0.095 \times \frac{5}{12} = \$94.21 \quad \text{Rounded}$$

with maturity value

$$M = \$2380 + \$94.21 = \$2474.21$$

Problem-Solving Hint When the length of the loan is given in months, do not convert the time to days in order to find the date due.

Expressing a promissory note in days instead of months is commonly done. For example, a loan might be signed on March 12, and be due in 90 days. To find the date due, use the exact number of days in a month. This can be done in either of two ways, as shown in Chapter 11.

One way is to look back at Table 11.1, which shows the number of each day. From the table, March 12 is the 71st day of the year. The loan is due 90 days after March 12. The number of the day on which the loan is due is found as follows.

$$\begin{array}{rl} 71 & \text{Number of the day loan is made.} \\ +\ 90 & \text{Number of days until due.} \\ \hline 161 & \text{Number of the day loan is due is 161.} \end{array}$$

From Table 11.1, day 161 is June 10. A 90-day loan made on March 12 is due on June 10.

As an alternate method, use the actual number of days in each month. The loan is made on March 12. Since March has 31 days, there are 19 more days in March.

$$\begin{array}{r} 31 \\ -\ 12 \\ \hline 19 \end{array}$$

There are 30 days in April, and 31 in May. Find the total as follows.

$$\begin{array}{rl} 19 & \text{Rest of days in March} \\ 30 & \text{Days in April} \\ +\ 31 & \text{Days in May} \\ \hline 80 & \end{array}$$

The loan is for 90 days, which is 10 more than 80, making the loan due on June 10.

The following table shows several more examples.

Date Made	Length of Loan	Due Date
January 9	60 days	March 10
May 28	120 days	September 25
November 21	100 days	March 1
October 9	180 days	April 7

Objective 3 ***Find the Face Value, Time, and Rate of a Note.*** The next examples show how to find the face value, time, or rate for a note. Each of these examples uses formulas from Chapter 11.

NOTE Face value for simple interest notes is the same as the principal P, as used in Chapter 11. ■

Example 2 Finding Face Value and Interest

Sheila Walker signed a 120-day note at 12% for funds she used for a vacation to Colorado. A single payment of \$1248 paid the note when it was due. Find the face value and interest for the note.

Solution The payment of \$1248 is the maturity value of the note. The maturity value is found with the following formula from Chapter 11.

$$M = P(1 + RT)$$

Since P is not known in this example, solve for P by dividing both sides by $1 + RT$.

$$P = \frac{M}{1 + RT}$$

Now substitute \$1248 for M, 0.12 for R, and $\frac{120}{360}$ for T.

$$\begin{aligned} P &= \frac{\$1248}{1 + (0.12 \times \frac{120}{360})} \\ &= \frac{\$1248}{1 + 0.04} \\ &= \frac{\$1248}{1.04} \\ &= \$1200 \end{aligned}$$

The face value of the note was \$1200; the interest charge was \$1248 − \$1200 = \$48.

Example 3 Finding the Time of a Note

A note is signed by the Movie Store, with a face value of \$820. The interest rate is 10%, with a maturity value of \$830.25. Find the time of the note, in days.

Solution Recall the formula for time in days from Chapter 11.

$$T \text{ (in days)} = \frac{M - P}{PR} \times 360$$

Find the time by substituting \$830.25 for M, \$820 for P, and 0.10 for R.

$$\begin{aligned} T \text{ (in days)} &= \frac{\$830.25 - \$820}{\$820 \times 0.10} \times 360 \\ &= \frac{\$10.25}{\$82} \times 360 \\ &= 0.125 \times 360 \\ &= 45 \text{ days} \end{aligned}$$

The loan was for 45 days. The same time in days can also be found using the equation from Chapter 11:

$$T \text{ (in days)} = \frac{I}{PR} \times 360$$

NOTE A common error when solving for time T is forgetting to multiply by 360. When solving for time in Example 3, the number 0.125 ($\frac{1}{8}$) is the time in years. Multiplying by 360 gives the time in days. ■

Example 4 Finding the Rate of a Note

Leslie Graham, owner of Creative Music, signed a note with a face value of \$12,500 on April 12. She needed the funds to pay her income taxes. On August 30, a maturity value of \$12,961.81 is to be repaid. Find the rate on the note.

Solution The interest on the loan is

$$\$12{,}961.81 - \$12{,}500 = \$461.81$$

Find the rate by starting with the formula for simple interest, $I = PRT$, and dividing both sides by PT.

$$R = \frac{I}{PT}$$

Now use \$461.81 for I, \$12,500 for P, and $\frac{140}{360}$ for T, since there are 140 days from April 12 to August 30. (Confirm this is 140 days.)

$$\begin{aligned} R &= \frac{\$461.81}{\$12{,}500 \times \frac{140}{360}} \\ &= 9.5\% \quad \text{Rounded.} \end{aligned}$$

The rate is 9.5%. The same rate can be found using the alternative form of the equation $R = \dfrac{M - P}{PT}$ found in Chapter 11.

NOTE Do not round before the final answer in problems such as Example 4. ■

12.1 Exercises

Identify each of the following from the note below.

1. maker
2. payer
3. payee
4. face value
5. term of loan
6. day loan made
7. day loan due
8. maturity value

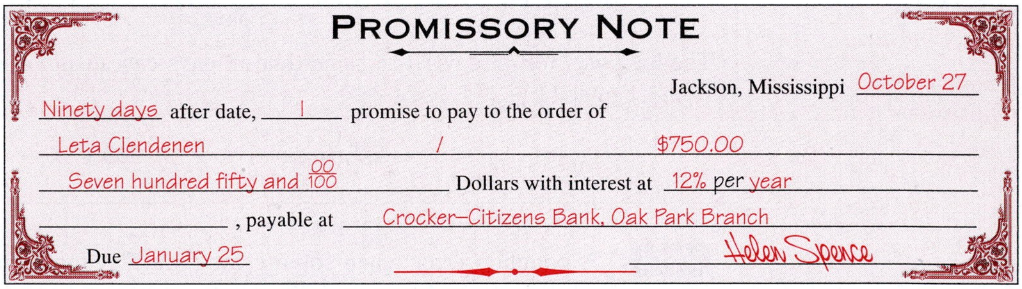

PROMISSORY NOTE

Jackson, Mississippi October 27

Ninety days after date, I promise to pay to the order of

Leta Clendenen / $750.00

Seven hundred fifty and 00/100 Dollars with interest at 12% per year

, payable at Crocker–Citizens Bank, Oak Park Branch

Due January 25 Helen Spence

Find the due date of each of the following.

	Date Made	Term of Loan
9.	June 20	6 months
10.	July 12	3 months
11.	December 31	6 months
12.	May 31	4 months
13.	January 6	70 days
14.	September 14	125 days
15.	November 24	150 days
16.	December 8	160 days

Find the date due and the maturity value for each of the following. Use Table 11.1 on page 389 in your textbook. Round money to the nearest cent.

	Date Made	Face Value	Term of Loan	Rate
17.	June 12	\$6000	150 days	9%
18.	March 18	\$9500	200 days	10%
19.	August 14	\$5000	300 days	$8\frac{1}{2}$%
20.	October 20	\$4500	180 days	12%

Complete the following table. Round money to the nearest cent, time to the nearest day, and rate to the nearest tenth of a percent.

	Principal	Rate	Time	Maturity Value
21.	______	10%	180 days	\$8,820
22.	______	9%	240 days	\$7,632
23.	\$3,600	8%	______	\$3,696
24.	\$9,000	11%	______	\$9,660
25.	\$8,400	______	140 days	\$8,759.33
26.	\$9,500	______	180 days	\$10,165
27.	\$12,240	$10\frac{1}{2}$%	200 days	________
28.	\$15,000	9%	150 days	________

29. State and explain the different parts of a simple interest promissory note. (See Objective 1.)

30. Name at least five reasons why a small firm such as a bicycle or boat shop might need to borrow money. List three reasons you might need to borrow money.

Solve each of the following application problems. Round to the nearest cent.

31. REAL ESTATE SIGNS Susan Eisenhammer and Pam Carlson manage the Nelison Real Estate Office. They bought a supply of "for sale" and "sold" signs by signing a 100-day, $8\frac{1}{4}$% note for \$1125 on October 1. Find (a) the due date and (b) the maturity value of the note.

32. SKI PURCHASE Jill Sample's Ski House bought \$15,900 worth of skis from Rossignol on September 19. The firm signed a 200-day, $11\frac{1}{2}$% note for the skis. Find (a) the due date and (b) the maturity value of the note.

33. FENCE COMPANY Jones Fence Company signed a 9-month, 11% note on March 13 for \$21,000. Find (a) the due date and (b) the maturity value.

34. SMALL ANIMAL CLINIC Jane Reeves just became a veterinarian and needs \$48,000 to start her small animal clinic. She borrows this amount at $12\frac{1}{8}$% for 8 months on a note dated March 31. Find (a) the due date and (b) the maturity value.

35. BAGEL SHOP Kyle Sleg wishes to remodel his bagel shop and signs a 320-day note at $10\frac{1}{2}$% with a maturity value of 46,466.67. Find the face value.

36. STARTUP COMPANY A 15-year-old computer whiz needs two computers and some software to start a home business. His parents help him by signing a note with a maturity value of \$5400, rate of 12%, and time of 240 days. Find the face value.

37. FLOWER SHOP Marie's Flowers uses the proceeds from an 11%, 7-month note with a maturity value of \$30,222.33 to build a garden area with fountain in front of her store. Given that the note is due on November 7, find (a) the date the loan was made and (b) the face value.

38. PHYSICAL THERAPIST A physical therapist signed an 11-month, $9\frac{1}{2}$% note with a maturity value of \$92,836.92 due on March 30. Find (a) the date the loan was made and (b) the face value.

39. STATIONERY SHOP West Stationery is getting ready to pay a note that is due. The interest rate is 12%, the face value is \$4000, and the interest is \$120. Find the time (in days) of the note.

40. SCUBA SHOP Lewin Scuba just paid the \$7536 maturity value on a note with a face value of \$7200. The interest rate was 8%. Find the time (in days) of the note.

41. SECURITY SYSTEM PURCHASE The manager of the local electronics plant bought a security system for \$12,000, paying for it with a 120-day note with a maturity value of \$12,320. Find the interest rate she paid on the note.

42. COFFEE SHIPMENT Dunn Brothers Coffee Company signed a \$17,000, 90-day note for a shipment of coffee. The maturity value of the note was \$17,382.50. Find the interest rate paid.

43. ALASKA CRUISE Bob and Sheryl Robinson sign a \$6800 note at 12% simple interest for 90 days and use the funds to help pay for a 2-week cruise to Alaska. If Sheryl receives an after-tax year-end bonus of \$5000 in 90 days, how much additional money will they need to pay off the note.

12.2 Simple Discount Notes

Objectives

1. *Define simple discount notes.*
2. *Find bank discount and proceeds.*
3. *Calculate proceeds if given face value, discount rate, and time.*
4. *Distinguish between discount rates and simple interest rates.*
5. *Find effective interest rates.*
6. *Understand T-bills.*
7. *Find the face value that produces the desired proceeds.*
8. *Find discount rate, face value, or time.*

Objective 1 *Define Simple Discount Notes.* The dollar amount written on the front of a simple interest note is called the face value (or principal). The face value of a simple interest note is the amount actually loaned to the borrower. However, face value is *defined differently* in the simple discount notes discussed in this section.

A **simple discount note** has the interest deducted *in advance* from the face value written on the note. In this type of note, the borrower *never receives* the face value—rather the borrower receives the face value *less interest.* These notes are sometimes called **interest-in-advance notes** since the interest is subtracted *before* any money is given to the borrower.

NOTE A simple discount note represents another method for calculating interest. ∎

The face value and the maturity value of a simple discount note are the same. The amount of interest charged is called the **bank discount**, or just the **discount**. Discount here is *not* the same as a discount received at a store when an item is on sale. Rather, the borrower receives a sum of money called the **proceeds**, which equals the face value of the note *less the discount.*

Type of Note	Amount Received by Borrower		Interest		Repayment Amount
Simple interest	Face value (Principal)	+	Interest	=	Maturity value
Simple discount	Proceeds	+	Discount (Interest)	=	Face value (Maturity value)

As an example of a simple discount note, suppose a borrower signs a note for $2000 with a bank discount of $150. The borrower receives proceeds of

$$\$2000 - \$150 = \$1850$$

The face value or maturity value of the note is $2000 and the interest charge is $150.

It is important to understand the difference between simple interest notes and simple discount notes.

Simple *interest* is calculated on the *principal*, while simple *discount* is calculated on the *maturity value.*

Objective 2 ***Find Bank Discount and Proceeds.*** The formula for finding the bank discount is similar to the formula for calculating simple interest. Different letters are used to emphasize that the loan is a *discount loan* and that interest charges are computed based on maturity value rather than on the amount received by the borrower.

Calculating Bank Discount

$$\text{Bank discount} = \text{Face value} \times \text{Discount rate} \times \text{Time or } B = MDT$$

where B = bank discount
M = face value (maturity value)
D = discount rate
T = time (in years)

Then, if P is the proceeds:

$$\text{Proceeds (loan amount)} = \text{Face value} - \text{Bank discount}$$

or $$P = M - B$$

The maturity value, or face value, M, is the sum of the discount and the proceeds.

$$M = P + B$$

NOTE Actually, the formula for simple interest, $I = PRT$, could be used with simple discount. However, the formula is written $B = MDT$ to emphasize that the loan is a *discount* loan. ■

Problem-Solving Hint When time is in months, use 12 in the denominator of T. The fraction can then be reduced.

Example 1 Finding Discount and Proceeds

Marie Gostowski borrowed \$12,000 for 10 months from Bank of America so that she could buy a new, larger commercial oven for her bakery. The banker discounted the note at 9%. Find the amount of the discount and the proceeds.

Solution Find the discount with the formula $B = MDT$, with $M = \$12{,}000$, $D = 9\%$, and $T = \frac{10}{12}$ or $\frac{5}{6}$.

$$B = MDT$$

$$= \$12{,}000 \times 0.09 \times \frac{5}{6} = \$900$$

The discount of \$900 is the interest charge on the loan. The proceeds that Gostowski actually received when making the loan is found using $P = M - B$.

$$P = M - B$$

$$= \$12{,}000 - \$900 = \$11{,}100$$

After signing the note for \$12,000, Gostowski will be given \$11,100. Then 10 months later, she must make a single payment of \$12,000 to the bank.

NOTE In Example 1, the proceeds of \$11,100 is the present value, and the maturity value of \$12,000 is the future value. ▬

Objective 3 *Calculate Proceeds If Given Face Value, Discount Rate, and Time.* The proceeds of a simple discount note is the face value (maturity value) *minus* the discount ($M - B$). The discount equals maturity value × discount rate × the time in years. Look at the following.

$$P = M - B$$

$$= M - MDT \qquad \text{Substitute } MDT \text{ for } B.$$

$$= M(1 - DT) \qquad \text{Use the distributive property.}$$

Example 2 Finding Proceeds

Adventure Travel borrows \$25,000 for 90 days at a discount rate of $10\frac{1}{2}\%$ to help cover operating expenses. Find the proceeds.

Solution There are two ways to find the proceeds. One way would be to use the formula $B = MDT$ to find the discount and then subtract the discount from the face value to find the proceeds.

$$B = MDT$$

$$= \$25{,}000 \times 0.105 \times \frac{90}{360} = \$656.25$$

The proceeds are then

$$P = M - D$$

$$= \$25{,}000 - \$656.25 = \$24{,}343.75$$

As a second method and a check for finding the proceeds, use the formula $P = M(1 - DT)$. Here, $M = \$25{,}000$, $D = 10\frac{1}{2}\%$, and $T = \frac{90}{360}$.

$$P = M(1 - DT)$$
$$= \$25{,}000\left[1 - \left(0.105 \times \frac{90}{360}\right)\right]$$

Inside the brackets, be sure to first multiply $0.105 \times \frac{90}{360}$, and then subtract.

$$P = \$25{,}000(1 - 0.02625)$$
$$= \$25{,}000(0.97375)$$
$$= \$24{,}343.75$$

By this method, it is not necessary to find the bank discount.

NOTE It is important *not to round before the final answer* in these problems, in order to avoid rounding errors.

Scientific *Calculator Approach*

For the calculator solution to Example 2, think of the problem as

$$\$25{,}000\left(1 - 0.105 \times \frac{90}{360}\right)$$

and enter the parentheses accordingly.

25000 [×] [(] 1 [−] .105 [×] 90 [÷] 360 [)] [=] 24343.75

Objective 4 ***Distinguish Between Discount Rates and Simple Interest Rates.*** A discount rate of 12% is not the same as a simple interest rate of 12%. The next example shows why.

Example 3 Comparing Discount Rates and Simple Interest Rates

Two different notes each have a face value of \$7500 and a time of 90 days. One has a simple interest rate of 12%, and the other has a discount rate of 12%.

(a) Find the interest owed on each.

Solution

For the Simple Interest Note	For the Simple Discount Note
$I = PRT$	$B = MDT$
$= \$7500 \times 0.12 \times \frac{90}{360}$	$= \$7500 \times 0.12 \times \frac{90}{360}$
$= \$225$	$= \$225$

In each case, the interest is \$225.

(b) Find the amount *actually received* by the borrower in each case.

Solution

For the Simple Interest Note	For the Simple Discount Note
Principal = Face value = \$7500	Proceeds = $M - B$ = \$7500 − \$225 = \$7275

With the simple interest note, the borrower has the use of \$7500, but only \$7275 is available with the simple discount note. In each case, the interest charge is the same, \$225, but *more money is available to the borrower with the simple interest note.*

(c) Find the maturity value of each note.

Solution

For the Simple Interest Note	For the Simple Discount Note
$M = P + I$ = \$7500 + \$225 = \$7725	Maturity value = Face value = \$7500

Notice the difference in the proceeds or loan amount and also the maturity value in (b) versus (c).

The differences between these two notes can be summarized as follows.

	Simple Interest	Simple Discount
Face value	\$7500	\$7500
Interest	\$225	\$225
Amount available to borrower	\$7500	\$7275
Maturity value	\$7725	\$7500

Clearly, a 12% simple interest note is better for the borrower than a 12% simple discount note.

Objective 5 *Find Effective Interest Rates.* Because of the possible confusion resulting from the different ways of calculating interest charges, the **Federal Truth in Lending Act** was passed in 1969. This act requires that all interest rates be given as comparable percents. While this law is discussed in more detail in Chapter 13, the next example shows how to find the simple interest rate corresponding to the given discount rate of Example 3.

Example 4 Finding Rate of Interest for Discount Notes

Find the rate of simple interest for the simple discount note in Example 3.

Solution The discount rate of 12% given in Example 3 is not the rate of simple interest, since the 12% applies to the maturity value of \$7500, *not* to the proceeds of \$7275. Since the borrower received only \$7275, the interest rate must be found from this amount. Do this with the formula for simple interest, $I = PRT$. Here $I = \$225$ (the discount), $P = \$7275$ (the proceeds), $T = \frac{90}{360}$ year, and R must be found.

Start with $I = PRT$, and divide both sides by PT.

$$R = \frac{I}{PT}$$

Then substitute the given numbers.

$$\begin{aligned} R &= \frac{\$225}{\$7275 \times \frac{90}{360}} \\ &= \frac{\$225}{\$1818.75} \\ &= 0.1237 \text{ or } 12.37\% \quad \text{Rounded} \end{aligned}$$

The rate of interest is 12.37% rounded to the nearest hundredth of a percent. This rate is called the **effective rate of interest** or the **true rate of interest**. By federal regulations, a person borrowing \$7500 for 90 days at a discount rate of 12% would have to be told that the **annual percentage rate** on the loan is 12.5% (instead of 12.37%, since the regulations allow rounding up to the nearest quarter of a percent).

Objective 6 *Understand T-Bills.* One common use of discount interest is that involved in the purchase of **U.S. Treasury bills** (often called just **T-bills**). The federal government uses T-bills to borrow money. T-bills are currently available with 13-week, 26-week, or 52-week maturity. An investor buys a T-bill at a price equal to the proceeds, which is the amount after the discount is subtracted. The investor then receives the full face value of the T-bill when it reaches maturity. Since T-bills are loans to the federal government, they are considered one of the safest of all possible investments.

Example 5 Finding Facts about T-Bills

An Italian bank is worried about devaluation of the national currency, so it purchases \$1 million in U.S. T-bills in order to place cash in a safe place for a short period of time. The T-bills are at a 4% discount rate for 26 weeks. Find (a) the total purchase price, (b) the total maturity value, (c) the interest earned, and (d) the effective rate of interest.

Solution

$M = \$1{,}000{,}000$; $D = 0.04$; $T = \frac{26}{52}$ (The denominator is 52 when time is in weeks.)

(a) Bank discount = Face value × Discount rate × Time

$$= \$1{,}000{,}000 \times 0.04 \times \tfrac{26}{52} = \$20{,}000$$

Purchase price = Face value − Bank discount

$$= \$1{,}000{,}000 - \$20{,}000 = \$980{,}000$$

(b) Maturity value $=$ Face value

$= \$1{,}000{,}000$

(c) Interest $=$ Bank discount

$= \$20{,}000$

(d) $$\text{Effective rate} = \frac{\text{Interest earned}}{\text{Purchase price (proceeds)} \times \text{Time}}$$

$$= \frac{\$20{,}000}{\$980{,}000 \times \frac{26}{52}} = 0.04081 = 4.08\% \quad \text{Rounded}$$

Objective 7 ***Find the Face Value That Produces the Desired Proceeds.*** Normally a borrower wants to borrow a certain amount of money. The next example shows how to find the face value of a simple discount note so that the proceeds will be the amount needed by the borrower.

Example 6 Find the Face Value That Produces the Desired Proceeds

Mike Collins needs \$4000 to repair his roof. Find the value of a note that will provide the \$4000 in proceeds if he plans to repay the note in 180 days and the bank charges a 12% discount rate.

Solution Start with the formula $P = M(1 - DT)$. Since M is not known, find a formula for M by dividing both sides by $1 - DT$.

$$M = \frac{P}{1 - DT}$$

Replace P with \$4000, D with 0.12, and T with $\frac{180}{360}$.

$$M = \frac{\$4000}{(1 - 0.12 \times \frac{180}{360})}$$

$$= \frac{\$4000}{1 - 0.06} = \$4255.32 \quad \text{Rounded.}$$

Collins must sign a note with a face value of \$4255.32 to get the \$4000 that he needs.

Scientific *Calculator Approach*

The calculator solution to Example 6 is

4000 [÷] [(] 1 [−] .12 [×] 180 [÷] 360 [)] [=] 4255.32.

Objective 8 ***Find Discount Rate, Face Value, or Time.*** Just as the formula for simple interest, $I = PRT$, can also be solved for P, R, or T, the formula for simple discount, $B = MDT$, can be solved for M, D, or T. Do this as shown in the next example.

Example 7 Finding Discount

Sheila Watts borrowed proceeds of \$4480 to help with college expenses. Her 240-day note had a maturity value of \$4800. Find the discount rate.

Solution To find the discount rate, start with the formula $B = MDT$ and divide both sides by MT.

$$D = \frac{B}{MT}$$

The discount is

$$B = \$4800 - \$4480 = \$320$$

also $M = \$4800$ and $T = \frac{240}{360}$. Now find D.

$$D = \frac{\$320}{\$4800 \times \frac{240}{360}} = 0.10$$

The discount rate is 10%.

12.2 Exercises

Find the discount and the proceeds. Round money to the nearest cent.

	Maturity Value	Discount Rate	Time in Days
1.	$4,600	9%	90
2.	$6,800	10%	180
3.	$6,200	$14\frac{1}{4}\%$	180
4.	$15,500	12%	200
5.	$8,400	$9\frac{1}{2}\%$	30
6.	$9,800	$8\frac{3}{4}\%$	50

Find the due date and the proceeds for the following. Round to the nearest cent.

	Maturity Value	Discount Rate	Date Made	Time in Days
7.	$8,400	11%	January 3	100
8.	$5,400	12%	March 25	90
9.	$12,000	$9\frac{1}{2}\%$	August 21	180
10.	$8,500	14%	November 12	240

Complete this table.

	Maturity Value	Discount Rate	Date Made	Due Date	Time in Days	Discount	Proceeds
11.	$24,000	$9\frac{1}{2}\%$	2/4	______	180	______	______
12.	$8,275	7%	______	9/10	120	______	______
13.	$14,400	______	______	1/4	150	$660	______
14.	$8,200	______	2/9	5/10	______	$205	______

	Maturity Value	Discount Rate	Date Made	Due Date	Time in Days	Discount	Proceeds
15.	______	______	11/12	______	90	$108	$7,092
16.	______	______	12/2	8/11	______	$1372	$18,228

Solve each of the following application problems. Round rate to the nearest tenth of a percent and money to the nearest cent.

17. BOAT PURCHASE An Exxon employee bought a boat with funds borrowed from First National Bank. He intends to repay the note with a Christmas bonus of $6000 to be paid to him in 120 days. Given an 11% discount, find the discount and the proceeds.

18. INCOME TAX PAYMENT LaTonya Barker must make a quarterly tax payment to the Internal Revenue Service. She decides to borrow money from her company (Bicycles Unlimited) and repay the loan in 100 days using an $8500 bonus she will receive at that time. Given an 8.5% discount rate, find the discount and the proceeds.

19. BUSINESS TRIP TO ALASKA Walter Bates needed money to go on a business trip to Alaska and he signed a note for $6000. At a $10\frac{1}{2}\%$ discount rate, the discount on the note was $210. Find the length of the loan in days.

20. CATERING SUPPLIES Margaret Jones signs a $4500 note at 12% simple discount to buy supplies for her catering company. Find the length of the loan in days if the discount is $135.

21. DISCOUNT NOTE A 60-day note for $14,000 was signed. Given a discount of $291.67, find the discount rate.

22. CONSTRUCTION IN MEXICO A construction firm in Mexico signed a 200-day note with a U.S. bank for $850,000. Find the discount rate given proceeds of $793,333.33.

23. LAPTOP COMPUTER Jane Peters frequently travels to Hong Kong and needs to purchase a special laptop computer costing $3200. Given a discount rate of 10% and a 140-day term, find the maturity value of the loan.

24. LAST YEAR AT COLLEGE Mary Gibb estimates that she needs $10,000 to finish her last year at college. She borrows the funds from her uncle for 12 months at a favorable 4% discount rate. Find the face value of the loan.

25. PERSONAL NOTE Marge Prullage signs a $4200 note at the bank. The bank charges an 11% discount rate. Find the proceeds if the note is for 10 months. Find the effective interest rate charged by the bank.

26. PERSONAL DISCOUNT NOTE Ed Foust goes to the bank and signs a note for $8400. The bank charges a 7% discount rate. Find the proceeds and the effective rate charged by the bank if the note is for 8 months.

27. CONSTRUCTION IN NEW ZEALAND An American contractor working on a project in New Zealand requires proceeds of $165,000 for 30 days to pay for supplies. A bank lends her the funds at a 15% discount rate. Find the face value of the note and the effective interest rate.

28. INTERNATIONAL FINANCE A Japanese electric company requires proceeds of $720,000 (local currency) and borrows from a bank in Thailand at 12% discount for 45 days. Find the face value of the note and the effective interest rate.

The following exercises apply to U.S. Treasury bills. (Assume 52 weeks per year for each exercise, and round to the nearest hundredth of a percent.)

29. PURCHASE OF T-BILLS A large Japanese investment firm purchases $25,000,000 in U.S. T-bills at a 6% discount for 13 weeks. Find (a) the purchase price of the T-bills, (b) the maturity value of the T-bills, (c) the interest earned, and (d) the effective rate.

30. T-BILLS Nina Horn buys a $50,000 T-bill at a 5.8% discount for 26 weeks. Find (a) the purchase price of the T-bill, (b) the maturity value, (c) the interest earned, and (d) the effective rate of interest.

31. Explain the difference between simple interest notes and simple discount notes.

32. Compare the formulas for simple interest rate and simple discount rate. Define all variables for both and explain the difference between simple interest rate and simple discount rate. (See Objective 4.)

33. As a borrower, would you prefer a simple interest note with a rate of 11% or a simple discount note at a rate of 11%? Explain using an example. (See Objective 4)

34. Why do you think banks and large corporations sometimes own T-bills?

12.3 Comparing Simple Interest and Simple Discount

Objectives

1. *Compare the differences between simple interest and simple discount notes.*
2. *Convert between simple interest and simple discount rates.*

If you work in the finance area of a firm, or if you ever own your own business, you will see *both* simple interest notes (introduced in Section 12.1) and simple discount notes (introduced in Section 12.2). We will compare these two types of notes in this section. Then we will discuss ways of comparing simple interest rates and simple discount rates.

First, let us list the key similarities between these two types of notes.

1. The borrower receives a lump sum of money at the beginning of each type of note.
2. Both types of notes are repaid with a single payment at the end of a stated period of time.
3. This length of time is generally one year or less.

Objective 1 ***Compare the Differences Between Simple Interest and Simple Discount Notes.*** Table 12.1 compares these two types of notes.

Objective 2 ***Convert Between Simple Interest and Simple Discount Rates.*** As shown earlier, a 12% simple interest rate is *not* the same as a 12% simple discount rate. The rest of this section shows formulas for conversion back and forth between simple interest rates and simple discount rates.

To find these formulas, start with the key formulas for simple interest and for simple discount.

Simple Interest		**Simple Discount**
$M = P(1 + RT)$	and	$P = M(1 - DT)$
↳ Principal		↳ Proceeds

Ταβλε +—⊳ A COMPARISON OF SIMPLE INTEREST AND SIMPLE DISCOUNT NOTES

Variables Used for Simple Interest	Variables Used for Simple Discount
I = Interest	B = Discount
P = Principal (face value)	P = Proceeds (amount received)
R = Rate of interest	D = Discount rate
T = Time in years, or fraction of a year	T = Time in years, or fraction of a year
M = Maturity value	M = Maturity value (face value)

Formulas

Elements	Simple Interest	Simple Discount
Face value	Stated on note, or $P = \frac{M}{1 + RT}$	Same as maturity value, or $M = \frac{P}{1 - DT}$
Interest charge	$I = PRT$	$B = MDT$
Maturity value	$M = P + I$ or $M = P(1 + RT)$	Same as face value, or $M = \frac{P}{1 - DT}$
Amount received by borrower	Face value or principal	Proceeds $P = M - B$ or $P = M(1 - DT)$
Identifying phrases	Interest at a certain rate Maturity value greater than face value Simple interest	Discounted at a certain rate Proceeds Maturity value equal to face value Simple discount rate
True annual interest rate	Same as stated rate, R	Greater than stated rate, D

Solve each of these formulas for P, the principal and proceeds, respectively. The principal or proceeds is the lump sum received by the borrower in each case.

Simple Interest	**Simple Discount**
$P = \frac{M}{1 + RT}$	$P = M(1 - DT)$

Since each right-hand side is equal to P, the two right-hand sides must be equal to each other, or

$$\frac{M}{1 + RT} = M(1 - DT)$$

Divide both sides by M to get

$$\frac{1}{1 + RT} = 1 - DT$$

By going through several more algebraic steps, this equation can be solved first for R and then for D, giving the results in the box.

The simple interest rate R and the simple discount rate D are calculated by the formulas

$$R = \frac{D}{1 - DT} \quad \text{and} \quad D = \frac{R}{1 + RT}$$

where T is time in years.

NOTE The simple interest rate corresponding to a simple discount rate is also called the *effective rate of interest*. As these formulas show, the dollar amounts play no part in converting between rates—only rate and time matter. ■

Example 1 Converting Interest and Discount Rates

John Patterson's lawn care business has two outstanding notes. (a) The first note is a 180-day simple discount note with a face value of \$16,800 and a discount rate of 10%. Convert this rate to a simple interest (effective interest) rate. (b) The second note is a 140-day simple interest note with a face value of \$24,600 and an interest rate of 11%. Find the corresponding simple discount rate.

Solution

(a) Find R with the following formula.

$$R = \frac{D}{1 - DT}$$

Again, notice that no dollar amounts, not even face value, are needed.

Here $D = 0.10$ and $T = \frac{180}{360}$.

$$R = \frac{0.10}{1 - (0.10 \times \frac{180}{360})} = \frac{0.1}{1 - 0.05} = \frac{0.1}{0.95} = 0.105263$$

Rounding to the nearest hundredth of a percent, the corresponding simple interest rate is 10.53%.

The answer can be checked by first finding the interest on the simple discount loan.

$$\begin{aligned} B &= MDT \\ &= \$16{,}800 \times 0.10 \times \frac{180}{360} \\ &= \$840 \end{aligned}$$

Then find the loan proceeds.

$$\begin{aligned} P &= M - B \\ &= \$16{,}800 - \$840 \\ &= \$15{,}960 \end{aligned}$$

Finally, use the formula for a simple interest calculation to find the simple interest rate:

$$\begin{aligned} R &= \frac{I}{PT} \\ &= \frac{\$840}{\$15{,}960 \times \frac{180}{360}} \\ &= 10.53\% \quad \text{Rounded} \end{aligned}$$

(b) Find the simple discount rate that corresponds to a simple interest rate of 11% for 140 days. Use the following formula.

$$D = \frac{R}{1 + RT}$$

Note that the face value is not needed.

Replace R with 0.11 and T with $\frac{140}{360}$. Then

$$D = \frac{0.11}{1 + (0.11 \times \frac{140}{360})}$$

$$= \frac{0.11}{1 + 0.042777778}$$

$$= 0.1055 \quad \text{Rounded}$$

or 10.55%, rounded to the nearest hundredth of a percent.

NOTE In both parts (a) and (b) in Example 1 above, the simple interest rate is larger than the equivalent simple discount rate. This is *always the case* when comparing the discount rate and interest rate of the same note. ▬

Scientific *Calculator Approach*

The calculator approach to Example 1 uses parentheses to group each denominator.

(a) .10 [÷] [(] 1 [−] .10 [×] 180 [÷] 360 [)] [=] 0.1053 Rounded.

(b) .11 [÷] [(] 1 [+] .11 [×] 140 [÷] 360 [)] [=] 0.1055 Rounded.

12.3 Exercises

Find the simple interest rate that corresponds to the given discount rate for the given time. Round to the nearest hundredth of a percent.

1. 9%, 120 days

2. 12%, 180 days

3. $14\frac{1}{4}$%, 200 days

4. $12\frac{1}{2}$%, 100 days

Find the simple discount rate that corresponds to the given simple interest rate for the given time. Round to the nearest hundredth of a percent.

5. 12%, 220 days

6. 8%, 240 days

7. 10%, 100 days

8. 9%, 180 days

9. Is the rate associated with simple discount notes the true annual interest rate? If not, explain why not and show how to calculate the true rate.

10. If you were offered a simple interest note at a rate of 12% and a simple discount note with a discount of 12%, which would you prefer? Why?

Solve each of the following application problems.

11. **OPERATION ON INVALID FATHER** Sherri Johnson needs to borrow money for an operation for her invalid father. She can borrow at a 13% simple interest rate or at a 12.8% simple discount rate. If she needs the money for 90 days, which loan should she take?

12. **WEDDING** Reann Kiang needs to borrow \$8600 to pay for her only daughter's wedding. One bank charges 12% simple interest and a second bank charges 11% simple discount. If she needs the money for 120 days, which loan should she take?

13. **CARIBBEAN CRUISE** Joan Boston wishes to borrow exactly \$4500 for 180 days so that she can go on a Caribbean cruise. How much less interest is paid with a 12% simple interest note compared to a 12% simple discount note?

14. **DRILLING FOR OIL** Alamo Energy wishes to borrow exactly \$900,000 for 9 months to drill three oil and gas wells. How much less interest is paid with a 14% simple interest note compared to a 14% simple discount note?

15. Show the algebraic steps needed to solve the following equation for D.

$$\frac{1}{1 + RT} = 1 - DT$$

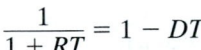

16. Which two variables are used when converting from a simple discount rate to an equivalent simple interest rate? Assume a simple discount rate of 12% and explore the effect of time on the simple interest rate. Do this by substituting three different times, each less than one year, into the formula. (See Objective 2.)

12.4 Discounting a Note

Objectives

1 *Find the discount and the proceeds of a note.*

2 *Find the proceeds of a rediscounted note.*

Businesses often accept notes, either simple interest notes or simple discount notes, in place of immediate payment for goods or services. For example, a manufacturer of ski equipment may deliver goods to ski shops in October and may agree not to collect from the shops until April. To secure its payment, the manufacturer may request promissory notes from the ski shops receiving the goods.

As a result, the manufacturer may have a considerable amount of cash tied up in promissory notes that will not be paid until April. To get cash earlier, *the manufacturer can sell the promissory notes* to a bank. The bank will give the manufacturer the maturity value of the notes, less a fee charged by the bank for this service. This fee is called the **bank discount**, or just **discount**. The process of receiving cash for a note, or selling a note, is called **discounting a note**.

The amount of cash actually received by the manufacturer is called the **proceeds**. The bank then collects the maturity value from the maker of the note when it becomes due. Normally, such notes are sold with **recourse**. This means that if the maker of the note does not pay for some reason, the bank collects from the seller of the note. This protects the bank against loss—you may have noticed that banks do not like to lose money!

Objective 1 ***Find the Discount and the Proceeds of a Note.*** Use the following procedure to discount a note.

1. Find the maturity value of the original note (if necessary).
2. Find the discount period.
3. Find the discount using the formula $B = MDT$.
4. The proceeds are found by $P = M - B$.

This method is shown in the next examples.

Example 1 Finding Proceeds

Blues Recording Studio holds a 200-day simple interest note from a rock music group that agreed to pay them to record an album and produce 1000 copies on compact disks. The note is dated March 24 and has a face value of $4800 with simple interest of 12%. Blues Recording wishes to convert the note to cash on August 15. Given a discount rate of 12.5%, find the proceeds to the Recording Studio.

Solution Go through the four steps of discounting a note.

Discounting a Note

Step 1. First find the interest on the simple interest note if held until maturity.

$$\text{Interest} = \$4800 \times 0.12 \times \frac{200}{360} = \$320$$

The **maturity value** is: $4800 + $320 = $5120.

Step 2. Find the **discount period**, which is the number of days remaining from August 15 until the note is due. The discount period is often found by using a diagram as shown in Figure 12.4.

In this example, the date of the note is March 24, the discount date is August 15, and the due date is October 10. August 15 to October 10 is 56 days, found as follows.

16	Days left in August
30	Days in September
+ 10	Days until note is due in October
56	Days of discount period

Verify 56 days using the chart on page 389.

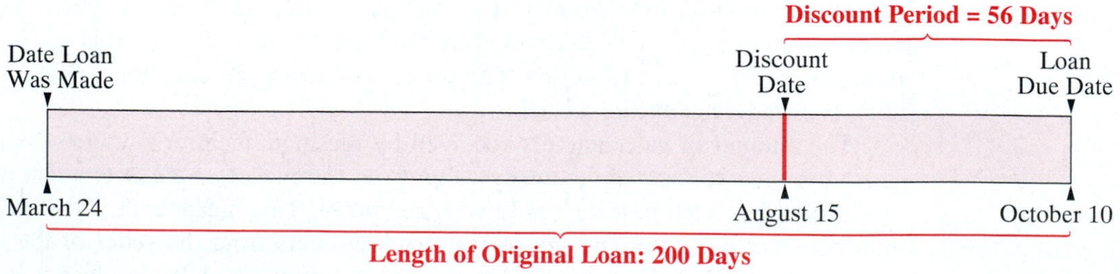

Figure 12.4

Step 3. Find the **bank discount** using the formula $B = MDT$, where $M = \$5120$ and T is $\frac{56}{360}$.

$$B = \$5120 \times 0.125 \times \frac{56}{360} = \$99.56 \quad \text{Rounded}$$

The bank discount is \$99.56.

Step 4. Find the **proceeds** using the formula $P = M - B$.

$$P = \$5120 - \$99.56 = \$5020.44$$

The bank purchases the note on August 15 for \$5020.44 in cash paid to Blues Recording Studio. Then, on the maturity date of October 10, the bank will collect \$5120 from the maker of the note. In summary:

Date	Transaction
March 24	The rock group signs 200-day simple interest note for \$4800.
August 15	Blues Recording Studio sells note to bank for \$5020.44.
October 10	The bank receives \$5120 from payer (rock group).

The recording studio in Example 1 earned \$220.44 (\$5020.44 − \$4800) in interest in 144 (200 − 56) days. Their effective interest rate is found as follows.

$$R = \frac{I}{PT}$$

$$= \frac{\$220.44}{\$4800 \times \frac{144}{360}}$$

$$= 11.5\% \quad \text{Rounded}$$

Their effective interest rate was a little less than the 12% rate stated on the note.

NOTE In discounting a note, the business receives less money than if it held the note to maturity, but it will receive the money sooner. ■

Example 2 Finding Proceeds and Discount

On February 27, Andrews Lincoln-Mercury receives a 150-day simple interest note with a face value of \$3500 at 8% interest per year. On March 27, the firm discounts the note at the bank. Find the proceeds if the discount rate is 12%.

Solution Again, go through the four steps in discounting a note.

Step 1 Find the interest and maturity value of the original loan. Find the interest with the formula $I = PRT$.

$$\text{Interest} = \$3500 \times 0.08 \times \frac{150}{360} = \$116.67$$

The interest on the note is \$116.67. The maturity value is found with the formula $M = P + I$.

$$\text{Maturity value} = \$3500 + \$116.67 = \$3616.67$$

Step 2 Find the discount period using the diagram in Figure 12.5.

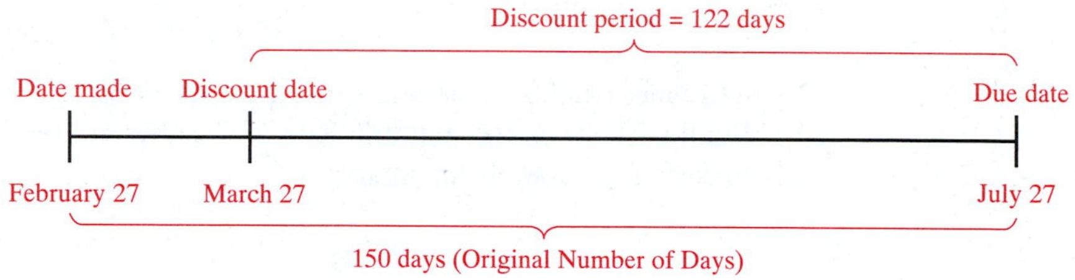

Figure 12.5

Problem-Solving Hint Remember, the discount period is calculated from the date of sale to the *due date* of the loan, not from the date the loan was made.

The discount period is 122 days.

Step 3 Find the bank discount, using the discount rate of 12%. Use the formula $B = MDT$.

$$\text{Discount} = \$3616.67 \times 0.12 \times \frac{122}{360} = \$147.08$$

The discount is \$147.08

Step 4 Find the proceeds.

$$\begin{aligned} P &= M - B \\ &= \$3616.67 - \$147.08 = \$3469.59 \end{aligned}$$

Andrews receives \$3469.59 from the bank on March 27.

Objective 2 ***Find the Proceeds of a Rediscounted Note.*** In the next example, a company borrows money from a bank by signing a simple discount note. The bank making the loan then sells the note to a finance company. In this case, with the same note being discounted twice, the note is said to have been **rediscounted**. Notice in the example that the rate charged by the finance company to the bank is lower than the rate charged to the public. This is very common in transactions between large financial institutions, with the lower rate a sort of "wholesale" rate.

Even though the note in Example 3 is rediscounted, the steps used are basically the same as those used in earlier examples. The only difference is that since the original note was a simple discount note instead of a simple interest note, it is not necessary to solve for maturity value. The face value *is* the maturity value.

Example 3 Finding Proceeds of a Rediscounted Note

Barbara Hanks, chief financial officer of Cole Springs Mattress, signed a 180-day note at the bank, with a face value of \$140,000. The bank discounted the note at 12%. Then, 54 days after the note was signed, the bank rediscounted the note to Century Finance which charged a 10% discount. Find the proceeds Century Finance must pay to the bank.

Solution Go through the steps given for discounting a note.

Step 1 The maturity value of the note is \$140,000.

NOTE The fact that the bank charged a 12% discount rate plays no part in the problem since only the maturity value of the note is needed.

Step 2 The discount period is $180 - 54 = 126$ days.

Step 3 The discount (at Century Finance) is

$$\text{Discount} = \$140{,}000 \times 0.10 \times \frac{126}{360} = \$4900$$

Step 4 The proceeds are

$$\$140{,}000 - \$4900 = \$135{,}100$$

This example can be summarized as follows.

1. Cole Springs Mattress signs a note with a maturity value of \$140,000. The original proceeds to Cole Springs Mattress is found as follows.

$$\begin{aligned} P &= M - MDT \\ &= \$140{,}000 - \$140{,}000 \times 0.12 \times \frac{180}{360} \\ &= \$131{,}600 \end{aligned}$$

Then, 54 days after the note is signed, the bank rediscounts the note to Century Finance, receiving proceeds of \$135,100. Finally, 180 days after signing the note (and 126 days after the note was discounted), Cole Springs Mattress pays \$140,000 to Century Finance.

2. The bank lends \$131,600 for 54 days and makes $\$135{,}100 - \$131{,}600 = \$3500$ in interest. The effective interest rate earned by the bank is found as follows.

$$\begin{aligned} R &= \frac{I}{PT} \\ &= \frac{\$3500}{\$131{,}600 \times \frac{54}{360}} \\ &= 17.7\% \quad \text{Rounded} \end{aligned}$$

Not only notes are discounted. It is very common for a business to sell part of its accounts receivable (money owed to the business—see Section 17.3) to a financial institution. This process is called **factoring**, and the people who buy the accounts receivable are called **factors**. The calculations involved with factoring are the same as those discussed in this section.

12.4 Exercises

Find the discount period for each of the following.

	Loan Made	Length of Loan	Date of Discount
1.	June 28	120 days	August 4
2.	March 13	180 days	June 1
3.	August 4	220 days	January 12
4.	November 5	60 days	December 18

Find the proceeds when each of the following simple discount notes are discounted. Round to the nearest cent.

	Maturity Value	Discount Period	Discount Rate
5.	\$4,800	90 days	$8\frac{1}{2}\%$
6.	\$8,000	200 days	9%
7.	\$15,000	180 days	12%
8.	\$18,200	240 days	11%

Each of the following simple interest notes was discounted at 12%. Find the discount period, the discount, and the proceeds for each. Round money to the nearest cent.

	Loan Made	Face Value	Length of Loan	Simple Interest Rate	Discounted
9.	January 9	\$3,500	120 days	9%	March 9
10.	April 23	\$4,000	150 days	7%	June 11
11.	May 5	\$6,800	130 days	$10\frac{1}{2}\%$	July 6
12.	May 29	\$4,500	80 days	8%	July 8
13.	September 18	\$10,000	220 days	10%	February 4
14.	October 11	\$17,500	100 days	11%	January 2

Solve each of the following application problems. Round money to the nearest cent.

15. PLUMBING COMPANY On May 10, Ace Plumbing accepted a 190-day, 10% interest note for \$12,000 from a contractor in lieu of a cash payment. The note is discounted on July 26 at a 12% rate. Find (a) the discount period, (b) the discount, and (c) the proceeds to Ace Plumbing.

16. ELECTRONIC EQUIPMENT Home Health signed a \$90,000 note at $11\frac{1}{2}\%$ interest for 180 days for electronic equipment, on October 1. On February 18 the note was sold to another firm at a discount rate of $12\frac{1}{2}\%$. Find (a) the discount period, (b) the discount, and (c) the proceeds.

17. **NOTE TO FOUNDRY** On April 18, Moline Foundry accepts a $4500 note in settlement of a bill for goods purchased by a customer. The note is for 150 days at 10% interest. If Moline sells the note at a 14% discount rate 30 days after receipt, find (a) the bank discount and (b) the proceeds.

18. **DISCOUNTING A NOTE** On August 7, Lane Plumbing accepted a 120-day note at 11% interest for $88,000. They sell the note to a bank on November 1 at a 12% discount rate. Find (a) the bank discount and (b) the proceeds.

19. **REDISCOUNTING A NOTE** Citizen's First Bank accepted a $24,000 simple discount note from a customer at a discount rate of $10\frac{1}{2}\%$ for 150 days. They rediscounted the note to Northside Bank at 11% exactly 90 days before its maturity date. Find (a) the proceeds to the customer, (b) the proceeds to Citizen's First Bank, and (c) the actual amount of interest earned by Citizen's First Bank on the note.

20. **SIMPLE DISCOUNT NOTE** Farmer's Bank accepted a $17,000 simple discount note from a customer. The note was for 120 days at an 11% discount rate. The bank then rediscounted the note at a second bank, at 8%, 15 days before the maturity date of the note. Find each of the following: (a) the proceeds to the customer of Farmer's Bank, (b) the proceeds to Farmer's Bank when it rediscounted the note, and (c) the actual amount of interest earned by Farmer's Bank on the note.

21. **DISCOUNT NOTE** State Bank accepts a simple discount note at 12% for 280 days on December 3, with a maturity value of $36,500. They rediscount the note at 10% on May 4 of the following year. Find (a) the proceeds to the maker of the note, (b) the proceeds to State Bank, and (c) the actual interest earned by State Bank.

22. **DISCOUNT NOTE** On December 9, the Sunrise Bank accepted a $27,000 simple discount note from a customer. The note is at a discount rate of 16% and is due on March 9. On February 12, the bank rediscounted the note at 10%. Find each of the following: (a) the proceeds to the customer of Sunrise Bank, (b) the proceeds to Sunrise Bank when it rediscounted the note, and (c) the actual amount of interest earned by Sunrise Bank on the note.

23. **CONSTRUCTION NOTE** Gilbert Construction Company signed a $78,000 simple interest note at 12% for 150 days with Union State Bank on November 20. On January 23, Union State Bank went bankrupt and sold all of their notes to National Bank effective 14 days later. Find the (a) maturity value of the note and (b) proceeds to Union State Bank given a discount rate of 13.5%.

24. **INVESTING IN A T-BILL** Tina Klein bought a $10,000, 7.5%, 52-week T-bill on June 29 and sold it 26 weeks later at a discount rate of 8%. Find Tina's (a) purchase price for the T-bill, (b) the discount at time of sale, (c) the proceeds to Tina, and (d) her effective interest rate.

25. Explain the purpose of discounting a note. What does the discounting procedure cost the original holder of the note?

26. Explain why banks or other financial institutions may need to rediscount notes. (See Objective 2.)

Chapter 12 Quick Review

Chapter Terms

Review the following terms to test your understanding of the chapter. For any terms you do not know, refer to the page number found next to that term.

annual percentage rate [p. 433]
bank discount [p. 428]
collateral [p. 422]
discount [p. 428]
discount period [p. 442]
discounting a note [p. 441]
effective rate of interest [p. 433]
face value [p. 421]
factors [p. 445]
factoring [p. 445]
Federal Truth in Lending Act [p. 433]
interest-in-advance notes [p. 428]
maker [p. 421]
maturity value [p. 421]
payee [p. 421]
payer [p. 421]
proceeds [p. 428]
promissory note [p. 420]
recourse [p. 441]
rediscounted [p. 444]
simple discount note [p. 428]
simple interest note [p. 421]
term of the note [p. 421]
true rate of interest [p. 433]
U.S. Treasury bills (T-bills) [p. 433]

Concepts / Example

12.1 Maturity value of a loan

Use $I = PRT$ to find interest; then use $M = P + I$ to find maturity value.

Example: A note with a face value of $12,000 is due in 60 days. The rate is 8%. Find the maturity value.

$$I = PRT$$
$$= \$12{,}000 \times 0.08 \times \frac{60}{360} = \$160$$
$$M = P + I$$
$$= \$12{,}000 + \$160 = \$12{,}160$$

12.1 Finding the due date, interest, and maturity value of a note with the term in months

1. Add the number of months to date of note to find due date.
2. Use $I = PRT$ to find interest.
3. Use $M = P + I$ to find maturity value.

Example: Find due date, interest, and maturity value of a $4800 loan made on May 12 for 9 months at 9%. Due date is 9 months from May 12, which is February 12 of the following year.

$$I = PRT$$
$$= \$4800 \times 0.09 \times \frac{9}{12} = \$324$$
$$M = P + I$$
$$= \$4800 + \$324 = \$5124$$

12.1 Determining the face value of a note, given rate, time, and maturity value

Use the formula

$$P = \frac{M}{1 + RT}$$

Example: A 90-day note has an interest rate of 9% and a maturity value of $1431.50. Find the face value.

$$P = \frac{M}{1 + RT}$$
$$= \frac{\$1431.50}{1 + (0.09 \times \frac{90}{360})} = \$1400$$

12.1 Finding the time of a note in days given maturity value, face value, and interest rate

Use $T = \frac{M - P}{PR} \times 360$ to find time in days. Alternatively, since $I = M - P$, use $T = \frac{I}{PR} \times 360$.

Example: A note has a face value of $4800, interest rate of 6%, and a maturity value of $5000. Find the time of the note in days.

$$T = \frac{\$5000 - \$4800}{\$4800 \times 0.06} \times 360$$
$$= 250 \text{ days}$$

12.1 Determining the rate of a note given the face value, time, and maturity value

Use $R = \frac{M - P}{PT}$ to find the rate. Alternatively, since $I = M - P$, use $R = \frac{I}{PT}$.

A 240-day note has a face value of $9000 and a maturity value of $9300. Find the interest rate.

$$R = \frac{\$9300 - \$9000}{\$9000 \times \frac{240}{360}} = 0.05$$

The rate is 5%.

12.2 Finding the discount and proceeds of a simple discount note

1. Calculate bank discount B using the formula $B = MDT$ with D = discount rate.
2. Calculate proceeds using the formula $P = M - B$.

Tom Jones borrows $5000 for 60 days at a discount rate of 9%. Find the bank discount and proceeds.

$$B = MDT$$
$$= \$5000 \times 0.09 \times \frac{60}{360} = \$75$$
$$P = M - B$$
$$= \$5000 - \$75 = \$4925$$

12.2 Determining the effective rate or true rate of interest given face value, time, and discount rate

First, find the discount (B) using the formula $B = MDT$. Find the proceeds using the formula $P = M - B$. Calculate the true rate of interest from either

$$R = \frac{M - P}{PT} \qquad \text{or} \qquad R = \frac{I}{PT}$$

A note has a face value of $9000, a time of 120 days, and a discount rate of 12%. Find the true rate of interest.

$$B = MDT$$
$$= \$9000 \times 0.12 \times \frac{120}{360} = \$360$$
$$P = M - B$$
$$= \$9000 - \$360 = \$8640$$
$$R = \frac{I}{PT}$$
$$= \frac{\$360}{\$8640 \times \frac{120}{360}} = 0.125$$

The effective rate is 12.5%.

12.2 Finding the face value of a simple discount note

Use the formula $M = \frac{P}{1 - DT}$ to find the face value M.

Find the face value of a note that will provide $38,000 proceeds if the note is repaid in 220 days and the bank charges an 8% discount rate.

$$M = \frac{P}{1 - DT}$$
$$= \frac{\$38{,}000}{1 - (0.08 \times \frac{220}{360})}$$
$$= \$39{,}953.27$$

12.2 Determining the discount rate of a note given face value and proceeds

Find the discount B from the formula $B = M - P$.

Find the rate from the formula $D = \frac{B}{MT}$

A 90-day note has a face value of $15,000 and proceeds of $14,568.75. Find the discount rate.

$$B = M - P$$
$$= \$15{,}000 - \$14{,}568.75$$
$$= \$431.25$$
$$D = \frac{B}{MT}$$
$$= \frac{\$431.25}{\$15{,}000 \times \frac{90}{360}} = 0.115$$

The discount rate is 11.5%.

12.3 Finding the simple interest rate that equates to a given discount rate.

Find the simple interest rate using the formula

$$R = \frac{D}{1 - DT}$$

A 180-day simple discount note is at 10%. Find the equivalent simple interest rate.

$$R = \frac{D}{1 - DT}$$

$$= \frac{0.10}{1 - 0.10 \times \frac{180}{360}} = 10.53\% \quad \text{Rounded}$$

12.3 Find the simple discount rate that equates to a given simple interest rate.

Find the discount rate using the formula

$$D = \frac{R}{1 + RT}$$

A 240-day simple interest note is at $10\frac{3}{4}\%$. Find the equivalent simple discount rate.

$$D = \frac{R}{1 + RT}$$

$$= \frac{0.1075}{1 + 0.1075 \times \frac{240}{360}} = 10.03\% \quad \text{Rounded}$$

12.4 Finding the proceeds, to an individual or firm, after discounting a note

1. Find I, the simple interest on the note, and add it to the face value to find the maturity value, M.
2. Find the discount period.
3. Find the bank discount using $B = MDT$.
4. Find the proceeds using $P = M - B$.

Moe's Ice Cream converts a 9%, 150-day simple interest note dated March 1 with a face value of $15,000 to cash on June 1. Assume a discount rate of 11% and find the proceeds.

1. $I = \$15{,}000 \times 0.09 \times \frac{150}{360} = \$562.50$

$M = \$15{,}000 + \$562.50 = \$15{,}562.50$

2. Find the discount period.

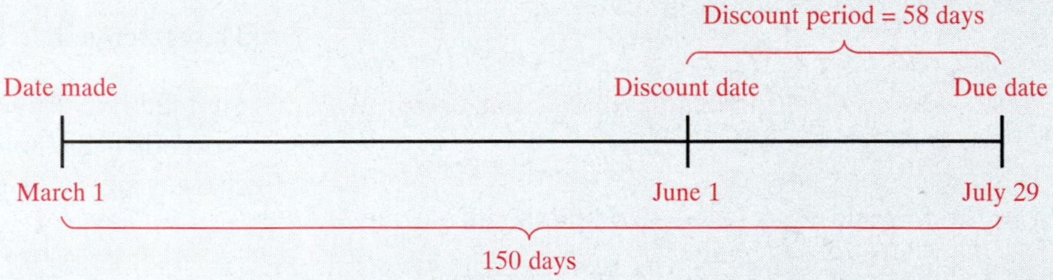

3. $B = \$15{,}562.50 \times 0.11 \times \frac{58}{360} = \$275.80$

4. $P = \$15{,}562.50 - \$275.80$

$= \$15{,}286.70$

The bank will pay $15,286.70 to Moe on June 1 and collect $15,562.50 on July 29 from the maker of the note.

Chapter 12 Review Exercises

The answer section includes answers to all Review Exercises.

Complete the following table for simple interest notes. Round money to the nearest cent. **[12.1]**

	Face Value	Rate	Time	Interest	Maturity Value
1.	$9,800	$8\frac{1}{2}\%$	200 days	________	________
2.	$3,000	________	90 days	$78.75	________
3.	$8,000	12%	________	$640	________
4.	________	10%	180 days	$615	________

In the following, find the due date and the maturity value. **[12.1]**

	Date Made	Face Value	Term of Loan	Interest Rate	Date Due	Maturity Value
5.	January 8	$12,000	120 days	9%	________	________
6.	June 19	$6,000	200 days	$12\frac{1}{2}\%$	________	________

Find the discount and the proceeds for the following discounted notes. **[12.2]**

	Face Value	Discount Rate	Time	Discount	Proceeds
7.	$18,000	12%	80 days	______	______
8.	$26,000	$10\frac{1}{2}\%$	180 days	______	______

Each of the following simple interest notes was discounted 12%. Find the (a) discount period, (b) discount, and (c) proceeds for the following discounted notes. **[12.4]**

	Loan Made	Face Value	Length of Loan	Rate	Date of Discount
9.	September 4	$12,000	150 days	9%	October 25
10.	December 20	$8,500	120 days	$11\frac{1}{8}\%$	February 28

Find the true simple interest rate corresponding to each of the following simple discount rates. Round to the nearest hundredth of a percent. **[12.3]**

	Time	Discount Rate
11.	150 days	11%
12.	170 days	9%

Solve the following application problems. Round rates to the nearest hundredth of a percent, time to the nearest day, and money to the nearest cent.

13. What simple discount rate corresponds to a simple interest rate of 12% on a 180-day note? **[12.3]**

14. Green Acres Pet Store borrowed $38,000 for 120 days at $11\frac{3}{4}\%$. Find the maturity value. **[12.1]**

15. The note in Exercise 14 was discounted 65 days before maturity at $12\frac{1}{2}\%$. Find the discount. **[12.2]**

16. Benito Maintenance signed a $45,000 simple interest note for 200 days and was charged $2250 in interest. Find the rate. **[12.1]**

17. A note for $9800 at 10% simple interest generated $490 interest for the payee. Find the term of the loan. **[12.1]**

18. A borrower signed a note with a face value of $50,000 at a $9\frac{1}{2}\%$ discount rate and received $47,361.11. Find the term of the note. **[12.2]**

19. A businesswoman signed a 50-day discount note with a face value of $35,000 and proceeds of $34,319.44. Find the discount rate. **[12.2]**

20. A 90-day note has a loan amount of $12,000 and a maturity value of $12,330. Find the simple interest and simple discount rate of this note. **[12.3]**

21. Bill Bates needed money for a new automobile and signed a note for $15,000 at a simple discount rate of 12% and a term of 240 days. Find the effective rate of interest. **[12.2]**

22. On December 8, Joan Jones signed a 100-day, 10% discount note at First Bank for $14,000. First Bank sold the note to a finance company on February 12 at a $9\frac{1}{2}\%$ discount. Find the proceeds to First Bank. **[12.4]**

23. West Stables must pay a note given to one of its suppliers. The interest rate on the note is 8%, the face value is $12,300, and the interest is $410. Find the time of the note. **[12.1]**

24. Martinez Cleaners borrows $42,000 for 200 days at $9\frac{1}{2}\%$ simple interest. Find the interest. **[12.1]**

25. A company signed a $79,000, 120-day note for a shipment of goods. The maturity value of the note is $83,187. Find the rate of interest paid on the note. **[12.1]**

26. Christina Barrett borrowed funds at a 11% discount rate for 180 days. Find the maturity value if the discount was $1012. **[12.2]**

27. Tom Watson Insurance accepted a 270-day, $8000 note on May 25. The interest rate on the note is 15%. The note was then discounted at 12% on August 7. Find the proceeds. **[12.4]**

28. A note with a face value of $6570 was discounted at 16%. If the discount was $788.40, find the length of the loan in days. **[12.4]**

29. Sylvia Tiboe signed a note with a discount of $1812 and proceeds of $28,400. Find the maturity value. **[12.2]**

30. A business owner signed a 120-day note with a face value of $26,000 and a discount rate of 10%. Find the bank discount and proceeds. **[12.2]**

31. On March 3, National Bank accepted a $25,000, 180-day note at a discount rate of 9%. They rediscount the note on May 26 at $9\frac{1}{2}\%$. Find the proceeds to National Bank. **[12.4]**

32. Linda Youngman accepted a $16,000, 120-day note from a customer. The note had a discount rate of 9% and was accepted on May 12. The note was then discounted at 11% on July 20. Find the proceeds to Youngman. **[12.4]**

33. Colonial Bank accepted a 200-day note at 10% simple interest with a face value of $83,000. They discount the note at a 10% simple discount rate 90 days before it matures. Find the proceeds to Colonial Bank. **[12.2]**

34. Diane Thompson needs $7580 to buy a computer. She signs a simple discount note at the bank, which charges a 14% discount rate. If the note is for 120 days, find the face value of the note. **[12.2]**

35. The Florist Wholesale Shop accepted a 210-day, $6420 note on December 12. Find the proceeds if the interest rate on the note is 12% and the note was discounted at 15% on January 19. **[12.4]**

36. Find the simple discount rate that corresponds to a simple interest rate of 15% on a 270-day note. **[12.3]**

37. Would you prefer a $15,000 note at 12% simple interest or one at 12% discount interest? Explain. **[12.3]**

38. Explain the differences between a simple interest note and a simple discount note. Which type of note (if either) is preferred? Explain. **[12.3]**

Chapter 12 Summary Exercise
Banking in a Global World: How Do Large Banks Make Money?

Bank of America borrowed $80,000,000 at 9% simple interest for 180 days from a Japanese investment house. At the same time, the bank made the following loans, each for the exact same 180-day period.

- An 11% simple interest note for $38,000,000 to a Canadian firm that extracts oil from Canadian tar sands
- A 12.8% simple discount note to a French contractor building a factory in South Africa for $27,500,000
- A 12%, simple discount note for $14,500,000 to a Louisiana company building minesweepers in New Orleans for the Egyptian government

(a) Find the difference between interest received and interest paid by the bank on these funds.

(b) The bank did not loan out all $80,000,000. Find the amount they actually loaned out.

Net Assets Business on the Internet

Nike

Statistics

- 1999: 22,800 employees
- Footwear revenue of $3.2 billion
- Apparel revenue of $1.4 billion

The founders of Nike, Phil Knight and Bill Bowerman, met at the University of Oregon in 1957 when Phil Knight was an athlete under Coach Bowerman. At the time, Bowerman was handcrafting running shoes for his runners who were breaking records around the country. In 1964, Knight and Bowerman each contributed $500 to form Nike. It wasn't until five years later that Knight was able to quit his day job as an accountant and go to work full time for the company. Since 1978, Nike has used air in the soles of shoes, thereby making shoes lighter. Great athletes that have used Nike shoes include: Steve Prefontaine, John McEnroe, Alberto Salazar, Carl Lewis, and Michael Jordan.

Nike designs shoes for performance but also for longevity since running shoes last anywhere from 250 to 1000 miles. The breakdown of the actual cost of a shoe is approximately 60% materials, 10% labor, 7% tooling, and 23% factory and other associated costs. Nike currently has over 5000 registered, qualified shoe testers in the United States alone covering every sport.

1. Estimate the number of pairs of shoes, to the nearest tenth of a million, that Nike has in inventory if the average cost to the company of a pair of shoes is $38.15 and inventory is $1.2 billion.

2. Find the average amount of debt per employee for Nike, given that they have $1446 billion in debt and 22,800 employees.

3. Assume that the company signs a 90-day simple interest note on November 11 at $9\frac{1}{2}\%$ with proceeds of $36,500,000. Find the interest, maturity value, and due date.

4. Assume that the company signs a simple discount note for 120 days at 10% and a face value of $12,000,000. Find the bank discount and proceeds.

13 Compound Interest

Simple interest is paid on the principal—not on any past interest. However, bank deposits and many other investments commonly earn **compound interest**. Compound interest is calculated on any interest previously credited (paid) to the account in addition to the original principal. This chapter is about compound interest, which can have a significant effect when applied over long periods of time. For example, assume $1 was invested in an account paying 3% compounded annually in 1492, the year Christopher Columbus arrived in the Americas. Figure 13.1 shows that the $1 investment would have grown to over $3,300,000 by 2000. If your parents invested $1 for you when you were born, it would grow to $79.06 by your 75th birthday assuming 6% per year compounded annually. Or, it would grow to $1271.90 by your 75th birthday assuming a growth rate of 10% per year compounded annually.

How can compound interest help you meet your personal financial goals?

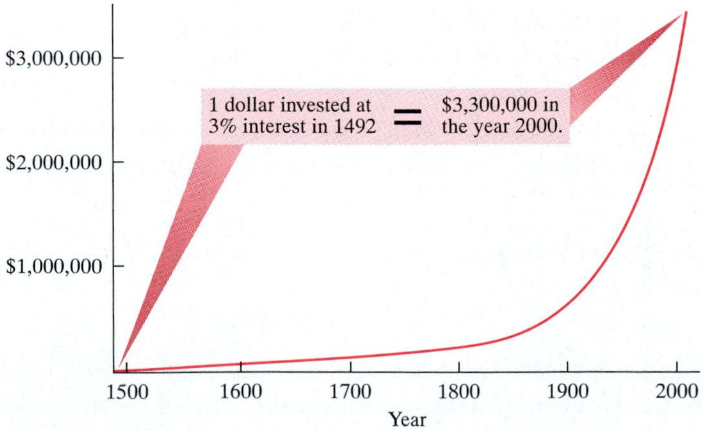

Figure 13.1

13.1 Compound Interest

Objectives

1. *Find compound interest and compound amount.*
2. *Determine the number of periods and rate per period.*
3. *Find values in the interest table.*
4. *Use the formula for compound interest to find the compound amount.*
5. *Find the effective rate of interest.*

Objective 1 *Find Compound Interest and Compound Amount.* With compound interest, the interest is found by calculating the interest on all past interest as well as on the original principal. For example, suppose \$1000 is deposited in a mutual fund that pays 10% interest, with interest calculated at the end of each year. Let us find the amount of compound interest that will be earned in 3 years assuming no interest is withdrawn during the 3 years. At the end of the first year, interest is paid on the original deposit of \$1000. Using the formula for simple interest, $I = PRT$.

$$\text{Interest} = \$1000 \times 0.10 \times 1 = \$100$$

At the end of the first year, the account contains

$$\text{Principal} + \text{Interest} = \$1000 + \$100 = \$1100$$

The interest for the second year is paid on the original deposit plus the first year's interest.

$$\text{Interest} = \$1100 \times 0.10 \times 1 = \$110$$

$$\text{Amount at end of second year} = \$1100 + \$110 = \$1210$$

The interest for the third year is paid on the amount at the end of the second year.

$$\text{Interest} = \$1210 \times 0.10 \times 1 = \$121$$

$$\text{Amount at end of third year} = \$1210 + \$121 = \$1331$$

This amount, \$1331, the final amount on deposit at the end of the three year investment period, is called the **compound amount** or **future amount**, and symbolized with the letter, M.

NOTE The compound amount is also referred to as maturity value, or as future value. ▬

The interest earned during the time of the investment is found by the following formula.

$$\text{Interest} = \text{Compound amount} - \text{Original principal}$$

In our example,

$$\text{Interest} = \$1331 - \$1000 = \$331$$

NOTE As a comparison, the *simple* interest on \$1000 at 10% for 3 years is \$300 ($I = PRT = \$1000 \times 0.10 \times 3$). The compound interest, therefore, is \$31 more than the simple interest (\$331 − \$300). ▬

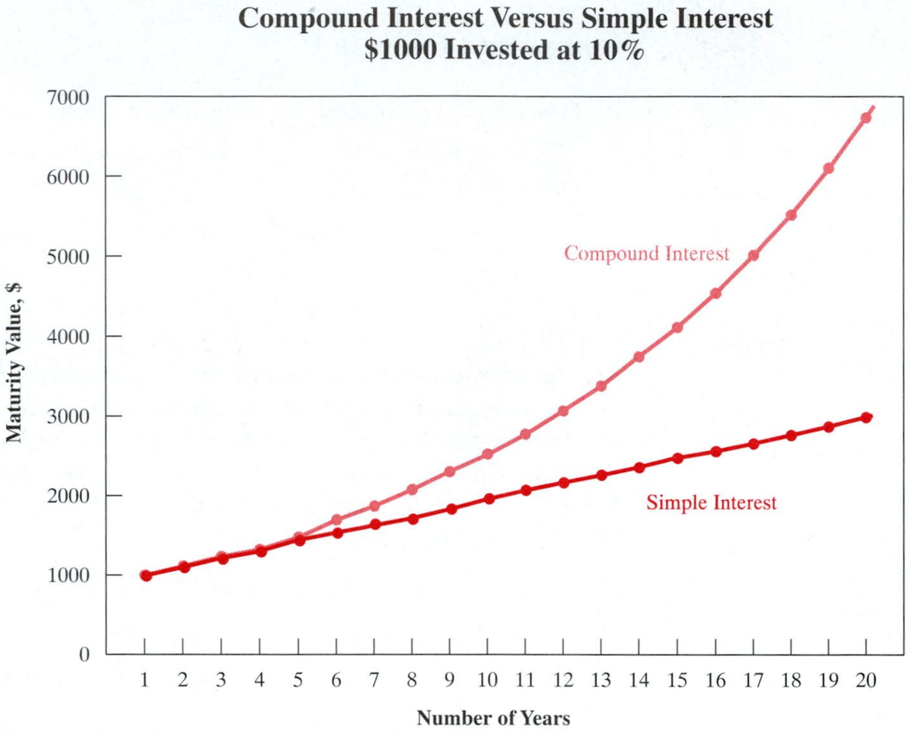

Figure 13.2

The advantage of compound interest over simple interest increases greatly as the number of years increase, as shown for a $1000 investment earning 10% in Figure 13.2.

The process used for finding the compound amount can be simplified by multiplying the principal by the expression

$$(1 + \text{Rate})$$

where rate is expressed as a decimal. This multiplication should be done as many times as the number of years in the problem. In the example above, the rate was 10%, or the decimal 0.10. The principal at the beginning of a year should be multiplied by $1 + 0.10 = 1.10$. Multiply by 1.10 three times to find the compound amount after 3 years. Work as follows.

	\$1000	Original principal
×	1.10	1 + rate as a decimal
	\$1100	Amount after 1 year
×	1.10	
	\$1210	Amount after 2 years
×	1.10	
	\$1331	Amount after 3 years

The total amount after 3 years is $1331; as mentioned earlier, this sum is the compound amount, or the future amount.

The compound amount has been found by multiplying the principal of \$1000 by 1.10 a total of three times.

$$\$1000 \times 1.10 \times 1.10 \times 1.10 = \$1331$$

In algebra, repeated multiplication with the same number is commonly written using exponents. An **exponent** is used to show how many times a number is used as a factor. The exponent is written using smaller type, and is placed to the upper right of the base number as follows.

$$\$1000 \times 1.10 \times 1.10 \times 1.10 = \$1000 \times (1.10)^{3} \quad \leftarrow \text{Exponent}$$

or \$1000 $(1.10)^3$.

NOTE The number 3 is the exponent and represents the number of years for which the interest is calculated.

Example 1 Finding Compound Interest

Tom and Gloria Peters hope to have \$5000 in 4 years for a down payment on their first home. They invest \$3800 in an account that pays 6% interest at the end of each year, on previous interest in addition to principal. (a) Find the excess of compound interest over simple interest after 4 years. (b) Will they have enough money at the end of 4 years to meet their goal of a down payment?

Solution

(a) Year	Interest	Compound Amount
1	$\$3800.00 \times 0.06 \times 1 = \$228.00$	$\$3800.00 + \$228.00 = \$4028.00 = \$3800\,(1.06)$
2	$\$4028.00 \times 0.06 \times 1 = \$241.68$	$\$4028.00 + \$241.68 = \$4269.68 = \$3800\,(1.06)^2$
3	$\$4269.68 \times 0.06 \times 1 = \$256.18$	$\$4269.68 + \$256.18 = \$4525.86 = \$3800\,(1.06)^3$
4	$\$4525.86 \times 0.06 \times 1 = \$271.55$	$\$4525.86 + \$271.55 = \$4797.41 = \$3800\,(1.06)^4$

$$\text{Compound Interest} = \$4797.41 - \$3800 = \$997.41$$

$$\text{Simple Interest (for comparison)} = \$3800 \times 0.06 \times 4 = \$912$$

$$\text{Difference} = \$997.41 - \$912 = \$85.41$$

(b) No, they will be short of their goal by \$5000 − \$4797.41 = \$202.59.

Objective 2 ***Determine the Number of Periods and Rate per Period.*** Interest is often credited to an account more than once a year when calculating compound interest. The interest rate used to find the amount of interest credited at the end of each compounding period is the **nominal**, or **stated**, annual rate divided by the number of compounding periods in a year.

Compounding Period	Interest Credited at the End of Each	Number of Times Interest Is Credited per Year	Number of Times Interest Would Be Credited over 5 Years	Rate per Compounding Period If R Is Rate per Year
Annual	year	1	$5 \times 1 = 5$	R
Semiannual	6 months	2	$5 \times 2 = 10$	$\frac{R}{2}$
Quarterly	quarter	4	$5 \times 4 = 20$	$\frac{R}{4}$
Monthly	month	12	$5 \times 12 = 60$	$\frac{R}{12}$

Example 2 Finding Number of Periods and Rate per Period

(a) A bank pays interest of 8%, compounded semiannually. This means that semiannually, or twice a year, interest of $8\% \div 2 = 4\%$ is added to all money that has been on deposit for 6 months or more.

(b) An interest rate of 12% per year, compounded quarterly, means that every 3 months (quarterly), interest of $12\% \div 4 = 3\%$ is added to all money that has been on deposit for at least a quarter.

NOTE In Example 2(a), the **period of compounding** is semiannual (every 6 months), while it is quarterly (every 3 months) in Example 2(b). ■

The formula for compound interest is expressed using exponents.

If P dollars are deposited at a rate of interest i per period for n periods, then the *compound amount* M, or the final amount on deposit, is

$$M = P(1 + i)^n$$

Interest earned is

$$I = M - P$$

where P is the original principal.

NOTE It is important to keep in mind that i is the interest rate *per compounding period*, and not per year, and n is the number of *compounding periods*, not the number of years. ■

Example 3 Finding Compound Interest

A savings account at Northstar Bank in Canada pays 7% per year compounded semiannually. If you initially deposit \$2500, (a) find the compound amount after 3 years and (b) find the compound interest.

Solution Both compound interest and compound amount can be estimated using simple interest calculations. Simple interest calculations will underestimate both compound amount and compound interest, but not by much, since the time involved is only 3 years.

$$I = PRT$$

$$I = \$2500 \times 0.07 \times 3 \quad \text{(Assume 7\% per year for 3 years.)}$$

$$= \$525$$

The future value estimate would be \$2500 + \$525 = \$3025. The compound interest should be somewhat larger than \$525, and the compound amount should be somewhat larger than \$3025.

(a) Every 6 months, $\frac{7\%}{2} = 3.5\%$ ($i = 3.5\%$) is added to all funds on deposit for 6 months or more. The number of 6-month periods in 3 years is 6. Thus, the annual rate of 7% is $i = \frac{7\%}{2} = 3.5\%$, and $n = 6$, the number of compounding periods. To find the compound amount, multiply \$2500 by (1 + 0.035) a total of 6 times.

$$\$2500 \times 1.035 \times 1.035 \times 1.035 \times 1.035 \times 1.035 \times 1.035 = \$3073.14$$

Using exponents, the compound amount can be written as follows.

$$\$2500(1 + \tfrac{0.07}{2})^6 = \$2500(1.035)^6$$

(b) Interest = compound amount − original principal

= \$3073.14 − \$2500

= \$573.14

Notice that compound amount and compound interest were a little larger than estimated.

Objective 3 ***Find Values in the Interest Table.*** The value of $(1 + i)^n$ can be found by direct calculation using scientific calculators or from tables. One such table is given in the compound interest column (column A) of the interest table in Appendix D. The interest rate per compounding period is on the upper left hand corner of each page.

Example 4 Using the Interest Table

Find the following values in the compound interest table in Appendix D.

(a) $(1 + 5\%)^{12}$ or $(1 + 0.05)^{12} = (1.\underline{05})^{12}$

(12 → Number of compounding periods; 05 → Interest rate per compounding period)

Find the 5% page of the interest table. Look in column A, for compound interest, and find 12 (or 12 periods) at the left side. You should find 1.79585633.

(b) $(1 + 8\%)^{27} = (1.\underline{08})^{27} = 7.98806147$

(27 → Number of compounding periods; 08 → Interest rate per compounding period)

Find the 8% page. Then look in column A and find 27 at the left.

Scientific *Calculator Approach*

For the calculator solution to Example 4,

(a) The value of $(1.05)^{12}$ can be found as follows.

1.05 [y^x] 12 [=] 1.795856326

(b) The value of $(1.08)^{27}$ is found as follows.

1.08 [y^x] 27 [=] 7.988061469

NOTE Some calculators have an a^x or x^y key instead of a y^x key. All of these keys are used in the same way. ■

Objective 4 ***Use the Formula for Compound Interest to Find the Compound Amount.*** The evaluation of $(1 + i)^n$ using tables or calculators can now be used to find the compound amount and interest.

Example 5 Finding Compound Interest

John Smith inherits $15,000, which he deposits in a retirement account that pays interest compounded semiannually. How much will he have after 25 years if the funds grow (a) at 6%, (b) at 8%, and (c) at 10%?

Solution In 25 years, there are $2 \times 25 = 50$ semiannual periods. The semiannual interest rates are (a) $\frac{6\%}{2} = 3\%$, (b) $\frac{8\%}{2} = 4\%$, and (c) $\frac{10\%}{2} = 5\%$. Using factors from the table or use the formula $M = P(1 + i)^n$:

(a) $\$15{,}000(1.03)^{50} = \$15{,}000 \times 4.38390602^* = \$65{,}758.59$
(b) $\$15{,}000(1.04)^{50} = \$15{,}000 \times 7.10668335 = \$106{,}600.25$
(c) $\$15{,}000(1.05)^{50} = \$15{,}000 \times 11.46739979 = \$172{,}011$

The $15,000 that John Smith inherits is the present value he has today. The future value is the amount he will have in 25 years.

Financial *Calculator Approach*

Present value, interest per compounding period, and number of compound periods are known. Future value is the unknown.

(a) −15000 [PV] 3 [i] 50 [n] [FV] 65758.59

(b) −15000 [PV] 4 [i] 50 [n] [FV] 106600.25

(c) −15000 [PV] 5 [i] 50 [n] [FV] 172011

Some calculators will not take all these digits. With these, round to as many digits as will fit. This procedure may make answers vary by a few cents, especially on large sums of money.

Problem-Solving Hint Simple interest rate calculations are usually indicated by phrases such as *simple interest*, *simple interest note*, or *discount rate*. Compound interest rate calculations are usually indicated by phrases such as *compounded annually*, *6% per quarter*, or *compounded daily*.

The more often interest is compounded, *the more interest is earned.* Use a financial calculator or a compound interest table more complete than the one in this text and use the compound interest formula to get the results shown in Table 13.1. (Leap years were ignored in finding daily interest.)

As suggested by Table 13.1, it makes a big difference whether interest is compounded or not. Interest over the 10 years differs by $1190.85 when simple interest is compared to interest compounded annually. However, increasing the frequency of compounding makes smaller and smaller differences in the amount of interest earned.

Table 13.1 INTEREST ON \$1000 AT 6% PER YEAR FOR 10 YEARS

Frequency of Compounding	Interest per Compounding Period	Number of Periods	Interest
Not at all (simple interest)	—	—	\$600.00
Annually	6%	10	\$1790.85
Semiannually	6%/2	20	\$1806.11
Quarterly	6%/4	40	\$1814.02
Monthly	6%/12	120	\$1819.40
Daily	6%/365	3650	\$1822.03
Hourly	6%/8760	87,600	\$1822.12
Every Minute	6%/525,600	5,256,000	\$1822.12

Example 6 Finding Compound Amount

Sarah Kline is comparing two different investment options. Find the compound amount earned on a deposit of \$6000 for 3 years at (a) 12% compounded quarterly and (b) 12% compounded monthly.

Solution

(a) Interest compounded quarterly is compounded 4 times a year. In 3 years, there are $3 \times 4 = 12$ compounding periods. Interest of $\frac{12\%}{4} = 3\%$ is credited at the end of each quarter.

From the table.

$$\$6000(1.03)^{12} = \$6000 \times 1.42576089 = \$8554.57$$

(b) Interest compounded monthly is compounded 12 times a year. In 3 years, there are $3 \times 12 = 36$ compounding periods. Interest of $\frac{12\%}{12} = 1\%$ is credited at the end of each month.

$$\$6000(1.01)^{36} = \$6000 \times 1.43076878 = \$8584.61$$

Scientific *Calculator Approach*

Solve Example 6 using a scientific calculator as follows.

(a) 6000 [×] 1.03 [y^x] 12 [=] 8554.57

(b) 6000 [×] 1.01 [y^x] 36 [=] 8584.61

Another use of compound interest calculations is to find the effect of inflation on real estate values.

Example 7 Finding the Effect of Inflation on Real Estate Values

Bill and Joy Lopez purchase a home for \$97,200. If the house goes up in value by 3% per year, find its value at the end of 4 years.

Solution The value of the house compounds at 3% for 4 years. Use the table to find 1.12550881.

$$\text{Future value of home} = \$97{,}200 \times 1.12550881$$
$$= \$109{,}399 \quad \text{Rounded.}$$

The home is expected to increase in value by $109,399 − $97,200 = $12,199 during the 4 years.

Objective 5 ***Find the Effective Rate of Interest.*** If interest is compounded *more often than annually*, then the actual rate of interest is greater than the nominal or stated rate of interest. For example, depositing $1000 for 1 year at 12% compounded quarterly produces a compound amount as follows.

$$M = \$1000(1.12550881)$$
$$= \$1125.51$$

The interest earned on this deposit is $125.51, which is 12.551% of the original deposit of $1000. This amount of interest is the same as if the $1000 were invested at simple interest of 12.551% for one year. Although the stated rate of interest was 12%, the actual increase in the investment was 12.551%, the **effective rate of interest**. It is important to know the effective rate of interest so that you can compare one investment or loan to another.

NOTE The effective rate of interest is sometimes called the **effective annual yield**.

Finding the Effective Rate of Interest

Step 1. Find the entry in column A of the interest table that corresponds to the proper rate per period and the proper number of periods.

Step 2. Subtract 1 from the number.

Step 3. Round to the nearest hundredth of a percent.

Example 8 Finding the Effective Rate of Interest

James Suhr is comparing two loans. Loan A is at a nominal rate of 10% compounded quarterly and Loan B is at a nominal rate of 9% compounded monthly. Find the effective rate of interest for both.

Solution Loan A: 10% ÷ 4 = 2.5% per quarter for 4 quarters. Look in column A of the interest table for $2\frac{1}{2}$% and 4 periods to find the effective interest rate as follows.

$$\begin{array}{r} 1.10381289 \\ -\ 1.00000000 \\ \hline 0.10381289 \end{array}, \text{ or } 10.38\%$$

Loan B: 9% ÷ 12 = 0.75% per month for 12 months. Look in column A of the interest table for $\frac{3}{4}$% and 12 periods to find the following.

$$\begin{array}{r} 1.09380690 \\ -\ 1.00000000 \\ \hline 0.09380690 \end{array}, \text{ or } 9.38\%$$

Problem-Solving Hint In Step 2, be sure to subtract 1 from the number in the table.

13.1 Exercises

Find the compound amount and the amount of interest earned for each of the following. Round to the nearest cent. Do not use tables in Exercises 1–4.

1. $12,000 at 8% compounded annually for 3 years
2. $8500 at 6% compounded semiannually for $4\frac{1}{2}$ years
3. $6000 at 12% compounded quarterly for 2 years
4. $10,500 at 10% compounded quarterly for 4 years

Find the compound amount when the following deposits are made. Round to the nearest cent.

5. $1000 at 8% compounded annually for 40 years
6. $925 at 5% compounded annually for 12 years
7. $470 at 12% compounded semiannually for 9 years
8. $8765.72 at 12% compounded monthly for 4 years

Find the amount of interest earned by the following deposits.

9. $8400 at 7% compounded annually for 8 years
10. $6200 at 11% compounded annually for 5 years
11. $12,600 at 8% compounded quarterly for $4\frac{3}{4}$ years
12. $23,000 at 12% compounded monthly for $2\frac{1}{2}$ years

Find simple interest, then use column A of the interest table and the same interest rate to find the interest compounded annually. Find the amount by which the compound interest is larger. Round to the nearest cent.

	Principal	Rate	Number of Years
13.	$1,000	6%	5
14.	$800	7%	6
15.	$7,908.42	5%	8
16.	$10,240	10%	11

Find the effective rate corresponding to each of the following nominal rates. Round to the nearest hundredth of a percent.

17. 8% compounded quarterly
18. 10% compounded quarterly
19. 15% compounded monthly
20. 12% compounded monthly

Solve each of the following application problems. Round to the nearest cent.

21. Jane Gates invested $2800 for $2\frac{1}{2}$ years at 12% interest. Find the compound amount if (a) the compounding is quarterly and (b) if the compounding is monthly.

22. FINDING COMPOUND AMOUNT Waters Electrical deposited $10,000 for 3 years at 10% interest. Find the compound amount if the compounding period is (a) semiannually and (b) quarterly.

23. COMPOUNDING PERIODS Find the interest earned on $10,000 for 4 years at 6% compounded (a) yearly, (b) semiannually, (c) quarterly, and (d) monthly. (e) Find the simple interest.

24. COMPOUNDING PERIODS Suppose $32,000 is deposited for 2 years at 10% interest. Find the interest earned on the deposit if the interest is compounded (a) yearly, (b) semiannually, (c) quarterly, and (d) monthly. (e) Find the simple interest.

25. LOAN TO A RESTAURANT Benjamin Moore loans $8800 to the owner of a new restaurant. He will be repaid at the end of 4 years, with interest at 8% compounded semiannually. Find how much he will be repaid.

26. COMPOUND INTEREST Glenda Wong deposits $8270 in a bank that pays 12% interest, compounded semiannually. Find the amount she will have at the end of 5 years.

27. COMPARING BANKS There are two banks in Citrus Heights. One pays interest of 4% compounded annually, and the other pays 4% compounded quarterly. If Stan deposits $10,000 in each bank, (a) how much will he have in each bank at the end of 3 years? (b) How much more would he have in the bank that paid more interest?

28. COMPARING YIELDS The two bond funds that Jessica Mendez is considering have yielded 10% compounded quarterly and 9% compounded monthly in the past. If she deposits $2500 in each, (a) how much will she have in each at the end of 2 years? (b) How much more will she have in the higher yielding fund?

29. GUARANTEED PENSION INCOME In order to guarantee a certain pension income, Mr. Watkins must have $125,000 in the bank 3 years from now. Suppose he deposits $80,000 today at 12% compounded monthly and leaves it there for 3 years. How much additional money would he then have to add to have the required amount?

30. GOVERNMENT CLAIM To settle a government claim about the company's pension plan, West Hardware must have $275,000 in an account $2\frac{1}{2}$ years from now. Suppose $200,000 is deposited today at 12% interest compounded quarterly. How much additional money would have to be added in $2\frac{1}{2}$ years to have the necessary amount?

31. COMPARING INVESTMENTS Joan Getz is comparing two investment opportunities, the first at 6% compounded semiannually and the second at 5% compounded monthly. Find the effective interest rates for each. Which would produce more income for Getz?

32. RAISING FUNDS A church plans to raise funds by selling bonds to individuals and businesses in the community. They are considering two options, the first at 8% compounded annually and the second at 6% compounded monthly. Find the effective interest rate for both. Which would produce the lower cost for the church?

Use either a scientific or financial calculator to solve the following for compound amount.

33. $12,800 at 12.3% compounded annually for 3 years

34. $24,500 at 8.9% compounded semiannually for 2 years

35. \$9500 at 10.3% compounded quarterly for $2\frac{1}{2}$ years

36. \$4300 at 10.2% compounded monthly for $2\frac{1}{3}$ years

37. Explain the basic difference between simple interest and compound interest. (See Objective 1.)

38. Do long-term investments grow more using simple interest or compound interest? Why is this type of interest to your advantage when saving for events several years in the future? (See Objective 1.)

39. Bill Baxter has \$25,000 to invest for a year. He can lend it to his sister who has agreed to pay 10% simple interest for the year. Or, he can invest it with a bank at 8% compounded quarterly for a year. How much additional interest would the simple interest loan to his sister generate?

40. **COMPARING YIELDS ON LOANS** Citizens Bank has \$850,000 to lend for 9 months. They can lend it to a local contractor at a simple interest rate of 12% or they can lend it to another bank that will pay 12% compounded monthly. How much additional interest would the compound interest loan to the bank generate?

41. Peter and Betsy Mueller inherit \$12,000 from Peter's aunt. Although they are only in their early thirties, the Muellers decide to set the money aside for their eventual retirement. They invest the funds in a Roth IRA account so that they will not have to pay any taxes on the earnings. Find the amount they will have in 25 years if the funds are invested in a Roth IRA that grows at 8% compounded semiannually for 25 years.

13.2 Daily and Continuous Compounding

Objectives

1. *Define passbook account.*
2. *Calculate interest compounded daily.*
3. *Find compound interest for time deposit accounts.*
4. *Determine the penalty for early withdrawal.*
5. *Find compound amount with continuous compounding.*

Objective 1 *Define Passbook Account.* **Savings accounts** or **passbook accounts** meet the daily money needs of a person or business. Money may be deposited in or withdrawn from a passbook account anytime, with no penalty, although you cannot write checks on funds in a savings account. Interest rates on savings accounts can vary from $2\frac{1}{2}$% to 6%. Many checking accounts now also pay interest but the banks charge a fee if the balance falls below a specific amount. The Truth in Savings Act of 1991 resulted in **Regulation DD**, which requires that interest on savings accounts be paid based on the *exact number of days*.

A savings account can be *one of the safest places* for money. Savings accounts at most, but not all, institutions are insured by the Federal Deposit Insurance Corporation (FDIC) on deposits up to \$100,000. Call your bank to see if your funds are federally insured.

Objective 2 *Calculate Interest Compounded Daily.* Interest on savings accounts is found using compound interest. It is common for banks to pay interest **compounded daily** so that interest is credited for every day that the money is on deposit.

The formula for daily compounding is *exactly* the same as that given in the last section. However, because the annual interest rate must be divided by 365 (for daily com-

Table 13.2 VALUES OF $(1 + i)^n$ FOR $3\frac{1}{2}\%$ COMPOUNDED DAILY

Number of Days n	Value of $(1 + i)^n$	n	Value of $(1 + i)^n$	n	Value of $(1 + i)^n$	n	Value of $(1 + i)^n$	n	Value of $(1 + i)^n$
1	1.000095890	19	1.001823491	37	1.003554076	55	1.005287650	73	1.007024219
2	1.000191790	20	1.001919556	38	1.003650307	56	1.005384048	74	1.007120783
3	1.000287699	21	1.002015631	39	1.003746548	57	1.005480454	75	1.007217357
4	1.000383617	22	1.002111714	40	1.003842797	58	1.005576870	76	1.007313939
5	1.000479544	23	1.002207807	41	1.003939056	59	1.005673296	77	1.007410531
6	1.000575480	24	1.002303909	42	1.004035324	60	1.005769730	78	1.007507132
7	1.000671426	25	1.002400021	43	1.004131602	61	1.005866174	79	1.007603742
8	1.000767381	26	1.002496141	44	1.004227888	62	1.005962627	80	1.007700362
9	1.000863345	27	1.002592271	45	1.004324184	63	1.006059089	81	1.007796990
10	1.000959318	28	1.002688410	46	1.004420489	64	1.006155560	82	1.007893628
11	1.001055300	29	1.002784558	47	1.004516803	65	1.006252041	83	1.007990276
12	1.001151292	30	1.002880716	48	1.004613127	66	1.006348531	84	1.008086932
13	1.001247293	31	1.002976882	49	1.004709460	67	1.006445030	85	1.008183598
14	1.001343303	32	1.003073058	50	1.004805802	68	1.006541538	86	1.008280273
15	1.001439322	33	1.003169243	51	1.004902153	69	1.006638056	87	1.008376958
16	1.001535350	34	1.003265438	52	1.004998513	70	1.006734583	88	1.008473651
17	1.001631388	35	1.003361641	53	1.005094883	71	1.006831119	89	1.008570354
18	1.001727435	36	1.003457854	54	1.005191262	72	1.006927665	90	1.008667067

Note: The value of $(1 + i)^n$ for $3\frac{1}{2}\%$ compounded daily for a quarter with 91 days is 1.008763788 and for a quarter with 92 days is 1.008860519.

pounding), the arithmetic is very tedious. To avoid this, use Table 13.2, which provides the necessary numbers for 1 to 92 days, as well as for one to four 90-day quarters, assuming $3\frac{1}{2}\%$ interest compounded daily. The table goes to only 92 days since interest is normally credited to the depositor's account only quarterly, even with daily compounding. The four quarters in a year begin on January 1, April 1, July 1, and October 1.

NOTE See Appendix A.2 for financial calculators that do not require the use of a table.

Example 1 Finding Interest Using Daily Compounding

Mr. Watson wants his 6-year-old son Billy to learn about savings and interest. He took Billy to a bank on September 12 and opened a savings account with \$500 to which Billy added \$23.50 he had received from doing chores. Find the amount of interest Billy has earned by his birthday, November 20, if he earns $3\frac{1}{2}\%$ compounded daily.

Solution There are 18 days remaining in September, 31 days in October and 20 days in November. The money was on deposit for $18 + 31 + 20 = 69$ days. The table value for 69 days is 1.006638056, so the compound amount is

$$\$523.50 \times 1.006638056 = \$526.98$$

and the interest earned is $\$526.98 - \$523.50 = \$3.48$. Billy is old enough that he isn't impressed with \$3.48. Mr. Watson, knowing the power of compound interest over the long term, insists that Billy continue to save part of his earnings.

Scientific *Calculator Approach*

Using a scientific calculator, Example 1 can be solved as follows.

523.50 [X] [(] 1 [+] .035 [÷] 365 [)] [y^x] 69 [=] 526.98

Financial *Calculator Approach*

Solve for the unknown future value using a financial calculator as follows.

−523.50 [PV] .009589041 [i] 69 [n] [FV] 526.98

(.009589041 ← 3.5 [÷] 365)

INTEREST BY QUARTER FOR $3\frac{1}{2}\%$ COMPOUNDED DAILY ASSUMING 90-DAY QUARTERS

Number of Quarters	Value of $(1 + i)^n$
1	1.008667067
2	1.017409251
3	1.026227205
4	1.035121585

Example 2 Finding Quarterly Interest

Tom Blackmore is a private investigator. On January 10, he deposited $2463 in a savings account paying $3\frac{1}{2}\%$ compounded daily. He deposits an additional $1320 on February 18 and $840 on March 3. Find the interest earned through April 10.

Solution Treat each of the three amounts separately. The $2463 was in the account for 21 days in January, 28 days in February, 31 days in March, and 10 days in April for a total of 90 days. The compound amount is found using 1.008667067 from the table.

Compound amount = $2463 × 1.008667067 = $2484.35 First deposit plus interest

A deposit of $1320 was made on February 18. This amount was on deposit for 10 days in February, 31 days in March and 10 days in April for a total of 51 days.

Compound amount = $1320 × 1.004902153 = $1326.47 Second deposit plus interest

The final deposit of $840 was on deposit for 28 days in March and 10 days in April for a total of 38 days.

Compound amount = $840 × 1.003650307 = $843.07 Final deposit plus interest

The total amount in the account is found by adding the three compound amounts together, and the interest earned is the total amount in the account minus deposits.

Total in account = $2484.35 + $1326.47 + $843.07 = $4653.89

Interest earned = $4653.89 − ($2463 + $1320 + $840) = $30.89

The interest earned is $30.89.

Example 3 Finding Quarterly Interest

Beth Gardner owns Blacktop Paving, Inc. She needs a place to keep extra cash, a place that will earn interest but that will allow her to get funds when needed. She opened a savings account on July 20 with a $24,800 deposit. She then withdrew $3800 on August 29 for an unexpected truck repair and she made another withdrawal for $8200 on September 29 for payroll. Find the interest earned through October 1, given interest at $3\frac{1}{2}\%$ compounded daily.

Solution Of the original $24,800, a total of $24,800 − $3800 − $8200 = $12,800 earned interest from July 20 to October 1, or for 274 − 201 = 73 days. Find the factor 1.007024219 from the table.

$$\text{Compound amount} = \$12{,}800 \times 1.007024219 = \$12{,}889.91$$
$$\text{Interest} = \$12{,}889.91 - \$12{,}800 = \$89.91$$

The withdrawn $3800 earned interest from July 20 to August 29, or for 241 − 201 = 40 days.

$$\text{Compound amount} = \$3{,}800 \times 1.003842797 = \$3814.60$$
$$\text{Interest} = \$3814.60 - \$3800 = \$14.60$$

Finally, the withdrawn $8200 earned interest from July 20 to September 29 or for 272 − 201 = 71 days.

$$\text{Compound amount} = \$8200 \times 1.006831119 = \$8256.02$$
$$\text{Interest} = \$8256.02 - \$8200 = \$56.02$$

The total interest earned is ($89.91 + $14.60 + $56.02) = $160.53. The total in the account on October 1 is found as follows.

$$\text{Deposits} + \text{Interest} - \text{Withdrawals} = \text{Balance on October 1}$$
$$\$24{,}800 + \$160.53 - (\$3800 + \$8200) = \$12{,}960.53$$

Objective 3 ***Find Compound Interest for Time Deposit Accounts.*** While passbook savings accounts are very useful for money needed for day-to-day living, the low interest rate paid by these accounts makes them undesirable for larger amounts of money, such as money being saved for retirement. These larger amounts are better off in a **time deposit account** where money must be left for a fixed period of time. For example, one local savings and loan pays $3\frac{1}{4}\%$ on money in a day-in and day-out passbook account, but pays 6% on money left with a **certificate of deposit** for 2 years. With a certificate of deposit (abbreviated **CD**), money must be left for some minimum time period, and a minimum amount of money must be deposited. Another popular certificate of deposit currently available pays interest depending on the interest rate paid by the U.S. government. While these CDs pay higher interest (as high as 6% at times) and require that money be left for only 6 months, a large minimum deposit (perhaps $10,000) *may be required.* A typical certificate of deposit is shown in Figure 13.3.

It is probably not a good idea to invest all of one's money in a certificate of deposit, since a penalty is charged for early withdrawal. At least some cash should be left in a passbook savings account (or perhaps in the money market accounts discussed later) for daily needs.

The higher interest paid by some certificates of deposit can be found using Table 13.3. (This table assumes daily compounding and 365 days per year.)

Example 4 Finding Compound Amount and Interest on Time Deposits

David Herren invests $20,000 in a certificate of deposit paying 6% compounded daily. Find the compound amount after 5 years. Also, find the interest earned.

Solution Look in Table 13.3 for 6% and 5 years, finding 1.34982552.

$$\text{Compound amount} = \$20{,}000(1.34982552)$$
$$= \$26{,}996.51$$

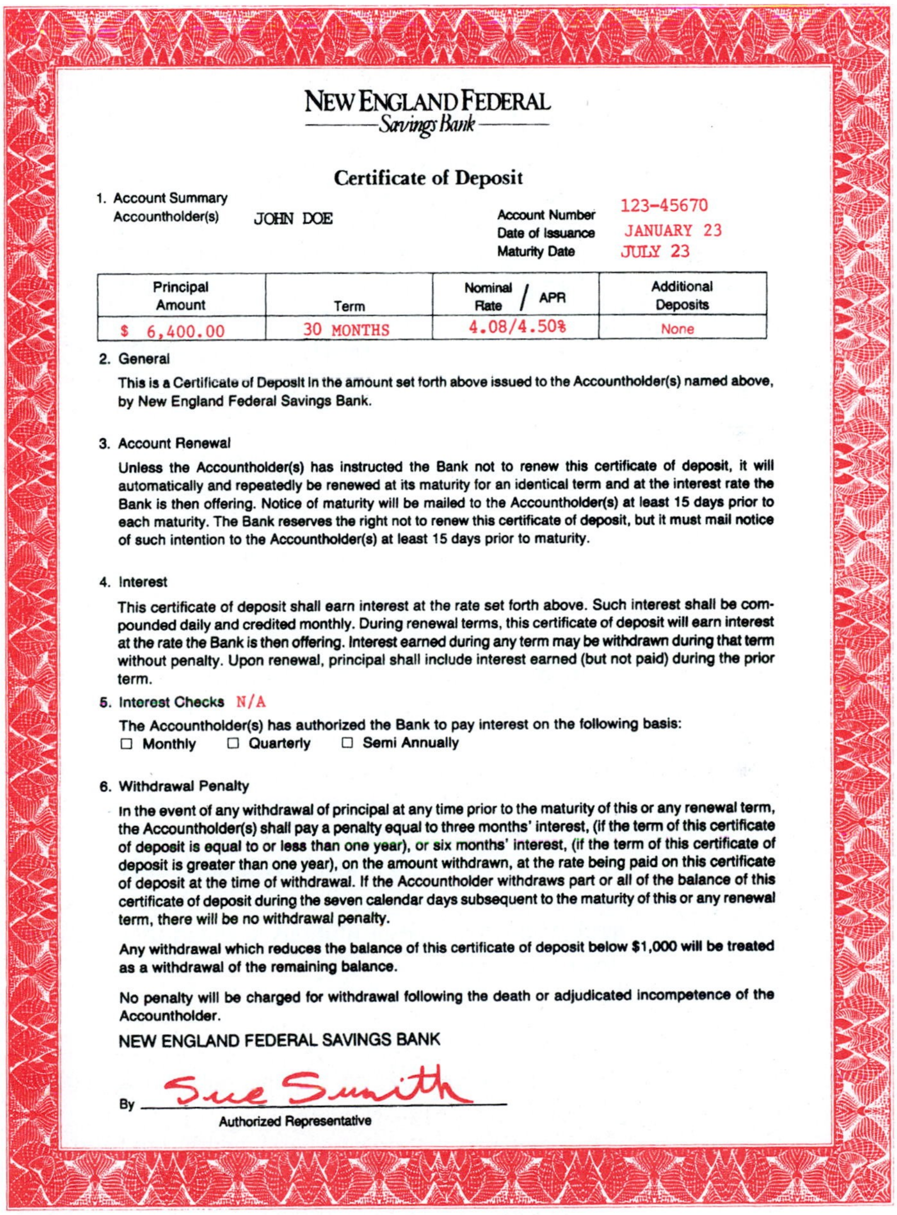

NEW ENGLAND FEDERAL
Savings Bank

Certificate of Deposit

1. Account Summary
   Accountholder(s) JOHN DOE

Account Number 123-45670
Date of Issuance JANUARY 23
Maturity Date JULY 23

Principal Amount	Term	Nominal Rate / APR	Additional Deposits
$ 6,400.00	30 MONTHS	4.08/4.50%	None

2. General

   This is a Certificate of Deposit in the amount set forth above issued to the Accountholder(s) named above, by New England Federal Savings Bank.

3. Account Renewal

   Unless the Accountholder(s) has instructed the Bank not to renew this certificate of deposit, it will automatically and repeatedly be renewed at its maturity for an identical term and at the interest rate the Bank is then offering. Notice of maturity will be mailed to the Accountholder(s) at least 15 days prior to each maturity. The Bank reserves the right not to renew this certificate of deposit, but it must mail notice of such intention to the Accountholder(s) at least 15 days prior to maturity.

4. Interest

   This certificate of deposit shall earn interest at the rate set forth above. Such interest shall be compounded daily and credited monthly. During renewal terms, this certificate of deposit will earn interest at the rate the Bank is then offering. Interest earned during any term may be withdrawn during that term without penalty. Upon renewal, principal shall include interest earned (but not paid) during the prior term.

5. Interest Checks N/A

   The Accountholder(s) has authorized the Bank to pay interest on the following basis:
   ☐ Monthly ☐ Quarterly ☐ Semi Annually

6. Withdrawal Penalty

   In the event of any withdrawal of principal at any time prior to the maturity of this or any renewal term, the Accountholder(s) shall pay a penalty equal to three months' interest, (if the term of this certificate of deposit is equal to or less than one year), or six months' interest, (if the term of this certificate of deposit is greater than one year), on the amount withdrawn, at the rate being paid on this certificate of deposit at the time of withdrawal. If the Accountholder withdraws part or all of the balance of this certificate of deposit during the seven calendar days subsequent to the maturity of this or any renewal term, there will be no withdrawal penalty.

   Any withdrawal which reduces the balance of this certificate of deposit below $1,000 will be treated as a withdrawal of the remaining balance.

   No penalty will be charged for withdrawal following the death or adjudicated incompetence of the Accountholder.

   NEW ENGLAND FEDERAL SAVINGS BANK

   By Sue Smith
   Authorized Representative

Figure 13.3

Table 13.3 COMPOUND INTEREST FOR TIME DEPOSIT ACCOUNTS (DAILY COMPOUNDING)

Number of Years	Interest Rate			
	5%	6%	7%	8%
1	1.05126750	1.06183131	1.07250098	1.08327757
2	1.10516335	1.12748573	1.15025836	1.17349030
3	1.16182231	1.19719965	1.23365322	1.27121572
4	1.22138603	1.27122407	1.32309429	1.37707948
5	1.28400343	1.34982552	1.41901993	1.49175931
10	1.64866481	1.82202895	2.01361755	2.22534585

Of this amount,

$$\$26{,}996.51 - \$20{,}000 = \$6996.51$$

is interest.

NOTE Many time deposits earn interest compounded monthly or quarterly rather than daily. In that event, calculate interest using the techniques of Section 13.1. ■

Objective 4 *Determine the Penalty for Early Withdrawal.* A depositor who agrees to leave money in an account for a certain period of time, but then withdraws it early, must pay a penalty. While it is not possible for the depositor to lose any of the *principal* involved as long as the institution is federally insured, it is very possible that all or part of the *interest* may be lost. The procedure for calculating this early withdrawal penalty is not standard. Many financial institutions use the following rules for calculating the **early withdrawal penalty**.

Calculating the Early Withdrawal Penalty

1. If money is withdrawn within 3 months of the deposit, no interest will be paid at all on the money withdrawn.
2. If money is withdrawn after 3 months but before the end of the term, then 3 months is deducted from the time the account has been open and regular passbook interest is paid on the account.

Example 5 Finding Interest When Early Withdrawal Occurs

Use the rules listed in the box to find the amount of interest earned on each of the following.

Solution

(a) On January 5, Raymond Hoyle deposited $5000 in a 1-year certificate of deposit paying 6% compounded daily. He withdrew the money on March 12 of the same year.

Since March 12 is within 3 months of January 5, no interest at all is earned. The bank will simply return the $5000.

(b) Bob Kashir deposited $6000 in a 4-year certificate of deposit paying 5% compounded daily. He withdrew the money 15 months later. The passbook rate at his bank is $3\frac{1}{2}\%$ compounded daily.

The money is withdrawn early, so 3 months interest is lost. The money was on deposit for 15 months, but only 15 − 3 = 12 months interest will be paid. Also, interest will be paid at the passbook rates, $3\frac{1}{2}\%$ compounded daily, instead of the more generous 5% compounded daily. The compound amount is found using the factor for four 90-day quarters (assume it is not a leap year) at $3\frac{1}{2}\%$. From Table 13.2, this is 1.035121585.

$$\$6000(1.035121585) = \$6210.73$$

The interest is found by subtracting the initial deposit of $6000 from $6210.73.

$$\text{Interest} = \$6210.73 - \$6000 = \$210.73$$

Table 13.4 SAMPLE IMMA ACCOUNT INTEREST RATES

Effective Date of Rate	Interest Rate
11/24	4.00%
11/30	3.75%
12/17	3.60%
12/18	3.50%
12/21	3.90%

Insured Money Market Accounts. Many savings institutions offer **insured money market accounts** (often called **IMMAs**). These accounts offer interest rates almost as high as those of certificates of deposit, while permitting funds to be withdrawn at any time. A typical IMMA account is insured by the federal government (up to a certain maximum), allows for balances as low as $2500, pays interest rates within 1% or so of CDs, and allows for three checks or more per month to be written without charge. Unlike certificates of deposit, however, the interest rate *may change weekly or even daily.* As an example, Table 13.4 lists the interest rates paid during one 30-day period by a local bank on an IMMA account. Notice how often the rates changed.

Insured money market accounts were set up to compete with the money market funds offered by many stock brokerage and mutual fund firms. IMMAs offer competitive interest rates with the benefit of federal insurance.

Objective 5 ***Find Compound Amount with Continuous Compounding.*** In the first section of this chapter, we discussed annual, semiannual, quarterly, and monthly compounding. Then we saw daily compounding at the beginning of this section. There is no reason that interest could not be compounded more frequently, such as every hour or even every minute. Table 13.1 on page 463 shows the interest earned on $1000 at 6% per year for 10 years, assuming various frequencies of compounding.

Clearly, more frequent compounding results in more interest. However, after a while increasing the frequency of compounding makes less and less difference. For example, going from compounding daily to compounding every minute produces only 9 cents additional interest on $1000 after 10 years.

It would be possible to extend Table 13.1 to include compounding every second, every half-second, or any desired small time interval. It would even be possible to think of compounding *every instant.* Compounding every instant is called **continuous compounding**. The formula for continuous compounding, given in the next box, uses the number e. The number e is approximately equal to 2.7182818. Some calculators have an

e^x button for working with this number.

If P dollars is deposited at a rate of interest r per year and *compounded continuously* for y years, the compound interest M is as follows.

$$M = Pe^{yr}$$

Continuous compounding is used by some financial institutions and various government agencies.

Example 6 Finding the Compound Amount for Continuous Compounding

Find the compound amount for the following deposits.

(a) \$1000 at 6% compounded continuously for 10 years
(b) \$48,906.11 at 9% compounded continuously for 6 years

Solution

(a) \$1000 at 6% compounded continuously for 10 years

Use the formula in the box with P = \$1000, $r = 0.06$, and $y = 10$. The compound amount is as follows.

$$M = Pe^{yr}$$
$$M = \$1000e^{10(0.06)}$$
$$= \$1000e^{0.6}$$

The value of $e^{0.6}$ is found in the table in Appendix C. Find 0.6 in the column labeled x and read across to the number 1.82211880 in the column labeled e^x.

$$M = \$1000(1.82211880) = \$1822.12$$

Of this amount,

$$\$1822.12 - \$1000 = \$822.12$$

is interest, the same as when interest is compounded every minute as shown in Table 13.1 in this section.

(b) \$48,906.11 at 9% compounded continuously for 6 years (use Appendix C)

$$M = \$48{,}906.11e^{6(0.09)}$$
$$= \$48{,}906.11e^{0.54}$$
$$= \$48{,}906.11(1.71600686)$$
$$= \$83{,}923.22$$

The compound amount is \$83,923.22, which includes \$35,017.11 of interest.

NOTE The interest rates paid by banks in some years are quite low compared to the returns from other investments such as stocks. However, in other years the returns from banks may equal or even exceed returns from stocks.

13.2 Exercises

Find the interest earned by the following. Assume $3\frac{1}{2}\%$ interest compounded daily, and use the exact number of days.

	Amount	Date Deposited	Date Withdrawn
1.	$6200	October 7	December 10
2.	$3850	January 5	February 9
3.	$6500	February 17	April 15
4.	$2830	May 4	June 23

Find the amount on deposit on the first day of the next quarter when the following sums are deposited as indicated. Assume $3\frac{1}{2}\%$ interest compounded daily.

	Amount	Date Deposited
5.	$7235.82	February 14
6.	$3018.25	April 7
7.	$2965.72	July 1
8.	$4031.46	April 1

Find the compound amount for each of the following certificates of deposit. Assume daily compounding.

	Amount Deposited	Interest Rate	Time in Years
9.	$5,000	6%	2
10.	$2,900	5%	10
11.	$14,000	7%	3
12.	$3,000	8%	4

Find the amount of interest earned by the following certificates of deposit. Assume daily compounding and use Table 13.3.

	Amount Deposited	Interest Rate	Time in Years
13.	$20,000	7%	3
14.	$6,800	5%	5
15.	$3,800	8%	4
16.	$1,000	6%	1

17. Explain the difference between passbook accounts, time deposit accounts, certificates of deposit, and insured money market accounts.

18. Which would you prefer on your own savings, semiannual compounding or continuous compounding? Why?

Solve each of the following application problems. Assume $3\frac{1}{2}\%$ interest compounded daily. Round to the nearest cent.

19. SAVINGS ACCOUNT Teresa Tabor had \$4300 in her savings account on July 1. She then deposited \$1000 on July 30 and \$500 on September 5. Find the balance in the account on October 1.

20. Vicki Phelps had \$8600 in her savings account on April 1. She then deposited \$800 on May 5 and an additional \$350 on June 20. Find the balance in the account on July 1.

21. SAVINGS ACCOUNT FOR EXTRA CASH On April 1, Action Sports opened a savings account with a deposit of \$17,500. A withdrawal of \$5000 was made 21 days later and another withdrawal of \$980 was made 12 days before July 1. Find (a) the interest earned through that date and (b) the balance on July 1.

22. EXCESS CASH The owner of Rondo's Magic Shop opened a savings account for the extra cash in the firm. The initial deposit of \$7800 was made on July 7. A withdrawal of \$1500 was made 46 days later, with an additional withdrawal of \$1000 made 30 days before October 1. Find (a) the interest earned through that date and (b) the balance on October 1.

For the following application problems on certificates of deposit, use the rules for finding the early withdrawal penalty as given in the text. Assume daily compounding of interest and use Table 13.2.

23. TIME DEPOSIT Honey Watts deposited \$4500 in a 5-year time deposit account paying 5% interest. She withdrew the funds 65 days later. How much interest did she receive?

24. TIME DEPOSIT Jana Box deposited \$6000 in a 3-year account paying 6% interest but needed the money 2 months later and withdrew it. How much interest did she receive?

25. TIME DEPOSIT Emma Gilger placed \$20,000 in a 10-year account paying 5% interest. The savings and loan where she placed her money pays a passbook rate of $3\frac{1}{2}\%$ compounded daily. She withdrew \$5000 of her money after 15 months. Find the interest that she earned on the \$5000.

26. TIME DEPOSIT Trish Hardison deposited \$18,680 in a 4-year account paying 6% interest. The bank where she put the money pays a passbook rate of $3\frac{1}{2}\%$ compounded daily. She withdrew \$2000 after 15 months. Find the interest that she earned on the \$2000.

Use the formulas and the value of e (Appendix C) to find the compound amount and the interest for the following deposits. Assume continuous compounding.

	Amount	Interest rate	Time
27.	\$8000	6%	2 years
28.	\$1200	8%	11 years

	Amount	Interest rate	Time
29.	\$4100.70	8%	9 months
30.	\$8008.43	6%	6 months

Solve the following application problems.

31. PARTIAL COLLATERAL FOR A LOAN An Italian firm deposited \$800,000 in a 2-year time deposit earning 6% compounded daily with a New York bank as partial collateral for a loan. At maturity, find (a) the compound amount and (b) the interest earned.

32. WELDING SHOP Joni Perez needs to borrow \$20,000 to open a welding shop but the bank will not lend her the money. Joni's uncle agrees to put up collateral for the loan with a \$20,000, 4-year certificate of deposit paying 7% compounded daily. This means that the bank will take all or part of his deposit if Joni should fail to repay the loan. Find (a) the compound amount earned by her uncle and (b) the interest earned if the funds remain at the bank for the full 4 years.

33. COMPARING INVESTMENTS Pat Metzger can choose between two investments: one pays 10% compounded semiannually, and the other pays $9\frac{1}{2}\%$ compounded continuously. If she wants to deposit \$17,000 for $1\frac{1}{2}$ years, which investment should she choose? How much extra interest will she earn by making the correct choice? (Hint: $e^{0.1425} = 1.15315308$.)

34. PENSION WITHDRAWAL Gary Orr wishes to invest \$62,904 he has withdrawn from his company's pension plan. One investment offered 12% compounded quarterly, while another offers 11.75% compounded continuously. Which investment should he choose if the money will be invested for 2 years? How much extra interest will he earn by making the proper choice? (Hint: $e^{0.235} = 1.26490877$.)

35. Is it important to you for your bank to have insurance on your savings account? Why or why not?

36. Go to your bank and find the current interest rates paid by them on savings accounts, IMMAs, and CDs.

13.3 Finding Time and Rate

Objectives

1. *Determine time given compound interest.*
2. *Determine rate.*
3. *Use more than one table to solve interest problems.*

We saw in Chapter 11 how to use the formula for simple interest, $I = PRT$, and find the value of any of the four variables when given the value of the other three. A similar thing can be done with the formula involving compound interest, $M = P(1 + i)^n$.

While basic algebra was used to solve $I = PRT$, more advanced algebra is needed for $M = P(1 + i)^n$. This different procedure is required because of the exponent n in the formula.

Objective 1 *Determine Time Given Compound Interest.* The next example shows how to find time for an investment.

Example 1 Finding Time

James Thompson needs \$45,000 to start a small music store but only has \$29,000. How long must he wait to start the business if he can earn 10% compounded annually?

Solution Use the formula $M = P(1 + i)^n$, with M = \$45,000, P = \$29,000, and i = 0.10. The value of n is unknown.

$$M = P(1 + i)^n$$
$$\$45{,}000 = \$29{,}000(1 + 0.10)^n$$
$$\$45{,}000 = \$29{,}000(1.1)^n$$
$$\frac{\$45{,}000}{\$29{,}000} = (1.1)^n \qquad \text{Divide by \$29,000.}$$
$$1.551724138 = (1.1)^n$$

Solving this equation would require advanced algebra. However, a good approximation can be found by looking down column A of the 10% page of the interest table in Appendix D. The number 1.61051000 in row 5 is very close to the number on the left side of the equation above. Since interest is compounded annually, he must wait about 5 years.

Example 2 Finding Time to Double

Suppose the general level of inflation in the economy averages 4% per year. Find the number of years it would take for the general level of prices to double.

Solution To find the number of years it will take for \$1 worth of goods and services to cost \$2, solve for n in the following equation.

$$2 = 1(1 + 4\%)^n$$

or

$$2 = 1(1 + 0.04)^n$$

where $M = 2$, $P = 1$, and $i = 4\%$. This equation simplifies as follows.

$$2 = (1.04)^n$$

As in the previous examples, look down column A of the 4% page of the interest table to find the number closest to 2. The closest number is 2.02581652. This number corresponds to 18 periods, so to the nearest year, the general level of prices would double in 18 years.

NOTE Prices would double about every $7\frac{1}{2}$ years if inflation were 10%. ■

Objective 2 ***Determine Rate.*** The same method used for finding time can be used to find rate. Find rate by looking in column A across from the proper number of periods. Then look through several pages corresponding to different rates, as necessary.

Example 3 Finding Rate

One of your classmates needs to borrow \$8000 for 1 year. He has been offered a loan with interest compounded monthly and a compound amount of \$8493.42. Find the rate.

Solution Use the equation $M = P(1 + i)^n$, with i unknown. Notice that n is the number of compounding periods, which is 12.

$$\$8493.42 = \$8000(1 + i)^{12}$$

Divide both sides by \$8000.

$$1.0616775 = (1 + i)^{12}$$

Find the row for 12 periods in column A of the interest tables in Appendix C. Check this number in column A under different rates as necessary until you find the number closest to 1.0616775. The closest number is 1.06167781 corresponding to an interest rate of $\frac{1}{2}\%$ per month, or

$$\frac{1}{2} \times 12 = 6\% \text{ per year}$$

Objective 3 *Use More than One Table to Solve Interest Problems.* Some problems require the use of more than one table, as in the next example.

Example 4 Using More than One Table

Jean King deposits \$2500 in an account paying 6% compounded quarterly. After 4 years the rate drops to 5% compounded semiannually. Find the amount in her account at the end of 7 years.

Solution First find the future value after 4 years. Look in column A of the interest rate table for $4 \times 4 = 16$ periods and $6\% \div 4 = 1.5\%$ interest to find 1.26898555.

$$\$2500(1.26898555) = \$3172.46$$

Now find the future value at the end of another 3 years. Look in column A of the interest rate table for $2 \times 3 = 6$ periods and $5\% \div 2 = 2.5\%$ interest to find 1.15969342.

$$\$3172.46(1.15969342) = \$3679.08$$

King will have \$3679.08 in her account at the end of 7 years.

Example 4 can also be solved using the equation for maturity value $M = P(1 + i)^n$. First find the value at the end of year 4 using $n = 16$ periods and $i = 1.5\%$. Use the y^x key on your scientific calculator and solve the following.

$$M = \$2500(1 + 0.015)^{16} = \$3172.46$$

This amount then grows at $i = 2.5\%$ every 6 months for $n = 6$ semiannual periods. Use the y^x key again.

$$M = \$3172.46(1 + 0.025)^6 = \$3679.08$$

Notice the future value is exactly the same as the value obtained using the tables.

13.3 Exercises

Complete this table. Round time to the nearest period, interest to the nearest whole percent per year, and money to the nearest cent. (As a review, some of these problems do not require the methods of this section.)

	Principal	Compound Amount	Interest Rate	Compounded	Time in Years
1.	\$6,200	\$7,384.30	6%	annually	______
2.	\$12,000	\$17,631.94	8%	annually	______
3.	\$3,600	\$4,824.34	10%	semiannually	______
4.	\$2,500	\$3,267.40	11%	semiannually	______
5.	______	\$11,082.73	8%	semiannually	7
6.	______	\$9,043.63	10%	quarterly	6
7.	\$12,000	\$15,149.72	______	annually	4
8.	\$13,200	\$22,680.06	______	annually	8
9.	\$8,500	\$13,403.64	______	quarterly	$5\frac{3}{4}$
10.	\$3,100	\$3,765.63	______	monthly	$3\frac{1}{4}$

11. During your working career of 40 years, approximately how many times can you double an investment at 6% interest and at 8% interest? (Hint: Use Column A in Appendix D.) What does this mean to you? (See Objective 2.)

12. Explain how to find the interest rate given principal, compound amount, compounding period, and time in years. (See Objective 2.)

Solve the following application problems. Round as in Exercises 1–10.

13. INVESTING FOR A NEW AUTO Betty Ford will need a new car and invests \$12,000 in a fund containing U.S. Treasury bills paying about 6% per year compounded quarterly. Find the compound amount in (a) 3 years and (b) 5 years.

14. INHERITANCE Jim Pierce inherits \$85,000 and places it in an account expected to yield 10% per year compounded semiannually. Find the compound amount in (a) 10 years and in (b) 25 years.

15. UNKNOWN INTEREST RATE Roy Bledsoe deposits \$46,000 in an account that has interest compounded semiannually. In $2\frac{1}{2}$ years the account contains \$58,708.95. Find the interest rate paid.

16. UNKNOWN INTEREST RATE Find the interest rate paid if a deposit of \$35,200 becomes \$61,723.41 in $4\frac{3}{4}$ years, with interest compounded quarterly.

17. STUDENT LOAN A student loan of \$5200 at 9% compounded semiannually resulted in a maturity value of \$5934.06. Find the term or length of the loan.

18. HOME CONSTRUCTION LOAN A home construction loan of \$80,000 at 12% compounded monthly resulted in a maturity value of \$92,877.52. Find the term or length of the loan.

Use the ideas of Example 2 in the text to answer the following questions. Find the time, to the nearest year, it would take for the general level of prices in the economy to double if the average annual inflation rate is as follows.

19. $2\frac{1}{2}\%$

20. 6%

21. $3\frac{1}{2}\%$

22. 4%

23. GASOLINE DEMAND Nationwide, demand for gasoline is increasing at a rate of about 2% per year. If it continues to increase at this rate indefinitely, find the number of years before the oil companies would need to double the supply of gasoline.

24. POPULATION GROWTH Assume that world population grows at a rate of 2% per year throughout the 21st century (Figure 13.4). Use column A in Appendix D to estimate the number of years it requires for world population to double. Approximately how many times would world population double during 100 years?

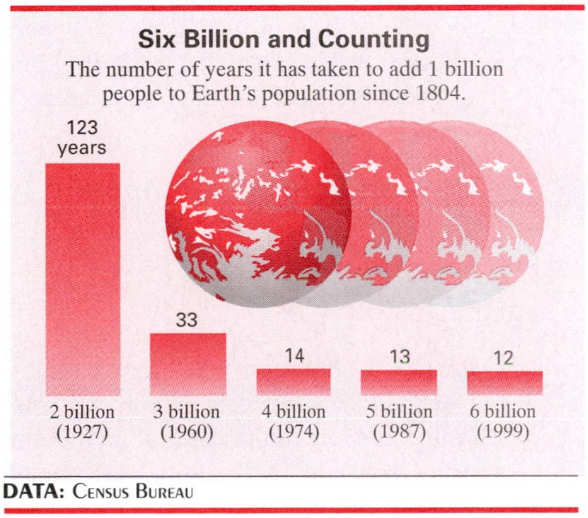

Figure 13.4

The remainder of these exercises may require the use of any of the tables presented so far in this chapter.

25. INVESTING Dawn Young deposits $10,000 at 8% compounded quarterly. Two years after she makes the first deposit, she adds another $20,000, also at 8%, compounded quarterly. What total amount will she have 5 years after her first deposit?

26. SAVING FOR COLLEGE Holly Crabtree has decided to help her grandson go to college. In 1997, she deposited $5000 in an account paying 6% compounded semiannually. Exactly two years later she deposited another $2000 in a different account paying 5% compounded quarterly. Find the total amount after an additional 5 years (in 2004).

For Exercises 27–30 use Table 13.3 for the certificates of deposit.

27. COMPOUND INTEREST Robert Jay deposits $4200 in an account paying 8% per year compounded quarterly. After $3\frac{3}{4}$ years, the compound amount is used to buy a

3-year certificate of deposit paying 8% compounded annually. Find the final amount Jay will have after $6\frac{3}{4}$ years.

28. LONG-TERM SAVINGS Lupe Gonzales puts $4000 in a 5-year certificate of deposit (CD) paying 6% compounded annually. After 5 years, she hopes to take the future value from the CD and invest it in a mutual fund yielding 8% per year compounded annually for 20 years. Find the final amount that Gonzales will have after 25 years.

29. COMPOUND INTEREST Jean Sides has $11,000 in a mutual fund paying 10% compounded semiannually. After the money has been on deposit for 4 years, $5000 is removed to buy a certificate of deposit paying 7% compounded daily for 3 years. (The remaining money stays at 10% compounded semiannually.) Find the total amount on deposit in both accounts, both principal and interest, after the 7 years.

30. COMPOUND INTEREST Frank Sabas has $45,000 in an account paying 8% compounded quarterly. After $2\frac{1}{4}$ years, Sabas removes $20,000 and buys a 5-year certificate of deposit paying 6% compounded daily. (The remaining money continues to earn 8% compounded quarterly.) Find the total amount on deposit in both accounts, both principal and interest, after the $7\frac{1}{4}$ years.

13.4 Present Value at Compound Interest

Objective
1 *Review the meaning of future value and present value.*
2 *Find the present value.*

Objective 1 *Review the Meaning of Future Value and Present Value.* In Sections 13.1 and 13.2, compound interest calculations were used to find the maturity value or **future value** (value at some future date) of an investment by using the formula $M = P(1 + i)^n$ or by using Appendix D. The initial deposit, interest rate, and term were given and future value was calculated.

In this section, the problem is reversed—the future value, interest rate, and term are given. The goal is to find the amount that must be deposited today to produce the desired future value at the specified future date. The amount that must be deposited today is called the **present value**; the amount to be accumulated at some specific future date is the future value (Figure 13.5).

Objective 2 *Find the Present Value.* Earlier, P was used for **principal**, which was the amount deposited at the beginning of the term. The phrase *present value* is often used instead of

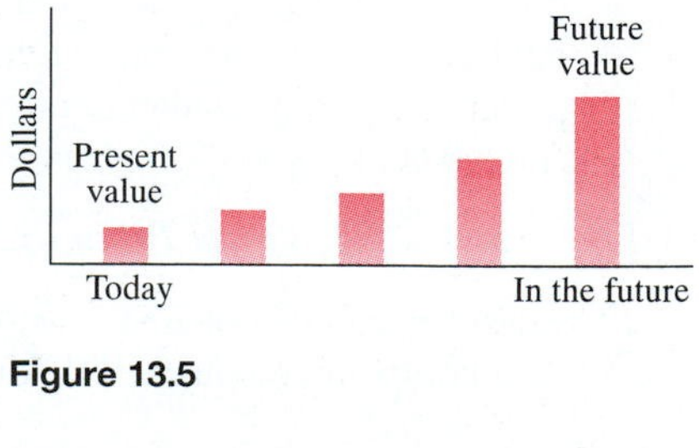

Figure 13.5

principal if the amount that must be deposited today is unknown. The letter P will also be used for present value. Find P as follows:

$$M = P(1 + i)^n$$

$$\frac{M}{(1 + i)^n} = \frac{P(1 + i)^n}{(1 + i)^n} \qquad \text{Divide by } (1 + i)^n.$$

$$P = \frac{M}{(1 + i)^n} \qquad \text{Use this formula for present value.}$$

The present value, P, is found by dividing the compound amount, M, by $(1 + i)^n$. The values of $(1 + i)^n$ are given in column A of the interest table. Another way to solve this problem is to rewrite the formula as shown below and use column B of the interest table.

From column A $\qquad\qquad$ From column B

$$\frac{M}{(1 + i)^n} \quad \text{as} \quad M \cdot \frac{1}{(1 + i)^n}$$

In summary, the present value, P, of the future amount, M, at an interest rate of i per period for n periods is shown here.

$$P = \frac{M}{(1 + i)^n} = M \cdot \frac{1}{(1 + i)^n}$$

(Use column B of the interest table in Appendix D for the value of $\frac{1}{(1 + i)^n}$).

Example 1 Finding Present Value

Shannon Walker estimates that her son will need \$3000 in 5 years in order to purchase a new, state-of-the-art, multimedia computer system, including software and a top-quality color printer. What lump sum deposited today at 8% compounded annually will produce the needed \$3000?

Solution Use the formula in the box with $M = \$3000$, $i = 8\%$, and $n = 5$.

$$P = M \cdot \frac{1}{(1 + i)^n}$$

$$P = \$3000 \cdot \frac{1}{(1 + 0.08)^5} = \$3000 \cdot \frac{1}{(1.08)^5}$$

Look on the 8% page of the interest table in Appendix D, in row 5, and the present value column (column B) to find that $\frac{1}{(1.08)^5} = 0.68058320$.

$$P = \$3000(0.68058320) = \$2041.75$$

A deposit of \$2041.75 today at 8% compounded annually will produce \$3000 in 5 years. Of this,

$$\$3000 - \$2041.75 = \$958.25$$

represents interest earned on the initial deposit of \$2041.75.

The same result could have been found by using the compound interest table, column A, and *dividing*. Looking up 8% and 5 periods in column A (or calculating $[1.08]^5$) gives 1.46932808. Now divide.

$$P = \frac{\$3000}{1.46932808} = \$2041.75$$

NOTE Here both methods gave the same answer. Sometimes the answers may differ by a few cents because of rounding error. ▬

Scientific
Calculator Approach

The calculator solution to this example uses parentheses to set the denominator apart.

3000 [÷] [(] 1 [+] .08 [)] [y^x] 5 [=] 2041.75

Financial
Calculator Approach

Future value, interest rate per compounding period, and number of compounding periods are known in this problem. Present value is the unknown.

3000 [*FV*] 8 [*i*] 5 [*n*] [*PV*] −2041.75

The minus sign indicates that Shannon Walker must pay out $2041.75 (the present value) now in order to earn the future value at a future date.

Example 2 Finding Present Value

Green Acres Sprinklers, Inc. will receive $185,000 in 3 years as part of a lawsuit settlement. Find the present value if money can be invested at 5% compounded quarterly.

Solution 3 years = 12 quarters and 5% compounded quarterly = $1\frac{1}{4}\%$ per quarter. Look at the $1\frac{1}{4}\%$ page in Appendix D and column B on the row for 12 periods to find 0.86150860. Multiply to find the present value.

$$\$185{,}000(0.86150860) = \$159{,}379.09$$

A deposit of $159,379.09 at 5% compounded quarterly for 3 years will result in $185,000.

Example 3 Finding the Value of a Business

Tom Fredrickson owns a men's clothing store worth $125,000. The business is doing well, with Fredrickson confident that the value of the business will increase at the rate of 16% per year, compounded semiannually, for the next 4 years. If he sells the business, he will invest the proceeds at 8% compounded quarterly. What sale price should he insist on?

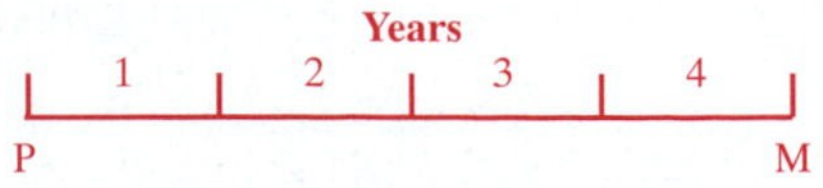

Solution This problem requires two steps. First decide on the future value of the business. This is done with the formula for compound amount, $M = P(1 + i)^n$. Here, $P =$ \$125,000, $i = 16\% \div 2 = 8\%$, and $n = 4 \times 2 = 8$ periods.

$$M = \$125{,}000(1 + 8\%)^8$$
$$= \$125{,}000(1.85093021) \quad \text{Use column A.}$$
$$= \$231{,}366.28$$

Now find the present value of this sum, assuming that money can be invested at 8% compounded quarterly. Use the formula

$$P = M \cdot \frac{1}{(1 + i)^n}$$

with $M = \$231{,}366.28$, $i = 8\% \div 4 = 2\%$, and $n = 4 \times 4 = 16$ periods.

$$P = \$231{,}366.28 \cdot \frac{1}{(1 + 0.02)^{16}}$$
$$= \$231{,}366.28(0.72844581) \quad \text{Use column B.}$$
$$= \$168{,}537.80$$

Fredrickson should not sell the business for less than \$168,537.80—an investment of this amount at 8% compounded quarterly for 4 years will produce the same future amount as the growth in the value of the business.

13.4 Exercises

Find the present value and the amount of interest earned for each of the following. Round to the nearest cent.

	Amount Needed	Time	Interest	Compounded
1.	\$4,800	3 years	8%	annually
2.	\$3,300	4 years	11%	annually
3.	\$12,200	$2\frac{1}{2}$ years	12%	quarterly
4.	\$7,000	$3\frac{1}{4}$ years	10%	quarterly
5.	\$8,500	1 year	12%	monthly
6.	\$18,000	$1\frac{1}{2}$ years	9%	monthly

7. Explain the difference between present value and future value. (See Objective 1.)

8. List three reasons why a business owner may need to place a value on a business. (See Objective 3.)

Solve each of the following application problems.

9. PACK MULES David Fontana Backpackers will need to buy some new pack mules in 5 years. These mules will cost a total of \$5000. What lump sum should the firm invest today at 7%, compounded annually, in order to be able to buy the mules? How much interest will be earned?

10. EXPANSION PLANS The Christmas Store needs $120,000 in 3 years to open a second store. Find the lump sum that should be invested today at 8% compounded quarterly to produce the needed funds. Find the interest earned.

11. RETAIL MEAT INDUSTRY The retail meat industry is gradually becoming more automated. Halmost Meat Market feels that it will need to invest $37,500 in 4 years for new machinery. What lump sum must be invested today at 12%, compounded quarterly so that the firm will have the necessary money?

12. ENCYCLOPEDIA A new *Encyclopedia of World Business* is being prepared that will cost $3675 for all volumes. The encyclopedia will be available in 2 years. What lump sum should be invested by the Manhattan Business Library at 8%, compounded quarterly, so that the library will have enough to buy the new books?

13. SMALL GROCERY STORE Jose Martinez, an immigrant from Mexico, estimates that he needs $9000 to start a small grocery store in 3 years. How much must he deposit today if his credit union will pay 8% compounded quarterly?

14. SAVING FOR COLLEGE Mrs. Jones wants all of her grandchildren to go to college and decides to help financially. How much must she give to each child at birth if they are to have $10,000 on entering college 18 years later, assuming 6% interest compounded annually?

15. TIME VALUE OF MONEY Assume that money can be invested at 8% compounded quarterly. Which is larger, $2500 now or $3800 in 5 years? (Hint: First find the present value of $3800, then compare present values.)

16. TIME VALUE OF MONEY Assume that money can be invested at 10% compounded semiannually. Which is larger, $3000 now or $7500 in 10 years? (Hint: First find the present value of $7500, then compare present values.)

17. FUTURE VALUE An investment of $30,000 earns interest of 10% compounded semiannually for $2\frac{1}{2}$ years. Find the future value of the investment. If money can be deposited elsewhere at 8% compounded quarterly, find the present value of the investment.

18. SELLING AN ESTATE Judith McGrath sold her deceased grandfather's home and accepted a note for $65,000 with interest of 9% compounded semiannually for 3 years. Find the future value. Then find the present value of this amount if funds can be invested elsewhere at 6% compounded monthly.

19. CAFE OF MODERN ART The owner of Cafe of Modern Art signs a note for $16,800 at 10% simple interest for 4 years. Find the maturity value of the note. What should the holder of the note be willing to accept in payment for the note today if funds can be invested at 6% per year compounded quarterly?

20. MATURITY VALUE In 6 years, Susan Hessney must pay off a note with a face value of $12,000, and simple interest of 9% per year. Find the maturity value of the note. What should the holder of the note be willing to accept in complete payment today, if money can be invested at 6% per year compounded quarterly?

21. VALUING A HAIR SALON Jessie Jones believes her hair salon is worth $20,000 and estimates that its value will grow at 10% per year for the next 3 years. If she sells the business, the funds will be invested at 8% compounded quarterly. (a) Find the future value if she holds onto the business. (b) What price should she insist on now if she sells the business?

22. BIKE SHOP Andy Sargent figures his bike shop is worth $88,000 if sold today and that it will grow in value at 8% per year for the next 6 years. If he sells the business, the funds will be invested at 5% compounded semiannually. (a) Find the future value of the shop. (b) What price should he insist on at this time if he sells the business?

Chapter 13 Quick Review

Chapter Terms

Review the following terms to test your understanding of the chapter. For any terms you do not know, refer to the page number found next to that term.

certificate of deposit (CD) [p. 470]
compound amount [p. 457]
compound interest [p. 456]
compounded daily [p. 467]
continuous compounding [p. 473]
early withdrawal penalty [p. 472]
effective annual yield [p. 464]
effective rate of interest [p. 464]
exponent [p. 459]
future amount [p. 457]
future value [p. 482]
inflation [p. 463]
insured money market account (IMMA) [p. 473]
nominal rate [p. 459]
passbook account [p. 467]
period of compounding [p. 460]
present value [p. 482]
principal [p. 482]
Regulation DD [p. 467]
savings account [p. 467]
simple interest [p. 456]
stated rate [p. 459]
time deposit account [p. 470]

Concept	Example
13.1 Finding the compound amount and the interest Determine the interest rate per period and the number of compounding periods. Use the formula $M = P(1 + i)^n$ and the interest table to calculate the compound amount. Then subtract the principal from the compound amount to obtain interest.	Find the compound amount and the interest if \$7200 is deposited at 9% compounded monthly for 4 years. $i = \frac{0.09}{12} = 0.0075$ $n = 4 \times 12 = 48$ $M = P(1 + i)^n$ $= \$7200(1.0075)^{48}$ $= \$7200(1.43140533) = \$10{,}306.12$ $I = M - P$ $= \$10{,}306.12 - \$7200 = \$3106.12$
13.1 Determining the effective rate of interest (also called the effective annual yield) Determine the value from column A of the interest table in Appendix D that corresponds to the rate and number of periods. Subtract 1 from this value and round as required.	Find the effective rate of interest to the nearest hundredth, if the annual rate is 8% compounded quarterly. $i = \frac{0.08}{4} = 0.02;\ n = 4$ 1.08243216 table value for $i = 2\%$, $n = 4$ $1.08243216 - 1.00000000 = 0.08243216$ The effective rate is 8.24%.
13.2 Interest compounded daily Determine the number of days and then find the value of $(1 + i)^n$ from a table or calculator. Multiply this value by the principal to obtain compound amount. Subtract principal from compound amount to obtain interest.	Janice deposited \$1535 on September 5 in an account paying $3\frac{1}{2}\%$ and withdrew everything on December 5. How much interest was earned? There are 25 additional days in September, 31 days in October, 30 days in November, and 5 days in December. $25 + 31 + 30 + 5 = 91$ days Compound amount = \$1535(1.008763788) = \$1548.45 Interest = \$1548.45 − \$1535 = \$13.45

13.2 Finding the interest and balance when withdrawals are made

Find the amount that earns interest for the entire quarter and determine compound amount and interest on this sum. Then, find compound amount and interest for the amount withdrawn. Find the final balance by adding the total interest to the original balance and subtracting the withdrawal.

Tom had \$1500 in his $3\frac{1}{2}\%$ savings account on October 1. He withdrew \$450 on October 15. How much interest had he earned by January 1 and what was his final balance?

\$1500 − \$450 = \$1050 earns interest for the entire 92-day quarter.

$$\text{Compound amount} = \$1050(1.008860519) = \$1059.30$$

$$\text{Interest} = \$1059.30 - \$1050 = \$9.30$$

The withdrawn \$450 earns interest for 14 days.

$$\text{Compound amount} = \$450(1.001343303) = \$450.60$$

$$\text{Interest} = \$450.60 - \$450 = \$0.60$$

$$\text{Total Interest} = \$9.30 + \$0.60 = \$9.90$$

$$\text{Final Balance} = \$1500 + \$9.90 - \$450 = \$1059.90$$

13.2 Finding the compound amount and interest earned on time deposits

Use Table 13.3 to find the value corresponding to the rate and number of years. Next multiply the table value by the initial amount to obtain compound amount. Then subtract initial investment from compound amount to obtain interest.

Mike invests \$15,000 in a certificate of deposit paying 7% compounded daily. Find the compound amount and interest after 4 years.

$$\text{Table value} = 1.32309429$$

$$\text{Compound amount} = \$15{,}000(1.32309429) = \$19{,}846.41$$

$$\text{Interest} = \$19{,}846.41 - \$15{,}000 = \$4{,}846.41$$

13.2 Finding the penalty for early withdrawal of funds from a time deposit account

Determine the number of months for which interest will be paid. Find value in table for rate and time. Then multiply this by initial deposit to obtain compound amount. Find the interest by subtracting initial deposit from compound amount.

Tom deposited \$5000 in a 3-year CD paying 7% compounded daily. He withdrew the money 15 months later. The passbook rate at his bank is $3\frac{1}{2}\%$ compounded daily. Find the amount of interest earned. Tom will get the passbook rate for 15 − 3 = 12 months, or 4 quarters. Passbook rate of $3\frac{1}{2}\%$ for 4 quarters days gives a table value of 1.035121585.

$$\text{Compound amount} = \$5{,}000 \times 1.0353121585 = \$5175.61$$

$$\text{Interest} = \$5175.61 - \$5{,}000 = \$175.61$$

13.2 Determining the interest and compound amount for continuous compounding

Use the formula

$$M = Pe^{yr}$$

Use the table in Appendix C to find the value of e^{yr}. Find interest from the formula $I = M - P$.

Find the interest and compound amount for \$2000 at 6% compounded continuously for 8 years.

$$M = Pe^{yr}$$

$$M = \$2000e^{(8)(0.06)} = \$2000e^{0.48} = \$2000(1.6160744) = \$3232.15$$

$$I = M - P$$

$$I = \$3232.15 - \$2000 = \$1232.15$$

13.3 Finding the time given interest rate, compound amount, and principal

Substitute values for M, P, and i in the formula $M = P(1 + i)^n$; then divide both sides by the value of P. Find the value in column A of the interest table (Appendix D) closest to the left-hand side of the equation for the correct value of i. The corresponding value of n is the number of periods required to obtain the compound amount.

How long would it take for \$1800 at 9% compounded semiannually to become \$7361.97?

$$i = 9\% \div 2 = 4.5\%$$
$$M = P(1 + i)^n$$
$$\$7361.97 = \$1800(1.045)^n$$
$$4.0899833 = (1.045)^n$$

In column A of the 4.5% page, the number 4.08998104 is *very close* to 4.0899833, so $n = 32$ and it would take 32 6-month periods, or 16 years.

13.3 Finding the rate, given principal, compound amount, and time

Divide the compound amount by the principal. Use the row of the interest table that corresponds to the number of interest periods and find the value of column A closest to the quotient obtained above. You may need to examine several pages. Then find the interest rate per year.

A principal of \$7000 is deposited for 18 months with interest compounded monthly. Find the rate if the compound amount is \$8373.03.

$$M = P(1 + i)^n$$
$$\$8373.03 = \$7000(1 + i)^{18}$$
$$1.196147 = (1 + i)^{18}$$

Find the row for 18 periods in the interest table. Check the table until the entry closest to 1.196147 is found. The closest number is 1.19614748 per month, or $1\% \times 12 = 12\%$ per year.

13.4 Finding the present value of a future amount

Use the formula

$$P = \frac{M}{(1 + i)^n}$$

where M = future value

i = interest rate per period

n = number of periods

Substitute values for i, n, and M and find the value of $\frac{1}{(1 + i)^n}$ from column B of the interest table.

What amount deposited today at 9% compounded monthly will produce \$8000 in 3 years?

$$M = \$8000$$
$$i = \frac{9\%}{12} = 0.75\%$$
$$n = 3 \times 12 = 36$$
$$P = \frac{M}{(1 + i)^n}$$
$$= M \cdot \frac{1}{(1 + i)^n}$$
$$P = \$8000 \times \frac{1}{(1 + 0.0075)^{36}}$$
$$= \$8000 \cdot \frac{1}{(1.0075)^{36}}$$
$$= \$8000(0.76414896)$$
$$= \$6113.19$$

Chapter 13 Review Exercises

The answer section includes answers to all Review Exercises.

Find the compound amount and the interest earned for each of the following. Round to the nearest cent. [13.1]

1. \$12,400 at $6\frac{1}{2}\%$ compounded annually for 6 years
2. \$7000 at 6% compounded semiannually for 4 years
3. \$4800 at 10% compounded quarterly for 3 years
4. \$18,000 at 7% compounded quarterly for 4 years
5. \$9000 at 9% compounded monthly for $2\frac{1}{2}$ years
6. \$12,000 at 6% compounded monthly for $3\frac{1}{4}$ years

Find the effective rate of interest corresponding to the following nominal rates. Round to the nearest hundredth of a percent. [13.1]

7. 7% compounded quarterly
8. 8% compounded quarterly
9. 7% compounded semiannually
10. 9% compounded monthly

Find the interest earned by the following. Assume $3\frac{1}{2}\%$ interest compounded daily. Round to the nearest cent. [13.2]

	Interest	Amount	Date Deposited	Date Withdrawn
11.	______	\$2900	March 4	May 30
12.	______	\$6000	May 20	July 1
13.	______	\$3020.80	July 15	October 1

Find the amount on deposit on the first day of the next quarter when the following sums are deposited as indicated. Assume $3\frac{1}{2}\%$ compounded daily. Round to the nearest cent. [13.2]

	Amount at End of Quarter	Amount	Date Deposited
14.	________	\$3500	January 22
15.	________	\$7200.35	April 22
16.	________	\$9600.40	August 10

Use Table 13.3 on page 472 to find the compound amount for each of the following certificates of deposit. Assume daily compounding. Round to the nearest cent. [13.2]

	Compound Amount	Amount Deposited	Interest Rate	Time in Years
17.	________	\$4000	7%	3
18.	________	\$6500	6%	2
19.	________	\$8800	8%	4

Find the compound amount and the interest for the following deposits. Assume continuous compounding. **[13.2]**

	Compound Amount	Interest	Amount	Interest Rate	Time
20.	________	________	$12,600	8%	7 years
21.	________	________	$5,000	7%	5 years

Complete the following table. Round time to the nearest period, interest to the nearest whole percent per year, and money to the nearest cent. **[13.3]**

	Principal	Compound Amount	Interest Rate	Compounded	Time in Years
22.	$4300	$5,754.37	______	annually	5
23.	$8600	$11,566.04	______	quarterly	3
24.	$7500	$9,914.25	______	quarterly	$3\frac{1}{2}$
25.	$6000	$7,986.00	10%	annually	______
26.	$8400	$10,357.20	6%	monthly	______

Find the present value for each of the following. Round to the nearest cent. Also find the interest earned. **[13.4]**

	Amount Needed	Time	Interest Rate	Compounded	Present Value	Interest Earned
27.	$14,300	3 years	$5\frac{1}{2}$%	annually	________	________
28.	$4,000	3 years	9%	semiannually	________	________
29.	$6,000	5 years	10%	quarterly	________	________
30.	$3,000	4 years	6%	monthly	________	________

Solve the following application problems.

31. Discount Auto Insurance deposited $1800 in a savings account paying $3\frac{1}{2}$% compounded daily on January 1 and deposited an additional $2300 in the account on March 12. Find the balance on April 1. **[13.2]**

32. Susan Chu opens a savings account on June 10 with a deposit of $4000 and makes an additional deposit of $1200 on July 6. Find the balance on September 1 assuming an interest rate of $3\frac{1}{2}$% compounded daily. **[13.2]**

33. Sam Cracker deposits $18,000 in an 8% certificate of deposit compounded daily for 10 years. Find the compound amount and the interest earned. (Use Table 13.3 on page 472.) **[13.2]**

34. Melissa Smith deposited $7350 in an 8%, 2-year certificate of deposit with daily compounding. She took out the money 15 months later. What was the interest Melissa received if her bank has a passbook rate of $3\frac{1}{2}$% compounded daily? (Use the rules for finding the early withdrawal penalty as given in the text.) **[13.2]**

35. Wooden Desk Company will need $47,500 in 4 years for capital improvements. What lump sum should be invested today at 12% compounded monthly to yield $47,500? **[13.4]**

36. Computers, Inc., accepted a 2-year note for $12,540 in lieu of immediate payment for computer equipment sold to a local firm. Find (a) the maturity value given a rate of 10% compounded annually and (b) the present value of the note at 6% per year compounded semiannually. **[13.4]**

37. Jack Taylor invests $15,000 in an account paying 6% per year compounded annually. In how many years will the compound amount become $18,937.15? **[13.3]**

38. The consumption of electricity in one area has increased historically at 6% per year. If it continued to increase at this rate indefinitely, find the number of years before the electric utility would need to double the amount of generating capacity. **[13.3]**

Chapter 13 Summary Exercise
Valuing a Chain of McDonald's Restaurants

James and Mary Watson own a small chain of McDonald's restaurants that is valued at $2,300,000. They believe that the chain will grow in value at 12% per year compounded annually for the next 5 years. If they sell the chain, the funds will be invested at a rate of 6% compounded semiannually. They expect inflation to be 4% per year for the next 5 years. Ignore taxes and answer the following; round answers to the nearest dollar.

(a) Find the future value of the chain after 5 years. Then find the price they should sell the chain for if they wish to have the same future value at the end of 5 years.

(b) Find the future value of the chain if it only grows at 2% per year for 5 years. Then find the price they should ask for the chain given a 2% growth rate per year.

(c) What future value would the chain be worth if it grew at their expected rate of inflation. Find the price they should ask for the chain if it grows at the rate of inflation.

(d) Complete the following table.

Growth Rate	Market Future Value	Value Today
2%	________	________
4% (inflation)	________	________
12%	________	________

NOTE The value of the chain varies by more than $1 million dollars depending on the rate of growth assumed for the business for the next 5 years.

Net Assets Business on the Internet

Bank of America

Statistics

- 1998: Revenues of $30.5 billion
- $357 billion combined total of loans and deposits
- 200,000 employees
- 2 million business customers

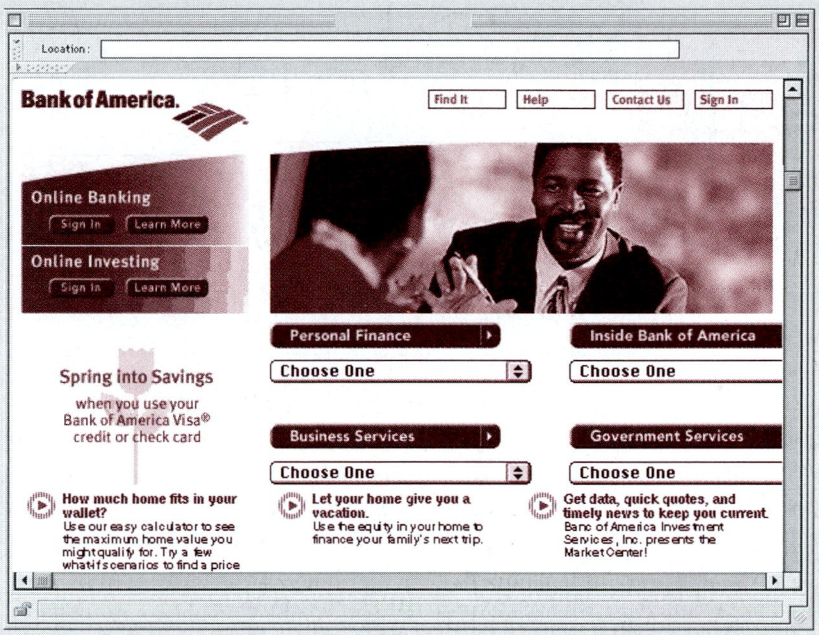

In 1998, NationsBank merged with Bank of America creating one of the world's largest banks. Bank of America has 30 million customers and does business in 190 countries around the world. Customers can obtain cash or make deposits at over 14,000 automated teller machines (ATMs) around the world. The new and enlarged Bank of America now has about 8% of all bank deposits in the United States, a call center that handles 38 million customers per month, and processing centers that process 37 million checks each day.

Using the World Wide Web, the bank allows customers to log in to the bank's computer system and balance their checking account, pay bills electronically, move money between checking and savings accounts, and make investment trades. Small businesses can also borrow money while on-line if they have the appropriate credit with the bank.

1. At 37 million checks each day, how many checks will the bank process per year assuming they process checks 5 days a week, 52 weeks per year?

2. Assume Bank of America borrows a total of $100,000,000 from several large customers and that the bank pays an average of 2.5% per year in interest. Further assume they can lend out $80,000,000 of this amount at 8% compounded quarterly for 1 year. (Use the compound interest table.) Find the difference between interest collected and interest paid out by the bank.

3. Assume that a large corporation deposits $8,400,000 in an interest-bearing checking account yielding $3\frac{1}{2}$% compounded daily for six days. Find the interest.

4. Assume one branch of Bank of America has $140,000,000 in assets. Given inflation of 4%, what amount of assets would they need after one year to have the same purchasing power?

14 Annuities and Sinking Funds

Chapters 11–13 involved **lump sums**, in which a sum of money was deposited or borrowed. However, few can afford to prepare for their children's college education or their own retirement by making one large deposit. Rather, as Figure 14.1 shows, people *start early and make periodic payments* to save up for major events such as retirement.

Why should people plan for retirement ***before*** *they retire?*

14.1 Amount of an Annuity

Objectives

1. *Identify the types of annuities.*
2. *Find the amount of an annuity.*
3. *Find the amount of an annuity payment.*
4. *Find the number of payments.*
5. *Find the amount of an annuity due.*
6. *Find the value of an individual retirement account (IRA).*

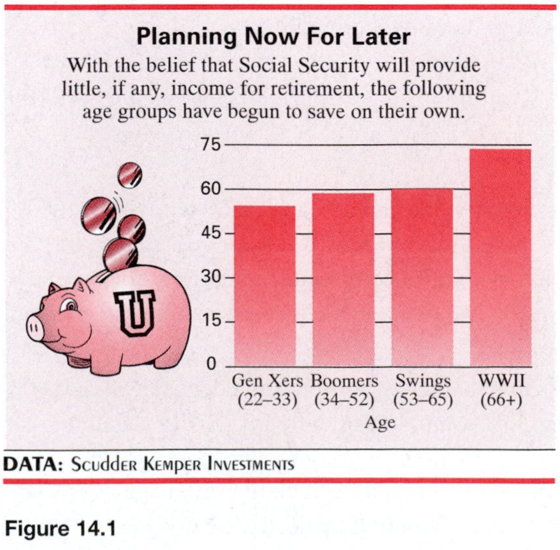

Planning Now For Later
With the belief that Social Security will provide little, if any, income for retirement, the following age groups have begun to save on their own.
75
60
45
30
15
0
U
Gen Xers (22–33)
Boomers (34–52)
Swings (53–65)
WWII (66+)
Age
DATA: Scudder Kemper Investments

Figure 14.1

An **annuity** is a sequence of *equal payments*. For example, a small firm may use an annuity to accumulate funds to buy a new truck. Assume the company makes deposits of \$2500 at the end of each year for 6 years into an account paying 8% per year. This is an annuity since it has a sequence of equal payments.

The time between the payments of an annuity is the **payment period**, with the time from the beginning of the first payment period to the end of the last called the **term of the annuity**. The **amount**, or **future value, of the annuity**, the final sum on deposit, is defined as the sum of the compound amounts of all the payments, compounded to the end of the term.

Objective 1 *Identify the Types of Annuities.* There are many kinds of annuities. A **contingent annuity** has a variable beginning or ending dates. For example, an insurance policy that pays a fixed amount per month beginning when a person is age 65 and lasting until that person's death has a variable ending date and is a contingent annuity. A person might prepare a will leaving a fixed annual sum to the surviving spouse for a fixed number of years. This is an example of a contingent annuity with a variable beginning date (because the annual payments do not start until the death of the person making the will). We will not discuss contingent annuities in this book. Instead, we will discuss only annuities certain.

An **annuity certain** has a specified beginning date and a specified ending date. The example of payments for 6 years is an annuity certain. Other examples would include the 48 monthly payments needed to pay off a car loan, or 18 annual payments into a college fund beginning at the birth of a child.

With an **ordinary annuity**, payments are made at the *end* of each period of time, while an **annuity due** has payments made at the *beginning* of each period. A **simple annuity** has payment dates matching the compounding period. For example, an annuity with payments made quarterly and having interest compounded quarterly is a simple annuity, while an annuity with payments made quarterly and interest compounded daily is not a simple annuity. While there are formulas available for annuities that are not simple (see a textbook on mathematics of finance), we discuss only simple annuities in this book.

Let us return to the annuity mentioned at the beginning of this section—payments of \$2500 at the end of each year for 6 years into an account paying 8% per year compounded annually. By the definitions, this annuity is *certain, ordinary, and simple*. To find the amount of the annuity, or the final sum on deposit, find the sum of the compound amounts of all the payments compounded to the end of the term.

The first deposit of \$2500 will produce a compound amount of

$$\$2500(1 + 0.08)^5 = \$2500(1.08)^5$$

Use 5 as the exponent instead of 6 since the money is deposited *at the end* of the first year, and earns interest *for only 5 years*. The second payment of \$2500 will produce a compound amount of $\$2500(1.08)^4$. Continuing in this way, the amount of the annuity is $\$2500(1.08)^5 + \$2500(1.08)^4 + \$2500(1.08)^3 + \$2500(1.08)^2 + \$2500(1.08)^1 + \$2500$. (The last payment earns no interest at all.) From column A of the interest table in Appendix D, this sum is calculated as follows.

$$\begin{aligned}&\$2500(1.46932808) + \$2500(1.36048896) + \$2500(1.25971200)\\&\quad + \$2500(1.16640000) + \$2500(1.08000000) + \$2500\\&= \$3673.32 + \$3401.22 + \$3149.28 + \$2916 + \$2700 + \$2500\\&= \$18{,}339.82\end{aligned}$$

The amount of the annuity is \$18,339.82. Since 6 deposits of \$2500 each were made, the interest earned is

$$\$18{,}339.82 - (6 \times \$2500) = \$3339.82$$

Figure 14.2 shows each payment into the annuity and the compound amount of each payment.

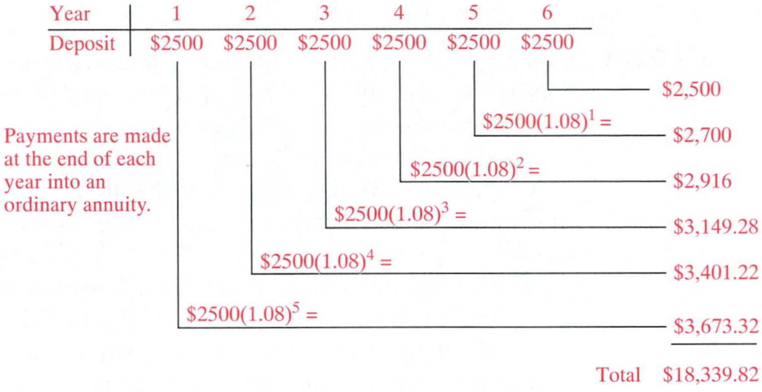

Figure 14.2

Scientific *Calculator Approach*

The value of \$2500 invested for 5 years, or \$2500(1.08)^5, is found as follows.

1.08 [y^x] 5 [×] 2500 [=] 3673.32

The other values on the right-hand side of Figure 14.2 can be found in a similar fashion.

Objective 2 ***Find the Amount of an Annuity.*** This method for finding the amount of the annuity of \$2500 at the end of each year for 6 years was tedious. With algebra, a formula for the amount of an annuity can be derived. This formula follows.

If a simple annuity is made up of payments of R dollars at the end of each period for n periods, at a rate of interest i per period, then the amount of the annuity S, is given by this equation.

$$S = R\left[\frac{(1+i)^n - 1}{i}\right]$$

The quantity in brackets is commonly written $s_{\overline{n}|i}$ (read "*s*-angle-*n*-at-*i*") so that

$$S = R \cdot s_{\overline{n}|i}.$$

The value of $s_{\overline{n}|i}$ for a specific n and i is found in Appendix D.

For our annuity, $R = \$2500$, $n = 6$, and $i = 0.08$. By the formula, the amount of the annuity is as follows.

$$S = R \cdot s_{\overline{n}|i}$$

$$S = \$2500 \cdot s_{\overline{6}|0.08}$$

$$= \$2500\left[\frac{(1 + 0.08)^6 - 1}{0.08}\right] \quad \text{Definition of } s_{\overline{6}|0.08}$$

$$= \$2500\left[\frac{(1.08)^6 - 1}{0.08}\right]$$

The value of $(1.08)^6$ can be found on the 8% page, line $n = 6$, of column A in the interest table in Appendix D or by using a calculator as shown below.

$$S = \$2500\left(\frac{1.58687432 - 1}{0.08}\right)$$

$$= \$2500(7.33592900)$$

$$= \$18{,}339.82$$

This is exactly the same result found earlier.

Scientific *Calculator Approach*

The expression $S = \$2500\left[\frac{(1 + 0.08)^6 - 1}{0.08}\right]$ above can be evaluated using a calculator as follows.

2500 [×] [(] [(] 1 [+] .08 [)] [y^x] 6 [−] 1 [)] [÷] .08 [=] 18339.82

Notice that the numerator in the brackets is entered as $((1 + 0.08)^6 - 1)$ in order to maintain the order of operations.

To save time, tables of values of $s_{\overline{n}|i}$ have been calculated using the expression given above for the amount of an annuity for different values of n and i. They appear in column C of the interest table in Appendix D. Look in this table for 8% and 6 periods to find

$$s_{\overline{6}|0.08} = 7.33592904$$

NOTE The number from the table is slightly different from the result found by using the formula. This variation is due to rounding. ▬

Example 1 Finding the Amount of an Annuity

The hospital where Tish Baker works pays \$150 at the end of each quarter into her retirement plan. If she chooses an investment fund that yields 6% per year compounded quarterly, how much will she have in 10 years? How much interest will she have earned?

Solution Baker's payments form an ordinary annuity, with $R = \$150$, $n = 10 \times 4 = 40$ compounding periods (quarters), and $i = 6\% \div 4 = 1.5\%$ per compounding period (per quarter). The amount, or future value, of the annuity is found as follows.

$$S = \$150(54.26789391) \quad \text{Using column C of table}$$
$$= \$8140.18$$

The interest earned is the future amount less all payments.

$$I = \$8140.18 - 40 \times \$150 = \$2140.18$$

The answer can be checked using the formula for calculating the amount of an annuity from page 496.

$$S = R\left[\frac{(1+i)^n - 1}{i}\right]$$

$$S = \$150\left[\frac{(1+0.015)^{40} - 1}{0.015}\right]$$

$$= \$150\left(\frac{1.814018409 - 1}{0.015}\right)$$

$$= \$150(54.26789391)$$

$$= \$8140.18$$

Notice that the factor 54.26789391 found using a scientific calculator is exactly the same as the factor from the table in Appendix D.

Financial *Calculator Approach*

Payment, number of compounding periods, and interest rate per compounding period are known. Future value is the unknown. Be sure to set your calculator to an ordinary annuity with payments at the end of each period, before keying in the following.

−150 [PMT] 40 [n] 1.5 [i] [FV] 8140.18

Example 2 Finding the Amount of an Annuity

At the birth of her grandson, Junella Smith decides to help pay for his college education. She commits to making deposits of \$600 into an account *at the end of each 6 months for 18 years*. Smith has narrowed her choices to an annuity at an insurance company paying 6% compounded semiannually, a CD at a bank paying 5% compounded semiannually, or a savings account at a credit union paying $3\frac{1}{2}\%$ compounded semiannually. (a) Find the amount of the annuity in each case. (b) Which is preferred?

Solution

(a) There are $2 \times 18 = 36$ semiannual periods in 18 years. The semiannual interest, value from Appendix D, and future amount is shown in the table on page 499. In each case, the future amount is found using either $S = R\left[\frac{(1+i)^n - 1}{i}\right]$ or $S = R \cdot s_{\overline{n}|i}$ and table values where $R = \$600$, $n = 36$, and i is the semiannual interest.

	Semiannual Interest	Value from Appendix D	Future Amount
Annuity	$\frac{6\%}{2} = 3\%$	63.27594427	\$37,965.57
CD	$\frac{5\%}{2} = 2.5\%$	57.30141263	\$34,380.85
Savings account	$\frac{3\frac{1}{2}\%}{2} = 1.75\%$	49.56612949	\$29,739.68

(b) The annuity with the insurance company is best as long as all three investment choices have equal risk in terms of loss of funds. Notice that an increase in the interest rate from $3\frac{1}{2}\%$ to 6% results in the accumulation of an extra \$8225.89 in 18 years.

NOTE Example 2 in Appendix A.2 shows how a financial calculator can be used to solve this same type of problem. ■

Objective 3 ***Find the Amount of an Annuity Payment.*** Sometimes the lump sum that will be needed at some time in the future is known, and the amount of each payment into an annuity that will guarantee the availability of the necessary amount must be found.

Example 3 Finding the Amount of an Annuity Payment

Wolf Films needs to buy a new model video camera 3 years from now. The firm wants to deposit an equal amount at the end of each quarter for 3 years in order to accumulate enough money to pay for the camera. The camera will cost \$2400, and the bank pays 8% interest compounded quarterly. Find the amount of each of the 12 deposits to be made.

Solution This example describes an ordinary annuity with $S = \$2400$, $i = 2\%$ ($8\% \div 4 = 2\%$), and $n = 3 \times 4 = 12$ periods. The unknown here is the amount of each payment, R. Use the formula for the amount of an annuity.

$$\$2400 = R \cdot s_{\overline{12}|0.02}$$

From column C of the interest table, with $n = 12$ and $i = 2\%$, find $s_{\overline{12}|0.02} = 13.41208973$, and

$$\$2400 = R(13.41208973)$$

Dividing both sides by 13.41208973 gives

$$\$178.94 = R$$

Each payment must be \$178.94. (As we shall see in Section 3 of this chapter, these payments form a sinking fund.)

Financial *Calculator Approach*

Future value, number of compounding periods, and interest rate per compounding period are known. Payment is the unknown.

2400 [FV] 12 [n] 2 [i] [PMT] −178.94

The negative sign in front of the payment indicates that the company must make payments into a fund, i.e. payments the company pays out are shown as a negative number.

Problem-Solving Hint The value of n is the number of compounding periods (*not years*) and i is the interest per compounding period (*not interest per year*).

Objective 4 ***Find the Number of Payments.*** The next example shows how to find the number of periods necessary to accumulate a certain amount of money.

Example 4 Finding the Number of Payments

The Dunlaps want to purchase their first home and need \$4000 for a down payment. They can save \$275 per quarter. If they deposit this amount in an account paying 3% interest compounded quarterly, how long will it be before they can purchase a home?

Solution Here the amount of the annuity (S), the amount of each payment (R), and the interest rate per period (i) are known. The unknown is n, the number of quarters for which deposits must be made. Start with the formula for the amount of an annuity.

$$S = R \cdot s_{\overline{n}|i}$$

Replace S with \$4000, R with \$275, and i with $3\% \div 4 = \frac{3}{4}\%$.

$$\$4000 = \$275 \cdot s_{\overline{n}|0.0075}$$

Divide both sides by \$275 to get the following.

$$14.54545455 = s_{\overline{n}|0.0075}$$

Go to the $\frac{3}{4}\%$ page in the interest table. Look down column C for the first number *larger* than 14.54545455. The first number that is larger is 14.70340370, which corresponds to 14 quarters. The family must save for 14 quarters, or $3\frac{1}{2}$ years; at the end of this period of time they will have

$$\$275(14.70340370) = \$4043.44.$$

NOTE The table value used is the one *larger* than the calculated value, not closest to it. ■

Objective 5 ***Find the Amount of an Annuity Due.*** As mentioned earlier, with ordinary annuities, payments are made at the *end of each time period.* With **annuities due** the payments are made at the *beginning of the time period.* The difference between an annuity and an annuity due is subtle, but both are used in the financial services industry. The following charts may help you understand the difference between an ordinary annuity and an annuity due.

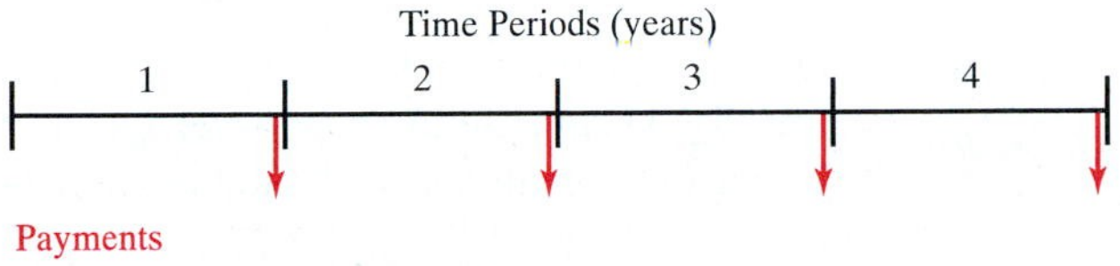

In an *ordinary annuity*, payments are made at the *end* of each period as shown above. Think of the payments being made in the *last* second of the *last* day of each period.

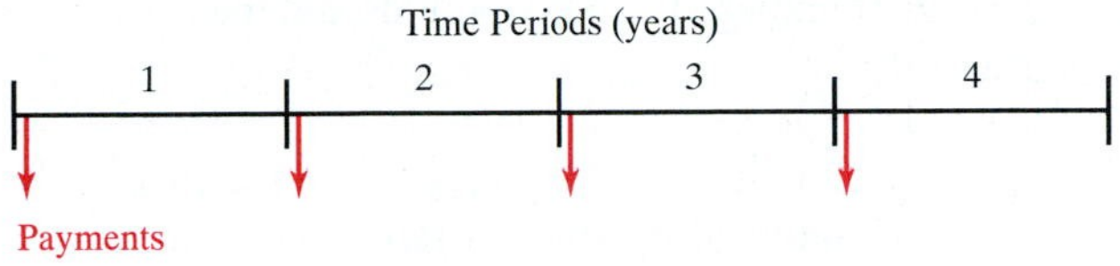

In an *annuity due*, payments are made at *beginning* of each period. Think of the payments being made on the *first* second of the *first* day of each period. If you subtract one from each time period in the chart below, the annuity due above is the same as the following ordinary annuity, *except for* the extra, last payment in the ordinary annuity.

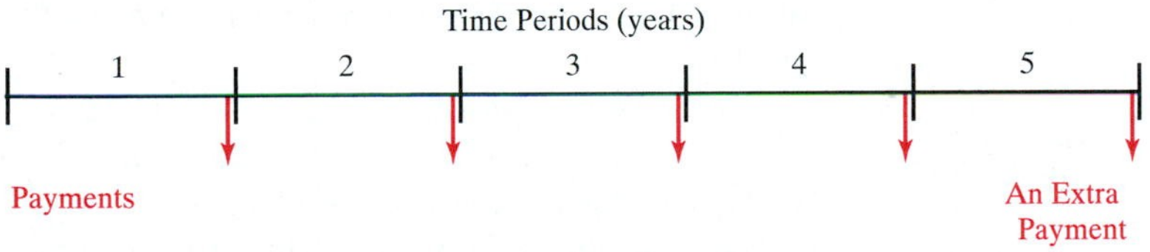

Therefore, the amount of an annuity due can be found using an ordinary annuity table and following the rule stated below.

> To find the amount of an annuity due, treat each payment as if it were made at the *end of the preceding period*. Then use column C of the interest table to find $s_{\overline{n}|i}$ for *one additional period*; to compensate for this, subtract the amount of one payment.

Example 5 Finding the Amount of an Annuity Due

Mr. and Mrs. Thompson set up an investment program using an *annuity due* with payments of \$500 *at the beginning of each quarter*. Find the amount of the annuity if they make payments for 7 years in a mutual fund paying 12% compounded quarterly.

Solution Look at column C of the 3% table in Appendix D and row 29 to find 45.21885020.

$$\$500(45.21885020) = \$22{,}609.43$$

Now subtract one payment.

$$S = \$22{,}609.43 - \$500 = \$22{,}109.43$$

The account will have \$22,109.43 in 7 years.

Financial
Calculator Approach

First change the calculator settings to annuity due (payments at the beginning of each period). Then enter known values for payment, number of periods, and interest rate per period. Finally, press the future value key to find the unknown.

−500 [PMT] 28 [n] 3 [i] [FV] 22109.43

Problem-Solving Hint When solving for the amount of an annuity due, be sure to add *one period* to the total number of annuity periods and then subtract *one payment* to find the total in the account.

Objective 6 ***Find the Value of an Individual Retirement Account (IRA).*** One great investment vehicle that working people can use to save for retirement is an **individual retirement account**, also called an **IRA**. There are two types of IRA accounts, regular IRAs and Roth IRAs.

Deposits to a **regular IRA** account are usually excluded from federal income taxes in the current year. Therefore, a contribution to a regular IRA may reduce the amount of income taxes *that you must pay this year*. Interest earned in a regular IRA is not subject to income taxes in the year earned. At retirement, the account holder withdraws funds from the regular IRA and pays taxes at that time. As a result, a regular IRA may allow you to save for retirement *without having to pay taxes* on the savings for years or even decades!

Deposits to a **Roth IRA** are not excluded from federal taxes in the year paid, so they *do not* reduce current income taxes. However, the deposit and interest do grow tax-free. In addition, one huge advantage is that the withdrawals from the Roth IRA at retirement are *not* subject to income taxes. This offers you a great opportunity to save money for retirement without having to pay taxes when you withdraw the funds.

Roth IRA

WHICH IRA IS BEST FOR YOU?
If you've been contributing to a non-deductible, traditional IRA, the new Roth IRA where earnings grow tax free, is by far the better option. If you're trying to decide between a deductible IRA and a Roth, here are some facts that may help.

	DEDUCTIBLE IRA	ROTH IRA
Tax deductible?	If you qualify	No
Taxable at withdrawal?	Yes	No
Penalty for early withdrawal?	Yes, prior to age 59.5	Yes*
Mandatory withdrawal age?	70.5	None
Penalty-free withdrawals?	$10,000 for first-time home buyers; unlimited for education	$10,000 for first-time home buyers, after five-year wait; unlimited for education

*Never any penalty for withdrawing your own contributions, but a penalty applies to withdrawal of any gains within five years of opening the account and/or before turning 59.5.

NOTE Early withdrawal from either a regular IRA or a Roth IRA may result in a penalty. ■

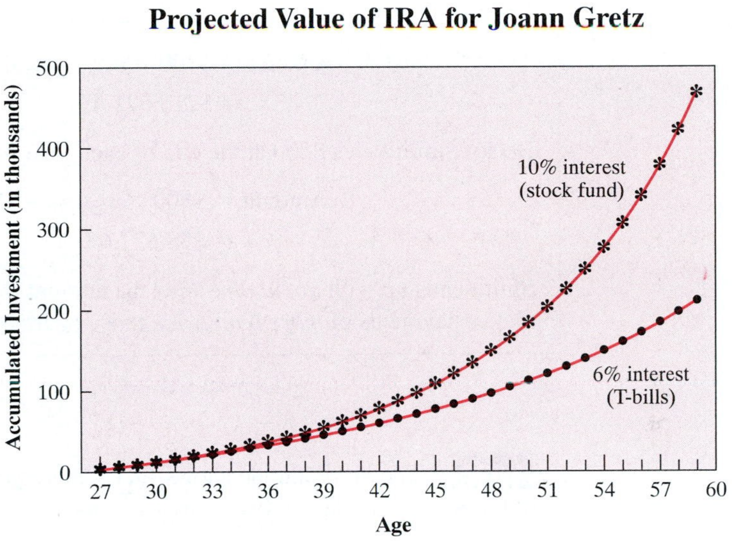

Figure 14.3

Example 6 Finding the Value of an IRA

At 27, Joann Gretz sets up an IRA with Merrill Lynch where she plans to deposit \$2000 at the end of each year until age 60. Find the amount of the annuity if she invests in (a) a Treasury bill fund, which has historically yielded 6% compounded annually versus (b) a stock fund, which has historically yielded 10% compounded annually. Assume that future yields equal historical yields.

Solution Age 60 is $60 - 27 = 33$ years away.

(a) Treasury bill fund: Look in column C of Appendix D with $n = 33$ and $i = 6\%$ to find 97.34316471.

$$\text{Amount} = \$2000 \times 97.34316471 = \$194{,}686.33$$

(b) Stock fund: Look in column C of Appendix D with $n = 33$ and $i = 10\%$ to find 222.25154420.

$$\text{Amount} = \$2000 \times 222.25154420 = \$444{,}503.09$$

Gretz can see the projected difference in the results of the Treasury bill fund and the stock fund using the graph in Figure 14.3. On this basis, she decides to try to maximize return on investment in the future—however, she plans to do so without taking too much risk.

Example 7 Saving for Retirement

At age 20, Tom Jones begins saving for retirement by making end-of-year payments of \$300 into an IRA account, until age 65 (a total of 45 years). Becky Smith doesn't begin saving for retirement until age 40, at which time she deposits \$800 at the end of each year until age 65 (a total of 25 years). Assume both funds earn 10% compounded annually and find the amount available at age 65 to both individuals.

Solution Tom Jones saves \$300 at the end of each year for 45 years at 10% per year.

$$\text{Amount} = \$300 \times s_{\overline{45}|0.10} = \$300 \times 718.90483685$$
$$= \$215{,}671.45$$

Becky Smith saves \$800 at the end of each year for 25 years at 10% per year.

$$\text{Amount} = \$800 \times s_{\overline{25}|0.10} = \$800 \times 98.34705943$$
$$= \$78{,}677.65$$

Smith ends up with *about one-third* the amount that Jones ends up with, *even though* she makes payments of *more than twice* those of Jones.

NOTE The last example shows that it is very important to begin saving early in life, but it is never too late to begin saving. ▬

14.1 Exercises

Find each of the following values using column C of the table in Appendix D.

1. $s_{\overline{15}|0.03}$
2. $s_{\overline{30}|0.08}$
3. $s_{\overline{10}|0.09}$
4. $s_{\overline{50}|0.12}$

Find the value of the following ordinary annuities. Interest is compounded annually. Find the total amount of interest earned.

5. $R = \$850$, $i = 0.06$, $n = 28$
6. $R = \$200$, $i = 0.10$, $n = 20$
7. $R = \$1000$, $i = 0.08$, $n = 25$
8. $R = \$2000$, $i = 0.065$, $n = 50$

Find the value of each of the following annuities and then find the total amount of interest earned.

9. $R = \$1400$, 10% per year compounded quarterly for 8 years.
10. $R = \$250$, 8% per year compounded quarterly for 10 years.
11. $R = \$800$, 9% per year compounded monthly for 4 years.
12. $R = \$500$, 6% per year compounded monthly for 3 years.

Find the amount of each of the following annuities due. Assume that interest is compounded annually. Find the amount of interest earned.

13. $R = \$1200$, $i = 0.075$, $n = 8$
14. $R = \$1400$, $i = 0.08$, $n = 10$
15. $R = \$17{,}544$, $i = 0.08$, $n = 6$
16. $R = \$64{,}715$, $i = 0.06$, $n = 12$

Assume that you are able to invest \$900 per year into an ordinary annuity in an individual retirement account. Find the amount accumulated after each of the time periods at each of the interest rates shown in the table on page 505.

	Annual Compounding Rate				
	5%	7%	9%	11%	13%
10 years	**17.** _______	**18.** _______	**19.** _______	**20.** _______	**21.** _______
20 years	**22.** _______	**23.** _______	**24.** _______	**25.** _______	**26.** _______
30 years	**27.** _______	**28.** _______	**29.** _______	**30.** _______	**31.** _______

32. Explain the difference in the methods used to calculate the amount of an annuity versus the amount of an annuity due. (See Objectives 2 and 5.)

33. Explain the difference between a regular IRA and a Roth IRA. Which is better for a young person? (See Objective 6.)

34. **BECOMING A MILLIONAIRE IS A CHOICE** A 22-year-old has decided to be a millionaire by investing \$2000 at the end of each year into an IRA with a mutual fund containing stocks. How long will it take him to reach his goal if his funds earn 10% compounded annually? If they earn 12% compounded annually? What does this mean to you? (See Objective 2.)

Find the amounts of each of the following annuities due. Find the amount of interest earned.

35. Payments of \$500 made at the beginning of each quarter for 10 years at 7% compounded quarterly

36. \$50 deposited at the beginning of each month for 4 years at 6% compounded monthly

37. \$100 deposited at the beginning of each quarter for 9 years at 8% compounded quarterly

38. \$1500 deposited at the beginning of each semiannual period for 11 years at 12% compounded semiannually

Find the periodic payment that will amount to the following sums under the given conditions.

39. S = \$20,000, interest at $7\frac{1}{2}$% compounded annually, payments made at the end of each year for 10 years

40. S = \$50,000, interest is 7% compounded semiannually, payments made at the end of each semiannual period for 8 years

41. S = \$50,000, interest is 12% compounded quarterly, payments made at the end of each quarter for 8 years

42. S = \$24,000, interest is 9% compounded monthly, payments are made at the end of each month for $3\frac{1}{4}$ years

Find the minimum number of payments that must be made to accumulate the given amount of money at the stated interest rate. Also, find the amount of the annuity.

43. \$15,000 needed, payments of \$450 are made at the end of each quarter at 6% compounded quarterly

44. \$21,000 needed, payments of \$1500 are made at the end of each 6-month period at 10% compounded semiannually

45. \$40,000 needed, payments of \$750 are made at the end of each month at 10% compounded monthly

46. \$37,500 needed, payments of \$1250 are made at the end of each month at 12% compounded monthly

Solve each of the following application problems.

47. INVESTING IN AN IRA At the end of each quarter, the Thompsons invest \$450 in an IRA earning 8% compounded quarterly. Find the amount of their annuity at the end of $3\frac{1}{2}$ years.

48. AMOUNT OF ANNUITY Sharon Stone deposits \$2000 at the end of each year in an account earning 10% compounded annually. Find the amount of the annuity after 25 years.

49. CHILD-SUPPORT PAYMENTS Becky Smith decides to put the child-support payments from her previous husband into an annuity for the education of her two children. At the end of each month for 4 years, she puts \$300 into a mutual fund that has yielded 12% compounded monthly. Find (a) the amount of the annuity and (b) the interest earned.

50. AMOUNT OF ANNUITY A large corporation deposits \$500,000 at the end of each year for 6 years in an account paying 10% interest compounded annually. Find (a) the amount of the annuity and (b) the interest earned, both to the nearest dollar.

51. FUTURE VALUE Pam Parker deposits \$2435 at the beginning of each year for 8 years in an account paying 6% compounded annually. She then leaves that money alone, with no further deposits, for an additional 5 years. Find the final amount on deposit after the entire 13-year period. Find the total amount of interest earned.

52. ANNUITY DUE Tim Bessner invests \$500 at the *beginning* of each quarter for 10 years in an account paying 7% compounded quarterly. He then leaves the money alone, with no further deposits, for an additional 6 years. Find the amount on deposit after the 16 years and find total interest earned.

53. RECREATIONAL VEHICLE Tom and Sandra Kip are trying to save \$4000 as the down payment on a new recreational vehicle. If they can deposit \$125 at the end of each month in an account paying 4% compounded monthly, how long will it take them to accumulate the necessary amount? How much will they actually accumulate?

54. NEW SHOWROOM Western Motors needs \$120,000 as a down payment on a new showroom. If the company deposits \$10,000 at the end of each quarter at 5% compounded quarterly, how long will it take them to get the needed money? How much will the company actually have?

55. DAUGHTER THROUGH COLLEGE In order to put his daughter through college, Bill Thomas plans to invest \$250 per quarter for 10 years in either (a) an annuity paying 8% compounded quarterly or (b) a CD yielding 6% compounded quarterly. Find the amount of the annuity for both choices.

56. COLLEGE FUND The Crockers have decided to invest \$1500 at the end of each year for 12 years in a college fund for their twin grandsons. They can choose either (a) a mutual fund of domestic stocks expected to yield 10% per year or (b) a savings account expected to yield about 4% per year. Find the amount of the annuity in both cases.

57. Complete Section 7 of the foldout.

14.2 Present Value of an Annuity

Objectives

1. *Calculate the present value of an annuity.*
2. *Use the formula for the present value of an annuity.*
3. *Find the equivalent cash price.*

The previous section discussed how to find the amount of an annuity after a series of equal, periodic payments. This section considers the present value of such an annuity. There are two ways to think of the **present value of an annuity**.

1. The *present value of an annuity* is a lump sum that can be deposited today that will amount to the same future amount as would the periodic *payments* of an annuity, (see Example 1 on page 508).
2. The present value of an annuity is a lump sum that could be deposited today so that equal periodic *withdrawals* could be made (see Example 2 on page 509).

Objective 1 *Calculate the Present Value of an Annuity.* As an example of this second way of looking at the present value of an annuity, let us find the amount that *must be deposited today*, at 10% compounded annually, so that \$1500 could be removed *at the end of each year for 6 years.*

The amount that must be deposited today is the sum of the present values of each of the separate withdrawals. In other words, today's deposit must include the present value of the withdrawal to be made at the end of the first year. The present value in 1 year of \$1500 at 10% compounded annually is found by looking in column B of the interest table in Appendix D for 10% and 1 period, finding 0.90909091. The present value is

$$\$1500(0.90909091) = \$1363.64$$

It is also necessary to deposit today the present value of the \$1500 to be withdrawn at the end of the second year. Again use column B of the interest table, this time with 10% and

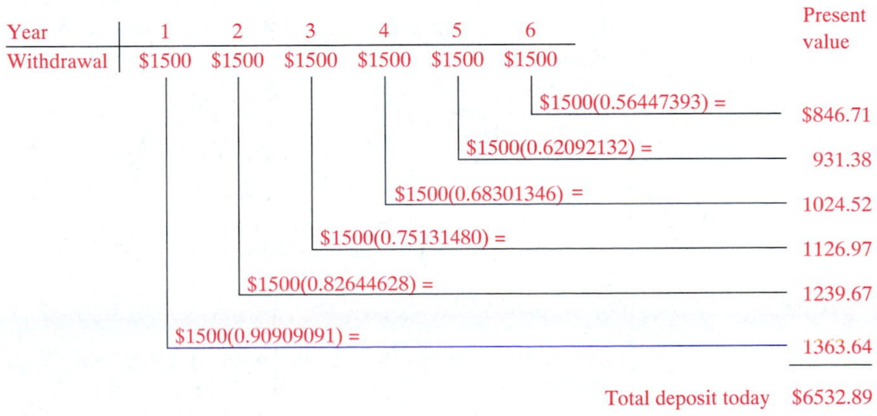

Figure 14.4

2 periods, finding 0.82644628. The present value for the withdrawal at the end of the second year is

$$\$1500(0.82644628) = \$1239.67$$

Continuing in this way for all six withdrawals gives the result shown in Figure 14.4.

A lump sum deposit today of \$6532.89 at 10% compounded annually would permit withdrawals of \$1500 at the end of each year for 6 years. Also, a lump sum deposit today of \$6532.89 left in an account for 6 years would produce *the same final total* as deposits of \$1500 at the end of each year for 6 years, with all deposits at 10% compounded annually.

Objective 2 ***Use the Formula for the Present Value of an Annuity.*** The formula for the present value of an annuity follows.

The present value of an annuity A, of R dollars at the end of each period for n periods, at a rate of interest i per period, follows.

$$A = R\left[\frac{(1+i)^n - 1}{i(1+i)^n}\right]$$

The expression in brackets is abbreviated $a_{\overline{n}|i}$.

$$A = R \cdot a_{\overline{n}|i}$$

$a_{\overline{n}|i}$ is read as *a*-angle-*n*-at-*i*.

Values of $a_{\overline{n}|i}$ are given in column D of the interest table. We can use the table to check the problem given on the preceding page. Look on the 10% page, for 6 payments at 10%, finding the number 4.35526070. Then multiply, which gives the following.

$$A = \$1500(4.35526070) = \$6532.89$$

This is the same answer found earlier.

Example 1 Finding the Present Value of an Annuity

Dion Martinez has decided to make annual payments of \$1200 at the end of each year for 15 years into an investment that she thinks will yield 8% compounded annually. What lump sum deposited today at 8% compounded annually will result in the same future value?

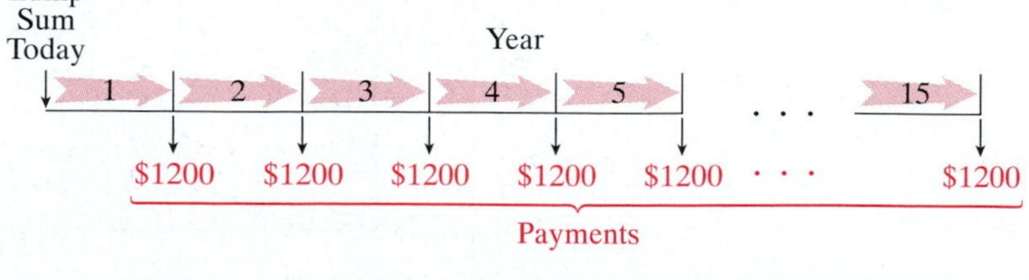

Solution Column D of the interest tables in Appendix D shows that $a_{\overline{15}|0.08} =$ 8.55947869.

Present value is $A = R \cdot a_{\overline{n}|i}$

$$A = \$1200 \cdot 8.55947869 = \$10{,}271.37$$

A lump sum of \$10,271.37 deposited today at 8% compounded annually will result in the same total after 15 years as year-end deposits of \$1200 for 15 years at 8%. This result can be checked by finding the future value of \$10,271.37 for 15 years at 8%. Use column A of Appendix D and the formula from Chapter 13 for compound amount M.

$$M = P(1 + i)^n$$
$$M = \$10{,}271.37 \cdot 3.17216911 = \$32{,}582.52$$

On the other hand, using the amount of an annuity that is column C of Appendix D, deposits of \$1200 at the end of each year for 15 years, at 8% produces the following.

$$S = R \cdot s_{\overline{n}|i}$$
$$S = R \cdot s_{\overline{15}|0.08}$$
$$= \$1200 \cdot 27.15211393 = \$32{,}582.54$$

The difference of 2 cents is due to rounding.

Financial
Calculator Approach

Set your calculator to an ordinary annuity (payments at the end of each period). Payment, number of compounding periods, and interest rate per compounding period are known. Present value is the unknown.

−1200 [PMT] 15 [n] 8 [i] [PV] 10271.37

NOTE There are two ways to produce \$32,582.54 in 15 years at 8% compounded annually—a single deposit of \$10,271.37 today, or payments of \$1200 at the end of each year for 15 years.

Example 2 Finding the Present Value

Fred and Sara Gonzales recently divorced. As a part of the divorce settlement, Fred must pay Sara \$2000 at the end of each year for 10 years. If money can be deposited at 6% compounded annually, find the lump sum he could deposit today to have enough money, with principal and interest, to make the payments.

Solution Look under $n = 10$ and $i = 0.06$ in column D of the interest tables to find $a_{\overline{10}|0.06} = 7.36008705$.

$$A = R \cdot a_{\overline{10}|0.06}$$
$$= \$2000 \cdot 7.36008705$$
$$= \$14{,}720.17$$

A deposit of \$14,720.17 today at 6% compounded annually is sufficient to make the 10 payments of \$2000 each. The difference between the sum of all payments, 10 × \$2000 = \$20,000, and the amount deposited today is the interest.

$$\text{Interest} = 10 \times \$2000 - \$14{,}720.17 = \$5279.83$$

NOTE Although the \$2000 withdrawals are at the end of each year, the original deposit must be made at the beginning of year 1.

Example 3 Finding the Present Value

An Australian engineering firm hires a new manager for their North American operations. The contract states that if the new manager works for 5 years, then he will receive a retirement benefit of $15,000 at the end of each semiannual period for 8 years. Find the lump sum the firm could deposit today to satisfy the retirement contract if funds can be invested at 10% compounded semiannually.

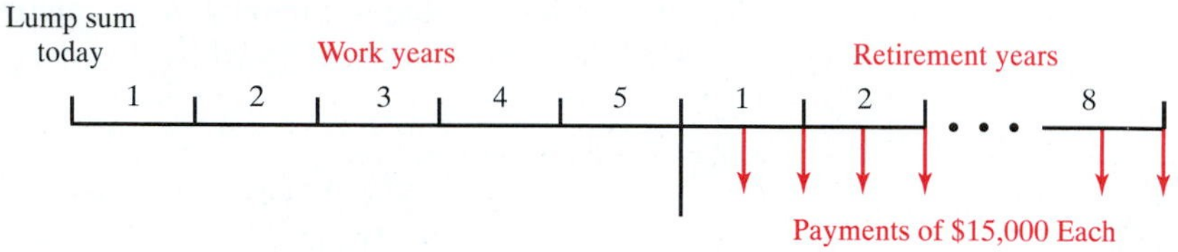

Solution First find the present value of an annuity with $2 \times 8 = 16$ periods at 5% per compounding period. Use column D of the interest tables to find $a_{\overline{16}|0.05} = 10.83776956$.

$$A = R \cdot a_{\overline{16}|0.05}$$
$$= \$15{,}000 \cdot 10.83776956$$
$$= \$162{,}566.54$$

The firm needs $162,566.54 at the end of the 5-year work period to satisfy the retirement benefits.

They can meet this liability today by depositing the present value of $162,566.54 given 10% compounded semiannually over 5 years. Use column B of the interest tables with $n = 5 \times 2 = 10$ and $i = \frac{10\%}{2} = 5\%$ and the formula from Chapter 13 to find the following.

$$P = \frac{M}{(1 + i)^n} = M\frac{1}{(1 + i)^n}$$
$$P = \$162{,}566.54 \cdot 0.61391325$$
$$= \$99{,}801.75$$

A lump sum of $99,801.75 deposited today will grow to $162,566.54 in 5 years which, with interest, is enough to make all 16 retirement payments of $15,000 each.

Example 4 Determining Retirement Income

Bill Jones wishes to retire at age 65 and withdraw $25,000 per year until age 90. (a) If money earns 8% per year compounded annually, how much will he need at age 65? (b) If Jones starts an IRA at age 32 by depositing $2000 per year in an account paying 8% per year compounded annually, will the IRA account contain enough money?

Solution

(a) The amount needed at age 65 is the present value of an annuity of $25,000 per year for $90 - 65 = 25$ years, with interest of 8% compounded annually. Use column D of the interest tables to find $a_{\overline{25}|0.08} = 10.67477619$.

$$A = R \cdot a_{\overline{n}|i}$$
$$A = \$25{,}000 \cdot 10.67477619$$
$$= \$266{,}869.40$$

Jones will need \$266,869.40 at age 65; this sum, at 8% compounded annually, will permit withdrawals of \$25,000 per year until age 90.

(b) Jones makes payments of \$2000 at the end of each year for 65 − 32 = 33 years, at 8% compounded annually. These payments form a regular annuity with $n = 33$ and $i = 0.08$. Use column C of the interest tables to find $s_{\overline{33}|0.08} = 145.95062044$.

$$S = R \cdot s_{\overline{n}|i}$$
$$S = \$2000 \cdot 145.95062044$$
$$= \$291{,}901.24$$

The value in the IRA account at age 65 (\$291,901.24) exceeds the amount needed to fund 25 yearly withdrawals of \$25,000 each (\$266,869.40). Therefore, Jones will have more than enough.

NOTE Example 7 in Appendix A.2 shows how a financial calculator can be used to solve a similar problem.

Objective 3 ***Find the Equivalent Cash Price.*** As we have seen, two sums of money *can look quite different and yet be equivalent* because the sums of money are available at different points in time. As in the next example, the present value of a future sum must be calculated to permit meaningful comparisons.

Example 5 Finding Equivalent Cash Price

Julia Smithers is an attorney trying to settle an estate. The estate owns a piece of property that is desired by two different developers. Developer A offers \$140,000 in cash, today, for the land. Developer B offers \$50,000 now as a down payment, with payments of \$8000 at the end of each quarter for 4 years. Money may be invested at 12% compounded quarterly. If Developer B offers a bank guarantee that the payment will be made, making each offer equally safe, which bid should the attorney accept?

Solution The bids can be compared *only by finding the present value* of the offer of Developer B. This offer is a down payment and an annuity of \$8000 at the end of each quarter for 4 years. The present value of this annuity is found with column D of the interest table, with $n = 4 \times 4 = 16$ periods, and, since money may be invested at 12% compounded quarterly, $i = 12\% \div 4 = 3\%$. The value of R is \$8000.

$$A = R \cdot a_{\overline{n}|i}$$
$$A = \$8000 \cdot a_{\overline{16}|0.03}$$
$$= \$8000(12.56110203)$$
$$= \$100{,}488.82$$

The present value of the annuity is \$100,488.82. Since Developer B also offers a down payment of \$50,000, the total cash price today of Developer B's offer is

$$\$50{,}000 + \$100{,}488.82 = \$150{,}488.82$$

This amount, \$150,488.82, is called the **equivalent cash price**. This exceeds the \$140,000 offered by Developer A, so the attorney should accept the bid of Developer B.

14.2 Exercises

Find each of the following values.

1. $a_{\overline{15}|0.075}$
2. $a_{\overline{20}|0.09}$
3. $a_{\overline{15}|0.12}$
4. $a_{\overline{40}|0.06}$

Find the present value of each of the following annuities. Round to the nearest cent.

	Amount per Payment	Payment at End of Each	No. of Years	Money Invested at
5.	$2400	year	9	8% compounded annually
6.	$2000	year	7	5% compounded annually
7.	$800	6 months	10	6% compounded semiannually
8.	$650	6 months	8	7% compounded semiannually
9.	$400	quarter	5	8% compounded quarterly
10.	$200	month	$3\frac{1}{2}$	5% compounded monthly

11. Explain the difference between future value and present value of an annuity. Illustrate the difference with an example.

12. An individual wins a state lottery that pays $50,000 a year for 20 years. Must the state have 1 million dollars to pay for this prize? If not, how can the state ensure that it has the necessary funds? (See Objective 1.)

Solve each of the following application problems. Round to the nearest cent.

13. STATE LOTTERY Gina Hardin was elated when she won a state lottery paying $85,480 at the end of each year for 20 years. Find the amount the state must set aside today to satisfy this annuity at 8% compounded annually.

14. DIVORCE SETTLEMENT Tom and Kitty Wysong recently divorced and Ms. Wysong agreed to pay Mr. Wysong $8000 every 6 months for 12 years since he had helped her through medical school. What amount should Ms. Wysong set aside today in an account earning 10% per year compounded semiannually in order to satisfy this obligation?

15. CARE OF DISABLED Mr. Roberts is retired and wishes to set up a 10-year annuity with quarterly payments of $8000 for the care of his disabled son. Find the amount he should deposit today at 6% interest compounded quarterly.

16. EQUIPMENT REPLACEMENT Physician's Pain Management plans to set aside an annual payment of $18,000 per year for 4 years for replacement of physical therapy equipment. Assuming 8% compounded annually, what lump sum deposited today would result in the same future value?

17. WITHDRAWALS What lump sum deposited today at 8% compounded semiannually would permit withdrawals of $1200 at the end of each 6-month period for 7 years? How much interest is earned?

18. WITHDRAWALS Find the amount that could be placed in a bank account today, at 8% compounded quarterly, to permit withdrawals of $1000 at the end of each quarter for 5 years.

19. RETIREMENT INCOME Lupé Garcia wishes to have a retirement income of \$15,000 at the end of each year for 25 years. Find the amount she must accumulate if she can earn (a) 8% per year versus (b) 12% per year.

20. RETIREMENT INCOME Kashundra Jones plans to make a lump sum deposit so that she can withdraw \$12,000 at the end of each year for 10 years. Find this lump sum if the money earns (a) 10% per year versus (b) 13% per year.

21. PAYING FOR COLLEGE Tom Potter estimates that his daughter's college needs, beginning in 8 years, will be \$3600 at the end of each quarter for 4 years. (a) Find the total amount needed in 8 years assuming 8% compounded quarterly. (b) Will he have enough money available in 8 years if he invests \$700 at the end of each quarter for the next 8 years at 8% compounded quarterly?

22. VAN PURCHASE In 4 years, Jennifer Videtto will need a delivery van, for her office supply store, that will require a down payment of \$10,000 with payments of \$1200 per month for 36 months. (a) Find the total amount needed in 4 years assuming 12% compounded monthly. (b) Will she have enough money available if she invests \$1000 at the end of each month for the next 4 years at 12% compounded monthly?

23. SALE OF COMMERCIAL BUILDING A small commercial building sells for a down payment of \$11,000, and payments of \$4000 at the end of each semiannual period for 20 years. Find the equivalent cash price of the building, if money may be invested at 10% compounded semiannually.

24. SALE OF FISHING BOAT A fishing boat is sold with a down payment of \$21,000 and payments of \$3500 at the end of each quarter for $7\frac{1}{2}$ years. Find the equivalent cash price of the boat if money may be invested at 12% compounded quarterly.

Based only on present value, which of the following bids should be accepted?

25. \$100,000 today, or \$51,000 down and \$8000 at the end of each year for 12 years; money may be invested at 10% compounded annually

26. \$140,000 today, or \$86,000 down and \$2000 at the end of each month for 3 years; money may be invested at 12% compounded monthly

27. \$420,000 today, or \$80,000 down and \$21,000 at the end of each quarter for 5 years; money may be invested at 10% compounded quarterly

28. \$4,000,000 today, or \$600,000 down and \$300,000 at the end of each 6-month period for 8 years; money may be invested at 10% compounded semiannually

Solve the following application problems.

29. COMMERCIAL FARMING A manager for a commercial farmer plans to retire in 5 years and is to receive \$10,000 at the end of each semiannual period for 15 years. Find the shortage in 5 years if the farmer deposits \$12,000 at the end of each semiannual period. Assume 8% compounded semiannually.

30. SEVERANCE PAY A company plans to lay off several workers in 3 years when a new automated assembly line is introduced. At that time, they anticipate a severance pay for the laid-off workers totaling \$22,000 per month for 2 years. Find their shortage in 3 years if they deposit \$6000 per month for 36 months. Assume a rate of 12% compounded monthly.

31. TERMINALLY ILL George Joyce is terminally ill and expects to live 5 years. He has a daughter who will need \$12,000 at the end of each year for 4 years beginning at the time of his death. Assume 8% per year and find the amount that must be deposited today to satisfy his daughter's needs.

32. **LIVE IN EUROPE** Beatrice Rice plans to live in Europe for 3 years beginning in 10 years. Assume an interest rate of 9% per year and assume she needs $35,000 at the end of each year while in Europe. Find the lump sum that must be deposited today to satisfy this need.

14.3 Sinking Funds

Objectives

1. *Find the amount of a sinking fund payment.*
2. *Set up a sinking fund table.*

Section 1 of this chapter showed how to find the amount of an annuity—the sum of money that will be in an account after making a series of equal periodic payments. Businesses often have a need to raise a certain amount of money at some fixed time in the future. In such a case, the problem is turned around. The businessperson knows how much money is needed in the future, and must find the amount of each periodic payment.

A **sinking fund** is a fund that is set up to receive these periodic payments. The periodic payments plus the interest on them produce the necessary lump sum needed in the future.

Sinking funds are set up to provide money to build new factories, buy equipment, and purchase other companies. Also, many corporations and governmental agencies set up sinking funds to cover the face value of **bonds** that must be paid off at some time in the future.

A bond has many similarities to a promissory note—a bond is *a promise to pay* a fixed amount of money at some stated time in the future. Bonds and notes both pay simple, *and not compound, interest.* However, interest on a bond is often paid periodically, while interest on a note is usually paid only on the maturity date of the note. Bonds are normally issued only by large corporations and by governments, including cities, states, and the federal government. A typical U.S. government bond is shown in Figure 14.5.

This section discusses only the mechanics of setting up a sinking fund to pay off a bond when it is due. In Chapter 18 the investment aspect of bonds is discussed in detail.

Figure 14.5

Objective 1 ***Find the Amount of a Sinking Fund Payment.*** The payments into a sinking fund are just the payments of an annuity. If S is the amount that must be accumulated, R is the amount of each periodic payment, n is the number of periods, and i is the interest rate per period, then

$$S = R \cdot s_{\overline{n}|i}.$$

For a sinking fund S, the amount needed, is known, and R, the amount of each periodic payment, is unknown. Divide both sides of the formula for S by $s_{\overline{n}|i}$ for R.

$$R = \frac{S}{s_{\overline{n}|i}}$$

The amount of each payment R, into a *sinking fund* that must contain S dollars after n payments, with interest of i per period, follows.

$$R = S \cdot \left(\frac{1}{s_{\overline{n}|i}}\right)$$

The right-hand side is read as capital S times 1 divided by s-angle-n-at-i.

Values of $1 \div s_{\overline{n}|i}$ are given in column E of the interest table. (These values also can be found by dividing the corresponding numbers from column C into 1.)

Example 1 Finding Periodic Payments

KidsToys, Inc. sold \$100,000 worth of bonds that must be paid off in 8 years. They now must set up a sinking fund to accumulate the necessary \$100,000 to pay off their debt. Find the amount of each payment going into a sinking fund if the payments are made at the end of each year and the fund earns 10% compounded annually.

Solution Look in column E of the interest table for 10% and 8 payments, finding 0.08744402. The amount of each payment is found as follows.

$$R = S \cdot \left(\frac{1}{s_{\overline{n}|i}}\right)$$

$$R = \$100{,}000\left(\frac{1}{s_{\overline{8}|0.10}}\right)$$

$$= \$100{,}000(0.08744402)$$

$$= \$8744.40$$

If the corporation deposits \$8744.40 at the end of each year for 8 years in an account paying 10% compounded annually, it will have the necessary \$100,000.

NOTE The 10% interest rate in Example 1 is what the corporation *earns* on money it has deposited. The *interest that the corporation pays* to the people who bought the \$100,000 in bonds plays no part in this calculation. ■

Objective 2 ***Set Up a Sinking Fund Table.*** To keep track of the various payments into a sinking fund, accountants often make up a sinking fund table, as shown in Example 2.

Example 2 Setting Up a Sinking Fund Table

KidsToys, Inc. in Example 1 deposited \$8744.40 at the end of each year in a sinking fund that earned 10% compounded annually. Set up a sinking fund table for these deposits.

Solution The sinking fund account contains no money until the end of the first year, when a single deposit of \$8744.40 is made. Since the deposit is made at the end of the year, no interest is earned that year.

At the end of the second year, the account contains the original \$8744.40, plus the interest earned by this money. This interest is found by the formula for simple interest.

$$I = \$8744.40(0.10)(1) = \$874.44$$

An additional deposit is also made at the end of the second year, so that the sinking fund then contains a total of

$$\$8744.40 + \$874.44 + \$8744.40 = \$18{,}363.24$$

Continue this work to get the following sinking fund table. A spreadsheet package on a personal computer is an excellent tool to calculate sinking fund tables. *Be sure to round interest to the nearest cent each time* before completing each row of the table.

SINKING FUND TABLE

	Beginning of Period		End of Period	
Period	Accumulated Amount	Periodic Deposit	Interest Earned	Accumulated Amount
1	\$0	\$8744.40	\$0	\$8,744.40
2	8,744.40	8744.40	874.44	18,363.24
3	18,363.24	8744.40	1836.32	28,943.96
4	28,943.96	8744.40	2894.40	40,582.76
5	40,582.76	8744.40	4058.28	53,385.44
6	53,385.44	8744.40	5338.54	67,468.38
7	67,468.38	8744.40	6746.84	82,959.62
8	82,959.62	8744.42	8295.96	100,000.00

NOTE In Example 2, the last payment differs by 2 cents due to rounding error. The accumulated amount must equal \$100,000. ■

In Example 1, a sinking fund was set up to pay off the principal due on some bonds. A sinking fund can be set up to pay off *both principal and interest* on a loan. The following example presents yet another application for a sinking fund in which funds are accumulated for a large purchase.

Example 3 Finding Periodic Payments and Interest Earned

Children's Hospital plans to purchase a new MRI machine in 3 years. The machine currently sells for \$2,100,000, but the price is expected to increase at 8% per year compounded semiannually. The hospital decides to set up a sinking fund to purchase the machine. Find the amount of each year-end payment into the fund, if annual payments are made and the money is expected to earn 6% compounded annually. Round each to the nearest dollar.

Solution First, find the future price of the MRI machine using $\frac{8\%}{2} = 4\%$ interest for $3 \times 2 = 6$ periods using column A in Appendix D.

$$\text{Future price} = \$2{,}100{,}000 \times 1.26531902 = \$2{,}657{,}170 \quad \text{Rounded}$$

This is the total amount that the sinking fund must accumulate. The required payment is found using column E in Appendix D with 3 periods and 6% interest to find 0.31410981.

$$\text{Payment} = \$2{,}657{,}170 \times 0.31410981 = \$834{,}643 \quad \text{Rounded}$$

Payments of $834,643 at the end of each year into a sinking fund paying 6% per year compounded annually will produce enough to pay cash for the MRI machine in 3 years.

Two different interest rates are involved in Example 3. The price is increasing at 8% per year compounded semiannually, but deposits in the sinking fund earn 6% compounded annually. **Interest rate spreads** such as this are common in business. For example, banks use an interest rate spread between what they pay for funds on deposit and what they charge on loans to customers.

14.3 Exercises

Find each of the following values.

1. $\dfrac{1}{s_{\overline{12}|0.075}}$

2. $\dfrac{1}{s_{\overline{20}|0.035}}$

3. $\dfrac{1}{s_{\overline{40}|0.09}}$

4. $\dfrac{1}{s_{\overline{30}|0.045}}$

Find the amount of each payment to be made into a sinking fund so that the indicated amount will be present. Round to the nearest cent.

5. $8500, money earns $5\frac{1}{2}\%$ compounded annually, 4 annual payments

6. $4850, money earns 7% compounded semiannually, 15 semiannual payments

7. $14,000, money earns 8% compounded quarterly, 20 quarterly payments

8. $28,000, money earns 10% compounded quarterly, 40 quarterly payments

9. Explain the purpose of a sinking fund. Why is it given a name other than annuity?

10. Define the phrase *interest rate spread*. Why do interest rate spreads exist? (See Objective 2.)

In Exercises 11–14, do the following. Find (a) the present value needed to fund the end of period retirement benefit using the interest rate given, and find (b) the end of period semiannual payment needed to accumulate the value found in part (a) assum-

ing regular investments for 25 years in an account yielding 8% compounded semiannually.

11. \$10,000 per year for 25 years, 8% per year

12. \$12,000 every 6 months for 25 years, 7% compounded semiannually

13. \$12,000 per quarter for 9 years, 8% compounded quarterly

14. \$10,000 per quarter for 12 years, 12% compounded quarterly

Solve each of the following application problems. Round to the nearest cent.

15. TRUCK PURCHASE Paul Pence needs \$28,000 to purchase a truck in 3 years. (a) Find the amount of each payment if payments are made at the end of each quarter with interest at 8% compounded quarterly. (b) Find the total amount of interest earned.

16. ALLIGATOR HUNTING Cajun Jack needs \$45,000 in 4 years for a boat used to hunt alligators which are used both for leather and for the meat. (a) Find the amount of each payment if payments are made at the end of each quarter with interest at 6% compounded quarterly. (b) Find the total amount of interest earned.

17. PURCHASE OF A CLEANING MACHINE Smith Dry Cleaning must buy a new cleaning machine in 9 years for \$110,000. The firm desires to set up a sinking fund for this purpose. Find the payment into the fund at the end of each year if money in the fund earns 6% compounded annually.

18. TALENT AGENCY Catriona Kaplan's Baby Beautiful Talent Agency needs \$79,000 in 6 years. To accumulate the necessary funds, the company sets up a sinking fund with payments made into the fund quarterly. Find the payment into this fund if money in the fund earns 5% compounded quarterly.

19. NEW JAIL A city sold \$4,000,000 worth of bonds to pay for a new jail. To pay off the bonds when they mature in 8 years, the city sets up a sinking fund. Find the amount of each payment into the fund if the city makes annual payments, and the money earns 6% compounded annually. Find the amount of interest earned by the deposits.

20. CHURCH FUND RAISING A small church-related college sold \$1,480,000 in bonds to pay for an addition to the administration building. The bonds must be paid off in 15 years. To accumulate the money to pay off the bonds, the college sets up a sinking fund. If the college makes payments at the end of each 6 months and the money earns 10% compounded semiannually, find the amount of each payment into the fund. Find the amount of interest earned by the deposits.

21. SINKING FUND Complete this sinking fund table. Assume 8% interest and a desired future value in 4 years of \$960,211.72.

	Beginning of period		End of Period	
Period	Accumulated Amount	Periodic Deposit	Interest Earned	Accumulated Amount
1	\$0	\$213,090.95	\$0	\$213,090.95
2	\$213,090.95	\$213,090.95	________	________
3	________	________	________	________
4	________	________	________	________

22. SINKING FUND Complete this sinking fund table for Exercise 19.

	Beginning of Period		End of Period	
Period	**Accumulated Amount**	**Periodic Deposit**	**Interest Earned**	**Accumulated Amount**
1	$0	$404,143.76	$0	$404,143.76
2	$404,143.76	$404,143.76	______	______
3	______	$404,143.76	______	______
4	______	$404,143.76	______	______
5	______	$404,143.76	______	______
6	______	$404,143.76	______	______
7	______	$404,143.76	______	______
8	______	______	______	______

23. A NEW SHOWROOM A Denver Ford dealership wants to build a new showroom costing $2,300,000. They set up a sinking fund with end of the month payments in an account earning 12% compounded monthly. Find the amount that should be deposited in this fund each month if they wish to build the showroom (a) in 3 years and (b) in 4 years.

24. AIRPORT IMPROVEMENTS A city near Chicago sold $9,000,000 in bonds to pay for improvements to an airport. They set up a sinking fund with end of the quarter payments in an account earning 8% compounded quarterly. Find the amount that should be deposited in this fund each quarter if they wish to pay off the bonds in (a) 7 years and (b) 12 years.

Chapter 14 Quick Review

Chapter Terms

Review the following terms to test your understanding of the chapter. For each term you do not know, refer to the page number found next to the term.

amount of an annuity [p. 495]
annuity [p. 495]
annuity certain [p. 495]
annuity due [p. 495]
bonds [p. 514]
contingent annuity [p. 495]
equivalent cash price [p. 511]
future value of an annuity [p. 495]
individual retirement account (IRA) [p. 502]
interest rate spread [p. 517]
lump sum [p. 494]
ordinary annuity [p. 495]
payment period [p. 495]
present value of an annuity [p. 507]
Roth IRA [p. 502]
simple annuity [p. 495]
sinking fund [p. 514]
stock fund [p. 503]
term of the annuity [p. 495]
Treasury bill fund [p. 503]

Concept	Example
14.1 Finding the amount of an annuity Use n, the number of periods, and i, the interest rate per period, to find $s_{\overline{n}\rvert i}$ in column C of the interest table in Appendix D. Find the amount S from the formula $S = R \cdot s_{\overline{n}\rvert i}$ with R the periodic payment into the annuity. $$S = R\left[\frac{(1+i)^n - 1}{i}\right]$$	Bill Thomas deposits \$4500 at the end of each year at 6% compounded annually. How much will Thomas have at the end of 40 years? $n = 40;\ i = 6\%$ From the table $s_{\overline{40}\rvert 0.06} = 154.76196562$ and $S = (\$4500)(154.76196562) = \$696{,}428.85$
14.1 Finding the amount of each payment into an annuity Determine the amount needed in the future S, the interest rate per period i, and the number of periods n. Use the interest table in Appendix D to find $s_{\overline{n}\rvert i}$, then find R from the formula $S = R \cdot s_{\overline{n}\rvert i}$.	Find the amount of the periodic payment that will produce \$18,000 if the interest is 6% compounded monthly and monthly payments are made for 4 years. $S = \$18{,}000$ $n = 4 \times 12 = 48$ $i = 0.06 \div 12 = 0.005$ $s_{\overline{48}\rvert 0.005} = 54.09783222$ $S = R \cdot s_{\overline{n}\rvert i}$ $\$18{,}000 = R(54.09783222)$ $\$332.73 = R$
14.1 Finding the number of payments Divide S, the amount of the annuity, by R, the amount of the payment. Find the page in Appendix D corresponding to the interest rate i. Look in column C for $S \div R$ (or the first larger number) and find n for this number.	George Gleine needs \$5000. He can put \$210 per quarter into an account paying 4% compounded quarterly. How long will it take to accumulate the money? $s_{\overline{n}\rvert 0.01} = \frac{S}{R} = \frac{\$5000}{\$210}$ $s_{\overline{22}\rvert 0.01} = 23.80952381$ Closest table value is 24.47158598 so $n = 22$ quarters; $22 \div 4 = 5\frac{1}{2}$ years.

14.1 Finding the amount of an annuity due

Add 1 to the number of periods and use this as the value of n. Use n and i, the interest rate per period, to find $s_{\overline{n}|i}$ in the interest table. To find the amount, use the formula

$$S = R \cdot s_{\overline{n}|i} - \text{one payment}$$

Find the amount of an annuity due if payments of \$700 are made at the beginning of each month for 3 years into an account paying 5% compounded monthly.

$$n = (12 \times 3) + 1 = 37$$

$$i = 5\% \div 12 = \frac{5}{12}\%$$

$$s_{\overline{37}|0.467} = 39.91480775$$

$$S = \$700(39.91480775) - \$700$$

$$= \$27{,}240.37$$

14.1 Finding the amount of an IRA

Use n, the number of periods, and i, the interest rate per period, to find $s_{\overline{n}|i}$ in column C of the interest table in Appendix D. Find the amount S from the formula $S = R \cdot s_{\overline{n}|i}$ with R the periodic payment into the annuity.

Deposits of \$1500 per year are made into an IRA account paying 10% per year. Find the accumulated amount in 30 years.

$$n = 30;\ i = 10\%$$

From the table, we have the following.

$$s_{\overline{30}|0.10} = 164.49402269$$

$$S = \$1500 \cdot 164.49402269$$

$$= \$246{,}741.03$$

14.2 Finding the present value of an annuity

Use the number of periods n, interest rate per period i, and column D of the interest table to find $a_{\overline{n}|i}$. Find the present value of the annuity A, using the formula

$$A = R \cdot a_{\overline{n}|i}$$

with R the payment per period.

What lump sum deposited today at 9% compounded annually will yield the same total as payments of \$2500 at the end of each year for 8 years?

$$R = \$2500;\ n = 8;\ i = 9\%$$

$$a_{\overline{8}|0.09} = 5.53401911$$

$$A = \$2500(5.53401911)$$

$$= \$13{,}837.05$$

14.2 Finding equivalent cash price

Use the number of periods n, interest rate per period i, and column D of the interest table to find $a_{\overline{n}|i}$. Find the present value using the formula

$$A = R \cdot a_{\overline{n}|i}$$

Add the value of A to the down payment to obtain the equivalent cash value.

A buyer offers to purchase a business for \$75,000 down and quarterly payments of \$4000 for 5 years. If money may be invested at 8% compounded quarterly, how much is the buyer actually offering?

$$R = \$4000$$

$$n = 4 \times 5 = 20;\ i = 8\% \div 4 = 2\%$$

$$a_{\overline{20}|0.02} = 16.3514334$$

$$A = \$4000 \times 16.3514334$$

$$= \$65{,}405.73$$

Equivalent cash value = \$75,000 + \$65,405.73
= \$140,405.73

14.3 Determining the payment into a sinking fund

Use the number of payments n, the interest rate per period i, and column E of the interest table to find the value of $1 \div s_{\overline{n}|i}$. Use the formula

$$R = P \cdot \frac{1}{s_{\overline{n}|i}}$$

to calculate the payment.

A company must set up a sinking fund to accumulate \$50,000 in 5 years. Find the amount of the payments if they are made at the end of each year and the fund earns 8%.

$$n = 5; i = 0.08$$

$$\frac{1}{s_{\overline{n}|i}} = 0.17045645$$

$$R = \$50{,}000(0.17045645)$$

$$= \$8522.82$$

14.3 Setting up a sinking fund table

Determine the payment R and the interest at the end of each payment. Then add the previous total, next payment, and interest to find the accumulated amount. Repeat for each period.

A company wants to set up a sinking fund to accumulate \$10,000 in 4 years. It wishes to make annual payments into the account, which pays 8% compounded annually. Set up a table.

$$n = 4; i = 0.08$$

$$\frac{1}{s_{\overline{4}|0.08}} = 0.22192080$$

$$R = \$10{,}000(0.22192080)$$

$$= \$2219.21 \quad \text{Rounded}$$

	Beginning of Period		End of Period	
Period	Accumulated Amount	Periodic Deposit	Interest Earned	Accumulated Amount
1	\$0	\$2219.21	\$0	\$2,219.21
2	2219.21	2219.21	177.54	4,615.96
3	4615.96	2219.21	369.28	7,204.45
4	7204.45	2219.19	576.36	10,000.00

Chapter 14 Review Exercises

The answer section includes answers to all Review Exercises.

Find the amount of each of the following annuities. Round to the nearest cent. **[14.1]**

1. \$1500 is deposited at the end of each year for 22 years, money earns $6\frac{1}{2}\%$ compounded annually.
2. \$1000 is deposited at the end of each semiannual period for 12 years, money earns 6% compounded semiannually.
3. \$3000 is deposited at the end of each quarter for $6\frac{3}{4}$ years, money earns 10% compounded quarterly.
4. \$1000 is deposited at the end of each month for $3\frac{1}{2}$ years, money earns 9% compounded monthly.
5. \$18,000 is deposited at the beginning of each year for 7 years, money earns $6\frac{1}{2}\%$ compounded annually.
6. \$3500 is deposited at the beginning of each quarter for $5\frac{1}{2}$ years, money earns 8% compounded quarterly.
7. Willa Burke deposits \$803.47 at the end of each quarter for $3\frac{3}{4}$ years, in an account paying 12% compounded quarterly. Find the final amount in the account. Find the amount of interest earned.
8. A firm of attorneys deposits \$7500 of profit sharing money at the end of each semiannual period for $7\frac{1}{2}$ years.

Find the final amount in the account if the deposit earns 5% compounded semiannually. Find the amount of interest earned.

Find the present value of each of the following ordinary annuities. Round to the nearest cent. **[14.2]**

9. Payments of $4200 are made annually at 7% compounded annually for 15 years.

10. Payments of $800 are made semiannually at 8% compounded semiannually for $4\frac{1}{2}$ years.

11. Payments of $450 are made quarterly for $5\frac{1}{4}$ years at 6% compounded quarterly.

12. Payments of $125 are made monthly for $4\frac{1}{6}$ years at 10% compounded monthly.

Find the amount of each payment to be made to a sinking fund so that enough money will be available to pay off the indicated loan. Round to the nearest cent. **[14.3]**

13. $85,000 loan, money earns 9% compounded annually, 6 annual payments

14. $42,000 loan, money earns 6% compounded quarterly, 26 quarterly payments

15. $100,000 loan, money earns 12% compounded semiannually, 9 semiannual payments

16. $35,000 loan, money earns 9% compounded monthly, 47 monthly payments

Solve each of the following application problems. Round to the nearest cent.

17. Bessie Smith invests $800 at the end of each quarter in a fund she hopes will average 10% compounded quarterly, for 10 years. Find the amount of the annuity. **[14.1]**

18. Bill Wild can save $600 at the end of each year in an account earning 10% per year. How many years are required for him to accumulate $100,000? **[14.1]**

19. Lupé Rivera owes her retired mother $45,000 on a piece of land. Find the required monthly payment into a sinking fund if Lupé pays it off in 4 years and the interest rate is 12% per year compounded monthly. **[14.1]**

20. In 3 years Ms. Thompson must pay a pledge of $7500 to her college's building fund. What lump sum can she deposit today, at 10% compounded semiannually, so that she will have enough to pay the pledge? **[14.2]**

21. According to the terms of a divorce settlement, one spouse must pay the other a lump sum of $28,000 in 17 months. What lump sum can be invested today, at 6% compounded monthly, so that enough will be available for the payment? **[14.2]**

22. A-1 Plumbing needs 48,000 in 4 years to replace two trucks. Find the amount they must deposit at the end of each quarter in a fund earning 6% compounded quarterly. **[14.3]**

23. A power company needs to replace some large generators at an estimated cost of $680,000 in $5\frac{1}{2}$ years. They expect to receive a payment of $240,000 from another source at that time. Find the payment required at the end of each semiannual period to accumulate the remaining funds needed to purchase the generators. Assume they can earn 8% compounded semiannually. **[14.3]**

24. Hilda Worth invests $500 at the *beginning* of each semiannual period into an IRA account paying 10% compounded semiannually. Given that she is 30 years old, find the amount she will have accumulated at age 54. **[14.1]**

25. Koplan Kitchens plans to purchase some land for expansion purposes in 3 years. They anticipate payments at that time of $11,546.48 at the end of each quarter for 5 years. Find the present value of the payments assuming 10% compounded quarterly. Find the lump sum they must deposit today in an account earning 12% compounded monthly to satisfy the debt. **[14.2]**

Prepare a sinking fund table for the following. **[14.3]**

26. A firm sets up a sinking fund to pay off a $100,000 note due in 4 years. The firm makes annual payments into an account earning 9% compounded annually.

	Beginning of Period		End of Period	
Period	Accumulated Amount	Periodic Deposit	Interest Earned	Accumulated Amount
1	$0	$21,866.87	$0	$21,866.87
2	________	________	________	________
3	________	________	________	________
4	________	________	________	________

Chapter 14 Summary Exercise
Planning for Retirement

At age 32, Tish Baker has decided to invest $1800 per year for 33 years until she is 65, in her retirement plan. She has also decided to place one-half of the funds in a stock index fund that roughly tracks the return on Standard and Poors. The other half of the funds will be placed in a mutual fund containing bonds.

(a) Find the amount she will have at 65 assuming the stock fund averages 12% per year compounded annually and the bond fund averages 8% per year compounded annually.

(b) The amount found in part (a) seems like a lot of money. But Baker knows that inflation will increase her cost of living. She wants to see the effect of 3% annual inflation on her financial goals. Find the income she needs at age 65 to have the same purchasing power as an income of $20,000 today. (*Hint:* Look at the section on inflation in Section 11.4 and use column A of the tables.)

(c) Baker wishes to fund her retirement for 20 years (to age 85). Find the present value of the annual income found in part (b) assuming that the funds earn 8% per year compounded annually.

(d) Will her expected savings from part (a) fund her retirement at a purchasing power of $20,000 per year in today's dollars?

Baker will also probably receive a monthly check from Social Security during her retirement. However, there is some discussion about the ability of Social Security to pay benefits, as you can see from the newspaper clipping.

Social Security Wins 3-Year Reprieve

WASHINGTON—The booming economy is bolstering Social Security and postponing the retirement programs projected insolvency by three years, to 2032, the system's trustees said Tuesday.

The number of workers for each retiree will fall from 3.4 now to two in 2030.

"Although there is no immediate financial crisis, the time to act is now to prevent a crisis," Social Security Commissioner Kenneth Apfel said.

Members of Congress warned against complacency.

"We should not have a false sense of security just because the Social Security program will go bankrupt a few years later than expected," said Sen. Judd Gregg, R-N.H.

This year's report on Social Security comes as both parties are debating ideas for altering the program, including proposals from conservatives and moderates to place some payroll taxes into privately controlled investments.

Source: USA Today, 4/29/98. Reprinted by permission.

Net Assets Business on the Internet

Tenet Health

Statistics

- 1998: Revenues of $9.9 billion
- 30,000 licensed beds
- 130,000 employees
- 130 acute-care facilities
- Over 900 member physicians

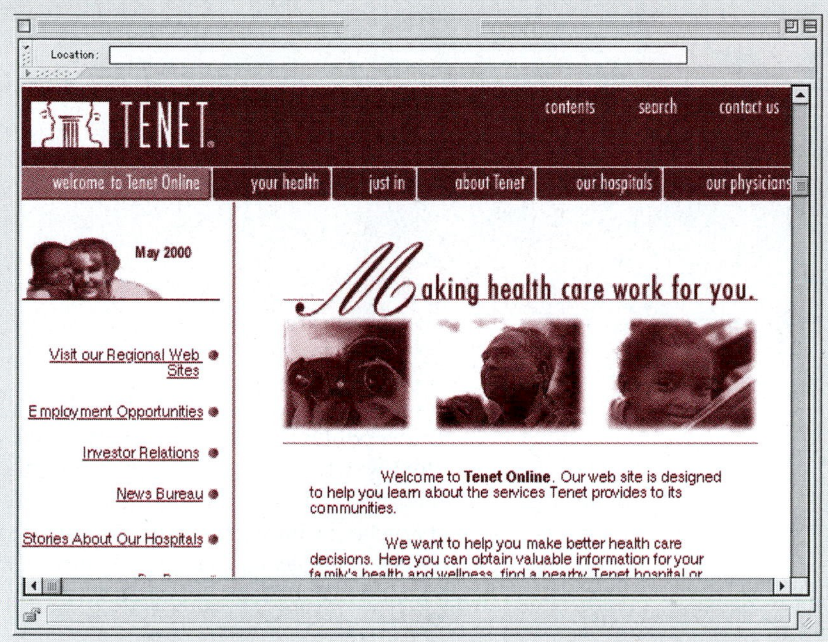

One way to save money while providing a high level of care is to use the emergency room for real emergencies. This is the reason that many of Tenet Health's hospitals have set up clinics that currently receive more than 50,000 visits per year from women seeking prenatal, maternity, and pediatric care. Patients get quality health care and the emergency room is kept free for emergency care.

Tenet Health also invests in technology to increase efficiency and allow health-care workers to handle medical problems. To quote one Tenet Health physician: "It's a phenomenal concept, to be able to do surgery without a knife. With our modern imaging tools ... we can see right into the body, down to the level of a millimeter." Technologies like these allow physicians to perform procedures only dreamed about 20 years ago in a manner that is both effective and cost efficient.

1. Tenet Health makes a contribution of \$800 each year into the retirement plan of employee Tish Baker. If the hospital continues to make this same contribution each year for 25 years and if the account earns 8% per year, find the future value in Baker's account at the end of 25 years.

2. Assume that Tenet Health deposits an average of \$800 into the retirement account each year for each of their 130,000 employees. What is the total amount of money that is deposited in retirement accounts each year?

3. Assume Tenet Health needs \$4,000,000 in 7 years to replace an MRI machine. What payments should they make into a sinking fund if the year-end payments grow at 9% per year?

4. Managers wish to build a new facility costing \$28,000,000 in 4 years. Assuming funds earn 8% interest compounded quarterly, will an investment of \$200,000 at the end of every quarter for 4 years generate enough money to pay for the facility? If not, what is the expected shortfall?

15 Business and Consumer Loans

Paying cash for everything is almost impossible. We use credit to pay for our phone, lights, and running water, for example, since we normally pay for these items at the end of a month. Many people buy on credit at the department store, use a credit card at the gas station, or buy cars or furniture on the installment plan. Occasionally it is necessary to borrow to pay taxes, medical expenses, or educational costs. Almost everyone borrows when buying a home. Recent data on household debt, excluding debt on the home itself, is shown in Figure 15.1.

What are some of the things you have borrowed money for in the past?

In addition, most businesses, including banks, borrow money from time to time. The person or firm doing the lending is called the **lender** or **creditor**. The person or firm doing the borrowing is the **borrower** or **debtor**. Even federal, state, and local govern-

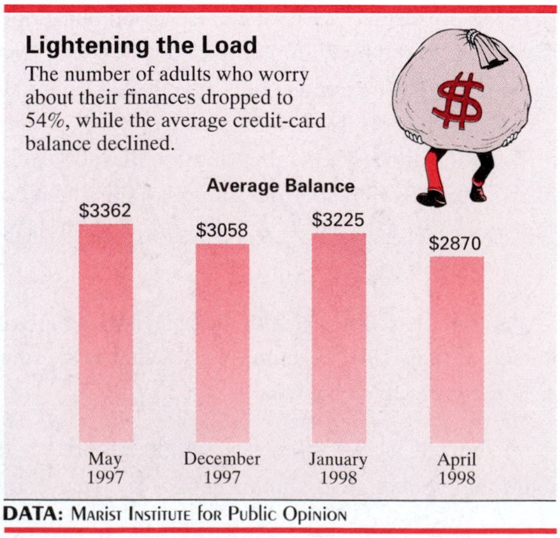

Figure 15.1

ments borrow money. Some of their debt is open-end credit but much of their debt is financed with bonds, which are discussed in Section 18.3 on page 680.

Credit card companies typically set the interest rates and minimum payments without consulting the borrower. However, *you may be able to negotiate a lower interest rate* on your credit cards with the lenders. Contact the companies that hold your credit cards and try it, *you may save money*! If they do not give you a good rate, shop around for a card with a lower interest rate or better terms.

Years ago, consumer credit was offered as an additional service to attract more customers and increase total sales. Today, consumer credit is not merely a service to the customer. For many large retail stores the interest or finance charge represents *a large portion* of company profit. This chapter examines the various methods used in determining interest charges.

There is a fundamental difference between the notes discussed in earlier sections and the loans discussed in this chapter. In the earlier situations, a sum of money was paid back, with interest, with a lump-sum payment at some future date. In this chapter, the principal and interest are paid back in periodic payments.

15.1 Open-End Credit

Objectives

1. *Define open-end credit.*
2. *Understand revolving charge accounts and credit accounts.*
3. *Use the average daily balance method to calculate finance charges.*
4. *Define loan consolidation.*

Objective 1 *Define Open-End Credit.* A common way of buying on credit, called **open-end credit**, has no fixed payments. The customer continues making payments until no outstanding balance is owed. With open-end credit, additional credit is often extended before the initial amount is paid off. Examples of open-end credit include most department store charge accounts and charge cards as well as MasterCard and Visa. Individuals are given a **credit limit** on these accounts based on their income and other factors. They are then allowed to charge up to this amount.

Objective 2 *Understand Revolving Charge Accounts and Credit Accounts.* Individuals may open **charge accounts** at certain stores such as department stores. This allows the customer to make frequent purchases from that store, or chain of stores, during the month without having to pay cash or write a check. Such accounts are often never paid off, although a minimum amount must be paid each month. Since the account may never be paid off, it is called a **revolving charge account**.

Credit cards such as MasterCard and Visa also allow for many purchases during a month at numerous stores and restaurants. Sometimes there is an annual membership fee or a minimum monthly charge for the use of this service. Figure 15.2 shows that more people are paying off their credit card debt each month. A sample copy of a receipt signed by a customer using a credit card is shown in Figure 15.3. Many credit cards also allow you to get cash immediately, although there may be an extra charge for this if you are at an automated teller machine (ATM). Whether you use an ATM, or even checks written on the credit card account, these are considered a cash advance and result in an immediate charge.

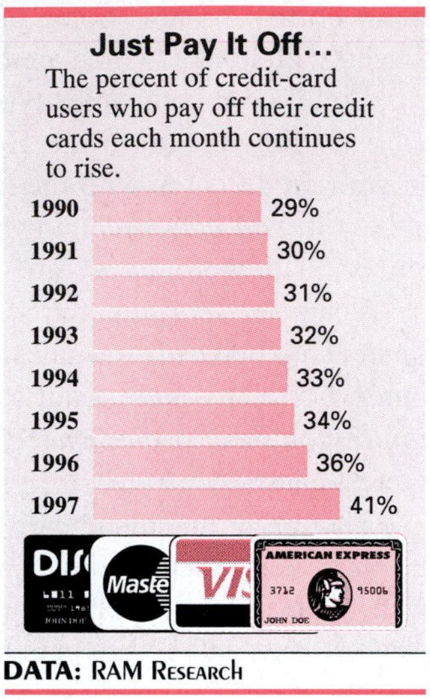

Figure 15.2

At the end of a billing period, the customer receives a statement of payments and purchases made. This statement typically takes one of two forms: **country club billing** provides a carbon copy of all original charge receipts, while **itemized billing**, becoming more and more common because of its lower cost to the credit card companies, provides an itemized listing of all charges, without copies of each individual charge. A typical itemized statement is shown in Figure 15.4.

Any charges beyond the cash price of an item are called **finance charges**. Finance charges include interest, credit life insurance, a time payment differential, and carrying charges. Many lenders charge **late fees** for payments that are received after the due date

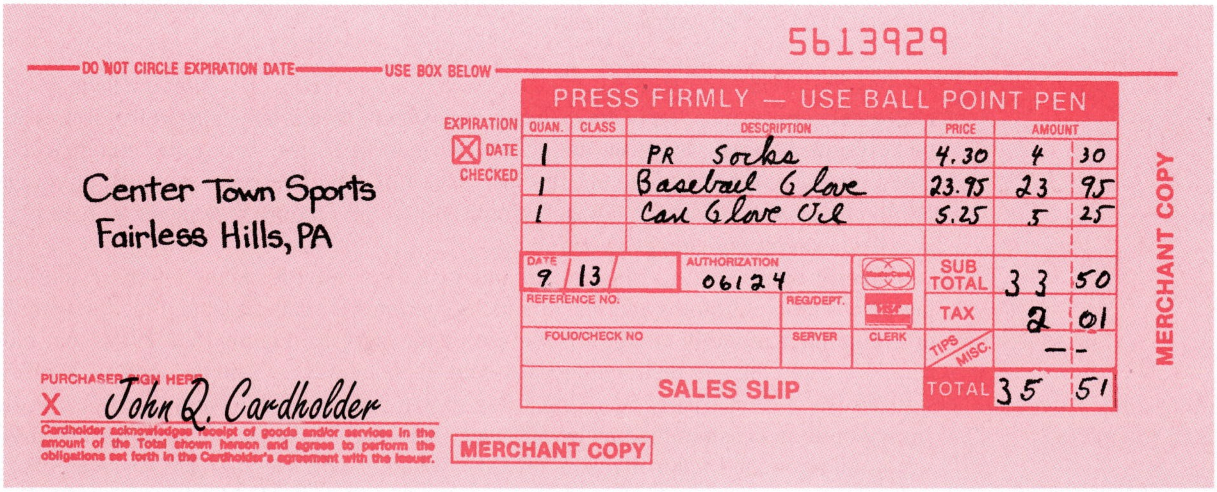

Figure 15.3

AMERICA'S CREDIT CARD

$

WRITE IN THE AMOUNT OF YOUR PAYMENT

IMPORTANT! **1.** Return this portion of your statement with your check: please do not fold or bend **2.** Write your account number on the face of your check **3.** Make checks payable to America's Charge Card

ACCOUNT NUMBER	BILLING DATE	PAYMENT DUE DATE	NEW BALANCE	MINIMUM PAYMENT DUE
5211-1234-5678	10-18	11-12	605.19	30.00

JOHN Q. CARDHOLDER
1000 MAIN STREET
ANYTOWN, USA 00000
521112345678 0060926003000

PLEASE BE SURE OUR MAILING ADDRESS ON THE REVERSE SIDE APPEARS IN THE WINDOW OF THE RETURN ENVELOPE

DO NOT INCLUDE INQUIRIES WITH PAYMENT. SEE REVERSE SIDE.

PLEASE PRINT

FOR CHANGE OF ADDRESS	NEW STREET ADDRESS		APT. P.O. BOX NO.
	CITY	STATE	ZIP CODE

RETAIN THIS PORTION FOR YOUR RECORDS

Account Number	Credit Line	Available Credit	Past Due Amount	BILLING DATE
5211-1234-5678	1,000.00	394.81	.00	10-18

TRAN DATE		POSTING DATE		REFERENCE NUMBER	TRANSACTION DESCRIPTION			TRANSACTION AMOUNT
09	06	09	09	12009230	TURNSTYLE 702	CHICAGO	IL	25.17CR
09	07	09	13	55190432	POLK BROS	CHICAGO	IL	150.00
09	13	09	21	86235671	CENTER TOWN SPORTS	NORRIDGE	IL	35.51
09	14	09	19	21345678	SPIEGELS, INC.	CHICAGO	IL	60.10
09	20	09	20	93703523	PAYMENT THANK YOU			110.00CR
09	21	09	26	86245671	MINNESOTA FABRICS 32	NORRIDGE	IL	42.64
09	25	09	28	12001051	TURNSTYLE 702	CHICAGO	IL	14.49

PREVIOUS BALANCE	AVERAGE DAILY BALANCE	FINANCE CHARGE	NEW BALANCE	Minimum Payment
432.64	498.44	4.98	605.19	30.00

For Customer Service Call	Charge Transactions	PAYMENTS	CREDITS	PAYMENT DUE DATE
	302.74	110.00	25.17CR	11-12

SEE ITEM NUMBER ON REVERSE SIDE FOR YOUR PERIODIC RATE(S), THE RANGE OF AVERAGE DAILY BALANC(S) TO WHICH IT APPLIES, AND THE CORRESPONDING **ANNUAL PERCENTAGE RATE**

Additional **FINANCE CHARGE** on purchases may be avoided if total NEW BALANCE is paid in full by PAYMENT DUE DATE

NOTICE SEE REVERSE SIDE FOR IMPORTANT INFORMATION

YOU MAY AT ANY TIME PAY ALL OR PART OF THE TOTAL BALANCE OWING ON YOUR ACCOUNT

Figure 15.4

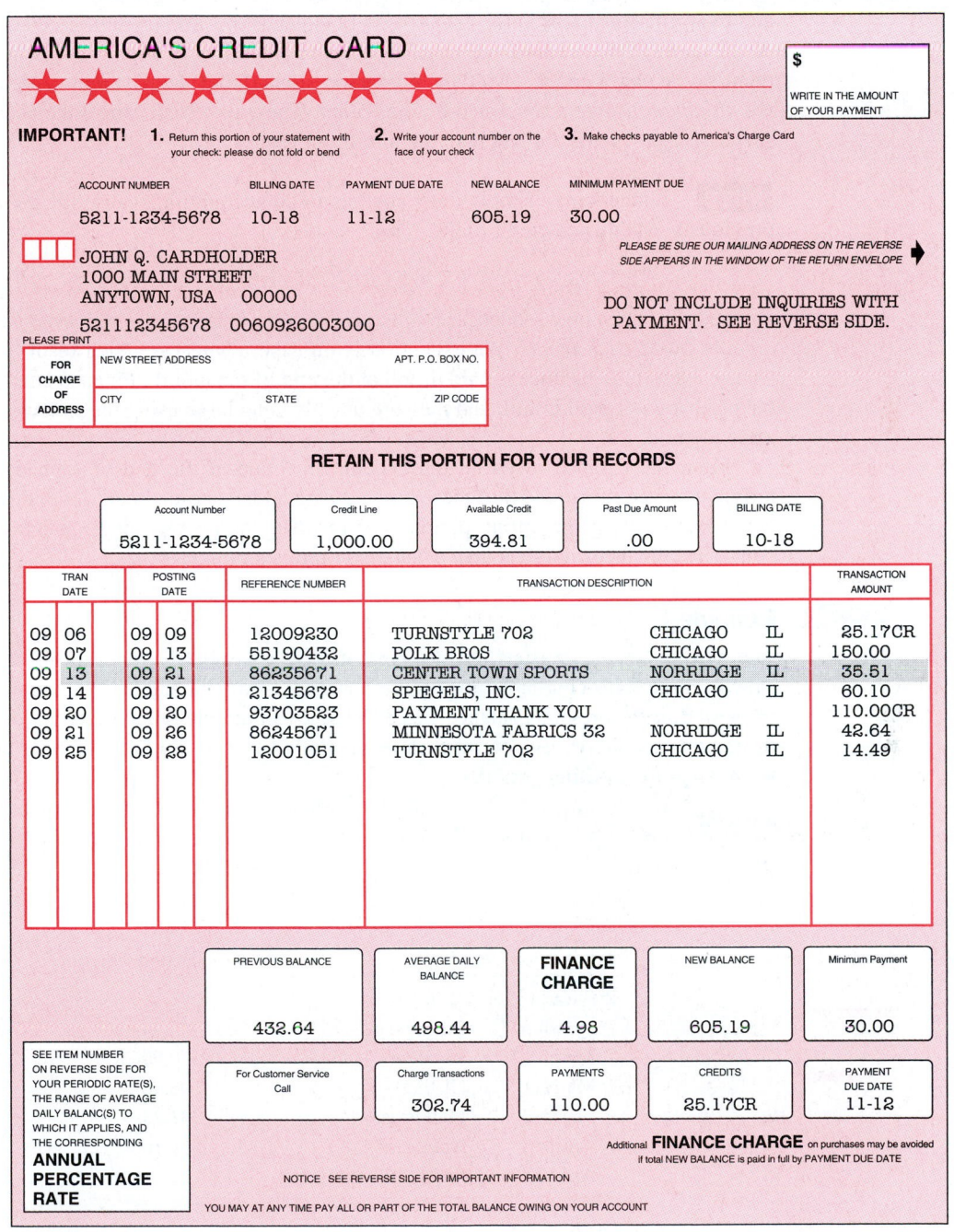

and **over-the-limit fees** in the event the debt exceeds the amount authorized by the issuer of the card. Both late fees and over-the-limit fees tend to be high, so it is best to avoid them if you can. Lenders also typically charge a late fee if the payment made is *less than the minimum payment* requested by the lender. They may even cancel the account if two successive payments are less than the minimum requested.

NOTE Lenders may simply deny charges to an account that cause the debt to exceed previously established credit limits. ■

Objective 3 *Use the Average Daily Balance Method to Calculate Finance Charges.* Finance charges on many open-ended accounts are zero as long as the borrower *pays the full amount owed each month.* An exception is on cash advances, which result in a finance charge even if the balance is paid in full at the end of the month. Finance charges apply if balances are not paid in full, and they are usually calculated using the **average daily balance** method.

First, the balance owed on the account is found at the end of each day during a month or billing period. All of these amounts are added and the total divided by the number of days during the billing period. The result is the average daily balance of the account. The finance charge is then calculated on this amount.

Example 1 Finding Average Daily Balance

The activity in the MasterCard account of Kay Chamberlin for one billing period is shown in the following table. (a) Find the average daily balance on the next billing date of April 3, if the previous balance was $209.46. (b) Find the finance charge for the month if the monthly charge is $1\frac{1}{2}\%$ of the average daily balance. (c) Find the total amount due at the end of the billing period.

Solution

(a)

Transaction Description		Transaction Amount
Previous balance, $209.46		
March 3	Billing date	
March 12	Payment	$50.00CR*
March 17	Clothes	$28.46
March 20	Mail order	$31.22
April 1	Auto parts	$59.10
April 3	End of billing cycle	

*CR represents "credit."

At the close of business on March 3, the unpaid balance was $209.46. This balance was the same for 9 days until March 12, when it changed to

$$\$209.46 - \$50 = \$159.46$$

This balance was the same for 5 days until March 17 when it became

$$\$159.46 + \$28.46 = \$187.92$$

In 3 days, on March 20, the balance became

$$\$187.92 + \$31.22 = \$219.14$$

which remained unchanged for 12 days, becoming, on April 1,

$$\$219.14 + \$59.10 = \$278.24$$

The new billing date is April 3, so the unpaid balance was $278.24 for 2 days. These results are summarized in the following table.

Date	Number of Days Until Balance Changes	Unpaid Balance
March 3	9	$209.46
March 12	5	$159.46
March 17	3	$187.92
March 20	12	$219.14
April 1	2	$278.24
April 3	31 (Total number of days in billing period)	

To find the average daily balance, weigh each unpaid balance according to the number of days for that balance, total the products, and then divide by the 31 days of this particular billing cycle. Do all this with the following shortcut procedure.

$$\begin{aligned}
\$209.46 \times 9 &= \$1885.14 \\
\$159.46 \times 5 &= 797.30 \\
\$187.92 \times 3 &= 563.76 \\
\$219.14 \times 12 &= 2629.68 \\
\$278.24 \times 2 &= \underline{556.48} \\
& \quad \$6432.36
\end{aligned}$$

(For example, $209.46 × 9 = $1885.14 represents the sum of the average daily balances at the end of the day from March 3 through March 11.)

Now divide the total by 31, since there are 31 days in this billing cycle.

$$\text{Average daily balance} = \$6432.36 \div 31 = \$207.50 \quad \text{Rounded.}$$

Chamberlin will pay a finance charge based on the average daily balance of $207.50.

Scientific *Calculator Approach*

The calculator solution to part (a) of Example 1 is as follows: multiply each unpaid balance times the respective number of days until the balance changes. Add these values together and divide by the number of days in the month (31 days).

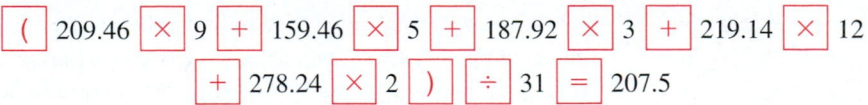

[( ] 209.46 [×] 9 [+] 159.46 [×] 5 [+] 187.92 [×] 3 [+] 219.14 [×] 12 [+] 278.24 [×] 2 [ )] [÷] 31 [=] 207.5

(b) Find the finance charge for the month by multiplying the monthly rate of $1\frac{1}{2}\%$ by the average daily balance.

$$\text{Finance charge} = 0.015 \times \$207.50 = \$3.11$$

The finance charge would be different for a month with a different average daily balance.

(c) The amount due at the end of the billing period is the previous balance minus any payments and credits plus any new charges, including the finance charge.

Previous Balance	−	Payment	+	Clothes	+	Mail Order	+	Auto Parts	+	Finance Charge
\$209.46	−	\$50	+	\$28.46	+	\$31.22	+	\$59.10	+	\$3.11

ending balance

The amount due is \$281.35.

Problem-Solving Hint The billing period in Example 2 is 31 days. Some billing periods are 30 days or less. Be sure to use the correct number of days.

NOTE Not all billings for open-end accounts occur on the same day of the month. ■

If finance charges are expressed on a per month basis, find the **annual percentage rate** by multiplying the monthly rate by 12, the number of months in a year. Table 15.1 shows typical monthly rates and the corresponding annual percentage rates.

Table 15.1 MONTHLY FINANCE CHARGES AND CORRESPONDING ANNUAL PERCENTAGE RATES

Quoted Monthly Finance Charge	Annual Percentage Rate
$\frac{1}{2}$ of 1%	6%
$\frac{2}{3}$ of 1%	8%
$\frac{3}{4}$ of 1%	9%
$\frac{5}{6}$ of 1%	10%
1%	12%
$1\frac{1}{4}\%$	15%
$1\frac{1}{2}\%$	18%
$1\frac{2}{3}\%$	20%
$1\frac{3}{4}\%$	21%
2%	24%

NOTE The monthly finance charge percent is only used to calculate the finance charge and not the minimum payment. ■

Objective 4 ***Define Loan Consolidation.*** Credit can be *so easy to obtain* for stable, employed individuals in our society *that it can be problematic*. Figure 15.5 shows students' use of credit cards. Individuals with many high-interest revolving account loans sometimes **consolidate their loans** into one lower-interest loan, frequently with a longer term. This allows

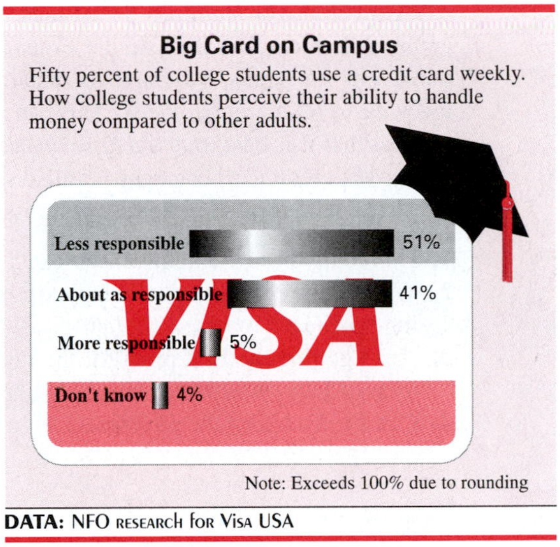
Big Card on Campus
Fifty percent of college students use a credit card weekly.
How college students perceive their ability to handle money compared to other adults.
Less responsible 51%
About as responsible 41%
More responsible 5%
Don't know 4%
VISA
Note: Exceeds 100% due to rounding
DATA: NFO RESEARCH FOR VISA USA

Figure 15.5

them to handle their monthly payments rather than defaulting on debt and ruining their credit history. Consolidating a loan may help someone afford the payments, but it does not extend the life of the items purchased that created the loans in the first place. Someone who consolidates debts and credit card balances, then makes additional loans, can develop serious financial problems and may end up with more debt than he or she can handle.

Example 2 Loan Consolidation

Between auto loans, a home loan, and the revolving account loans shown below, teachers Bill and Cynthia Taylor have more debt than they can handle. They have gone to George Willis at Glaston Credit Union for help.

Revolving Account	Debt	Annual Percentage Rate	Minimum Payment
Sears	$3880.54	18%	$150
Dillards	1620.13	16%	50
MasterCard	3140.65	14%	100
Visa	4920.98	20%	135
Totals	$13,562.30		$435

Solution George Willis (1) put the Taylors on a *strict budget*, (2) consolidated the revolving account debts into one longer-term low-interest loan at the credit union, and (3) decreased the payment on one auto by refinancing it at a lower rate. In all, Willis reduced the Taylors' monthly payments by about $280 per month. The Taylors should be okay *as long as they stay on the budget*, do not make additional credit purchases, and pay off some of the debt.

Unfortunately, families sometimes owe so much on high-interest credit cards and automobile loans that they simply cannot handle the debt even by consolidating loans. In that event the family may have to declare bankruptcy or get second jobs, pushing workloads up to 60 hours per week. This can make life *very* difficult. Most financial planners suggest that it is best to *avoid high interest charges on credit cards*. This can be done by paying off credit card balances in full at the end of each monthly billing cycle.

In general, consider the following issues *before* borrowing:

1. Do I really need the item *now*, or can I delay the purchase until I have the cash to buy it?
2. What is the real cost, including cash price, sales tax, and finance charges, of purchasing this item?
3. Am I getting the best price for the item and best loan terms (such as interest rate and term of loan) possible or should I shop around for a better deal?
4. Can I truly afford the payments?

15.1 Exercises

Find the finance charge for each of the following revolving charge accounts. Assume interest is calculated on the average daily balance of the account. (See Example 1.)

	Average Daily Balance	Monthly Interest Rate
1.	$836.15	1.2%
2.	$8431.10	1.4%
3.	$389.95	$1\frac{1}{4}\%$
4.	$2235.46	1.6%

Find the average daily balance for each of the following credit card accounts. Assume there is one month between billing dates (using the proper number of days in the month). Then find the finance charge if interest is 1.5% per month on the average daily balance. (See Example 1.) Finally, find the balance at the end of the billing cycle.

5. Previous balance $139.56

September 12	Billing date	
September 20	Payment	$45
September 21	CD's	$37.25

6. Previous balance $412.48

November 5	Billing date	
November 18	Payment	$150
November 30	Dinner and play	$84.50

7. Previous balance $684.32

May 4	Billing date	
May 15	Payment	$50
May 19	Theater tickets	$75.75

8. Previous balance $228.95

January 27	Billing date	
February 9	Cheese	$11.08
February 13	Returns	$26.54
February 20	Payment	$29
February 25	Repairs	$71.19

9. Previous balance $312.78

June 11	Billing date	
June 15	Returns	$106.45
June 20	Jewelry	$115.73
June 24	Car rental	$74.19
July 3	Payment	$115

10. Previous balance $355.72

March 29	Billing date	
March 31	Returns	$209.53
April 2	Auto parts	$28.76
April 10	Gasoline	$14.80
April 12	Returns	$63.54
April 13	Returns	$11.71
April 20	Payment	$72
April 21	Flowers	$29.72

11. Previous balance $714.58

August 17	Billing date	
August 21	Mail order	$26.94
August 23	Returns	$25.41
August 27	Beverages	$31.82
August 31	Payment	$128
September 9	Returns	$71.14
September 11	Plane ticket	$110
September 14	Cash advance	$100

12. Previous balance $412.42

March 10	Billing date	
March 13	Returns	$28.18
March 15	Gasoline	$16
March 20	Payment	$200
April 3	Restaurant	$28.45
April 5	Clothing	$86.80

Solve the following application problems.

13. CREDIT CARD DEBT Jerry Jasper has an average daily balance of $2800.35 for the entire month on a credit card that charges 1.5% on the average daily balance. (a) Find the monthly finance charge. (b) Find the finance charge if he can change the loan to a credit card that charges 1% on the average unpaid balance. (c) Find the savings.

14. REDUCING FINANCE CHARGES Maria Estefan has an average daily balance of \$4509.66 for the entire month on a credit card that charges 1.8% on the average daily balance. (a) Find the monthly finance charge. (b) Find the finance charge if she can change the loan to a credit card that charges 1% on the average unpaid balance. (c) Find the savings.

15. DREAM VACATION After taking his dream vacation to Europe, Benjamin Thompson has an average daily balance of \$4850.39 on a credit card that charges 1.5% per month. (a) Find the interest charges. (b) Find the interest charges if he can change the loan to a credit card that charges 0.8% per month. (c) Find the savings.

16. CREDIT CARD DEBT Jerry Johnson has been ill and has an average daily balance of \$16,432.51 on a credit card charging 1.6% interest. (a) Find the monthly finance charge. (b) Find the finance charge if he can change the loan to a credit card that charges 0.5% on the average daily balance. (c) Find the savings.

17. Explain how the average daily balance is determined. (See Objective 3.)

18. Explain how consolidating loans may be of some advantage to a credit card holder. What disadvantages can you think of? (See Objective 4.)

19. Assume your brother owes \$1200, \$1500, and \$440 on credit cards charging 15%, 18%, and 12%, respectively. What would you recommend to your brother? Why? (See Objective 4.)

20. Make a list of all of your debts and write down the payments, amounts owed, and interest rates. Explain how you can rearrange your loans to your advantage.

21. Complete Section 5 of the foldout.

15.2 Installment Loans

Objectives

1 *Define installment loan and annual percentage rate.*
2 *Find the total installment cost and finance charge.*
3 *Use a table to find the APR.*

Objective 1 *Define Installment Loan and Annual Percentage Rate.* A loan is **amortized** if both principal and interest are paid off by a sequence of periodic payments. An example is a payment of \$250 per month for 48 months on a car loan. This type of loan is called an **installment loan**. Installment loans are used for cars, boats, home improvements, furniture, and even for consolidating several smaller loans into one affordable payment.

Since the enactment of the **Federal Truth-in-Lending Act (Regulation Z)** in 1969, lenders must report their **finance charge** (the charge for credit) and their **annual percentage rate (APR)** on installment loans. The Truth-in-Lending Act does *not* regulate interest rates or credit charges but merely requires a standardized and truthful report of what they are to within one-eighth of a percent of the actual APR. In practice, many institutions round to the nearest one-quarter of a percent. Each state sets the allowable interest rates and charges. Lenders normally give borrowers a document such as the one in Figure 15.6, showing all credit charges.

Note and Disclosure Statement

BORROWER NAME (Last – First – Middle Initial) AND ADDRESS (Street – City – State – Zip Code) | DATE | MEMBER NUMBER | NOTE NUMBER

Smith, John Q.
10123 Fair Oaks Blvd.
Fair Oaks, CA 95628

CONTRACT NUMBER	REFERENCE NUMBER	MATURITY DATE
012-2719-6	XXXXXXXXX	XXXXXXX

In this agreement "you" and "your" mean each person who signs this agreement. The "credit union" means the credit union whose name appears above and anyone to whom the credit union transfers its rights under this agreement. The terms on the reverse side are part of this agreement. Boxes checked below apply to this agreement.

TRUTH IN LENDING DISCLOSURE

ANNUAL PERCENTAGE RATE The cost of your credit as a yearly rate.	FINANCE CHARGE The dollar amount the credit will cost you.	Amount Financed The amount of credit provided to you or on your behalf.	Total of Payments The amount you will have paid when you have made all payments as scheduled.	Prepayment: If you pay off early you will not have to pay a penalty. e means an estimate
13.99%	$ 1,605.64	$ 6,800.00	$ 8,405.64	

Your Payment Schedule will be:	Number of Payments	Amount of Payments	When Payments Are Due
	36	$233.49	monthly beginning May15

Property Insurance: You may obtain property insurance from anyone you want that is acceptable to the credit union. XXXXXXXXXXXXXXXXXXXXXXXXXXXXXXXXXX $

Security: Collateral securing other loans with the credit union will also secure this loan. You are giving a security interest in your shares and/or deposits in the credit union; and [X] the goods/property being purchased; [ ] Other (Describe)

Late Charge: N/A

Filing Fees $ N/A **Non-Filing Insurance** $ N/A

See your contract documents for any additional information about nonpayment, default, and any required repayment in full before the scheduled date.

ITEMIZATION OF THE AMOUNT FINANCED

ITEMIZATION OF AMOUNT FINANCED OF $ | AMOUNT GIVEN TO YOU DIRECTLY $ | AMOUNT PAID ON YOUR ACCOUNT $ | PREPAID FINANCE CHARGE $

AMOUNT PAID TO OTHERS ON YOUR BEHALF $ To $ To
$ To $ To

NOTE AND SECURITY AGREEMENT CONTINUED ON REVERSE SIDE

Promise to Pay: You promise to pay $ to the credit union plus interest on the unpaid balance at % per year until what you owe has been repaid.

Collection Costs: You promise to pay all costs of collecting the amount you owe under this agreement including court costs and reasonable attorney fees

Security Offered:	MODEL	YEAR	I.D. NUMBER	TYPE	VALUE
	Chevrolet Caprice		1NA1G1H96XE6811		

Other (Describe):

You Pledge Shares and/or Deposits of $ in account number Key No. This Note is governed by the laws of Illinois

SIGNATURE: If you agree to make and be bound by the terms of this Note and Security Agreement sign below. *If you are not a borrower but an owner of the collateral for this loan, sign below and check the box for "Owner of Collateral". By doing so you agree only to the terms of the Security Agreement.*

CAUTION: IT IS IMPORTANT THAT YOU THOROUGHLY READ THIS CONTRACT BEFORE YOU SIGN IT.

Borrower	Date	Borrower ☐ Owner of Collateral (other than a Borrower)	Date
X (SEAL)		X (SEAL)	
Borrower ☐ Owner of Collateral (other than a Borrower)	Date	Witness	Date
X (SEAL)		X (SEAL)	

CREDIT INSURANCE APPLICATION

"You" or "Your" means the member and the joint insured (if applicable).

Credit insurance **is voluntary and not required in order to obtain this loan.** You may select any insurer of your choice. You can get this insurance only if you check the "yes" box below and sign your name and write in the date. The rate you are charged for the insurance is subject to change. You will receive written notice before any increase goes into effect. You have the right to stop this insurance by notifying your credit union in writing. Your signature below means you agree that:

- If you elect insurance, you authorize the credit union to add the charges for insurance to your loan each month.
- You are eligible for disability insurance only if you are working for wages or profit for 25 hours a week or more on the date of any advance. If you are not, that particular advance will not be insured until you return to work. If you are off work because of temporary layoff, strike or vacation, but soon to resume, you will be considered at work. Are you working for wages or profit for 25 hours a week or more? ☐ Yes ☐ No
- You are eligible for insurance up to the Maximum Age for Insurance. Insurance will stop when you reach that age.

NOTE: THE LIFE AND DISABILITY INSURANCE CONTAINS CERTAIN BENEFIT EXCLUSIONS, INCLUDING A PRE-EXISTING CONDITION EXCLUSION. PLEASE REFER TO YOUR CERTIFICATE FOR DETAILS.

YOU ELECT THE FOLLOWING INSURANCE COVERAGE(S)	YES	NO	PREMIUM SCHEDULE	INSURANCE MAXIMUMS	DISABILITY	LIFE
CREDIT DISABILITY		X	$ e	MONTHLY TOTAL BENEFIT	$ 600	N/A
				INSURABLE BALANCE PER LOAN ACCOUNT	$50,000	N/A
				MAXIMUM AGE FOR INSURANCE	66	N/A

If you are totally disabled for more than 30 days, then the Disability Benefit will begin with the 31st day of disability. SECONDARY (If you desire to name one) BENEFICIARY

DATE	DATE OF BIRTH	DATE	DATE OF BIRTH
April 3			

SIGNATURE OF BORROWER ELIGIBLE TO BE INSURED (Be sure to check the boxes above.)
X John Q. Smith

SIGNATURE OF JOINT INSURED (CO-BORROWER) (Only required if JOINT CREDIT LIFE coverage is selected)
X N/A

Figure 15.6

The Cost of Immediate Money

A "fee" of 20% of a check's face value for a loan until payday may seem small. But when compounded over six weeks, it adds up, compared to a cash advance on a credit card.

$200 Loan for Six Weeks

Payday advance loan	
"Interest rate" (20% of face value)	$40
Setup charge (weeks 1 and 2)	$15
First rollover (weeks 3 and 4)	$55
Second rollover (weeks 5 and 6)	$55
TOTAL COST	**$165**
Effective APR	**715%**
Credit-card cash advance	
TOTAL COST	**$5**
APR	**21%**

DATA: *FORTUNE MAGAZINE*

Figure 15.7

The **nominal**, or **stated**, interest rate can differ from the APR. The APR is the *true effective annual interest rate* for a loan. For example, a $1000 loan for 1 year with $120 in interest charges has an APR of 12%.

$$R = \frac{I}{PT} = \frac{\$120}{\$1000 \times 1}$$

A loan of $1000 for 9 months with interest charges of $120 has an APR of 16%.

$$R = \frac{I}{PT} = \frac{\$120}{\$1000 \times \frac{9}{12}}$$

Interest charges *vary significantly* from one loan source to another and can be quite high as you can see from Figure 15.7. The finance charge for a loan depends on the borrower's past credit record, income, down payment, and other factors. *It pays to shop and make comparisons based on APR figures before borrowing!*

Objective 2 *Find the Total Installment Cost and Finance Charge.* Find the annual percentage rate by first finding the **total installment cost** (or the **deferred payment price**) and the **finance charge** on the loan. Do this with the following steps.

Finding Amount Financed

Step 1. Find the total installment cost.

$$\text{Total installment cost} = \text{Down payment} + (\text{Amount of each payment} \times \text{Number of payments})$$

Step 2. Find the finance charge.

$$\text{Finance charge} = \text{Total installment cost} - \text{Cash price}$$

Step 3. Finally, find the amount financed.

$$\text{Amount financed} = \text{Cash price} - \text{Down payment}$$

NOTE Lenders frequently *allow the borrower to choose the day of the month* that installment payments are due. This allows the borrower to pay the bill after a payday for example. ■

Example 1 Finding Total Installment Cost, Amount Financed, and Finance Charge

Ed Chamski makes his living by teaching piano both at home and at a nearby community college where he is a professor of music. He purchased a new baby grand piano for \$8500 with \$1500 down and 48 monthly payments of \$184.34 each. Find (a) the total installment cost, (b) the finance charge, and (c) the amount financed.

Solution

(a) Find the total installment cost by adding the down payment to the total amount of all payments.

$$\text{Total installment cost} = \$1500 + 48 \times \$184.34 = \$10{,}348.32$$

(b) The finance charge is the total installment cost less the cash price.

$$\text{Finance charge} = \$10{,}348.32 - \$8500 = \$1848.32$$

(c) The amount financed is \$8500 − \$1500 = \$7000.

NOTE In determining the total installment cost, the down payment is added to the total of the monthly payments. ■

Students frequently borrow money using a **Stafford loan**, which is a type of installment loan. The government pays the interest on a *subsidized* Stafford loan while the student borrower is in school on at least half-time basis. On the other hand, the student is responsible for interest on *unsubsidized* Stafford loans. Repayment of a loan begins 6 months after the borrower ceases at least half-time enrollment. You can find information about Stafford loans at the financial aid office of your college or at a bank.

Objective 3 *Use a Table to Find the APR.* Annual percentage rate tables, available from the nearest federal reserve bank or the Board of Governors of the Federal Reserve System, Washington, D.C. 20551, are used for APR rates that *are accurate enough* to satisfy federal law. Table 15.2 is a portion of these tables, which incidentally consist of two volumes.

The APR is found using the annual percentage rate table as follows.

Finding Annual Percentage Rate Using Table

Step 1. Divide the finance charge by the amount financed, and multiply by \$100.

$$\frac{\text{Finance charge}}{\text{Amount financed}} \times \$100$$

The result is the finance charge per \$100 of the amount financed.

Table 15.2 ANNUAL PERCENTAGE RATE TABLE FOR MONTHLY PAYMENT PLANS

Number of Payments	Annual Percentage Rate (Finance Charge per $100 of Amount Financed)															
	10.00%	10.25%	10.50%	10.75%	11.00%	11.25%	11.50%	11.75%	12.00%	12.25%	12.50%	12.75%	13.00%	13.25%	13.50%	13.75%
1	0.83	0.85	0.87	0.90	0.92	0.94	0.96	0.98	1.00	1.02	1.04	1.06	1.08	1.10	1.12	1.15
2	1.25	1.28	1.31	1.35	1.38	1.41	1.44	1.47	1.50	1.53	1.57	1.60	1.63	1.66	1.69	1.72
3	1.67	1.71	1.76	1.80	1.84	1.88	1.92	1.96	2.01	2.05	2.09	2.13	2.17	2.22	2.26	2.30
4	2.09	2.14	2.20	2.25	2.30	2.35	2.41	2.46	2.51	2.57	2.62	2.67	2.72	2.78	2.83	2.88
5	2.51	2.58	2.64	2.70	2.77	2.83	2.89	2.96	3.02	3.08	3.15	3.21	3.27	3.34	3.40	3.46
6	2.94	3.01	3.08	3.16	3.23	3.31	3.38	3.45	3.53	3.60	3.68	3.75	3.83	3.90	3.97	4.05
7	3.36	3.45	3.53	3.62	3.70	3.78	3.87	3.95	4.04	4.12	4.21	4.29	4.38	4.47	4.55	4.64
8	3.79	3.88	3.98	4.07	4.17	4.26	4.36	4.46	4.55	4.65	4.74	4.84	4.94	5.03	5.13	5.22
9	4.21	4.32	4.43	4.53	4.64	4.75	4.85	4.96	5.07	5.17	5.28	5.39	5.49	5.60	5.71	5.82
10	4.64	4.76	4.88	4.99	5.11	5.23	5.35	5.46	5.58	5.70	5.82	5.94	6.05	6.17	6.29	6.41
11	5.07	5.20	5.33	5.45	5.58	5.71	5.84	5.97	6.10	6.23	6.36	6.49	6.62	6.75	6.88	7.01
12	5.50	5.64	5.78	5.92	6.06	6.20	6.34	6.48	6.62	6.76	6.90	7.04	7.18	7.32	7.46	7.60
13	5.93	6.08	6.23	6.38	6.53	6.68	6.84	6.99	7.14	7.29	7.44	7.59	7.75	7.90	8.05	8.20
14	6.36	6.52	6.69	6.85	7.01	7.17	7.34	7.50	7.66	7.82	7.99	8.15	8.31	8.48	8.64	8.81
15	6.80	6.97	7.14	7.32	7.49	7.66	7.84	8.01	8.19	8.36	8.53	8.71	8.88	9.06	9.23	9.41
16	7.23	7.41	7.60	7.78	7.97	8.15	8.34	8.53	8.71	8.90	9.08	9.27	9.46	9.64	9.83	10.02
17	7.67	7.86	8.06	8.25	8.45	8.65	8.84	9.04	9.24	9.44	9.63	9.83	10.03	10.23	10.43	10.63
18	8.10	8.31	8.52	8.73	8.93	9.14	9.35	9.56	9.77	9.98	10.19	10.40	10.61	10.82	11.03	11.24
19	8.54	8.76	8.98	9.20	9.42	9.64	9.86	10.08	10.30	10.52	10.74	10.96	11.18	11.41	11.63	11.85
20	8.98	9.21	9.44	9.67	9.90	10.13	10.37	10.60	10.83	11.06	11.30	11.53	11.76	12.00	12.23	12.46
21	9.42	9.66	9.90	10.15	10.39	10.63	10.88	11.12	11.36	11.61	11.85	12.10	12.34	12.59	12.84	13.08
22	9.86	10.12	10.37	10.62	10.88	11.13	11.39	11.64	11.90	12.16	12.41	12.67	12.93	13.19	13.44	13.70
23	10.30	10.57	10.84	11.10	11.37	11.63	11.90	12.17	12.44	12.71	12.97	13.24	13.51	13.78	14.05	14.32
24	10.75	11.02	11.30	11.58	11.86	12.14	12.42	12.70	12.98	13.26	13.54	13.82	14.10	14.38	14.66	14.95
25	11.19	11.48	11.77	12.06	12.35	12.64	12.93	13.22	13.52	13.81	14.10	14.40	14.69	14.98	15.28	15.57
26	11.64	11.94	12.24	12.54	12.85	13.15	13.45	13.75	14.06	14.36	14.67	14.97	15.28	15.59	15.89	16.20
27	12.09	12.40	12.71	13.03	13.34	13.66	13.97	14.29	14.60	14.92	15.24	15.56	15.87	16.19	16.51	16.83
28	12.53	12.86	13.18	13.51	13.84	14.16	14.49	14.82	15.15	15.48	15.81	16.14	16.47	16.80	17.13	17.46
29	12.98	13.32	13.66	14.00	14.33	14.67	15.01	15.35	15.70	16.04	16.38	16.72	17.07	17.41	17.75	18.10

30	13.43	13.78	14.13	14.48	14.83	15.19	15.54	15.89	16.24	16.60	16.95	17.31	17.66	18.02	18.38	18.74
31	13.89	14.25	14.61	14.97	15.33	15.70	16.06	16.43	16.79	17.16	17.53	17.90	18.27	18.63	19.00	19.38
32	14.34	14.71	15.09	15.46	15.84	16.21	16.59	16.97	17.35	17.73	18.11	18.49	18.87	19.25	19.63	20.02
33	14.79	15.18	15.57	15.95	16.34	16.73	17.12	17.51	17.90	18.29	18.69	19.08	19.47	19.87	20.26	20.66
34	15.25	15.65	16.05	16.44	16.85	17.25	17.65	18.05	18.46	18.86	19.27	19.67	20.08	20.49	20.90	21.31
35	15.70	16.11	16.53	16.94	17.35	17.77	18.18	18.60	19.01	19.43	19.85	20.27	20.69	21.11	21.53	21.95
36	16.16	16.58	17.01	17.43	17.86	18.29	18.71	19.14	19.57	20.00	20.43	20.87	21.30	21.73	22.17	22.60
37	16.62	17.06	17.49	17.93	18.37	18.81	19.25	19.69	20.13	20.58	21.02	21.46	21.91	22.36	22.81	23.25
38	17.08	17.53	17.98	18.43	18.88	19.33	19.78	20.24	20.69	21.15	21.61	22.07	22.52	22.99	23.45	23.91
39	17.54	18.00	18.46	18.93	19.39	19.86	20.32	20.79	21.26	21.73	22.20	22.67	23.14	23.61	24.09	24.56
40	18.00	18.48	18.95	19.43	19.90	20.38	20.86	21.34	21.82	22.30	22.79	23.27	23.76	24.25	24.73	25.22
41	18.47	18.95	19.44	19.93	20.42	20.91	21.40	21.89	22.39	22.88	23.38	23.88	24.38	24.88	25.38	25.88
42	18.93	19.43	19.93	20.43	20.93	21.44	21.94	22.45	22.96	23.47	23.98	24.49	25.00	25.51	26.03	26.55
43	19.40	19.91	20.42	20.94	21.45	21.97	22.49	23.01	23.53	24.05	24.57	25.10	25.62	26.15	26.68	27.21
44	19.86	20.39	20.91	21.44	21.97	22.50	23.03	23.57	24.10	24.64	25.17	25.71	26.25	26.79	27.33	27.88
45	20.33	20.87	21.41	21.95	22.49	23.03	23.58	24.12	24.67	25.22	25.77	26.32	26.88	27.43	27.99	28.55
46	20.80	21.35	21.90	22.46	23.01	23.57	24.13	24.69	25.25	25.81	26.37	26.94	27.51	28.08	28.65	29.22
47	21.27	21.83	22.40	22.97	23.53	24.10	24.68	25.25	25.82	26.40	26.98	27.56	28.14	28.72	29.31	29.89
48	21.74	22.32	22.90	23.48	24.06	24.64	25.23	25.81	26.40	26.99	27.58	28.18	28.77	29.37	29.97	30.57
49	22.21	22.80	23.39	23.99	24.58	25.18	25.78	26.38	26.98	27.59	28.19	28.80	29.41	30.02	30.63	31.24
50	22.69	23.29	23.89	24.50	25.11	25.72	26.33	26.95	27.56	28.18	28.80	29.42	30.04	30.67	31.29	31.92
51	23.16	23.78	24.40	25.02	25.64	26.26	26.89	27.52	28.15	28.78	29.41	30.05	30.68	31.32	31.96	32.60
52	23.64	24.27	24.90	25.53	26.17	26.81	27.45	28.09	28.73	29.38	30.02	30.67	31.32	31.98	32.63	33.29
53	24.11	24.76	25.40	26.05	26.70	27.35	28.00	28.66	29.32	29.98	30.64	31.30	31.97	32.63	33.30	33.97
54	24.59	25.25	25.91	26.57	27.23	27.90	28.56	29.23	29.91	30.58	31.25	31.93	32.61	33.29	33.98	34.66
55	25.07	25.74	26.41	27.09	27.77	28.44	29.13	29.81	30.50	31.18	31.87	32.56	33.26	33.95	34.65	35.35
56	25.55	26.23	26.92	27.61	28.30	28.99	29.69	30.39	31.09	31.79	32.49	33.20	33.91	34.62	35.33	36.04
57	26.03	26.73	27.43	28.13	28.84	29.54	30.25	30.97	31.68	32.39	33.11	33.83	34.56	35.28	36.01	36.74
58	26.51	27.23	27.94	28.66	29.37	30.10	30.82	31.55	32.27	33.00	33.74	34.47	35.21	35.95	36.69	37.43
59	27.00	27.72	28.45	29.18	29.91	30.65	31.39	32.13	32.87	33.61	34.36	35.11	35.86	36.62	37.37	38.13
60	27.48	28.22	28.96	29.71	30.45	31.20	31.96	32.71	33.47	34.23	34.99	35.75	36.52	37.29	38.06	38.83

(*continued*)

Table 15.2 (CONTINUED) ANNUAL PERCENTAGE RATE TABLE FOR MONTHLY PAYMENT PLANS

Number of	Annual Percentage Rate (Finance Charge per $100 of Amount Financed)															
Payments	14.00%	14.25%	14.50%	14.75%	15.00%	15.25%	15.50%	15.75%	16.00%	16.25%	16.50%	16.75%	17.00%	17.25%	17.50%	17.75%
1	1.17	1.19	1.21	1.23	1.25	1.27	1.29	1.31	1.33	1.35	1.37	1.40	1.42	1.44	1.46	1.48
2	1.75	1.78	1.82	1.85	1.88	1.91	1.94	1.97	2.00	2.04	2.07	2.10	2.13	2.16	2.19	2.22
3	2.34	2.38	2.43	2.47	2.51	2.55	2.59	2.64	2.68	2.72	2.76	2.80	2.85	2.89	2.93	2.97
4	2.93	2.99	3.04	3.09	3.14	3.20	3.25	3.30	3.36	3.41	3.46	3.51	3.57	3.62	3.67	3.73
5	3.53	3.59	3.65	3.72	3.78	3.84	3.91	3.97	4.04	4.10	4.16	4.23	4.29	4.35	4.42	4.48
6	4.12	4.20	4.27	4.35	4.42	4.49	4.57	4.64	4.72	4.79	4.87	4.94	5.02	5.09	5.17	5.24
7	4.72	4.81	4.89	4.98	5.06	5.15	5.23	5.32	5.40	5.49	5.58	5.66	5.75	5.83	5.92	6.00
8	5.32	5.42	5.51	5.61	5.71	5.80	5.90	6.00	6.09	6.19	6.29	6.38	6.48	6.58	6.67	6.77
9	5.92	6.03	6.14	6.25	6.35	6.46	6.57	6.68	6.78	6.89	7.00	7.11	7.22	7.32	7.43	7.54
10	6.53	6.65	6.77	6.88	7.00	7.12	7.24	7.36	7.48	7.60	7.72	7.84	7.96	8.08	8.19	8.31
11	7.14	7.27	7.40	7.53	7.66	7.79	7.92	8.05	8.18	8.31	8.44	8.57	8.70	8.83	8.96	9.09
12	7.74	7.89	8.03	8.17	8.31	8.45	8.59	8.74	8.88	9.02	9.16	9.30	9.45	9.59	9.73	9.87
13	8.36	8.51	8.66	8.81	8.97	9.12	9.27	9.43	9.58	9.73	9.89	10.04	10.20	10.35	10.50	10.66
14	8.97	9.13	9.30	9.46	9.63	9.79	9.96	10.12	10.29	10.45	10.62	10.78	10.95	11.11	11.28	11.45
15	9.59	9.76	9.94	10.11	10.29	10.47	10.64	10.82	11.00	11.17	11.35	11.53	11.71	11.88	12.06	12.24
16	10.20	10.39	10.58	10.77	10.95	11.14	11.33	11.52	11.71	11.90	12.09	12.28	12.46	12.65	12.84	13.03
17	10.82	11.02	11.22	11.42	11.62	11.82	12.02	12.22	12.42	12.62	12.83	13.03	13.23	13.43	13.63	13.83
18	11.45	11.66	11.87	12.08	12.29	12.50	12.72	12.93	13.14	13.35	13.57	13.78	13.99	14.21	14.42	14.64
19	12.07	12.30	12.52	12.74	12.97	13.19	13.41	13.64	13.86	14.09	14.31	14.54	14.76	14.99	15.22	15.44
20	12.70	12.93	13.17	13.41	13.64	13.88	14.11	14.35	14.59	14.82	15.06	15.30	15.54	15.77	16.01	16.25
21	13.33	13.58	13.82	14.07	14.32	14.57	14.82	15.06	15.31	15.56	15.81	16.06	16.31	16.56	16.81	17.07
22	13.96	14.22	14.48	14.74	15.00	15.26	15.52	15.78	16.04	16.30	16.57	16.83	17.09	17.36	17.62	17.88
23	14.59	14.87	15.14	15.41	15.68	15.96	16.23	16.50	16.78	17.05	17.32	17.60	17.88	18.15	18.43	18.70
24	15.23	15.51	15.80	16.08	16.37	16.65	16.94	17.22	17.51	17.80	18.09	18.37	18.66	18.95	19.24	19.53
25	15.87	16.17	16.46	16.76	17.06	17.35	17.65	17.95	18.25	18.55	18.85	19.15	19.45	19.75	20.05	20.36
26	16.51	16.82	17.13	17.44	17.75	18.06	18.37	18.68	18.99	19.30	19.62	19.93	20.24	20.56	20.87	21.19
27	17.15	17.47	17.80	18.12	18.44	18.76	19.09	19.41	19.74	20.06	20.39	20.71	21.04	21.37	21.69	22.02
28	17.80	18.13	18.47	18.80	19.14	19.47	19.81	20.15	20.48	20.82	21.16	21.50	21.84	22.18	22.52	22.86
29	18.45	18.79	19.14	19.49	19.83	20.18	20.53	20.88	21.23	21.58	21.94	22.29	22.64	22.99	23.35	23.70
30	19.10	19.45	19.81	20.17	20.54	20.90	21.26	21.62	21.99	22.35	22.72	23.08	23.45	23.81	24.18	24.55

31	19.75	20.12	20.49	20.87	21.24	21.61	21.99	22.37	22.74	23.12	23.50	23.88	24.26	24.64	25.02	25.40
32	20.40	20.79	21.17	21.56	21.95	22.33	22.72	23.11	23.50	23.89	24.28	24.68	25.07	25.46	25.86	26.25
33	21.06	21.46	21.85	22.25	22.65	23.06	23.46	23.86	24.26	24.67	25.07	25.48	25.88	26.29	26.70	27.11
34	21.72	22.13	22.54	22.95	23.37	23.78	24.19	24.61	25.03	25.44	25.86	26.28	26.70	27.12	27.54	27.97
35	22.38	22.80	23.23	23.65	24.08	24.51	24.94	25.36	25.79	26.23	26.66	27.09	27.52	27.96	28.39	28.83
36	23.04	23.48	23.92	24.35	24.80	25.24	25.68	26.12	26.57	27.01	27.46	27.90	28.35	28.80	29.25	29.70
37	23.70	24.16	24.61	25.06	25.51	25.97	26.42	26.88	27.34	27.80	28.26	28.72	29.18	29.64	30.10	30.57
38	24.37	24.84	25.30	25.77	26.24	26.70	27.17	27.64	28.11	28.59	29.06	29.53	30.01	30.49	30.96	31.44
39	25.04	25.52	26.00	26.48	26.96	27.44	27.92	28.41	28.89	29.38	29.87	30.36	30.85	31.34	31.83	32.32
40	25.71	26.20	26.70	27.19	27.69	28.18	28.68	29.18	29.68	30.18	30.68	31.18	31.68	32.19	32.69	33.20
41	26.39	26.89	27.40	27.91	28.41	28.92	29.44	29.95	30.46	30.97	31.49	32.01	32.52	33.04	33.56	34.08
42	27.06	27.58	28.10	28.62	29.15	29.67	30.19	30.72	31.25	31.78	32.31	32.84	33.37	33.90	34.44	34.97
43	27.74	28.27	28.81	29.34	29.88	30.42	30.96	31.50	32.04	32.58	33.13	33.67	34.22	34.76	35.31	35.86
44	28.42	28.97	29.52	30.07	30.62	31.17	31.72	32.28	32.83	33.39	33.95	34.51	35.07	35.63	36.19	36.76
45	29.11	29.67	30.23	30.79	31.36	31.92	32.49	33.06	33.63	34.20	34.77	35.35	35.92	36.50	37.08	37.66
46	29.79	30.36	30.94	31.52	32.10	32.68	33.26	33.84	34.43	35.01	35.60	36.19	36.78	37.37	37.96	38.56
47	30.48	31.07	31.66	32.25	32.84	33.44	34.03	34.63	35.23	35.83	36.43	37.04	37.64	38.25	38.86	39.46
48	31.17	31.77	32.37	32.98	33.59	34.20	34.81	35.42	36.03	36.65	37.27	37.88	38.50	39.13	39.75	40.37
49	31.86	32.48	33.09	33.71	34.34	34.96	35.59	36.21	36.84	37.47	38.10	38.74	39.37	40.01	40.65	41.29
50	32.55	33.18	33.82	34.45	35.09	35.73	36.37	37.01	37.65	38.30	38.94	39.59	40.24	40.89	41.55	42.20
51	33.25	33.89	34.54	35.19	35.84	36.49	37.15	37.81	38.46	39.12	39.79	40.45	41.11	41.78	42.45	43.12
52	33.95	34.61	35.27	35.93	36.60	37.27	37.94	38.61	39.28	39.96	40.63	41.31	41.99	42.67	43.36	44.04
53	34.65	35.32	36.00	36.68	37.36	38.04	38.72	39.41	40.10	40.79	41.48	42.17	42.87	43.57	44.27	44.97
54	35.35	36.04	36.73	37.42	38.12	38.82	39.52	40.22	40.92	41.63	42.33	43.04	43.75	44.47	45.18	45.90
55	36.05	36.76	37.46	38.17	38.88	39.60	40.31	41.03	41.74	42.47	43.19	43.91	44.64	45.37	46.10	46.83
56	36.76	37.48	38.20	38.92	39.65	40.38	41.11	41.84	42.57	43.31	44.05	44.79	45.53	46.27	47.02	47.77
57	37.47	38.20	38.94	39.68	40.42	41.16	41.91	42.65	43.40	44.15	44.91	45.66	46.42	47.18	47.94	48.71
58	38.18	38.93	39.68	40.43	41.19	41.95	42.71	43.47	44.23	45.00	45.77	46.54	47.32	48.09	48.87	49.65
59	38.89	39.66	40.42	41.19	41.96	42.74	43.51	44.29	45.07	45.85	46.64	47.42	48.21	49.01	49.80	50.60
60	39.61	40.39	41.17	41.95	42.74	43.53	44.32	45.11	45.91	46.71	47.51	48.31	49.12	49.92	50.73	51.55

Step 2. Read down the left column of Table 15.2 to the proper number of payments. Go across to the number closest to the number found in Step 1. Read up the column to find the annual percentage rate.

Example 2 Finding Annual Percentage Rate

An automobile costing \$24,000 was financed at \$473.48 per month for 48 months after a down payment of \$5500 including the value of the trade-in. The total finance charge was \$4227.04 and the amount financed was \$18,500. Find the annual percentage rate.

Solution Divide the finance charge (\$473.48 × 48 − \$18,500 = \$4227.04) by the total amount financed (\$24,000 − \$5500 = \$18,500) and multiply by \$100.

$$\frac{\$4227.04}{\$18{,}500} \times \$100 = \$22.85$$

This gives the finance charge per \$100 of amount financed.

Read down the left column of Table 15.2 to the line for 48 months (the *actual* number of monthly payments). Follow across the right to find the number *closest to* \$22.85. Here, find \$22.90. Look at the top of this column of figures to find the annual percentage rate, 10.50%.

In this example, 10.5% is the annual percentage rate that must be disclosed to the buyer of the car.

Financial *Calculator Approach*

In solving Example 2, a more precise value of the APR can be found using a financial calculator. Payment, number of compounding periods, and present value (loan amount) are known. The interest rate per compounding period is unknown.

−473.48 [*PMT*] 48 [*n*] 18500 [*PV*] [*i*] 0.873297346

Multiply this number by 12 (12 months in a year) to get 10.48% rounded to the nearest hundredth.

15.2 Exercises

Find the finance charge and the total installment cost for the following. (See Example 1.)

	Amount Financed	Down Payment	Cash Price	Number of Payments	Amount of Payment	Total Installment Cost	Finance Charge
1.	\$12,400	\$5000	\$17,400	60	\$264.94	______	______
2.	\$650	\$125	\$775	24	\$32	______	______
3.	\$150	None	\$150	12	\$15	______	______
4.	\$1200	None	\$1200	20	\$70	______	______
5.	\$65	\$25	\$90	10	\$7.50	______	______
6.	\$490	None	\$490	12	\$43.30	______	______

Find the annual percentage rate using Table 15.2. (See Example 2.)

	Amount Financed	Finance Charge	No. of Monthly Payments	APR
7.	$1850	$157.30	15	______
8.	$345	$24.62	12	______
9.	$442	$28.68	14	______
10.	$4690	$1237.22	48	______
11.	$145	$13.25	18	______
12.	$650	$73.45	24	______

13. Explain the difference between open-end credit and installment loans. (See Section 15.1 and Objective 1 of this section.)

14. Find the APR on one of your personal loans. (See Objective 3.)

Solve the following application problems.

15. **TV PURCHASE** Holly Canon bought a portable color television set manufactured in Ontario, Canada, for $375. She paid nothing down but agreed to payments of $33.16 per month for 12 months. Find the annual percentage rate for the loan.

16. **REFRIGERATOR PURCHASE** Sears offers a refrigerator for $1600 with no down payment, $294.06 in interest charges and 30 equal payments. Find the annual percentage rate.

17. **FINANCING A TV** House Television and Appliance wants to advertise a table model color television for $400, with 25% down and monthly payments of only $39.32 per month for 8 months. They must also include the annual percentage rate in the ad. Find the annual percentage rate.

18. **SOFA ON SALE** A department store has a sofa on sale for $900. A buyer paying 20% down may finance the balance by paying $34.30 per month for 2 years. Find the annual percentage rate that the store must include in its advertising.

19. **COMPUTER SYSTEM PURCHASE** A contractor in Mexico City purchases a computer system for 650,000 pesos. After making a down payment of 100,000 pesos he agrees to payments of 26,342.18 pesos per month for 24 months. Find (a) the total installment cost and (b) the annual percentage rate.

20. **TRUCK PURCHASE** An electrical contractor in Hiroshima, Japan, purchases a truck costing 2,700,000 yen. He makes a down payment of 1,000,000 yen and agrees to monthly payments of 54,855 yen for 36 months. Find (a) the total installment cost and (b) the annual percentage rate.

21. **STORE EXPANSION** Dillon Sporting Goods borrows $180,000 to expand their store to include winter sporting supplies including skis and snowmobiles. They make no down payment but agree to monthly payments of $6974.66 for 30 months. Find (a) the total installment cost and (b) the annual percentage rate.

22. **NEW STORE** Bob Zombo starts a small stereo store with $60,000 borrowed from his uncle. Zombo agrees to repay his uncle with 24 monthly payments of $2820. Find (a) the total installment cost and (b) the annual percentage rate.

23. When are people more likely to use installment loans rather than open-end credit? (*Hint:* Search your local newspaper for examples.)

24. Which type of loan tends to have higher interest rates—installment loans or credit loans? Document this with several examples of each type of loan.

15.3 Early Payoffs of Loans

Objectives

1. *Use the United States Rule when prepaying a loan.*
2. *Find the amount due on the maturity date using the United States Rule.*
3. *Use the Rule of 78 when prepaying a loan.*

Objective 1 ***Use the United States Rule When Prepaying a Loan.*** It is common for a payment to be made on a loan *before it is due*. This may occur when a person receives extra money or when a debt is refinanced elsewhere. Prepayments of loans are discussed in this section.

The first method for calculating early loan payment is the **United States Rule**, which is used by the U.S. government as well as most states and financial institutions. Under the United States Rule, any payment is first applied to any interest owed. The balance of the payment is then used to reduce the principal amount of the loan.

The United States Rule

Step 1. Find the simple interest due from the date the loan was made until the date the partial payment is made. Use the formula $I = PRT$.

Step 2. Subtract this interest from the amount of the payment.

Step 3. Any difference is used to reduce the principal.

Step 4. Treat additional partial payments in the same way, always finding interest on *only* the unpaid balance after the last partial payment.

The remaining principal plus interest on this unpaid principal is then due on the due date of the loan.

Objective 2 ***Find the Amount Due on the Maturity Date Using the United States Rule.*** If the partial payment is not large enough to pay the interest due, the payment is held in suspension until enough money is available to pay the interest due; or, the payment is rejected and returned. Either way, the borrower is contacted. This means that a partial payment smaller than the interest due *offers no advantage* to the borrower—the lender just holds the partial payment until enough money is available to pay the interest owed.

Example 1 Finding the Amount Due

On August 14, Jane Ficker signed a 180-day note for $28,500 for an X-ray machine for her dental office. The note has an interest rate of 10%. On October 25, a payment of $8500 is made. (a) Find the balance owed on the principal. (b) If no additional payments are made, find the amount due at maturity of the loan.

Solution

(a) First find the interest from August 14 to October 25 (use the table near the inside back cover of the text; this shows 298 − 226 = 72 days.) using $I = PRT$.

$$\text{Interest} = \$28{,}500 \times 0.10 \times \frac{72}{360} = \$570$$

Subtract the interest from the October 25 payment to find the amount of the payment to be applied to principal. Then reduce the original principal by this amount.

$$\text{Applied to principal} = \$8500 - \$570 = \$7930$$
$$\text{New principal} = \$28{,}500 - \$7930 = \$20{,}570$$

(b) The note was originally for 180 days with the partial payment made after 72 days. Thus, interest on the new principal of $20,570 will be charged for 180 − 72 = 108 days.

$$\text{Interest} = \$20{,}570 \times 0.10 \times \frac{108}{360} = \$617.10$$

If no additional partial payments are made, the amount due at the maturity date is the remaining principal plus interest on that remaining principal.

$$\text{Amount due at maturity} = \$20{,}570 + \$617.10 = \$21{,}187.10$$

In order to find the total interest paid when partial payments are made, the individual interest payments are added.

Example 2 Finding Total Interest Paid

A lawn furniture manufacturer signs a 140-day note on February 5. The note, for $45,600, is to a supplier of aluminum tubing for the furniture and carries a simple interest rate of 12%. On March 19, the manufacturer receives an unexpected payment from one of its customers and applies $16,000 toward the note. A further early payment permits a second $13,250 partial payment on April 23. Find the interest paid on the note and the amount paid on the due date of the note.

Solution The first partial payment was made on March 19, which is 23 + 19 = 42 days after the loan is made. In 42 days, the interest on the note is found as follows.

$$I = PRT$$
$$I = \$45{,}600 \times 0.12 \times \frac{42}{360} = \$638.40$$

A partial payment of $16,000 was made on March 19. Of this amount, $638.40 is applied to interest.

$16,000.00	Amount of payment
− 638.40	Interest owed
$15,361.60	Applied to principal

After March 19, the balance on the loan is as follows.

\$45,600.00	Original amount of loan
− 15,361.60	Applied to principal
\$30,238.40	New amount owed

After March 19, the balance on the note is \$30,238.40. A second partial payment is made on April 23, which is 12 + 23 = 35 days later. Interest on \$30,238.40 for 35 days is

$$I = \$30{,}238.40 \times 0.12 \times \frac{35}{360} = \$352.78$$

A payment of \$13,250 is made on April 23. Of this, \$352.78 applies to interest, leaving

$$\$13{,}250 - \$352.78 = \$12{,}897.22$$

to reduce the principal. After April 23, the principal is as follows.

\$30,238.40	Previous balance
− 12,897.22	Applied to principal
\$17,341.18	New principal

The first partial payment was made 42 days after the note was signed, with the second payment made 35 days after that. The second payment was made 42 + 35 = 77 days after the note was signed. Since the note was for 140 days, the note is due

$$140 - 77 = 63$$

days after the second partial payment. Interest on the new balance of \$17,341.18 for 63 days is

$$I = \$17{,}341.18 \times 0.12 \times \frac{63}{360} = \$364.16$$

On the date the loan matures, a total of

$$\$17{,}341.18 + \$364.16 = \$17{,}705.34$$

must be paid. The total interest paid over the life of the loan is

$$\$638.40 + \$352.78 + \$364.16 = \$1355.34$$

All this work can be summarized in the following table.

Date Payment Made	Amount of Payment	Applied to Interest	Applied to Principal	Remaining Balance
March 19	\$16,000	\$638.40	\$15,361.60	\$30,238.40
April 23	13,250	352.78	12,897.22	17,341.18
Date of maturity (June 25)	17,705.34	364.16	17,341.18	0
Totals		\$1355.34	\$45,600	

Objective 3 *Use the Rule of 78 When Prepaying a Loan.* A variation of the United States Rule, called the **Rule of 78**, is still used by many lenders. The rule of 78 is sometimes called the **sum-of-the-balances method** when the length of the contract is other than 1 year. This rule allows a lender to earn more of the finance charge during the early months of

the loan compared to the United States Rule. Lenders typically use this rule to protect against early payoffs on *small loans*. Effectively, the lender will earn a higher rate of interest in the event of an early payoff under the Rule of 78 than under the United States Rule. The Rule of 78 *favors the lender* in the event of an early payoff.

The Rule of 78 gets its name from a loan of 12 months—the sum of the months $1 + 2 + 3 + \cdots + 12 = 78$. The finance charge for the first month is $\frac{12}{78}$ of the total charge, with $\frac{11}{78}$ in the second month, $\frac{10}{78}$ in the third month, and $\frac{1}{78}$ in the final month. The Rule of 78 can be applied to loans with terms other than 12 months. For example, the sum of the months in a 6-month contract is $1 + 2 + 3 + 4 + 5 + 6 = 21$. The finance charge for the first month would be $\frac{6}{21}$, $\frac{5}{21}$ for the second month, and so on.

The **unearned interest**, or interest not earned by the lender under the Rule of 78 for a loan of 12 months, depends on the month in which the loan is paid off. The unearned interest is returned to the borrower. If the loan is paid off at the end of 2 months, then the interest earned by the lender is $\frac{12}{78} + \frac{11}{78} = \frac{23}{78}$ of the finance charge. Thus, the interest not earned by the lender is

$$1 - \frac{23}{78} \qquad \text{or equivalently} \qquad \frac{1}{78} + \frac{2}{78} + \cdots + \frac{10}{78}$$

which is $\frac{55}{78}$ of the finance charges. The process is similar for loans of lengths other than 12 months.

The unearned interest is given by

$$U = F\left(\frac{N}{P}\right)\left(\frac{1 + N}{1 + P}\right)$$

where U = unearned interest, F = finance charge, N = number of payments remaining, and P = total number of payments.

Example 3 Finding Payoff Value

Richard Buck borrowed \$600 that he is paying back in 24 monthly payments of \$29.50 each. With 9 payments remaining, he decides to repay the loan in full. Find (a) the amount of unearned finance charge and (b) the amount necessary to repay the loan in full.

Solution

(a) Buck is scheduled to make 24 payments of \$29.50 each, for a total repayment of

$$24 \times \$29.50 = \$708$$

His finance charge is

$$\$708 - \$600 = \$108$$

Find the amount of unearned interest as follows. The finance charge is \$108, the scheduled number of payments is 24, and the loan is paid off with 9 payments left. Use the formula.

$$\begin{aligned}\text{Unearned interest} &= \$108 \times \left(\frac{9}{24}\right) \times \frac{(1 + 9)}{(1 + 24)} \\ &= \$108 \times \left(\frac{9 \times 10}{24 \times 25}\right) \\ &= \$16.20\end{aligned}$$

Paying off the loan 9 months early produces a savings of \$16.20 in interest.

(b) When Buck decides to pay off the loan, he has 9 payments of \$29.50 left. These payments total

$$9 \times \$29.50 = \$265.50$$

By paying the loan early, Buck saves the unearned interest of \$16.20, so

$$\$265.50 - \$16.20 = \$249.30$$

is needed to pay the loan in full.

Scientific
Calculator Approach

The calculator solution to Example 3 is as follows.

(a) first think of the problem as $\dfrac{(108 \times 9 \times (1 + 9))}{(24 \times (1 + 24))}$ or as $\dfrac{108 \times 9 \times 10}{24 \times 25}$ then solve.

[( ] 108 [×] 9 [×] [( ] 1 [+] 9 [ )] [ )] [÷]

[( ] 24 [×] [( ] 1 [+] 24 [ )] [ )] [=] 16.2

(b) 9 [×] 29.5 [−] 16.2 [=] 249.3

15.3 Exercises

Find the balance due on the maturity date of the following notes. Find the total amount of interest paid on the note. Use the United States Rule.

	Principal	Interest	Time in Days	Partial Payments
1.	\$6,500	8%	100	\$1500 on day 56
2.	\$5,800	10%	120	\$2500 on day 60
3.	\$8,500	12%	150	\$5000 on day 45
4.	\$6,000	$10\frac{1}{2}\%$	130	\$3000 on day 100
5.	\$10,000	$8\frac{1}{4}\%$	180	\$6000 on day 120
6.	\$12,600	$11\frac{1}{2}\%$	140	\$8000 on day 100

Each of the following loans is paid in full before their date of maturity. Find the amount of unearned interest. Use the Rule of 78.

	Finance Charge	Total Number of Payments	Remaining Number of Payments When Paid in Full
7.	\$975	48	30
8.	\$325	36	6
9.	\$460	20	4

	Finance Charge	Total Number of Payments	Remaining Number of Payments When Paid in Full
10.	$325	24	22
11.	$3653.82	48	9
12.	$3085.54	60	15

13. Explain the concept of the United States Rule. When is the United States Rule likely to be applied? (See Objective 1.)

14. Explain the concept of the Rule of 78. When is the Rule of 78 likely to be applied? (See Objective 3.)

15. Why is it of no advantage to the borrower to make a partial payment smaller than the amount of interest due? What would the lender do with this payment? (See Objective 2.)

16. List some advantages of paying loans off faster than required.

Solve the following application problems.

17. **COMPUTER CONSULTANT** The computer system at Genome Therapy crashed several times last year. On January 10, the company borrows $125,000 at 11% for 250 days to pay a consultant to work on their Novell network. However, they decide to pay the loan in full on July 1. Find (a) the interest due and (b) the total amount due using the United States Rule.

18. **LOAN REPAYMENT** Titus Czech signed a 100-day note for $2800 at 15% compounded annually, but repaid the loan in full in 80 days instead. Find (a) the interest paid and (b) the amount due in 80 days using the United States Rule.

19. **TRAVEL AGENCY LOAN** Sheila Goshorn decides to use her income tax refund to pay her travel agency loan in full. She finds that her 36-month loan includes $240 in interest and that she will be paying the loan in full with 21 months remaining. Calculate the amount of unearned interest using the Rule of 78.

20. **EARLY PAYOFF** George Duda has decided that making the small monthly payment on a 6-month loan is a nuisance. The total finance charge is only $34, and the loan is paid in full with 3 months remaining. Find the unearned interest using the Rule of 78.

21. **EASY PAYMENT PLAN** Anne Kelly purchased a refrigerator on the "easy payment plan" with only $100 down and 18 equal monthly payments of $45.20. The total cash price of the refrigerator was $800. After making 6 monthly payments, she decided to pay the loan in full. Use the Rule of 78 and find (a) the amount of unearned interest and (b) the amount necessary to pay the loan in full.

22. **FINANCING A USED CAR** A used car costs $8850. After a down payment of $2000, the balance is financed with 48 payments of $194.25 each. Suppose the loan is paid off with 15 payments left. Use the Rule of 78 and find (a) the amount of unearned interest and (b) the amount necessary to pay the loan in full.

23. **PAPER GOODS INVENTORY** To save on freight charges, Wholesale Paper orders large quantities of basic paper goods every 4 months. For their last order, the firm signed a note on February 18 that will mature on May 15. The face value of the note was $104,500, with interest of 11%. The firm made a partial payment of $38,000 on March 20, with a second partial payment of $27,200 on April 16. Find the amount due on the maturity date of the note and the amount of interest paid on the note using the United States Rule.

24. TESTING EQUIPMENT Mid-City Electronics bought new testing equipment, paying for it with a note for $32,000. The note was made on July 26 and is due on November 20. The interest rate is 13%. The firm made a partial payment of $6000 on August 31, with a second partial payment of $11,700 on October 4. Find the amount due on the maturity date of the note and the amount of interest paid on the note using the United States Rule.

15.4 Personal Property Loans

Objectives

1. *Define personal property and real estate.*
2. *Use the formula for amortization to find payment.*
3. *Set up an amortization table.*
4. *Find monthly payments.*

Objective 1 *Define Personal Property and Real Estate.* Items that can be moved from one location to another such as an automobile, a boat, or a stereo are called **personal property**. In contrast, buildings, land, and homes cannot be moved and are called **real estate** or **real property**. Personal property loans are discussed in this section and real estate loans are discussed in the next section.

As the following headline suggests, people can end up with more debt than they can afford. In that event, individuals are sometimes forced to return personal property such as an automobile to the lender. When this happens, the property is said to be **repossessed** by the lender.

Consumers Cautioned to Pay Up Debts Before Trouble

Source: The Tyler-Courier-Times Telegraph. Reprinted by permission.

Objective 2 *Use the Formula for Amortization to Find Payment.* A loan is **amortized** if both the principal and interest are paid off by *a sequence of equal payments* made at regular intervals in time. One example of a loan that is amortized is a car loan with 48 equal monthly payments.

There is no loan and no need to make any payments for an item purchased with cash. However, many larger, more expensive items are purchased on credit and require a series of regular payments. The amount of each payment can be calculated using the formula, given in Section 14.2, for the present value A of an annuity with payment R, interest rate per period i, and n periods (the formula is repeated here).

$$A = R\left[\frac{(1+i)^n - 1}{i(1+i)^n}\right]$$

The unknown in Section 14.2 was the present value A. Now the present value (loan amount) is known along with the interest rate per compounding period and number of periods—the payment R is the unknown. The equation above can be solved for R and a scientific calculator can be used to calculate payments. Alternately, you can use the tables in Appendix D and solve the equation $A = R \cdot a_{\overline{n}|i}$ for R.

The periodic payment R, needed to amortize a loan of A dollars, with interest of i per period for n periods follows.

$$R = A\left(\frac{1}{a_{\overline{n}|i}}\right)$$

The notation $a_{\overline{n}|i}$ is read "a-angle–n-at-i." Values of $\frac{1}{a_{\overline{n}|i}}$ for different n's and i's can be found in column F of Appendix D. Thus, the periodic payment R is found by multiplying the loan amount A times a number taken from the table in Appendix D.

Example 1 Finding Amortization Information

Pablo Valdez recently received his college degree and accepted a job as a designer of World Wide Web pages. Valdez and his spouse, a teacher and avid water skier, decide to borrow \$15,000 to purchase a new ski boat. They go to Teachers Credit Union where George Willis checks their credit and authorizes a 36-month loan at 12% per year. Find (a) the monthly payment, (b) the portion of the first payment that is interest, (c) the balance due after 1 payment, (d) the interest owed for the second month, and (e) the balance after the second payment.

Solution

(a) Use $\frac{12\%}{12} = 1\%$ per period for i and 36 periods for n in the amortization table in Appendix D to find 0.03321431. Payment = \$15,000 × 0.03321431 = \$498.21 at the end of each month.

(b) Interest for the month is found using the simple interest formula. One month is $\frac{1}{12}$ of a year which is used for T.

$$I = PRT$$

$$I = \$15{,}000 \times 0.12 \times \frac{1}{12} = \$150$$

The amount of the first payment that is applied to reduce the loan is

$$\$498.21 - \$150 = \$348.21$$

(c) The debt after the first payment is

$$\$15{,}000 - \$348.21 = \$14{,}651.79$$

(d) Interest owed for the second month is found using the loan balance after the first payment.

$$\text{Interest} = \$14{,}651.79 \times 0.12 \times \frac{1}{12} = \$146.52$$

A payment of \$498.21 is made at the end of period 2 and a total of \$498.21 − \$146.52 = \$351.69 is applied to the debt.

(e) The balance after the second payment follows.

$$\$14{,}651.79 - \$351.69 = \$14{,}300.10$$

Financial *Calculator Approach*

The unknown payment in Example 1 can also be found using a financial calculator. Enter the known values for present value, number of compounding periods, and interest rate per compounding period.

15000 [PV] 36 [n] 1 [i] [PMT] −498.21

↳ 12% ÷ 12 months

Objective 3 ***Set Up An Amortization Table.*** An **amortization table** or **schedule** shows the amount of each payment that goes to interest and to principal. It also shows the debt remaining after each payment. One excellent way to calculate an amortization table is to use a spreadsheet package on a personal computer. Sometimes, amortization tables are referred to as **loan repayment tables**.

Problem-Solving Hint Be sure to round interest to the nearest cent each time before proceeding.

Example 2 Finding Payments, Interest, and Loan Balances

A contractor agrees to pay \$15,000 for a new computer system. This amount will be repaid in 3 years with semiannual payments at an interest rate of 8% compounded semiannually. Set up an amortization schedule for this loan.

Solution An interest rate of $\frac{8\%}{2} = 4\%$ for i is applied to the $3 \times 2 = 6$ semiannual payments. A payment of \$2861.43 is found using column F in Appendix D. Interest is calculated using $I = PRT$, with $P =$ the balance at the end of the previous period, $R =$ annual interest of 8%, and $T = \frac{1}{2}$ of a year. The interest rate per compounding period is $\frac{8\%}{2} = 4\%$ and the number of compounding periods is 1.

Payment Number	Amount of Payment	Interest for Period	Portion of Principal	Balance at End of Period
0	—	—	—	\$15,000.00
1	\$2861.43	\$600.00	\$2261.43	12,738.57
2	2861.43	509.54	2351.89	10,386.68
3	2861.43	415.47	2445.96	7,940.72
4	2861.43	317.63	2543.80	5,396.92
5	2861.43	215.88	2645.55	2,751.37
6	2861.42	110.05	2751.37	0

Table 15.3 LOAN PAYOFF TABLE

APR \ Months	18	24	30	36	42	48	54	60
8%	0.059138	0.045229	0.036887	0.031336	0.027376	0.024413	0.022113	0.020277
9%	0.0596	0.045683	0.037347	0.0318	0.027845	0.024885	0.022589	0.020758
10%	0.060056	0.046146	0.03781	0.032267	0.028317	0.025363	0.023072	0.021247
11%	0.060516	0.046608	0.038277	0.032739	0.028793	0.025846	0.023561	0.021742
12%	0.060984	0.047075	0.038747	0.033214	0.029276	0.026333	0.024057	0.022245
13%	0.06145	0.047542	0.03922	0.033694	0.029762	0.026827	0.024557	0.022753
14%	0.061917	0.048013	0.0397	0.034178	0.030252	0.027327	0.025065	0.023268
15%	0.062383	0.048488	0.04018	0.034667	0.03075	0.027831	0.025578	0.02379
16%	0.062855	0.048963	0.040663	0.035159	0.03125	0.02834	0.026096	0.024318
17%	0.063328	0.049442	0.04115	0.035653	0.031755	0.028854	0.026620	0.024853
18%	0.063806	0.049925	0.04164	0.036153	0.032264	0.029369	0.027152	0.025393
19%	0.064283	0.050408	0.042133	0.036656	0.032779	0.0299	0.027687	0.02594
20%	0.064761	0.050896	0.04263	0.037164	0.033298	0.030431	0.02823	0.026493
21%	0.065244	0.051388	0.04313	0.037675	0.033821	0.030967	0.028776	0.027053

NOTE As in the last example, the final payment will frequently vary slightly from the regular payments due to rounding errors. ■

Objective 4 ***Find Monthly Payments.*** Table 15.3 can also be used to find the monthly payment. Determine the table value corresponding to the annual percentage rate (APR) and the number of monthly payments. Then find the monthly payment by multiplying the number from the table by the amount to be financed.

Example 3 Finding Monthly Payments

After a down payment, Linda Dean owes \$8700 on a Ford Taurus. She wishes to pay the loan off in 60 monthly payments. Find the amount of each payment and the finance charge, if the APR on her loan is 18%. Her interest rate is high because she has a poor credit history.

Solution Multiply the amount to be financed, \$8700, and the number from Table 15.3 for 60 months and 18%, 0.025393.

$$\text{Payment} = (\$8700)(0.025393) = \$220.92$$

The total amount repaid in 60 months is

$$60(\$220.92) = \$13{,}255.20$$

The finance charge is

$$\$13{,}255.20 - \$8700 = \$4555.20$$

15.4 Exercises

Find each of the following.

1. $\frac{1}{a_{\overline{15}|0.075}}$

2. $\frac{1}{a_{\overline{40}|0.075}}$

3. $\frac{1}{a_{\overline{36}|0.10}}$

4. $\frac{1}{a_{\overline{48}|0.12}}$

Find the payment necessary to amortize each of the following loans. Use column F of the interest table in Appendix D and the formula of this section.

	Amount of Loan	Interest Rate	Payments Made	Number of Years	Payment
5.	\$1,850	$7\frac{1}{2}\%$	annually	2	________
6.	\$3,500	9%	annually	4	________
7.	\$4,500	8%	semiannually	$7\frac{1}{2}$	________
8.	\$1,900	16%	semiannually	$4\frac{1}{2}$	________
9.	\$96,000	8%	quarterly	$7\frac{3}{4}$	________
10.	\$210,000	12%	quarterly	8	________
11.	\$4,876	12%	monthly	3	________
12.	\$6,400	9%	monthly	$3\frac{1}{2}$	________

Use Table 15.3 to find the monthly payment and then find the finance charge for each loan.

	Amount Financed	Number of Months	APR	Monthly Payment	Finance Charge
13.	\$4,800	36	9%	______	________
14.	\$4,800	24	12%	______	________
15.	\$12,000	48	13%	______	________
16.	\$8,102	48	8%	______	________

Solve each of the following application problems. Use column F of the interest table in Appendix D.

17. OPENING A RESTAURANT Chuck and Judy Nielson opened a restaurant at a cost of \$340,000. They paid \$40,000 of their own money and agreed to pay the remainder in quarterly payments over 7 years at 12% compounded quarterly. Find the quarterly payment and the total amount of interest paid over 7 years.

18. NEW LIMOUSINE Midtown Limousine Service bought a new Lincoln limousine for \$57,000. The company agreed to pay 10% down and pay off the rest with monthly payments for 36 months at 9%. Find the amount of each monthly payment necessary to amortize the loan. Find the total amount of interest paid over 3 years.

19. NEW PRINTER An insurance firm pays $4000 for a new high-speed printer for its computer. It amortizes the loan for the printer in 4 annual payments at 8%. Prepare an amortization schedule for this printer.

Payment Number	Amount of Payment	Interest for Period	Portion to Principal	Principal at End of Period
0	—	—	—	$4000.00
1	______	______	______	______
2	______	______	______	______
3	______	______	______	______
4	______	______	______	______

20. TRUCK FINANCING Long Haul Trucking purchases a used tractor for pulling large trailers on interstate highways at a cost of $72,000. They agree to pay for it with a loan that will be amortized over 9 annual payments at 8% interest. Prepare an amortization schedule for the truck.

Payment Number	Amount of Payment	Interest for Period	Portion to Principal	Principal at End of Period
0	—	—	—	$72,000.00
1	______	______	______	______
2	______	______	______	______
3	______	______	______	______
4	______	______	______	______
5	______	______	______	______
6	______	______	______	______
7	______	______	______	______
8	______	______	______	______
9	______	______	______	______

Solve the following application problems. Use Table 15.3.

21. ENGINEERING WORK STATIONS An engineering firm purchases 7 new work stations for $3500 each. They make a down payment of $10,000 and amortize the balance with monthly payments at 11% for 4 years. Prepare an amortization schedule showing the first 5 payments.

Payment Number	Amount of Payment	Interest for Period	Portion to Principal	Principal at End of Period
0	—	—	—	$14,500.00
1	______	______	______	______
2	______	______	______	______
3	______	______	______	______
4	______	______	______	______
5	______	______	______	______

22. SETTING UP A LAW OFFICE Denise Sullivan purchased $14,000 worth of law books and $7200 worth of office furniture when she opened her law office. She paid $1200 down and agreed to amortize the balance with monthly payments for 5 years at 12%. Prepare an amortization schedule for the first 5 payments.

Payment Number	Amount of Payment	Interest for Period	Portion to Principal	Principal at End of Period
0	—	—	—	$20,000.00
1	______	______	______	______
2	______	______	______	______
3	______	______	______	______
4	______	______	______	______
5	______	______	______	______

23. Identify and explain three important items that you will see in an installment loan. (See Objectives 2 and 3.)

24. Explain how people can get in over their heads in terms of monthly payments. How can this be avoided?

Prepare an amortization schedule for the following loans. Use a spreadsheet package on a computer if you have one available. Do not forget to round interest to the nearest cent each month before finding the portion of each payment that goes to principal.

25. Tim Gates financed $8000 on a used car at 12% for 15 months.

Payment Number	Amount of Payment	Interest for Period	Portion to Principal	Principal at End of Period
0	—	—	—	$8,000.00
1	______	______	______	______
2	______	______	______	______
3	______	______	______	______
⋮	⋮	⋮	⋮	⋮
14	______	______	______	______
15	______	______	______	______

26. NETWORK LINK Caribbean Tours, Inc. financed $108,000 to set up a network link between their offices in Canada, the United States, and Mexico. They agree to repay the 10% loan with 16 quarterly payments.

Payment Number	Amount of Payment	Interest for Period	Portion to Principal	Principal at End of Period
0	—	—	—	$108,000.00
1	______	______	______	______
2	______	______	______	______
3	______	______	______	______
⋮	⋮	⋮	⋮	⋮
15	______	______	______	______
16	8271.79	201.77	8070.02	0

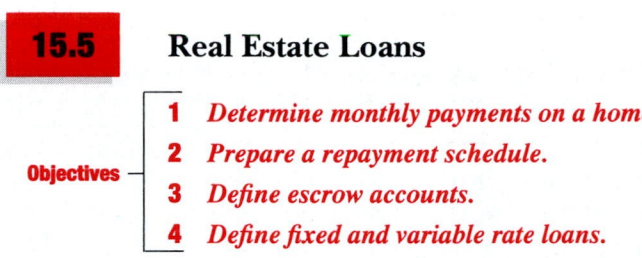

15.5 Real Estate Loans

Objectives

1. *Determine monthly payments on a home.*
2. *Prepare a repayment schedule.*
3. *Define escrow accounts.*
4. *Define fixed and variable rate loans.*

Objective 1 *Determine Monthly Payments on a Home.* A home is *one of the most expensive purchases* made by the average person. The amount of the monthly payment is a major concern of prospective buyers. The size of this payment is found by the exact same methods and formulas used in Section 15.4, but because of the many different interest rates and repayment periods, special tables are often used. Figure 15.8 shows that 30-year mortgage rates have varied considerably depending on when you took out a home loan. In particular, rates were very high in 1980–1984. As shown in Figure 15.8, the 1990s saw rates that were more moderate. The World Wide Web can be used to find current mortgage rates, or call a mortgage lender near you.

The real estate amortization table (Table 15.4) shows the monthly payment necessary to repay a $1000 loan for differing interest rates and lengths of repayment. This table is used to find the monthly payment by multiplying the loan amount, in thousands of dollars, times the correct factor from the table. Higher interest rates mean higher borrowing costs, as you can clearly see from the next example.

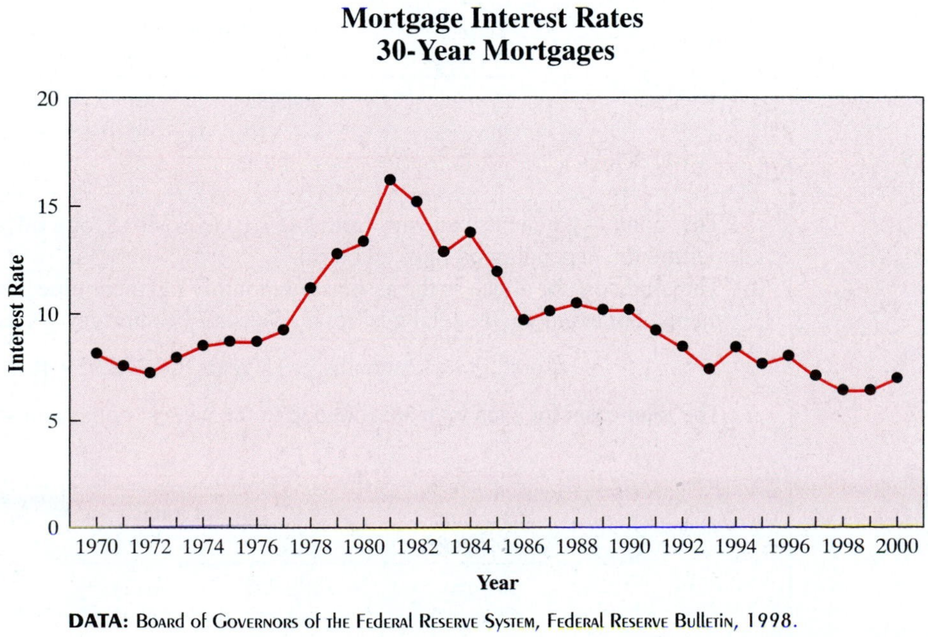

Figure 15.8

Table 15.4 AMORTIZATION (PRINCIPAL AND INTEREST PER THOUSAND DOLLARS)

Term in Years	7%	$7\frac{1}{4}$%	$7\frac{1}{2}$%	$7\frac{3}{4}$%	8%	$8\frac{1}{4}$%	$8\frac{1}{2}$%	$8\frac{3}{4}$%	9%	$9\frac{1}{4}$%	$9\frac{1}{2}$%	$9\frac{3}{4}$%
10	11.62	11.75	11.88	12.01	12.14	12.27	12.40	12.54	12.67	12.81	12.94	13.08
15	8.99	9.13	9.27	9.42	9.56	9.71	9.85	10.00	10.15	10.30	10.45	10.60
20	7.76	7.91	8.06	8.21	8.37	8.53	8.68	8.84	9.00	9.16	9.33	9.49
25	7.07	7.23	7.39	7.56	7.72	7.89	8.06	8.23	8.40	8.57	8.74	8.92
30	6.66	6.83	7.00	7.17	7.34	7.52	7.69	7.87	8.05	8.23	8.41	8.60

Example 1 Finding Payments to Amortize a Loan

Bob and Mary McArthur need to borrow $65,300 to purchase a home and are trying to decide whether they should finance it for 15 years or for 30 years. They are also looking at the effect of interest rates on total costs. (a) Find the monthly payment for both 15 years and 30 years at both $7\frac{1}{2}$% and $9\frac{3}{4}$%. (b) Then find the total cost of financing the home for each case.

Solution

(a) First, find the amount to be financed in thousands ($65,300 ÷ 1000 = 65.3 thousands). Then, multiply this number times the factor from the table for each interest rate. For example, to find the payment for a 15-year amortization at $7\frac{1}{2}$%, multiply 65.3 × 9.27 = $605.33, which is rounded to the nearest cent. The monthly payment at $9\frac{3}{4}$% for 15 years is 65.3 × 10.60 = $692.18. The monthly payments are as follows.

Term of Mortgage	Interest Rate	
	$7\frac{1}{2}$%	$9\frac{3}{4}$%
15 years	$605.33	$692.18
30 years	$457.10	$561.58

The monthly payments can vary from $457.10 to $692.18, or a difference of $235.08 per month, depending on rates and term.

(b) The total cost for a loan is the associated monthly payment times the number of payments. For example, the total cost for a 15-year $7\frac{1}{2}$% mortgage is as follows.

$605.33 × 12 months × 15 years = $108,959.40

The total costs for each loan are rounded to the nearest dollar and shown below.

Term of Mortgage	Interest Rate	
	$7\frac{1}{2}$%	$9\frac{3}{4}$%
15 years	$108,959	$124,592
30 years	$164,556	$202,169

Notice that the total costs to finance the \$65,300 loan range from about \$109,000 to about \$202,000 depending on the option chosen. The monthly payment on a 15-year loan is higher, but the total cost of financing the home *is considerably lower*. Of course, it is always best to get the lowest interest rate that you can when borrowing.

Problem-Solving Hint Be sure to divide the loan amount by \$1000 before calculating the monthly payment.

For many years, mortgage payoffs of 25 or 30 years have been the most common. The last few years, however, have seen **accelerated mortgages**, with payoffs of 15 or 18 years. As you can see from the last example, an accelerated mortgage results in significantly less interest over the life of the loan.

Example 2 Finding Loan Amount

The Mocks wish to purchase the largest, nicest home they can but are limited to a monthly payment of \$900, not including insurance and taxes. If the current mortgage rate is $8\frac{1}{2}\%$ and they would like a 20-year mortgage, what is the most they can finance?

Solution The relationship outlined in the first part of this section is:

$$\begin{matrix}\text{Number of thousands} \\ \text{of debt}\end{matrix} \times \begin{matrix}\text{Factor from} \\ \text{Table 15.4}\end{matrix} = \text{Monthly payment}$$

Previously the monthly payment was unknown. Now the payment (\$900) and the factor from Table 15.4 (20 years, $8\frac{1}{2}\%$ interest yields a factor of 8.68) are known but the debt is unknown. Using algebra, the equation above can be solved for debt.

$$\begin{aligned}\begin{matrix}\text{Number of thousands} \\ \text{of debt}\end{matrix} &= \frac{\text{Monthly payment}}{\text{Factor from Table 15.4}} \\ &= \frac{\$900}{8.68} \\ &= 103.687 \text{ (rounded)}\end{aligned}$$

Therefore, they can afford a mortgage of \$103,687.

Objective 2 ***Prepare a Repayment Schedule.*** The interest on real estate loans is computed on *the decreasing balance* of the loan. Each equal monthly payment is first applied toward the interest for the previous month. The balance is then applied toward reduction of the amount owed. Payments in the *early years* of a real estate loan are mostly interest; only a small amount goes toward reducing the principal. The amount of interest gradually decreases each month, so that larger and larger amounts of the payment apply to the principal. During the last years of the loan, most of the monthly payment is applied toward the principal.

Many lenders supply an **amortization schedule**, also called a **repayment schedule** or **loan reduction schedule**, showing the amount of payments for interest, the amount for principal, and the principal balance for each month over the entire life of the loan. These calculations can be done by hand as shown in the next example, but they are commonly done on computers.

Example 3 Preparing a Repayment Schedule

Prepare a repayment schedule for the first 2 months on a loan of $60,000 at 8% interest for 30 years. The monthly payment on this loan is $440.40.

Solution Find the interest for the first month using the formula for simple interest.

$$I = PRT$$

$$\text{Interest} = \$60{,}000 \times 0.08 \times \frac{1}{12} = \$400$$

Subtract to find the amount of the first payment that reduces the debt.

$$\$440.40 - \$400 = \$40.40$$

Find the remaining debt after the first payment by subtracting.

$$\$60{,}000 - \$40.40 = \$59{,}959.60$$

Use the simple interest formula with the new loan balance to find the interest for month 2.

$$\text{Interest} = \$59{,}959.60 \times 0.08 \times \frac{1}{12} = \$399.73$$

Subtact to find the amount of the second payment that reduces the debt.

$$\$440.40 - \$399.73 = \$40.67$$

Find the remaining debt after the second payment by subtracting.

$$\$59{,}959.60 - \$40.67 = \$59{,}918.93$$

The following repayment schedule shows the first and second months of this loan.

REPAYMENT SCHEDULE

Payment Number	Interest Payment	Principal Payment	Balance of Principal
1	$400.00	$40.40	$59,959.60
2	$399.73	$40.67	$59,918.93

Table 15.5 shows the interest payment, principal payment, and loan balance for the first 24 payments, payments numbered 256–270, and payments numbered 355–360 of the loan in Example 3. Notice how slowly the remaining balance falls during the first 12 months. In fact, during the first 12 months a total of $4781.83 is paid to interest and the remaining balance is reduced by only $502.97.

The 257th payment is the first payment in which a larger amount goes to principal than to interest. The remaining balance drops below one-half of the original loan of $60,000 only after 269 payments (22 years, 5 months). In other words, it takes almost $22\frac{1}{2}$ years to cut the loan balance in half and then about $7\frac{1}{2}$ years more to pay off the other half of the loan. The final payment is $230.21.

Objective 3 *Define Escrow Accounts.* Many lenders require **escrow accounts** (also called **impound accounts**) for people taking out a mortgage. With an escrow account, buyers pay $\frac{1}{12}$ of the total estimated property tax and insurance each month. The lender holds these

Table 15.5 LOAN REDUCTION SCHEDULE*

Payment Number	Interest Payment	Principal Payment	Remaining Balance	Payment Number	Interest Payment	Principal Payment	Remaining Balance
1	$400.00	$40.40	$59,959.60	256	$220.50	$219.90	$32,854.74
2	399.73	40.67	59,918.93	257	219.03	221.37	32,633.37
3	399.46	40.94	59,877.99	258	217.55	222.85	32,410.53
4	399.19	41.21	59,836.78	259	216.07	224.33	32,186.20
5	398.91	41.49	59,795.29	260	214.57	225.83	31,960.37
6	398.64	41.76	59,753.53	261	213.07	227.33	31,733.04
7	398.36	42.04	59,711.49	262	211.55	228.85	31,504.19
8	398.08	42.32	59,669.17	263	210.03	230.37	31,273.82
9	397.79	42.61	59,626.56	264	208.49	231.91	31,041.91
10	397.51	42.89	59,583.67	265	206.95	233.45	30,808.46
11	397.22	43.18	59,540.49	266	205.39	235.01	30,573.45
12	396.94	43.46	59,497.03	267	203.82	236.58	30,336.87
13	396.65	43.75	59,453.28	268	202.24	238.16	30,098.72
14	396.36	44.04	59,409.24	269	200.66	239.74	29,858.98
15	396.06	44.34	59,364.90	270	199.06	241.34	29,617.64
16	395.77	44.63	59,320.27				
17	395.47	44.93	59,275.34				
18	395.17	45.23	59,230.11	355	15.87	424.53	1,955.31
19	394.87	45.53	59,184.58	356	13.04	427.36	1,527.95
20	394.56	45.84	59,138.74	357	10.19	430.21	1,097.74
21	394.26	46.14	59,092.60	358	7.32	433.08	664.66
22	393.95	46.45	59,046.15	359	4.43	435.97	228.69
23	393.64	46.76	58,999.39	360	1.52	228.69	0.00
24	393.33	47.07	58,952.32				
					$98,331.96	$60,000	

*Interest rate 8%, loan amount $60,000, monthly principal and interest payment $440.40, term in years 30, total number of payments 360.

funds until the taxes and insurance fall due and then pays the bills for the borrower. Many consumer groups oppose this practice, since the lender earns interest on the money while waiting for payments to come due. In fact, a few states require that interest be paid to the homeowner on escrow accounts on any homes located in those states.

Example 4 Finding Total Monthly Payment

Bob Jones used a $75,000 loan for 25 years at 8% to purchase a summer cabin. Annual insurance and taxes on the property are $654 and $1329 respectively. Find the monthly payment.

Solution Use Table 15.4 to find 7.72. Add monthly insurance and taxes to the loan amortization.

$$\text{Payment} = 75 \times 7.72 + \frac{\$654 + \$1329}{12}$$

$$= \$579 + \$165.25$$

$$= \$744.25$$

Scientific *Calculator Approach*

The calculator solution to Example 4 is to first find the monthly payment on the principal by multiplying the number of thousands (75) times the number from the table. Add to this the monthly payment due to insurance and taxes.

75 [×] 7.72 [+] [(] 654 [+] 1329 [)] [÷] 12 [=] 744.25

Objective 4 *Define Fixed and Variable Rate Loans.* Home loans with fixed, stated interest rates are called **fixed-rate loans**. These loans help borrowers during times of rising interest rates since monthly payments remain fixed. Conversely, fixed rate loans hurt lenders during times of rising interest rates. The lenders may have to pay higher interest rates on deposits than they receive on loans. This can result in large losses to the financial institutions.

As a result, some lenders now use **variable-interest-rate loans**, also called **adjustable-rate mortgages**. The interest rates on these loans can vary up or down at stated periods over the life of the loan depending on the movement of interest rates in general. The monthly payments will change as interest rates are changed on the loan. Thus a borrower's monthly payment is not fixed for 15 or 30 years, but may go up or down every few years. In Canada and Britain, variable-rate mortgages are the only ones available.

When filing an annual income tax return, a person with a home loan must decide whether to take the standard deduction or to itemize deductions. Both the interest paid on the loan and the property taxes may currently be used in itemizing deductions. If interest and taxes are itemized, the result is frequently more than the standard deduction. The resulting tax savings are often *a major reason* to buy instead of rent. However, because of the standard deduction given to everyone, these deductions do not usually result in a dollar-for-dollar gain. While the tax advantages of owning are often praised greatly by zealous real estate sales people, don't forget that because of the standard deduction the savings *are often not as great as claimed*, and the savings may not show up until a refund is received next April, while the house payments begin immediately.

15.5 Exercises

Use Table 15.4 to find the monthly payment for the following loans.

	Amount of Loan	Interest Rate	Term of Loan	Monthly Payment
1.	$78,000	$7\frac{1}{4}\%$	30 years	______
2.	$69,500	8%	25 years	______
3.	$112,800	$8\frac{1}{2}\%$	15 years	______
4.	$27,000	7%	20 years	______
5.	$96,500	$9\frac{3}{4}\%$	15 years	______
6.	$38,200	$7\frac{1}{4}\%$	15 years	______

7. Explain how different interest rates can make a large difference in interest charges over a 30-year loan. (See Objective 1.)

8. Explain how changing the term of a loan can result in significantly different interest charges over the life of the loan. (See Objective 2.)

Find the total monthly payment including taxes and insurance for the following loans. Round to the nearest cent.

	Amount of Loan	Interest Rate	Term of Loan	Annual Taxes	Annual Insurance	Monthly Payment
9.	\$69,000	7%	30 years	\$1850	\$450	______
10.	\$75,400	$8\frac{1}{2}\%$	20 years	\$1177	\$520	______
11.	\$58,600	8%	30 years	\$745	\$380	______
12.	\$68,400	9%	30 years	\$1256	\$350	______
13.	\$91,580	$8\frac{1}{4}\%$	25 years	\$1326	\$489	______
14.	\$64,750	$7\frac{3}{4}\%$	30 years	\$1101	\$342	______

Solve the following application problems.

15. **HOME PURCHASE** Peter and Betsy Mueller hope to buy a home costing \$145,000 with annual insurance and taxes of \$780 and \$2950 respectively. They have \$20,000 saved for a down payment and their real estate agent has told them they can get a $7\frac{1}{2}\%$ rate for a 15-year mortgage. They are qualified for a home loan as long as the total monthly payment does not exceed \$1500. Are they qualified for the loan?

16. **CONDOMINIUM PURCHASE** Mr. and Mrs. Ariz wish to buy a condominium costing \$95,000 with annual insurance and taxes of \$680 and \$2278 respectively. They have \$6000 to pay down and plan to amortize the balance at 9% for 25 years. They are qualified for a home loan as long as the total monthly payment does not exceed \$850. Are they qualified for the loan?

17. **REPAYMENT SCHEDULE** June and Bill Able borrow \$122,500 at $7\frac{1}{2}\%$ for 15 years. Prepare a repayment schedule for the first two payments.

Payment Number	Total Payment	Interest Payment	Principal Payment	Balance of Principal
0	—	—	—	\$122,500.00
1	______	______	______	________
2	______	______	______	________

18. **REPAYMENT SCHEDULE** Tom Ajax purchases a home for his elderly mother. He finances \$44,300 at $7\frac{1}{4}\%$ for 10 years. Prepare a repayment schedule for the first two payments.

Payment Number	Total Payment	Interest Payment	Principal Payment	Balance of Principal
0	—	—	—	\$44,300.00
1	______	______	______	________
2	______	______	______	________

19. **FINANCING A DUPLEX** Jane Bickers purchased a duplex for \$110,000 with a down payment of 20%. She financed the balance at $8\frac{1}{2}\%$ for 20 years. Find the monthly payment and the total interest charges. (See Example 2.)

20. BORROWING PESOS Raul Aguinaga borrowed 8,000,000 pesos for 15 years at $9\frac{3}{4}\%$. Find the monthly payment and the sum of the interest charges.

21. AFFORDABLE MORTGAGE PAYMENT Paul Shingle can afford a mortgage payment of \$650 per month not including insurance and taxes. Given a 30-year loan with a rate of 9%, find the maximum mortgage (to the nearest thousand) that he can afford.

22. AFFORDABLE MORTGAGE PAYMENT Jessie Baker can spend \$780 per month not including insurance and taxes. Given a 15-year loan with a rate of $9\frac{1}{2}\%$, find the maximum mortgage (to the nearest hundred) that she can afford.

23. Take the amount you currently pay each month for rent and subtract \$150 (for insurance and taxes). Use this amount to estimate the home loan you can afford to buy assuming an 8%, 30-year mortgage. Are there homes in your area selling for this amount? (See Example 2.)

24. Mortgage companies tend to require a 5% to 20% down payment, a good credit history, and a steady job before financing a home for you. First calculate 10% of the amount calculated in Exercise 23 (this approximates your required down payment). How long will it take you to save this amount if you can save \$200 per month in a fund yielding 6% compounded monthly (ignore taxes)? (See Section 14.6).

Chapter 15 Quick Review

Chapter Terms

Review the following terms to test your understanding of the chapter. For any terms you do not know, refer to the page number found next to that term.

accelerated mortgages [p. 561]
adjustable-rate mortgages [p. 564]
amortization schedule [p. 554]
amortization table [p. 554]
amortized [p. 536]
annual percentage rate (APR) [p. 532]
average daily balance [p. 530]
borrower [p. 526]
charge accounts [p. 527]
consolidate loans [p. 532]
country club billing [p. 528]
credit card [p. 527]
credit limit [p. 527]
creditor [p. 526]
debtor [p. 526]
deferred payment price [p. 538]
escrow accounts [p. 562]
Federal Truth-in-Lending Act [p. 536]
finance charges [p. 528]
fixed-rate loans [p. 564]
impound accounts [p. 562]
installment loan [p. 536]
itemized billing [p. 528]
late fees [p. 528]
lender [p. 526]
loan reduction schedule [p. 561]
loan repayment tables [p. 554]
nominal rate [p. 538]
open-end credit [p. 527]
over-the-limit fees [p. 530]
personal property [p. 552]
real estate [p. 552]
real property [p. 552]
Regulation Z [p. 536]
repayment schedule [p. 561]
repossessed [p. 552]
revolving charge account [p. 527]
Rule of 78 [p. 548]
Stafford loan [p. 539]
stated rate [p. 538]
sum-of-the-balances method [p. 548]
total installment cost [p. 538]
unearned interest [p. 549]
United States Rule [p. 546]
variable-interest-rate loans [p. 564]

Concepts	Examples
15.1 Finding the finance charge on a revolving charge account using the average daily balance method 1. Find the unpaid balance on each day. 2. Add the daily unpaid balances and divide by the number of days in the billing period. 3. Calculate the finance charge by multiplying the average daily balance by the interest rate.	Previous balance, \$115.45; November 1, Billing date; November 15, Payment of \$35; November 22, Jacket \$45. Find the finance charge if interest is 1% per month on the average daily balance. 1. 14 days at \$115.45 = \$1616.30 7 days at (\$115.45 − \$35 = \$80.45) = \$563.15 9 days at (\$80.45 + \$45 = \$125.45) = \$1129.05 2. 14 + 7 + 9 = 30 days \$1616.30 + \$563.15 + \$1129.05 = \$3308.50 $\text{Average daily balance} = \frac{\$3308.50}{30} = \$110.28$ 3. Finance charge = \$110.28 × 0.01 = \$1.10
15.2 Finding the total installment cost, finance charge, and amount financed Total installment cost = Down payment + (Amount of each payment × Number of payments) Finance charge = Total installment cost − Cash price amount financed = Cash price − Down payment	Joan Taylor bought a fur coat for \$1580. She put \$350 down and then made 12 payments of \$115 each. Find the total installment cost, the finance charge, and the amount financed. Total installment cost = \$350 + (12 × \$115) = \$1730 Finance charge = \$1730 − \$1580 = \$150 Amount financed = \$1580 − \$350 = \$1230

15.2 Finding APR using a table

1. Determine the finance charge per $100.

$$\frac{\text{Finance charge}}{\text{Amount financed}} \times \$100$$

2. Find the number of payments in the leftmost column of Table 15.2; then go across to the number closest to the number found in Step 1 and read up to find APR.

Use the table to find the APR for the following example.

$$\text{Finance charge} = \$992$$

$$\text{Amount financed} = \$8200$$

$$\text{Finance charge per } \$100 = \frac{992}{8200} \times \$100 = \$12.10$$

Number of payments = 24; table value closest to $12.10 is $12.14; APR = 11.25%.

15.3 United States Rule for repayment of loans

1. Find interest from date of loan to date of partial payment.
2. Subtract interest from partial payment.
3. Reduce principal by any difference.
4. Find additional interest from date of partial payment to next partial payment or maturity date and add this interest to unpaid principal.

Sam Spade signed a 120-day note for $3000 at 11% on February 1. Spade made a partial payment of $1200 on March 18. What is the amount due at maturity?

1. There are 27 + 18 = 45 days from February 1 to March 18.

$$I = PRT$$

$$I = \$3000(0.11)\left(\frac{45}{360}\right) = \$41.25$$

2. $1200 − $41.25 = $1158.75 is applied to reduction of principal.
3. $3000 − $1158.75 = $1841.25 is balance owed.
4. There are 120 − 45 = 75 days until maturity, so additional interest is

$$I = \$1841.25(0.11)\left(\frac{75}{360}\right) = \$42.20.$$

Then $1841.25 + $42.20 = $1883.45 is due at maturity.

15.3 Finding the unearned interest using the Rule of 78

Find the unearned interest:

$$U = F\left(\frac{N}{P}\right)\left(\frac{1+N}{1+P}\right)$$

where U = unearned interest,

F = finance charge,

N = number of payments remaining, and

P = total number of payments

Then find the amount left to pay and subtract the unearned interest to find balance remaining.

Tom Fish borrows $1500, which he is paying back in 36 monthly installments of $52.75 each. With 10 payments remaining he decides to pay the loan in full. Find the amount of unearned interest and the amount necessary to pay the loan in full.

$$\text{Installment cost} = 36 \times \$52.75 = \$1899$$

$$\text{Finance charge} = \$1899 - \$1500 = \$399$$

$$\text{Unearned interest} = \$399 \times \frac{10}{36} \times \frac{(1+10)}{(1+36)} = \$32.95$$

The 10 payments of $52.75 that are left amount to $52.75 × 10 = $527.50.

$$\text{Balance} = \$527.50 - \$32.95 = \$494.55$$

15.4 Finding the periodic payment to amortize a loan

Use the number of periods for the loan n, interest rate per period i, and column F of the interest table to find $1 \div a_{\overline{n}|i}$. Then use the formula

$$R = A \cdot \left(\frac{1}{a_{\overline{n}|i}}\right)$$

to calculate the payment.

Bob agrees to pay \$18,600 for a car. The amount will be repaid in monthly payments over 4 years at an interest rate of 9%. Find the amount of each payment.

$$n = 12 \times 4 = 48$$

$$i = \frac{9\%}{12} = \frac{3}{4}\%$$

$$\frac{1}{a_{\overline{48}|0.0075}} = 0.02488504$$

$$R = \$18{,}600 \times 0.02488504$$
$$= \$462.86$$

15.4 Setting up an amortization schedule

Find the periodic payment R; then find the interest for the first period using $I = PRT$. Subtract I from R and reduce the original debt by this amount, D. Find the balance by subtracting the debt reduction D from the original amount A. Repeat until original debt is amortized.

Teri Meyer borrows \$1800 for 2 years at 10%. She will repay this amount with semiannual payments. Set up an amortization schedule.

$$n = 4 \qquad i = \frac{10\%}{2} = 5\%$$

$$\frac{1}{a_{\overline{4}|0.05}} = 0.28201183$$

$$R = \$1800 \times 0.28201183$$
$$= \$507.62$$
$$I = PRT$$
$$I = \$1800 \times 0.10 \times \frac{1}{2}$$
$$= \$90$$
$$D = R - I$$
$$D = \$507.62 - \$90$$
$$= \$417.62$$
$$A - D = \$1800 - \$417.62$$
$$= \$1382.38$$

Continue to get table shown below.

Payment Number	Amount of Payment	Interest for Period	Portion to Principal	Principal at End of Period
0	—	—	—	\$1800.00
1	\$507.62	\$90.00	\$417.62	1382.38
2	507.62	69.12	438.50	943.88
3	507.62	47.19	460.43	483.45
4	507.62	24.17	483.45	0

15.4 Finding monthly payments, total amount paid, and finance charge

Multiply the amount to be financed by the number from Table 15.3. This is the periodic payment. Find the total amount repaid by multiplying the periodic payment by the number of payments. Subtract the amount financed from the total amount repaid to obtain the finance charge.

Nick owes $9600 on a Ford Taurus. He wishes to pay the car off in 48 monthly payments. Find the amount of each payment and the finance charge if the APR on his loan is 12%.

Amount financed = $9600

Table value (Table 15.3) for 48 payments and 12% is 0.026333.

Payment = $9600 × 0.026333 = $252.80

Total amount repaid = $252.80 × 48 = $12,134.40

Finance charge = $12,134.40 − $9600 = $2534.40

15.5 Finding the amount of the monthly mortgage payments and total interest charges over the life of a mortgage

Use the number of years and the interest rate to find the amortization value per thousand dollars from Table 15.4. Then multiply the table value by the number of thousands in the principal to obtain monthly payment. Find total amount of payments and subtract original amount owed from total payments to obtain interest paid.

Lou and Rose buy a house at the shore. After a down payment, they owe $75,000. Find the monthly payment at $7\frac{3}{4}\%$ and the total charges over the life of a 25-year mortgage.

$$n = 25 \qquad i = 7\frac{3}{4}\%$$

Table value (Table 15.4) = 7.56.

There are $\frac{\$75{,}000}{1000} = 75$ thousands in $75,000.

Monthly payment = 75 × 7.56 = $567

There are 25 × 12 = 300 payments.

Total payments = 300 × $567 = $170,100

Interest paid = $170,100 − $75,000 = $95,100

Chapter 15 Review Exercises

The answer section includes answers to all Review Exercises.

Find the finance charge for each of the following revolving charge accounts. Assume interest is calculated on the average daily balance of the account. **[15.1]**

	Average Daily Balance	Monthly Interest Rate
1.	$243	$1\frac{1}{2}\%$
2.	$3240.60	$1\frac{1}{4}\%$
3.	$875.12	1.62%

Find the average daily balance for each of the following credit card accounts. Assume 1 month between billing dates (using the proper number of days in the month). Then find the finance charge if interest is $1\frac{1}{2}\%$ per month on the average daily balance. Finally, find the amount due at the end of the billing cycle. **[15.1]**

4. Previous balance $634.25

March 9	Billing date	
March 17	Payment	$125
March 30	Lunch	$34.26

5. Previous balance $236.26

July 10	Billing date	
July 15	Athletic shoes	$28.25
July 20	Payment	$75
July 31	Pillow cases	$35
August 5	Returns	$24.36

Find the total cash price, total installment cost, and the finance charge for each of the following. **[15.2]**

	Amount Financed	Down Payment	Cash Price	Number of Payments	Amount of Payment	Total Installment Cost	Finance Charge
6.	$2300	$400	______	18	$139.73	______	______
7.	$3800	$800	______	20	$212	______	______
8.	$6500	$1500	______	36	$225	______	______

Find the annual percentage rate using Table 15.2. **[15.2]**

	Amount Financed	Finance Charge	Number of Monthly Payments
9.	$4,100	$435	18
10.	$5,600	$698	20
11.	$8,800	$766.48	15
12.	$10,270	$1065	20

Find the balance due on the maturity date and the total amount of interest on the following notes. Use the United States Rate. **[15.3]**

	Principal	Interest	Time in Days	Partial Payments
13.	$12,400	$8\frac{1}{2}\%$	180 days	$2000 on day 62
14.	$9,000	12%	120 days	$2000 on day 40
15.	$6,000	11%	120 days	$3200 on day 30 $2000 on day 90
16.	$9,000	9%	90 days	$3000 on day 30 $1500 on day 45

Find the amount of each payment needed to amortize each of the following loans. Round to the nearest cent. **[15.4]**

17. $24,000 loan, repaid at 9% in 8 semiannual payments

18. $18,500 loan, repaid at 10% in 10 semiannual payments

19. $12,400 loan; repaid at 10% in 20 quarterly payments

20. $8600 loan, repaid at 9% in 24 monthly payments

Find the monthly payment and finance charge for each of the following loans using Table 15.3. Round to the nearest cent. [15.4]

21. \$9400 financed for 24 months at 12%

22. \$7500 financed for 42 months at 16%

23. \$9000 financed for 48 months at 14%

24. \$15,000 financed for 30 months at 11%

Find the monthly payment for each of the following real estate loans. [15.5]

25. \$74,000 loan at $8\frac{3}{4}\%$ for 30 years

26. \$65,000 loan at $7\frac{1}{2}\%$ for 20 years

27. \$100,000 loan at 8% for 15 years

28. \$120,000 loan at $7\frac{1}{4}\%$ for 30 years

Solve each of the following application problems.

29. Sandmeyer Concrete Company was the maker of an 11% note for \$5600 dated June 20 for 150 days. The firm made partial payments of \$1330 on July 20 and \$1655 on September 3. (a) What payment is required when the note is due? (b) What was the total interest paid on the note? [15.2]

30. Debbie Blaisdell has a revolving charge at a department store. Her monthly statement contained the following information. [15.1]

6–10	Billing date; previous balance	\$52.45
6–20	Payment	\$15 CR
6–25	Craft department	\$17.40
7–2	Shoe department	\$23

Find the average daily balance on the next billing date of July 10. Then, find the finance charge if the interest is 1.5% per month on the average daily balance.

31. Ben Franklin purchased a 1-year-old Ford Taurus with a cash price of \$15,780 with a down payment of \$2780 and agreed to make 48 monthly payments of \$332.84. Find the total installment cost and the annual percentage rate. [15.2]

32. ABC Plumbing borrows \$28,100 to buy a new truck. They agree to make quarterly payments for 3 years at 12% per year. Find the amount of the quarterly payment. [15.4]

33. Mr. and Mrs. Zagorin plan to buy a \$90,000 home, paying 20% down and financing the balance at 8% for 30 years. The taxes are \$960 per year, with fire insurance costing \$252 per year. Find the monthly payment (including taxes and insurance). [15.5]

34. Jerome Watson, owner of Watson Welding, purchases a building for his business and makes a \$25,000 down payment. He finances the balance of \$122,500 for 20 years at 8%. (a) Find the total monthly payment given taxes of \$3200 per year and insurance of \$1275 per year. (b) Assume that insurance and taxes do not increase and find the total cost of owning the building for 20 years (include the down payment). [15.5]

Chapter 15 Summary Exercise
Consolidating Loans

John and Kathy MacGruder are struggling to make their monthly payments. Kathy works one job and takes care of their two small children. John has had to take on a second job to make the payments.

(a) Find the monthly payments on each of the following purchases and the total monthly payment. Use column F in the interest tables in Appendix D and Table 15.4.

Purchase	Original Loan Amount	Interest Rate	Term of Loan	Monthly Payment
Honda Accord	$18,800	12%	4 years	_______
Ford Truck	$14,300	18%	4 years	_______
Home	$96,500	$8\frac{1}{2}$%	15 years	_______
2nd mortgage on home	$4,500	12%	3 years	_______
Total				_______

(b) These monthly expenses do not include car insurance ($215 per month), health insurance ($120 per month), or taxes on their home ($2530 per year), among other expenses. Find their total monthly outlay for all of these expenses.

Expense	Monthly Outlay
Payments on debt from (a)	_______
Car insurance	_______
Health insurance	_______
Taxes on home	_______
Total	_______

(c) The MacGruders learn that they can (1) refinance the remaining $14,900 amount on the Honda Accord at 12% over 4 years, (2) refinance the remaining $8600 loan amount on the Ford Truck at 12% over 3 years, (3) refinance the remaining $94,800 loan amount on their home at 8% over 30 years, and (4) reduce their car insurance payments by $28 per month. Complete the table below. Use column F in the interest tables in Appendix D and Table 15.4.

Item	New Loan Amount	New Interest Rate	New Term of Loan	New Monthly Payment
Honda Accord	$14,900	12%	4 years	_______
Ford Truck	8,600	12%	3 years	_______
Home	94,800	8%	30 years	_______
2nd mortgage on home	4,500	12%	3 years	_______
Car insurance				_______
Health insurance				_______
Taxes on home				_______
Total				_______

(d) Find the reduction in their monthly payments.

NOTE Part of the savings in the monthly payment came from reducing the interest rates; the remainder of the savings came from extending the loans further into the future. ■

Net Assets Business on the Internet

York Tracktown Credit Union

Statistics

- 1999: 6000 customers
- Nonprofit
- FDIC insured up to $100,000 per member
- Up to 72-month financing of a new automobile

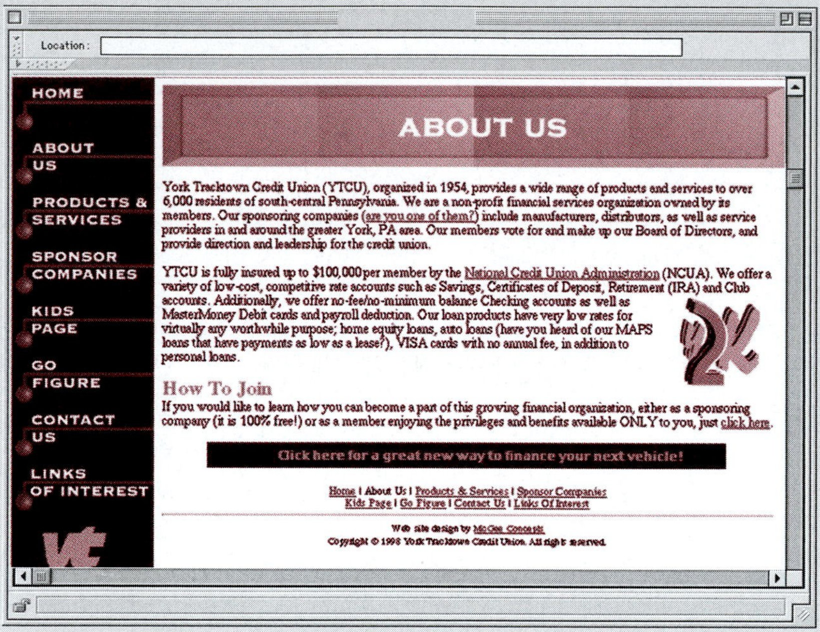

York Tracktown Credit Union (YTCU) provides a wide range of products and services to residents located in south-central Pennsylvania. Only individuals who are employees of companies sponsoring the credit union can be members. The company offers a variety of competitive-rate accounts such as savings, certificates of deposit, and retirement accounts such as IRAs. In addition, the firm offers no-fee, no-minimum-balance checking accounts as well as debit cards and payroll deduction. YTCU will also arrange for you to lease an automobile for those who prefer to lease over purchasing.

Credit unions are similar to banks in that you can have a savings or checking account with them and in turn you can borrow money for virtually any worthwhile purpose such as home-equity loans and auto loans. Interest rates paid on savings are typically higher at a credit union and interest rates paid on loans are typically lower than area banks—check out any credit union that you are eligible to join.

1. Loan officer George Willis loaned Betty Faber $14,000 for a 2-year-old Honda Accord. Given that the interest rate was 12% per year compounded monthly and the loan was for 4 years, find the monthly payment rounded to the nearest dollar.

2. Betty Faber also borrowed $84,000 from the credit union to purchase a home. Find the payment on the 30-year loan at 7% and at 9% interest rates.

3. Betty Faber recommended her credit union to her cousin who made the loans listed below. Find the monthly payment for each.

	Loan Amount	Term	Interest Rate	Monthly Payment
Boat	$11,300	3 years	12% comp. monthly	________
Automobile	$21,500	4 years	9% comp. monthly	________

Part 4 Cumulative Review Chapters 11–15

Round money amounts to the nearest cent, time to the nearest day, and rates to the nearest tenth of a percent.

Find the value of the unknown quantity using simple interest. Use banker's interest. **[11.1–11.2]**

	Interest	Principal	Rate	Time
1.	______	\$4,500	8%	5 months
2.	______	\$6,200	9.7%	250 days
3.	\$46.67	________	7%	100 days
4.	\$302.60	________	12.5%	70 days
5.	\$50.93	\$2,100	______	90 days
6.	\$306	\$6,800	______	120 days
7.	\$202.22	\$9,100	10%	______
8.	\$915	\$18,300	12%	______

Find the discount and the proceeds. **[12.2]**

	Face Value	Discount Rate	Time (Days)
9.	\$9000	12%	90
10.	\$875	$6\frac{1}{2}\%$	210

11. Convert a simple discount rate of 10% for 90 days to a simple interest rate and round to the nearest tenth of a percent. **[12.3]**

12. Convert a simple interest rate of 12% for 180 days to a simple discount rate and round to the nearest tenth of a percent. **[12.3]**

Find the compound amounts for the following. **[13.1]**

13. \$1000 at 4% compounded annually for 17 years

14. \$3520 at 8% compounded annually for 10 years

Find the compound amounts and interest earned for each of the following. Assume $3\frac{1}{2}\%$ interest compounded daily. **[13.2]**

	Amount	Date Deposited	Date Withdrawn
15.	\$12,600	March 24	June 3
16.	\$7500	November 20	February 14

Find the present value and the amount of interest earned for the following. Round to the nearest cent. **[13.4]**

	Amount Needed	Time (Years)	Interest	Compounded
17.	\$1000	7	8%	Annually
18.	\$19,000	9	5%	Semiannually

Find the amount of each of the following ordinary annuities. **[14.1]**

	Amount of Each Deposit	Deposited	Rate	Time (Years)
19.	$1000	Annually	4%	8
20.	$2000	Annually	6%	6

Find the present value of the following annuities. **[14.2]**

	Amount per Payment	Payment at End of Each	Time (Years)	Rate of Investment	Compounded
21.	$925	6 months	11	8%	Semiannually
22.	$27,235	Quarter	8	8%	Quarterly

Find the required payment into a sinking fund. **[14.3]**

	Future Value	Interest Rate	Compounded	Time (Years)
23.	$3600	8%	Annually	7
24.	$4500	10%	Quarterly	7

Solve the following application problems. Use 360-day years where applicable.

25. Walter Bates sets aside $5000 today to help pay his son's college expenses, which begin in 9 years. Given a rate of 8% compounded semiannually, find the future value of this investment. **[13.1]**

26. According to the terms of a divorce settlement, one spouse must pay the other a lump sum of $2800 in 17 months. What lump sum can be invested today, at 18% compounded monthly, so that enough will be available for the payment? **[13.4]**

27. Bill Jones borrowed $12,000 on a simple interest note at 8% for 40 days. Find (a) the maturity value and (b) the interest. **[12.1]**

28. Sherie Whatly borrowed $15,000 on a simple discount note at 10% for 100 days. Find (a) the interest and (b) the proceeds. **[12.2]**

29. Tom Davis owes $7850 to a relative. He has agreed to pay the money in 5 months, at an interest rate of 6%. One month before the loan is due, the relative discounts the loan at the bank. The bank charges a 7.92% discount rate. How much money does the relative receive? **[12.4]**

30. The Pool family earned $23,500 last year. Find the gain in purchasing power for a year in which they receive a raise of 4.1% but inflation as measured by the CPI was 3.7%. **[11.4]**

31. Thomas Wood has an average daily balance of $6327.12 on a credit card that charges 1.5% per month. Find (a) the monthly finance charge, (b) the finance charge if you could change the loan to a credit card that charges 0.8% on the average daily balance, and (c) the savings. **[15.1]**

32. A used automobile costing $10,500 was purchased with $1000 down and payments of $315 per month for 36 months. Find (a) the total installment cost, (b) the finance charge, (c) the amount financed, and (d) the annual percentage rate. **[15.2]**

33. On July 7, Gene Harper borrows \$5000 at 9% interest to pay for surgery to correct his nearsightedness. Find the balance owed immediately after a partial payment on September 30 of \$1200. Assume the United States Rule applies for partial payments. **[15.3]**

34. A public utility needs \$60 million in 5 years for a major capital expansion. What annual payment must it place in a sinking fund earning 10% per year in order to accumulate the required funds? **[14.3]**

35. At 58, Thomas Jones knows that he must start saving for his retirement. He decides to invest \$300 per quarter in an account paying 10% compounded quarterly. Find the accumulated amount (a) at age 65 and (b) at age 70. **[14.1]**

36. Ben Torres borrows \$18,500 at 10% compounded monthly for 48 months to purchase a farm tractor. Find the monthly payment using Table 15.3. **[15.4]**

37. The owner of Jessica's Cookies has an extra \$3200 that he puts into a savings account paying $3\frac{1}{2}$% per year compounded daily. Find the interest if the funds are left there for 65 days. **[13.2]**

38. Cheryl Crow wishes to purchase a home costing \$140,000. Given a down payment of 20%, find the monthly payment needed to finance the balance for 30 years at $8\frac{3}{4}$%. **[15.5]**

Part 5 Accounting and Other Applications

16 Depreciation

A business finds its net income (profit) by subtracting all expenses from the amount of money received by the business (revenues). Major expenses include the cost of goods sold, salaries, rent, and utilities. Other expenses include the cost of assets such as machines, buildings, and fixtures. These assets usually last several years, so it would not show the true income of the business if the entire cost of an asset were considered an expense in the year of purchase. Instead, a method called **depreciation** is used to spread the cost of the asset over the length of its useful life, which is measured in years. Depreciation is used with assets having a useful life of more than one year. The asset to be depreciated must have a predictable life: a truck can be depreciated because its useful life can be estimated, but land cannot be depreciated because its life is indefinite. For example, Figure 16.1 shows the remaining value of a Kenworth truck as it is depreciated over its useful life. The truck in this example has a 5 year life.

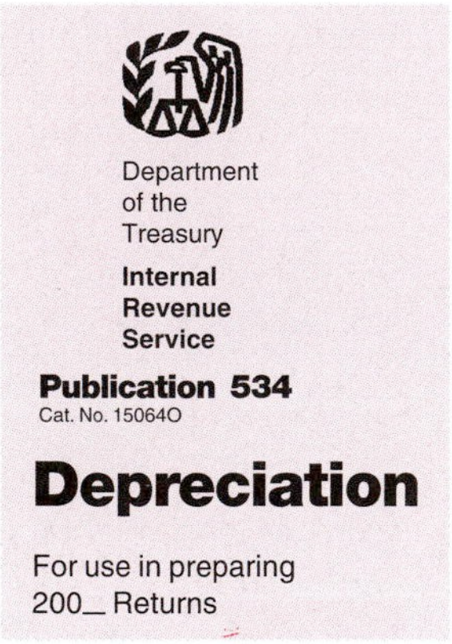
Department of the Treasury
Internal Revenue Service
Publication 534
Cat. No. 15064O
Depreciation
For use in preparing 200_ Returns

Which company assets would have a life greater than 5 years?

Over the years, several methods of computing depreciation have been used, including *straight-line*, *declining-balance*, *sum-of-the-years'-digits*, and *units-of-production*. These methods are used in keeping company accounting records, and, in many states, for preparing state income tax returns. Assets purchased after 1981 are depreciated for federal tax returns with the accelerated cost recovery system or the modified accelerated cost recovery system, which will be discussed later. The use of depreciation for federal in-

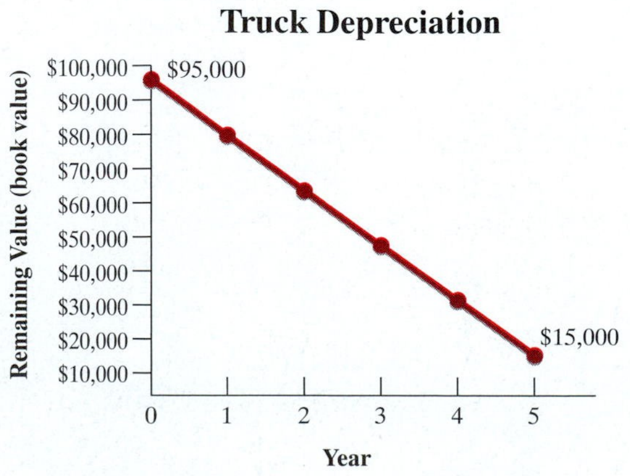

Figure 16.1

come tax purposes is detailed in an Internal Revenue Service publication. The complete title of this publication is shown because of this.

A company need not use the same method of depreciation for all of its various assets. For example, the straight-line method of depreciation might be used on some assets and the declining-balance method on other assets. Furthermore, the depreciation method used in preparing a company's financial statements may be different from the method used in preparing income tax returns.

16.1 Straight-Line Method

Objectives

1. *Understand the terms used in depreciation.*
2. *Use the straight-line method to find the annual depreciation.*
3. *Use the straight-line method to find the book value of an asset.*
4. *Determine the book value of an asset after several years.*
5. *Calculate the accumulated depreciation.*
6. *Use the straight-line method to prepare a depreciation schedule.*

Objective 1 *Understand the Terms Used in Depreciation.* Physical assets such as machinery, cars, or buildings are *tangible assets.* A tangible asset may be depreciated as long as its useful life can be estimated. The key terms used in depreciation are summarized below.

Cost, the basis for determining depreciation, is the total amount paid for the asset.

Useful life is the period of time during which the asset will be used.

The Internal Revenue Service (IRS) has guidelines as to the estimated life of assets used in a particular trade or business. However, useful life depends on the use of the asset, repair policy, replacement policy, obsolescence, and other factors.

Salvage value or **scrap value** (sometimes called **residual value**) is the estimated value of an asset when it is retired from service, traded in, disposed of, or exhausted. An asset may have a salvage value of zero or **no salvage value**.

Accumulated depreciation is the amount of depreciation taken so far, a running balance of depreciation to date.

Book value is the cost of an asset minus the total depreciation to date (cost minus accumulated depreciation). The book value at the end of an asset's life is equal to the salvage value. Book value can never be less than salvage value.

Objective 2 *Use the Straight-Line Method to Find the Annual Depreciation.* The simplest method of depreciation, **straight-line depreciation**, assumes that assets lose an equal amount of value during each year of life. For example, suppose a heavy-equipment trailer is purchased by Village Nursery and Landscaping at a cost of $9400. The trailer has an estimated useful life of 8 years, and a salvage value of $1400. Find the amount to be depreciated (**depreciable amount**) using the following formula.

$$\text{Amount to be depreciated} = \text{Cost} - \text{Salvage value}$$

Here the amount to be depreciated over the 8-year period is figured as follows.

$9400	Cost
− 1400	Salvage value
$8000	Amount to be depreciated

With the straight-line method, an equal amount of depreciation is taken each year of the 8-year life of the trailer. The annual depreciation for this trailer is

$$\$8000 \div 8 = \$1000$$

Or, use the following formula.

The annual depreciation by the *straight-line method* for an item having a cost of c, a salvage value s, and a life of n years, is d, where

$$d = \frac{c - s}{n}$$

Use the formula for the example by substituting \$9400 for c, \$1400 for s, and 8 for n.

$$d = \frac{\$9400 - \$1400}{8} = \frac{\$8000}{8} = \$1000$$

The annual straight-line depreciation is \$1000. Each year during the 8-year life of the trailer, \$1000 will be treated as an expense by the company owning the trailer. The annual depreciation of \$1000 is $\frac{1}{8}$ of the depreciable amount. The annual rate of depreciation is often given as a percent, in this case, $12\frac{1}{2}\%$ $\left(\frac{1}{8} = 12\frac{1}{2}\%\right)$.

Objective 3 *Use the Straight-Line Method to Find the Book Value of an Asset.* The **book value**, or remaining value of an asset at the end of a year, is the original cost minus the depreciation up to and including that year (the **accumulated depreciation**). In the example, the book value at the end of the first year is \$8400.

\$9400	Cost
− 1000	First-year's depreciation
\$8400	Book value at the end of the first year

Book value is found with the following formula.

Book value = Cost − Accumulated depreciation

Example 1 Finding First-Year Depreciation and Book Value

Dependable Insurance Company purchased a high-end personal computer at a cost of \$2650. The estimated life of the computer is 5 years, with a salvage value of \$350. Find the (a) annual rate of depreciation, (b) annual amount of depreciation, and (c) book value at the end of the first year.

Solution

(a) The annual rate of depreciation is 20%, since a 5-year life means $\frac{1}{5}$ or 20% per year.
(b) First find the depreciable amount.

\$2650	Cost
− 350	Salvage value
\$2300	Depreciable amount

This \$2300 will be depreciated evenly over the 5-year life for an annual depreciation of \$460 (\$2300 × 20% = \$460, or \$2300 ÷ 5 = \$460). The annual depreciation can also be found by the formula

$$d = \frac{c - s}{n}$$

Substitute \$2650 for c, \$350 for s, and 5 for n.

$$d = \frac{\$2650 - \$350}{5} = \frac{\$2300}{5} = \$460$$

The annual depreciation by the straight-line method is \$460.

(c) Since the annual depreciation is \$460, the book value at the end of the first year will be found as follows.

\$2650	Cost
− 460	First year's depreciation
\$2190	Book value after 1 year

The book value of the personal computer after 1 year is \$2190.

Scientific *Calculator Approach*

To solve Example 1 using a calculator, first use parentheses to find the depreciable amount. Next, divide to find depreciation. Finally, find the book value.

(2650 − 350) ÷ 5 = 460

2650 − 460 = 2190

If an asset has **no salvage value** at the end of the expected life, then the entire cost will be depreciated over the life of the asset. In Example 1, if the personal computer was expected to have no salvage value at the end of 5 years, the annual amount of depreciation would have been \$530 (since \$2650 ÷ 5 = \$530).

NOTE Find the book value at the end of any year by multiplying the annual amount of straight-line depreciation by the number of years and subtract this result, the accumulated depreciation to date, from the cost. ■

Objective 4 *Determine the Book Value of an Asset after Several Years.*

Example 2 Finding the Book Value at the End of Any Year

A lighted display case at the Bead Works costs \$3400 and has an estimated life of 10 years and a salvage value of \$800. Find the book value at the end of 6 years.

Solution The annual rate of depreciation is 10% (10-year life leads to $\frac{1}{10}$ or 10%).

\$3400	Cost
− 800	Salvage value
\$2600	Depreciable amount

Since \$2600 is depreciated evenly over the 10-year life of the case, the annual depreciation is \$260 (\$2600 × 10% = \$260, or \$2600 ÷ 10 = \$260).

Objective 5 *Calculate the Accumulated Depreciation.* The accumulated depreciation over the 6-year period is

\$260 × 6 (years) = \$1560 Accumulated depreciation (6 years)

Find the book value at the end of 6 years by subtracting the accumulated depreciation from the cost.

\$3400	Cost
− 1560	Accumulated depreciation (6 years)
\$1840	Book value at the end of 6 years

After 6 years, this display case would be carried "on the books" with a value of \$1840.

NOTE This book value helps the owner of a business estimate the value of the business, which is important when the owner is borrowing money or trying to sell the business. ■

Objective 6 *Use the Straight-Line Method to Prepare a Depreciation Schedule.* A **depreciation schedule** is often used to show the annual depreciation, accumulated depreciation, and book value over the useful life of an asset. As an aid in comparing the three methods of depreciation, the depreciation schedule of Example 3 and the schedules shown in the double-declining-balance method (see Section 16.2) and the sum-of-years'-digits method (see Section 16.3) use the same asset.

Example 3 Preparing a Depreciation Schedule

Village Nursery and Landscaping bought a new pickup truck for \$18,500. It is estimated that the truck will have a useful life of 5 years, at which time it will have a salvage value (trade-in value) of \$3500. Prepare a depreciation schedule using the straight-line method of depreciation.

Solution The annual rate of depreciation is 20% (5-year life = $\frac{1}{5}$ per year = 20%). Find the depreciable amount as follows.

\$18,500	Cost
− 3,500	Salvage value
\$15,000	Depreciable amount

This \$15,000 will be depreciated evenly over the 5-year life, giving an annual depreciation of \$3000 (\$15,000 × 20% = \$3000).

This depreciation schedule includes a year 0 to show the initial cost of the truck.

Year	Computation	Amount of Depreciation	Accumulated Depreciation	Book Value
0	—	—	—	\$18,500
1	(20% × \$15,000)	\$3000	\$3,000	15,500
2	(20% × \$15,000)	3000	6,000	12,500
3	(20% × \$15,000)	3000	9,000	9,500
4	(20% × \$15,000)	3000	12,000	6,500
5	(20% × \$15,000)	3000	15,000	3,500

The depreciation is \$3000 each year, the accumulated depreciation at the end of 5 years is equal to the depreciable amount, and the book value at the end of 5 years is equal to the salvage value.

Problem-Solving Hint If the rate is a repeating decimal, use the fraction that is equivalent to the decimal. Instead of 33.3%, use the fraction $\frac{1}{3}$; instead of 16.7%, use the fraction $\frac{1}{6}$.

16.1 Exercises

Find the annual straight-line rate of depreciation, given each of the following estimated lives.

1. 5 years	**2.** 4 years	**3.** 8 years	**4.** 10 years
5. 20 years	**6.** 25 years	**7.** 15 years	**8.** 30 years
9. 80 years	**10.** 40 years	**11.** 50 years	**12.** 100 years

Find the annual amount of depreciation for each of the following, using the straight-line method.

13.	Cost:	$9000	**14.**	Cost:	$3400
	Estimated life:	20 years		Estimated life:	4 years
	Estimated scrap value:	None		Estimated scrap value:	$800
15.	Cost:	$2700	**16.**	Cost:	$8100
	Estimated life:	3 years		Estimated life:	6 years
	Estimated scrap value:	$300		Estimated scrap value:	$750
17.	Cost:	$4200	**18.**	Cost:	$12,200
	Estimated life:	5 years		Estimated life:	10 years
	Estimated scrap value:	None		Estimated scrap value:	$3200

Find the book value at the end of the first year for each of the following, using the straight-line method.

19.	Cost:	$3200	**20.**	Cost:	$35,000
	Estimated life:	8 years		Estimated life:	10 years
	Estimated scrap value:	$400		Estimated scrap value:	$2500
21.	Cost:	$5400	**22.**	Cost:	$4500
	Estimated life:	12 years		Estimated life:	5 years
	Estimated scrap value:	$600		Estimated scrap value:	None

Find the book value at the end of 5 years for each of the following, using the straight-line method.

23.	Cost:	$4800	**24.**	Cost:	$16,000
	Estimated life:	10 years		Estimated life:	20 years
	Estimated scrap value:	$750		Estimated scrap value:	$2000
25.	Cost:	$80,000	**26.**	Cost:	$660
	Estimated life:	50 years		Estimated life:	8 years
	Estimated scrap value:	$10,000		Estimated scrap value:	$100

27. Develop a single formula that will show how to find annual depreciation using the straight-line method of depreciation. (See Objective 2.)

28. Explain the procedure used to calculate depreciation when there is no salvage value. Why will the book value always be zero at the end of the asset's life?

Solve each of the following application problems.

29. MACHINERY DEPRECIATION Dallas Tool and Diecasting Company selects the straight-line method of depreciation for a lathe costing $12,000 with a 3-year life and an expected scrap value of $3000. Prepare a depreciation schedule.

Year	Computation	Amount of Depreciation	Accumulated Depreciation	Book Value
0	—	—	—	$12,000
1	______	______	______	______
2	______	______	______	______
3	______	______	______	______

30. TRUCK DEPRECIATION Village Nursery and Landscaping paid $25,600 for a $1\frac{1}{2}$-ton dual axle flatbed truck with an estimated life of 6 years and a salvage value of $7000. Prepare a depreciation schedule using the straight-line method of depreciation.

Year	Computation	Amount of Depreciation	Accumulated Depreciation	Book Value
0	—	—	—	$25,600
1	______	______	______	______
2	______	______	______	______
3	______	______	______	______
4	______	______	______	______
5	______	______	______	______
6	______	______	______	______

31. OFFICE EQUIPMENT DEPRECIATION Shippers' Express paid $9400 for office equipment with an estimated life of 6 years and a salvage value of $2200. Prepare a depreciation schedule using the straight-line method of depreciation.

32. BUSINESS FIXTURES Dorothy Sargent buys fixtures for her shop at a cost of $7800 and estimates the life of the fixtures as 10 years, after which they will have no salvage value. Prepare a depreciation schedule, calculating depreciation by the straight-line method.

33. BARGE DEPRECIATION A Dutch petroleum company purchased a barge for $1,300,000. The estimate life is 20 years, at which time it will have a salvage value of $200,000. (a) Use the straight-line method of depreciation to find the annual amount of depreciation. (b) Find the book value at the end of 5 years.

34. DEPRECIATING COMPUTER EQUIPMENT The new computer equipment at Leisure Travel has a cost of $14,500, an estimated life of 8 years, and scrap value of $2100. Find (a) the annual depreciation and (b) the book value at the end of 4 years using the straight-line method of depreciation.

35. BOAT-RAMP DEPRECIATION A freshwater boat ramp has an estimated life of 15 years and no scrap value. If the boat ramp cost $37,900 and the straight-line method is used, find (a) the annual depreciation and (b) the book value after 7 years. (Round depreciation to the nearest dollar.)

36. BAKERY PACKAGING EQUIPMENT The packaging equipment line at Rainbow Bakery has a cost of $132,400, an estimated life of 10 years, and a salvage value of $35,000. Find the book value of the packaging equipment line after 8 years using the straight-line method of depreciation.

37. FURNITURE DEPRECIATION A bookcase costs $880, has an estimated life of 8 years, and has a scrap value of $160. Use the straight-line method of depreciation to find (a) the annual rate of depreciation, (b) the annual amount of depreciation, and (c) the book value at the end of the first year.

38. FORKLIFT DEPRECIATION Levinson Supply purchased a new forklift for $12,500. The estimated life is 10 years, with a salvage value of $2500. Use the straight-line method of depreciation to find (a) the annual rate of depreciation, (b) the annual amount of depreciation, (c) the book value at the end of 5 years, and (d) the accumulated depreciation at the end of 8 years.

16.2 Declining-Balance Method

Objectives

1. ***List the declining-balance methods.***
2. ***Find the declining-balance rate.***
3. ***Find depreciation and book value using the double-declining-balance method.***
4. ***Use the formula to find depreciation by the double-declining-balance method.***
5. ***Prepare a depreciation schedule using double-declining-balance method.***

The straight-line method of depreciation assumes that the cost of an asset is spread equally and evenly over each year of its life. This is not realistic for many assets. For example, a new utility van loses much more value during its first year of life than during its fifth year. Using straight-line depreciation for such assets would give a book value higher than the actual value of the asset during the early years.

Methods of accelerated depreciation are used to more accurately reflect the rate at which most assets actually lose value. **Accelerated depreciation** produces larger amounts of depreciation in the earlier years of the life of an asset and smaller amounts in the later years. The total amount of depreciation taken over the life of an asset is the same as with the straight-line method (the difference of cost and salvage value), but the distribution of the annual amounts is different.

One of the more common accelerated methods of depreciation is called the **declining-balance method**; with this method a *declining-balance rate* is first established. This rate is multiplied by the previous year's book value to get the current year's depreciation. *Since the book value declines from year to year, the annual depreciation will also decline*; this explains the method's name.

Objective 1 *List the Declining-Balance Methods.*

Three Common Declining-Balance Methods

200%, or double-declining-balance, method. With this method, 200% of the straight-line rate, or twice the straight-line rate, is used.
150% declining-balance method. With this method, $1\frac{1}{2}$ times the straight-line rate is used.
125% declining-balance method. With this method, $1\frac{1}{4}$ times the straight-line rate is used.

Objective 2 | ***Find the Declining-Balance Rate.*** Calculate depreciation using the declining-balance method by first finding the straight-line rate of depreciation. Then adjust the straight-line rate to the desired declining-balance rate (200%, 150%, or 125% as desired). The following examples of declining-balance depreciation show the **200% method**, or double-declining-balance method.

Example 1 Finding the 200% Declining-Balance Rate

Find the straight-line rate and the double-declining-balance (200%) rate for each of the following years of life.

Solution

Years of Life	Straight-Line Rate	Double-Declining-Balance Rate
3	33.33% ($\frac{1}{3}$)	66.67% ($\frac{2}{3}$)
4	25%	50%
5	20%	40%
8	12.5%	25%
10	10%	20%
20	5%	10%
25	4%	8%
50	2%	4%

NOTE Throughout the remainder of this chapter, money amounts will be rounded to the nearest dollar, a common practice when dealing with depreciation. The rounded value is then used in further calculations. ▬

Objective 3 | ***Find Depreciation and Book Value Using the Double-Declining-Balance Method.***

Example 2 Finding Depreciation and Book Value Using Double-Declining-Balance

Northridge Golf and Country Club purchased a portable storage building for \$8100. It is expected to have a life of 10 years, at which time it will have no salvage value. Using the double-declining-balance method of depreciation, find the first and second years' depreciation and the book value at the end of the first and second year.

Solution Start by finding the double-declining-balance rate; find this rate by doubling the straight-line rate. The straight-line rate for a life of 10 years is $\frac{1}{10}$, or 10%, and 10% doubled is 20%. The double-declining-balance rate (20%) is then multiplied by the book value, or in year 1 the *original cost*. The depreciation in year 1 is

$$\text{Original cost} \times \text{Double-declining rate}$$

or

$$\$8100 \times 0.20 = \$1620$$

Depreciation in year 1 is \$1620. Therefore, the book value at the end of the first year is as follows.

\$8100	Cost
− 1620	Depreciation to date
\$6480	Book value at the end of the first year

The second year's depreciation is 20% of \$6480 (last year's book value or declining balance).

\$6480 (declining balance) × 0.20 = \$1296   Depreciation in second year, rounded

The book value at the end of the second year is \$8100 − \$2916 (depreciation year 1 and year 2) = \$5184.

Objective 4 ***Use the Formula to Find Depreciation by the Double-Declining-Balance Method.*** Declining-balance depreciation can also be found with the following formula.

The annual depreciation, d, by the declining-balance method, is given by

$$d = r \times b$$

where r is the declining-balance rate and b is the book value in the previous year. (In year 1, b is the original cost of the asset.)

Example 3 Using a Formula to Find Depreciation and Book Value

Pioneer Beverage buys a bottling machine for \$59,400. The expected life of the bottling machine is 5 years, at which time it will have no salvage value. Using the double-declining-balance method of depreciation, find the first and second years' depreciation and the book value at the end of the first and second year.

Solution The straight-line depreciation rate for a 5-year life is 20%. The double-declining rate is 40% (20% times 2). The first year's depreciation is 40% of the declining balance or, in the first year, 40% of the cost. Use the formula, substituting 0.40 for r and \$59,400 for b.

$$d = r \times b$$
$$d = 0.40 \times \$59{,}400 = \$23{,}760$$

The depreciation in the first year is \$23,760. The book value at the end of the first year is as follows.

\$59,400	Cost
− 23,760	Depreciation to date
\$35,640	Book value at the end of the first year

The second year's depreciation is 40% of \$35,640 (last year's book value or declining balance) or, again with the formula,

$$d = r \times b$$
$$d = 0.40 \times \$35{,}640 = \$14{,}256$$

The depreciation in the second year is $14,256. At the end of the second year the book value is figured as shown.

$59,400	Cost
− 38,016	Depreciation to date ($23,760 + $14,256)
$21,384	Book value at the end of the second year

NOTE The total amount of depreciation taken over the life of the asset is the same using either the straight-line or the double-declining-balance methods of depreciation, but the distribution of the annual amounts is different. ▬

Objective 5 *Prepare a Depreciation Schedule Using the Double-Declining-Balance Method.* The next example shows a depreciation schedule for the same pickup truck used in Example 3 of Section 16.1. Here, the declining-balance method is used, where the same rate is used each year, and the rate is multiplied by the declining balance (last year's book value). This example shows that the amount of depreciation in a given year may have to be adjusted so that book value is never less than salvage value.

Example 4 Preparing a Depreciation Schedule

Village Nursery and Landscaping bought a new pickup truck at a cost of $18,500. It is estimated that the truck will have a useful life of 5 years, at which time it will have a salvage value (trade-in value) of $3500. Prepare a depreciation schedule using the double-declining-balance method of depreciation.

Solution the annual rate of depreciation is 40% (20% straight-line × 2 = 40%). *Do not subtract salvage value from cost before calculating depreciation. In year 1, the full cost is used to calculate depreciation.*

Year	Computation	Amount of Depreciation	Accumulated Depreciation	Book Value
0	—	—	—	$18,500
1	(40% × $18,500)	$7400	$7,400	$11,100
2	(40% × $11,100)	4440	11,840	$6,660
3	(40% × $6,660)	2664	14,504	3,996
4		496	15,000	3,500
5		0	15,000	3,500

NOTE In year 4, 40% of $3996 is $1598. If this amount were subtracted from $3996, the book value would drop below the salvage value of $3500. Since book value may never be less than salvage value, depreciation of $496 is taken in year 4 so that book value equals salvage value. No further depreciation remains for year 5. The total amount of depreciation taken over the life of the asset is the same using either the straight-line or the double-declining-balance method of depreciation. ▬

16.2 Exercises

Find the annual double-declining-balance (200% method) rate of depreciation, given each of the following estimated lives.

1. 5 years **2.** 20 years **3.** 8 years

4. 25 years **5.** 15 years **6.** 4 years

7. 10 years **8.** 30 years **9.** 6 years

10. 40 years **11.** 50 years **12.** 100 years

Find the first year's depreciation for each of the following by using the double-declining-balance method of depreciation.

13. Cost: $18,000
Estimated life: 10 years
Estimated scrap value: $3000

14. Cost: $4950
Estimated life: 20 years
Estimated scrap value: None

15. Cost: $10,500
Estimated life: 5 years
Estimated scrap value: $500

16. Cost: $38,000
Estimated life: 40 years
Estimated scrap value: $5000

17. Cost: $3800
Estimated life: 4 years
Estimated scrap value: None

18. Cost: $1140
Estimated life: 6 years
Estimated scrap value $350

Find the book value at the end of the first year for each of the following by using the double-declining-balance method of depreciation. Round to the nearest dollar.

19. Cost: $4200
Estimated life: 10 years
Estimated scrap value: $1000

20. Cost: $2500
Estimated life: 6 years
Estimated scrap value: $400

21. Cost: $1620
Estimated life: 8 years
Estimated scrap value: None

22. Cost: $5640
Estimated life: 5 years
Estimated scrap value: $800

Find the book value at the end of 3 years for each of the following by using the double-declining-balance method of depreciation. Round to the nearest dollar.

23. Cost: $16,200
Estimated life: 8 years
Estimated scrap value: $1500

24. Cost: $8500
Estimated life: 10 years
Estimated scrap value: $1100

25. Cost: $6000
Estimated life: 3 years
Estimated scrap value: $750

26. Cost: $75,000
Estimated life: 50 years
Estimated scrap value: None

27. Another name for the double-declining-balance method of depreciation is the 200% method. Explain why the straight-line method of depreciation is often called the 100% method.

28. Explain why the amount of depreciation taken in the last year of an asset's life may be zero when using the double-declining-balance method of depreciation.

Solve each of the following application problems. Round to the nearest dollar.

29. WEIGHT-TRAINING EQUIPMENT Gold's Gym selects the double-declining-balance method of depreciation for some weight training equipment costing $14,400. If the estimated life of the equipment is 4 years and the salvage value is zero, prepare a depreciation schedule.

Year	Computation	Amount of Depreciation	Accumulated Depreciation	Book Value
0	—	—	—	$14,400
1	______	______	______	______
2	______	______	______	______
3	______	______	______	______
4	______	______	______	______

30. STUDIO SOUND SYSTEM A studio sound system costing $11,760 has a 3-year life and a scrap value of $1400. Prepare a depreciation schedule using the double-declining-balance method of depreciation.

Year	Computation	Amount of Depreciation	Accumulated Depreciation	Book Value
0	—	—	—	$11,760
1	______	______	______	______
2	______	______	______	______
3	______	______	______	______

31. ELECTRONIC ANALYZER Neilo Lincoln-Mercury decides to use the double-declining-balance method of depreciation on a Barnes Electronic Analyzer that was acquired at a cost of $25,500. If the estimated life of the analyzer is 8 years and the estimated scrap value is $3500, prepare a depreciation schedule.

32. CONVEYOR SYSTEM Prepare a depreciation schedule for the installation of a conveyor system using the double-declining-balance method of depreciation. Cost = $14,000; estimated life = 5 years; estimated scrap value = $2500.

33. CARPET CLEANING EQUIPMENT John Walker, owner of the Carpet Solution, purchased some truck-mounted carpet-cleaning equipment at a cost of $8200. The estimated life of the equipment is 8 years and the expected salvage value is $1250. Use the double-declining-balance method of depreciation to find the depreciation in the third year.

34. HARBOR BOATS A harbor boat costs $478,000 and has an estimated life of 10 years and a salvage value of $150,000. Find the depreciation in the second year using the double-declining-balance method of depreciation.

35. JUICE EXTRACTOR Nature's Products purchased a new juice extractor. The cost of the extractor is $1090 and it has an estimated life of 5 years with no salvage value.

Use the double-declining-balance method of depreciation to find the book value at the end of the third year.

36. COMMUNICATION EQUIPMENT Karen Guardino purchased some communication equipment for her public relations firm at a cost of $19,700. She estimates the life of the equipment to be 8 years, at which time the salvage value will be $1000. Use the double-declining-balance method of depreciation to find the book value at the end of 5 years.

37. CONSTRUCTION POWER TOOLS West Construction purchased some power tools for the shop at a cost of $5800. They have a life of 8 years and a scrap value of $1000. Use the double-declining-balance method of depreciation to find (a) the annual rate of depreciation, (b) the amount of depreciation in the first year, (c) the accumulated depreciation at the end of the fifth year, and (d) the book value at the end of the fifth year.

38. PRIVATE SCHOOL EQUIPMENT Gale Klein bought some white boards for her reading clinic at a cost of $3620. The estimated life of the white boards is 5 years, with a salvage value of $400. Use the double-declining-balance method of depreciation to find (a) the annual rate of depreciation, (b) the amount of depreciation in the first year, (c) the accumulated depreciation at the end of the third year, and (d) the book value at the end of the third year.

16.3 Sum-of-the-Years'-Digits Method

Objectives

1. *Understand the sum-of-the-years'-digits method.*
2. *Find the depreciation fractions for the sum-of-the-years'-digits method.*
3. *Use the formula to calculate the sum-of-the-years'-digits depreciation.*
4. *Prepare a depreciation schedule using the sum-of-the-years'-digits method.*

Objective 1 *Understand the Sum-of-the-Years'-Digits Method.* The **sum-of-the-years'-digits method** of depreciation is another accelerated depreciation method. The double-declining-balance method of depreciation produces more depreciation than the straight-line method in the early years of an assets' life and less in the later years. The sum-of-the-years'-digits method, however, produces results in between the straight-line and the double-declining-balance method—more depreciation than straight-line at the beginning and more depreciation than double-declining at the end.

Objective 2 *Find the Depreciation Fractions for the Sum-of-the-Years'-Digits Method.* The use of the sum-of-the-years'-digits method requires a **depreciation fraction**, instead of the depreciation rate used earlier. When this depreciation fraction, which decreases annually, is multiplied by the depreciable amount (cost minus salvage value) the result is the annual depreciation.

To find the depreciation fraction, first find the denominator, which remains constant for every year of the life of the asset. The denominator is the sum of all the years of the estimated life of the asset (sum-of-the-years'-digits). For example, if the life is 6 years, the denominator is 21 (since $1 + 2 + 3 + 4 + 5 + 6 = 21$). The numerator of the fraction changes each year, and represents the years of life remaining at the beginning of that year.

Example 1 Finding the Depreciation Fraction

Find the depreciation fraction for each year if the sum-of-the-years'-digits method of depreciation is to be used for an asset with a useful life of 6 years.

Solution First determine the denominator of the depreciation fraction. The denominator is 21 ($1 + 2 + 3 + 4 + 5 + 6 = 21$). Next determine the numerator for each year. The number of years of life remaining at the beginning of any year is the numerator.

Year	Depreciation Fraction
1	$\frac{6}{21}$
2	$\frac{5}{21}$
3	$\frac{4}{21}$
4	$\frac{3}{21}$
5	$\frac{2}{21}$
6	$\frac{1}{21}$
21 sum of the year's digits	$\frac{21}{21}$

As the table shows, when using the sum-of-the-years'-digits method, an asset having a life of 6 years is assumed to lose $\frac{6}{21}$ of its value the first year, $\frac{5}{21}$ the second year, and $\frac{4}{21}$ the third year, and so on. The sum of the six fractions in the table is $\frac{21}{21}$, or 1, so that the entire depreciable amount is used over the 6-year life.

NOTE It is common not to write these fractions in lowest terms, so that the year in question can be seen. ■

A fast method of finding the sum-of-the-years'-digits is to use the formula

$$\text{Sum-of-the-years'-digits} = \frac{n(n + 1)}{2}$$

where n is the estimated life of the asset.

For example, if the life is 6 years, 6 is multiplied by 6 plus 1, resulting in 6×7, or 42. Then 42 is divided by 2, giving 21, the same denominator used in Example 1. This method eliminates adding digits and is especially useful when the life of an asset is long.

Objective 3 *Use the Formula to Calculate the Sum-of-the-Years'-Digits Depreciation.* The depreciation fraction in any year is multiplied by the depreciable amount (as in the straight-line method) to calculate the amount of depreciation in any one year.

The formula for sum-of-the-years'-digits depreciation is

$$d = r \times (c - s)$$

where r is the depreciation fraction, c is the cost of the asset, and s is the salvage value.

Example 2 Finding Depreciation Using the Sum-of-the-Years'-Digits Method

A Ditch Witch 1220 Trencher is purchased by Village Nursery and Landscaping at a cost of \$8940. It has a useful life of 8 years and an estimated salvage value of \$1200. Find the first and second years' depreciation using the sum-of-the-years'-digits method.

Solution The depreciation fraction has a denominator of 36 (or $1 + 2 + 3 + 4 + 5 + 6 + 7 + 8$). The numerator for the first year of useful life is 8. The first year fraction $\frac{8}{36}$ is multiplied by the depreciable amount, \$7740 (\$8940 cost − \$1200 salvage value). Substitute the proper numbers into the following formula

$$\begin{aligned} d &= r \times (c - s) \\ &= \frac{8}{36} \times (\$8940 - \$1200) \\ &= \frac{8}{36} \times \$7740 = \$1720 \end{aligned}$$

The first year's depreciation is \$1720.

The book value at the end of the first year is \$8940 (original cost) − \$1720 (first year's depreciation) = \$7220. For the second and succeeding years, always go back to the *original* depreciable amount and not to book value, as in the declining-balance method. The depreciation fraction for the second year, $\frac{7}{36}$, is multiplied by the original depreciable amount, \$7740 (\$8940 cost − \$1200 salvage value). This gives

$$\frac{7}{36} \times \$7740 = \$1505$$

The second year's depreciation is \$1505. (In this example the depreciation for any year is always found by multiplying the appropriate fraction by \$7740.)

Scientific *Calculator Approach*

The calculator solution to Example 2 finds $\frac{1}{36}$ of the depreciation amount and stores the result [STO] in memory. This amount is then recalled [RCL] to find future depreciation amounts.

8940 [−] 1200 [=] [÷] 36 [=] [STO] [×] 8 [=] 1720 [RCL] [×] 7 [=] 1505

Example 3 Finding Depreciation When There Is No Salvage Value

City Saturn purchased an electronic smog analyzer for \$9000. Using the sum-of-the-years'-digits method of depreciation, find the first and second years' depreciation if the analyzer has an estimated life of 4 years and no salvage value.

Solution The depreciation fraction has a denominator of 10 (or $1 + 2 + 3 + 4$). The numerator in the first year is 4. The first year depreciation fraction is $\frac{4}{10}$, which is multiplied by the amount to be depreciated, or \$9000 (\$9000 cost because there is no salvage value). Use the formula as follows.

$$d = r \times (c - s)$$
$$= \frac{4}{10} \times (\$9000 - \$0)$$
$$= \frac{4}{10} \times \$9000$$
$$= \$3600$$

The first year's depreciation is \$3600.

The fraction for finding the depreciation in the second year is $\frac{3}{10}$, which is multiplied by the depreciable amount, or \$9000.

$$\frac{3}{10} \times \$9000 = \$2700$$

The second year's depreciation is \$2700.

Objective 4 ***Prepare a Depreciation Schedule Using the Sum-of-the-Years'-Digits Method.*** For comparison, the next example uses the same pickup truck discussed in Sections 16.1 and 16.2.

Example 4 Preparing a Depreciation Schedule

Village Nursery and Landscaping bought a new pickup truck for \$18,500. It is estimated that the truck will have a useful life of 5 years, at which time it will have a salvage value (trade-in value) of \$3500. Prepare a depreciation schedule using the sum-of-the-years'-digits method of depreciation.

Solution Using the formula on page 592, the depreciation fraction has a denominator of (5 × 6) ÷ 2, or 15.

Year	Computation	Amount of Depreciation	Accumulated Depreciation	Book Value
0	—	—	—	\$18,500
1	$(\frac{5}{15} \times \$15{,}000)$	\$5000	\$5,000	13,500
2	$(\frac{4}{15} \times \$15{,}000)$	4000	9,000	9,500
3	$(\frac{3}{15} \times \$15{,}000)$	3000	12,000	6,500
4	$(\frac{2}{15} \times \$15{,}000)$	2000	14,000	4,500
5	$(\frac{1}{15} \times \$15{,}000)$	1000	15,000	3,500

NOTE The sum-of-the-years'-digits method of depreciation allows rapid depreciation in the early years of the asset's life and also provides some depreciation during the last years. ■

The three methods of depreciation can be compared visually by graphing the amounts of depreciation in each year and the book values at the end of each year. Figures

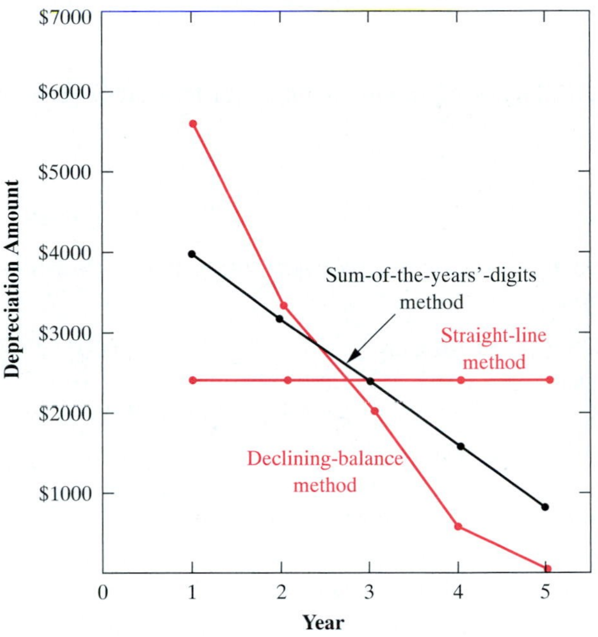

Figure 16.2 Comparison of depreciation on Village Nursery and Landscaping pickup truck using three depreciation methods

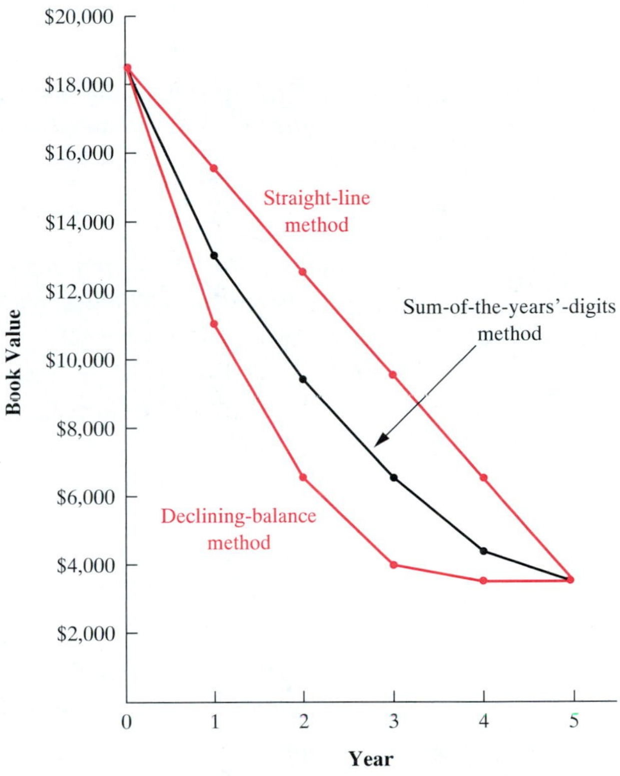

Figure 16.3 Comparison of book value on Village Nursery and Landscaping pickup truck using three depreciation methods

16.2 and 16.3 show the annual depreciation and book value for the pickup truck owned by Village Nursery and Landscaping.

Choosing a Method of Depreciation

While there is no simple way to decide which method of depreciation is preferable in a given case, the following considerations may help to arrive at an answer.

- Will a larger deduction in the first year or years help pay for the asset with the tax dollars saved?
- Is it expected that earnings during the first years will be larger than in the following years? Larger depreciation deductions help to reduce taxes.
- Will a steady deduction over the life of the asset be advantageous?
- Accelerated deductions in early years mean little or no deductions in later years.
- Is profit expected to increase in the coming years during the life of the asset? If this is the case, a steady deduction over the life of the asset will allow depreciation in later years.
- Will repair bills be more in later years? Since these are also deductions, they might offset lower depreciation amounts in later years.

16.3 Exercises

Find the sum-of-the-years'-digits depreciation fraction for the first year given each of the following estimated lives.

1. 4 years **2.** 3 years **3.** 6 years **4.** 5 years

5. 7 years **6.** 8 years **7.** 10 years **8.** 20 years

Find the first year's depreciation for each of the following using the sum-of-the-years'-digits method of depreciation.

9.	Cost:	$4800	**10.**	Cost:	$5600
	Estimated life:	4 years		Estimated life:	5 years
	Estimated scrap value:	$700		Estimated scrap value:	$800
11.	Cost:	$60,000	**12.**	Cost:	$1440
	Estimated life:	10 years		Estimated life:	8 years
	Estimated scrap value:	$5000		Estimated scrap value:	None
13.	Cost:	$1350	**14.**	Cost:	$9500
	Estimated life:	3 years		Estimated life:	8 years
	Estimated scrap value:	$150		Estimated scrap value:	$1400

Find the book value at the end of the first year for each of the following using the sum-of-the-years'-digits method of depreciation. Round to the nearest dollar.

15.	Cost:	$9500	**16.**	Cost:	$25,000
	Estimated life:	8 years		Estimated life:	10 years
	Estimated scrap value:	$1400		Estimated scrap value:	None
17.	Cost:	$3800	**18.**	Cost:	$15,650
	Estimated life:	5 years		Estimated life:	6 years
	Estimated scrap value:	$500		Estimated scrap value:	$2000

Find the book value at the end of 3 years for each of the following using the sum-of-the-years'-digits method of depreciation.

19.	Cost:	$2240	**20.**	Cost:	$27,500
	Estimated life:	6 years		Estimated life:	10 years
	Estimated scrap value:	$350		Estimated scrap value:	None
21.	Cost:	$4500	**22.**	Cost:	$6600
	Estimated life:	8 years		Estimated life:	5 years
	Estimated scrap value:	$900		Estimated scrap value:	$1500

23. Write a description of how the depreciation fraction is determined in any year of an asset's life when using the sum-of-the-years'-digits method of depreciation. (See Objective 2.)

24. If you were starting your own business, what type of business would it be? Which of the three depreciation methods, straight-line, double-declining-balance, or sum-of-the-years'-digits, would you decide to use? why?

Solve each of the following application problems. Round to the nearest dollar.

25. COMMERCIAL FREEZER Big Town Market has purchased a new freezer case at a cost of $10,800. The estimated life of the freezer case is 6 years, at which time the salvage value is estimated to be $2400. Complete a depreciation schedule using the sum-of-the-years'-digits method of depreciation.

Year	Computation	Amount of Depreciation	Accumulated Depreciation	Book Value
0	—	—	—	$10,800
1	______	______	______	______
2	______	______	______	______
3	______	______	______	______
4	______	______	______	______
5	______	______	______	______
6	______	______	______	______

26. RESTAURANT EQUIPMENT Old South Restaurant has purchased a new steam table for $14,400. The expected life of the unit is 4 years, at which time the salvage value is estimated to be $2400. Complete a depreciation schedule using the sum-of-the-years'-digits method of depreciation.

Year	Computation	Amount of Depreciation	Accumulated Depreciation	Book Value
0	—	—	—	$14,400
1	______	______	______	______
2	______	______	______	______
3	______	______	______	______
4	______	______	______	______

27. OFFICE FURNITURE DEPRECIATION Sunset Real Estate Company has purchased office furniture at a cost of $2700. The estimated life of the furniture is 6 years, at which time the salvage value is estimated to be $600. Complete a depreciation schedule using the sum-of-the-years'-digits method of depreciation.

28. FORKLIFT DEPRECIATION Prepare a depreciation schedule for the following fork lift, using the sum-of-the-years'-digits method of depreciation. Cost is $15,000, estimated life is 10 years, and estimated scrap value is $4000.

29. LANDSCAPE EQUIPMENT Village Nursery and Landscaping purchased a new Ditch Witch 3500 at a cost of $32,000. The expected life of the unit is 8 years, and the salvage value is expected to be $5000. Use the sum-of-the-years'-digits method of depreciation to determine the first year's depreciation.

30. HOSPITAL EQUIPMENT Orangevale Rents uses the sum-of-the-years'-digits method of depreciation on all hospital rental equipment. If they purchase new hospital beds at a cost of $12,800 and estimate the life of the beds to be 10 years with no scrap value, find the book value at the end of the fourth year.

31. LIGHT-RAIL TOOLING Electro-car, a light-rail manufacturer, purchased a power-wheel and axle jig from a German manufacturer for $31,880. The jig has an expected life of 20 years and an estimated salvage value of $5000. Find the book value at the end of the third year.

32. SOLAR COLLECTOR Find the depreciation in the third year for a solar collector, using the sum-of-the-years'-digits method of depreciation. Cost is $23,000, estimated life is 8 years, and estimated scrap value is $5000.

33. COMMERCIAL CARPET The cost to install new carpeting at Norma's Art Studio is $6360. It has a useful life of 5 years and no salvage value. Find (a) the first and (b) the second year's depreciation using the sum-of-the-years'-digits method.

34. ASSET DEPRECIATION Using the sum-of-the-years'-digits method of depreciation, find (a) the first and (b) the second year's depreciation for an asset that has a cost of $3375, an estimated life of 5 years and no salvage value.

35. FAST-FOOD RESTAURANTS In-N-Out Burgers purchased a new deep fry unit at a cost of $12,420. The expected life of the unit is 8 years, with a scrap value of $1800. Use the sum-of-the-years'-digits method of depreciation to find (a) the first year's depreciation fraction, (b) the amount of depreciation in the first year, (c) the accumulated depreciation at the end of the eighth year, and (d) the book value at the end of the fourth year.

36. SEWER DRAIN SERVICE Armour Drain bought a new sewer de-rooter for $6725. The life of the machine is 10 years and the scrap value is $1500. Use the sum-of-the-years'-digits method of depreciation to find (a) the first year's depreciation fraction, (b) the amount of depreciation in the first year, (c) the accumulated depreciation at the end of the tenth year, and (d) the book value at the end of the sixth year.

16.4 Units-of-Production Method and Partial-Year Depreciation

Objectives

1. *Describe the units-of-production method of depreciation.*
2. *Use the units-of-production method to find the depreciation per unit.*
3. *Calculate annual depreciation by the units-of-production method.*
4. *Prepare a depreciation schedule using the units-of-production method.*
5. *Calculate partial-year depreciation by the straight-line method.*
6. *Calculate partial-year depreciation by the double-declining-balance method.*
7. *Calculate partial-year depreciation by the sum-of-the-years'-digits method.*

Objective 1 *Describe the Units-of-Production Method of Depreciation.* An asset often has a useful life given in terms of *units of production* or *units of output*, such as hours of use or miles of service. For example, an airliner or truck may have a useful life given as hours of air time or miles of travel. A steel press or stamping machine may have a life given as the total number of units that it can produce. With these assets, the **units-of-production method** of depreciation is used. Just as with the straight-line method of depreciation, a constant amount of depreciation is taken with the units-of-production method. With the straight-line method a constant amount of depreciation is taken each year, while the units-of-production method depreciates a constant amount per unit of use or production.

Objective 2 *Use the Units-of-Production Method to Find the Depreciation per Unit.* Find the depreciation per unit with the following formula.

$$\text{Depreciation per unit} = \frac{\text{Depreciable amount}}{\text{Units of life}}$$

For example, suppose a stump chipper owned by Brent's Tree Service costs \$15,000, has a salvage value of \$3000, and is expected to operate 700 hours. Find the depreciation per hour by dividing the depreciable amount by the number of hours of life.

The depreciable amount is \$15,000 − \$3000 = \$12,000. Use the formula to find the depreciation per unit.

$$\frac{\$12{,}000 \text{ depreciable amount}}{700 \text{ hours of life}} = \$17.14 \text{ Depreciation per hour (rounded)}$$

Objective 3 *Calculate Annual Depreciation by the Units-of-Production Method.* Multiply the depreciation per unit by the number of units produced during the year to find the annual depreciation.

Example 1 Using Units-of-Production Depreciation

North American Trucking purchased a new Kenworth truck for \$95,000. The truck has a salvage value of \$15,000 and an estimated life of 500,000 miles. Find the depreciation for a year in which the truck is driven 128,000 miles.

Solution First find the depreciable amount.

\$95,000	Cost
− 15,000	Scrap value
\$80,000	Depreciable amount

Next find the depreciation per unit.

$$\frac{\$80{,}000 \text{ depreciable amount}}{500{,}000 \text{ miles of life}} = \$0.16 \text{ depreciation per mile}$$

Multiply to find the depreciation for the year.

128,000 miles × \$0.16 = \$20,480 depreciation for the year

Scientific *Calculator Approach*

The calculator solution to Example 2 uses parentheses to find the depreciable amount and then chain calculations to find the depreciation amount.

[( ] 95000 [−] 15000 [ )] [÷] 500000 [=] [×] 128000 [=] 20480

Example 2 Preparing a Depreciation Schedule

Global Electronics purchased a shrink-wrapping machine that costs \$52,300, has an estimated salvage value of \$4000, and has an expected life of 690,000 units. Prepare a depreciation schedule using the units-of-production method of depreciation. Use the following packaging schedule.

Year 1	240,000 units
Year 2	150,000 units
Year 3	90,000 units
Year 4	120,000 units
Year 5	90,000 units

Solution The depreciable amount is \$48,300 (\$52,300 − \$4000). The depreciation per unit is found as follows.

$$\frac{\$48{,}300}{690{,}000 \text{ units}} = \$0.07 \text{ per unit}$$

The annual depreciation is found by multiplying the number of units packaged each year by the depreciation per unit.

Year 1	240,000 units × \$0.07 =	\$16,800
Year 2	150,000 units × \$0.07 =	\$10,500
Year 3	90,000 units × \$0.07 =	\$6,300
Year 4	120,000 units × \$0.07 =	\$8,400
Year 5	90,000 units × \$0.07 =	\$6,300
Total	690,000 units	\$48,300 *depreciable amount*

Objective 4 *Prepare a Depreciation Schedule Using the Units-of-Production Method.* These results were used to help prepare the following depreciation schedule.

Year	Computation	Depreciation	Accumulated Depreciation	Book Value
0	—	—	—	\$52,300
1	(240,000 × \$0.07)	\$16,800	\$16,800	35,500
2	(150,000 × \$0.07)	10,500	27,300	25,000
3	(90,000 × \$0.07)	6,300	33,600	18,700
4	(120,000 × \$0.07)	8,400	42,000	10,300
5	(90,000 × \$0.07)	6,300	48,300	4,000

NOTE In Example 2, the book value at the end of year 5 (\$4000) is the amount of the salvage value. This is true because the total number of units of life (690,000) has been used up by the machine during the 5 years. The machine may continue in use producing more units; however, no additional depreciation may be taken. ■

Objective 5 *Calculate Partial-Year Depreciation by the Straight-Line Method.* So far, each of the examples in depreciation assumed that the depreciable asset was purchased at the beginning of a year. If the asset is purchased during the year, only a fraction of the first year's depreciation may be taken. For example, if an asset is acquired on June 1, only $\frac{7}{12}$ (since 7 months remain) of the first year's depreciation may be taken. For an asset acquired on April 1, only $\frac{9}{12}$ or $\frac{3}{4}$ of the first year's depreciation may be taken. If the asset is purchased at other than the beginning of the month, then count that month for depreciation if purchased on or before the 15th of the month. This **partial-year depreciation** is explained in the following examples.

Each of the methods of depreciation studied so far (except the units-of-production method) is affected differently by a partial year. The units-of-production method is not affected by the partial first year since actual use determines depreciation.

Example 3 Finding Partial-Year Depreciation with Straight-Line

A mountain bike display rack purchased on October 1 at a cost of $6750 has an estimated salvage value of $750 and an expected life of 5 years. Using the straight-line method of depreciation, find the depreciation in the year of purchase.

Solution The depreciation for a full year is

$$\frac{\$6750 - \$750}{5 \text{ years}} = \frac{\$6000}{5} = \$1200$$

Since the purchase date is October 1, 3 months remain in the year. This means that only $\frac{3}{12}$ or $\frac{1}{4}$ of the $1200 may be taken in the year of purchase.

$$\$1200 \times \frac{1}{4} = \$300 \quad \text{Depreciation allowed in the year of purchase}$$

The next 4 years are depreciated at the full $1200 per year and the last 9 months ($\frac{3}{4}$) of the fifth year has depreciation figured as follows.

$$\$1200 \times \frac{3}{4} = \$900 \quad \text{Partial-year depreciation}$$

As before, the total depreciation over all the years is the depreciable amount, or $6000.

Objective 6 *Calculate Partial-Year Depreciation by the Double-Declining-Balance Method.*

Example 4 Finding Partial Year Depreciation with Double-Declining Balance

A boat dock with a life of 10 years is installed on April 12 at a cost of $18,000. If the double-declining-balance method is used, find the depreciation for the first partial year and the next full year.

Solution The double-declining-balance rate is 20% ($\frac{1}{10} \times 2 = \frac{2}{10} = \frac{1}{5} = 20\%$). Calculate the first year's depreciation.

$$\$18{,}000 \times 0.2 = \$3600 \quad \text{Depreciation first full year}$$

Now find the depreciation for 9 months or $\frac{3}{4}$ year. (Since April 12 is on or before April 15 count all of April.)

$$\$3600 \times \frac{3}{4} = \$2700 \quad \text{Partial-year depreciation}$$

The book value (declining balance) is $15,300 ($18,000 − $2700). Use this book value to find the first full year depreciation.

$$\$15{,}300 \text{ declining balance} \times 0.2 = \$3060$$

The depreciation for the first full year is \$3060. Depreciation for the following years would be calculated as usual. (Make sure that book value does not go below any scrap value.)

Objective 7 ***Calculate Partial-Year Depreciation by the Sum-of-the-Years'-Digits Method.***

Example 5 Finding Partial-Year Depreciation with Sum-of-the-Years'-Digits

An industrial air conditioning compressor is purchased on July 1 at a cost of \$3800. It is estimated that the compressor will have a life of 4 years and a scrap value of \$600. Use the sum-of-the-years'-digits method to find the depreciation for each year of the life of the asset.

Solution The partial year here is $\frac{1}{2}$ (6 months). The depreciation fraction in year 1 is $\frac{4}{10}$ and the depreciable amount is \$3200 (\$3800 − \$600).

$$\frac{4}{10} \times \$3200 = \$1280 \qquad \text{Depreciation year 1}$$

$$\$1280 \times \frac{1}{2} = \$640 \qquad \text{Partial-year depreciation}$$

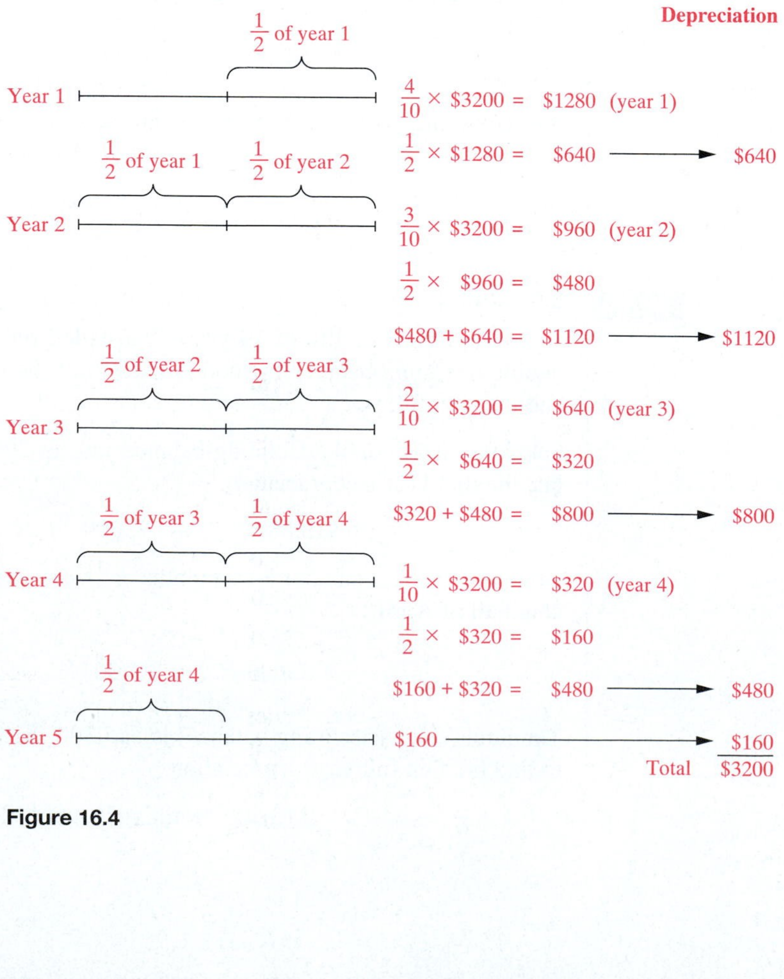

Figure 16.4

Partial-year depreciation in the sum-of-the-years'-digits method requires depreciation in the first full year (year 2) to be the sum of the second half of depreciation in year 1 and the first half of depreciation in year 2. Depreciation in year 3 will be the sum of the second half of year two and the first half of year three and so on for the remaining life. See Figure 16.4.

NOTE If an asset is purchased on or before the 15th of the month, then count that month for depreciation. However, if an asset is purchased on the 16th of the month or after, begin depreciation with the following month. ■

16.4 Exercises

Find the depreciation per unit in each of the following. Round to the nearest thousandth of a dollar.

	Cost	Salvage Value	Estimated Life
1.	$22,500	$1,500	60,000 units
2.	$5,000	$400	10,000 units
3.	$3,750	$250	120,000 units
4.	$7,500	$500	15,000 units
5.	$300,000	$25,000	4,000 hours
6.	$600,000	$50,000	2,000 hours
7.	$175,000	$25,000	5,000 hours
8.	$125,000	$20,000	500,000 miles

Find the amount of depreciation in each of the following.

	Depreciation per Unit	Units Produced		Depreciation per Unit	Units Produced
9.	$0.23	78,000	**10.**	$0.15	380,000
11.	$0.54	32,000	**12.**	$0.73	16,500
13.	$0.185	15,000	**14.**	$0.032	73,000
15.	$0.40	17,400	**16.**	$0.015	180,000

17. In your own words, describe the conditions under which the units-of-production method of depreciation is most applicable. (See Objective 1.)

18. Use an example of your own to demonstrate how the annual depreciation amount is found using the units-of-production method of depreciation. (See Objective 3.)

Find the depreciation in the first partial year and the next full year for each of the following. Round to the nearest dollar.

	Cost	Salvage Value	Life	Depreciation Method	Date Acquired
19.	$9,700	$700	4 years	Straight-line	June 1
20.	$5,600	$1000	10 years	Straight-line	Oct. 1
21.	$20,000	$1000	20 years	Double-declining	Mar. 1
22.	$5,250	$550	10 years	Double-declining	May 1
23.	$3,150	None	6 years	Sum-of-years'-digits	July 12
24.	$9,600	$600	5 years	Sum-of-years'-digits	Sept. 10
25.	$6,300	$900	8 years	Sum-of-years'-digits	Mar. 28
26.	$14,375	$2000	10 years	Sum-of-years'-digits	Apr. 19

Solve each of the following application problems. Round to the nearest dollar.

27. COMMERCIAL DEEP FRYER Dunkin Donuts purchased a new deep fryer at a cost of $6800. The expected life is 5000 hours of production, at which time it will have a salvage value of $500. Prepare a depreciation schedule, using the units-of-production method, given the following production: year 1: 1350 hours; year 2: 1820 hours; year 3: 730 hours; year 4: 1100 hours.

Year	Computation	Amount of Depreciation	Accumulated Depreciation	Book Value
0	—	—	—	$6800
1	________	______	______	______
2	________	______	______	______
3	________	______	______	______
4	________	______	______	______

28. TRUCKING BUSINESS Jack Armstrong purchased a Kenworth truck at a cost of $87,000. He estimates that it is good for 300,000 miles and will have a salvage value of $15,000. Use the units-of-production method to prepare a depreciation schedule given the following production: year 1: 108,000 miles; year 2: 75,000 miles; year 3: 117,000 miles.

Year	Computation	Amount of Depreciation	Accumulated Depreciation	Book Value
0	—	—	—	$87,000
1	________	______	______	______
2	________	______	______	______
3	________	______	______	______

29. **SOFT DRINK BOTTLING** A small soft drink bottler purchased an automatic filling and capping machine for $185,000. The machine has a scrap value of $30,000 and an estimated life of 20,000 hours. Find the depreciation for a year in which the machine was in operation 3400 hours.

30. **CARDBOARD RECYCLING** A cardboard crusher costs $13,800, has an estimated salvage value of $2400, and has an expected life of 300,000 units. Find (a) the amount of depreciation each year and (b) the book value at the end of each year. Use the units-of-production method of depreciation given the following production schedule.

Year 1	36,000 units
Year 2	42,000 units
Year 3	39,000 units

31. **PACKAGING EQUIPMENT** Action Packing bought a new packing machine for $156,000 and estimates the life of the machine to be 10 years with a salvage value of $10,000. If the straight-line method of depreciation is used and the system is purchased on October 1, find the first partial-year depreciation and the following full year's depreciation.

32. **SMALL TRACTOR** Village Nursery and Landscaping bought a new Bobcat Tractor for $19,450. The tractor has a life of 7 years and a scrap value of $3000. If the straight-line method of depreciation is used and the tractor was purchased on March 1, find the first partial-year depreciation and the following full year's depreciation.

33. **CELLULAR TELEPHONES** Tri-State Insurance Company purchased cellular phones for all outside sales people at a cost of $4500. The phones were purchased on October 8, the estimated life of the phones is 5 years, and they are expected to have no salvage value. Use the double-declining-balance method of depreciation to find the first partial-year depreciation and the following full year's depreciation.

34. **TELEPHONE PAGING** Action Paging Service purchased a $14,200 switching system on June 25. The estimated life of the system is 8 years and the salvage value is $3000. Use the double-declining-balance method of depreciation to find the first partial-year's depreciation and the following full year's depreciation.

35. **COMMERCIAL REROOFING** A warehouse has a new roof installed on June 27 at a cost of $44,400. The estimated life is 15 years and there is no scrap value. Use the sum-of-the-years'-digits method to find the first partial-year's depreciation and the following full year's depreciation.

36. **WAITING ROOM FURNITURE** On September 14, Dr. Umeda purchased new furniture for his office. The cost of the furniture was $7850, the salvage value is $500, and expected life is 6 years. Use the sum-of-the-years'-digits method to find the first partial-year's depreciation and the following full year's depreciation.

16.5 Modified Accelerated Cost Recovery System

Objectives

1. *Understand the modified accelerated cost recovery system (MACRS).*
2. *Determine the recovery period of different types of property.*
3. *Find the depreciation rate given the recovery period and recovery year.*
4. *Use the MACRS method to find the amount of depreciation.*
5. *Prepare a depreciation schedule using the MACRS.*

Objective 1 *Understand the Modified Accelerated Cost Recovery System (MACRS).* A depreciation method known as the **accelerated cost recovery system (ACRS)** originated as part of the Economic Recovery Tax Act of 1981. It was later modified by the Tax Equity and Fiscal Responsibility Act of 1982 and again by the Tax Reform Act of 1984. The Tax Reform Act of 1986 brought the most recent and significant overhaul to the accelerated cost recovery system (ACRS), and applies to all property placed in service after 1986. This new method is known as the **modified accelerated cost recovery system (MACRS)**. The result is that there are now three systems for computing depreciation for *federal tax purposes*.

Federal Tax Depreciation Methods

1. The MACRS method of depreciation is used for all property placed in service after 1986.
2. The ACRS method of depreciation will continue to be used for all property placed in service from 1981 through 1986.
3. The straight-line, declining-balance and sum-of-the-years'-digits methods continue to be used if the property was placed in service before 1981.

NOTE The units-of-production method of depreciation is still allowed under the MACRS. ▬

Keep two things in mind about MACRS. First, the system is designed, really, for tax purposes (it is sometimes called the **income tax method**), and businesses often use some alternate method of depreciation (in addition to MACRS) for financial accounting purposes. Second, many states do not allow the modified accelerated cost recovery system of depreciation for finding state income tax liability. This means businesses must use the MACRS on the federal tax return and one of the previous methods on the state tax return.

Objective 2 *Determine the Recovery Period of Different Types of Property.* Under the modified accelerated cost recovery system, assets are placed in one of nine recovery classes, depending on whether the law assumes a 3-, 5-, 7-, 10-, 15-, 20-, 27.5-, 31.5-, or 39-year life for the asset. These lives, or **recovery periods**, are determined as follows.

MACRS Recovery Classes

3-year property	Tractor units for over-the-road use, any racehorse that is over 2 years old, or any other horse that is over 12 years old
5-year property	Automobiles, taxis, trucks, buses, computers and peripheral equipment, office machinery (typewriters, calculators), copiers, and research equipment
7-year property	Office furniture and fixtures (desks, files, safes), and any property not designated by law to be in any other class
10-year-property	Vessels, barges, tugs, and similar water transportation equipment
15-year property	Improvements made directly to land, such as shrubbery, fences, roads, and bridges
20-year property	Certain farm buildings
27.5-year property	Residential rental real estate such as rental houses and apartments

continued

31.5-year property	Nonresidential rental real estate such as office buildings, stores, and warehouses if placed in service before May 13, 1993
39-year property	Nonresidential property placed in service after May 12, 1993

Example 1 Finding the Recovery Period for Property

Rancher's Supply owns the following assets. Determine the recovery period for each of them.

(a) Computer equipment
(b) An industrial warehouse (after May 12, 1993)
(c) A pickup truck
(d) Office furniture
(e) A farm building (storage shed)

Solution Use the list just given.

(a) 5 years
(b) 39 years
(c) 5 years
(d) 7 years
(e) 20 years

Objective 3 ***Find the Depreciation Rate Given the Recovery Period and Recovery Year.*** With MACRS, salvage value is ignored, so that depreciation is based on the entire original cost of the asset. The depreciation rates are determined by applying the double-declining-balance (200%) method to the 3-, 5-, 7-, and 10-year class properties, the 150% declining-balance method to the 15- and 20-year class properties, and the straight-line (100%) method to the 27.5-, 31.5-, and 39-year class properties. Since these calculations are repetitive and require additional knowledge, the Internal Revenue Service provides tables that show the depreciation rates expressed as percents. The rates are shown in Table 16.1. To determine the rate of depreciation for any year of life, find the recovery year in the left-hand column, and then read across to the allowable recovery period.

Notice that the number of recovery years is one greater than the class life of the property. This is because only a half-year of depreciation is allowed for the first year the property is placed in service, regardless of when the property is placed in service during the year. This is known as the **half-year convention** and is used by most taxpayers. A complete coverage of depreciation, including all depreciation tables, is included in Internal Revenue Service **Publication 534**, *Depreciation*, and may be obtained by contacting the IRS Forms Distribution Center. Publication 534 lists several items that the taxpayer or tax preparer might find useful and is shown at the side.

Modified Accelerated Cost Recovery System (MACRS)

Useful Items
You may want to see:

Publication

- ☐ **225** Farmer's Tax Guide
- ☐ **463** Travel, Entertainment, and Gift Expenses
- ☐ **544** Sales and Other Dispositions of Assets
- ☐ **581** Basis of Assets
- ☐ **583** Taxpayers Starting a Business
- ☐ **587** Business Use of Your Home
- ☐ **917** Business Use of a Car

Source: IRS

Problem-Solving Hint

MACRS is the income tax method of depreciation and several important points should be remembered.

1. No salvage value is used.
2. The life of the asset is determined by using the recovery periods assigned to different types of property.
3. A depreciation rate is usually found for each year by referring to a MACRS table of depreciation rates.

Table 16. 1 MACRS DEPRECIATION RATES

Recovery Year	Applicable Percent for the Class of Property								
	3-year	5-year	7-year	10-year	15-year	20-year	27.5-year	31.5-year	39-year
1	33.33	20.00	14.29	10.00	5.00	3.750	1.818	1.587	2.568
2	44.45	32.00	24.49	18.00	9.50	7.219	3.636	3.175	2.564
3	14.81	19.20	17.49	14.40	8.55	6.677	3.636	3.175	2.564
4	7.41	11.52	12.49	11.52	7.70	6.177	3.636	3.175	2.564
5		11.52	8.93	9.22	6.93	5.713	3.636	3.175	2.564
6		5.76	8.92	7.37	6.23	5.285	3.636	3.175	2.564
7			8.93	6.55	5.90	4.888	3.636	3.175	2.564
8			4.46	6.55	5.90	4.522	3.636	3.175	2.564
9				6.56	5.91	4.462	3.637	3.175	2.564
10				6.55	5.90	4.461	3.636	3.174	2.564
11				3.28	5.91	4.462	3.637	3.175	2.564
12					5.90	4.461	3.636	3.174	2.564
13					5.91	4.462	3.637	3.175	2.564
14					5.90	4.461	3.636	3.174	2.564
15					5.91	4.462	3.637	3.175	2.564
16					2.95	4.461	3.636	3.174	2.564
17						4.462	3.637	3.175	2.564
18						4.461	3.636	3.174	2.564
19						4.462	3.637	3.175	2.564
20						4.461	3.636	3.174	2.564
21						2.231	3.637	3.175	2.564
22							3.636	3.174	2.564
23							3.637	3.175	2.564
24							3.636	3.174	2.564
25							3.637	3.175	2.564
26							3.636	3.174	2.564
27							3.637	3.175	2.564
28							3.636	3.174	2.564
29								3.175	2.564
30								3.174	2.564
31								3.175	2.564
32								3.174	2.564
33–39									2.564

Example 2 Finding the Rate of Depreciation with MACRS

Use Table 16.1 to find the rate of depreciation given the following recovery year and recovery period.

	(a)	(b)	(c)	(d)
Recovery year	4	2	3	9
Recovery period	5 years	3 years	10 years	27.5 years

Solution

(a) 11.52% **(b)** 44.45% **(c)** 14.40% **(d)** 3.637%

Objective 4 *Use the MACRS Method to Find the Amount of Depreciation.* No salvage value is subtracted from the cost of property. Under the MACRS method, the depreciation rate multiplied by the original cost determines the depreciation amount. Depreciation by the MACRS method is given by

$$d = r \times c$$

where d is the depreciation for a rate r (from Table 16.1) and an original cost c.

Example 3 Finding the Amount of Depreciation with MACRS

Village Nursery and Landscaping purchased a pickup truck. Find the amount of depreciation in the third year if the truck had a cost of $18,500.

Solution A pickup truck has a recovery period of 5 years. From Table 16.1, the depreciation rate in the third year of recovery of 5-year property is 19.20%. Multiply this rate by the full cost of the property to determine the amount of depreciation.

$$19.20\% \times \$18{,}500 = \$3552$$

The amount of depreciation is $3552.

Objective 5 *Prepare a Depreciation Schedule Using the MACRS.*

Example 4 Preparing a Depreciation Schedule with MACRS

Omaha Insurance Company has purchased desks and chairs at a cost of $24,160. Prepare a depreciation schedule using the modified accelerated cost recovery system.

Solution No salvage value is used with MACRS. Office desks and chairs have a 7-year recovery period. The annual depreciation rates for 7-year property from Table 16.1 are as follows.

Recovery Year	Recovery Percent (Rate)
1	14.29%
2	24.49%
3	17.49%
4	12.49%
5	8.93%
6	8.92%
7	8.93%
8	4.46%

Multiply the appropriate percents by $24,160 to get the results shown in the following depreciation schedule.

Year	Computation	Amount of Depreciation	Accumulated Depreciation	Book Value
0	—	—	—	$24,160
1	(14.29% × $24,160)	$3452	$3,452	20,708
2	(24.49% × $24,160)	5917	9,369	14,791
3	(17.49% × $24,160)	4226	13,595	10,565
4	(12.49% × $24,160)	3018	16,613	7,547
5	(8.93% × $24,160)	2157	18,770	5,390
6	(8.92% × $24,160)	2155	20,925	3,235
7	(8.93% × $24,160)	2157	23,082	1,078
8	(4.46% × $24,160)	1078	24,160	0

The MACRS method of depreciation allows a rapid rate of investment recovery and at the same time results in a less complicated computation. By eliminating the necessity of estimating the life of an asset and the need for using a salvage value, the tables provide a more direct method of calculating depreciation.

16.5 Exercises

Use Table 16.1 to find the recovery percent (rate) given the following recovery year and recovery period.

	Recovery Year	Recovery Period		Recovery Year	Recovery Period
1.	3	5 years	**2.**	2	3 years
3.	4	10 years	**4.**	1	7 years
5.	1	5 years	**6.**	2	20 years
7.	14	27.5 years	**8.**	10	31.5 years
9.	6	5 years	**10.**	4	27.5 years
11.	14	39 years	**12.**	4	31.5 years

Find the first year's depreciation for each of the following using the MACRS method of depreciation. Round to the nearest dollar.

13. Cost: $12,250
Recovery period: 7 years

14. Cost: $8790
Recovery period: 5 years

15. Cost: $9680
Recovery period: 3 years

16. Cost: $72,300
Recovery period: 20 years

17. Cost: $48,000
Recovery period: 10 years

18. Cost: $786,400
Recovery period: 31.5 years

Find the book value at the end of the first year for each of the following using the MACRS method of depreciation. Round to the nearest dollar.

19. Cost: $9380
Recovery period: 3 years

20. Cost: $32,750
Recovery period: 10 years

21. Cost: $18,800
Recovery period: 10 years

22. Cost: $137,000
Recovery period: 27.5 years

Find the book value at the end of 3 years for each of the following using the MACRS method of depreciation. Round to the nearest dollar.

23. Cost: $9570
Recovery period: 5 years

24. Cost: $6500
Recovery period: 3 years

25. Cost: $87,300
Recovery period: 27.5 years

26. Cost: $390,800
Recovery period: 31.5 years

27. The same business asset may be depreciated using two or more different methods. Explain why a business would do this. (See Objective 1.)

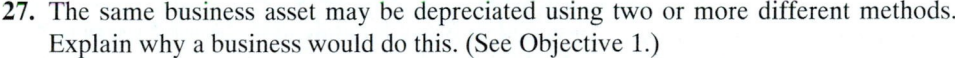

28. After learning about MACRS, what three features stand out to you as being unique to this method? (See Objective 3.)

Solve the following application problems. Use Table 16.1. Round to the nearest dollar.

29. **RACE HORSE** Rocking Horse Ranch purchased a race horse for $10,980. Prepare a depreciation schedule using the MACRS method of depreciation (3-year property).

Year	Computation	Amount of Depreciation	Accumulated Depreciation	Book Value
0	—	—	—	$10,980
1				
2				
3				
4				

30. **COMPANY VEHICLES** Village Nursery and Landscaping purchased a pickup truck at a cost of $18,500. Prepare a depreciation schedule using the MACRS method of depreciation (5-year property).

Year	Computation	Amount of Depreciation	Accumulated Depreciation	Book Value
0	—	—	—	$18,500
1				
2				
3				
4				
5				
6				

31. **OFFSHORE DRILLING** Gulf Drilling purchased a tugboat for $122,700. Prepare a depreciation schedule using the MACRS method of depreciation.

32. **RESIDENTIAL RENTAL PROPERTY** Andy Kirkpatrick purchased some nonresidential rental real estate before May 12, 1993, for $415,000. Prepare a depreciation schedule for the first ten years using the MACRS method of depreciation (31.5-year property).

33. **LAPTOP COMPUTERS** Susan Raymond purchased a laptop computer for her office for $3700. Find the book value at the end of the third year using the MACRS method of depreciation.

34. **OFFICE FURNITURE** Delta Dental purchased new office furniture at a cost of $13,800. Find the book value at the end of the fifth year using the MACRS method of depreciation.

35. **BOOKKEEPING BUSINESS** Jim Bralley, the owner of Bralley's Bookkeeping, purchased an office building at a cost of $480,000. Find the amount of depreciation for each of the first 5 years using the MACRS method of depreciation (39-year property).

36. **INDEPENDENT BOOKSTORE OWNERSHIP** Maretha Roseborough, owner of the Barnstormer Bookstore, bought a building to use for her business. The cost of the building was $220,000. Find the amount of depreciation each of the first 5 years using the MACRS method of depreciation (39-year property).

Chapter 16 Quick Review

Chapter Terms

Review the following terms to test your understanding of the chapter. For each term you do not know, refer to the page number found next to that term.

200% method [p. 586]
accumulated depreciation [p. 579]
accelerated cost recovery system (ACRS) [p. 606]
accelerated depreciation [p. 585]
book value [p. 579]
cost [p. 579]
declining balance method [p. 585]
depreciable amount [p. 579]
depreciation [p. 578]
depreciation fraction [p. 591]
depreciation schedule [p. 582]
double-declining balance method [p. 586]
half-year convention [p. 607]
income tax method [p. 606]
intangible assets [p. 579]
modified accelerated cost recovery system (MACRS) [p. 606]
no salvage value [p. 579]
partial-year depreciation [p. 601]
Publication 534 [p. 607]
recovery classes [p. 606]
recovery periods [p. 606]
residual value [p. 579]
salvage value [p. 579]
scrap value [p. 579]
straight-line method [p. 579]
sum-of-the-year's-digits method [p. 591]
tangible assets [p. 579]
units-of-production method [p. 598]
useful life [p. 579]

Concepts / Examples

16.1 Straight-line method of depreciation

The depreciation is the same each year. Use the formula

$$d = \frac{c - s}{n}$$

with c = cost, s = salvage value, and n = life (in years).

Cost, \$500; scrap value, \$100; life of 8 years; find the annual amount of depreciation.

$$d = \frac{\$500 - \$100}{8} = \frac{\$400}{8} = \$50$$

16.1 Book value

Book value is the value remaining at the end of the year. Cost minus accumulated depreciation equals book value.

Cost, \$400; scrap value, \$100; life of 3 years; find the book value at the end of the first year.

$$d = \frac{\$400 - \$100}{3} = \frac{\$300}{3} = \$100$$

Book value = \$400 − \$100 = \$300

16.2 Declining-balance rate

First, find the straight-line rate, then adjust it as follows:

200%: Multiply by 2.
150%: Multiply by $1\frac{1}{2}$.
125%: Multiply by $1\frac{1}{4}$.

Life of an asset is 10 years. Find the double-declining-balance (200%) rate.

$$10 \text{ years} = 10\% \left(\frac{1}{10}\right) \text{ straight-line}$$

$$2 \times 10\% = 20\%$$

16.2 Double-declining-balance depreciation method

First, find the double-declining-balance rate; then use the formula

$$d = r \times b$$

where b is total cost in the first year and the declining balance in the following years.

Cost, \$1400; life of 5 years; find the depreciation in years 1 and 2 (2 × 20% [straight-line rate] = 40%.)

Year 1:

$$\text{Depreciation} = 40\% \times \$1400 = \$560$$

$$\text{Book value} = \$1400 - \$560 = \$840$$

Year 2:

$$\text{Depreciation} = 40\% \times \$840 = \$336$$

16.3 Sum-of-years'-digits depreciation fraction

Add the years' digits together to get the denominator. The numerator is the number of years of life remaining at the beginning of the year. The denominator shortcut is as follows.

$$\text{Sum-of-the-years'-digits} = \frac{n(n+1)}{2}$$

Useful life is 4 years; find the depreciation fraction for each year.

$$1 + 2 + 3 + 4 = 10$$

Year	Depreciation Fraction
1	$\frac{4}{10}$
2	$\frac{3}{10}$
3	$\frac{2}{10}$
4	$\frac{1}{10}$

16.3 Sum-of-years'-digits depreciation method

First find the depreciation fraction, and then use the formula.

$$d = r \times (c - s)$$

Cost, \$2500; salvage value, \$400; life of 6 years; find depreciation in year 1.

$$\text{Depreciation fraction} = \frac{6}{21}$$

$$d = \frac{6}{21} \times (\$2500 - \$400)$$

$$= \frac{6}{21} \times \$2100 = \$600$$

16.4 Units-of-production depreciation amount

Use the following formula.

$$\text{Depreciation per unit} = \frac{\text{Depreciable amount}}{\text{Units of life}}$$

Cost, \$10,000; salvage value, \$2500; useful life of 15,000 units; find depreciation per unit.

$$\text{Depreciation per unit} = \frac{\$7500 \text{ depreciable amount}}{15{,}000 \text{ units of life}} = \$0.50$$

16.4 Units-of-production depreciation method

Multiply the depreciation per unit (per hour) by the number of units (hours) of production.

In the preceding example: the first year's production is 3800 units. Find depreciation in year 1.

$$3800 \times \$0.50 = \$1900$$

16.4 Partial-year depreciation

If an asset is purchased during the year, take only a fraction of the year's depreciation.

Depreciable amount, \$4500; life, 5 years; date purchased, June 7. Find first partial-year depreciation by the straight-line method.

$$7 \text{ months} = \frac{7}{12} \text{ year}$$

$$\frac{\$4500}{5} \times \frac{7}{12} = \$525$$

16.5 Modified accelerated cost recovery system (MACRS)

Established in 1986 for federal tax. No salvage value. Recovery periods are:

3-year	5-year	7-year
10-year	15-year	20-year
27.5-year	31.5-year	39-year

Find the proper rate from the table and then multiply by the cost to find depreciation.

Use the table, finding the recovery period column at the top of the table and the recovery year in the left-hand column. Cost: \$4850; recovery period 5 years; recovery year, 3; find the depreciation.

Find 5-year recovery period column at top of Table 16.1, and recovery year 3 in leftmost column; rate is 19.20%.

$$d = 19.20\% \times \$4850 = \$931.20$$

Chapter 16 Review Exercises

The answer section includes answers to all Review Exercises.

Find the annual straight-line and double-declining-balance rates (percents) of depreciation and the sum-of-the-years'-digits fractions in the first year for each of the following estimated lives. [16.1–16.3]

1. 5 years **2.** 6 years **3.** 4 years
4. 10 years **5.** 20 years **6.** 8 years

Use Table 16.1 to find the recovery percent (rate) given the following recovery year and recovery period. [16.5]

	Recovery Year	Recovery Period		Recovery Year	Recovery Period
7.	4	5-year	**8.**	1	3-year
9.	8	20-year	**10.**	3	7-year
11.	20	39-year	**12.**	20	31.5-year

Solve the following application problems. Round to the nearest dollar if necessary.

13. Cloverdale Creamery purchased an ice cream machine at a cost of $12,400. The machine has an estimated life of 10 years, and a scrap value of $3000. Use the straight-line method of depreciation to find the annual depreciation. [16.1]

14. The water filtration system at Micro Brew costs $74,000, has an estimated life of 20 years, and an estimated scrap value of $12,000. Use the straight-line method of depreciation to find the book value of the machinery at the end of 10 years. [16.1]

15. Sunset Swimming pools purchased a dump truck for $38,000. If the estimated life of the dump truck is 8 years, find the book value at the end of 2 years using the double-declining-balance method of depreciation. [16.2]

16. Star Bushing Company bought some Belgian manufactured drill presses at a total cost of $18,500. The estimated life of the drill presses is 5 years, and there is no scrap value. Find the depreciation in the first year using the double-declining-balance method of depreciation. [16.2]

17. The Feather River Youth Camp has purchased a diesel generator for $8250. Use the sum-of-the-years'-digits method of depreciation to determine the amount of depreciation to be taken during *each of the 4 years* on the diesel generator that has a 4-year life and scrap value of $1500. [16.3]

18. Murray's Delicatessen purchased a new display case for $7375. It has an estimated life of 10 years, and a salvage value of $500. Use the sum-of-the-years'-digits method of depreciation to find the book value at the end of the third year. [16.3]

19. A grape press costs $11,000, has a scrap value of $2500, and an estimated life of 5000 hours. Use the units-of-production method to find the depreciation for a year in which the machine was in operation 900 hours. [16.4]

20. Table Fresh Foods purchased a machine to package their pre-sliced garden salads. The machine costs $20,100, has an estimated life of 30,000 hours, and a salvage value of $1500. Use the units-of-production method of depreciation to find (a) the annual amount of depreciation and (b) the book value at the end of each year, given the following use information: year 1: 7800 hours; year 2: 4300 hours; year 3: 4850 hours; year 4: 7600 hours. [16.4]

21. The Fashion Express purchased new clothing racks at a cost of $22,400. Using the straight-line method of depreciation, a 5-year life, and a scrap value of $3500, find the accumulated depreciation at the end of the fourth year. [16.1]

22. Using the sum-of-the-years'-digits method of depreciation, find the amount of depreciation to be charged off each year on a tour bus purchased by Bayside Tours. The bus has a cost of $85,000, an estimated life of 5 years, and a scrap value of $13,000. [16.3]

23. A recycled glass crusher costs $48,000, has an estimated life of 5 years, and a scrap value of $4500. If the straight-line method of depreciation is used and the system is purchased on September 10, find the first partial-year depreciation. **[16.4]**

24. A parking lot is resurfaced on June 1 at a cost of $9720. The estimated life is 8 years and there is no scrap value. Use the sum-of-the-years'-digits method to find the first partial-year depreciation and the following full year's depreciation. **[16.4]**

25. Instant Copy Service purchased a new copy system at a cost of $28,400 on June 19. The estimated life of the system is 10 years and the salvage value is $6000. Use the double-declining-balance method of depreciation to find the first partial-year depreciation and the following full year's depreciation. **[16.4]**

26. Karl Schmidt, owner of Toy Train Hobby, has added paging and intercom features to the communication systems of his 4 stores at a cost of $2800 per store. The estimated life of the systems is 10 years, with no expected salvage value. Using the sum-of-the-years'-digits method of depreciation, find the total book value of all the systems at the end of the third year. **[16.3]**

27. A job printer purchased an automatic composer at a cost of $15,000. If the estimated life of the composer is 8 years, find the book value at the end of 4 years, using the double-declining-balance method of depreciation. Assume no scrap value. **[16.2]**

28. Reef Fisheries has purchased a commercial fishing boat for $74,125. The life of the boat is estimated to be 10 years and the scrap value $15,000. Use the sum-of-the-years'-digits method of depreciation to find the book value of the fishing boat at the end of the fourth year. **[16.3]**

29. King's Table bought new dining room tables at a cost of $14,750. If the estimated life of the tables is 8 years, at which time they will be worthless, and the double-declining-balance method of depreciation is used, find the book value at the end of the third year. **[16.2]**

30. A private road costs $56,000 and has a 15-year recovery period. Find the depreciation in the third year using the MACRS method of depreciation. **[16.5]**

31. Kathy Woodward purchased a new refrigeration system (7-year property) for her flower shop at a cost of $8100. Find the amount of depreciation for each of the first five years using the MACRS method of depreciation. **[16.5]**

32. The Rice Growers paid $2,800,000 to build a drying plant in 1998. The recovery period is 39 years. Use MACRS method of depreciation to find the book value of the drying plant at the end of the fifth year. **[16.5]**

Chapter 16 Summary Exercise *Comparing Depreciation Methods: A Business Application*

TRADER JOE'S purchased freezer cases at a cost of $285,000. The estimated life of the freezer cases is 5 years, at which time they will have no salvage value. The company would like to compare allowable depreciation methods and decides to prepare depreciation schedules for the fixtures using the straight-line, double-declining-balance, and the sum-of-the-years'-digits methods of depreciation. Using these depreciation schedules, find the answer to these questions for Trader Joe's.

(a) What is the book value at the end of 3 years using the straight-line depreciation method?

(b) Using the double-declining-balance method of depreciation, what is the book value at the end of the third year?

(c) With the sum-of-the-year's-digits method of depreciation, what is the accumulated depreciation at the end of 3 years?

(d) What amount of depreciation will be taken in year 4 with each of the methods?

Net Assets Business on the Internet

GMC

Statistics

- 1902: First GMC truck built
- 1999: 647,000 employees
- Facilities in 50 countries
- Named Top 100 Products by *Construction Equipment* magazine

Founded in 1902, General Motors Corporation has grown into the world's largest industrial corporation and full-line vehicle manufacturer. In the late 1990s, the company had partnered with over 30,000 companies worldwide. As the world's largest U.S. exporter of cars and trucks with manufacturing, assembly, or component operations in 50 countries, General Motors has a global presence in over 190 countries.

West Construction purchased a new GMC Sierra pickup truck for $22,500. The useful life of the truck is 5 years and the estimated salvage value is $4500. Use this information to solve the following.

1. Find the book value of the pickup truck after 3 years using the straight-line method of depreciation.

2. Find the book value of the pickup truck at the end of 3 years using the double-declining balance method of depreciation.

3. If the pickup truck has an expected life of 100,000 miles, find **(a)** the depreciation per mile and **(b)** the book value of the pickup truck after 75,000 miles.

4. If the MACRS depreciation rates for the pickup truck were 20%, 32%, and 19.2% in the first three years of the truck's life, find the book value of the pickup truck at the end of 3 years.

17 Financial Statements and Ratios

McDonald's had nearly 14 billion customer visits last year, which is equivalent to serving lunch and dinner to every person on earth (Figure 17.1). Even so, on any given day, McDonald's only serves about 1% of the world's population.

Why has McDonald's expanded their operations to Europe, Asia, Latin America, and Canada?

Business owners and managers must keep careful records of the expenses and income of their business. They need these records to help them manage the business, to inform other owners of current operations, to provide required information to lenders from whom they wish to borrow money, and for tax purposes.

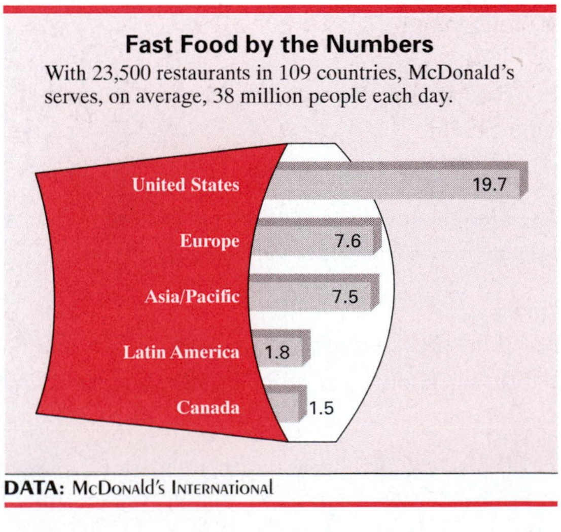

Figure 17.1

Bookkeepers and accountants keep track of *the income and expenses of the firm*. Accountants are concerned with issues such as meeting payroll, paying suppliers on time, estimating and paying taxes, as well as estimating and handling revenues. This chapter covers some of the tools used by managers, accountants, lenders, and investors when looking at the financial health of a firm.

17.1 The Income Statement

Objectives
1. *Learn the terms used with income statements.*
2. *Understand the income statement of McDonald's Corporation.*
3. *Complete an income statement.*

Objective 1 *Learn the Terms Used with Income Statements.* An **income statement** is prepared for the management and owners of a business to summarize all income and expenses for a given period of time, such as a month or a year. As a first step, find the **gross sales**, the total amount of money received from customers for the goods or services sold by the firm. Then subtract the value of any **returns** from customers to arrive at **net sales**, the value of the goods and services bought and kept by customers. Use the following formula.

Net sales = Gross sales − Returns

After finding net sales, look at company records to find the **cost of goods sold**, the amount paid by the firm for the items sold to customers during the period of time covered by the income statement. Then subtract the cost of goods sold from the net sales to find the gross profit. Gross profit is often called *gross profit on sales*. The **gross profit** is the amount of money left over after the business pays for the goods it sells. This money is used to pay the expenses involved in running the business and anything remaining after expenses goes to either taxes or profit.

Gross profit = Net sales − Cost of goods sold

Operating expenses represent the amount paid by the firm in an attempt to sell its goods. Common expenses include rent, salaries and wages, advertising, utilities, losses from uncollectible accounts, and taxes on inventory and payroll. Operating expenses are sometimes called **overhead**. Finally, **net income before taxes** is the actual amount earned by the firm during the given time period.

Net income before taxes = Gross profit − Operating expenses

Net income, also called **net income after taxes**, is found by subtracting income taxes from net income before taxes.

Net income = Net income before taxes − Income taxes

Banks do not wish to lend money to firms that may not be able to repay the loans. As a result, a bank will usually want to see all of the values identified in this section so far, in addition to other data, before making a loan to a firm. The values introduced here are typically included in an income statement, which we will now examine.

Objective 2 *Understand the Income Statement of McDonald's Corporation.* A portion of McDonald's Corporation 1998 income statement is shown in Example 1. Shares of the company are publicly traded on the New York Stock Exchange under the symbol MCD. Check the

New York Stock Exchange page in a newspaper or on the World Wide Web for a current stock price and dividend. The income statement is **consolidated** since it shows the total results of all of the subsidiary companies, or companies owned by McDonald's Corporation.

Example 1 Finding the Net Income

In 1998, McDonald's Corporation had gross sales of approximately $12,421,400,000. They had no returns to speak of due to the nature of their business, food and packaging costs of $2,997,400,000, operating expenses of $7,116,600,000, and income taxes of $757,300,000. Find the net income after taxes for the year.

Solution Use the formulas from the previous page. For convenience, work with tenths of millions of dollars by dropping the last five digits; 2 hence, $12,421,400,000 becomes $12,421.4 million.

Net sales = Gross sales − Returns
= $12,421.4 − 0 = $12,421.4 In millions of dollars

Gross sales equal net sales since they have no returns. Now find gross profit.

Gross profit = Net sales − Cost of goods sold
= $12,421.4 − $2997.4 = $9424.0 In millions of dollars

The gross profit is used to pay the expenses of running the business. Now find the net income before taxes.

Net income before taxes = Gross profit − Operating expenses
= $9424.0 − $7116.6 = $2307.4 In millions of dollars

Finally, calculate the net income after income taxes.

Net income after taxes = Net income before taxes − Income taxes
= $2307.4 − $757.3 = $1550.1

In 1998, McDonald's Corporation paid approximately $757,300,000 in *income taxes* and still had an *after-tax profit* of $1,550,100,000. The results are summarized in the following table.

McDONALD'S CORPORATION
CONSOLIDATED STATEMENTS OF INCOME
YEAR ENDING DECEMBER 31, 1998
(MILLIONS OF DOLLARS)

Gross sales	$12,421.4
Returns	− 0
Net sales	$12,421.4
Cost of goods sold	− 2,997.4
Gross profit	$ 9,424.0
Operating expenses	− 7,116.6
Net income before taxes	$ 2,307.4
Taxes	− 757.3
Net income after taxes	$ 1,550.1

Check the results shown on the income statement with the following fundamental formula.

Cost of goods sold	+	Expenses	+	Taxes	+	Net income after taxes	=	Net sales
$2997.4	+	$7116.6	+	$757.3	+	$1550.1	=	$12,421.4

The income statement checks.

The value of a company's stock is based on *financial results* in addition to *perceived opportunities*. As a publicly held corporation, McDonald's Corporation must publish financial results. The firm's stock price generally rises when profits are rising and generally falls when company profits are falling. You can obtain McDonald's Corporation current financial statements mailed to you *free* by calling their home office in Oak Brook, Illinois. You can also get financial information on the company using the World Wide Web.

Objective 3 *Complete an Income Statement.* Example 1 gives the value for the cost of goods sold, whereas this amount would normally need to be calculated. The cost of goods sold can be found using the formula below. **Initial inventory** is the at-cost value of all goods on hand for sale at the beginning of the period, and **ending inventory** is the at-cost value of all goods on hand for sale at the end of the period.

Initial (or beginning) inventory
+ Cost of goods purchased during time period
+ Freight
− Ending inventory
Cost of goods sold

Example 2 Preparing an Income Statement

Josie's Clothing had gross sales of $159,000 during the past year, with returns of $9000. Inventory on January 1 of last year was $47,000. A total of $104,000 worth of goods was purchased last year, with freight on the goods totaling $2000. Inventory on December 31 of last year was $56,000. Wages paid to employees totaled $18,000. Rent was $9000, advertising was $1000, utilities totaled $2000, and taxes on inventory and payroll totaled $4000. Miscellaneous expenses totaled $6000 and income taxes were $500. Complete an income statement for the store.

Solution Go through the steps that follow, which refer to the income statement in Figure 17.2.

Working Through an Income Statement

Step 1. Enter gross sales and sales returns. Subtract sales returns from gross sales to find net sales. Net sales in this example were $150,000.

Step 2. Enter the cost of goods purchased and the freight. Add these two numbers.

Step 3. Add the inventory on January 1 and the total cost of goods purchased.

Step 4. Subtract the inventory on December 31 from the result of Step 3. This gives the cost of goods sold.

JOSIE'S CLOTHING
INCOME STATEMENT
FOR THE YEAR ENDING DECEMBER 31

Gross sales				$159,000
Returns			Subtract ①	$9,000
Net sales				$150,000
Inventory, January 1		$47,000		
Cost of goods purchased	$104,000	Add ②	Add ③	
Freight	$2,000			Subtract ⑤
Total cost of goods purchased		$106,000		
Total of goods available for sale		$153,000	Subtract ④	
Inventory, December 31		$56,000		
Cost of goods sold				$97,000
Gross profit				$53,000
Expenses				
Salaries and wages		$18,000		
Rent		$9,000		
Advertising		$1,000	Add ⑥	
Utilities		$2,000		
Taxes on inventory, payroll		$4,000		
Miscellaneous expenses		$6,000		
Total expenses				$40,000
Net income before taxes			Subtract ⑦	$13,000
Income taxes			Subtract ⑧	$ 500
Net income				$12,500

Figure 17.2

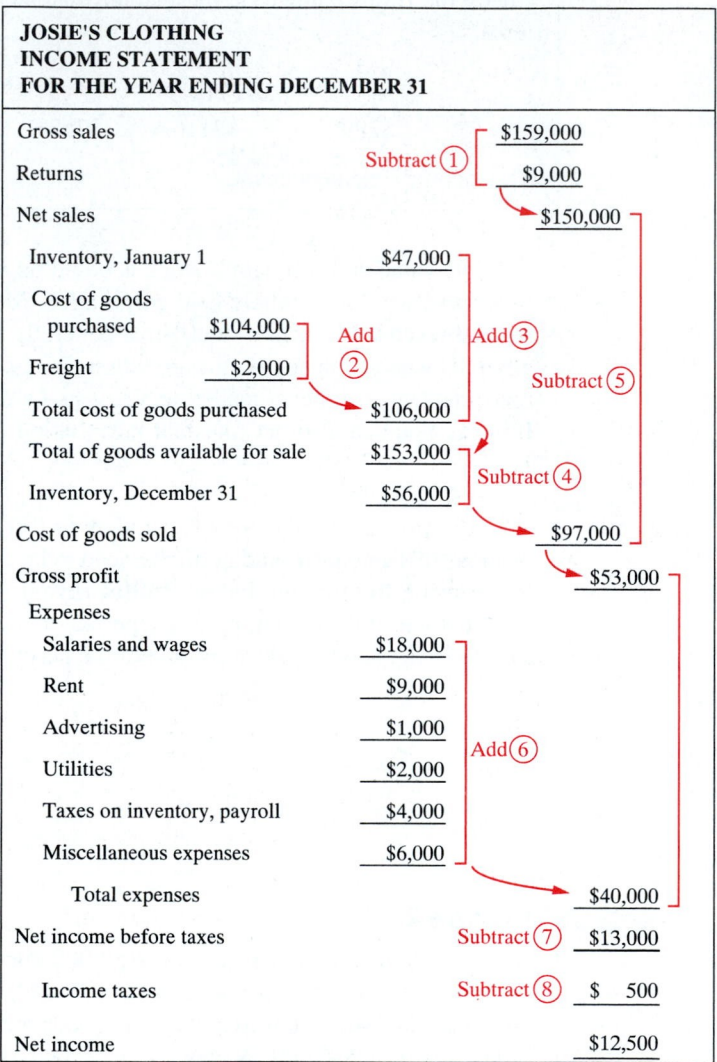

Step 5. Subtract the cost of goods sold from net sales, which were found in Step 1. The result is the gross profit.

Step 6. Enter all expenses and add them to get the total expenses.

Step 7. Subtract the total expenses from the gross profit to find the net income before taxes.

Step 8. Subtract taxes from net income before taxes to find net income after taxes.

NOTE Be sure to check the results of your income statement by adding the cost of goods sold, expenses, net income before taxes, and taxes. This total should equal net sales. ▬

17.1 Exercises

Find (a) the gross profit, (b) the net income before taxes, and (c) the net income after taxes for each firm.

1. **CLOTHING STORE** Jerome Buchanan opened a franchise clothing store in the mall three years ago. Last year, the cost of goods sold was $367,200, operating expenses were $228,300, income taxes were $22,700, gross sales were $685,900, and returns were $2350.

2. **CIGAR STORE** A cigar store in Ohio had net sales of $289,300, operating expenses of $68,200, cost of goods sold of $165,000, and paid $9900 in taxes.

Find the cost of goods sold for the following firms.

3. **PAINT INVENTORY** At Rainbow Paint, the inventory on January 1 was $263,400, with $343,500 worth of goods purchased during the year. Freight was $4800 and the inventory on December 31 was $287,500.

4. **AUTO INVENTORY** Inventory at Southside Auto was $428,340 on January 1 and $387,708 on December 31. They paid freight of $8400 during the year on purchases of $548,200.

5. **COMPUTER COMPANY** Future Tech Computing had gross sales of $284,000 last year, with returns of $6000. The inventory on January 1 was $58,000. A total of $232,000 worth of goods was purchased, with freight of $3000. The inventory on December 31 was $69,000. Wages and salaries were $15,000, rent was $6000, advertising was $2000, utilities were $1000, taxes on inventory and payroll totaled $3000, miscellaneous expenses totaled $4000, and income taxes amounted to $2400. Complete the income statement.

FUTURE TECH COMPUTING
INCOME STATEMENT
YEAR ENDING DECEMBER 31

Gross sales			________	
Returns			________	
Net sales				________
Inventory, January 1		________		
Cost of goods purchased	________			
Freight	________			
Total cost of goods purchased		________		
Total of goods available for sale		________		
Inventory, December 31		________		
Cost of goods sold			________	
Gross profit				________
Expenses				
Salaries and wages		________		
Rent		________		
Advertising		________		
Utilities		________		

Taxes on inventory, payroll	________	
Miscellaneous expenses	________	
Total expenses		________
Net Income before Taxes		________
Income taxes		________
Net Income		________

6. DENTAL SUPPLY New England Dental Supply is a regional wholesaler with gross sales last year of $2,215,000. Returns totaled $26,000. Inventory on January 1 was $215,000. Goods purchased during the year totaled $1,123,000. Freight was $4000. Inventory on December 31 was $265,000. Salaries and wages were $154,000, rent was $59,000, advertising was $11,000, utilities were $12,000, taxes on inventory and payroll totaled $10,000, and miscellaneous expenses were $9000. In addition, taxes for the year amounted to $287,400. Complete the following income statement for the firm.

NEW ENGLAND DENTAL SUPPLY
INCOME STATEMENT
YEAR ENDING DECEMBER 31

Gross sales			________	
Returns			________	
Net sales				________
Inventory, January 1		________		
Cost of goods purchased	________			
Freight	________			
Total cost of goods purchased		________		
Total of goods available for sale		________		
Inventory, December 31		________		
Cost of goods sold			________	
Gross profit				________
Expenses				
Salaries and wages		________		
Rent		________		
Advertising		________		
Utilities		________		
Taxes on inventory, payroll		________		
Miscellaneous expenses		________		
Total expenses				________
Net Income before Taxes				________
Income taxes				________
Net Income				________

7. INCOME STATEMENT FOR SELF-EMPLOYED Kathy Gilmore is self-employed as a computer consultant. She sells her services to customers and has no inventory, no returns, no freight, and no cost of goods sold. Gross sales were $170,500, salaries and wages were $63,000, rent was $28,000, advertising was $12,000, utilities were $4000, taxes on payroll were $3,800, and miscellaneous office expenses were $9400. Complete the income statement for this firm given that income taxes were $6800.

KATHY GILMORE, CONSULTANT
INCOME STATEMENT FOR THE
YEAR ENDING DECEMBER 31

Gross sales			________	
Returns			________	
Net sales				________
Inventory, January 1		________		
Cost of goods purchased	________			
Freight	________			
Total cost of goods purchased		________		
Total of goods available for sale		________		
Inventory, December 31		________		
Cost of goods sold			________	
Gross profit				________
Expenses				
Salaries and wages		________		
Rent		________		
Advertising		________		
Utilities		________		
Taxes on inventory, payroll		________		
Miscellaneous expenses		________		
Total expenses				________
Net Income before Taxes				________
Income taxes				________
Net Income				________

8. What are the different factors that control gross profit? Is it important for management to continue to watch their gross profit? Why? (See Objective 1.)

9. List the items on a family's income statement that a bank would need to look at when thinking of lending you money for a new automobile. (See Objective 2.)

10. Name several companies for which returns might not be significant, as was the situation with McDonald's Corporation.

11. Explain why a lender might like to see an income statement before making a loan. (See Objective 1.)

12. (a) Discuss the purpose of an income statement. (b) Outline the basic structure of an income statement. (See Objective 2.)

13. Complete item 9 in the foldout and prepare a financial statement for yourself. If you do not currently work, use numbers you believe will be reasonable for your situation 6 months after you graduate.

17.2 Analyzing the Income Statement

Objectives

1. *Compare income statements using vertical analysis.*
2. *Calculate percents of net sales.*
3. *Compare an income statement with published charts.*
4. *Prepare a horizontal analysis.*

By going through the steps presented in the previous section, a firm can find its net income for a given period of time. A question that might be asked is, "What happened to each part of the sales dollar?" The first step toward answering this question is to *list each of the important items* on the income statement as a percent of net sales in a process called a **vertical analysis** of the income statement.

Objective 1 *Compare Income Statements using Vertical Analysis.* A vertical analysis of an income statement is another application of the fundamental formula for percent from Chapter 3, $P = RB$. Since a percent is needed, the formula must be solved for the rate R as follows. In a vertical analysis, each item is found as a percent of net sales.

$$\left(R = \frac{P}{B}\right) \quad \text{or} \quad \left(R = \frac{\text{Particular item}}{\text{Net sales}}\right)$$

A comparison of the results for two different years can be made by calculating a vertical analysis for each year. This results in a **comparative income statement**.

Example 1 Performing a Vertical Analysis

First, perform a vertical analysis of the summary 1997 and 1998 income statements shown below for McDonald's Corporation. Then construct a comparative income statement by showing the results in a table.

McDONALD'S CORPORATION
CONSOLIDATED STATEMENTS OF INCOME
(IN MILLIONS OF DOLLARS)

Year Ending December 31	1997	1998
Gross sales	$11,408.8	$12,421.4
Returns	− 0	− 0
Net sales	$11,408.8	$12,421.4
Cost of goods sold	− 2,772.6	− 2,997.4
Gross profit	$ 8,636.2	$ 9,424.0
Operating expenses	− 6,228.9	− 7,116.6
Net income before taxes	$ 2,407.3	$ 2,307.4
Income taxes	− 764.8	− 757.3
Net income	$ 1,642.5	$ 1,550.1

Solution Calculate each value in the column labeled 1998 as a percent of 1998 net sales. Then do the same using 1997 data. Percents are rounded to the nearest tenth.

COMPARATIVE INCOME STATEMENT

	1997	1998
Percent cost of goods sold	$\frac{\$2772.6}{\$11{,}408.8} = 24.3\%$	$\frac{\$2997.4}{\$12{,}421.4} = 24.1\%$
Percent gross profit	$\frac{\$8636.2}{\$11{,}408.8} = 75.7\%$	$\frac{\$9424.0}{\$12{,}421.4} = 75.9\%$
Percent operating expenses	$\frac{\$6228.9}{\$11{,}408.8} = 54.6\%$	$\frac{\$7116.6}{\$12{,}421.4} = 57.3\%$
Percent net income before taxes	$\frac{\$2407.3}{\$11{,}408.8} = 21.1\%$	$\frac{\$2307.4}{\$12{,}421.4} = 18.6\%$
Percent income taxes	$\frac{\$764.8}{\$11{,}408.8} = 6.7\%$	$\frac{\$757.3}{\$12{,}421.4} = 6.1\%$
Percent net income after taxes	$\frac{\$1642.5}{\$11{,}408.8} = 14.4\%$	$\frac{\$1550.1}{\$12{,}421.4} = 12.5\%$

Notice that the operating expenses increased from 54.6% of net sales in 1997 to 57.3% of net sales in 1998. This is *not a good sign* since it contributed to a decrease in net income before taxes from 21.1% of net sales in 1997 to only 18.6% of net sales in 1998. As an investor, you may wish to find out why operating expenses increased before you buy stock in the company.

Objective 2 ***Calculate Percents of Net Sales.*** The formula given in the previous section for checking income statements is just as valid for percents as for dollar amounts.

$$\text{Cost of goods sold} + \text{Operating expenses} + \text{Income taxes} + \text{Net income after taxes} = \text{Net sales}$$

Use the information from Example 1 to verify that this is true for 1998 data. Notice that net sales equals gross sales for McDonald's Corporation since they have no returns.

$$24.1\% + 57.3\% + 6.1\% + 12.5\% = 100\%$$

These figures may not always add to 100% due to rounding.

The next example shows how to calculate these percents when there are returns.

Example 2 Finding Percents of Net Sales

Write each of the following items as a percent of net sales. The salaries and wages for this firm are low because it is a small business and the owner currently is not taking any income out of the business.

Gross sales	$209,000	Salaries and wages	$11,000
Returns	$9,000	Rent	$6,000
Cost of goods sold	$145,000	Advertising	$11,000

Solution Use the formula for net sales, with gross sales = \$209,000 and returns = \$9000.

$$\text{Net sales} = \text{Gross sales} - \text{Returns}$$
$$= \$209{,}000 - \$9000 = \$200{,}000$$

Now find all the desired percents.

$$\text{Percent gross sales} = \frac{\$209{,}000}{\$200{,}000} = 104.5\%$$

NOTE This percent is more than 100% because returns are included in it, but not in net sales. ■

$$\text{Percent returns} = \frac{\$9000}{\$200{,}000} = 4.5\%$$
$$\text{Percent cost of goods sold} = \frac{\$145{,}000}{\$200{,}000} = 72.5\%$$
$$\text{Percent salaries and wages} = \frac{\$11{,}000}{\$200{,}000} = 5.5\%$$
$$\text{Percent rent} = \frac{\$6000}{\$200{,}000} = 3\%$$
$$\text{Percent advertising} = \frac{\$11{,}000}{\$200{,}000} = 5.5\%$$

Objective 3 ***Compare an Income Statement with Published Charts.*** Once the percent of net sales for each item on the income statement has been found, they can be compared to the percents for similar businesses. To do this, consult published charts that have the required data. One such chart is shown in Table 17.1. These charts are compiled by averaging statistics from many similar firms.

Table 17.1 TYPICAL PERCENTS

Type of Business	Cost of Goods	Gross Profit	Total Operating Expenses*	Net Income	Wages	Rent	Advertising
Supermarkets	82.7%	17.3%	13.9%	3.4%	6.5%	0.8%	1.0%
Men's and women's apparel	67.0%	33.0%	21.2%	11.8%	8.0%	2.5%	1.9%
Women's apparel	64.8%	35.2%	23.4%	11.7%†	7.9%	4.9%	1.8%
Shoes	60.3%	39.7%	24.5%	15.2%	10.3%	4.7%	1.6%
Furniture	68.9%	31.2%	21.7%	9.6%†	9.5%	1.8%	2.5%
Appliances	66.9%	33.1%	26.0%	7.2%†	11.9%	2.4%	2.5%
Drugs	67.9%	32.1%	23.5%	8.6%	12.3%	2.4%	1.4%
Restaurants	48.4%	51.6%	43.7%	7.9%	26.4%	2.8%	1.4%
Service station	76.8%	23.2%	16.9%	6.3%	8.5%	2.3%	0.5%

*This column represents the total of all expenses involved in running the firm. Total operating expenses include, but are not limited to, wages, rent, and advertising.

†These numbers cannot be found by subtraction because of rounding.

Example 3 Compare Business Ratios

Gina Burton wishes to compare the business ratios of her shoe store to industry averages. Figures from her store and industry averages for shoe stores are shown below. Burton sees that her expenses are higher and her net income is lower than the industry averages. What might Ms. Burton do to decrease total expenses and increase net income?

	Cost of Goods	Gross Profit	Total Expense	Net Income	Wages	Rent	Advertising
Burton's Shoes	58.2%	41.8%	28.3%	13.5%	11.7%	5.6%	2.8%
Shoes (from chart)	60.3%	39.7%	24.5%	15.2%	10.3%	4.7%	1.6%

Solution Burton's wages, rent, and advertising all seem to be above the averages. If she can decrease any of these, or increase total sales without increasing any of these, then her net income will improve. Perhaps she can reschedule some employees to reduce overtime or shift more work to lower-wage employees. It may be that she can try to renegotiate her rent with her landlord or purchase her own building and move the store. Perhaps she can get the same advertising exposure by changing her advertising strategy to one that costs less.

On the other hand, her store may never compare favorably with national averages. That is fine as long as she makes an adequate profit.

Objective 4 *Prepare a Horizontal Analysis.* Another way to analyze an income statement is to prepare a **horizontal analysis**. A horizontal analysis finds percents of change (either increases or decreases) between the current time period and a previous time period. This comparison can expose *unusual changes*, such as a rapid increase in expenses or decline in net sales or profits.

A horizontal analysis is done by finding the amount of any change from the previous year to the current year, both in dollars and as a percent. For example, the income statement for McDonald's Corporation given earlier shows that net sales increased from \$11,408.8 (in millions) in 1997 to \$12,421.4 (in millions) in 1998, an increase of \$12,421.4 − \$11,408.8 = \$1012.6 (in millions). Find the percent of increase by comparing the increase to 1997 sales.

$$\frac{\$1012.6}{\$11{,}408.8} = 8.9\% \quad \text{Rounded}$$

$$\text{Percent of change} = \frac{\text{Change}}{\text{Last year's amount}}$$

Always use last year as the base.

This *relatively slow growth rate in total sales* may be part of the reason that McDonald's stock has been hot and cold, as shown in Figure 17.3.

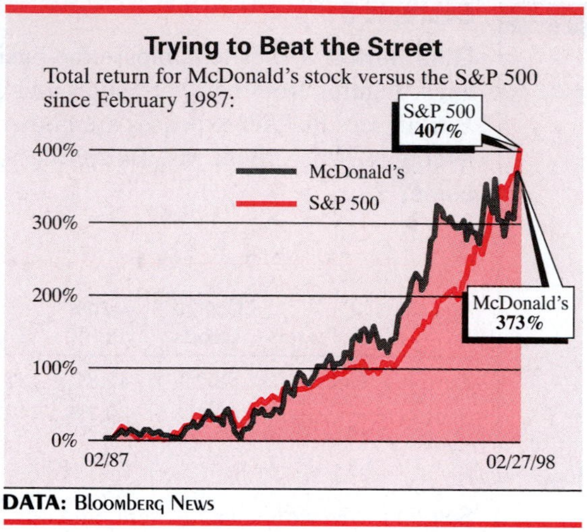

Figure 17.3

Example 4 Performing a Horizontal Analysis

Perform a horizontal analysis for the 1997 and 1998 income statements for McDonald's Corporation shown in Example 1.

Solution Find the increase by subtracting the 1997 figure from the 1998 figure, then divide by the 1997 figure to find the percent increase and finally round to the nearest tenth of a percent.

MCDONALD'S CORPORATION
CONSOLIDATED STATEMENTS OF INCOME (MILLIONS OF DOLLARS)
YEAR ENDING DECEMBER 31

	1997	1998	Increase	Percent
Net sales	$11,408.8	$12,421.4	$1,012.6	8.9%
Gross profit	8,636.2	9,424.0	787.8	9.1%
Income before taxes	2,407.3	2,307.4	(99.9)	(4.1%)

NOTE The parentheses indicate a *negative increase* in profits. In other words, the 1998 profits were less than 1997 profits. ■

Net sales increased by almost 9%, which is a slow growth rate, but perhaps not that bad for a company as large as McDonald's Corporation. A large growth rate probably cannot be maintained, and a small growth rate suggests that the company's profits may never increase much above current levels. Of course, the managers also want profits to grow, but income before taxes actually decreased by 4.1%. McDonald's Corporation hopes to maintain a good growth rate.

1. By encouraging existing customers to consume more of their products
2. By finding new customers for their products in international markets such as China and Russia

17.2 Exercises

Prepare a vertical analysis for each of the following firms. Round percents to the nearest tenth of a percent.

1. **GUITAR SHOP** In 1999, Classic Guitars had a cost of goods sold (guitars) of $243,570, operating expenses of $140,450 and net sales of $480,300.

2. **SHOE SHOP** Monica's Shoe Shop had net sales of $294,380, operating expenses of $68,650, and a cost of goods sold of $163,890.

The following charts show some figures from the income statements of several companies. In each case, prepare a vertical analysis by expressing each item as a percent of net sales. Then write in the appropriate average percent from the table in the book.

3. CAPITAL APPLIANCE CENTER

	Amount	Percent	Percent from Table 17.1
Net sales	$900,000	100.0%	100.0%
Cost of goods sold	617,000	______	______
Gross profit	283,000	______	______
Wages	108,900	______	______
Rent	20,700	______	______
Advertising	27,000	______	______
Total expenses	216,000	______	______
Net income before taxes	67,000	______	______

4. ELLIS RESTAURANT

	Amount	Percent	Percent from Table 17.1
Net sales	$600,000	100.0%	100.0%
Cost of goods sold	280,000	______	______
Gross profit	320,000	______	______
Wages	160,000	______	______
Rent	15,000	______	______
Advertising	8,000	______	______
Total expenses	255,000	______	______
Net income before taxes	65,000	______	______

Complete the following comparative income statement. Round percents to the nearest tenth of a percent.

5. BEST TIRES, INC.
COMPARATIVE INCOME STATEMENT

	This Year		Last Year	
	Amount	Percent	Amount	Percent
Gross sales	$1,856,000	_______	$1,692,000	_______
Returns	6,000	_______	12,000	_______
Net sales	_______	100.0%	_______	100.0%
Cost of goods sold	1,202,000	_______	1,050,000	_______
Gross profit	648,000	_______	630,000	_______
Wages	152,000	_______	148,000	_______
Rent	82,000	_______	78,000	_______
Advertising	111,000	_______	122,000	_______
Utilities	32,000	_______	17,000	_______
Taxes on inv., payroll	17,000	_______	18,000	_______
Miscellaneous expenses	62,000	_______	58,000	_______
Total expenses	456,000	_______	441,000	_______
Net income	_______	_______	_______	_______

Complete the following horizontal analysis for Best Tires, Inc. comparative income statement given above. Round percents to the nearest tenth of a percent.

6. BEST TIRES, INC.
COMPARATIVE INCOME STATEMENT

			Increase (Decrease)	
	This Year	Last Year	Amount	Percent
Gross sales	$1,856,000	$1,692,000	_______	_______
Returns	6,000	12,000	_______	_______
Net sales	1,850,000	1,680,000	_______	_______
Cost of goods sold	1,202,000	1,050,000	_______	_______
Gross profit	648,000	630,000	_______	_______
Wages	152,000	148,000	_______	_______
Rent	82,000	78,000	_______	_______
Advertising	111,000	122,000	_______	_______
Utilities	32,000	17,000	_______	_______
Taxes on inv., payroll	17,000	18,000	_______	_______
Miscellaneous expenses	62,000	58,000	_______	_______
Net income	192,000	189,000	_______	_______

The following tables give the percents for various items from the income statements of firms in various businesses. Complete these tables by including the appropriate percents from Table 17.1. Identify any areas that might require attention by management. Also list suggestions for improving any problem area.

	Type of Store	Cost of Goods	Gross Profit	Total Operating Expenses	Net Income	Wages	Rent	Advertising
7.	Supermarkets	84.5%	15.5%	14.4%	1.1%	6.4%	2.1%	0.9%
		___	___	___	___	___	___	___
8.	Shoes	60.5%	39.5%	24.3%	15.2%	10.4%	4.8%	1.5%
		___	___	___	___	___	___	___
9.	Drug store	71.2%	28.8%	26.5%	2.3%	12.9%	5.3%	2.0%
		___	___	___	___	___	___	___
10.	Restaurant	57.8%	42.2%	38.6%	3.6%	28.1%	2.4%	1.0%
		___	___	___	___	___	___	___

11. Compare a vertical analysis to a horizontal analysis. What are the strengths of each? (See Objectives 1 and 4.)

12. Explain the purpose of comparing percent of net sales on an income statement to percents for similar businesses. (See Objective 2.)

13. The average net income before taxes for supermarkets is 3.4%. Go to the supermarket where you normally shop and list 10 items that you normally buy and the price of each. Total these and find 3.4% of this total. Approximately what is the net income for the store for selling these 10 items?

14. Why would a lender want to use both vertical and horizontal analyses before making a long-term loan to a firm? (See Sections 17.1 and 17.2.)

17.3 The Balance Sheet

Objectives

1. *Identify the terms used with balance sheets.*
2. *Prepare a balance sheet.*
3. *Analyze the balance sheet of McDonald's Corporation.*

An income statement summarizes the financial affairs of a business firm for a given period of time, such as a year. On the other hand, a **balance sheet** describes the financial condition of a firm *at one point in time*, such as the last day of a year. A balance sheet shows the worth of a business at a particular time by listing its **assets**, which are the things it owns, such as property, equipment, and money owed to the business, as well as its **liabilities**, which are amounts owed by the business to others. The difference of these two amounts gives the **owner's equity** in the business.

Objective 1 *Identify the Terms Used with Balance Sheets.* Both assets and liabilities are divided into two categories, **long-term** and **current** (**short-term**). Long-term generally applies when the time involved is more than one year, whereas short-term applies when the time involved is less than one year. The following items appear as assets on balance sheets.

ASSETS

Current assets: cash or items that can be converted into cash within a short period of time such as a year

Cash: cash in checking and savings accounts
Marketable securities: stocks, bonds, and other securities that can quickly be converted to cash
Accounts receivable: funds owed by customers of the firm
Notes receivable: value of all notes owed to the firm
Inventory: cost of merchandise that the firm has for sale

Plant and equipment: assets that are expected to be used for more than one year (also called **fixed assets** or **plant assets**)

Land: book value of any land owned by the firm
Buildings: book value of any building owned by the firm
Equipment: book value of equipment, store fixtures, furniture, and similar items owned by the firm

The following items, which must be paid by the firm, appear as liabilities on balance sheets.

LIABILITIES

Current liabilities: items that must be paid by the firm within a short period of time, usually one year

Accounts payable: amounts that must be paid to other firms
Notes payable: value of all short-term notes owed by the firm

Long-term liabilities: items that will be paid after one year

Mortgages payable: total due on all mortgages
Long-term notes payable: total of all long-term notes

The difference between the total of all assets and the total of all liabilities is called the **owner's equity** in the firm. That is

$$\text{Owner's equity} = \text{Assets} - \text{Liabilities}$$

Owner's equity is also called **proprietorship**, **net worth**, or, for a corporation, **stockholder's equity**. The formula for owner's equity is equivalent to the following *fundamental formula for accounting.*

$$\text{Assets} = \text{Liabilities} + \text{Owner's equity}$$

The fundamental formula for accounting makes intuitive sense. Your net worth is the value of all of your assets minus your debts.

Objective 2 | ***Prepare a Balance Sheet.*** Now that all the terms have been defined, a balance sheet can be prepared.

Example 1 Preparing a Balance Sheet

The Farmersville Market lists its current cash assets as \$8000. Notes receivable total \$11,000, accounts receivable total \$15,000, and inventory is \$51,000. Plant assets include land worth \$24,000, buildings worth \$22,000, and fixtures worth \$18,000. Current liabilities include notes payable, which are \$8000, and accounts payable, which are \$26,000. Mortgages total \$39,000 and long-term notes payable total \$24,000. Owner's equity is \$52,000. Complete a balance sheet for the market, and use the fundamental formula for accounting to find owner's equity.

Solution To prepare a balance sheet, go through the following steps. Refer to the balance sheet in Figure 17.4.

Step 1. Enter all current assets. On the balance sheet cash is \$8000, notes receivable are \$11,000, accounts receivable are \$15,000, and inventory is \$51,000.

Step 2. Add the current assets. The total in the example is \$85,000.

Step 3. Enter all plant assets. In the example, land is \$24,000, buildings are \$22,000, and fixtures are \$18,000.

Step 4. Add the plant assets of Step 3. In the example, the total is \$64,000.

Step 5. Add the results from Steps 2 and 4. This gives the total value of all assets owned by the firm. Total assets in the example are \$149,000.

Step 6. Enter all current liabilities. In the example, notes payable are \$8000 and accounts payable are \$26,000.

Step 7. Add all current liabilities. The sum in the example is \$34,000.

Step 8. Enter long-term liabilities. In the example, mortgages total \$39,000, and long-term notes payable total \$24,000.

Step 9. Add all long-term liabilities. The sum in the example is \$63,000.

Step 10. Add the results of Steps 7 and 9. This gives the total of all liabilities owed by the firm. The total liabilities in the example is \$97,000.

Step 11. Enter owner's equity. In the example, owner's equity is \$52,000.

Step 12. Add owner's equity to total liabilities. The total in the example is \$97,000 + \$52,000 = \$149,000.

Step 13. Use the fundamental formula of accounting:

$$\text{Assets} = \text{Liabilities} + \text{Owner's equity}$$

In the example, assets = \$149,000, liabilities = \$97,000, and owner's equity = \$52,000. To check, see that

$$\$149{,}000 = \$97{,}000 + \$52{,}000$$

which is correct.

Problem-Solving Hint Always make sure that the total assets equal *total liabilities plus owner's equity*.

Objective 3 *Analyze the Balance Sheet of McDonald's Corporation.*

Example 2 Preparing a Balance Sheet

The assets and liabilities of McDonald's Corporation on December 31, 1998 in millions of dollars are: cash and marketable securities, \$299.2; accounts receivable, \$609.4; inventories, \$77.3; other current assets, \$323.5; subsidiaries and other assets, \$2433.4; property and equipment, \$16,041.6; accounts payable, \$621.3; loans and notes payable, \$686.8; other payables and accrued taxes, \$1189.0; long-term debt \$6188.6, and other liabilities \$1634.0. Complete a balance sheet for the company.

FARMERSVILLE MARKET
BALANCE SHEET
DECEMBER 31

		Assets		
Current Assets				
Cash	(1) Enter these figures	$8,000		
Notes Receivable		$11,000		
Accounts Receivable		$15,000		
Inventory	(2) Add all current assets	$51,000		
	Total Current Assets		$85,000	
Plant Assets				
Land	(3) Enter these figures	$24,000		
Buildings		$22,000		
Fixtures	(4) Add all plant assets	$18,000		
	Total Plant Assets		$64,000	
(5) Add (2) and (4) **TOTAL ASSETS**				$149,000
		Liabilities		
Current Liabilities	(6) Enter these figures			
Notes Payable		$8,000		
Accounts Payable	(7) Add all current liabilities	$26,000		
	Total Current Liabilities		$34,000	
Long-term Liabilities	(8) Enter these figures			
Mortgages Payable		$39,000		
Long-term Notes Payable		$24,000		
	Total Long-term Liabilities (9) Add all long-term liabilities		$63,000	
(10) Add (7) and (9) **Total Liabilities**				$97,000
		Owner's Equity		
Owner's Equity	(11) Enter this figure		$52,000	
(12) Add (11) and (10) **TOTAL LIABILITIES AND OWNER'S EQUITY**	(13) Check			$149,000

Figure 17.4

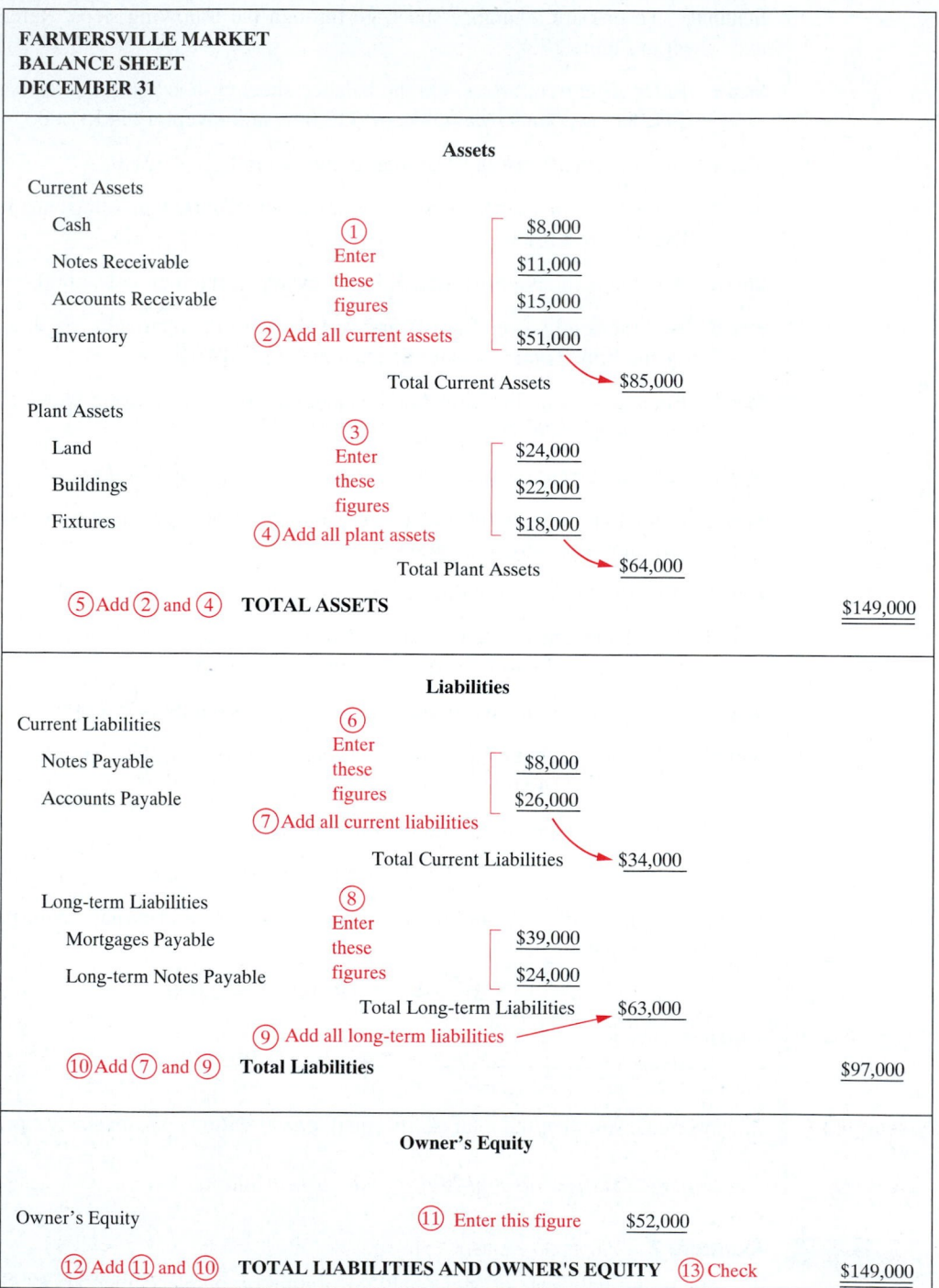

Solution

McDONALD'S CORPORATION
CONSOLIDATED BALANCE SHEET
DECEMBER 31, 1998 (IN MILLIONS OF DOLLARS)

Current Assets		
Cash and equivalents	$ 299.2	
Accounts receivable	609.4	
Inventories	77.3	
Other current assets	323.5	
Total current assets	$ 1,309.4	Sum of all current assets
Other Assets		
Subsidiaries and other assets	$ 2,433.4	
Property and equipment	16,041.6	
Total assets	$19,784.4	Sum of current assets plus all other assets
Current Liabilities		
Accounts payable	$ 621.3	
Loans and notes payable	686.8	
Other payables and accrued taxes	1,189.0	
Total current liabilities	$ 2,497.1	Sum of all current liabilities
Other Liabilities		
Long-term debt	$ 6,188.6	Sum of current liabilities plus all other liabilities
Other liabilities	1,634.0	
Total liabilities	$10,319.7	
Stockholder's Equity	$ 9,464.7	
Total liabilities and equity	$19,784.4	Total assets − total liabilities

NOTE The sum of the liabilities and the stockholder's equity must equal the assets. ■

A balance sheet shows the position of a company *at one point in time*—the figures could be very different a month later. For example, McDonald's Corporation had approximately $299,200,000 in cash and equivalents on December 31, 1998. This balance sheet *does not* provide any information about cash and equivalents held on any date other than that specific date.

17.3 Exercises

Complete the corresponding balance sheets for the following business firms.

1. CONSTRUCTION OF BRIDGES Apple Construction (all figures in millions): fixtures, $28; buildings, $290; land, $466; cash, $273; notes receivable, $312; accounts receivable, $264; inventory, $180; notes payable, $312; mortgages payable, $212; accounts payable, $63; long-term notes payable, $55.

APPLE CONSTRUCTION
BALANCE SHEET FOR DECEMBER 31 (IN MILLIONS)

	Assets		
Current assets			
Cash	______		
Notes receivable	______		
Accounts receivable	______		
Inventory	______		
Total current assets		______	
Plant assets			
Land	______		
Buildings	______		
Fixtures	______		
Total plant assets		______	
Total assets			______
	Liabilities		
Current liabilities			
Notes payable	______		
Accounts payable	______		
Total current liabilities		______	
Long-Term liabilities			
Mortgages payable	______		
Long-term notes payable	______		
Total long-term liabilities		______	
Total liabilities			______
	Owner's Equity		
Owner's equity	______		
Total liabilities and owner's equity			______

2. FENCE COMPANY A-1 Fence Company has land valued at \$8750. Its accounts payable total \$49,230; notes receivable are \$2600; accounts receivable are \$37,820; cash is \$14,800; buildings are \$21,930; notes payable are \$3780; long-term notes payable total \$18,740; mortgages total \$26,330; inventory is \$49,680; and fixtures are \$16,820. Use the fundamental formula of accounting to find owner's equity.

A-1 FENCE COMPANY
BALANCE SHEET FOR DECEMBER 31

	Assets	
Current assets		
Cash	______	
Notes receivable	______	
Accounts receivable	______	
Inventory	______	
Total current assets		______

Plant assets			
Land	______		
Buildings	______		
Fixtures	______		
Total plant assets		______	
Total Assets			______
	Liabilities		
Current liabilities			
Notes payable	______		
Accounts payable	______		
Total current liabilities		______	
Long-term liabilities			
Mortgages payable	______		
Long-term notes payable	______		
Total long-term liabilities		______	
Total Liabilities			______
	Owner's Equity		
Owner's equity	______		
Total Liabilities and Owner's Equity			______

3. Compare a balance sheet to an income statement. What are the similarities and the differences? How is each used? (See Sections 17.1 and 17.3.)

4. Explain how the fundamental formula for accounting is used to check a balance sheet. (See Objective 1.)

5. Prepare a balance sheet for your own family or for your parents.

6. List several reasons why a family needs to have some cash. List several reasons why a corporation needs to have some cash.

17.4 Analyzing the Balance Sheet; Financial Ratios

Objectives

1. ***Compare balance sheets by vertical analysis.***
2. ***Prepare a horizontal analysis.***
3. ***Find the current ratio.***
4. ***Find the acid-test ratio.***
5. ***Find the ratio of net income after taxes to average owner's equity.***
6. ***Find the accounts receivable turnover.***
7. ***Find the average age of accounts receivable.***
8. ***Find the debt-to-equity ratio.***

Objective 1 *Compare Balance Sheets by Vertical Analysis.* A balance sheet can be analyzed in much the same way as an income statement. In a **vertical analysis**, each item on the balance sheet is expressed as a percent of total assets. A **comparative balance sheet** shows the vertical analysis for two different years. You may want to look at a firm's comparative balance sheet before buying their stock or before accepting a job offer.

Example 1 Comparing Balance Sheets

First, do a vertical analysis for both the 1997 and 1998 balance sheets for McDonald's Corporation by calculating each value as a percent of the total assets for the year. Then compare the percents to identify changes from 1997 to 1998.

Solution

McDONALD'S CORPORATION
CONSOLIDATED BALANCE SHEET
DECEMBER 31, 1998 (IN MILLIONS OF DOLLARS)

	1997		1998	
Assets	Amount	Percent	Amount	Percent
Current Assets				
Cash and equivalents	$ 341.4	1.9%	$ 299.2	1.5%
Accounts receivable	$ 483.5	2.7%	$ 609.4	3.1%
Inventories	$ 70.5	0.4%	$ 77.3	0.4%
Other current assets	$ 246.9	1.4%	$ 323.5	1.6%
Total current assets	$ 1,142.3	6.4%	$ 1,309.4	6.6%
Other Assets:				
Subsidiaries and other assets	$ 2,137.8	11.7%	$ 2,433.4	12.3%
Property and equipment	$14,961.4	+82.0%	$16,041.6	+81.1%
Total assets	$18,241.5	100.1%	$19,784.4	100.0%
Liabilities	Amount	Percent	Amount	Percent
Current Liabilities				
Accounts payable	$ 650.6	3.6%	$ 621.3	3.1%
Loans and notes payable	$ 1,293.8	7.1%	$ 686.8	3.5%
Other payables and accrued taxes	$ 1,040.1	5.7%	$ 1,189.0	6.0%
Total current liabilities	$ 2,984.5	16.4%	$ 2,497.1	12.6%
Other liabilities				
Long-term debt	$ 4,834.1	26.5%	$ 6,188.6	31.3%
Other liabilities	$ 1,571.3	8.6%	$ 1,634.0	8.3%
Total liabilities	$ 9,389.9	51.5%	$10,319.7	52.2%
Stockholder's equity	$ 8,851.6	48.5%	$ 9,464.7	47.8%
Total liabilities and owner's equity	$18,241.5	100.0%	$19,784.4	100.0%

NOTE Each value in the table above, including column subtotals, was divided by the respective value for total assets. Thus the amounts add up to the appropriate subtotal, which cannot be said for all of the percent figures. ■

Problem-Solving Hint The tax year for McDonald's Corporation ends on December 31. Other companies use tax years that end *on a different day of the year*, such as April 30 or July 31.

Notice that current liabilities, those expected to be paid within 1 year, decreased from 16.4% of total assets in 1997 to 12.6% of total assets in 1998. This corresponds to a significant decrease in current liabilities.

$$\$2,984,500,000 - \$2,497,100,000 = \$487,400,000 \text{ decrease}$$

Simultaneously, current assets, cash or items that can be converted to cash quickly, increased significantly.

$$\$1,309,400,000 - \$1,142,300,000 = \$167,100,000 \text{ increase}$$

It is a good sign when current liabilities go down and current assets go up; however, long-term debt went up significantly.

$$\$6,188,600,000 - \$4,834,100,000 = \$1,354,500,000$$

Apparently, managers used funds from long-term debt to pay down current liabilities and increase current assets. Managers might do this when they feel that the cost of long-term debt is particularly low, as it was in 1998.

Objective 2 *Prepare a Horizontal Analysis.* Perform a **horizontal analysis** by finding the change, both in dollars and in percent, for each item on the balance sheet from one year to the next. As before, always use the *previous year* as a base when finding the percents.

Example 2 Using Horizontal Analysis

According to the balance sheet for McDonald's Corporation, cash and equivalents on December 31, 1998 were $299.2 (in millions) compared to $341.4 (in millions) on December 31, 1997. This represents a decrease of $299.2 − $341.4 = ($42.2) million. The percent decrease is as follows.

$$\frac{(\$42.2)}{\$341.4} = (12.4\%) \quad \text{Rounded}$$

Similarly, complete a horizontal analysis of the current assets portion of McDonald's Corporation's balance sheet.

Solution

COMPARATIVE ANALYSIS OF CONSOLIDATED BALANCE SHEETS
McDONALD'S CORPORATION (IN MILLIONS OF DOLLARS)

			Increase or (Decrease)	
Current Assets—December 31	**1997**	**1998**	**Amount**	**Percent**
Cash and equivalents	$341.4	$299.2	($42.2)	(12.4%)
Accounts receivable	483.5	609.4	125.9	26.0%
Inventories	70.5	77.3	6.8	9.6%
Other current assets	246.9	323.5	76.6	31.0%
Total Current Assets	$1142.3	$1309.4	$167.1	14.6%

Cash and Equivalents actually *went down* by 12.4% from 1997 to 1998.

Objective 3 *Find the current ratio.* The **current ratio**, also known as **banker's ratio**, is found by dividing current assets by current liabilities.

$$\text{Current ratio} = \frac{\text{Current assets}}{\text{Current liabilities}}$$

Example 3 Finding the Current Ratio

According to the 1998 balance sheet for McDonald's Corporation, current assets were \$1309.4 (in millions) and current liabilities were \$2497.1 (in millions). Find the current ratio.

Solution Use the formula for current ratio.

$$\text{Current ratio} = \frac{\$1309.4}{\$2497.1} = 0.524$$

This ratio is often expressed as .524 to 1 or as .524 : 1. Many lending institutions calculate current ratio in the process of determining if a firm will get a loan. A conservative rule of thumb, not necessarily applicable to all businesses or at all times, is that the current ratio should be at least 2 : 1. A firm with a current ratio much less than 2 : 1 may have an increased risk of financial difficulty and may have difficulty borrowing money. This rule of thumb probably does not apply to McDonald's Corporation since they are able to turn over their inventory so rapidly during the year.

One disadvantage of the current ratio is that inventory is included in current assets. In a period of financial difficulty, a firm might have trouble disposing of the inventory at a reasonable price. Some accountants feel that the "acid test" for a firm's financial health is to consider only **liquid assets**: assets that are either cash or that can be converted to cash quickly, such as securities and accounts and notes receivable.

Objective 4 *Find the Acid-Test Ratio.* The **acid-test ratio**, also called the **quick ratio**, is defined as follows.

$$\text{Acid-test ratio} = \frac{\text{Liquid assets}}{\text{Current liabilities}}$$

NOTE As a general rule, the acid-test ratio should be at least 1 to 1, with the idea that liquid assets are at least enough to cover current liabilities. ∎

Example 4 Finding the Acid-Test Ratio

Find the 1998 acid-test ratio for McDonald's Corporation.

Solution Liquid assets are made up of cash and equivalents (short-term, liquid investments) in addition to accounts receivable. Using data from the 1998 balance sheet:

$$\text{Liquid assets} = \$299.2 + \$609.4 = \$908.6$$

Since current liabilities are \$2497.1, the acid-test ratio follows.

$$\frac{\$908.6}{\$2497.1} = 0.364 \quad \text{Rounded}$$

The rule of thumb probably does not apply to McDonald's Corporation since they are a large, stable corporation with many customers, and a company that can turn their inventory over very quickly.

Objective 5 ***Find the Ratio of Net Income After Taxes to Average Owner's Equity.*** A company with a large amount of capital invested should have a higher net income than a company that has only a relatively small amount invested. To check on this, accountants often find the **ratio of net income after taxes to average owner's equity**. The average owner's equity is found by adding the owner's equity at the beginning and end of the year and dividing by 2.

$$\text{Average owner's equity} = \frac{\text{Owner's equity at beginning} + \text{Owner's equity at end}}{2}$$

Then the ratio of net income after taxes to average owner's equity is found as follows.

$$\text{Ratio of net income after taxes to average owner's equity} = \frac{\text{Net income after taxes}}{\text{Average owner's equity}}$$

NOTE The ratio of net income after taxes to average owner's equity is also called "return on equity."

Example 5 Finding the Return on Average Equity

Find the 1998 ratio of net income after taxes to average owner's equity for McDonald's Corporation.

Solution At the end of 1997, which is the same as the beginning of 1998, the firm had a stockholder's equity of \$8851.6 (in millions) and at the end of 1998, the stockholder's equity was \$9464.7. The average owner's equity was

$$\frac{\$8851.6 + \$9464.7}{2} = \$9158.2 \quad \text{Rounded}$$

Using the net income after taxes for 1998 from the income statement in Section 17.1 (\$1550.1) results in the following ratio of net income after taxes to average owner's equity.

$$\frac{\$1550.1}{\$9158.2} = 16.9\% \quad \text{Rounded}$$

This ratio should be at least as much as the interest paid on savings accounts by banks. Otherwise the capital represented by these assets should be deposited in a bank account. After all, savings accounts should have less risk than an investment in a company. Increased risk should bring a *higher return* to the investor than that of a risk-free savings account. Notice that this ratio is considerably higher than savings account yields!

NOTE The ratio of net income after taxes to average owner's equity is the only ratio of the three we have looked at that requires you to look at both the income statement and the balance sheet.

Objective 6 *Find the Accounts Receivable Turnover.* The accounts receivable of a firm represent credit sales—goods sold on credit and later billed to the customer. The **accounts receivable turnover** is an indication of how fast the firm is collecting its bills. If this ratio *starts to decline*, then the firm may well need to be more aggressive in collecting its receivables from customers. By collecting receivables promptly, the firm will need to borrow less money, thus cutting its interest charges.

To find the accounts receivable turnover ratio, first find the average accounts receivable.

$$\text{Average accounts receivable} = \frac{\text{Accounts receivable at beginning} + \text{Accounts receivable at end}}{2}$$

Then,

$$\text{Accounts receivable turnover} = \frac{\text{Net sales}}{\text{Average accounts receivable}}$$

Example 6 Finding the Accounts Receivable Turnover

Find the accounts receivable turnover for McDonald's Corporation for 1998.

Solution Accounts receivable at the end of 1997 were the same as those at the beginning of 1998 or \$483.5. Accounts receivable at the end of 1998 were \$609.4.

$$\text{Average accounts receivable} = \frac{\$483.5 + \$609.4}{2} = \$546.5 \quad \text{In millions}$$

From the income statement on page 628, net sales for 1998 were \$12,421.4.

$$\text{Accounts receivable turnover} = \frac{\$12{,}421.4}{\$546.5} = 22.7$$

This ratio should be watched over time. If the ratio were to decline it could indicate some fundamental problems within the company.

Objective 7 *Find the Average Age of Accounts Receivable.* As mentioned, a firm must collect its accounts receivable promptly to minimize its own need to borrow money. To tell how well accounts are being collected, the firm can find the **average age of accounts receivable**, or the average number of days needed to collect its receivables. This is found by this formula.

$$\text{Average age of accounts receivable} = \frac{365}{\text{Accounts receivable turnover}}$$

Example 7 Finding the Average Age of Accounts Receivable

Find the average age of accounts receivable for McDonald's Corporation during 1998.

Solution Divide 365 by the accounts receivable turnover found in Example 6.

$$\text{Average age of accounts receivable} = \frac{365}{22.7} = 16.1 \text{ days} \quad \text{Rounded}$$

Again, this ratio is one that should be watched over time. A decline in this ratio would suggest that the company is improving their collection of receivables.

Objective 8 *Find the Debt-to-Equity Ratio.* Companies borrow money to expand and take advantage of business opportunities. This is fine as long as total debt *does not* become excessive. One common measure used to see if debt is reasonable is the **debt-to-equity ratio** which is the ratio of all liabilities to all owner's equity.

$$\text{Debt-to-equity ratio} = \frac{\text{Current liabilities} + \text{Long-term liabilities}}{\text{Owner's equity}}$$

Example 8 Finding the Debt-to-Equity Ratio

Find the debt-to-equity ratio for McDonald's Corporation at the end of 1998.

Solution Divide the sum of the current liabilities and all other liabilities by the owner's equity.

$$\text{Debt-to-equity ratio} = \frac{\$2497.1 + \$7822.6}{\$9464.7} = 1.09 \text{ or } 109\% \quad \text{Rounded}$$

NOTE An acceptable level for debt-to-equity ratios varies drastically from industry to industry. For some, a ratio of 150% is reasonable, while for others a 25% ratio might make the firm's bankers nervous. Companies that can turn inventory into cash very quickly, such as McDonald's Corporation, are said to be **liquid**. These types of companies can usually justify a higher debt-to-equity ratio than other companies. ■

17.4 Exercises

Complete the following.

1. RUBBER SUPPLY COMPANY Complete this balance sheet using vertical analysis. Round to the nearest tenth of a percent.

INTERSTATE RUBBER SUPPLY
COMPARATIVE BALANCE SHEET

	This Year		Last Year	
	Amount	**Percent**	**Amount**	**Percent**
Assets				
Current assets				
Cash	$52,000	______	$42,000	______
Notes receivable	$8,000	______	$6,000	______
Accounts receivable	$148,000	______	$120,000	______
Inventory	$153,000	______	$120,000	______
Total current assets	______	______	______	______
Plant assets				
Land	$10,000	______	$8,000	______
Buildings	$14,000	______	$11,000	______
Fixtures	$15,000	______	$13,000	______
Total plant assets	______	______	______	______
Total assets	______	100.0%	______	100.0%

	This Year		Last Year	
	Amount	Percent	Amount	Percent
Liabilities				
Current liabilities				
Accounts payable	$3,000	______	$4,000	______
Notes payable	$201,000	______	$152,000	______
Total current liabilities	______	______	______	______
Long-term liabilities				
Mortgages payable	$20,000	______	$16,000	______
Long-term notes payable	$58,000	______	$42,000	______
Total long-term liabilities	______	______	______	______
Total liabilities	______	______	______	______
Owner's equity	$118,000	______	$106,000	______
Total liabilities and owner's equity	______	______	______	______

2. **RUBBER SUPPLY COMPANY** Complete the horizontal analysis for a portion of the balance sheet for Interstate Rubber Supply.

INTERSTATE RUBBER SUPPLY
HORIZONTAL ANALYSIS

	This Year	Last Year	Increase (Decrease)	
			Amount	Percent
Assets				
Current assets				
Cash	$52,000	$42,000	______	______
Notes receivable	$8,000	$6,000	______	______
Accounts receivable	$148,000	$120,000	______	______
Inventory	$153,000	$120,000	______	______
Total current assets	$361,000	$288,000	______	______
Plant assets				
Land	$10,000	$8,000	______	______
Buildings	$14,000	$11,000	______	______
Fixtures	$15,000	$13,000	______	______
Total plant assets	$39,000	$32,000	______	______
Total assets	$400,000	$320,000	______	______

In Exercises 3–6, find (a) the current ratio and (b) the acid-test ratio. Round each ratio to the nearest hundredth. (c) Do the ratios suggest that the company is financially healthy using the guidelines given in the text?

3. **INTERSTATE RUBBER SUPPLY** Use data from Exercises 1 and 2.
4. **MUSIC STORE** Virginia Music has current assets of $216,750, current liabilities of $213,000, cash of $25,400, notes and accounts receivable of $42,500 and an inventory of mostly electric pianos and various electronic equipment valued at $148,850.

5. CADILLAC DEALER Wagner Cadillac has current assets of $2,210,350, current liabilities of $1,232,500, total cash of $480,500, notes and accounts receivable of $279,050, and an inventory of $1,450,800.

6. MUSIC SHOP Best Music Shoppe has cash of $48,000; current liabilities total $72,500; notes and accounts receivable of $3800; and inventory valued at $42,800.

A portion of a comparative balance sheet is shown below. First complete the chart. Then find the current ratio and the acid-test ratio for the indicated year. Round each ratio to the nearest hundredth.

7. This year

8. Last year

	This Year		Last Year	
	Amount	**Percent**	**Amount**	**Percent**
Current assets				
Cash	$12,000	________	$15,000	________
Notes Receivable	4,000	________	6,000	________
Accounts Receivable	22,000	________	18,000	________
Inventory	26,000	________	24,000	________
Total Current Assets	64,000	80.0%	63,000	84.0%
Total Plant and Equipment	16,000	________	12,000	________
Total assets	________	100.0%	________	100.0%
Total current liabilities	$30,000	________	$25,000	________

Find the debt-to-equity ratio for each of the following firms. Round to the nearest hundredth.

9. INTERSTATE RUBBER SUPPLY Use data from Exercises 1 and 3 for this year.

10. VIRGINIA MUSIC Use data from Exercise 4; owner's equity of $265,000 and long-term liabilities of $174,300.

11. WAGNER CADILLAC Use data from Exercise 5; owner's equity of $1,280,000 and long-term debt of $650,000.

12. BEST MUSIC SHOPPE Use data from Exercise 6; owner's equity of $64,500 and long-term debt of $8000.

13. Explain the meaning to the manager of a firm of an increase in the age of accounts receivable. (See Objective 2.)

14. List three ratios that can be used to analyze a balance sheet and the purpose of each ratio.

Find the ratio of net income after taxes to average owner's equity for the following firms. Round to the nearest tenth of a percent.

15. SMALL INTERNATIONAL AIRLINE The stockholders equity in TNA Airline, a small international airline that uses small airplanes to move freight to and from Mexico, is $845,000 at the beginning of the year and $928,500 at the end of the year. Net income after taxes for the year was $54,400.

16. **PIPELINE COMPANY** TransCanada Pipe is an international company that does business both in Canada and in the United States. In thousands of dollars, owner's equity at the beginning of the year was $48,340 and $62,842 at the end of the year. Net income after taxes for the year was $6838.

Find the accounts receivable turnover rate and the average age of the accounts receivable for the following firms. Round each rate to the nearest tenth.

17. **MEN'S FACTORY WAREHOUSE** Accounts receivable at the beginning of the year of $320,000; accounts receivable at the end of the year, $450,000; and net sales $6,500,000.

18. **PLATES AND PLANTS GALORE** Accounts receivable at the beginning of the year of $110,000; accounts receivable at the end of the year, $80,000; and net sales of $875,000.

19. Explain why the acid-test ratio is a better measure of the financial health of a firm than the current ratio. (See Objectives 3 and 4.)

20. Explain why increased risk requires a higher return on investment. (See Objective 8.)

Chapter 17 Quick Review

Chapter Terms

Review the following terms to test your understanding of the chapter. For any terms you do not know, refer to the page number found next to that term.

accounts receivable turnover [p. 646]
acid-test ratio [p. 644]
assets [p. 635]
average age of accounts receivable [p. 646]
balance sheet [p. 635]
banker's ratio [p. 644]
comparative balance sheet [p. 641]
comparative income statement [p. 628]
consolidated [p. 622]
cost of goods sold [p. 621]
current assets/liabilities [p. 635, 636]
current ratio [p. 644]
debt-to-equity ratio [p. 647]
ending inventory [p. 623]
fixed assets [p. 636]
gross profit [p. 621]
gross sales [p. 621]
horizontal analysis [p. 631]
income statement [p. 621]
initial inventory [p. 623]
liabilities [p. 636]
liquid assets [p. 644]
long-term assets/liabilities [p. 635, 636]
net income [p. 621]
net income after taxes [p. 621]
net income before taxes [p. 621]
net sales [p. 621]
net worth [p. 636]
operating expenses [p. 621]
overhead [p. 621]
owner's equity [p. 636]
plant and equipment [p. 636]
plant assets [p. 636]
proprietorship [p. 636]
quick ratio [p. 644]
ratio of net income after taxes to average owner's equity [p. 645]
returns [p. 621]
short-term assets/liabilities [p. 635, 636]
stockholder's equity [p. 636]
vertical analysis [p. 628]

Concepts / Examples

17.1 Gross profit, net income before taxes, and net income after taxes

Gross profit = Net sales − Cost of goods sold

Net income before taxes = Gross profit − Operating expenses

Net income after taxes = Net income before taxes − income taxes

Example: Cost of goods sold, \$156,000; operating expenses, \$35,000; net sales, \$210,000; taxes, \$4000. Find gross profit, net income before taxes, and net income after taxes.

$$\text{Gross profit} = \$210{,}000 - \$156{,}000 = \$54{,}000$$

$$\text{Net income before taxes} = \$54{,}000 - \$35{,}000 = \$19{,}000$$

$$\text{Net income after taxes} = \$19{,}000 - \$4000 = \$15{,}000$$

17.2 Finding the percent of net sales of individual items

Subtract returns from gross sales to determine net sales. Then divide the value of an item by net sales to obtain percent of net sales.

Example: Express the following items for Mr. Bill's Appliance Store as percents of net sales: gross sales, \$340,000; returns, \$15,000; cost of goods sold, \$210,000; wages, \$19,000; gross profit.

$$\text{Net sales} = \$340{,}000 - \$15{,}000 = \$325{,}000$$

$$\text{Gross profit} = \$325{,}000 - \$210{,}000 = \$115{,}000$$

$$\text{Percent gross sales} = \frac{\$340{,}000}{\$325{,}000} = 104.6\%$$

$$\text{Percent returns} = \frac{\$15{,}000}{\$325{,}000} = 4.6\%$$

$$\text{Percent cost of goods sold} = \frac{\$210{,}000}{\$325{,}000} = 65\%$$

$$\text{Percent wages} = \frac{\$19{,}000}{\$325{,}000} = 5.8\%$$

$$\text{Percent gross profit} = \frac{\$115{,}000}{\$325{,}000} = 35.4\%$$

17.2 Comparing income statements to published charts

In one chart, list the percents of net sales for each item from a published chart and a particular company.

Prepare a vertical analysis of Monica's Shoe Store.

Cost of Goods	Gross Profit	Wages	
65%	35%	6.4%	Monica's Shoes
60.3%	39.7%	10.3%	From chart

Her cost of goods is high, gross profit is low, and wages are low.

17.2 Preparing a horizontal analysis

List last year's and this year's values for each item. Then calculate the amount of the increase or decrease for each item and express as a percent of previous year.

The results of a horizontal analysis of the portion of a business is given. Calculate the percent increases or decreases in each item.

	This Year	Last Year	Increase (Decrease) Amount	Percent
Gross sales	$735,000	$700,000	$35,000	5%
Returns	5,000	10,000	(5,000)	(50%)
Net sales	730,000	690,000	40,000	5.8%
Cost of goods sold	530,000	540,000	(10,000)	(1.9%)
Gross profit	200,000	150,000	50,000	33.3%

17.3 Fundamental formula for accounting

$$\text{Assets} = \text{Liabilities} + \text{Owner's equity}$$

or

$$\text{Assets} - \text{Liabilities} = \text{Owner's equity}$$

Stemco Tubing has assets of $842,300 and liabilities of $625,100. Find the owner's equity.

$$\$842{,}300 - \$625{,}100 = \$217{,}200$$

17.4 Current ratio; acid-test ratio; ratio of net income after taxes to the average owner's equity

$$\text{Current ratio} = \frac{\text{Current assets}}{\text{Current liabilities}}$$

$$\text{Acid-test ratio} = \frac{\text{Liquid assets}}{\text{Current liabilities}}$$

$$\text{Ratio} = \frac{\text{Net income after taxes}}{\text{Average owner's equity}}$$

The Circle Towne Agency has current assets of $250,000, current liabilities of $110,000, cash of $45,000, and accounts receivable of $80,000. The agency had owner's equity of $140,000 at the beginning of the year, and $180,000 at the end of the year. The net income after taxes was $25,000. Calculate the following ratios.

$$\text{Current ratio} = \frac{\$250{,}000}{\$110{,}000} = 2.27$$

$$\text{Acid-test ratio} = \frac{\$45{,}000 + \$80{,}000}{\$110{,}000} = 1.14 \quad \text{Rounded}$$

$$\text{Average owner's equity} = \frac{\$140{,}000 + \$180{,}000}{2} = \$160{,}000$$

Ratio of net income after taxes to average owner's equity:

$$\frac{\$25{,}000}{\$160{,}000} = 0.156 = 15.6\%$$

17.4 Accounts receivable turnover

$$\text{Accounts receivable turnover} = \frac{\text{Net sales}}{\text{Average accounts receivable}}$$

A firm has net sales of \$793,750; accounts receivable of \$50,000 at the beginning of the year; \$75,000 at the end of the year; current liabilities of \$15,000; long-term liabilities of \$10,000; and owner's equity of \$80,000. Find the accounts receivable turnover, the average age of accounts receivable, and the debt-to-equity ratio.

$$\text{Average accounts receivable} = \frac{\$50{,}000 + \$75{,}000}{2} = \$62{,}500$$

$$\text{Accounts receivable turnover} = \frac{\$793{,}750}{\$62{,}500} = 12.7$$

$$\text{Average age of accounts receivable} = \frac{365}{12.7} = 28.7 \text{ days}$$

$$\text{Debt-to-equity ratio} = \frac{\$15{,}000 + \$10{,}000}{\$80{,}000} = 0.31$$

Average age of accounts receivable

$$\text{Average age of accounts receivable} = \frac{365}{\text{Accounts receivable turnover}}$$

Debt-to-equity ratio

$$\text{Debt-to-equity ratio} = \frac{\text{Current liabilities} + \text{Long-term liabilities}}{\text{Owner's equity}}$$

Chapter 17 Review Exercises

Find the gross profit and net income before taxes for each of the following. **[17.1]**
The answer section includes answers to all Review Exercises.

	Cost of Goods	Operating Expenses	Net Sales
1.	$124,800	$89,200	$312,200
2.	$379,520	$124,800	$643,250
3.	$300,900	$98,400	$442,500
4.	$606,520	$212,300	$842,400

Find the cost of goods sold for each of the following firms. **[17.1]**

	Initial Inventory	Cost of goods Purchased	Freight	Final Inventory
5.	$215,400	$422,000	$26,300	$247,100
6.	$125,400	$94,300	$8,200	$101,400
7.	$84,000	$52,400	$4,300	$98,000
8.	$184,200	$245,000	$18,300	$165,400

Find the net income after taxes for each of the following. Then find the ratio of net income after taxes to average owner's equity. **[17.1]**

9. Blues Recording Studio has net sales of $660,500; no inventory and thus no cost of goods sold; total expenses of $412,900; average owner's equity of $340,000; and taxes of $58,800.

10. Buy Rite Hardware has net sales of $894,200; cost of goods sold of $462,800; expenses of $304,100; average owner's equity of $389,700; and taxes of $36,700.

Complete the accompanying income statements for the following firms. **[17.1]**

11. Lori's Boutique had gross sales of $175,000 last year, with returns of $8000. The inventory on January 1 was $44,000. A total of $126,000 worth of goods was purchased with freight of $2000. The inventory on December 31 was $52,000. Salaries and wages were $9000, rent was $4000, advertising was $1500, utilities were $1000, taxes on inventory and payroll totaled $2000, and miscellaneous expenses totaled $3000.

LORI'S BOUTIQUE
INCOME STATEMENT
FOR THE YEAR ENDING DECEMBER 31

Gross sales			________	
Returns			________	
Net sales				________
Inventory, January 1		________		
Cost of goods purchased	________			
Freight	________			
Total cost of goods purchased		________		
Total of goods available for sale		________		
Inventory, December 31		________		

Cost of goods sold				
Gross profit				
Expenses				
Salaries and wages				
Rent				
Advertising				
Utilities				
Taxes on inventory, payroll				
Miscellaneous expenses				
Total expenses				
Net income before taxes				

12. The Guitar Warehouse had gross sales of $2,215,000 with returns of $26,000. The inventory on January 1 was $215,000. A total of $1,123,000 worth of goods was purchased with freight of $4000. The inventory on December 31 was $265,000. Salaries and wages were $154,000, rent was $59,000, advertising was $11,000, utilities were $12,000, taxes on inventory and payroll totaled $10,000, and miscellaneous expenses totaled $9000. Income taxes for the year were $242,300.

THE GUITAR WAREHOUSE
INCOME STATEMENT
FOR THE YEAR ENDING DECEMBER 31

Gross sales				
Returns				
Net sales				
Inventory, January 1				
Cost of goods purchased				
Freight				
Total cost of goods purchased				
Total of goods available for sale				
Inventory, December 31				
Cost of goods sold				
Gross profit				
Expenses				
Salaries and wages				
Rent				
Advertising				
Utilities				
Taxes on inventory, payroll				
Miscellaneous expenses				
Total expenses				
Net income before taxes				
Income taxes				
Net income				

Prepare a vertical analysis for each of the following firms. Round to the nearest tenth of a percent. **[17.2]**

	Cost of Goods Sold	Operating Expenses	Net Sales
13.	$485,800	$104,300	$812,200
14.	$813,200	$387,100	$1,329,400

Complete the following chart. Express each item as a percent of net sales and then write in the appropriate average percent from Table 17.1. Round each percent to the nearest tenth of a percent. **[17.2]**

15. ANDY'S STEAK HOUSE

	Amount	Percent	Percent from Table 17.1
Net sales	$300,000	100%	100%
Cost of goods sold	$125,000	______	______
Gross profit	$175,000	______	______
Wages	$72,000	______	______
Rent	$12,000	______	______
Advertising	$5,700	______	______
Total expenses	$123,000	______	______
Net income	$52,000	______	______

Complete the following balance sheet. **[17.3]**

16. Gaskets, Inc. manufactures gaskets for gasoline engines. They have been under intense financial pressure recently due to foreign competition. The firm has notes payable of $410,000; accounts receivable of $460,000; cash is $240,000; long-term notes payable of $194,000; buildings worth $260,000; inventory of $225,000; fixtures are $48,000; notes receivable of $180,000; land worth $180,000; accounts payable of $882,000; and mortgages payable of $220,000. Use the fundamental formula of accounting to find owner's equity.

GASKETS, INC.
BALANCE SHEET
FOR DECEMBER 31

Assets

Current assets		
Cash	______	
Notes receivable	______	
Accounts receivable	______	
Inventory	______	
Total current assets		______
Plant assets		
Land	______	
Buildings	______	
Fixtures	______	
Total plant assets		______
Total Assets		______

	Liabilities		
Current liabilities			
Notes payable	______		
Accounts payable	______		
Total current liabilities		______	
Long-term liabilities			
Mortgages payable	______		
Long-term notes payable	______		
Total long-term liabilities		______	
Total liabilities			______
	Owner's Equity		
Owner's equity	______		
Total Liabilities and Owner's Equity			______

***Calculate the current ratio, acid-test ratio, and debt-to-equity ratio for each of the following. Round to the nearest hundredth.* [17.4]**

	Current Assets	Current Liabilities	Long-Term Liabilities	Owner's Equity	Liquid Assets
17.	$342,000	$260,000	$140,000	$225,000	$120,000
18.	$95,000	$115,000	$85,000	$48,000	$5,000
19.	$160,000	$205,000	$0	$185,000	$145,000

***Find the accounts receivable turnover rate and the average age of the accounts receivable in each of the following firms. Round to the nearest tenth.* [17.4]**

20. Accounts receivable at beginning of year $875,400
Accounts receivable at end of year $962,300
Net Sales $4,612,000

21. Accounts receivable at beginning of year $126,800
Accounts receivable at end of year $92,400
Net Sales $942,500

22. Complete the following comparative balance sheet. [17.4]

	This Year		Last Year	
	Amount	Percent	Amount	Percent
Current assets				
Cash	$28,000	______	$22,000	______
Notes receivable	$12,000	______	$15,000	______
Accounts receivable	$39,000	______	$31,500	______
Inventory	$22,000	______	$20,000	______
Total current assets	$101,000	______	$88,500	______
Total plant and equipment	$48,000	______	$16,000	______
Total Assets	______	______	______	______
Total current liabilities	$38,000	______	$36,000	______

Chapter 17 Summary Exercise
Owning Your Own Small Business

Tom Walker wants to expand his bicycle shop and has gone to a bank for a loan. The commercial loan officer asks Walker for his most recent income statement and balance sheets based on the following data.

Gross sales	$212,000	Salaries and wages	$37,000
Returns	$12,500	Rent	$12,000
Inventory on January 1	$44,000	Advertising	$2,000
Cost of goods purchased	$75,000	Utilities	$3,000
Freight	$8,000	Taxes on inventory, payroll	$7,000
Inventory on December 31	$26,000	Miscellaneous expenses	$4,500

(a) Prepare an income statement.

WALKER BICYCLE SHOP
INCOME STATEMENT
YEAR ENDING DECEMBER 31

Gross sales			________
Returns			________
Net sales			________
Inventory, January 1		________	
Cost of goods purchased	________		
Freight	________		
Total cost of goods purchased		________	
Total of goods available for sale		________	
Inventory, December 31		________	
Cost of goods sold			________
Gross profit			________
Expenses			
Salaries and wages		________	
Rent		________	
Advertising		________	
Utilities		________	
Taxes on inventory, payroll		________	
Miscellaneous expenses		________	
Total expenses			________
Net income before taxes			________

(b) Express the following items as a percent of net sales. Round to tenths of a percent.

Gross sales	________	Salaries and wages	________
Returns	________	Rent	________
Cost of goods sold	________	Utilities	________

(c) After the year is completed, Walker has $62,000 in cash, $2500 in notes receivable, $8200 in accounts receivable, and $26,000 in inventory. He has land worth $7600, buildings valued at $28,000, and fixtures worth $13,500. He also has $4500 in notes payable and $27,000 in accounts payable, mortgages for $15,000, long-term notes payable of $8000, and owner's equity of $93,300. Prepare a balance sheet.

WALKER BICYCLE SHOP
BALANCE SHEET
DECEMBER 31

Assets			
Current assets			
Cash	________		
Notes receivable	________		
Accounts receivable	________		
Inventory	________		
Total current assets		________	
Plant assets			
Land	________		
Buildings	________		
Fixtures	________		
Total plant assets		________	
Total assets			________
Liabilities			
Current liabilities			
Notes payable	________		
Accounts payable	________		
Total current liabilities		________	
Long-term liabilities			
Mortgages payable	________		
Long-term notes payable	________		
Total long-term liabilities		________	
Total liabilities			________
Owner's Equity			
Owner's equity	________		
Total liability and owner's equity			________

(d) Find the current ratio and the acid-test ratio for Walker's business. Round to nearest hundredth.

(e) If you were the commercial loan officer, would you approve Walker's requested loan? Why or why not?

Net Assets Business on the Internet

McDonald's

Statistics

- 1998: Revenues of $12 billion
- 267,000 employees
- 24,500 restaurants in 115 countries

In 1954, a 52-year-old salesman named Ray Kroc mortgaged his home and invested his entire life savings to become the exclusive distributor of a five-spindled milk shake maker called the Multimixer. Hearing about a McDonald's hamburger stand in California that was running eight Multimixers at a time, he packed up his car and headed west. Kroc made a deal with the owners of the restaurant in California and opened his first restaurant in 1955.

McDonald's has a global reputation for quick service, quality food at low prices, and cleanliness, as indicated by the 15,000 people who lined up when the first McDonald's opened in Kuwait. Even so, the company serves less than 1% of the world's population. McDonald's is also quite active in helping others, as shown by their aid to families of critically ill children through Ronald McDonald House.

1. Given an annual revenue of $12 billion and 24,500 restaurants around the world, find the average sales per restaurant. Why might this information be important to someone thinking of buying a McDonald's restaurant?

2. Peter Voight owns a McDonald's restaurant with current assets of $124,500, total assets of $289,350, current liabilities of $82,654, and total liabilities of $147,320. Find the current ratio.

3. Assume McDonald's currently feeds 1% of the world's population with a resulting annual revenue of $12 billion. What would McDonald's revenue be if they expanded their market share to the point of serving 1.25% of the world's population? What assumptions did you make to arrive at your answer?

18 Securities and Distribution of Profit and Overhead

The graph on the left in Figure 18.1 shows that people have more in their 401(k) retirement plans now than they did in 1994. The graph on the right shows that a higher percent of the retirement fund are in equities (stocks). This chapter discusses the way profits are distributed and will help you understand more about publicly held companies and stock markets.

Why are people investing much of their 401(k) money in equity funds and stocks?

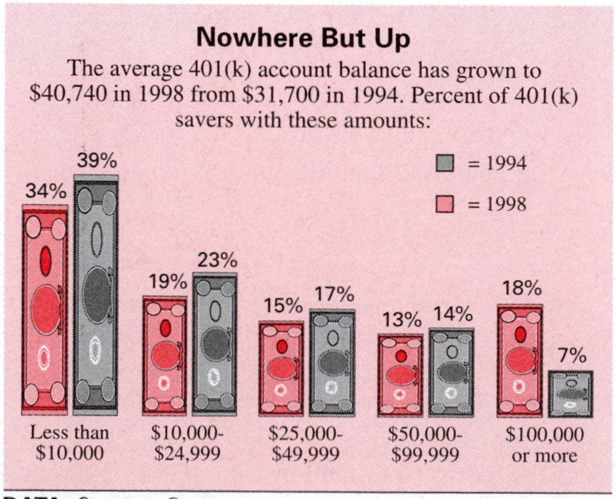

DATA: Spectrum Group

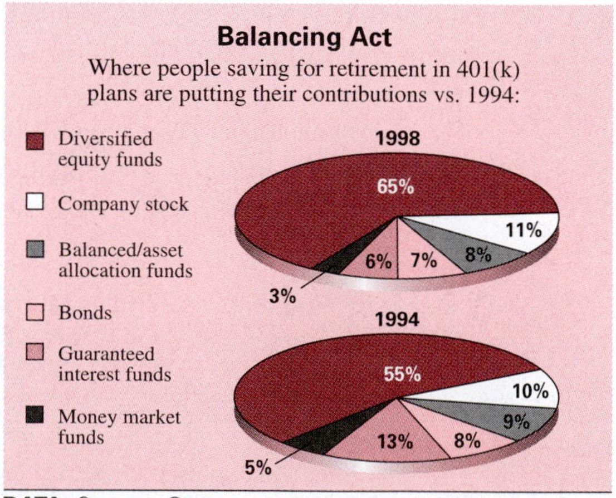

DATA: Spectrum Group

Figure 18.1

The distribution of profits depends on the way a company is formed. If a business is a **sole proprietorship**, that is, if it is owned by one person, then all the profits go to that person. Sole proprietorships are most common among the smallest businesses, such as a small retail store, restaurant, hair salon, and the like. With a typical sole proprietorship, there is no division of assets between personal assets and business assets, so that personal property such as an automobile, could be **attached** (seized under court order) to pay debts associated with the business. Business assets such as inventory, could also be attached to pay personal debts.

A **partnership** is a business formed by two or more parties. Since partnerships involve the resources of several people, they are common for somewhat larger businesses, such as a small factory or perhaps a small chain of retail stores. Unless they have a prior agreement setting up a different method, profits in a partnership are divided equally among the partners.

Partnerships are often set up by individuals with different interests and abilities—one partner might actually run the business on a day-to-day basis, while the other partner makes a substantial financial investment but does not actively take part in the business. A partner who makes only a financial investment is called a **silent partner**.

The law often says that partners are responsible "jointly and separately" for partnership debts, which means that a partner with money may have to pay all debts of the partnership. The methods of dividing the profits in a partnership are discussed in Section 18.4.

To get around the problem of the lack of protection for personal assets, most larger businesses are set up as **corporations**. The people investing money in a corporation have **limited liability**—they can lose no more money than they have invested in the corporation. Distribution of profits in a corporation is discussed in the first section.

18.1 Distribution of Profits in a Corporation

Objectives

1. *Compare preferred and common stocks.*
2. *Distribute profits to shareholders and calculate dividend per share.*
3. *Find earnings per share.*

A corporation is set up with money, or **capital**, raised by selling shares of **stock**. A share of stock represents partial ownership in a corporation. If 1 million shares of stock are sold to establish a new firm, the owner of one share will hold 1/1,000,000 of the corporation. For a publicly held corporation, the ownership of stock is shown by **stock certificates**, such as the one shown in Figure 18.2.

In most states, corporations are required to have an **annual meeting**. At this meeting, open to all owners of the stock (or **stockholders**), the management of the firm is open to questions from stockholders. The stockholders also elect a **board of directors**—a group of people who represent the stockholders. The board of directors hires the **executive officers** of the corporation, such as the president, vice presidents, and so on. The board of directors also typically authorize the distribution of some of the profits in the form of dividends. Dividends are covered in more detail later in this section.

Objective 1 *Compare Preferred and Common Stocks.* Corporations normally issue two types of stock, **preferred stock** and **common stock**. As the name implies, preferred stock has certain rights over common stock. Owners of preferred stock must be paid dividends be-

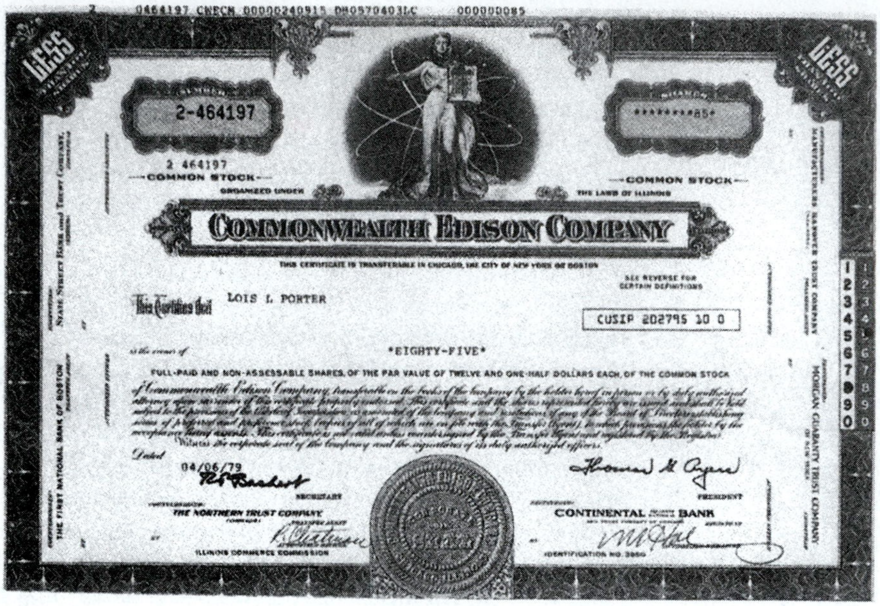

Figure 18.2

fore any dividends can be paid to owners of common stock. In the case of bankruptcy of the corporation, owners of preferred stock must be paid off completely before common stock owners get any money. (Preferred stockholders receive no money, however, until all other debts of the corporation are paid.)

Each share of preferred stock has a **par value**—the amount printed on the stock certificate that must be paid to its owner before common stockholders receive any money. Each share also has a stated **dividend**, often given as a percent of par value. These dividends must be paid by the company to the preferred stockholders before any dividends can be paid to the common stockholders. (However, there is no guarantee that any dividends will be paid at all.)

Sometimes, preferred stock is given additional features to make it more attractive to potential buyers. For example, the stock may be **cumulative preferred stock**, which means that any dividends not paid in the past must be paid before common stockholders receive any money. The stock might also be **convertible preferred stock**, which means that one share is convertible into a stated number of shares of common stock at some future date.

Holders of preferred stock usually are not able to vote at the annual meeting of the corporation. Also, most preferred stock is **nonparticipating**, which means the corporation will never pay dividends above the stated rate. (Holders of **participating** preferred stock could share in any good success of the corporation by an increase in the dividend.)

NOTE *Common stock carries no guarantees.* Its holders are last in line when profits are distributed or when the corporation is dissolved. However, common stockholders are able to vote at the annual meeting. Common stock may or may not have a par value, but since nothing is guaranteed, par value for common stock is of little importance. ■

Objective 2 | ***Distribute Profits to Shareholders and Calculate Dividend Per Share.*** The next examples show how the profits of a corporation might be distributed.

Example 1 Calculating Dividend per Share

Thornton Electronics had a net income of \$1,200,000 last year. The board of directors decides to reinvest \$500,000 in the business and distribute the remaining \$700,000 to stockholders. The company has 40,000 shares of \$100 par value, 6% preferred stock, and 350,000 shares of common stock. Find (a) the amount paid to holders of preferred stock, and (b) the amount per share given to holders of common stock.

Solution

(a) Each share of preferred stock has a par value of \$100 and pays a dividend of 6%. The dividend per share is

$$\text{Dividend per share} = \$100 \times 6\% = \$100 \times 0.06 = \$6$$

A dividend of \$6 must be paid for each share of preferred stock. Since there are 40,000 shares of preferred, a total of

$$\$6 \times 40{,}000 = \$240{,}000$$

will be paid to owners of preferred shares.

(b) A total of \$700,000 is available for stockholders, with \$240,000 going to the owners of preferred shares, leaving

\$700,000	Total
− 240,000	Preferred
\$460,000	

available for the common stockholders. There are 350,000 shares of common stock outstanding, with each share being paid a dividend of

$$\frac{\$460{,}000}{350{,}000} = \$1.31 \text{ per share}$$

Example 2 Finding Cumulative Dividends

Due to a steep drop in the price of oil, Alamo Energy paid no dividend last year. The company has done much better this year and the board of directors has set aside \$175,000 for the payment of dividends. The company has outstanding 12,500 shares of cumulative preferred stock having par value of \$50, with an 8% dividend. The company also has 40,000 shares of common stock. What dividend will be paid to the owners of each type of stock?

Solution The dividend per share of preferred stock is

$$\$50 \text{ par value} \times \underset{\text{Dividend rate}}{8\%} = \$4$$

Dividends have not been paid for 2 years (last year and this year), so each share of preferred stock must be paid

$$\$4 \text{ per share} \times 2 \text{ years} = \$8$$

before any dividends can be paid to holders of common stock. Since there are 12,500 shares of preferred stock outstanding, a total of

$$\$8 \times 12{,}500 = \$100{,}000$$

must be paid to the owners of the preferred stock. This leaves

$$\begin{array}{rl} \$175{,}000 & \text{Total} \\ -\ 100{,}000 & \text{Preferred} \\ \hline \$75{,}000 & \end{array}$$

to be divided among owners of common stock. Since there are 40,000 shares of common stock, each share will be paid a dividend of

$$\frac{\$75{,}000}{40{,}000} = \$1.88 \quad \text{Rounded}$$

NOTE In Example 2, the dividend for owners of preferred stock was paid for each of the last 2 years before any common stock dividends were paid. ■

Objective 3 *Find Earnings per Share.* One way to measure the financial success of a corporation is by finding the **earnings per share** made by the corporation. Earnings per share is found with the following formula.

$$\text{Earnings per share} = \frac{\text{Net income} - \text{Dividends on preferred}}{\text{Number of shares of common outstanding}}$$

Example 3 Finding Earnings per Share

Ben Martinez started a home construction company in 1982 as a sole proprietorship. He took on a partner who remains with the company today as part owner. The business continued to grow rapidly, and in 2000 the two partners incorporated the company under the name of Martinez Construction, Inc.

The two then gave stock to their children. They have also allowed three of their key employees to earn bonuses paid in the stock of the company. The two founders hope to convince these key employees to remain with the firm.

(a) Martinez Construction, Inc. made \$420,000 last year. They had 500,000 shares of common stock outstanding and no preferred stock.

$$\text{Earnings per share} = \frac{\$420{,}000 - 0}{500{,}000} = \$0.84 \text{ per share}$$

(b) This year, the company issued preferred stock and paid \$85,000 in dividends to preferred shareholders out of a net income of \$544,000. The earnings per share are

$$\text{Earnings per share} = \frac{\$544{,}000 - \$85{,}000}{500{,}000} = \$0.92 \quad \text{Rounded}$$

(c) Find Mr. Martinez's total dividend income this year if he owns 300,000 shares of common stock and one-fourth of the preferred stock. His total dividend income is the sum of the dividends from common stock plus those from preferred stock.

$$(300{,}000 \times \$0.92) + \left(\frac{1}{4} \times \$85{,}000\right) = \$297{,}250$$

(d) Find Mr. Martinez's total dividends in a bad year with no profits, no dividends on common stock, and only a $10,000 dividend to preferred stock owners.

$$\frac{1}{4} \times \$10{,}000 = \$2500$$

18.1 Exercises

Find the dividend that will be paid for each share of preferred stock and common stock.

	Common Stock (in shares)	Preferred Stock (in shares)	Total Dividends
1.	80,000	None	$0
2.	200,000	None	$16,000
3.	175,000	50,000, $10 par value, 3%	$40,000
4.	200,000	30,000, $5 par value, 5%	$80,000
5.	1,000,000	100,000, $1000 par value, 2%	$2,000,000
6.	500,000	40,000, $50 par value, 3%	$44,800

Solve the following application problems.

7. ANIMAL HOSPITAL Reeves Animal Hospital had a net income of $320,000 last year. If $280,000 was reinvested in an expansion of the business, and the remainder was paid in dividends, what is the dividend per share on the 400,000 shares of common stock outstanding?

8. TUGBOAT COMPANY Fleet Operations, Inc. had a net income of $1,480,000 last year. The board of directors decided to purchase another tugboat for $850,000 and pay the remainder of the profits out as dividends. Find the dividend per share given 504,000 common shares of stock outstanding.

9. AUTO LEASING Jenson Auto Leasing has 10,000 shares of preferred stock outstanding with a $100 par value at 8%. They also have 200,000 shares of common stock outstanding. Find the dividend per share for both preferred and common stock given total dividends of $592,000.

10. DISHES Plate & Platter has 100,000 shares of $10 par value, 6% preferred stock, and 280,000 shares of common stock outstanding. Find the dividend per share for the preferred and common stock if the board allocates $589,200 for dividends.

11. PAPER PRODUCTS COMPANY Northern Maine Paper Products has sold 25,000 shares of $40 par value, 4% preferred stock to help remodel its mill. It already had 300,000 shares of common stock. Find the dividend per share if the total profit is $850,000, and 35% was distributed to shareholders.

12. **RETAIL BOAT SALES** The manager of Boat & Ski sold 30,000 shares of \$80 par value, 9% preferred stock for funds to build a new building. The company already had 1,000,000 shares of common stock outstanding. Find the dividend per share for both types of shares if 40% of the \$790,000 in profits is distributed to shareholders.
13. **COMPUTER COMPANY** Since the computer business is booming, a small new company has decided to pay no dividend for 3 years, and instead, reinvest the profits in expansion. During the fourth year, the board decides to pay a dividend of \$2,675,000. Find the dividend per share if the company has outstanding 20,000 cumulative preferred shares of \$100 par value at 5%, and 450,000 common shares.
14. **RUBBER GOODS COMPANY** To pay for a new factory, Wilson Rubber Goods has paid no dividend for 2 years. Now, at the end of the third year, a dividend of \$1,200,000 was declared. Find the dividend per share if the company has 10,000 cumulative preferred shares outstanding of \$100 par value at 10%, and 400,000 common shares.
15. **EARNINGS PER SHARE** A corporation had net income of \$7,000,000, 1,200,000 common shares, and no preferred shares. Find the earnings per share.
16. **EARNINGS PER SHARE** Suppose a corporation with 750,000 common shares outstanding and no preferred shares had a net income of \$2,350,000 last year. Find the earnings per share.
17. **FAR EAST IMPORTS** Far East Imports has 300,000 common shares, and 12,500 preferred shares, of \$150 par value, 10%. The company last year had a net income of \$1,500,000, with 45% of this going to shareholders. Find the earnings per share.
18. **DRY CLEANERS** Boyle Cleaners had a net income of \$600,000 last year, and distributed 55% to shareholders. The firm has 50,000 common shares, and 10,000 preferred shares, of \$100 par value, 12%. Find the earnings per share.
19. **EARNINGS PER SHARE** Arckat has 250,000 shares of common stock and 40,000 shares of preferred stock (\$50 par, 4%) outstanding. Net income last year and this year were, respectively, \$680,000 and \$765,000. The board of directors disbursed 40% of net income as dividends in both years. Find the dividend per share of common stock (a) last year and (b) this year. (c) Find the percent increase in the *earnings per share* of this year over last year.
20. **RECORDING STUDIO** Rhythm Recording Studio has 60,000 shares of \$1000 par value, 4% preferred stock. They also have 2,000,000 shares of common stock outstanding. This year the company increased their net income by 20% over last year's \$4,300,000. In both years the board of directors authorized the payment of 60% of net income for payment of dividends. Find the dividend per share of common stock for (a) last year and (b) this year. (c) Find the percent increase in *earnings per share* of this year over last year.
21. **PAVING COMPANY** Bill Baker owns a paving company with 1,000,000 shares of common stock. He wishes to help provide for his six grandchildren, so he establishes 20,000 shares of preferred stock with a \$100 par value, 8%, and donates the preferred stock to a trust for his grandchildren. Given a year with a net income of \$620,000 and a payment of 30% of net income for dividends, find the dividend to be paid on behalf of *each* grandchild.
22. **PAVING COMPANY** In Exercise 21, find the dividend to be paid on behalf of *each* grandchild if net income is \$85,000, and 30% of net income is used for dividends.

23. Explain the difference between common stock and preferred stock. (See Objective 1.)

24. List several reasons why a corporation would issue preferred stock in addition to common stock. (See Objective 1.)

18.2 Buying Stock

Objectives

1. *Know the basics of stock ownership.*
2. *Read stock tables.*
3. *Find the commission for buying or selling stocks.*
4. *Find the total price of a stock purchase.*
5. *Find the current yield on a stock.*
6. *Find the PE ratio of a stock.*
7. *Define the Dow Jones Industrial Average.*
8. *Define a mutual fund.*

Objective 1 *Know the Basics of Stock Ownership.* As mentioned in the previous section, buying stock in a corporation makes the stockholder a *part owner* of the corporation. In return for the money a person invests in stock, he or she shares in any profits the company makes. Hopefully, the company will do well and prosper. If this happens, many other people will want the company's stock, and they will be willing to pay a good price for its shares. If this happens, the stockholders can sell at a profit.

THE WALL STREET JOURNAL.

What's News—

On the other hand, if the company does not do well, then fewer people will want its stock and the price will fall. The price of the shares of most large, publicly held firms, is set by supply and demand at institutions called **stock exchanges**. The largest stock exchange is the New York Stock Exchange, located on Wall Street in New York City. Many foreign countries, including Japan, Taiwan, England, Canada, and Mexico, have their own stock exchanges. One of the most widely circulated financial newspapers in the world is *The Wall Street Journal*, which is published daily by Dow Jones & Company, Inc. This newspaper provides daily prices for stocks and bonds in several different markets. *The Wall Street Journal* also maintains a home page on the World Wide Web (Internet).

The public does not go directly to the exchange to buy and sell stock. Instead, members of the public buy their stock through **stockbrokers**, people who have access to the exchanges. As discussed below, stockbrokers charge a fee for buying or selling stock.

Most people buy stock in hopes of making a profit. Obviously, there is no way to know for sure which stocks will increase in value and which will decrease. There are many people in the financial world offering to give you financial advice for a fee, but this advice is not guaranteed to produce a profit. Individuals can now purchase stock using the World Wide Web, without talking to a stockbroker.

Objective 2 *Read Stock Tables.* Find the current price of a stock by looking in many daily local newspapers or in *The Wall Street Journal*. A portion of the stock market page from the *Journal* is shown in Figure 18.3.

NOTE In reading the stock table, the following should be noted.

1. Price increases or declines of 5% or more are highlighted with boldface type.

NEW YORK STOCK EXCHANGE COMPOSITE TRANSACTIONS

-C-C-C-

52 Weeks Hi	52 Weeks Lo	Stock	Sym	Div	Yld %	PE	Vol 100s	Hi	Lo	Close	Net Chg
10 7/8	9	♣ColonlIntmk	CMK	.89	9.8	...	390	9 1/8	9	9 1/16	− 1/16
7	5 3/4	♣ColonlHigh	CIF	.68a	10.8	...	479	6 3/8	6 1/4	6 5/16	...
11 13/16	9	♣ColonlInvMun	CXH	.56	6.1	...	121	9 1/4	9 3/16	9 1/4	+ 1/8
8 7/16	6 1/4	♣ColonlMuni	CMU	.43	6.7	...	178	6 7/16	6 3/8	6 7/16	+ 1/16
29	24	♣ColonlProp	CLP	2.32	8.6	17	170	27 3/8	26 7/8	26 15/16	− 1/8
25 1/8	22 3/8	♣ColonlProp pfA		2.19	9.6	...	4	22 15/16	22 7/8	22 7/8	− 1/8
64 11/16	43 7/8	ColumEngy	CG	.90f	1.5	19	875	61 1/2	60 3/8	61 1/2	+ 15/16
27 1/2	16 3/16	ColumHCA	COL	.08	.4	27	11638	22 11/16	22 1/4	22 9/16	+ 1/8
26 3/8	24 3/8	ClmbSo A	CSJ	2.09	8.5	...	11	24 3/4	24 1/2	24 1/2	− 1/8
26 1/4	24	ClmbsSo pfB		1.98	8.1	...	151	24 1/2	24 1/8	24 1/2	+ 1/8
30 7/8	10 3/4	Comdisco	CDO	.10	.4	66	3146	22 7/16	22 1/8	22 7/16	...
26 3/8	24 5/8	ComEd TOPrS		2.12	8.7	...	39	24 9/16	24 1/2	24 1/2	− 1/8
70	46 1/2	Comerica	CMA	1.44	2.6	14	4007	56 1/8	54 1/2	54 7/8	− 1 11/16
21 3/8	11 3/8	CmfrtSysUSA	FIX		...	11	1036	13 5/16	12 3/4	12 15/16	− 7/16
50 1/2	31 5/8	ComrcBcpNJ	CBH	.88b	2.0	20	531	43 13/16	43 1/8	43 1/2	− 5/16
27 1/4	24	CommCap pfT		2.19	8.8	...	6	25	25	25	− 1/16
36 1/2	21 9/16	CommrcGpInc	CGI	1.12f	4.4	10	161	25 1/2	25 1/8	25 1/2	− 1/8
28 3/16	18 3/4	ComrclFed	CFB	.26	1.1	16	543	24 1/4	23 15/16	24	− 1/4
21 7/8	11 1/8	ComrclIntech	TEC	.60	4.6	8	76	13 1/4	13	13 1/16	− 1/16
34 3/16	19 11/16	CmrclMtls	CMC	.52	1.6	11	66	33 3/16	32 5/8	32 15/16	− 1/16
15 7/8	11 1/16	CmrclNetRlty	NNN	1.24	10.5	10	573	12	11 13/16	11 13/16	− 3/8
44 3/8	28 13/16	CmwlthEngy	CES	1.66	4.0	16	494	41 11/16	41	41 11/16	+ 1/2
40	8 3/4	Commscope	CTV		...	36	1474	37 3/8	36 1/2	36 7/8	− 7/16
33 5/8	22 5/8	CmntyBkSys	CBU	.92	3.7	12	2	24 3/4	24 3/4	24 3/4	+ 1/16
10 1/16	**2 15/16**	**Copel**	**ELP**	**.28e**	**4.4**	**...**	**8370**	**6 3/4**	**5 3/4**	**6 5/16**	**+ 3/8**
30 3/4	8 3/8	CompnhiaSidr	SID	2.14e	9.9	...	456	21 5/8	21	21 5/8	− 5/8
51 1/4	20	♣Compaq	CPQ	.08	.3	43	88769	24 1/8	23 1/8	24	+ 1/2
20 7/8	5 9/16	CompUSA	CPU		...	dd	8083	6 9/16	6 1/8	6 7/16	+ 1/16
58	26	CptrAssoc	CA	.08	.2	41	8800	48 3/4	47 1/4	48 1/2	+ 1 1/16
74 7/8	46 1/4	CptrSci	CSC		...	29	3732	65 3/16	63 7/8	64 1/16	− 1 1/16
34 1/2	15 3/8	♣CptrTask	TSK	.05	.3	12	139	17 7/8	17 1/2	17 1/2	− 3/8
26 11/16	12 1/8	CompxInt A	CIX		...	...	2	16 15/16	16 15/16	16 15/16	...
39 5/8	21 3/4	Comsat	CQ	.20	.5	47	3452	37 1/16	36	36 3/8	...
25 11/16	23 1/2	ComsatCap pfA		2.03	8.4	...	40	24 7/16	24 1/16	24 3/16	− 1/4
8	**2 3/16**	**ComstkRes**	**CRK**		**...**	**dd**	**978**	**5 1/4**	**4 7/8**	**5 3/16**	**+ 5/16**
34 3/8	22 9/16	ConAgra	CAG	.71	2.9	32	7869	24 11/16	24 1/8	24 1/4	− 1/8
25 11/16	24 15/16	ConagraCap pfA		2.25	8.9	...	34	25 5/16	25 1/4	25 5/16	...
21 9/16	18 7/8	ConagraCap pfB		1.30e	6.5	...	163	19 15/16	19 13/16	19 15/16	+ 3/16
26 1/4	25 1/4	ConagraCap pfC		2.34	9.2	...	41	25 3/8	25 5/16	25 5/16	...
8 1/8	3 13/16	♣ConeMills	COE		...	dd	215	6 1/8	5 15/16	6 1/8	+ 3/16
25 1/2	19 3/8	♣Conectiv	CIV	.88m	4.2	11	1640	21 3/16	20 7/16	21 1/8	+ 3/16

	52 Weeks Hi	52 Weeks Lo	Stock	Sym	Div	Yld %	PE	Vol 100s	Hi	Lo	Close	Net Chg
	24 13/16	19 5/16	♣CrownPac	CRO	2.26	10.0	21	144	22 1/2	22 1/4	22 1/2	+ 3/16
	15 3/4	9 3/16	CryoLife	CRY		...	25	709	13	12 5/8	12 5/8	− 5/16
s	28 15/16	20 7/16	♣CullnFrst	CFR	.70	2.6	16	464	26 9/16	26 1/4	26 9/16	− 1/16
	11 1/16	5 1/8	Culp	CFI	.14	1.6	15	81	8 3/4	8 1/2	8 3/4	...
	65 11/16	28 5/16	CumminsEng	CUM	1.10	1.8	89	1596	60 1/4	59 3/16	59 5/8	− 7/8
	12 11/16	10 5/16	♣CurIncoShrs	CUR	.82e	7.8	...	22	10 11/16	10 9/16	10 9/16	...
	48 3/8	31	♣CurtWrght	CW	.52	1.5	12	13	34 5/8	34 1/2	34 5/8	+ 3/16
	24 3/4	**5 1/2**	**CyprsSemi**	**CY**		**...**	**cc**	**29838**	**24 9/16**	**22 9/16**	**24**	**+ 1 1/4**
	16 1/16	**9**	**♣CyprusAmax**	**CYM**	**.20**	**1.4**	**dd**	**7059**	**14 1/2**	**13 1/2**	**14 1/2**	**+ 13/16**
	31 15/16	14 7/8	♣CytecInd	CYT		...	9	3847	24 5/8	22 1/2	23 5/8	− 1 1/8
			-D-D-D-									
	39 15/16	12 3/8	DBT Online	DBT		...	87	265	32	31 1/8	31 3/8	− 3/4
	27	16 1/2	DECS Tr II	RYD	1.81	7.6	...	1129	23 11/16	23 5/8	23 11/16	+ 15/64
n	**45 5/8**	**15 13/16**	**DLJdirect**	**DIR**		**...**	**cc**	**3501**	**20 1/8**	**18 9/16**	**18 3/4**	**− 1 3/4**
	10	8 3/8	DLJ HiYld	DHY	.99	11.6	...	449	8 1/2	8 3/8	8 1/2	+ 1/16
	22	16 5/16	♣DPL Inc	DPL	.94	5.0	15	1497	18 15/16	18 3/16	18 13/16	+ 5/16
	44 1/4	35 1/8	♣DQE	DQE	1.52	4.0	15	578	38 1/4	37 11/16	38 1/4	+ 3/16
	23	10 5/8	♣DR Horton	DHI	.12	.8	6	12819	14 13/16	13 7/8	14 7/16	− 3/8
	32 5/8	**4 5/8**	**♣DSP Comm**	**DSP**		**...**	**33**	**5661**	**26 5/8**	**25 1/8**	**25 7/16**	**− 1 3/8**
	70 1/4	34	DST Sys	DST		...	46	1702	69 7/16	67 9/16	69	+ 1/16
	49 1/4	37 15/16	DTE Engy	DTE	2.06	5.3	12	2177	39	38 9/16	38 15/16	+ 3/16
	22 1/2	9 1/2	♣DVI Inc	DVI		...	12	27	15 15/16	15 7/8	15 7/8	...
	106 1/2	65	DamlrBnz 2002	DAJ	4.33e	5.5	...	5	78 3/4	78 3/4	78 3/4	+ 3/4
n	108 5/8	71 9/16	DmlrChrylr	DCX	2.50e	3.3	...	4570	76 3/8	75 7/8	76 1/8	− 1 3/16
	59 1/8	25 9/16	DainRaus	DRC	.88	1.6	19	143	54 1/4	53 3/16	53 1/2	− 13/16
	13 1/8	6	DalTile	DTL		...	12	593	10 13/16	10 1/4	10 7/16	− 1/2
	57 1/16	22 11/16	DallasSemi	DS	.20	.4	25	798	49 1/8	48 1/4	49 1/8	..
	14 15/16	6 3/8	DanRivr A	DRF		...	6	261	7 1/2	7 1/4	7 1/2	+ 5/16
	54 1/16	31 5/16	♣Dana Cp	DCN	1.24	3.0	12	4580	42 1/4	40 11/16	41 15/16	+ 7/16
	69	28	Danaher	DHR	.06	.1	38	2819	57 7/16	55 3/8	55 15/16	− 1 3/8
	60 1/4	47 3/8	Danone ADS	DA	.64e	1.2	...	102	51 3/4	51 3/8	51 3/4	+ 1/4
	23 3/8	14 3/16	Darden	DRI	.08	.4	19	4131	18 3/4	17 15/16	18 9/16	+ 3/8
	21 13/16	7	DataGen	DGN		...	dd	3979	18 15/16	18 9/16	18 13/16	− 3/16
	29 3/8	10 1/2	DaveBusters	DAB		...	21	122	23 3/4	23 3/8	23 5/8	− 1/8
	76	31 7/16	DaytonHud	DH	.40	6	28	11167	61 15/16	60	61 7/8	+ 9/16
	23 1/2	14 3/8	DaytonSuper A	DSD		...	10	3	18 13/16	18 11/16	18 13/16	− 1/16
	8 1/2	3 1/4	DeRigo ADR	DER	.10p	...	...	426	5 5/16	5 1/16	5 5/16	+ 3/16
	50 7/16	32 15/16	DeanFood	DF	.88f	2.1	11	1671	41 5/8	40 13/16	41 9/16	+ 5/8
x	9 9/16	7	DebtStratFd	DBS	.88e	12.4	...	507	7 3/16	7	7 1/8	+ 1/16
x	9 1/4	7 7/8	DebtStratFd II	DSU	.89e	11.0	...	681	8 1/8	7 15/16	8 1/8	+ 1/16

Figure 18.3

2. Up or down arrowheads next to a stock indicate a new 52-week high or low.
3. Stocks with unusual volume activity are underlined. ▬

The first step in reading this table is to find the line corresponding to the company whose stock is of interest. Company names are usually abbreviated in newspapers. For example, Compaq is abbreviated CPQ. Look in the stock tables to find the following information on Compaq.

52 Weeks Hi	52 Weeks Lo	Stock	Sym	Div	Yld %	PE	Vol 100s	Hi	Lo	Close	Net Chg
$51\frac{1}{4}$	20	Compaq	CPQ	.08	.3	43	88769	$24\frac{1}{8}$	$23\frac{1}{8}$	24	$+\frac{1}{2}$

The numbers $51\frac{1}{4}$ and 20 in front of the company's name show that $51\frac{1}{4}$ ($51.25) was the highest price the stock reached in the previous 52 weeks, while 20 ($20) was the lowest. After the name of the company and its symbol, find .08, which means that the company pays $0.08 per year per share of stock as a dividend to the owners of the stock. Dividends typically go up when the company is making good profits and down when business is bad. However, a small dividend does not mean that a company is doing poorly. Rather, a company may pay a small dividend and use remaining profits for expansion purposes. The .3 in the next column is the current yield on the company's stock, in percent. The dividend of $0.08 per share is 0.3% of the current purchase price of the stock. After the .3 is 43, the price-earnings ratio, discussed later. Then comes 88769, the number of shares traded for the day in hundreds of shares. On the day reported, a total of

$$88769 \times 100 = 8{,}876{,}900$$

shares of Compaq were sold. After 88769 comes a list of prices reached by the stock during the day. The first number, $24\frac{1}{8}$, was the highest price reached by the stock during the day while $23\frac{1}{8}$ was the lowest. The next number, 24, says the stock closed at 24 ($24) at the end of the day. The last number is $+\frac{1}{2}$, indicating that the stock price increased by $\frac{1}{2}$ ($0.50) compared to the previous day.

Example 1 Reading a Stock Table

Use the stock table to find (a) the highest price for the last 52 weeks for the large international moving company Crown Pacific (CRO) and (b) the dividend for Culp.

(a) First, find the correct line in the stock table for Crown Pacific.

52 Weeks Hi	52 Weeks Lo	Stock	Sym	Div	Yld %	PE	Vol 100s	Hi	Lo	Close	Net Chg
$24\frac{13}{16}$	$19\frac{5}{16}$	Crown Pac	CRO	2.26	10.0	21	144	$22\frac{1}{2}$	$22\frac{1}{4}$	$22\frac{1}{2}$	$+\frac{3}{16}$

The highest price during the last 52 weeks was $24\frac{13}{16}$, or $24.8125, the first number on the left.

(b) Going back to the stock table, we can see that the dividend for Culp is given just after the company symbol, CFI. The dividend is $0.14 per share per year.

NOTE The letters "pf" appear after some of the company names in the stock table. These letters represent "preferred stock," which was discussed in Section 18.1. ▬

Example 2 Reading a Stock Table

Find the cost for 100 shares of Copel at the high price for the day.

Solution Find the total cost of a stock purchase by multiplying the price per share by the number of shares. From the stock table, the high price of Copel was $6\frac{3}{4}$ per share. First change $6\frac{3}{4}$ to the decimal 6.75, then multiply.

$$100 \times 6.75 = \$675$$

The cost of 100 shares of this stock is $675 (plus any broker's fee).

Objective 3 | ***Find the Commission for Buying or Selling Stocks.*** Individuals usually use a broker to buy or sell a stock. The broker has representatives at the exchange who will execute a buyer's order. (It is sometimes possible to buy stock directly from a bank, which might sell it with no sales charge.) The broker will charge a fee, or **commission**, for executing an order. Commission rates formerly were set by stock exchange rules and did not vary from broker to broker. Now, however, commissions are competitive and vary considerably among brokers.

In the last few years, several **discount brokers** have become popular. These brokers merely buy and sell stock and offer no additional services, such as research or recommendations on stocks to buy or sell. Several of these discount operations allow individuals to trade stock very cheaply over the World Wide Web, usually for a flat fee for each trade.

The rates charged depend on whether the order is for a **round lot** of shares (multiples of 100) or an **odd lot** (fewer than 100 shares). Odd-lot orders often involve an **odd-lot differential**, an extra charge of $\$\frac{1}{8}$, or 12.5 cents, a share. Recently, competition has driven the commission cost of buying and selling stocks down substantially. Typical expenses in buying and selling stock are as follows. (There is often a minimum commission charge.)

Buying Stock	**Selling Stock**
Broker's commission of 1.5% of purchase price	Broker's commission of 1.5% of selling price
+	+
Any odd-lot differential	Any odd-lot differential
	+
	SEC fee (see below)
	+
	Any transfer taxes (see below)

The SEC fee applies *only when stock is sold*, and is set by the Securities and Exchange Commission, a federal government agency that regulates stock markets. The fee is currently 1 cent ($0.01) per $300 in value (or any fraction of $300). For example, to find the fee for a sale of $1100, first divide $1100 by $300.

$$\frac{\$1100}{\$300} = 3.666$$

Since the fee is 1¢ per $300, or *fraction thereof*, round 3.666 *up* to 4. The fee is then 4 × 1¢ = 4¢.

NOTE Some state and local governments charge a **transfer tax** when stock is sold. The amount of this tax would be subtracted by the broker before turning over the balance of the money to the seller. ■

Objective 4 | ***Find the Total Price of a Stock Purchase.*** The next two examples show the total cost of a stock purchase and the cost of selling a stock.

Example 3 Finding Total Cost of a Purchase

Marie Wilson bought 75 shares of stock, paying $\$26\frac{3}{4}$ per share. Find her total cost for the purchase.

Solution The basic cost of the stock is found by multiplying the price per share and the number of shares.

First, we estimate the total cost to insure our final answer is reasonable. Commissions and an odd-lot differential will make the cost to Wilson somewhat higher than $\$26\frac{3}{4}$ per share, say \$27 per share. Her total cost should be about

$$\$27 \times 75 \text{ shares} = \$2025$$

Now we find the exact cost.

$$\begin{aligned}\text{Price per share} \times \text{Number of shares} &= \$26\frac{3}{4} \times 75 \\ &= \$26.75 \times 75 \\ &= \$2006.25\end{aligned}$$

The broker's commission is 1.5% of this amount, or

$$\begin{aligned}\text{Broker's commission} &= 1.5\% \times \$2006.25 \\ &= 0.015 \times \$2006.25 \\ &= \$30.09\end{aligned}$$

Since Wilson bought 75 shares (and not a multiple of 100 shares), she must pay an odd-lot differential of $\$\frac{1}{8}$, or 12.5 cents, per share. For 75 shares, this odd-lot differential amounts to

$$75 \text{ shares} \times \$\frac{1}{8} \text{ per share} = \$9.38 \quad \text{Rounded}$$

The total cost of the shares of stock follows.

\$2006.25	Basic cost
30.09	Broker's commission
+ 9.38	Odd-lot differential
\$2045.72	

This figure is close to our estimate of \$2025, and therefore is reasonable.

Example 4 Finding Total Cost of a Stock Sale

Find the amount received by a person selling 700 shares of a stock at $\$63\frac{5}{8}$.

Solution The basic price of the stock is

$$700 \times \$63\frac{5}{8} = 700 \times \$63.625 = \$44{,}537.50$$

Find the SEC fee as described.

$$\frac{\$44{,}537.50}{\$300} = 148.458$$

Round 148.458 up to 149. The SEC fee is then

$$149 \times 1¢ = \$1.49$$

The broker's commission is 1.5% of the basic price, or

$$0.015 \times \$44{,}537.50 = \$668.06$$

Finally, the seller receives the following.

	\$44,537.50	Basic price
	\$668.06	Broker's fee
−	\$1.49	SEC fee
	\$43,867.95	

There is no odd-lot differential since the number of shares traded is a multiple of 100.

Most people buy stock because they hope that the price will go up. If the price does go up, the investor can then sell the stock at a profit. Some people, however, buy a stock because of the dividend that it pays. (A dividend is usually paid quarterly.)

While there is no certain way of choosing stocks that will go up, there are several **stock ratios** that can be looked at when considering a stock purchase. Two useful ratios are the current yield and the price-earnings ratio.

Objective 5 ***Find the Current Yield on a Stock.*** The **current yield** on a stock is used to compare the dividends paid by stocks selling at different prices. Find current yield with the following formula.

$$\text{Current yield} = \frac{\text{Annual dividend per share}}{\text{Current price per share}}$$

This result usually is converted to a percent (rounded to the nearest tenth). The annual dividend rate and the current yield can be found in the stock tables in daily newspapers.

Example 5 Finding Current Yield

Find the current yield for each of the following stocks: (a) Dana Corporation, with a dividend of \$1.24 per share per year and (b) DVI Incorporated, with no dividend.

Solution

(a) The closing price of Dana Corporation, symbol DCN, was $41\frac{15}{16}$. Use the formula for current yield to get the following.

$$\text{Current yield} = \frac{\$1.24}{\$41.9375} = 3.0\% \quad \text{Rounded}$$

(b) The closing price of DVI Incorporated, symbol DVI, was $15\frac{7}{8}$. The following is found using the formula for current yield.

$$\text{Current yield} = \frac{0}{\$15.875} = 0\%$$

NOTE A stock pays no dividend when the company has been going through bad times or is investing in research or new plants that promise a long-term payoff. Sometimes a small new company will pay no dividends during its early years, preferring to reinvest the money for long-term growth.

Objective 6 *Find the PE Ratio of a Stock.* One number that some people use to help decide which stock to buy is the **price-earnings ratio** (abbreviated **PE ratio**). This ratio is found with the following formula.

$$\text{PE ratio} = \frac{\text{Price per share}}{\begin{array}{c}\text{Annual net income per share}\\ \text{(Earnings per share)}\end{array}}$$

NOTE Annual net income per share is the same thing as earnings per share discussed in Section 18.1.

Example 6 Finding the PE Ratio

Find the PE ratio for the following stocks: (a) Coca-Cola Co., price per share \$58.75, annual earnings per share \$1.28 and (b) Microsoft, price per share \$83.50, annual earnings per share \$1.42.

Solution

(a) For Coca-Cola,

$$\text{PE ratio} = \frac{\$58.75}{\$1.28} = 45.9$$

The PE ratio is 45.9, or 46 after rounding.

(b) For Microsoft,

$$\text{PE ratio} = \frac{\$83.50}{\$1.42} = 58.8$$

The PE ratio is 58.8, or 59 after rounding.

Problem-Solving Hint Be sure to round the PE ratio to the nearest whole number.

Sometimes a low PE ratio may indicate that the stock is a "sleeper," which has not been found by other investors. On the other hand, a high PE ratio may indicate that the stock's price is too high; as other people notice this, the stock's price could fall.

Unfortunately, *the PE ratio is not a perfect guide* to future stock market behavior. The PE ratio may be low, not because the stock is an undervalued sleeper but because investors see a poor future for the company. A PE ratio may be high, not because the stock is overpriced but because investors feel that company earnings will increase rapidly over the next few years.

Many daily newspapers now give the PE ratio. If no PE ratio is given for a particular stock, most probably that company has lost money during the previous year.

Objective 7 *Define the Dow Jones Industrial Average.* The **Dow Jones Industrial Average** is frequently used as an indicator of overall trends in stock prices. Actually, Dow Jones &

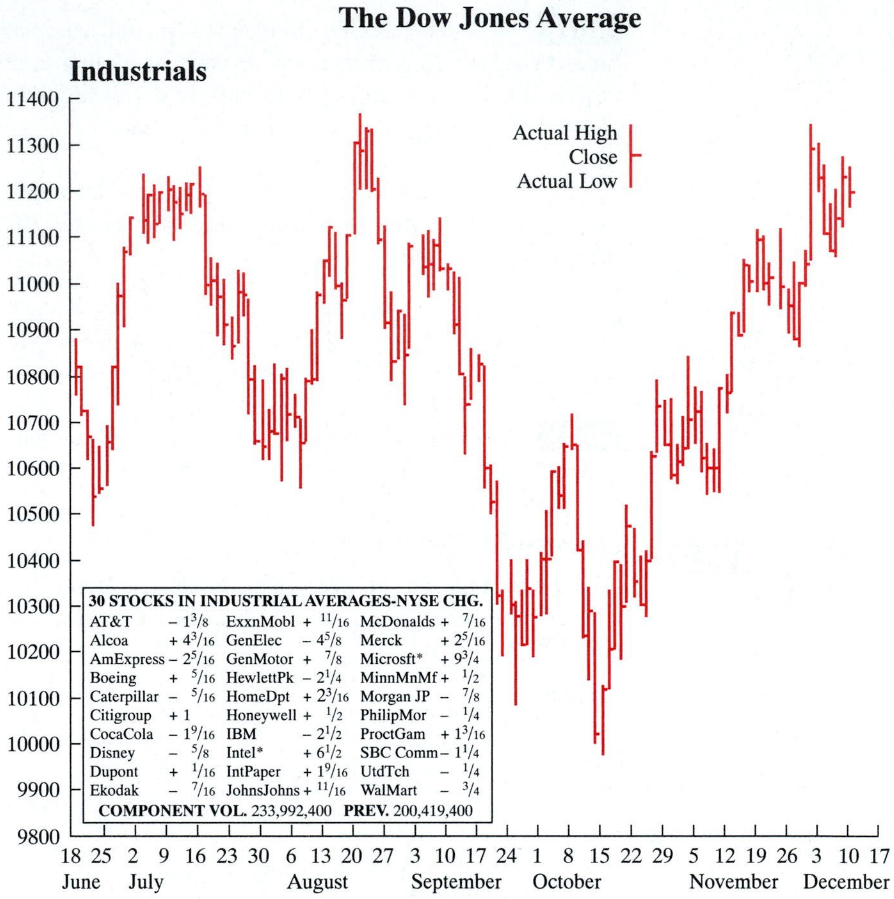

30 STOCKS IN INDUSTRIAL AVERAGES-NYSE CHG.					
AT&T	$-1\frac{3}{8}$	ExxnMobl	$+\frac{11}{16}$	McDonalds	$+\frac{7}{16}$
Alcoa	$+4\frac{3}{16}$	GenElec	$-4\frac{5}{8}$	Merck	$+2\frac{5}{16}$
AmExpress	$-2\frac{5}{16}$	GenMotor	$+\frac{7}{8}$	Microsft*	$+9\frac{3}{4}$
Boeing	$+\frac{5}{16}$	HewlettPk	$-2\frac{1}{4}$	MinnMnMf	$+\frac{1}{2}$
Caterpillar	$-\frac{5}{16}$	HomeDpt	$+2\frac{3}{16}$	Morgan JP	$-\frac{7}{8}$
Citigroup	$+1$	Honeywell	$+\frac{1}{2}$	PhilipMor	$-\frac{1}{4}$
CocaCola	$-1\frac{9}{16}$	IBM	$-2\frac{1}{2}$	ProctGam	$+1\frac{3}{16}$
Disney	$-\frac{5}{8}$	Intel*	$+6\frac{1}{2}$	SBC Comm	$-1\frac{1}{4}$
Dupont	$+\frac{1}{16}$	IntPaper	$+1\frac{9}{16}$	UtdTch	$-\frac{1}{4}$
Ekodak	$-\frac{7}{16}$	JohnsJohns	$+\frac{11}{16}$	WalMart	$-\frac{3}{4}$

Source: Wall Street Journal, reprinted by permission of Dow Jones, Inc. via Copyright Clearance Center,

Figure 18.4

Company publishes several different averages. The one most commonly used refers to an average of the stock prices of 30 large industrial companies listed in Figure 18.4. The movement of this average is typically quoted on television news and in the newspapers. As you can see in Figure 18.5, the Dow Jones Industrial Average has increased greatly over the past 90 years.

Objective 8 *Define a Mutual Fund.* Ownership of shares in a single company *can be risky*—the company may suffer poor financial results, causing the stock price to fall. This risk can be reduced by simultaneously investing in the stocks of several different companies. One way of doing this is to invest with a mutual fund that buys stocks. A **mutual fund** that invests in stocks receives investment funds from many different investors and uses the money to purchase stock in several different companies. For example, a $1000 investment in a typical mutual fund owning stock means that you own a very small piece of perhaps 100 different companies or more.

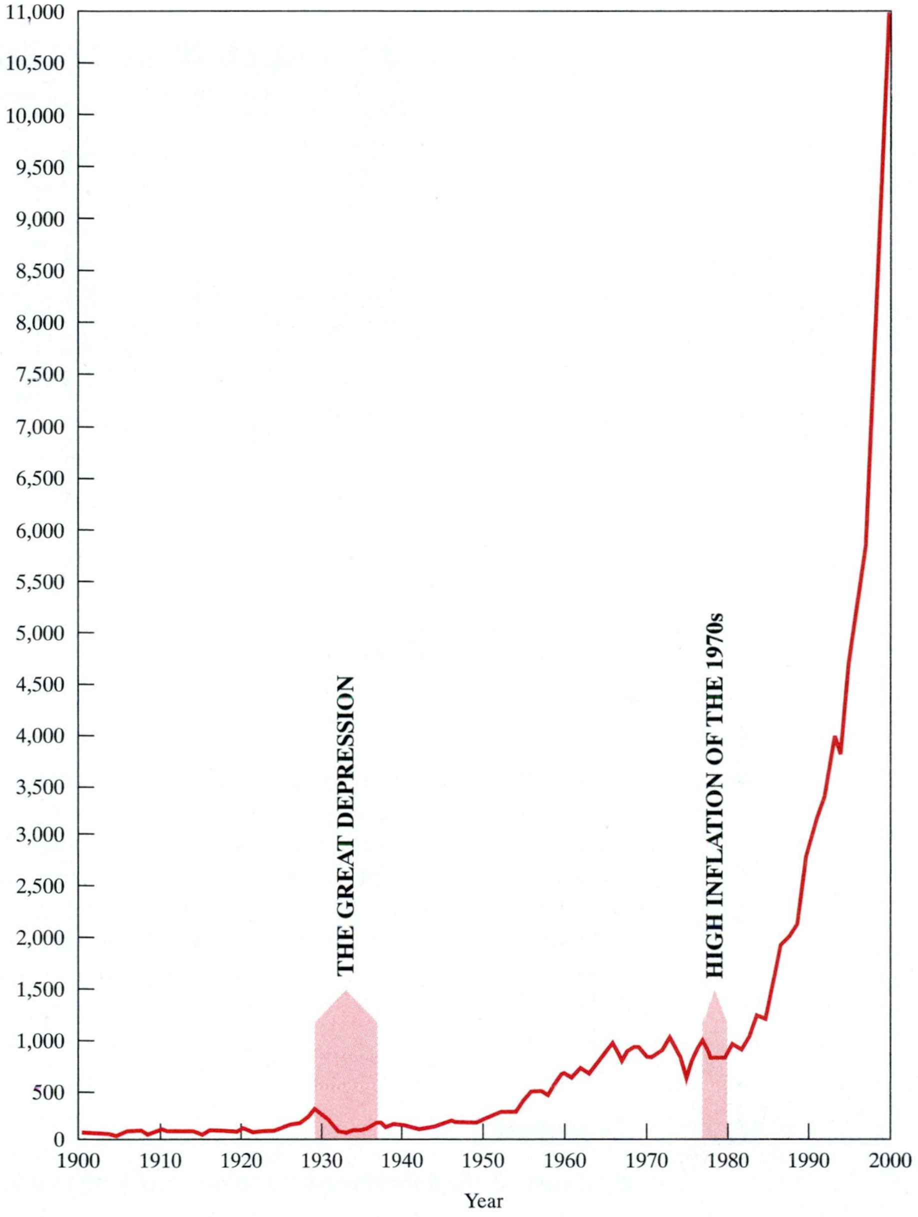

Data: Dow Jones & Co., Inc.

Figure 18.5

Figure 18.6 shows mutual fund quotes as listed in *The Wall Street Journal*. These are only a few of the hundreds of mutual funds available to individual investors. From the figure, the Janus Worldwide fund (JANUS: WrldW) had a net asset value (NAV) of 52.49, which means that each share was worth $52.49 at the close of business that day. The share price fell by $0.31 from the previous day, and the year-to-date percent return on investment (YTD % ret) was 10.8%.

MUTUAL FUND QUOTATIONS

Name	NAV	Net Chg	YTD %ret
JANUS:			
Balanced	20.80	−0.18	+ 7.0
Enterprise	46.12	−0.67	+27.3
EqInc	21.18	−0.27	+13.5
FedTE	6.84	...	− 3.1
FlxInc	9.44	−0.01	− 1.2
Fund	38.78	−0.37	+15.2
Gl LifeSci	12.32	+0.04	+23.2
GlTech	15.83	−0.19	+58.3
GrInc	32.66	−0.32	+12.4
HiYld	10.25	−0.01	+ 1.9
Mercury	30.82	−0.51	+27.8
Olympus	33.79	−0.31	+22.5
Ovrseas	22.30	−0.05	+11.1
ShTmBd	2.84	...	+ 1.1
SplSitu	21.87	−0.18	+25.6
Twen	60.25	−0.63	+13.0
Ventur	70.75	−0.20	+23.8
WrldW	52.49	−0.31	+10.8
Janus Aspen:			
AggGr	33.88	−0.46	+27.9
Balanced	24.30	−0.19	+ 9.1
BalRet p	24.37	−0.18	+ 8.8
CapAppr	23.25	−0.28	+16.8
FixInc	11.55	...	− 0.7
GrInc	14.12	−0.12	+18.1
Growth	26.59	−0.21	+13.7
IntlGr	23.49	−0.03	+10.7
WrldWGr	32.18	−0.18	+10.8
WrldwRet p	32.11	−0.18	+10.5
JapanFd	13.13	+0.10	+57.6

Name	NAV	Net Chg	YTD %ret
Legg Mason:			
AmerLdTr p	19.28	−0.15	+ 6.1
BalTr p	11.90	−0.06	− 2.0
EmgMktTr p	9.77	+0.02	+40.4
FocusTr	23.07	−0.28	+ 4.9
GblGovt p	9.20	+0.06	− 6.0
GvInt p	10.02	−0.02	− 1.5
HiYld p	14.91	−0.06	+ 5.1
IntEqTr p	13.13	+0.02	+ 4.3
InvGr p	9.83	−0.02	− 2.8
MdTF	15.53	...	− 2.7
PATF	15.69	...	− 2.4
SpInv p	34.18	−0.06	+11.6
TxFrInt p	15.14	...	− 1.5
TotRt p	21.38	−0.08	+ 3.6
USSmCap p	8.86	+0.01	+ 3.4
ValTr p	66.77	−0.90	+12.4
NvgSpInv	36.12	−0.06	+12.4
NvgValTr	68.35	−0.92	+13.1
LepIstel p	19.79	−0.26	− 0.6
LeutholdCl	11.28	−0.01	+ 4.1
Lexington Grp:			
CLdr	17.68	−0.01	+13.4
GNMA	8.28	−0.04	+ 0.5
Globl	10.55	−0.09	+11.5
Goldfd	3.01	−0.01	− 0.7
GthInc	22.19	−0.13	+ 1.3
Intl	13.53	−0.01	+16.5
Ram Glb	9.83	+0.02	− 1.2
Russia p	4.51	+0.04	+70.8

Figure 18.6

Over periods of many years, stocks have consistently provided a greater return on investment than savings accounts, certificates of deposit, or bonds. Most financial planners agree that stock ownership through mutual funds should be a part of any long-range retirement or investment plan. Many financial counselors suggest that *the first mutual fund one should invest in* is one that tracks the Standard and Poor's 500. The Standard and Poor's 500 is an index of the stock prices of 500 leading companies in leading industries, chosen to reflect the U.S. stock market. Many mutual fund houses such as Vanguard, Fidelity, and Janus have a specific fund that is designed to track the Standard and Poor's 500 index.

18.2 Exercises

Find each of the following from the stock table shown in Figure 18.3.

1. High for the day for Danaher (DHR)
2. Low for the day for Daimler Chrysler (DCX)
3. Closing price for the day for Cummins Engines (CUM)
4. Change from the previous day for Cone Mills (COE)
5. 52-week high for Comdisco (CDO)
6. 52-week low for DTE Energy (DTE)
7. Dividend for Comerica (CMA)
8. Dividend yield for Data General (DGN)
9. Volume for the day for CompUSA (CPU)
10. Closing price for Columbia HCA (COL)

Find the price for each of the following stock purchases. Ignore any broker's fees.

	Stock	Number of Shares	Transaction
11.	Dean Food (DF)	1000	Close
12.	Dallas Semi (DS)	100	High
13.	Cytec Ind (CYT)	200	Low
14.	Copel (ELP)	300	Close
15.	Colon Prop (CLP)	100	Low
16.	Dan River A (DRF)	300	High

Find the price for each of the following stock purchases. Use the broker's charges given in the text on page 672. (Hint: In Exercises 23 and 24, the odd-lot differential is charged on only 40 shares.)

17. 200 shares at $27\frac{5}{8}$

18. 700 shares at $19\frac{1}{8}$

19. 1200 shares at $37\frac{3}{4}$

20. 1400 shares at $16\frac{7}{8}$

21. 60 shares at $35\frac{5}{8}$

22. 90 shares at $53\frac{3}{4}$

23. 540 shares at $69\frac{1}{4}$

24. 740 shares at $32\frac{3}{8}$

Find the amount received by the sellers of the following stocks. Use the broker's charges listed in the text.

25. 300 shares at $30\frac{1}{2}$

26. 500 shares at $10\frac{1}{4}$

27. 100 shares at $40\frac{3}{4}$

28. 200 shares at $13\frac{7}{8}$

29. 830 shares at $52\frac{3}{4}$

30. 360 shares at $72\frac{1}{8}$

Find the current yield for each of the following stocks. Round to the nearest tenth of a percent.

	Stock	Current Price per Share	Annual Dividend
31.	Nike	$\$52\frac{11}{16}$	$0.48
32.	Seagram	$\$54\frac{1}{16}$	$0.66
33.	Ford	$\$51\frac{11}{16}$	$1.84
34.	Office Depot	$\$15\frac{1}{8}$	$0
35.	Travelers	$\$36\frac{1}{4}$	$0.50
36.	Exxon	$\$82\frac{9}{16}$	$1.64

Find the PE ratio for each of the following. Round all answers to the nearest whole number.

	Stock	Price per Share	Net Income per Share
37.	Caterpillar	$\$60\frac{5}{16}$	$3.10
38.	Dupont	$\$72\frac{3}{16}$	$3.84

	Stock	Price per Share	Net Income per Share
39.	General Motors	$\$63\frac{1}{8}$	\$7.36
40.	Hilton	$\$12\frac{5}{8}$	\$1.14

Solve each of the following application problems. Exercises 43 and 44 require concepts from Chapter 14.

41. STOCK PURCHASES Becka Jones purchased 200 shares of General Motors at $\$60\frac{3}{8}$ and 300 shares of Microsoft at $\$120\frac{1}{2}$. Find the total cost if she was charged a commission of 1%.

42. STOCK PURCHASES Theresa Tabor purchased 200 shares of Ford at $\$51\frac{11}{16}$ and 100 shares of Schwab at $\$42\frac{11}{16}$. Find the total amount paid if her broker charges 1.5%.

43. STOCK PURCHASES Leslie Toombs purchased 250 shares of Microsoft at \$105 per share and paid a 2% commission. She later sold the same stock at $\$118\frac{1}{4}$ but paid a 1.5% sales commission. Find her profit ignoring any dividends paid.

44. STOCK PURCHASES Gerome Smith bought 100 shares of Cendant at $\$15\frac{1}{4}$ per share and was charged a commission of 1.5%. He sold the 100 shares later at $\$22\frac{1}{8}$ and had a flat commission of \$25 through a discount broker. Find the profit, ignoring any dividends.

45. Explain current yield and the PE ratio. Why might investors be interested in these numbers? (See Objectives 5 and 6.)

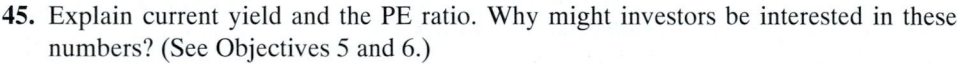

46. List four advantages of mutual funds. Name any particular mutual fund(s) that you have read or heard about. (See Objective 8.)

47. TWO INVESTMENT ALTERNATIVES Toni Chavez is comparing two different investment alternatives for his \$500 semiannual contribution to a retirement plan. Plan A uses certificates of deposits currently yielding 8% compounded semiannually. Plan B uses a global stock mutual fund with Merrill Lynch that has historically yielded 12% compounded semiannually. Find the amount of an annuity after 20 years of (a) plan A and (b) plan B. (c) Find the difference between the two. (See Section 14.1.)

48. RETIREMENT ACCOUNT INVESTMENT Laurie Zimms is trying to decide whether she should place her quarterly \$2000 retirement account investment in a certificate of deposit yielding 6% compounded quarterly or in a stock mutual fund that has historically yielded 10% compounded quarterly. Find the amount of an annuity after 12 years using (a) the CDs and (b) the mutual fund. (c) Find the difference between the two. (See Section 14.1.)

18.3 Bonds

Objectives

1. *Know the basics of bonds.*
2. *Read bond tables.*
3. *Find the cost of bonds, including commission.*
4. *Understand how mutual funds containing bonds can be used for monthly income.*

A corporation can raise money by selling additional shares of stock. The purchasers of this stock become part owners of the business. However, company management may feel that the sale of additional stock would excessively dilute the ownership rights of current stockholders. If so, management might decide to raise money by borrowing.

Objective 1 *Know the Basics of Bonds.* For short-term money needs, a corporation might borrow from a bank or an insurance company. For longer-term borrowing (such as 5 years or more), the corporation might borrow money from the public. The corporation borrows this money by selling **bonds**. A bond is *a promise to repay* the borrowed money at some specified time. The issuer of the bond promises to pay interest at a certain annual rate.

However, there is one type of bond, called a **zero-coupon bond**, that does not pay annual interest—it only pays the face value at maturity. An example of a zero-coupon bond is a bond bought for \$600 in 1998 with no annual interest payments, but it pays the bondholder \$1000 at maturity in 2005. In this case, a present value of \$600 would grow to \$1000 in 7 years, resulting in a yield of just over 7.5%.

National governments, states and provinces, and some cities also issue bonds. Bonds issued by a state or local governmental authority are called **municipal bonds**. Bonds issued by a corporation are called **corporate bonds**. One advantage of municipal bonds is that the income from these bonds is usually *not* subject to federal income taxes but may be subject to a state income tax.

Suppose a person buys a bond for \$1000 from Chase Manhattan due in 2006. This bond is a promise to repay the \$1000 in the year 2006 and until then to pay interest of $6\frac{1}{4}\%$ per year. For this bond, the investor would receive annual interest payments of

$$\$1000 \times 6\tfrac{1}{4}\% = \$1000 \times 0.0625 = \$62.50$$

In this example, \$1000 is the amount that the company promises to repay to the investor. This amount is called the **face value** or **par value** of the bond. Almost all corporations issue bonds with a par value of \$1000.

Objective 2 *Read Bond Tables.* The Chase Manhattan bond will be redeemed by the company in 2006 for \$1000. Suppose, however, that the bond's owner needs money now. The bond can be sold quickly through a bond dealer. However, the price for the bond is set, not by Chase Manhattan, but by market conditions. To find the selling price of a bond, look in the "Corporate Bond" section of the daily newspaper. A portion of *The Wall Street Journal* corporate bond page is reproduced in Figure 18.7.

For these Chase Manhattan bonds, the table gives the following.

Bonds	Cur Yld	Vol	Close	Net Chg
ChaseM $6\frac{1}{4}$ 06	6.4	20	$97\frac{1}{2}$	—

After the name of the company, which may be abbreviated, comes $6\frac{1}{4}$ 06. The $6\frac{1}{4}$ says that the bonds pay $6\frac{1}{4}\%$ interest on their face value of \$1000, while 06 is an abbreviation for the year 2006. (An "s" indicates that interest is paid by the company every 6 months instead of annually.) The number 6.4 is the current yield. The 20 shows that 20 bonds with a face value of \$1000 each were sold that day. The next number gives the closing price of the bonds. The bond prices in the table represent *not dollar amounts but percents*. Here $97\frac{1}{2}$ says that the bond was selling at 97.5% of its par value of \$1000. An investor who sold these bonds at this price would receive

$$\$1000 \times 97.5\% = \$1000 \times 0.975 = \$975$$

NEW YORK EXCHANGE BONDS

CORPORATION BONDS
Volume, $9,753,000

Bonds	Cur Yld.	Vol.	Close	Net Chg.
AES Cp 4½05	cv	7	125	+ 6⅛
AES Cp 8s8	8.6	235	93¼	− 1¼
ATT 5⅛01	5.2	228	97⅞	+ ⅛
ATT 7⅛02	7.0	35	101⅞	...
ATT 6¾04	6.7	25	100½	− ¼
ATT 7½06	7.2	10	103⅝	− ¼
ATT 6s09	6.4	50	93⅛	− ⅛
ATT 8⅛22	7.9	54	103¼	+ ⅛
ATT 8⅛24	7.9	14	103	+ ⅞
ATT 8.35s25	7.8	25	107	+ 1½
ATT 6½29	7.3	35	89½	+ ¼
ATT 8⅝31	8.1	40	107	+ ⅛
Aames 10½02	13.3	15	79	− 1
AlldC zr2000	...	2	93½	...
Alza 5s06	cv	36	133	− 1
Alza zr14	...	17	62⅛	...
AForP 5s30	7.5	5	66⅝	+ 1½
Amresco 10s03	11.8	230	84¾	− ¼
Amresco 10s04	11.8	806	85	+ 2
BellsoT 6¼03	6.3	25	99⅜	− ⅛
BellsoT 6⅜04	6.4	10	100	− ⅜
BellsoT 6½05	6.5	25	99½	+ ⅜
BellsoT 5⅞09	6.3	5	93½	− ¼
BellsoT 8¼32	8.0	60	103½	− ⅜
BellsoT 7⅞32	7.7	26	103	+ ¼
BellsoT 7½33	7.7	10	97¾	+ ¼
BellsoT 6¾33	7.4	1	91¾	...
BellsoT 7⅝35	7.7	75	99	...
BethSt 8.45s05	8.5	51	99	− ½
Bevrly 9s06	9.7	31	92½	− ½
Bluegrn 8¼12	cv	25	93½	+ 1
Bordn 8⅜16	8.7	18	96⅜	− ⅝
BrnSh 9½06	9.3	27	102	− ¾
CaterpInc 9⅜01	8.9	2	105⅛	− 4
CentrTrst 7½01	cv	6	97½	+ 2
ChaseM 6¼06	6.4	20	97½	...
ChaseM 6¾08	6.9	15	98⅛	− ⅛
CPoM 7¼12	7.3	25	100	...
DukeEn 7s00	7.0	10	100 17/32	− 1/32
DukeEn 6¼04	6.5	38	96½	− ⅛
DukeEn 6⅞23	7.4	7	93⅛	− 1⅜
FedDS 10s01	9.6	45	104⅛	− ⅞
FedDS 8⅛02	8.0	25	102⅛	+ ⅛
FnclFed 4½05	cv	5	82⅝	− 2⅜
FUnRE 8⅞03	8.9	35	99½	− ¾
Florsh 12¾02	12.0	2	105⅞	+ 1¼
FordCr 6⅜08	6.8	25	94	− ⅛
GBCB 8⅜07	8.8	120	95⅛	− 1⅞
GMA 5½01	5.7	70	96¾	− ¼
GMA 5⅝01	5.7	30	98	+ ⅜
GMA 7s02	7.0	11	100⅝	− ¼
GMA 6⅝02	6.6	16	99¾	+ ⅛
GMA zr12	...	17	366¾	+ ¾
GenesisH 9¾05	13.6	35	71½	− 2
Hlthso 9½01	9.3	85	101⅝	...
Hexcel 7s03	cv	100	82½	+ 1⅜
Hilton 5s06	cv	160	88¼	+ ¼
Hollngr 8⅝05	8.6	5	100	− ½
Hollgnr 9¼07	9.2	2	100⅞	− 1
Hollngr 9¼06	9.2	50	100⅛	− 1⅜
HomeDpt 3¼01	cv	7	264¼	− 2¾
HuntPly 11¾04	11.6	3	101⅝	− 1⅜
IBM 6⅜00	6.4	25	100 1/32	− 3/32
IBM 7¼02	7.1	210	102⅜	− ¼
IBM 5⅜09	6.0	5	90⅜	+ 2⅞
IBM 6½28	7.2	1	90½	+ ⅛
IPap dc5⅛12	6.3	20	81⅝	− 1⅝
JCPL 7⅛04	7.2	30	99⅜	− ¼
KaufB 7¾04	8.2	55	95	...
KentE 4½04	cv	13	79	...
KerrM 7½14	cv	15	98⅛	− ¾
Kolmrg 8¾09	cv	13	99¾	+ ¼
LehmnBr 8¾02	8.2	5	106⅝	− 1⅜
Leucadia 7¾13	8.0	15	96⅜	− ⅛
Loews 3⅛07	cv	200	89	+ ¼
LglsLt 9s22	8.6	7	105¼	− ⅞
Lucent 6.9s01	6.8	4	101⅛	− ¼
MSC Sf 7⅞04	cv	10	87½	+ ½

Figure 18.7

NOTE This selling price of \$975 *is lower than* the face value of \$1000, since general interest rates are higher than the $6\frac{1}{4}\%$ of this bond. ▬

Example 1 Finding the Selling Price of Bonds

Find the current selling price of the following bonds: (a) Duke Energy, $6\frac{1}{4}\%$ bonds of 2004, (b) ATT, $7\frac{1}{2}\%$ bonds of 2006 and (c) Florsh, $12\frac{3}{4}\%$ bonds of 2002.

Solution

(a) The listing for Duke Energy provides the following.

Bonds	Cur Yld	Vol	Close	Net Chg
DukeEn $6\frac{1}{4}$ 04	6.5	38	$96\frac{1}{2}$	$-\frac{1}{8}$

The price information for these bonds comes from the number $96\frac{1}{2}$. Notice that only 38 of these bonds were sold that day. This number represents 38 bonds, not 38 hundreds, as with stocks.

Since the price of these bonds for the day was $96\frac{1}{2}$, or $96\frac{1}{2}\%$ of par value, the selling price for one bond was

$$\$1000 \times 96\tfrac{1}{2}\% = \$1000 \times 0.965 = \$965$$

(b) Find the listing for ATT $7\frac{1}{2}$ 06. The price for the day is given as $103\frac{5}{8}$, with one bond selling for

$$\$1000 \times 103\tfrac{5}{8}\% = \$1000 \times 1.03625 = \$1036.25$$

(c) The selling price of a Florsh $12\frac{3}{4}\%$ bond of 2002 is given in the bond table as $105\frac{7}{8}$. The selling price of one bond is

$$\$1000 \times 105\tfrac{7}{8}\% = \$1000 \times 1.05875 = \$1058.75$$

Objective 3 ***Find the Cost of Bonds, Including Commission.*** Commissions charged on bond sales vary among brokers. A common charge is \$5 per bond, either to buy or to sell.

Example 2 Finding the Cost of Purchasing Bonds

Find the charge to purchase 15 bonds of ATT $6\frac{1}{2}\%$ of 2029. Use the closing price for the day and assume a sales charge of \$5 per bond.

Solution From the bond table in Figure 18.7, the price is found to be $89\frac{1}{2}$, or $89\frac{1}{2}\%$. The cost to buy one bond of \$1000 par value is

$$\$1000 \times 89\tfrac{1}{2}\% = \$1000 \times 0.895 = \$895$$

Fifteen of these bonds cost

$$15 \times \$895 = \$13{,}425$$

Commission is \$5 per bond, with a total commission of

$$15 \times \$5 = \$75$$

The purchase price of the 15 bonds is

$$\$13{,}425 + \$75 = \$13{,}500.$$

Bonds are a debt; a corporation owes money to its bondholders. As such, *bondholders have first claim*, after bankruptcy lawyers, taxing authorities, and wage earners, on the assets of the corporation if it goes into bankruptcy. (Stockholders have the last claim.) Even so, bonds may pay off only a few cents on the dollar in the event of bankruptcy. Bondholders have lost substantial sums in recent bankruptcies. Some investors like to buy the bonds of bankrupt and troubled companies—such **junk bonds** have been known to pay off handsomely when and if a company regains its financial health. We might mention that bond salespeople do not like the term "junk bonds"; they prefer "high-yield securities." Junk bonds are usually very risky and best left for professional investors.

Objective 4 ***Understand How Mutual Funds Containing Bonds Can Be Used for Monthly Income.*** A mutual fund can invest in stocks, in bonds, or in both. Typically, people and companies invest in stocks when they have a long period of time (at least a few years) during which they are accumulating funds. They tend to invest in bonds when they need the income from their investments. Many financial planners suggest that people invest in both stocks and bonds throughout their adult lifetime—*perhaps a higher percent of stocks when young and a higher percent of bonds when approaching retirement.*

Example 3 Using a Bond Fund for Income

Ada Clen needs income and is worried about investing her money in the bonds of a single company since that company might go bankrupt. Therefore she chooses to invest in a bond fund that invests in the bonds of many different companies and that currently yields 7% per year. Find the annual interest if she invests \$75,000 in the bond fund.

Solution Use the formula for simple interest $I = PRT$.

$$\text{Annual interest} = \$75{,}000 \times 0.07 \times 1 = \$5250 \text{ per year}$$

18.3 Exercises

Find the following information from the bond table in Figure 18.7. Use the listing for Genesis H.

1. Closing price

2. Number of bonds sold

3. Interest paid on the par value of the bonds, both as a percent and in dollars

4. Year when the bonds will be paid off by the company

5. Change since the previous day in the price of one bond

6. Price to buy 25 such bonds at the closing price with sales charges of \$5 per bond

Find the cost, including sales charges of \$5 per bond, for each of the following transactions.

Bond	Number Purchased
7. Alza, 5% bonds of 2006	30
8. Hilton, 5% bonds of 2006	20
9. FordCr, $6\frac{3}{8}$% bonds of 2008	100
10. Bordn, $8\frac{3}{8}$% bonds of 2016	40
11. IBM, $5\frac{3}{8}$% bonds of 2009	50
12. GenesisH, $9\frac{3}{4}$% bonds of 2005	10
13. Bellso T, $8\frac{1}{4}$% bonds of 2032	25
14. ATT, $6\frac{3}{4}$% bonds of 2004	35

15. Explain the purpose of bonds. (See Objective 1.)

16. Explain the differences between common stock, preferred stock, and bonds. (See Sections 18.1 and 18.3.)

Solve each application problem. Assume sales commissions of \$5 per bond. Use the table of bond prices in Figure 18.7, and don't forget the sales commission.

17. STOCK PURCHASES Onita Fields purchased 20 bonds of Bluegrn. Find the total cost.

18. STOCK PURCHASES Mary Dunlap bought 15 bonds of KaufB. Find the total cost.

19. STOCK PURCHASES Linda Cypert purchased 10 bonds of IPap and 10 bonds of Kolmrg. Find the total cost.

20. **PURCHASING BONDS** Jake Botswana bought 10 bonds of AForP and 15 bonds of Hithso. Find the total cost.

21. **PURCHASING BONDS** Pey Tang purchased 10 ATT $6\frac{1}{2}$29 bonds and 10 IBM $5\frac{3}{8}$09 bonds. Find (a) the total cost and (b) the annual interest.

22. **CARE OF DISABLED SON** Kitty Wysong purchased 20 Kolmrg $8\frac{3}{4}$09 bonds and 20 Loews $3\frac{1}{8}$07 bonds and put them in a trust for her disabled son's future care. Find (a) the total cost and (b) the annual interest.

23. **INTEREST** A wealthy couple places $100,000 in a municipal bond fund that allows them to reduce their taxes. Find their tax-free income from this investment if the fund earns $4\frac{1}{2}\%$ per year.

24. **ANNUAL INTEREST** Helmut Schmidt places all $150,000 of his retirement funds in a mutual fund containing only high-quality corporate bonds. Find his annual interest if the fund earns 8% per year.

25. **INCOME FROM A DIVORCE** Bill and Jane Fickland recently divorced. As a part of the divorce settlement, Mr. Fickland agrees to pay Ms. Fickland $7000 per year for several years. Find the amount that must be invested in a mutual fund containing bonds that is expected to yield 7.5% annually to generate the needed income.

26. **LAWSUIT AWARD** Joan Klein lost a lawsuit to Milton Freeman. As a result, she must make annual payments of $28,500 to Freeman for several years. Find the amount that must be invested in a mutual fund containing bonds that is expected to yield 7.8% annually to generate the needed income.

18.4 Distribution of Profits and Losses in a Partnership

Objectives

1. *Divide profits by equal shares.*
2. *Divide profits by agreed ratio.*
3. *Divide profits by original investment.*
4. *Divide profits by salary and agreed ratio.*
5. *Divide profits by interest on investment and agreed ratio.*

In a partnership, a business is owned by two or more people. These partners may have invested equal amounts of money to start the business, or one may have invested money while another invested specialized knowledge. The partners must agree on the relative amounts of money and time that will be invested in the business. They must also agree on the method by which any profits will be distributed. This section considers the various methods by which partnership profits may be distributed.

Objective 1 *Divide Profits by Equal Shares.* The partners may simply agree to share all profits and losses equally. (In fact, if there is no formal agreement stating the terms under which profits are to be divided, most states require that profits be divided equally.)

Example 1 Dividing Profits by Equal Shares

Three partners open a men's wear store. Each agrees to put up one-third of the investment funds needed and to work 40 hours per week in the business for no salary at first. They also agree to split profits equally. In a year with $96,000 in profits, each partner would get

$$\frac{1}{3} \times \$96{,}000 = \$32{,}000$$

Objective 2 *Divide Profits by Agreed Ratio.* Partners may agree to divide the profits using some given rule. For example, two partners might agree that profits will be divided so that 60% goes to one partner and 40% to the other. Profit divisions are sometimes given as a ratio; this division could be written 60:40, or in a reduced form, 3:2, with profits said to be divided in an **agreed ratio**.

Example 2 Dividing Profits by Agreed Ratio

Aaron Ortego and Gary Wayne purchase the Ter Marche Apartments in New Orleans, Louisiana, and agree to split profits in the ratio of 1:3. How would profits of $18,000 be divided?

Solution The ratio 1:3 says that profits will be divided into 1 + 3 = 4 equal shares. Ortego gets 1 and Wayne gets 3 of the equal shares.

$$\$18{,}000 \div 4 = \$4500$$

Ortego's share:	$4500 × 1 =	$ 4,500
Wayne's share:	$4500 × 3 =	$13,500
Total		$18,000

Objective 3 *Divide Profits by Original Investment.* A common way of dividing the profits is on the basis of the **original investments** made by each partner. The fraction of the total original investment supplied by each partner is used to find the fraction of the profit that each partner receives.

Example 3 Dividing Profits by Original Investment

Bob Huffman and Gary White form a partnership to drill for oil. Huffman contributes $80,000 and White contributes $40,000. They sign an agreement that says that each partner will lose the money each invested if they drill a dry hole, and that any profits will be distributed based on the original investment. If they discover an oil field and sell it for $600,000, find the share received by each.

Solution The total amount contributed to start the venture was

$$\$80{,}000 + \$40{,}000 = \$120{,}000$$

Of this total, Huffman contributed

$$\frac{\$80{,}000}{\$120{,}000} = \frac{2}{3}$$

and White contributed $\frac{1}{3}$. Therefore, Huffman is entitled to $\frac{2}{3}$ of the price received for the oil field, or

$$\frac{2}{3} \times \$600{,}000 = \$400{,}000$$

and White is entitled to

$$\frac{1}{3} \times \$600{,}000 = \$200{,}000$$

NOTE Each partner's fraction or percent of the total investment must be determined before the profit distribution is calculated. ■

Example 4 Dividing Losses by Original Investment

Suppose the oil company in Example 3 had a loss of $40,000 the first year. Find the share of the loss that each partner must pay.

Solution Just as partners share profits, they may be called on to share losses. In this case, Huffman must pay $\frac{2}{3}$ of the loss, or

$$\frac{2}{3} \times \$40{,}000 = (\$26{,}666.67)$$

and White must pay $\frac{1}{3}$ of the loss.

$$\frac{1}{3} \times \$40{,}000 = (\$13{,}333.33)$$

Losses are usually indicated with parentheses.

Objective 4 *Divide Profits by Salary and Agreed Ratio.* Sometimes one partner contributes money to get a business started, while a second partner contributes money and also operates the business on a daily basis. In such a case, the partner operating the business may be paid a salary out of profits, with any additional profits divided in some agreed-upon ratio, called dividing profits by **salary and agreed ratio**. As mentioned in the introduction to this chapter, a partner who makes only a financial investment but takes no part in running the business is a **silent partner**.

Example 5 Dividing Profits by Salary and Original Investment

Ben Walker has managed a restaurant for 8 years and would like to open his own restaurant except that he does not have enough money. He finally decides to form a partnership with Herma Gonzalez to open a restaurant that will be managed by Walker at a guaranteed annual salary of $18,000. Walker invests $20,000 and Gonzalez invests $60,000. They agree to divide any profits based on the original investment. Find the amount each partner would receive from a profit of $70,000.

Solution The first $18,000 is used for Walker's salary.

$70,000	Profit
− 18,000	Salary
$52,000	Profit to be divided

Walker would also receive

$$\frac{1}{4} \times \$52{,}000 = \$13{,}000$$

and Gonzalez would receive

$$\frac{3}{4} \times \$52{,}000 = \$39{,}000$$

Thus Walker receives $18,000 + $13,000 = $31,000 and Gonzalez receives $39,000.

Objective 5 ***Divide Profits by Interest on Investment and Agreed Ratio.*** Sometimes one partner will put up a large share of the money necessary to start a firm, while other partners may actually operate the firm. In such a case, an agreement to divide profits by **interest on investment and agreed ratio** may be reached by which the partner putting up the money gets interest on the investment before any further division of profits.

Example 6 Dividing Profits by Interest and Agreed Ratio

Laura Cameron, Jay Davis, and Donna Friedman opened a tool rental business. Cameron contributed \$250,000 to the opening of the business, which will be operated by Davis and Friedman. The partners agree that Cameron will first receive a 10% return on her investment before any further division of profits. Additional profits will be divided in the ratio 1:2:2. Find the amount that each partner would receive from a profit of \$75,000.

Solution Cameron is first paid a 10% return on her investment of \$250,000. This amounts to

$$\$250{,}000 \times 10\% = \$250{,}000 \times 0.10 = \$25{,}000$$

This leaves an additional

\$75,000	Total profit
− 25,000	Amount to Cameron
\$50,000	

to be divided. The additional profit of \$50,000 is to be divided in the ratio 1:2:2, respectively. First divide this amount into 1 + 2 + 2 = 5 equal shares. Cameron gets 1 of these 5 shares, or

$$\frac{1}{5} \times \$50{,}000 = \$10{,}000$$

Davis and Friedman each get 2 of the 5 shares, or

$$\frac{2}{5} \times \$50{,}000 = \$20{,}000$$

In summary, Cameron gets the following amount.

\$25,000	Return on investment
+ 10,000	Share of profit
\$35,000	Total

Both Davis and Friedman get \$20,000.

NOTE The return on investment is paid first and then the remaining profit is divided among the partners. ■

Example 7 Dividing Losses by Agreed Rates

Suppose the tool rental business in Example 6 had a profit of only \$15,000. What would be the distribution of this amount?

Solution The partners agreed to give Cameron a 10% return, or \$25,000. The profits were only \$15,000, leaving a loss of \$10,000.

$$\begin{array}{r} \$25{,}000 \\ -\ 15{,}000 \\ \hline (\$10{,}000) \end{array}$$

This loss of \$10,000 will be shared in the ratio 1:2:2, just as were the profits. Cameron's share of the loss is

$$\frac{1}{5} \times \$10{,}000 = (\$2000)$$

while the share of both Davis and Friedman is

$$\frac{2}{5} \times \$10{,}000 = (\$4000)$$

Losses are shown in parentheses.

Cameron gets \$25,000, minus her share of the loss.

	\$25,000	Due to Cameron
−	2,000	Her share of loss
	\$23,000	Actually received by Cameron

Both Davis and Friedman must each *contribute* \$4000 toward the loss. The \$23,000 that Cameron actually receives is made up as follows.

	\$15,000	Profit
	4,000	From Davis
+	4,000	From Friedman
	\$23,000	Total to Cameron

NOTE Cameron does absorb her share of the loss by accepting \$23,000 instead of the \$25,000 return on investment initially agreed upon. ▬

18.4 Exercises

Divide the following profits. Round all answers to the nearest dollar.

	Partners	Investment	Method	Profits
1.	1	\$40,000	Equal shares	\$58,000
	2	\$40,000		
2.	1	\$20,000	Ratio 1:3:4	\$120,000
	2	\$50,000		
	3	\$50,000		
3.	1	\$80,000	Ratio of investment	\$46,000
	2	\$20,000		
4.	1	\$20,000	\$22,000 salary to partner 3;	\$72,000
	2	\$20,000	balance divided 3:3:2	
	3	\$5,000		

	Partners	Investment	Method	Profits
5.	1	$40,000	12% return to partner 1;	$25,000
	2	$10,000	balance in ratio 3:2	
6.	1	$40,000	10% return to partner 2;	$111,000
	2	$60,000	balance in ratio of	
	3	$50,000	investment	
7.	1	$20,000	$10,000 salary to partner 1;	$22,000
	2	$15,000	$12,000 salary to partner 2;	
	3	$40,000	balance in ratio of investment	
8.	1	$40,000	$20,000 salary to partner 3;	$133,000
	2	$40,000	10% return to partners 1	
	3	$5,000	and 2; balance divided	
	4	$35,000	2:2:1:2	

Solve the following application problems.

9. **CIRCUS ACT** Two partners train horses for a circus act. Find the amount of profit each receives if they have no agreement for dividing profits in a year with a profit of $84,500.

10. **SURGEONS** A firm with five surgeons has agreed to split the profits equally. Find the profit to each if the firm has a profit of $1,400,000 one year.

11. **MEXICAN RESTAURANT** Three brothers invest $10,000, $20,000, and $20,000, respectively, to start up a Mexican restaurant. Find the share that each receives if profit the first year is $15,600 and profits are divided in the ratio of the original investments.

12. **FLORIST SHOP** Kathy Bates and Marion Tomlin have started a new florist shop. Bates contributed $25,000 to the firm and Tomlin contributed $15,000. Find the division of a profit of $30,000 if profits are divided in the ratio of the original investments.

13. **DIVIDING PROFITS** Four partners have agreed to divide profits in the ratio 3:5:7:9. Find the division of a profit of $180,000.

14. Suppose the partners in Exercise 13 have a loss of $96,000. How much of the loss would be paid by each partner?

15. **TRAVEL AGENCY** Mary Finch and Pete Renz have started a new travel agency. Finch will run the agency. She gets a $15,000 salary, with any additional profits distributed in the ratio 1:4. Find the distribution of a profit of $57,000.

16. **TOY STORE** June Thomas, Ben Walker, and his sister Sara Walker start a toy store to be managed by Thomas at a salary of $20,000 per year. They agree that any additional profits will be divided in the ratio of 1:3:2. Find the division of a profit of $86,000.

17. **HARDWARE STORE** Bob Coker has invested $80,000 in a new hardware store. His partner, Will Toms, will actually run the store. The partners agree that Coker will get a 10% return on his investment, with any additional profits divided in the ratio 1:3. Find the division of a profit of (a) $60,000 and (b) $6000.

18. ELECTRONICS Wilma Dickson has invested \$350,000 in a small electronics plant, to be run by her partner, John Ardery. They agree that she will receive a 10% return on her investment, and that any additional profits will be divided in the ratio 2:3. Divide a profit of (a) \$90,000 and (b) \$30,000.

19. PARTNERSHIP Three partners invest \$15,000, \$25,000, and \$30,000 in a business. The partners agree that partner 1 will receive a 10% return on investment, with partner 2 receiving a salary of \$12,000. Any additional profits will be divided in the ratio of the original investments. Divide a profit of \$110,000.

20. WHOLESALE PLUMBING A wholesale plumbing business has three partners. Partner 1 invested \$50,000 in the business and is given a 10% return on investment. Partner 2 invested \$75,000 and earns a 6% return on investment plus a salary of \$21,000. Partner 3 invested \$100,000 and earns a 12% return on investment. Any additional profits are divided in the ratio 2:1:2. Divide a profit of \$180,000.

21. State the approach used to divide profits among 3 partners if partner 1 gets a fixed salary and the remaining profits are divided according to a ratio of 2:3:2. (See Objective 4.)

22. Explain the difference between the distribution of profits and loses in a partnership versus those in a corporation. (See Sections 18.1 and 18.4.)

18.5 Distribution of Overhead

Objectives

1. *Allocate overhead by floor space.*
2. *Allocate overhead by sales value.*
3. *Allocate overhead by number of employees.*

Businesses have many expenses in addition to the cost of the materials and labor that are actually used to make a product. Rent on the factory building must be paid, insurance premiums and executive salaries must be paid, office supplies must be ordered, and so on. These general expenses, which cannot be avoided but which do not go directly for the production of goods and services, are called **overhead**. The cost of an office computer would come under overhead, while the cost of sheet metal used to actually make a product would not.

A company can usually decide on the total overhead expenses fairly quickly; however, a problem often comes up in dividing the overhead among the various products or lines of business in a company. This is one area in which **cost accounting** is used. Various methods are used by different firms to divide overhead. The choice of a method often depends on industry practice.

In any case, the **allocation of overhead** is usually done by forming a ratio of each product or department to the total firm. There are several ways of forming this ratio.

Objective 1 *Allocate Overhead by Floor Space.* Overhead can be allocated by department according to the **floor space** used by each department of the company.

Example 1 Allocating Overhead by Floor Space

Clover Printing has three departments, with floor space as shown.

Department	Floor Space
Magazine printing	50,000 square feet
Book printing	30,000 square feet
Catalog printing	20,000 square feet
Total	100,000 square feet

Allocate an overhead of $275,000.

Solution The magazine printing department has a floor space of 50,000 square feet out of a total of 100,000 square feet. Therefore, this department is allocated

$$\frac{50{,}000}{100{,}000} = \frac{1}{2}$$

of the overhead, or

$$\frac{1}{2} \times \$275{,}000 = \$137{,}500$$

When finding the expenses of this department, the company accountants would assign an overhead expense of $137,500 for the department.

The book printing department uses

$$\frac{30{,}000}{100{,}000} = \frac{3}{10}$$

of the floor space, and so would be allocated $\frac{3}{10}$ of the overhead, or

$$\frac{3}{10} \times \$275{,}000 = \$82{,}500$$

Finally, catalog printing would be allocated

$$\frac{20{,}000}{100{,}000} = \frac{1}{5}$$

of the overhead, or

$$\frac{1}{5} \times \$275{,}000 = \$55{,}000$$

NOTE Check your answer by adding the individual departmental allocations. This sum should equal the total overhead. ▬

Objective 2 *Allocate Overhead by Sales Value.* It is common to allocate overhead according to the **sales value** of each department or product, as shown in the next example.

Example 2 Allocating Overhead by Sales Value

Bales Manufacturing produces four products with monthly production and values as shown.

Product	Production	Value of Each
Wheelbarrows	2,500	$40
Ladders	5,000	25
Shovels	6,000	10
Hammers	25,000	3

Allocate an overhead of $50,000.

Solution First find the total value of each item.

Product	Production	Value of Each	Total Value
Wheelbarrows	2,500	$40	2,500 × $40 = $100,000
Ladders	5,000	25	5,000 × 25 = 125,000
Shovels	6,000	10	6,000 × 10 = 60,000
Hammers	25,000	3	25,000 × 3 = 75,000
			Total $360,000

Wheelbarrows amounted to $100,000 of the total value of $360,000. Therefore, the fraction

$$\frac{\$100{,}000}{\$360{,}000} = \frac{5}{18}$$

of the total overhead must be applied to wheelbarrows. The total overhead is $50,000, so

$$\frac{5}{18} \times \$50{,}000 = \$13{,}888.89$$

must be applied to wheelbarrows. Also,

$$\frac{\$125{,}000}{\$360{,}000} \times \$50{,}000 = \frac{25}{72} \times \$50{,}000 = \$17{,}361.11$$

of overhead will be applied to ladders, and

$$\frac{\$60{,}000}{\$360{,}000} \times \$50{,}000 = \frac{1}{6} \times \$50{,}000 = \$8333.33$$

to shovels. Finally,

$$\frac{\$75{,}000}{\$360{,}000} \times \$50{,}000 = \$10{,}416.67$$

is applied to hammers. Check that the sum of the various allocated overheads is the total overhead of $50,000.

Objective 3 *Allocate Overhead by Number of Employees.* Overhead can also be allocated by the **number of employees** associated with a department or product.

Example 3 Allocating Overhead by Number of Employees

An insurance office has an overhead charge of $80,000. Allocate this overhead expense based on the number of employees.

Solution Allocate this expense based on a ratio of the number of employees in the department to the total number of employees.

Department	Number of Employees	Ratio of Employees	Overhead of Department
Commercial	5	$\frac{5}{8}$	$\frac{5}{8} \times \$80,000 = \$50,000$
Personal	2	$\frac{2}{8}$	$\frac{2}{8} \times \$80,000 = \$20,000$
Life	1	$\frac{1}{8}$	$\frac{1}{8} \times \$80,000 = \$10,000$
Total	8		$80,000

18.5 Exercises

Allocate overhead as indicated. Round to the nearest dollar.

1.

Department	Floor Space
1	4,000 square feet
2	8,000 square feet
3	10,000 square feet

Overhead; $330,000

2.

Department	Floor Space
A	8,000 square feet
B	14,000 square feet
C	10,000 square feet

Overhead: $288,000

3.

Department	Floor Space
1	2400 square feet
2	3600 square feet
3	4000 square feet
4	6000 square feet

Overhead: $120,000

4.

Department	Floor Space
1	4,200 square feet
2	13,500 square feet
3	21,800 square feet
4	3,600 square feet

Overhead: $420,000

5.

Product	Number Produced	Value of Each
M	15,000	$8.00
N	20,000	$3.50
P	35,000	$2.00

Overhead: $140,000

6.

Product	Number Produced	Value of Each
X	5000	$20
Y	8000	$25
Z	4000	$50

Overhead: $62,500

7.

Product	Number Produced	Value of Each
1	140	$100
2	2000	$15
3	150	$20
4	1000	$22

Overhead: $48,000

8.

Product	Number Produced	Value of Each
1	150	$6
2	200	$12
3	75	$3
4	125	$8

Overhead: $10,000

9.

Department	Number of Employees
X	110
Y	60
Z	80

Overhead: $650,000

10.

Department	Number of Employees
J	1000
K	7000
L	2000

Overhead: $900,000

11.

Department	Number of Employees
1	100
2	120
3	140
4	60

Overhead: $800,000

12.

Department	Number of Employees
1	90
2	20
3	50
4	40

Overhead: $120,000

13. AUTO PARTS STORE Dayton Auto Parts allocates its $360,000 overhead by the floor space used by each department. Allocate the overhead for the following departments.

Department	Floor Space
Hoses	2000 square feet
Carburetors	8000 square feet
Water pumps	6000 square feet
Fuel pumps	9000 square feet
Gaskets	1000 square feet
Filters	4000 square feet

14. OFFICE SUPPLIES Savon Office and School Supplies wishes to allocate its $110,000 overhead among its various departments by floor space. Allocate the overhead for the following departments.

Department	Floor Space
Typing paper	750 square feet
Copy machine paper	600 square feet
Copy machines	1000 square feet
Office furniture	1500 square feet
Filing cabinets	500 square feet
Calculators	650 square feet

15. LUMBER COMPANY Salvage Lumber wishes to allocate its overhead of $68,000 by the sales value of each product. Allocate overhead for the following products.

Product	Number Produced	Value per Unit
Construction 2 × 4's	150	$200
Plywood	200	$400
Veneers	100	$600
Wood chips	500	$75
Furniture wood	300	$150

16. MEAT PRODUCTS Allocate the $8732 monthly overhead of Victor Meats by sales value of products, using the information in this chart.

Product	Number Produced	Value per Unit
Beef	10	$800
Lamb	7	$300
Pork	5	$750
Chicken	14	$120
Sausage	12	$150
Luncheon meats	15	$300

17. OVERHEAD TEA COMPANY Boston Teas has a weekly overhead of $6000 that they wish to allocate by the number of employees per department. Use the following chart.

Department	Number of Employees
Office	6
Marketing	8
Distribution	28
Accounting	8

18. DRUG RESEARCH Drug Research, Inc. wishes to allocate a weekly overhead of $11,400 among its departments according to the number of employees per department. Use the following chart.

Department	Number of Employees
Headache remedies	15
Pain killer	20
Cold remedies	20
Foot powder	5
Eye wash	8
Skin lotion	8

19. Define the term *overhead.* List at least three expenses that would be included in overhead. Explain why a manager needs to allocate overhead. (See Objective 1.)

20. Compare and contrast the three ways of allocating overhead that were discussed in this section. Give circumstances under which each of the three might be appropriate. (See Objectives 1–3.)

Chapter 18 Quick Review

Chapter Terms

Review the following terms to test your understanding of the chapter. For any terms you do not know, refer to the page number found next to that term.

agreed ratio [p. 686]
allocation of overhead [p. 691]
annual meeting [p. 663]
attached [p. 663]
board of directors [p. 663]
bond [p. 681]
capital [p. 663]
commission [p. 672]
common stock [p. 663]
convertible preferred stock [p. 664]
corporate bonds [p. 681]
corporations [p. 663]
cost accounting [p. 691]
cumulative preferred stock [p. 664]
current yield [p. 674]
discount broker [p. 672]
dividend [p. 664]
Dow Jones Industrial Average [p. 675]
earnings per share [p. 666]
executive officers [p. 663]
face value [p. 681]
floor space [p. 691]
interest on investment and agreed ratio [p. 688]
junk bond [p. 683]
limited liability [p. 663]
municipal bond [p. 681]
mutual fund [p. 676]
nonparticipating [p. 664]
number of employees [p. 694]
odd lot [p. 672]
odd-lot differential [p. 672]
original investments [p. 686]
overhead [p. 691]
par value [p. 664, 681]
participating [p. 664]
partnership [p. 663]
preferred stock [p. 663]
price-earnings ratio (PE ratio) [p. 675]
round lot [p. 672]
sales value [p. 692]
salary and agreed ratio [p. 687]
silent partner [p. 663, 687]
sole proprietorship [p. 663]
stock [p. 663]
stockbroker [p. 669]
stock certificate [p. 663]
stock exchange [p. 669]
stockholder [p. 663]
stock ratios [p. 674]
transfer tax [p. 672]
zero coupon bond [p. 681]

Concepts

18.1 Determining the amounts paid to holders of preferred and common stock

To find total paid to owners of preferred stock, multiply par value by dividend rate to obtain dividend per share, then multiply by number of shares.

To find the dividend paid to owners of common stock, subtract total paid to owners of preferred stock from total available to stockholders, then divide by number of shares of common stock.

Examples

A company distributes \$750,000 to stockholders. It has 15,000 shares of \$100 par value 4% preferred stock and 150,000 shares of common stock. Find (a) amount paid to holders of preferred stock and (b) amount per share to holders of common stock.

(a) Dividend per share 5 \$100 3 0.04 5 \$4

$$\text{Total to preferred} = \$4 \times 15{,}000 = \$60{,}000$$

(b) Dividend to common

$$= \frac{\$750{,}000 - \$60{,}000}{150{,}000} = \$4.60$$

18.1 Finding earnings per share

Subtract dividends on preferred stock from net income, then divide by the number of shares of common stock outstanding.

A company made \$500,000 last year. The company has 750,000 shares of common stock outstanding and paid \$75,000 to owners of preferred stock. Find the earnings per share.

$$EPS = \frac{\$500{,}000 - \$75{,}000}{750{,}000} = \$0.57$$

18.2 Reading the stock table

Locate the stock involved and determine the various quantities required.

Use the stock table to find the following information for Comsat: dividend; high for day, low for day, total sales; yearly high; yearly low.

52 Weeks Hi	Lo	Stock	Sym	Yld Div.	%	Vols PE	100s	Hi	Lo	Close	Net Chg
$39\frac{5}{8}$	$21\frac{3}{4}$	Comsat	CQ	.20	.5	47	3452	$37\frac{1}{16}$	36	$36\frac{3}{8}$	—

Dividend is \$0.20, high is $37\frac{1}{16}$ or \$37.0625, low is 36 or \$36; total sales are 345,200; yearly high is $39\frac{5}{8}$ or \$39.625; yearly low is $21\frac{3}{4}$ or \$21.75; and net change is 0.

18.2 Finding the current yield on a stock

$$\text{Current yield} = \frac{\text{Annual dividend}}{\text{Current price}}$$

Find the current yield of a stock if the purchase price is $\$112\frac{1}{8}$ per share and the annual dividend is \$1.40.

$$\text{Current yield} = \frac{\$1.40}{\$112.125}$$

$$= 0.012 = 1.2\%$$

18.2 Selling shares of a stock

Find the basic price of the stock from the table. Subtract the SEC fee and the broker's commission from the basic price of the stock.

Find the amount received by a person selling 500 shares of a stock at $53\frac{3}{8}$.

$$\text{Basic price} = 500 \times \$53.375 = \$26,687.50$$

Broker's commission

$$= 0.015 \times \$26,687.50 = \$400.31$$

$$\text{SEC fee: } \$26,687.50 \div \$300 = 89 \text{ (rounded up)}$$

$$\text{SEC fee} = 89 \times \$0.01 = \$0.89$$

Seller's proceeds

$$\begin{array}{r} \$26,687.50 \\ -400.31 \\ -\quad 0.89 \\ \hline \$26,286.30 \end{array}$$

18.2 Finding the price to earnings ratio (PE ratio)

To find the price to earnings ratio use the formula.

$$\text{PE ratio} = \frac{\text{Price per share}}{\text{Annual net income per share}}$$

Price per share, \$42.50; annual net income per share, \$2.75.

$$\text{PE ratio} = \frac{\$42.50}{\$2.75} = 15.45$$

18.3 Determining the cost of purchasing bonds

Locate the bond in the table, then multiply the price of the bond by 1000 and the number of bonds purchased. Then, add \$5 per bond to the total cost.

Find the cost, including sales charges, of 20 Bevrly bonds.

$$20 \times (1000 \times 0.925) + (20 \times \$5) = \$18,600$$

18.3 Determining the amount received from the sale of bonds

Locate the bond in the table, then multiply the price of the bond by 1000 and the number of bonds sold. Subtract $5 per bond from the total selling price.

Find the amount received from the sale of 15 JCPL bonds.

$$15 \times (1000 \times 0.99375) - (15 \times \$5) = \$14{,}831.25$$

18.3 Finding the annual income from a mutual fund containing bonds

Multiply the amount invested in the fund by the yield rate.

Find the annual income from $120,000 invested in a bond fund yielding $7\frac{1}{2}\%$.

$$\$120{,}000 \times 0.075 = \$9000$$

18.4 Dividing profits in a partnership

Use one of the following methods to determine each partner's ratio of the profits.

1. Equal shares
2. Agreed ratio
3. Original investment
4. Salary and agreed ratio
5. Interest on investment and agreed ratio
6. Multiply total profits by each partner's ratio.

Divide profits of $75,000 among three investors by the original investment if each partner invests the following amount.

Partner	Investment
1	$12,000
2	$15,000
3	$18,000

Total initial investment = $12,000 + $15,000 + $18,000 = $45,000

Ratios for each partner:

1. $\frac{12{,}000}{45{,}000} = \frac{4}{15}$
2. $\frac{15{,}000}{45{,}000} = \frac{1}{3}$
3. $\frac{18{,}000}{45{,}000} = \frac{2}{5}$

Profit for each partner:

1. $\frac{4}{15}(\$75{,}000) = \$20{,}000$
2. $\frac{1}{3}(\$75{,}000) = \$25{,}000$
3. $\frac{2}{5}(\$75{,}000) = \$30{,}000$

18.5 Allocating overhead by floor space

Determine the percent of floor space each department occupies. Multiply the percent by the amount of overhead to be allocated.

Department	Floor Space
Printing	40,000 sq. ft.
Cutting	25,000 sq. ft.
Binding	55,000 sq. ft

Allocate $330,000 overhead.

$$\text{Printing: } \frac{40{,}000}{120{,}000} \times \$330{,}000 = \$110{,}000$$

$$\text{Cutting: } \frac{25{,}000}{120{,}000} \times \$330{,}000 = \$68{,}750$$

$$\text{Binding: } \frac{55{,}000}{120{,}000} \times \$330{,}000 = \$151{,}250$$

18.5 Allocating overhead by sales value

Determine the percent of sales for each department. Multiply this percent by the amount of overhead to be allocated.

Allocate $120,000 overhead.

Product	Number Produced	Value of Each
A	5,000	$12
B	8,000	$5
C	10,000	$10

Total Value =

$$(5000)(\$12) + (8000)(\$5) + (10{,}000)(\$10) = \$200{,}000$$

$$\text{A: } \frac{(5000)(\$12)}{\$200{,}000} \times \$120{,}000 = \$36{,}000$$

$$\text{B: } \frac{(8000)(\$5)}{\$200{,}000} \times \$120{,}000 = \$24{,}000$$

$$\text{C: } \frac{(10{,}000)(\$10)}{\$200{,}000} \times \$120{,}000 = \$60{,}000$$

18.5 Allocating overhead by number of employees

Find the percent of total employees in each department and multiply by the overhead to be allocated.

Allocate $180,000 overhead.

Dept.	Number of Employees
A	40
B	20

$$\text{A: } \frac{40}{40 + 20} \times \$180{,}000 = \$120{,}000$$

$$\text{B: } \frac{20}{40 + 20} \times \$180{,}000 = \$60{,}000$$

Chapter 18 Review Exercises

The answer section includes answers to all Review Exercises.

In each of the following find the amount per share paid to holders of (a) preferred stock and (b) common stock. [18.1]

	Net Income	Reinvested Funds	Par Value	Rate	Number of Preferred Stockholders	Number of Common Stockholders
1.	$460,000	$317,000	$50	$6\frac{1}{2}\%$	8,000	180,000
2.	$2,375,000	$750,000	$150	8%	15,000	200,000
3.	$2,640,000	$425,000	$125	7%	22,750	750,000

In each of the following, find the earnings per share. Round to the nearest cent. [18.1]

	Net Income	Dividends on Preferred Stock	Number of Shares of Common Stock
4.	$127,500	$0	82,000
5.	$1,425,000	$675,000	275,000
6.	$2,750,000	$900,000	500,000

Use the stock table to find each of the following. Give money answers to the nearest thousandth of a dollar. [18.2]

7. High for the day for Comerica (CMA)
8. Low for the day for CurtWright (CW)
9. Closing price for CryoLife (CRY)
10. Change from the previous day for Copel (ELP)
11. Dividend for Danaher (DHR)
12. Sales for the day for Darden (DRI)
13. Yield for Comsat (CQ)
14. 52-week high for Data General (DGN)
15. 52-week low for Comdisco (CDO)

Find the price for each of the following stock purchases. Ignore any broker's fees. [18.2]

	Stock	Number of Shares	Transaction
16.	DanRivr A (DRF)	200	Close
17.	DPL Inc (DPL)	300	Low
18.	DST Sys (DST)	800	High

Find the price for each of the following stock purchases. Use the broker's fees in the text. [18.2]

19. 200 shares at $41\frac{5}{8}$
20. 340 shares at $73\frac{1}{8}$

Find the amount received from each of the following sales of stocks. Use the broker fees in the text. [18.2]

21. 100 shares at $\$30\frac{5}{8}$

22. 180 shares at $\$47\frac{1}{4}$

Find the current yield for the following. Round to the nearest tenth of a percent. [18.2]

23. Hewlett Packard at $\$104\frac{7}{8}$ per share and a dividend of $0.64

24. Intel at $\$19\frac{15}{16}$ per share and a dividend of $0.04

Find the PE ratio for each of the following. Round all answers to the nearest whole number. [18.2]

25. Microsoft at $\$83\frac{3}{8}$ per share with a net income of $1.42

26. General Electric at $\$112\frac{13}{16}$ per share with a net income per share of $2.99

Use the bond table to find each of the following. Use the listing for Bellsouth $6\frac{3}{4}$ 33. [18.3]

27. Closing price

28. Number of bonds sold

29. Year when the bonds will be paid off by the company

30. Change since the previous day in the price of one bond

31. Price to buy 30 such bonds at the closing price, with sales charges of $5 per bond.

32. Interest paid on the par value of each bond, both as a percent and in dollars.

Find the cost, including sales charges of $5 per bond, for each of the following. [18.3]

	Bond	Number Purchased
33.	IBM $6\frac{1}{2}$% of 2028	50
34.	ChaseM $6\frac{3}{4}$% of 2008	100

In problems 35–37, divide the profits based on the indicated method. Round all answers to the nearest dollar. [18.4]

	Partners	Investment	Method	Profits
35.	1	$8,500	Equal shares	$48,000
	2	$7,000		
	3	$10,500		
36.	1	$16,000	Ratio 2:3	$120,000
	2	$25,000		
37.	1	$9,000	Ratio of investment	$90,000
	2	$12,000		

In problems 38–40, allocate the overhead to each department of the company. Round to the nearest dollar. [18.5]

38.

Department	Floor Space
A	3000 square feet
B	5000 square feet
C	4000 square feet

Overhead: $100,000

39.

Product	Number Produced	Value of Each
1	8,000	$12
2	40,000	$6
3	20,000	$9

Overhead: $125,000

40.

Department	Number of Employees
A	70
B	55
C	45
D	60

Overhead: $85,000

Solve each of the following application problems. **[18.3]**

41. Ralph Toombs invested his retirement funds of $225,000 in a bond fund currently paying $8\frac{1}{4}\%$. Find his annual income.

42. George and Wanda Joyce invested their life's savings of $320,000 in a bond fund. If the fund charged them a 1% sales commission, find the annual income if the fund is currently paying $7\frac{5}{8}\%$.

Chapter 18 Summary Exercise *Partnership Interests*

Dougherty Educational Services, Inc. was formed several years ago with an investment of $15,000 from Trish Shields, $10,000 from Katie Abbot, and $25,000 from Beth Dougherty. Every $1 of their initial contribution resulted in the purchase of 2 shares of common stock. They also sold 10,000 preferred shares of $50 par value at 8% to other investors. Last year the firm had a net income after taxes of $250,000, and the board of directors allocated 45% of net income to the shareholders.

(a) Find the number of common shares of stock owned by Shields, Abbot, and Dougherty.

(b) Find the dividend per share.

(c) Shields, Abbot, and Dougherty respectively earn salaries of $32,000, $40,000 and $48,000. Find the sum of salary plus dividend for each of the three.

(d) Twenty-five percent of the net income after dividends is paid to a profit sharing plan which invests in a mutual fund containing both stocks and bonds. Recently the fund has yielded 12%. Find the contribution to the pension plan and one year's return on this investment assuming the current yield continues.

(e) An overhead of $142,000 has to be allocated to three departments of the company based on the following number of employees. Find the allocation to each department. Round to the nearest dollar.

Dept	No. of Employees
Math	7
English	5
Reading	4

(f) Would the profits represent a good return on their original investment if they formed the firm 5 years ago? If the company had been formed 30 years ago, would they be better off if they had placed their original investments in the stock market?

Net Assets Business on the Internet

Morrill Lynch

Statistics

- 1998: Revenues $17,547,000,000
- Total client assets of $1.5 trillion
- Presence in 43 countries across six continents
- 64,000 employees; 3000 in Japan alone
- World's third largest active fund manager

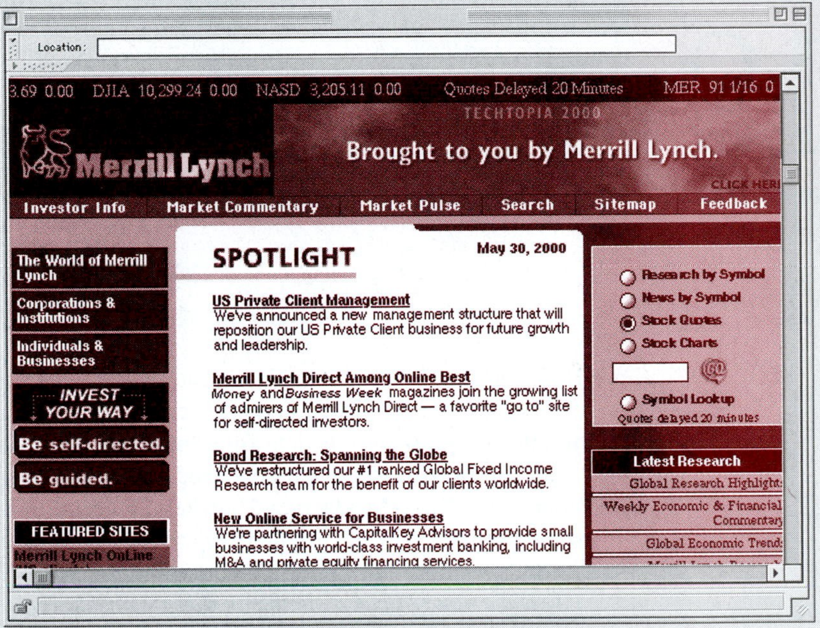

Charles E. Merrill founded Charles E. Merrill & Co. in 1914 and 16 months later the name was changed to Merrill, Lynch & Co. Charles Merrill was the first on Wall Street to have the vision of helping individuals who were not necessarily wealthy participate in stock and bond ownership. In 1928, just before the Great Depression, Charles Merrill predicted that "the financial skies are not clear," and prepared his firm's clients for financial survival. The company has long had an international presence as witnessed by their opening of offices in London and Tokyo in 1960 and 1961, respectively. Today, Merrill Lynch is a leading global financial management and advisory company that serves the needs of both individual and institutional clients.

1. A college professor nearing retirement bought 300 shares of McDonald's stock at $\$42\frac{3}{8}$ and 200 shares of Merrill Lynch stock at $\$79\frac{5}{8}$ through his long-time broker at Merrill Lynch. Find the total cost if commission charged was 1.5%.

2. Merrill Lynch underwrote and placed (sold) $125,000,000 in 30-year bonds, each yielding $7\frac{1}{4}\%$ per year, for a U.S. corporation that builds large diesel engines for ships. Find the commission if Merrill Lynch charged 2.6% for their services. Find the annual income to an investor who purchased 100 of these $1000 bonds.

3. Merrill Lynch helped a German construction company borrow $82,500,000 in U.S. dollars at $8\frac{3}{8}\%$ per year simple interest. Find the interest for 220 days. Find the commission earned by Merrill Lynch if they were paid $\frac{3}{16}$ of a percent of the amount borrowed.

4. Use either your local library or the World Wide Web to find the current price for Merrill Lynch stock (symbol MER). Also, find values for several of the stock ratios discussed in this chapter including PE ratio, current ratio and acid ratio.

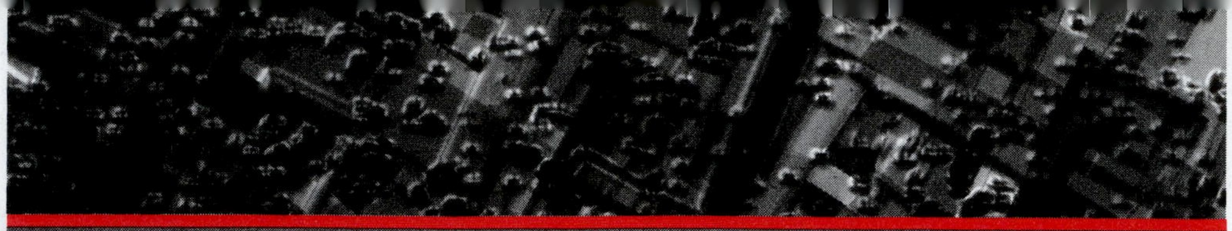

19 Business Statistics

The word **statistics** originally meant *state numbers*, data gathered by the government on the number of births, deaths, etc. Today the word "statistics" is used in a much broader sense to include data from business, economics, and many other fields. Statistics is a powerful and commonly used tool in business. For example, Figure 19.1 shows some data relevant to farmers, butchers, feed-lot operators, and supermarket managers among others: a 25-year trend of decreasing beef consumption per person in the United States.

Have you noticed a similar trend in your family?

Today, many companies use statistics on a regular basis. In this chapter, only a few of the basic ideas of statistics are introduced. We encourage you to take a class in statistics at your college if you wish to know more about this important subject.

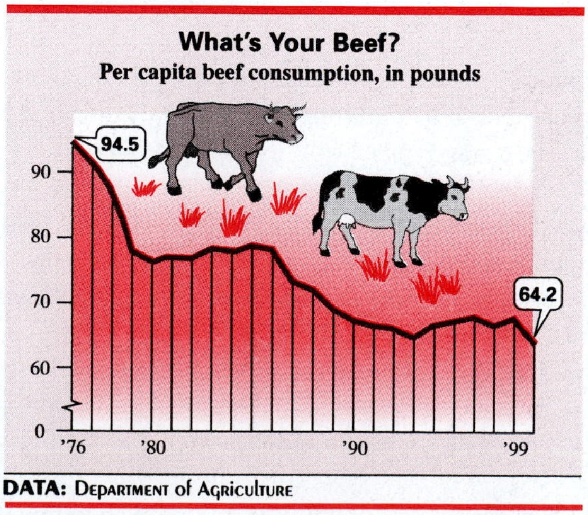

Figure 19.1

19.1 Frequency Distributions and Graphs

Objectives

1. *Construct a frequency distribution.*
2. *Make a bar graph.*
3. *Make a line graph.*
4. *Make a circle graph.*

Objective 1 *Construct a Frequency Distribution.* It can be difficult to interpret or find patterns in a large group of numbers called **raw data**. One way of analyzing the numbers is to organize them into a table that shows the frequency of occurrence of the various numbers. This type of table is called a **frequency distribution**.

Example 1 Constructing a Frequency Distribution

Tina McCartle is analyzing sales activity over the past 24 weeks at her new restaurant, Big-n-Juicy Hamburgers. The weekly sales data has been rounded to the nearest thousand dollars. Read down the columns, beginning with the left column, for successive weeks of the year.

\$3.9	\$4.0	\$4.3	\$4.6	\$5.1	\$5.6
\$3.2	\$4.2	\$4.8	\$4.9	\$4.8	\$4.8
\$3.3	\$4.1	\$4.1	\$5.2	\$5.0	\$5.3
\$3.5	\$3.9	\$4.8	\$5.0	\$5.3	\$5.3

Construct a table that shows each value of sales. Then go through the data and place a tally mark (|) next to each corresponding value, thereby creating a frequency distribution table.

Solution

Table 19.1 shows that the most common weekly sales amount was \$4800, although there were 3 weeks with sales of \$5300.

The frequency distribution given in the previous example contains a great deal of information, perhaps more than is needed. It can be simplified by combining weekly sales into groups, forming the grouped data shown in Table 19.2.

Problem-Solving Hint The number of groups in the left column of Table 19.2 is arbitrary and usually varies between 5 and 15.

Table 19.1 FREQUENCY DISTRIBUTION TABLE

Sales (thousands)	Tally	Frequency	Sales (thousands)	Tally	Frequency	Sales (thousands)	Tally	Frequency
\$3.2	|	1	\$4.2	|	1	\$5.1	|	1
\$3.3	|	1	\$4.3	|	1	\$5.2	|	1
\$3.5	|	1	\$4.6	|	1	\$5.3	|||	3
\$3.9	||	2	\$4.8	||||	4	\$5.6	|	1
\$4.0	|	1	\$4.9	|	1			
\$4.1	||	2	\$5.0	||	2			

Table 19.2 GROUPED DATA

Sales (thousands)	Frequency (number of weeks)
\$3.1–\$3.5	3
\$3.6–\$4.0	3
\$4.1–\$4.5	4
\$4.6–\$5.0	8
\$5.1–\$5.5	5
\$5.6–\$6.0	1

Example 2 Analyzing a Frequency Distribution

Based on the data from Big-n-Juicy Hamburgers, answer the following questions.

(a) McCartle can take no salary and the business still loses money when sales are less than or equal to \$4000 per week. During how many weeks did this occur?

(b) McCartle can take a small salary out of the company once sales go above \$5000 per week. During how many weeks did this occur?

Solution

(a) The first two classes in Table 19.2 represent weeks in which sales were equal to or less than \$4000. Thus, McCartle took no salary and the restaurant lost money for 6 weeks.

(b) The last two classes in Table 19.2 are the number of weeks during which sales were above \$5000, or 6 weeks. Therefore, McCartle took a small salary for 6 weeks.

Objective 2 *Make a Bar Graph.* The next step in analyzing this information is to use it to make a **graph**. In statistics, a graph is a visual presentation of numerical data. One of the most common graphs is a **bar graph**, where the height of a bar represents the frequency of a particular value. A bar graph for the sales data is shown in Figure 19.2.

The information from the grouped data is shown in Figure 19.3. This graph shows that weekly sales of between \$4600 and \$5000 were the most common. Notice that this graph *does not* show any trend that may be occurring over time.

Figure 19.2

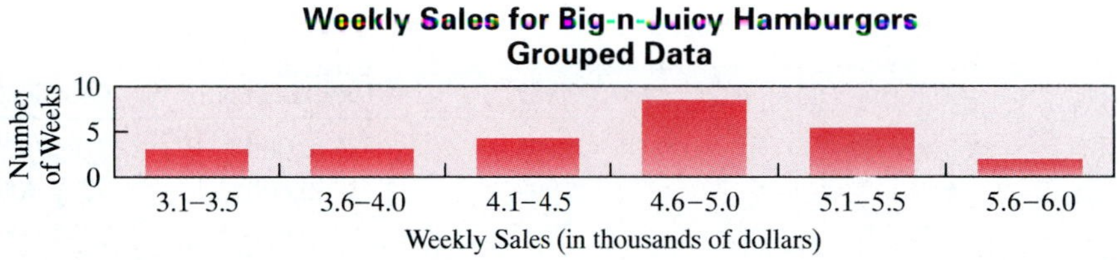

Figure 19.3

Objective 3 *Make a Line Graph.* Bar graphs show which numbers occurred and how many times, but do not necessarily show the order in which the numbers occurred. To discover any trends that may have developed, draw a **line graph** of the data over time.

Example 3 Draw a Line Graph

Show the progression of weekly sales at Big-n-Juicy Hamburgers through the year using a line graph. Do this by totaling the first 4 weeks (the first column) of data in Example 1 for the first data point. Similarly, total the second 4 weeks (second column) of data for the next data point, and so on.

Solution The total for the first four weeks is \$3.9 + \$3.2 + \$3.3 + \$3.5 = \$13.9 or \$13,900 in sales for the first 4 weeks. The total for the second 4 weeks is \$4.0 + \$4.2 + \$4.1 + \$3.9 = \$16.2 or \$16,200 in sales. The six data points of the graph are \$13.9, \$16.2, \$18, \$19.7, \$20.2 and \$21 in thousands of dollars.

It is apparent from Figure 19.4 that weekly sales are growing. Tina McCartle is excited about this trend and she is determined to continue the trend since her livelihood depends on the restaurant. She plans to work very hard in the restaurant over the next few months.

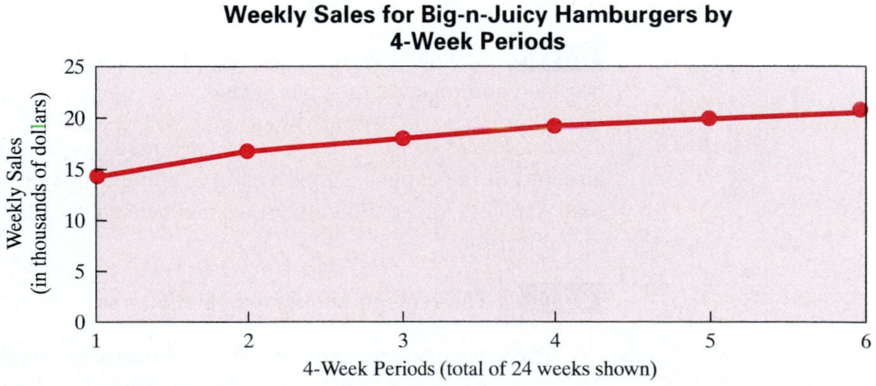

Figure 19.4

One advantage of line graphs is that two or more sets of data can be shown on the same graph. For example, suppose the managers of a company called Eastside Tire Sales want to compare total sales, profits, and overhead. Assume that they have extracted the following data from their records (Table 19.3).

Table 19.3 EASTSIDE TIRE SALES, 1998–2001

Year	Total Sales	Overhead	Profit
1998	\$740,000	\$205,000	\$83,000
1999	\$860,000	\$251,000	\$102,000
2000	\$810,000	\$247,000	\$21,000
2001	\$1,040,000	\$302,000	\$146,000

Separate lines can be made on a line graph for each category so that necessary comparisons can be made. A graph such as this is called a **comparative line graph** (see Figure 19.5).

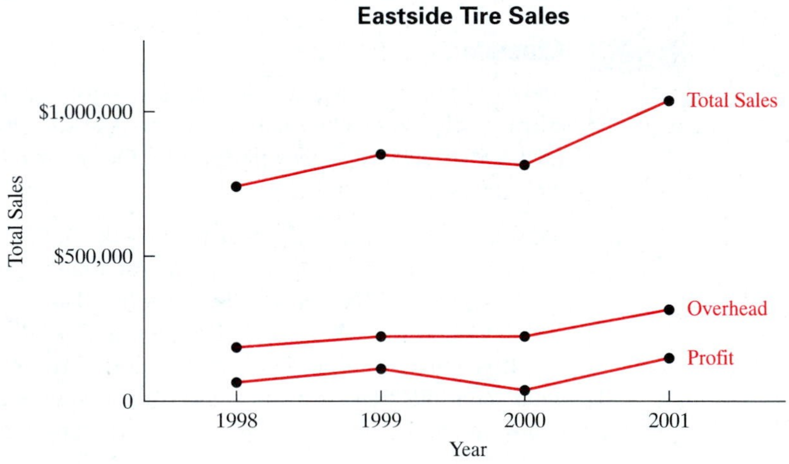

Figure 19.5

NOTE Including zero on one or both scales of a line graph *may help* the viewer understand and relate to the data. ■

Objective 4 ***Make a Circle Graph.*** Suppose a sales manager for Novel Recording Company makes a record of the expenses involved in keeping a sales force on the road. After finding the total expense, she could convert each expense into a percent of the total, with the results in Table 19.4.

NOTE The percents should total 100%. ■

The sales manager can show these percents by using a **circle graph**. Every circle has 360 degrees (written 360°). The 360° represents the total expenses. Since entertainment is 10% of the total expense, she used

$$360° \times 10\% = 360° \times 0.10 = 36° \text{ slice of circle}$$

to represent her entertainment expense. Since lodging is 25% of the total expenses, she used

$$360° \times 25\% = 90° \text{ slice of circle}$$

to represent lodging. After she found the degrees that represent each of her expenses, she drew the circle graph shown in Figure 19.6.

Table 19.4 NOVEL RECORDING COMPANY EXPENSES

Item	Percent of Total
Car and plane	30%
Lodging	25%
Food	15%
Entertainment	10%
Sales meetings	10%
Other	10%

Circle graphs are used to show comparisons when one item represents a small portion compared to another. In the circle graph shown here, an item representing 1% of the total could be drawn as a small but noticeable slice; such a small item would hardly show up in a line graph.

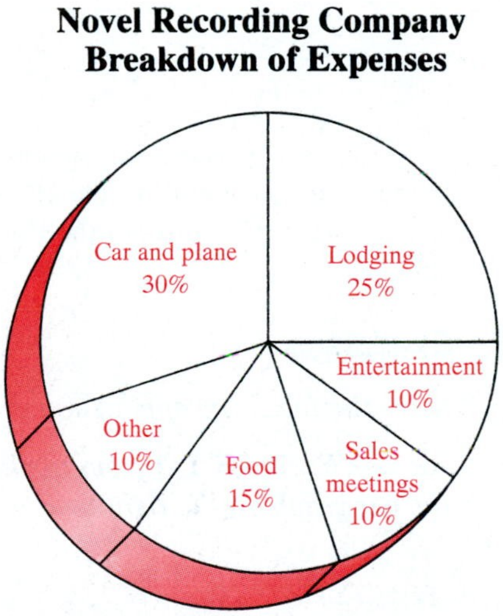

Figure 19.6

Example 4 Interpreting a Circle Graph

Based on the circle graph of expenses, answer the following questions.

(a) What percent of expenses was spent on travel and entertainment?
(b) What percent of expenses was spent on food and lodging?

Solution

(a)	Travel is	30%	Car and plane
	Entertainment is	+ 10%	
	Total spent	40%	
(b)	Food is	15%	
	Lodging is	+ 25%	
	Total spent	40%	

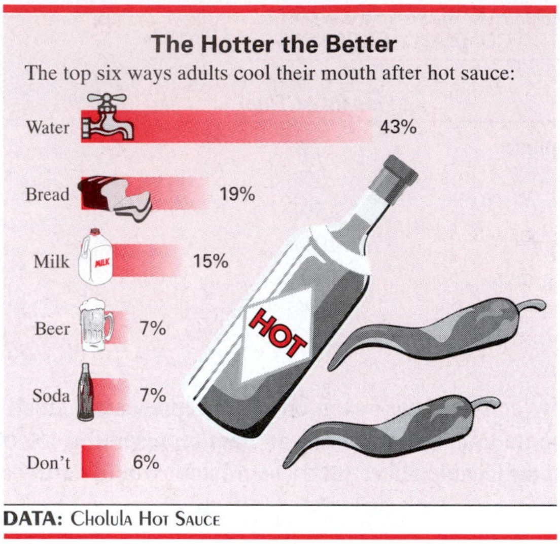

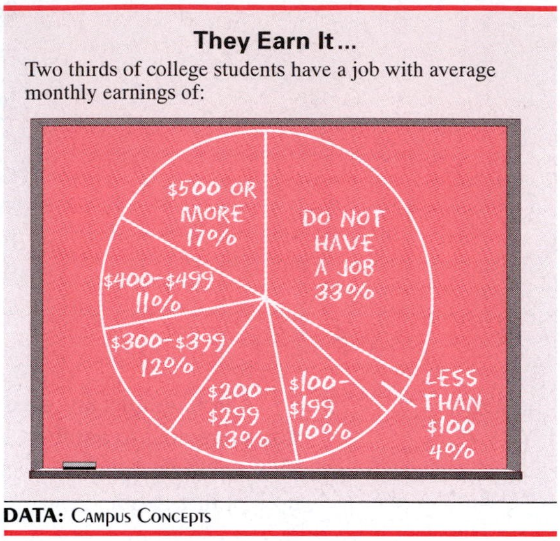

Figure 19.7

Graphs can show information in many interesting ways. Figure 19.7 shows two graphs. The bar graph to the left shows how people cool their mouths after eating hot sauce. The circle graph to the right shows average monthly income of working college students.

19.1 Exercises

Solve the following application problems using the information provided.

Answer Exercises 1–3 from the line graph on the left and answer Exercises 4–6 from the bar graph on the right.

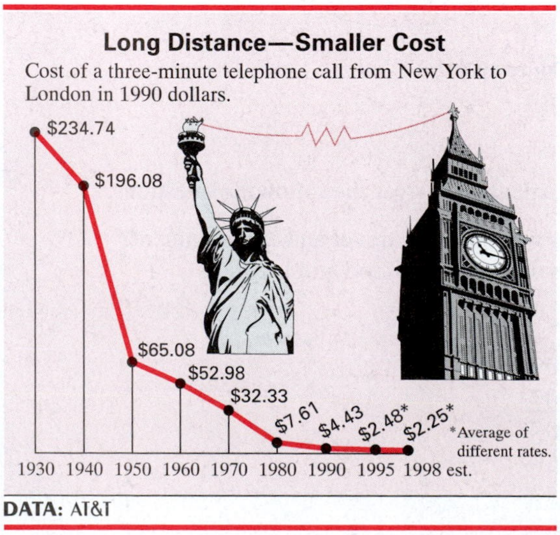

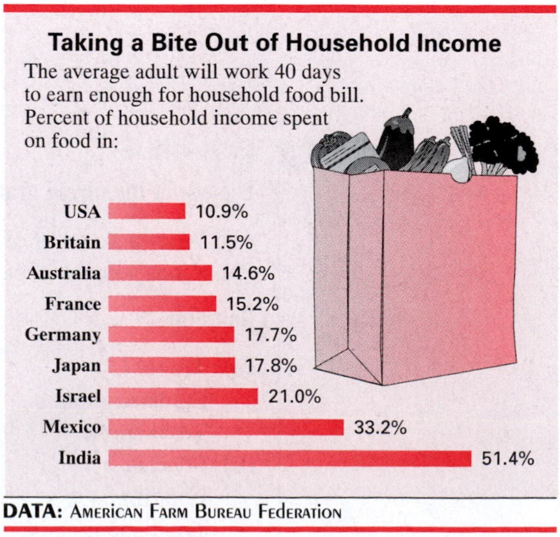

1. COST OF LONG DISTANCE Find the cost of a 3-minute telephone call from New York to London in 1950.

2. COST OF LONG DISTANCE Estimate the cost of a 3-minute telephone call from New York to London in 1985 by finding the average of the cost in 1980 and 1990.

3. COST OF LONG DISTANCE By what percent did the cost of a 3-minute telephone call from New York to London fall from 1950 to 1998. Round to the nearest percent.

4. FOOD COSTS Find the share of household income spent for food in the United States.

5. FOOD COSTS Find the share of household income spent for food in India.

6. FOOD COSTS List all countries in the graph in which less than 15% of household income is spent, on average, for food.

COLLEGE CREDITS ***The following list shows the number of college credits completed by 30 employees of the Franklin Bank.***

74	133	4	127	20	30
103	27	139	118	138	121
149	132	64	141	130	76
42	50	95	56	65	104
4	140	12	88	119	64

Use these numbers to complete the following table.

	Number of Units	Frequency
7.	0–24	______
8.	25–49	______
9.	50–74	______
10.	75–99	______
11.	100–124	______
12.	125–149	______

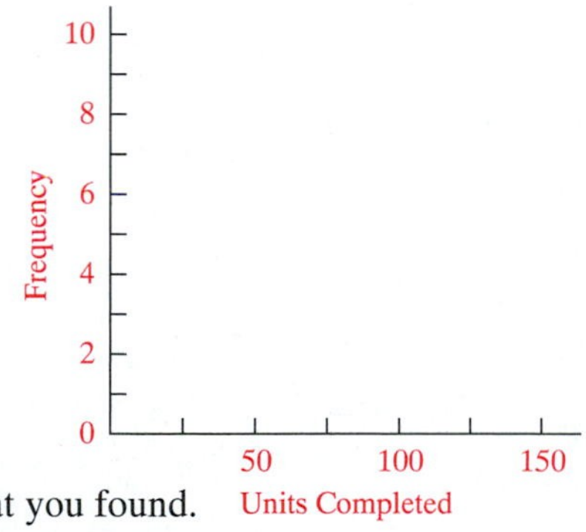

13. Make a line graph using the frequencies that you found.

14. How many employees completed fewer than 25 credits?

15. How many employees completed 50 or more credits?

16. How many employees completed from 50 to 124 credits?

17. How many employees completed from 0 to 49 credits?

18. What percent of the employees have completed at least 100 credits? Round to the nearest tenth of a percent.

19. Compare bar graphs, line graphs, and circle graphs. (See Objectives 1–3.)

20. Explain the purpose of a comparative line graph. (See Objective 3.)

21. FARM TRACTOR The following graph is used to estimate the acreage covered by a farm implement per hour, when its width and speed of travel are known. For example, a $7\frac{1}{2}$ foot (90-inch) mower blade moving 4 miles per hour would cover about $3\frac{5}{8}$

acres per hour. This is found by going across the graph from the working width (90 inches) to the diagonal line for speed (4 mph), then down to the bottom to find acreage per hour.

(a) What is the acreage per hour for a 36-inch implement moving $2\frac{1}{2}$ miles per hour?

(b) What is the acreage per hour for an 8-foot-wide combine moving 4 miles per hour?

(c) How fast must a tractor pull a 48-inch plow in order to plow one acre per hour?

(d) How wide a spray pattern is needed in order to spray $4\frac{1}{2}$ acres per hour at a speed of $4\frac{1}{2}$ miles per hour?

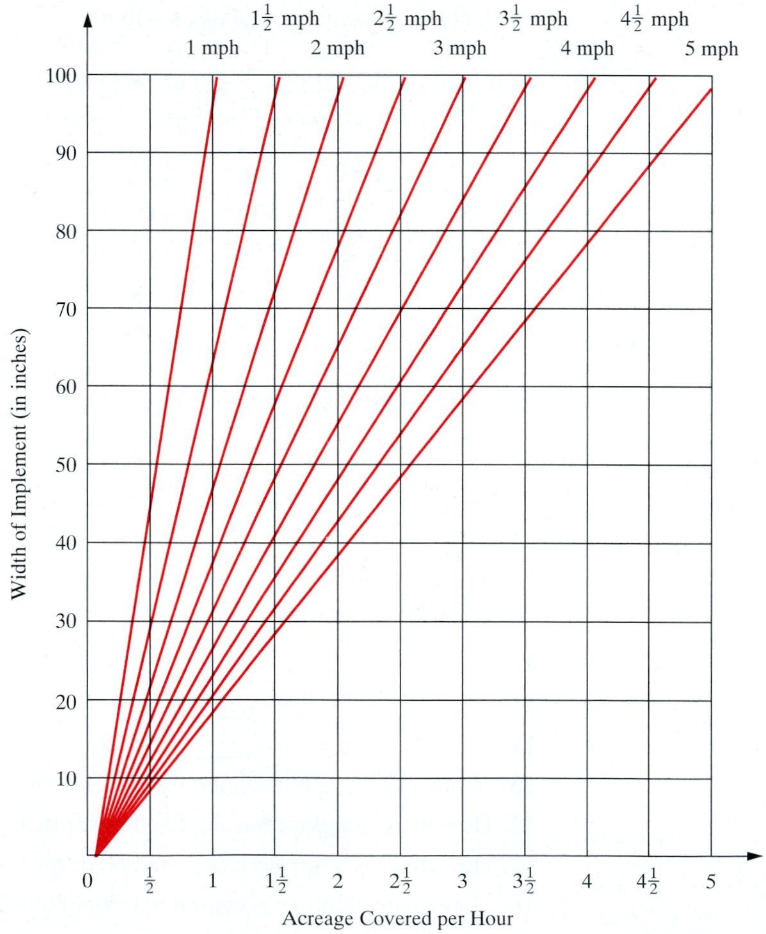

22. AVERAGE WEIGHT The comparative line graph shows the change in average weight (in pounds) for a recent 20-year period for various categories of American adults. Find the change in weight for each of the following groups of people.

(a) Men aged 20–24 who are 5 feet 10 inches tall (*Hint:* first find 5′10″ on the horizontal line in the center of the graph.)

(b) Women aged 40–49 whose height is 5 feet 8 inches

(c) 5-foot-tall women aged 20–24

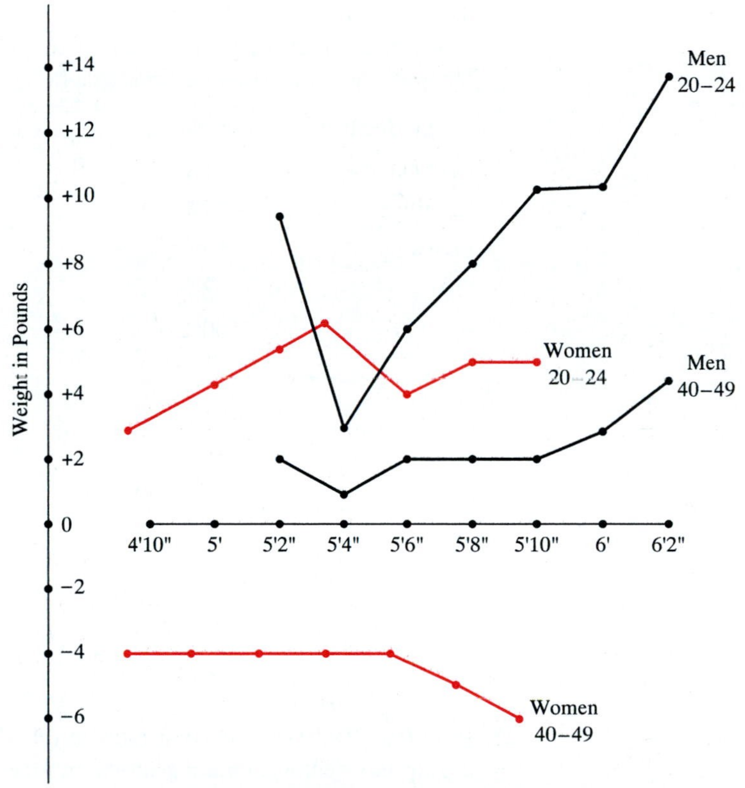

20-Year Change in Average Weight of Americans

23. COMPUTER SKILLS In today's business environment it is important for administrative and office-support professionals to possess computer skills. A recent survey asked the question of employers: "What does your company use as the primary method for evaluating a job candidate's software proficiency?" Draw a circle graph using the data.

The interview only	47%
Reference checking	26%
A test on a computer	8%
Other	19%
	100%

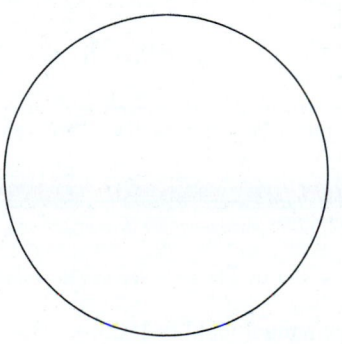

24. SOURCES OF RETIREMENT INCOME The Social Security Administration recently found that the average retired couple has the following source of retirement income. Show the data in a circle graph.

Social Security	40%
Personal assets	21%
Earnings	17%
Pensions	19%
Other	3%
	100%

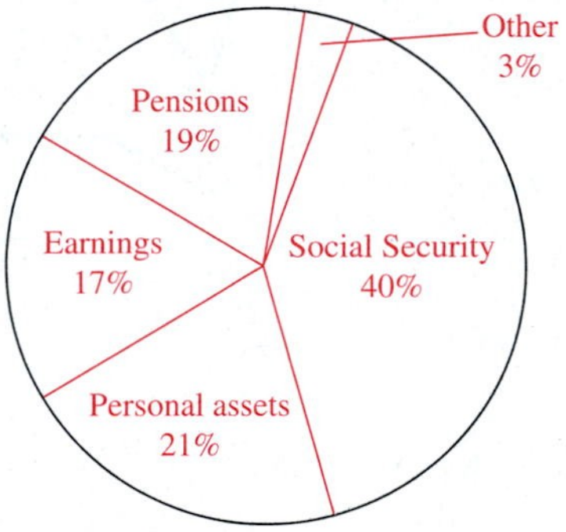

25. INTEREST RATES Interest rates in the United States have changed considerably during this century, creating an environment that either nourishes business growth and associated stock market rallies or stunts their growth. Use the data in the line chart to answer the questions that follow.

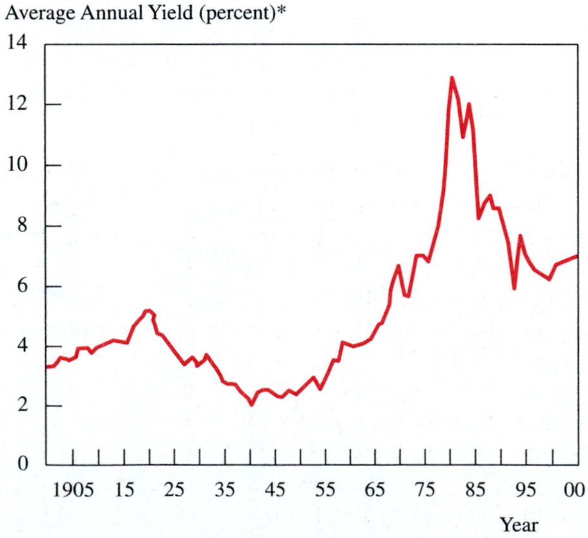

**1900–1919, average long-term corporate bonds; 1920–1995, average ten-year Treasury bonds.*

Source: Moody's Investor Service. Reprinted by permission.

(a) Find the average annual yield in 1925. 3.5%

(b) Over what range did interest rates fluctuate between 1905 and 1965? 2% to 5%

(c) Over what range have interest rates fluctuated since 1965?
(d) Approximate the increase in interest rates from 1941 (World War II) to 1980.

26. INJURED PEDESTRIANS The following circle graph, based on data from the National Safety Council, shows how pedestrians were injured. (a) Find the number of pedestrians who were killed or injured when crossing or entering an intersection on foot. (b) Find the number of pedestrians who were killed or injured while standing or playing in the road.

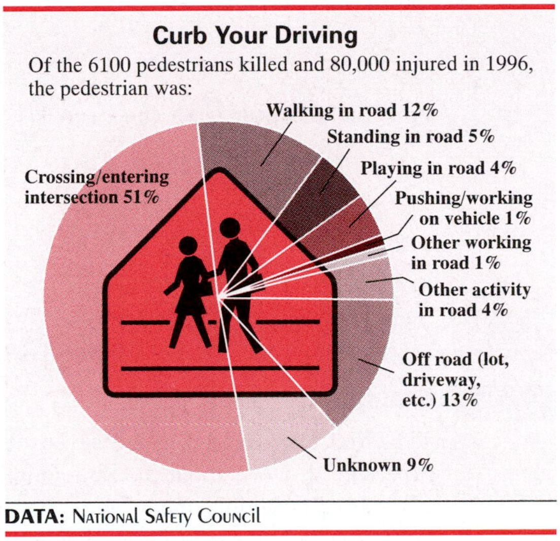

27. If you have not already done so, complete item 9 of the foldout, including the pie chart.

19.2 The Mean

Objectives

1. *Understand the difference between population and sample.*
2. *Find the mean of a set of data.*
3. *Find a weighted mean.*
4. *Find the mean for grouped data.*

Objective 1 *Understand the Difference Between Population and Sample.* In statistics, it is important to distinguish between the concepts of population and sample. **Population** is the entire group being studied, whereas **sample** is a portion of the entire group. Samples should be chosen randomly, meaning that no one individual in the population is more likely to be chosen than is another.

An administrator, for example, might be interested in the grade point average (GPA) of all freshmen at a community college. The GPA of the *entire population* of freshmen can be obtained, but this might be time-consuming to compile. Or, if an estimate of the GPA is adequate, the administrator can randomly choose a sample of perhaps 50 freshmen and find their grade point averages. The administrator might then *assume* that the grade point average of *this sample of students* is close to that of the entire population of all freshmen.

Objective 2 ***Find the Mean of a Set of Data.*** Businesses are often faced with the problem of analyzing a mass of raw data. Reports come in from many different branches of a company, or salespeople may send in a large number of expense claims, for example. In analyzing all the data, one of the first things to look for is a **measure of central tendency**—a single number that is designed to represent the entire list of numbers. One such measure of central tendency is the **mean**, which is just the common **average** used in everyday life.

For example, suppose the sales of carnations at Tom's Flower Shop for each of the days last week were \$86, \$103, \$118, \$117, \$126, \$158, and \$149. To find a single number that is representative of this list, use the following formula.

$$\text{Mean} = \frac{\text{Sum of all values}}{\text{Number of values}}$$

For Tom's Flower Shop, the mean is

$$\text{Mean} = \frac{\$86 + \$103 + \$118 + \$117 + \$126 + \$158 + \$149}{7}$$

$$= \frac{\$857}{7}$$

$$= \$122.43 \quad \text{Rounded to the nearest cent}$$

Example 1 Finding the Mean

Tina McCartle has promised seven of her employees at Big-n-Juicy Hamburgers that they will all work about the same number of hours. One employee complained that she worked considerably more hours than the other employees last month. The number of hours worked by each of the seven employees during the past month are given. Find the mean to the nearest hour.

Hours worked: 75, 63, 76, 82, 70, 81, 149

Solution The mean can be *estimated* by first rounding each number to the tens position. Therefore, 75 rounds to 80, 63 to 60, and so on, providing the following.

80, 60, 80, 80, 70, 80, 150

Add these numbers and divide by 7 to find an estimate of the mean.

$$600 \div 7 = 86 \quad \text{Rounded to the nearest whole number}$$

Now find the exact value by adding the original numbers and dividing by 7. Check that the sum of numbers is 596.

$$\text{Mean} = \frac{596}{7} = 85 \quad \text{Rounded}$$

The mean of 85 seems a bit large since one employee worked a lot more than the other six employees. The mean without this value of 149 is the sum of the remaining 6 hours worked divided by 6.

$$\text{Mean} = \frac{447}{6} = 75 \quad \text{Rounded}$$

This value seems more in line with the average number of hours worked. Perhaps there was an unusual reason the one employee worked 149 hours (someone else was sick, etc.).

Notice that our estimate of 86 is close to the mean of 85. Estimating an answer before calculating the exact value helps minimize errors.

Averages are used in many different places. For example, the graph in Figure 19.8 shows that the average total stopping distance on a dry road for a 3000-pound car depends on the speed of the automobile.

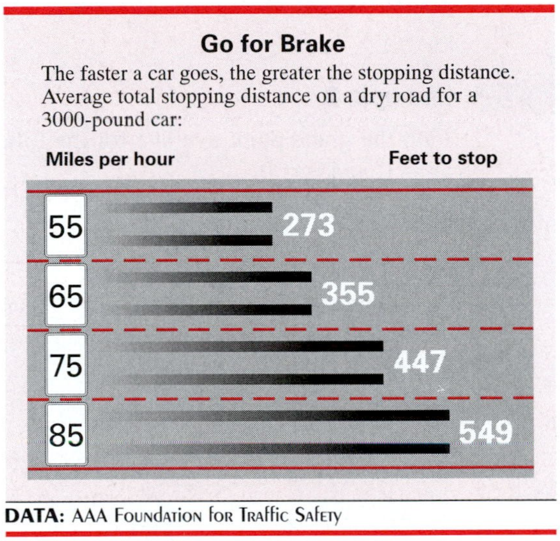

Figure 19.8

Objective 3 ***Find a Weighted Mean.*** Table 19.5 shows a frequency distribution of annual salaries received by the management employees of a medium-sized corporation.

As Table 19.5 shows, 27 management employees were paid $32,000 each, 16 were paid $40,000 each, and so on. The average salary paid to these employees cannot be found by just adding the salaries, since different salaries are earned by different numbers of employees. To find the mean of the salaries, it is necessary to first multiply each annual salary by the number of employees receiving that salary. This process produces a **weighted mean**, where each number (a salary here) is "weighted" by multiplying it by the number of times it occurs.

Table 19.5 SAMPLE SALARIES OF MANAGEMENT EMPLOYEES

Salary	Number of Managers (Frequency)
$32,000	27
40,000	16
50,000	11
72,000	6
90,000	4
96,000	4
110,000	3
160,000	2
296,000	1

Salary	Number of Managers	Salary × Number of Managers
$32,000	27	$864,000
40,000	16	640,000
50,000	11	550,000
72,000	6	432,000
90,000	4	360,000
96,000	4	384,000
110,000	3	330,000
160,000	2	320,000
296,000	1	296,000
	Totals 74	$4,176,000

By adding the numbers in the "Number of Managers" column, we find that the corporation has a total of 74 management employees. Find the mean by dividing the total of all salaries, \$4,176,000, by the total number of employees, 74.

$$\text{Mean salary} = \frac{\$4{,}176{,}000}{74} = \$56{,}432 \qquad \text{Rounded}$$

The mean salary of a management employee is \$56,432.

As you can see in the next example, weighted means are also used to find grade point averages.

Example 2 Finding Grade Point Average

Find the grade point average for the following student. Assume A = 4, B = 3, C = 2, D = 1, and F = 0.

Solution

Course	Units	Grade	Grade × Units
Business mathematics	3	A (= 4)	4 × 3 = 12
Retailing	4	C (= 2)	2 × 4 = 8
English	3	B (= 3)	3 × 3 = 9
Computer science	2	A (= 4)	4 × 2 = 8
Lab for computer science	2	D (= 1)	1 × 2 = 2
	Totals 14		39

The grade point average for this student is

$$\frac{39}{14} = 2.79$$

NOTE Grade point averages are frequently rounded to the nearest hundredth.

Objective 4 ***Find the Mean for Grouped Data.*** The mean can also be found for data that has been grouped into a frequency distribution. To do so, find the midpoint of each interval or class. This midpoint is found by averaging the highest and lowest numbers that can go into a class. For example, the midpoint, or **class mark**, of the interval 100–109 is the mean of 100 and 109.

$$\text{Class midpoint} = \frac{100 + 109}{2} = \frac{209}{2} = 104.5$$

NOTE The intervals in a frequency table should not overlap with one another.

Example 3 Finding the Mean for Grouped Data

A researcher surveyed a number of farmers on their use of a particular brand of fertilizer. Although the recommended usage is 135 pounds per acre, she found that the usage varied from that. Find the mean for the frequency distribution shown.

Intervals	Frequency
100–109	9
110–119	12
120–129	17
130–139	28
140–149	21
150–159	16
160–169	4

Solution Begin by finding the class mark for each class of fertilizer use. As explained, the class mark or midpoint of the first class is 104.5, while the midpoint of the second class is

$$\frac{110 + 119}{2} = \frac{229}{2} = 114.5$$

Find the other midpoints in a similar way. Then multiply the frequencies and the class marks, completing the column at the right (labeled "Frequency × Class Mark"). Next, find the totals in the "Frequency" column and the "Frequency × Class Mark" column.

Intervals	Frequency	Class Mark	Frequency × Class Mark
100–109	9	104.5	9 × 104.5 = 940.5
110–119	12	114.5	12 × 114.5 = 1,374.0
120–129	17	124.5	17 × 124.5 = 2,116.5
130–139	28	134.5	28 × 134.5 = 3,766.0
140–149	21	144.5	21 × 144.5 = 3,034.5
150–159	16	154.5	16 × 154.5 = 2,472.0
160–169	4	164.5	4 × 164.5 = 658.0
	Totals 107		14,361.5

Finally, the mean is the quotient of these totals, or

$$\text{Mean} = \frac{14{,}361.5}{107} = 134.2 \qquad \text{To the nearest tenth}$$

The mean usage rate of 134.2 pounds per acre is close to the recommended usage rate of 135 pounds per acre.

NOTE When a set of data is divided up into classes, it is no longer possible to tell where a particular item falls in a class. For this reason, *a mean found from grouped data is only approximate.* ■

19.2 Exercises

Find the mean for each of the following sets of data. Round to the nearest tenth.

1. 128, 240, 164, 380

2. 60, 65, 67, 62, 59, 58, 70

3. 3800, 3625, 3904, 3296, 3400, 3650, 3822, 4020

4. 10.3, 11.7, 12.4, 8.6, 9.9, 12.1, 13.2, 10.8, 9.6, 8.8

Round to the nearest whole number in Exercises 5–10.

5. RIVER FERRY A ferry is carrying four cars across a river. The weights of the cars are 4220 pounds, 3840 pounds, 3640 pounds, and 4080 pounds. Find the mean.

6. MILK PRODUCTION The yield of milk at a dairy for four successive days was 410 gallons, 440 gallons, 520 gallons, and 423 gallons. Find the mean.

7. COST OF DINING ROOM SET A young couple is shopping for a dining room set. The costs of the ones they have looked at so far are \$1280, \$2650, \$870, \$940 and \$760. Find the mean.

8. LIFE INSURANCE SALES Life insurance sold last year by the six new agents of National Insurance Agency totaled \$294,000, \$580,000, \$722,000, \$463,000, \$814,000, and \$1,785,000. Find the mean total of life insurance sold.

9. CAR SALES Last year, the value of the new cars sold by the eight salespeople at the Autoplex was \$385,000, \$495,000, \$873,000, \$1,210,000, \$611,000, \$802,000, \$173,000, and \$708,000. Find the mean value of the cars sold per salesperson.

10. TELEPHONE SALES Telesales employs nine people to make telephone calls to sell magazines. Last year, these people produced total sales of \$492,811, \$763,455, \$901,852, \$179,806, \$244,193, \$382,574, \$591,873, \$1,003,058, and \$473,902. Find the mean sales per employee.

Find the weighted mean for each of the following. Round to the nearest tenth.

11.

Value	Frequency
30	5
40	3
45	2
48	4

12.

Value	Frequency
10	8
12	5
15	1
18	1

13.

Value	Frequency
125	6
130	4
150	5
190	3
220	2
230	5

14.

Value	Frequency
25	1
26	2
29	5
30	4
32	3
33	5

15. Explain the difference between population and sample. (See Objective 1.)

16. Think of an example in which the mean really does not give a good understanding of the situation involving the data. (See Objective 2.)

In the following problems, find the weighted mean salary for the employees. Round to the nearest thousand dollars.

17.

Salary	Number of Employees
$18,000	8
$21,000	10
$28,000	8
$29,000	6
$38,000	4
$41,000	3
$53,000	2
$162,000	1

18.

Salary	Number of Employees
$15,000	8
$20,000	15
$22,000	13
$25,000	9
$30,000	4
$42,000	3
$57,000	2
$260,000	1

Find the grade point average for each of the following students. Assume A = 4, B = 3, C = 2, D = 1, and F = 0. Round to the nearest hundredth.

19.

Credits	Grade
4	D
2	A
3	C
1	F
3	D

20.

Credits	Grade
4	B
3	B
3	C
3	C
3	D

Find the mean for the following grouped data. Round to the nearest tenth.

21.

Interval	Frequency
50–59	15
60–69	20
70–79	21
80–89	27
90–99	18
100–109	2

22.

Interval	Frequency
320–339	7
340–359	9
360–379	12
380–399	11
400–419	6
420–439	5

23.

Interval	Frequency
25–49	18
50–74	15
75–99	30
100–124	18
125–149	32
150–174	14
175–199	7

24.

Interval	Frequency
150–154	4
155–159	7
160–164	9
165–169	12
170–174	16
175–179	8
180–184	3

Solve the following application problems.

25. **MINIMUM GRADE TO PASS** The final grade in a history class is the average of three tests taken during the semester. Wanda Kroll made 74% and 68% on the first two tests. Find the minimum score she can make on the third test and still have a 70% average.

26. **LOADING A TRAILER** A trailer can hold 4200 pounds. The front of the trailer is loaded with a piece of scrap steel weighing 1300 pounds and an engine block weighing 840 pounds. Find the number of cows with a mean weight of 660 pounds that can be safely added to the trailer.

19.3 The Median and the Mode

Objectives

1. *Find the median of a set of data.*
2. *Find the mode of a set of data.*
3. *Find the median and the mode of data in a frequency table.*

In everyday life, the word "average" usually refers to the mean. However, there are two other "averages" in common use, the **median** and the **mode**. Median and mode are discussed in this section.

Suppose the owner of a small company pays five employees annual salaries of \$12,500, \$13,000, \$13,200, \$14,000, and \$15,000. The average, or mean, salary paid to the employees is

$$\text{Mean} = \frac{\$12{,}500 + \$13{,}000 + \$13{,}200 + \$14{,}000 + \$15{,}000}{5}$$

$$= \frac{\$67{,}700}{5} = \$13{,}540$$

Now suppose that the employees go on strike and demand a raise. To get public support, they appear on television to talk about their low salaries, which average only \$13,540 per year.

The television station sends a reporter to interview the owner of the company. Before the interviewer arrives, the owner decides to find the average salary of *all employees*, including the five on strike, plus his own. To do this, he adds the five salaries given to the employees, plus his salary of \$127,000. This gives an average of

$$\text{Mean} = \frac{\$12{,}500 + \$13{,}000 + \$13{,}200 + \$14{,}000 + \$15{,}000 + \$127{,}000}{6}$$

$$= \frac{\$194{,}700}{6} = \$32{,}450$$

When the television reporter arrives, the owner is prepared to state that there is no reason for the employees to be on strike, since "the average salary of all company employees is \$32,450."

There are two points to this story. First, both averages are correct, depending on what is being measured. *This shows how easily statistics can be manipulated.* Second, *the mean is often a poor indicator of the "middle"* of a list of numbers. In fact, when the mean was computed by the owner, it was greater than 5 of the 6 employees' salaries. The mean may be greatly affected by extreme values, such as the owner's salary of \$127,000.

Objective 1 *Find the Median of a Set of Data.* To avoid such a misleading result, use a different measure of the "middle" of a list of numbers, the **median**. The median divides a list of numbers in half: one-half of the numbers lie at or above the median and one-half lie at or below the median.

Since the median divides a list of numbers in half, the first step in finding a median is to rewrite the list of numbers as an **ordered array, or list**, with the numbers going from *smallest to largest*. For example, the list of numbers 9, 6, 11, 17, 14, 12, 8 would be written in order as the ordered array

6, 8, 9, 11, 12, 14, 17

The median is found from the ordered array as explained in the following box. Notice that the procedure for finding the median depends on whether the number of numbers in the list is even or odd.

1. If the ordered array has an *odd* number of values, divide the number of values by 2. The next higher whole number gives the *location* of the median.
2. If the ordered array has an *even* number of values, there is no single middle number. Find the median by first dividing the number of values by 2. The median is the average (mean) of the number in this position and the number in the next higher position.

Example 1 shows lists of numbers having an *odd* number of values. Example 2 shows an *even* number of values.

Example 1 Finding the Median (Odd Number)

Find the median for the annual salaries of the 5 employees introduced earlier in this section ($12,500, $13,000, $13,200, $14,000, and $15,000).

Solution First, make sure to list the numbers from smallest to largest.

$12,500, $13,000, $13,200, $14,000, $15,000

There are five numbers in the list. Divide 5 by 2 to get $\frac{5}{2} = 2.5$. The *next higher number* is 3 so that the median is the third number or $13,200. Two numbers are larger than $13,200 and two are smaller.

Example 2 Finding the Median (Even Number)

Find the median of the salaries of the employer and the five employees introduced earlier in this section ($127,000, $12,500, $13,000, $13,200, $14,000, and $15,000).

Solution First arrange the numbers from smallest to largest.

$12,500, $13,000, $13,200, $14,000, $15,000, $127,000

There are six numbers (an even number of numbers) in the list. Divide 6 by 2 to get 3. The median is the *mean of the numbers in the third and fourth positions.*

$$\text{Median} = \frac{\$13{,}200 + \$14{,}000}{2} = \$13{,}600$$

The median of this set of numbers is $13,600 and the mean is $32,450. The median is probably a better measure of central tendency for this set of numbers than is the mean, which is distorted by one large number, i.e., $127,000.

Objective 2 *Find the Mode of a Set of Data.* The last important statistical measure of central tendency is called the **mode**. The mode is the number which occurs *the most often*. For example, 10 students earned the following scores on a business law examination.

74, 81, 39, 74, 82, 80, 100, 92, 74, 85

The mode is 74, since more students obtained this score than any other. (It is not necessary to form an ordered array when looking for the mode.)

Problem-Solving Hint A bar graph can be used to find the mode.

Example 3 Finding the Mode

Professor Miller gave the same test to both his day and evening sections of business math at American College. Find the mode of the tests given in each class. Which class had the lower mode?

(a) Day class: 85, 92, 81, 73, 78, 80, 83, 80, 74, 69, 80, 65, 71, 65, 80, 93, 54, 78, 80, 45, 70, 76, 73, 80, 71, 68
(b) Evening class: 68, 73, 59, 76, 79, 73, 85, 90, 73, 69, 73, 75, 93, 73, 76, 70, 73, 68, 82, 84, 77

Solution

(a) The number 80 is the mode for the day because it occurs more often than any other number.

(b) The number 73 is the mode for the evening class because it occurs more often than any other number.

The evening class has the lower mode.

NOTE It is not necessary to place the numbers in numerical order when looking for the mode, but it helps with a large array of numbers. ▬

A set of data is called **bimodal** if it has two different modes, i.e., two numbers that occur the same number of times with each occurring more than any other number in the data set. A data set in which every number occurs the same number of times is said to have *no mode*.

NOTE The mean, median, and mode are different ways of estimating the middle or center of a list of numbers. Each of these three ways is a measure of central tendency. ▬

Objective 3 *Find the Median and the Mode of Data in a Frequency Table.* The same basic ideas of median and mode are applied as shown in the next example.

Example 4 Finding the Mean, Median, and Mode

The diameter of a part coming out of a machining process is measured regularly. The diameters vary some, as shown in the frequency table. Find (a) the mean, (b) the median, and (c) the mode.

Diameter (inches)	Frequency
0.720–0.729	3
0.730–0.739	12
0.740–0.749	8
0.750–0.759	9
0.760–0.769	2

Solution

(a) The weighted mean is found using the technique shown in the previous section. The class mark for the first class is

$$\frac{0.720 + 0.729}{2} = 0.7245$$

and so on for the other classes.

Class Mark	Frequency × Class Mark
0.7245	3 × 0.7245 = 2.1735
0.7345	12 × 0.7345 = 8.814
0.7445	8 × 0.7445 = 5.956
0.7545	9 × 0.7545 = 6.7905
0.7645	2 × 0.7645 = 1.529
	Totals 34 = 25.263

The group average is

$$\frac{25.263}{34} = 0.743 \qquad \text{Rounded}$$

(b) There are 34, or an even number, of values. Divide 34 by 2 to get 17. Thus the 17th and 18th numbers from the smallest are averaged to find the median. The numbers have already been arranged from smallest to largest in the table. Three numbers fall in the first class (0.720–0.729) and 15 numbers fall in the first and second classes combined (0.720–0.729 and 0.730–0.739). The 17th and 18th numbers fall in the third class (0.740–0.749). The median is the class mark of the class containing the 17th and 18th values, or 0.7445.

(c) The mode is the class mark of the class with the highest frequency of occurrences, or 0.7345.

Thus, we have the following.

Mean = 0.743 Median = 0.7445 Mode = 0.7345

19.3 Exercises

Find the median for each of the following sets of data.

1. 37, 14, 65, 12, 32

2. 46, 27, 91, 34, 68, 53, 39

3. 95, 98, 75, 81

4. 6.8, 9.7, 5.2, 6.0, 6.8, 6.3

5. 0.81, 0.82, 0.86, 0.84

6. 900, 860, 840, 880, 920, 940

Find the mode or modes for each of the following sets of data.

7. 60, 50, 60, 40, 70, 60, 40

8. 12, 13, 10, 13, 14, 13

9. 65, 60, 68, 72, 56, 70, 85

10. 180, 195, 162, 173, 184, 195, 186, 170

11. 6, 4, 8, 4, 6, 9, 3, 2

12. 5.8, 5.6, 5.8, 5.5, 5.3, 5.4, 5.6, 5.2

13. Compare mean, median, and mode. (See Objectives 1–3.)

14. Give an example in which values for the mean, median, and mode of a set of data still do not provide a good understanding of the situation involving the data. (See Objectives 1–3.)

Solve the following application problems.

15. **SCUBA DIVING** A scuba diver used his log book to review the depths of his last five dives in the waters of the Caribbean.

68 feet, 90 feet, 56 feet, 82 feet, 110 feet

Find the mean and median.

16. **SKY DIVING** The altitudes of a sky diver's recent jumps are given in feet.

4300, 5000, 3800, 4800, 3400, 3600

Find the mean and median.

17. If you want to avoid a single extreme value having a large effect on the average, would you use the mean or the median? Explain.

18. **SHOE INVENTORY** A manager trainee at a shoe store is trying to figure out how many pairs of each size shoe of a particular style to order. Should she use mean, median, or mode? Or, are none of these values adequate? Explain. (See Section 19.2 and 19.3.)

Find the mean, median, and mode for the following grouped data sets. Round to the nearest tenth.

19.

Interval	Frequency
100–109	10
110–119	12
120–129	8
130–139	2

20.

Interval	Frequency
10–14	2
15–19	1
20–24	5
25–29	7

Solve the following application problems.

21. CHEMICAL PRODUCTION A biochemical company tries to place exactly 3 pounds of a particular enzyme in each bag that they produce. However, a tolerance of ± 0.1 pound is considered acceptable. A recent sample of 56 bags of enzyme showed the following.

Pounds of Enzyme	Frequency
2.76–2.85	0
2.86–2.95	3
2.96–3.05	48
3.06–3.15	5
3.16–3.25	0

Find the weighted mean. Based on the data available, does the company appear to have a successful bagging operation?

22. ROBOT-DRILLED HOLES A robot designed to drill a hole to a depth of 1.5 centimeters in a block of steel is being tested with the following results.

Depth of Hole (centimeters)	Frequency
1.16–1.25	0
1.26–1.35	1
1.36–1.45	18
1.46–1.55	20
1.56–1.65	0

Find the weighted mean. Does the robot appear to be working as desired?

19.4 Range and Standard Deviation

Objectives

1. *Find the range for a set of data.*
2. *Find the standard deviation.*
3. *Use the normal curve to estimate data.*

The mean is a good indicator of the middle, or central tendency, of a set of data values, but *it does not give the whole story* about the data. To see why, compare distribution A with distribution B in Table 19.6 on page 730.

Both distributions of numbers have the same mean (and the same median also), but beyond that they are quite different. In the first, 7 is a fairly typical value; but in the second, most of the values differ quite a bit from 7. In order to show this difference some measure of the **dispersion**, or spread, of the data is required.

Table 19.6 COMPARISON OF DISTRIBUTIONS A AND B

	A	B
	5	1
	6	2
	7	7
	8	12
	9	13
Mean	7	7
Median	7	7

Objective 1 *Find the Range for a Set of Data.* Two of the most common measures of dispersion, the range and the standard deviation, are discussed here.

The **range** for a set of data is defined as the *difference between the largest value and the smallest value in the set.* The range is *never* a negative number. In distribution A in Table 19.6, the largest value is 9 and the smallest is 5. The range is

$$\text{Highest} - \text{Lowest} = \text{Range}$$
$$9 - 5 = 4$$

In distribution B, the range is $13 - 1 = 12$.

The range can be misleading if it is interpreted unwisely. For example, suppose three executives rate two employees, Mark and Myrna, on five different jobs as shown in the following table.

Job	Mark	Myrna
1	28	27
2	22	27
3	21	28
4	26	6
5	18	27
Mean	23	23
Median	22	27
Range	10	22

By looking at the range for each person, we might be tempted to conclude that Mark is a more consistent worker than Myrna. However, by checking more closely, we might decide that Myrna is actually more consistent with the exception of one very poor score, which may be due to some special circumstance. Myrna's median score is not affected much by the single low score and is more typical of her performance as a whole than is her mean score.

One of the most useful measures of dispersion, the standard deviation, is based on *deviation from the mean* of the data. To find how much each value deviates from the mean, first find the mean, and then subtract the mean from each data value.

Example 1 Finding Deviations from the Mean

Find the deviations from the mean for the data values 32, 41, 47, 53, 57.

Solution Add these numbers and divide by 5. *The mean is 46.* To find the deviations from the mean, subtract 46 from each data value. (Subtracting 46 from a smaller number produces a negative result.)

Data value	32	41	47	53	57
Deviation	**−14**	**−5**	**1**	**7**	**11**

NOTE To check the work in Example 1, add the deviations. The sum of the deviations for a set of data is always 0, as long as the mean was not rounded. ■

Objective 2 *Find the Standard Deviation.* To find the measure of dispersion, it might be tempting to find the mean of the deviations. However, this number always turns out to be 0 no matter how much the dispersion in the data is because the positive deviations simply cancel out the negative ones.

Get around this problem of adding positive and negative numbers by *squaring* each deviation. (The square of a negative number is positive.) Take Example 1 one step further.

Data value	32	41	47	53	57
Deviation from mean	−14	−5	1	7	11
Square of deviation	**196**	**25**	**1**	**49**	**121**

We can now define the **standard deviation**: it is *the square root of the mean of the squares of the deviation.* The square root of a number n is written $\sqrt{n}$. It is the number that when multiplied by itself equals n. Thus, $\sqrt{144} = 12$ since $12 \times 12 = 144$, $\sqrt{1} = 1$ since $1 \times 1 = 1$, and so on. Therefore, the standard deviation is defined as follows.

$$\sqrt{\frac{\text{Sum of the squared deviations from the mean}}{\text{Number of observations}}}$$

Continuing the example on the preceding page we have the following, where s is used for standard deviation.

$$\begin{aligned} s &= \sqrt{\frac{196 + 25 + 1 + 49 + 121}{5}} \\ &= \sqrt{\frac{392}{5}} \\ &= \sqrt{78.4} \\ &= 8.9 \quad \text{Rounded to the nearest tenth} \end{aligned}$$

Scientific *Calculator Approach*

The calculator solution to this problem is as follows.

(196 + 25 + 1 + 49 + 121) ÷ 5 = $\sqrt{x}$ 8.854377448

The algebraic expression for the standard deviation is

$$s = \sqrt{\frac{\Sigma d^2}{n}}$$

where d = a deviation from the mean, n is the *number of data points in the group of numbers*, and the Greek letter Σ (sigma) represents a "sum of."

NOTE Some calculators have statistical function keys that can be used to calculate means and standard deviations. Check your calculator manual to see if your calculator has these keys.

Example 2 Finding Standard Deviation

The number of attendees at a recent children's soccer game is given: 7, 9, 18, 22, 27, 29, 32, 40. Find the standard deviation.

Solution

Step 1. Find the mean of the values.

$$\frac{7 + 9 + 18 + 22 + 27 + 29 + 32 + 40}{8} = 23$$

Step 2. Find the deviations from the mean.

Data values	7	9	18	22	27	29	32	40
Deviations	−16	−14	−5	−1	4	6	9	17

Step 3. Square each deviation.

Squares of deviations: 256 196 25 1 16 36 81 289

These numbers are the d^2 values in the formula.

Step 4. Find the sum of the d^2 values.

$$\Sigma d^2 = 256 + 196 + 25 + 1 + 16 + 36 + 81 + 289 = 900$$

Now divide Σd^2 by n, which is 8 in this example.

$$\frac{\Sigma d^2}{n} = \frac{900}{8} = 112.5$$

Step 5. Take the square root of the answer in Step 4. The standard deviation of the given list of numbers is

$$s = \sqrt{112.5} = 10.6$$

NOTE The **variance**, which is used in some textbooks, is the square of the standard deviation.

$$\text{Variance} = (\text{Standard deviation})^2$$

Similarly, the standard deviation is the square root of the variance.

$$\text{Standard deviation} = \sqrt{\text{Variance}}$$

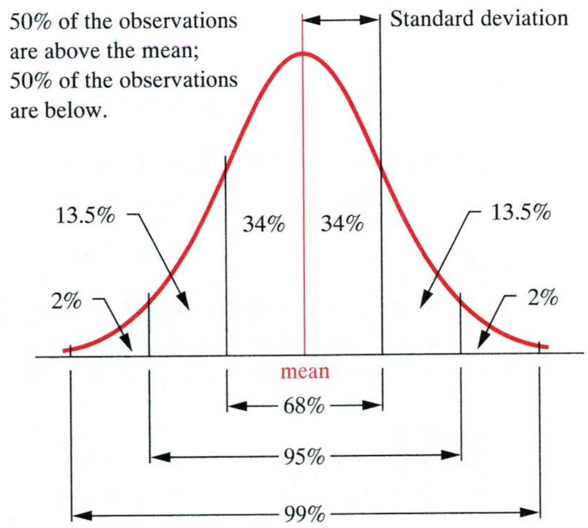

Figure 19.9

Objective 3 *Use the Normal Curve to Estimate Data.* Soft-drink bottlers such as the Coca-Cola Company wish to deliver a quality product to their customers. Of course, a primary measure of quality is taste. The level of liquid placed in each container during the bottling process affects the level of carbonation which, in turn, influences taste. Thus, these bottlers try to put a very specific amount of the liquid drink into each container. Some even X-ray each bottle as it rapidly moves down the production line in order to make sure that the amount of liquid is "close enough" to the desired level.

If the bottling process is working well, the amount of liquid in successive containers will be centered around the desired amount. Further, there will be very few containers with a fluid level that is either too low or too high. (This problem continues in Example 3.) Experience has shown that the output in this case (or in similar situations) can be described using the **normal curve**, which is shown in Figure 19.9.

It turns out that if a group of data is very closely approximated by a normal curve, approximately 68% of the data will lie *within 1 standard deviation* of the mean or between (mean − one standard deviation) and (mean + one standard deviation). Approximately 95% of the data will lie within 2 standard deviations of the mean; about 99% will lie within 3 standard deviations of the mean.

Example 3 Using the Normal Distribution

A worker uses statistics to look at the quality, in terms of fluid level in the containers, of 300 containers of a beverage bottled at his workstation. Suppose that the fluid level in the containers is approximated by the normal curve with a mean of 6 inches and a standard deviation of 0.04 inches. (a) Find the mean minus 3 standard deviations and the mean plus 3 standard deviations. (b) Use Figure 19.9 to find the number of containers out of the 300 that are expected to fall within the two limits found in part (a).

Solution

(a) Mean − 3 standard deviations $= 6 - (3 \cdot 0.04) = 5.88$ inches
Mean + 3 standard deviations $= 6 + (3 \cdot 0.04) = 6.12$ inches

(b) Simply add the percents within 3 standard deviations of the mean in Figure 19.9 to find the total percent of the containers that should fall within the limits found in (a) above.

$$2\% + 13.5\% + 34\% + 34\% + 13.5\% + 2\% = 99\%$$

Now multiply this percent times the 300 containers to find the number of containers that should fall within the limits found in (a).

$$99\% \text{ of } 300 = 0.99 \times 300 = 297 \text{ containers}$$

If a random sample results in more than 3 containers, out of 300, outside the acceptable range, then the production line is stopped and equipment is adjusted.

NOTE Workers from all over the world have been trained in statistics so they can monitor the quality of their own work. ▬

19.4 Exercises

Find the mean and standard deviation for each set of data. Round answers to the nearest tenth.

1. 15, 18, 20, 19

2. 6.8, 5.4, 3.7, 7.2, 6.4

3. 20, 22, 23, 18, 21, 22

4. 120, 118, 109, 115, 112, 110

5. 55, 58, 54, 52, 51, 59, 58, 60

6. 7.5, 7.3, 7.2, 7.5, 7.8, 7.1, 7.4, 8.0, 7.2, 7.6

Find the range for each set of data.

7. 18, 24, 60, 42, 51, 61

8. 10, 15, 12, 17, 21, 13

9. 500, 274, 361, 295, 112

10. 10.3, 7.4, 8.1, 6.5, 9.7

11. When can the range of a set of numbers be misleading? (See Objective 1.)

12. Explain the meaning of range and standard deviation. (See Objectives 1 and 2.)

The weight of two hundred 12-year-old students was measured. Obviously, not all of the students had exactly the same weight. However, the results can be approximated by a normal curve with a mean of 80 pounds and a standard deviation of 0.5 pound. Use the following graph to find the number of students weighing as indicated in Exercises 13–20.

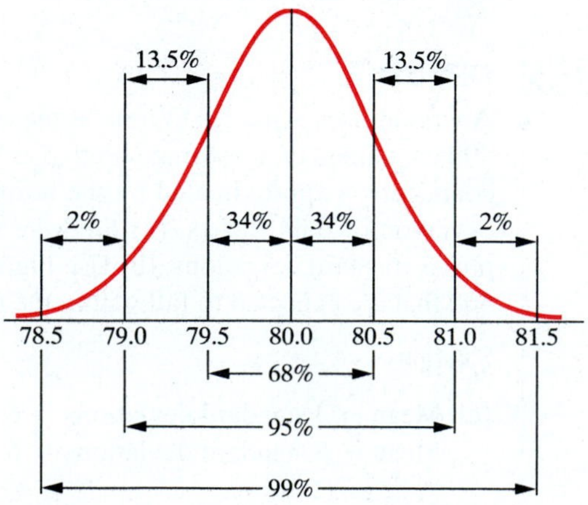

13. More than 80 pounds

14. More than 79.5 pounds

15. Between 79.5 and 80.5 pounds

16. Between 79 and 81 pounds

17. Between 80.5 and 81 pounds

18. Between 79.5 and 81 pounds

19. Within 1 pound of the mean

20. More than $1\frac{1}{2}$ pounds away from the mean

The number of contacts per telemarketing employee has recently averaged 120 per day, with a standard deviation of 20. Use the graph of the normal curve used for Exercises 13–20 to estimate the likelihood of the following number of contacts.

21. Less than 80

22. More than 160

23. Between 80 and 160

24. Between 100 and 140

On standard IQ tests, the mean is 100 and the standard deviation is 15. The results are very close to fitting a normal curve. Suppose an IQ test is given to a very large group of people. Find the percent of people whose IQ score is:

25. More than 100

26. Less than 100

27. Greater than 115

28. Between 85 and 115

29. Between 70 and 130

30. Between 55 and 145

31. Less than 55

32. More than 145

After a 30,000-mile road test, the wear on a particular brand of tire averaged 0.2 inches, with a standard deviation of 0.015 inches. The tire wear fits a normal curve closely. What percent of the tire wear is:

33. 0.185 inches or more?

34. Less than 0.185 inches?

35. 0.23 inches or less?

36. More than 0.23 inches?

The average amount of time taken by runners to complete a 10-kilometer race was 47.6 minutes, with a standard deviation of 2.7 minutes. Given that the times are approximately normally distributed, what percent of the runners required:

37. At least 44.9 minutes?

38. No more than 44.9 minutes?

39. No more than 53 minutes?

40. At least 53 minutes?

41. Between 44.9 minutes and 50.3 minutes?

42. Between 42.2 minutes and 53 minutes?

43. CLAIM SETTLEMENT The time required to handle a claim by one claim adjuster at an insurance company is approximated by the normal curve with mean 12.5 minutes and standard deviation 2.2 minutes. Out of 200 claims, how many would you expect to take more than 19.1 minutes?

44. INFLATING AUTOMOBILE TIRES A machine automatically inflates automobile tires at a factory. Results can be approximated by a normal curve with mean 30.8 pounds per square inch (psi) and, a standard deviation of 0.4 psi. Find the number of tires, out of 2000 filled, that will likely have less than 30 psi.

19.5 Index Numbers

Objectives

1 *Find the price relative.*

2 *Use the Consumer Price Index to compare costs.*

Objective 1 *Find the Price Relative.* A house that cost \$55,000 10 years ago now sells for \$115,000. Tuition at one state university has gone from \$320 10 years ago to \$2000 today. Both items increased in price. To find out which increased by the higher percent, find the *price relative* for each item. A **price relative** is the quotient of the current price and the price in some past year with the quotient multiplied by 100. The past year is called the **base year**. The formula for the price relative is as follows.

$$\text{Price relative} = \frac{\text{Price this year}}{\text{Price in base year}} \times 100$$

Example 1 Finding the Price Relative

(a) Using the prices just given, the price relative for the house follows.

$$\text{Price relative} = \frac{\$115{,}000}{\$55{,}000} \times 100 = 209.1 \quad \text{Rounded}$$

The price of the house today is 209.1% of its price 10 years ago. Note that a price relative is really just a percent—it gives the percent that this year's price is of the price in the base year.

(b) The price relative for the university tuition, using the prices given, is as follows.

$$\text{Price relative} = \frac{\$2000}{\$320} \times 100 = 625$$

The price of tuition at this particular university is 625% of its price 10 years ago.

Compare the price relatives of Example 1 to find that the price of the university tuition has increased at a faster rate than that of the house.

Example 2 Finding the Price Relative

The price of a particular computer chip dropped from \$1100 2 years ago to \$220 today. The price relative is given here.

$$\frac{\$220}{\$1100} \times 100 = 20$$

The computer chip sells today for 20% of its previous selling price. This example shows that a price relative can be less than 100; this occurs when the price of an item drops over time. A few things go down in price over the years *but most go up in price with time.* In particular, almost all price relatives are over 100 during inflationary periods.

NOTE The price relative gives a way of comparing the two prices and showing a percent increase. ■

Objective 2 *Use the Consumer Price Index to Compare Costs.* The **Consumer Price Index (CPI)**, published by the Bureau of Labor Statistics, can be given in terms of price relatives. The bureau keeps track of the costs of a great many items in different cities throughout the country and publishes its findings monthly. A recent portion of the report for one month is included in Table 19.7.

Table 19.7 URBAN PRICES INDEX*

	Chicago	Dallas	Los Angeles	New York	Philadelphia
All items	161.7	151.4	160.0	170.8	166.5
Food and beverages	160.5	157.0	163.2	163.0	154.3
Housing	160.2	138.8	156.6	171.7	167.3
Apparel and upkeep	122.3	137.2	121.0	130.3	104.0
Transportation	140.2	142.9	145.2	154.1	152.8
Medical	235.5	225.3	232.0	244.5	247.8
Entertainment	175.3	168.9	147.6	167.6	181.8

*All figures are expressed as a percent of the 1987 base of 100.

The numbers in this table represent price relatives, with a base year of 1987. For example, in Chicago the cost of food and beverages is now 160.5% of the cost in 1987, or, in other words, it now costs $160.50 to buy the food and beverages that could have been bought with $100 in 1987. As a further example, medical care that cost $100 in Philadelphia in 1987 now costs $247.80, and so on.

The price relatives in Table 19.7 can only be used to compare prices for the given city—they *cannot* be used for comparisons between cities. For example, the cost of housing in Los Angeles is now 156.6% of what it was in 1987, while the cost of housing in Chicago is now 160.2% of what it was in 1987. However, it cannot be said that housing in Chicago is more expensive than in Los Angeles from the data in the table—only that housing costs have increased at a faster rate in Chicago. It is possible that housing in Chicago was less expensive to begin with, so even with a greater percent increase it may still be less expensive than Los Angeles.

Example 3 Using the CPI

Suppose that a specific surgical operation costs $3200 in Dallas in 1987. Estimate its cost today.

Solution From Table 19.7, medical costs in Dallas today are 225.3% of what they were in 1987. A $3200 operation in 1987 should cost about

$$\$3200 \times 2.253 = \$7209.60$$

or approximately $7200 today.

19.5 Exercises

Find the price relatives for the following items. Round to the nearest tenth.

	Item	Price Then	Price Now
1.	Rent	$225 per month	$550 per month
2.	Ski boat	$2600	$4800
3.	Computer	$2000	$1200

	Item	Price Then	Price Now
4.	Jeans	$25 per pair	$40 per pair
5.	Natural gas	$1.25 per thousand cubic feet	$2.10 per thousand cubic feet
6.	House	$60,000	$95,000

Use the Urban Prices Index in Table 19.7 to complete the following chart.

	Urban Area	Item	$100 Worth in 1987 Will Cost Today
7.	Philadelphia	Food and beverages	______
8.	Chicago	Transportation	______
9.	New York	Medical	______
10.	Los Angeles	Entertainment	______
11.	Dallas	All items	______
12.	New York	Apparel and upkeep	______

Solve the following application problems using Table 19.7.

13. HOUSING COSTS Suppose a house near New York costs $180,000 in 1987. Estimate its cost today. Round to the nearest $1000.

14. MEDICAL OPERATION Estimate the cost today of a medical operation in Chicago that cost $1850 in 1987. Round to the nearest $500.

15. FOOD AND BEVERAGE COST A family of five in Dallas spent an average of $7200 on food and beverages in 1987. Estimate to the nearest $100 what a family of five would spend today.

16. BUDGET A family in Philadelphia budgeted $28,000 for all expenditures other than taxes and vacations in 1987. Find the amount they would need to budget today to the nearest $100.

17. INCREASING PRICES Which expense item increased most rapidly over the past 10 years?

18. INCREASING PRICES In which area in Table 19.7 did prices rise most rapidly over the past 10 years?

19. Explain price relative and inflation. Are they identical to one another? Explain (See Objectives 1 and 2.)

20. Go to the library and find the government's estimate of the CPI for the past 3 years. Compare the average inflation of these 3 years to inflation during the early 1930s, when inflation was actually negative. Also compare to inflation in the late 1970s, when inflation was very high. (See Objective 3.)

Chapter 19 Quick Review

Chapter Terms

Review the following terms to test your understanding of the chapter. For any terms you do not know, refer to the page number found next to that term.

Concept	Example

19.1 Constructing a frequency distribution from raw data

1. Construct a table listing each value and the number of times this value occurs.
2. For a distribution with grouped data, combine the data into classes.

For the following data, construct a frequency distribution:

12, 15, 15, 14, 13, 20, 10,
12, 11, 9, 10, 12, 17, 20, 16,
17, 14, 18, 19, 13.

Data	Tally	Frequency
9	|	1
10	||	2
11	|	1
12	|||	3
13	||	2
14	||	2
15	||	2
16	|	1
17	||	2
18	|	1
19	|	1
20	||	2

Classes	Frequency
9–11	4
12–14	7
15–17	5
18–20	4

19.1 Constructing a bar graph from a frequency distribution

Draw a bar for each class using the frequency of the class as the height of the bar.

Construct a bar graph from the frequency distribution of the previous example.

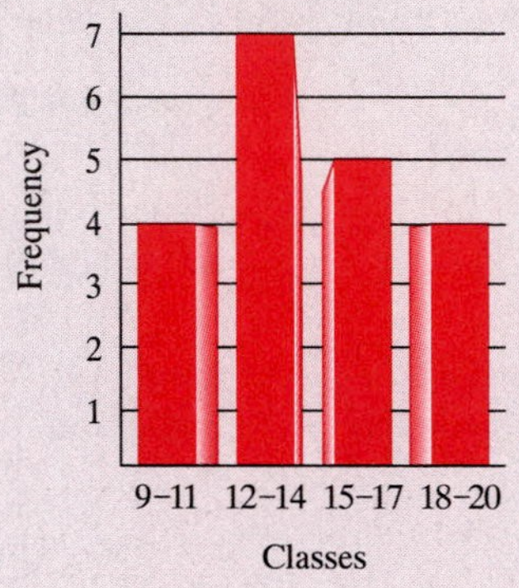

19.1 Constructing a line graph

1. Plot each year on the horizontal axis.
2. For each year, find the values for that year and plot a point for each value.
3. Connect all points with straight lines.

Construct a line graph for the following data.

Year	Value
1998	$850,000
1999	920,000
2000	875,000
2001	975,000

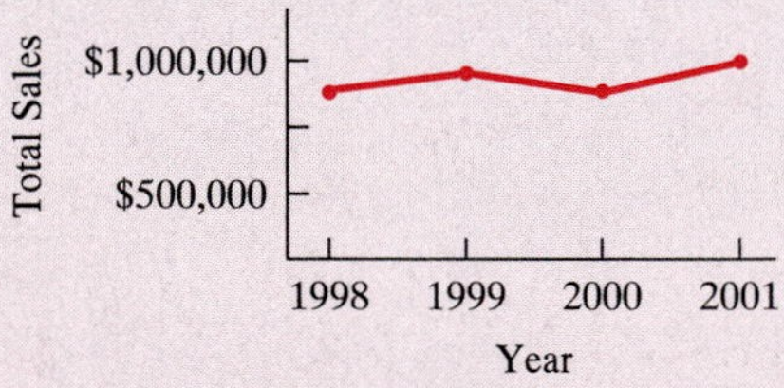

19.1 Constructing a circle graph

1. Determine the percent of the total for each item.
2. Find the number of degrees represented by each percent.
3. Draw the circle.

Construct a circle graph for the following expenses.

Item	Amount	Percent
Car	$200	20%
Lodging	300	30%
Food	250	25%
Entertainment	150	15%
Other	100	10%

Multiply each percent by 360° to find the angle for each sector.

Car: $360° \times 0.2 = 72°$

Lodging: $360° \times 0.3 = 108°$

Food: $360° \times 0.25 = 90°$

Entertainment: $360° \times 0.15 = 54°$

Other: $360° \times 0.1 = 36°$

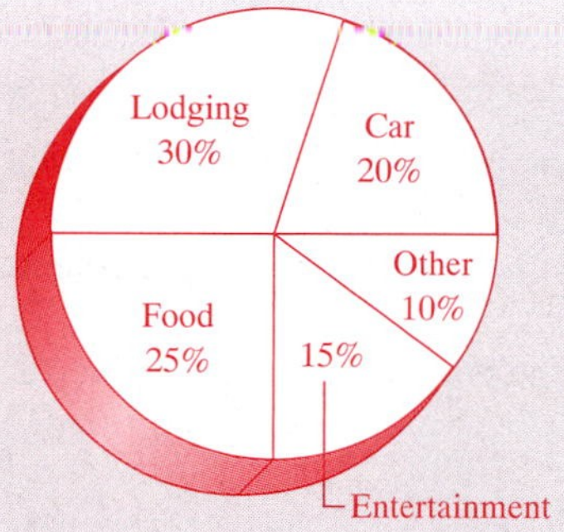

19.2 Finding the mean of a set of data

Divide the sum of the data by the number of data points.

The test scores for Pat Phelan in her business math course were 85, 76, 93, 91, 78, 82, 87, and 85. Find Phelan's test average.

$$\text{Mean} = \frac{85 + 76 + 93 + 91 + 78 + 82 + 87 + 85}{8}$$

$$= \frac{677}{8} = 84.63$$

19.2 Finding the weighted mean

1. Multiply frequency by value.
2. Add all products obtained in Step 1.
3. Divide the sum in Step 2 by the total number of data points.

No. of School-Age Children	Frequency
0	12
1	6
2	7
3	3
4	2
Total families	30

Find the mean number of school-age children per family.

No.	Frequency	No. × Frequency
0	12	0
1	6	6
2	7	14
3	3	9
4	2	8
		37

$$\text{Mean} = \frac{37}{30} = 1.23$$

19.2 Finding the mean of a grouped frequency distribution

1. Determine the class mark (the midpoint) for each class.
2. Multiply the class mark by the frequency of each class.
3. Add all the products obtained in Step 2.

Find the mean.

Intervals	Frequency
3–5	2
6–8	10
9–11	12
12–14	9
15–17	7

Intervals	Freq.	Class Mark	Frequency × Class Mark
3–5	2	4	8
6–8	10	7	70
9–11	12	10	120
12–14	9	13	117
15–17	7	16	112
	40		427

$$\text{Mean} = \frac{427}{40} = 10.68$$

19.3 Finding the median of a set of data

1. Arrange the data from lowest to highest.
2. If there are an *odd* number of numbers, the median is the number in the middle.
3. If there are an *even* number of numbers, the median is the average of the two in the middle.

Find the median for Pat Phelan's grades from an earlier example.

The data arranged from lowest to highest is 76, 78, 82, 85, 85, 87, 91, and 93.

The average of the middle two values is $\frac{85 + 85}{2} = 85$.

19.3 Determining the mode of a set of data

The mode is the most frequently occurring value.

Find the mode for Phelan's grades in the previous example.

The mode is 85, which occurs most frequently (twice).

19.3 Finding the median of a grouped frequency distribution

The median is the class mark associated with the middle number or with the average of the middle numbers.

Intervals	Frequency
10–19	5
20–29	8
30–39	2

Divide 15 by 2 to find 7.5. Thus, the class mark associated with the eighth value from the smallest, or 24.5, is the median.

19.3 Finding the mode of a grouped frequency distribution

The mode is the class mark of the class with the highest frequency.

Intervals	Frequency
10–19	5
20–29	8
30–39	2

The second class has the highest frequency. The mode is the associated class mark of 24.5.

19.4 Finding the range of a set of data

$$\text{Range} = \text{Highest} - \text{Lowest}$$

Find the range of the values 7, 6, 10, 7, 9, 5, 2, 8, and 9.

$$\text{Range} = \text{Highest} - \text{Lowest}$$
$$= 10 - 2 = 8$$

19.4 Finding the standard deviation

1. Determine the mean of the data.
2. Subtract the mean from each value to obtain individual deviations, d.
3. Square each deviation.
4. Sum all the squared deviations.
5. Divide the sum by the number of data.
6. Take the square root of the number obtained in Step 5.

Find the standard deviation of the values, 7, 6, 10, 7, 9, 5, 2, 8, and 9.

$$\text{Mean} = \frac{7 + 6 + 10 + 7 + 9 + 5 + 2 + 8 + 9}{9}$$
$$= 7$$

Data Value	Deviation d	Deviation Squared, d^2
7	0	0
6	−1	1
10	3	9
7	0	0
9	2	4
5	−2	4
2	−5	25
8	1	1
9	2	4

$$\frac{\Sigma d^2}{n} = \frac{48}{9} = 5.33$$
$$s = \sqrt{5.33} = 2.31$$

19.5 Finding the price relative

Divide the price this year by the price in a base year and multiply by 100.

A car cost \$16,000 10 years ago. It now costs \$22,500. Find the price relative for the car.

$$\text{Price relative} = \frac{\text{Price this year}}{\text{Price in base year}} \times 100$$
$$= \frac{\$22{,}500}{\$16{,}000} \times 100 = 140.6$$

19.5 Finding cost today using the Consumer Price Index (CPI)

Multiply CPI by cost in base year to estimate current cost.

A typical house in Philadelphia cost \$88,000 in 1987. Estimate its cost today.

$$\text{Cost today} = \$88{,}000 \times 167.3\%$$
$$= \$88{,}000 \times 1.673$$
$$= \$147{,}224$$

Chapter 19 Review Exercises

The answer section includes answers to all Review Exercises.

Work the following application problems. The following numbers are the number of gallons of gasoline sold at a convenience store by week for the past 20 weeks. **[19.1]**

12,450	11,300	12,800	10,850	14,100
14,900	12,300	11,600	12,400	12,900
13,300	12,500	13,390	12,800	12,500
15,100	13,700	12,200	11,800	12,600

1. Use these numbers to complete the following table.

Gallons of Gasoline	Number of Weeks
10,000–10,999	______
11,000–11,999	______
12,000–12,999	______
13,000–13,999	______
14,000–14,999	______
15,000–15,999	______

2. How many weeks had sales of 13,000 gallons or more?

3. Use the numbers in Exercise 1 to draw a bar graph. Be sure to put a heading and labels on the graph.

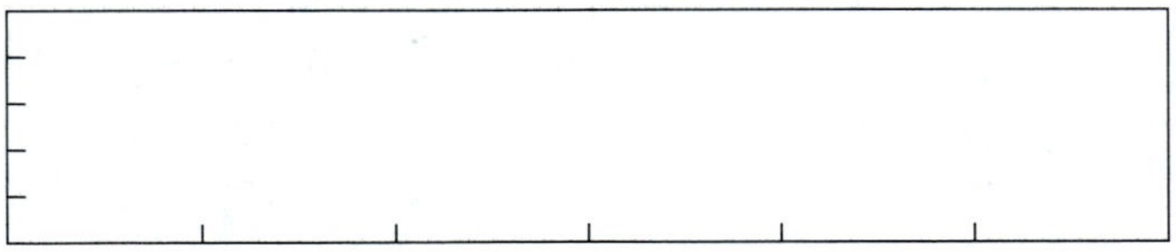

4. During a 1-year period, the campus newspaper at Comfort Community College had the following expenses. Find all numbers missing from the table.

Item	Dollar Amount	Percent of Total	Degrees of a Circle
Newsprint	$12,000	20%	______
Ink	$6,000	______	36°
Wire Service	$18,000	30%	______
Salaries	$18,000	30%	______
Other	$6,000	10%	______

5. Draw a circle graph using the information in Exercise 4.

6. What percent of the expenses were for newsprint, ink, and wire service?

Calculate the mean, median, and mode for the following sets of data. Round answers to the nearest tenth. **[19.2–19.3]**

7. 25, 20, 18, 35, 19

8. 85, 80, 82, 82, 88, 90, 92

9. 21, 20, 20, 18, 21, 19, 21, 22

10. 42, 44, 41, 44, 45, 44

11. 8, 7, 6, 6, 7, 7, 5, 9

12. 2.5, 2.4, 2.4, 2.3, 2.4, 2.6, 2.0, 2.2

Calculate the mean for the frequency distribution given in Exercises 13 and 14. Round answers to the nearest tenth. **[19.2]**

13.

Intervals	Frequency
10–14	6
15–19	3
20–24	5
25–29	7
30–34	5
35–39	9

14.

Intervals	Frequency
10–19	6
20–29	5
30–39	9
40–49	4
50–59	7

Calculate the median and the mode for the frequency distribution given in Exercises 15 and 16. **[19.3]**

15.

Intervals	Frequency
1–5	20
6–10	12
11–15	14
16–20	10
21–25	5

16.

Intervals	Frequency
50–59	3
60–69	5
70–79	18
80–89	12
90–99	4

Find the range and standard deviation for Exercises 17–20. Round answers to the nearest hundredth. **[19.4]**

17. 62, 24, 38, 91, 56

18. 5, 7, 12, 10, 7, 12, 18

19. 82, 86, 78, 74, 65

20. 150, 145, 130, 120, 162, 158

Find the price relative for Exercises 21–24. Round to the nearest tenth. **[19.5]**

Item	Price in Base Year	Price in Current Year
21. Automobile	$22,500	$32,300
22. Television	$250	$410
23. Personal computer	$1,800	$950
24. Lawn mower	$185	$300

Solve the following application problems.

25. The Dallas Chamber of Commerce advertised in 1987 that a family of five could live comfortably in their community with an income of $33,500 per year. Estimate the amount that a family of five would need today to live as comfortably to the nearest $100. **[19.5]**

26. The thickness of an alloy coating on a military weapon can be approximated by a normal curve with mean 7.2 millimeters and standard deviation 0.08 millimeters. Find the percent of items expected to have a coating thicker than 7.12 millimeters. **[19.4]**

27. Okba Asad, the owner of Current's Restaurant, has monitored weekly sales for the past 6 weeks.

Week	Sales	Week	Sales
1	$4227	4	$5009
2	$4806	5	$4198
3	$4559	6	$5126

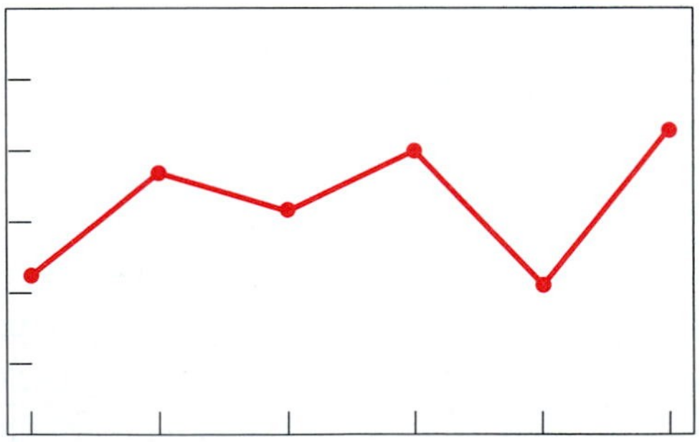

Plot these data in a line graph. Do sales appear to be increasing or can you tell from the data given? **[19.1]**

28. Marie Dee wishes to calculate her grade point average for the semester. Round your answers to the nearest hundredth. (A = 4.0, B = 3.0, C = 2.0, D = 1.0, F = 0.0)

Course	Semester Hours	Grade
Business math	3	A
English	3	C
Biology	4	B
Spanish	3	D

Also, find her grade point average if she had been unable to pull her Spanish grade up from the F she had in the course prior to the final exam. **[19.2]**

29. The average price of a condominium in a resort area in Colorado increased from $110,000 in 1990 to $195,000 today. Find the price relative. Use this value to estimate the cost today of a small home that sold in the resort area for $90,000 in 1990. **[19.5]**

30. Ted Smith sells stocks and bonds at Merrill Lynch. His wife developed a serious illness at the beginning of 2000 and her condition slowly improved through the balance of the year. Do you think his personal problems may have influenced his work performance?

Year	Quarterly Commissions			
1998	$14,250	$12,375	$15,750	$13,682
1999	$13,435	$14,230	$11,540	$15,782
2000	$8,207	$7,350	$10,366	$11,470

Support your view by drawing a line graph. Be sure to label the quarter in which Mrs. Smith became ill.

Sales Commission for Ted Smith
By Quarter for 1998–2000

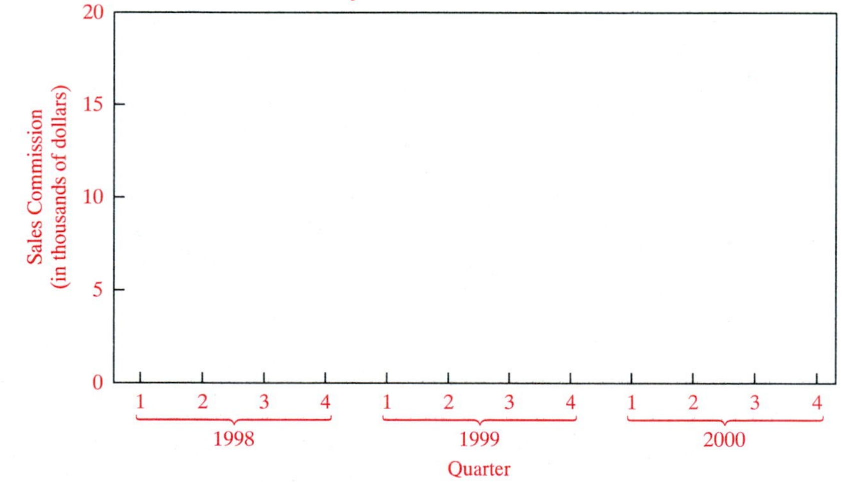

Chapter 19 Summary Exercise
Watching the Growth of a Small Business

Pat Gutierrez expanded her company, Christian Books Unlimited, this year by opening stores in different cities. One store was opened in February, and a second was opened in June. Sales, in thousands of dollars, for the two stores are given below.

	Feb.	Mar.	Apr.	May	June	July	Aug.	Sep.	Oct.
Store 1	6.5	6.8	7.0	6.9	7.5	7.8	8.0	7.6	8.2
Store 2	—	—	—	—	8.2	6.2	8.2	8.7	9.6

(a) Find the median, mean, and mode sales for each store to the nearest tenth.
(b) Plot sales for both stores on the same line graph with month on the horizontal axis and sales on the vertical axis.

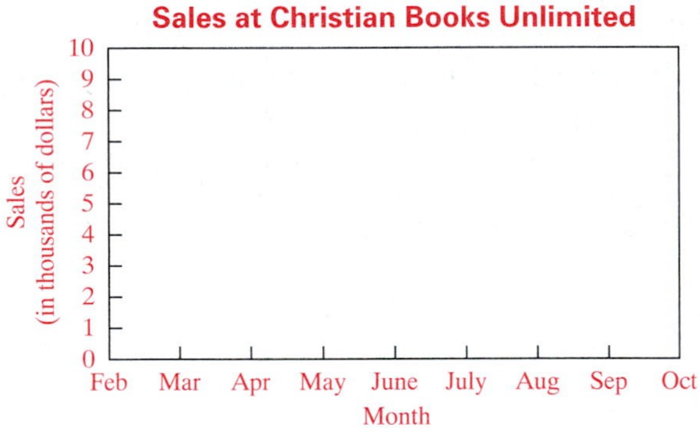

(c) What trends are apparent from the line graph above?

Net Assets Business on the Internet

Burger King

Statistics

- 1998: Revenues of $10.9 billion
- 300,000 employees
- 10,896 restaurants worldwide
- 14,000,000 customers served daily

Burger King Corporation was founded in 1954 in Miami, Florida. The company introduced the Whopper sandwich in 1957. In 1975, the firm began using drive-through service to satisfy customers who were on the go. The drive-through now accounts for about 60% of Burger King's business. In 1985, average restaurant sales passed the $1 million mark and a European training center opened in London to service overseas company and franchise employees. Burger King serves approximately 1404 customers per restaurant per day and the company notes with pride that a customer can order a Whopper in 1024 different possible ways.

1. Sales on consecutive days at a Burger King restaurant rounded to the nearest dollar were: Monday, $1475; Tuesday, $1456; Wednesday, $1278; Thursday, $1503; Friday, $1895; Saturday, $1753; and Sunday, $1298. Find the average daily sales (mean) and also the median.

2. Find the range and the standard deviation for the data in question 1. Round the standard deviation to the nearest whole number.

3. Use the average daily sales from question 1 and the standard deviation from question 2. Assume that daily sales are normally distributed and find the mean minus 2 standard deviations and the mean plus 2 standard deviations. Daily sales will fall between these two numbers 95% of the time.

Part 5 Cumulative Review Chapters 16–19

Solve the following application problems.

1. The Fashion Shoppe purchased new shelving for its store at a cost of $4840. The estimated life of the shelving is 4 years while the salvage value is estimated to be $200. Assuming straight-line depreciation, find (a) the annual rate of depreciation, (b) the annual amount of depreciation, and (c) the book value at the end of the *second* year. *[16.1]*

2. The manager of a gas station purchased new gasoline pumps at a cost of $16,400. The pumps are expected to have a useful life of 10 years, but will then have no salvage value. Find (a) the first year's annual depreciation and (b) the second year's annual depreciation using the 200% declining-balance method. *[16.2]*

3. To control the rate at which the bananas ripen, Wholesale Bananas, Inc. bought a new walk-in freezer for $12,820. It is estimated to have a salvage value of $400 after 8 years. Find the first and second year's depreciation using the sum-of-the-years'-digits method. Then find the book value at the end of the second year. *[16.3]*

4. A garbage truck costing $42,250 is expected to operate for 11,500 hours and to then have a salvage value of $2000. Use the units-of-production method to find the depreciation for the first year if the truck is operated 12 hours per day for all 365 days of the year. Then find the book value at the end of the first year. *[16.4]*

5. An automobile manufacturing company bought a new stamping press to stamp out automobile fenders for $115,800 and placed it in service on April 1. The salvage value after an expected useful life of 5 years is $3800. Find the first year's partial depreciation using, (a) the straight-line method, (b) the double-declining-balance method and (c) the sum-of-the-years'-digits method. *[16.4]*

6. Find the recovery period under MACRS for (a) a tugboat, (b) a delivery van, (c) a 4-year-old race horse, (d) a 32-unit apartment house, and (e) a couch for the waiting room in a dentist's office. *[16.5]*

7. Atlas Moving and Storage purchased a tractor unit for moving people's household goods long distances. The cost of the truck was $96,000. Prepare a depreciation schedule using the MACRS method of depreciation. *[16.5]*

Year	Computation	Amount of Depreciation	Accumulated Depreciation	Book Value
0	—	—	—	$96,000
1	______	______	______	______
2	______	______	______	______
3	______	______	______	______
4	______	______	______	______

8. The Fashion Shoppe had gross sales of $240,800 with returns of $4300. Inventory at the beginning of the year was $48,300 and by the end of the year inventory was $41,500. A total of $102,000 worth of goods was purchased last year and freight charges were $2900. Wages were $32,400, rent was $15,000, advertising was $2200, utilities were $3100, and taxes on inventory and payroll totaled $6100. Miscellaneous expenses totaled $8900 and income taxes were $11,400. Complete an income statement for the company. *[17.1]*

THE FASHION SHOPPE
INCOME STATEMENT
YEAR ENDING DECEMBER 31

Gross sales			______	
Returns			______	
Net sales				______
Inventory, January		______		
Cost of Goods Purchased	______			
Freight	______			
Total cost of goods purchased		______		
Total of goods available for sale		______		

Inventory, December 31	______	
Cost of goods sold		______
Gross profit		______
Expenses		
Salaries and wages	______	
Rent	______	
Advertising	______	
Utilities	______	
Taxes on inventory, payroll	______	
Miscellaneous expenses	______	
Total expenses		______
Net Income Before Taxes		______
Income taxes		______
Net Income		______

9. Analyze the income statement of Exercise 8 by finding cost of goods sold, gross profit, net income before taxes, and net income as a percent of net sales. Round to the nearest tenth of a percent. **[17.2]**

10. The Fashion Shoppe (see Exercise 8) has $28,400 cash, $8400 in notes receivable, $3800 in accounts receivable, and inventory of $41,500. They do not own the land or building that they use, but fixtures have an estimated value of $12,200. The company has notes payable of $4800 and accounts payable of $32,500, but has neither mortgages nor long-term notes payable. Complete a balance sheet for this company. **[17.3]**

THE FASHION SHOPPE
BALANCE SHEET FOR DECEMBER 31

	Assets		
Current assets			
Cash	______		
Notes receivable	______		
Accounts receivable	______		
Inventory	______		
Total current assets		______	
Plant assets			
Land	______		
Buildings	______		
Fixtures	______		
Total plant assets		______	
Total Assets			______
	Liabilities		
Current liabilities			
Notes payable	______		
Accounts payable	______		
Total current liabilities		______	
Long-term liabilities			
Mortgages payable	______		
Long-term notes payable	______		
Total long-term liabilities		______	
Total Liabilities			______

Owner's Equity	
Owner's equity	______
Total Liabilities and Owner's Equity	______

11. Find the current ratio, acid-test ratio, and ratio of net income after taxes to average owner's equity assuming owner's equity one year previous to the balance sheet of Exercise 10 was $42,800 and using the income statement from Exercise 8. **[17.4]**

12. Noel Saturn owns 50 shares of preferred stock with a par value of $80 and a dividend of 5%. Find his total dividend payment. **[18.1]**

13. Jackson Brewery paid no dividend last year but has set aside $80,750 for the payment of dividends this year. It has 10,000 shares of cumulative preferred stock outstanding with a par value of $60 and a 6% dividend. Find this year's dividend per share of common stock given 25,000 shares of common stock outstanding. **[18.1]**

14. ABC, Inc. was incorporated in 1990 and now has 200,000 shares of common stock outstanding. Given a net income of $85,000 and preferred dividends of $8000, find the earnings per share of common stock. **[18.1]**

15. José Torres purchased 80 shares of Microsoft at $107 $\frac{1}{4}$ per share. Find his total cost assuming a broker's commission of 1%. *Hint:* Be sure to include the odd-lot differential. **[18.2]**

16. A publicly held company that writes computer software had net income this year of $2.06 per share and paid a dividend this year of $0.40 per share. Given a stock price of $62 $\frac{1}{2}$, find (a) the current yield and (b) the PE ratio. **[18.2]**

17. Jessica Thomas sold 100 shares of Intel at $85 $\frac{1}{4}$ per share. She paid a commission of $15. Find the amount Thomas received after expenses. **[18.2]**

18. Use the stock table on page 670 (Figure 18.3) to find the following values for Compaq: low for the year, dividend yield, and high for the day. **[18.2]**

19. Use the bond table on page 682 (Figure 18.7) to find the current yield, volume, and close for IBM 5 $\frac{3}{8}$ 09 bonds. **[18.3]**

20. Find the net asset value and year-to-date return on investment for Janus Mercury mutual fund using Figure 18.6 on page 678. **[18.2]**

21. Find the cost of purchasing 25 bonds selling at 97 $\frac{1}{4}$ assuming a commission of $5 per bond. **[18.3]**

22. Find the net proceeds after a $5 commission per bond from the sale of 40 bonds at a market price of $92 $\frac{3}{8}$. **[18.3]**

23. Two partners invested $40,000 and $60,000, respectively, in a business venture. Given a profit of $48,000, find the distribution of profit to the two partners using (a) equal shares, (b) an agreed ratio of 3:1 to the first partner and 25% to the second partner, and (c) the original investment. **[18.4]**

24. Alan Padgett and Gina Harden own a retail business that Harden manages at a salary of $28,000. Originally, Padgett invested $60,000 and Harden invested $20,000 to start the business. Find the amount each partner would receive from a profit of $85,000 if they agree to divide profits based on their original investment. **[18.4]**

25. Allocate an overhead of $340,000 by floor space. **[18.5]**

Department	Floor Space
Machining	30,000 square feet
Stamping	15,000 square feet
Assembly	35,000 square feet
Total	80,000 square feet

26. Allocate an overhead of $180,000 by number of employees. **[18.5]**

Department	Number of Employees
A	20
B	12
C	18
	Total 50

27. Allocate an overhead of $140,000 by sales value. **[18.5]**

Product	Number Produced	Value of Each
1	12,800	$25
2	14,200	$15
3	180	$200

28. Use the line graph below to find the weekly pay for production workers at the beginning of 1998 and at the beginning of 1999. Then find the percent of increase to the nearest tenth. **[19.1]**

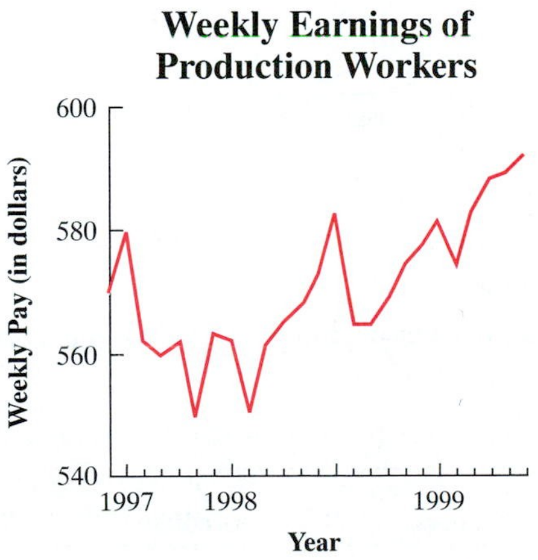

Data: Wall Street Journal, reprinted by permission of Dow Jones, Inc. via Copyright Clearance Center, Inc. © 1999 Dow Jones & Co., Inc. All rights reserved worldwide.

29. Use the figure on page 754 to estimate the percent of decrease in timber revenue for Alpine County, California, from 1991–92 to 1998–99. **[19.1]**

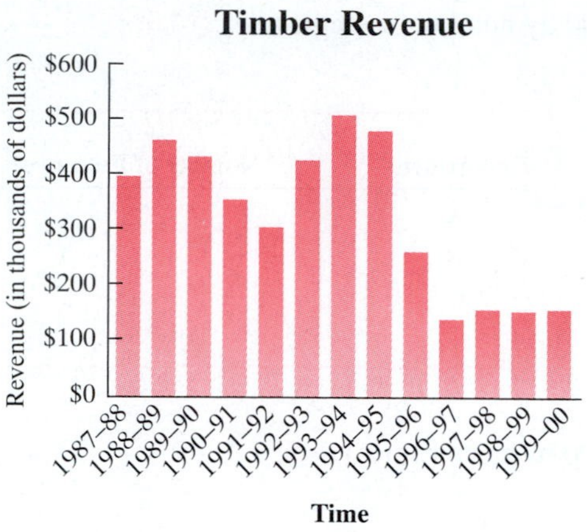

DATA: Alpine County School District

30. Benson's Medical Supplies had quarterly sales of $85,200, $102,000, $92,000, and $124,000 last year. Show this data using both a bar chart and a pie chart. **[19.1]**

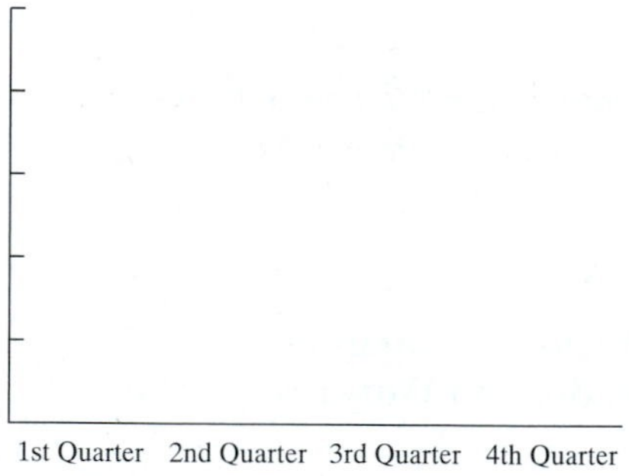

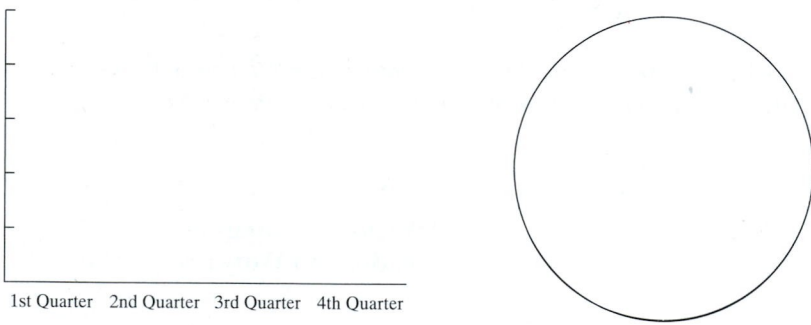

31. Find the mean, median, and mode for each data set: (a) 18, 14, 16, 17, 16, 19; (b) 24, 32, 25, 24, 31, 28, 27. Round to the nearest tenth, if needed. **[19.2–19.3]**

32. Glenda Hickey made the grades shown below. Find her grade point average to the nearest hundredth. (Note: A = 4, B = 3, C = 2, D = 1, and F = 0.) **[19.2]**

Course	Credits	Grade
Math	3	B (= 3)
English	4	C (= 2)
Sociology	3	C (= 2)
Music	3	A (= 4)
Total	13	

33. Find the mean, range, and standard deviation for the following data: 78, 82, 71, 69. Round to the nearest tenth where applicable. **[19.2 and 19.4]**

34. The diameter of a metal transmission shaft fits a normal curve, with a mean of 0.625 inches and a standard deviation of 0.045 inches. Find the percent of the shafts with a diameter of between 0.625 inches and 0.715 inches assuming a normal curve. **[19.4]**

35. The price of a network card for a computer network dropped from $1200 one year ago to $680 today. Find the price relative, rounded to the nearest whole number. **[19.5]**

36. Assume that a house cost $68,000 in Chicago in 1987. Use Table 19.7 on page 737 to find the price relative and estimate its cost today. Round to the nearest thousand. **[19.5]**

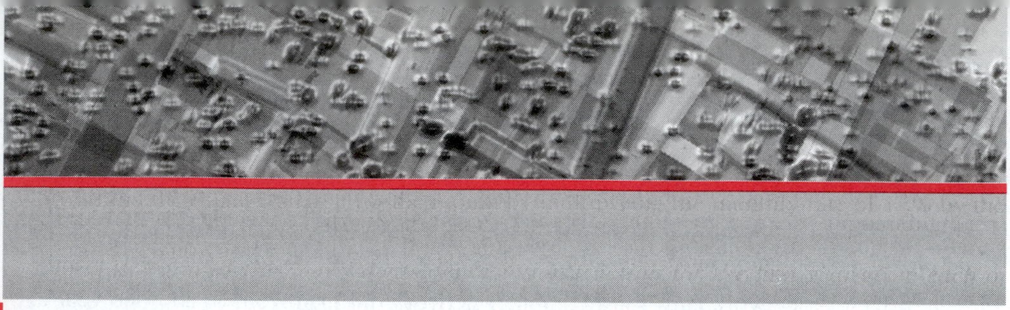

Appendix

A Calculator Basics

Calculators are among the more popular inventions of the last three decades. Each year better calculators are developed and costs drop. The first all-transistor desktop calculator was introduced to the market in 1966; it weighed 55 pounds, cost $2500, and was slow. Today, these same calculations are performed quite well on a calculator costing less than $10. And today's $200 pocket calculators have more ability to calculate than some of the early computers.

Many colleges allow students to use calculators in business mathematics courses. Some courses require calculator use. Although you can still purchase a basic four-function calculator, you're probably better off spending $10 to $20 on either a **scientific** or a **financial calculator**. These calculators allow you to do a lot more than the basic four-function calculators. A **graphing calculator** allows you to graph functions and visualize data, however, it is beyond the scope of this text.

In Section A.1, we discuss the common scientific calculator including percent key, reciprocal key, exponent keys, square root key, memory function, order of operations, and parentheses keys. In Section A.2, the financial calculator with its associated financial keys is discussed.

NOTE The various calculator models differ significantly. *Use the booklet that came with your calculator* for specifics about that calculator if your answers differ from those in this section. ■

A.1 Scientific Calculators

Objectives

1. *Learn the basic calculator keys.*
2. *Understand the* [C], [CE], *and* [ON/C] *keys.*
3. *Understand the floating decimal point.*
4. *Use the* [%] *and* [1/x] *keys.*
5. *Use the* [y^x] *and* [$\sqrt{\ }$] *keys.*
6. *Use the* [$a^b/_c$] *key.*
7. *Solve problems with negative numbers.*
8. *Use the calculator memory function.*
9. *Solve chain calculations using order of operations.*
10. *Use the parentheses keys.*
11. *Use the calculator for problem solution.*

Objective 1 *Learn the Basic Calculator Keys.* Most calculators use **algebraic logic**. Some problems can be solved by entering number and function keys in the same order as you would solve a problem by hand, but many others require a knowledge of the order of operations when entering the problem.

Example 1 Using the Basic Keys

(a) 12 + 25 **(b)** 456 ÷ 24

Solution

(a) The problem 12 + 25 would be entered as

12 [+] 25 [=]

and 37 would appear as the answer.

(b) Enter 456 ÷ 24 as

456 [÷] 24 [=]

and 19 appears as the answer.

Objective 2 *Understand the* [C], [CE], *and* [ON/C] *Keys.* All calculators have a [C] key. Pressing this key erases everything in most calculators and prepares them for a new problem. Some calculators have a [CE] key. Pressing this key erases only the number displayed, thus allowing for correction of a mistake without having to start the problem over. Many calculators combine the [C] key and the [CE] key and use an [ON/C] key. This key is used both to turn the calculator on and to erase the calculator display. If [ON/C] is pressed after the [=] or after one of the operation keys ([+], [−], [×], [÷]), everything in the calculator is erased. If the wrong operation key is pressed, simply press the correct key and the error is corrected. For example, in 7 [+] [−] 3 [=] 4, pressing the [−] key cancels out the previous [+] key entry.

NOTE Be sure to look at the directions that come with your calculator in terms of clearing the memory since keys and the operational sequence vary from calculator to calculator. ■

Objective 3 *Understand the Floating Decimal Point.* Most calculators have a **floating decimal**, which locates the decimal point in the final result.

Example 2 Calculating with Decimal Numbers

A contractor purchased 55.75 square yards of vinyl floor covering at a cost of $18.99 per square yard. Find her total cost.

Solution Proceed as follows.

55.75 [×] 18.99 [=] 1058.6925

The decimal point is automatically placed in the answer. Since money answers are usually rounded to the nearest cent, the answer is $1058.69.

In using a machine with a floating decimal, enter the decimal point as needed. For example, enter $47 as

47

with no decimal point, but enter $0.95 as follows.

[.] 95

One problem in using a floating decimal is shown by the following example.

Example 3 Placing the Decimal Point in Money Answers

Add $21.38 and $1.22.

Solution

21.38 [+] 1.22 [=] 22.6

The final 0 is left off. Remember that the problem deals with dollars and cents, so write the answer as $22.60.

Objective 4 *Use the* [%] *and* [1/x] *Keys.* The [%] key moves the decimal point two places to the left when used following multiplication or division.

Example 4 Using the [%] Key

Find 8% of $4205.

Solution

4205 [×] 8 [%] [=] 336.4 = $336.40

The [1/x] key replaces a number with the reciprocal of that number.

Example 5 Using the [1/x] Key

Find the multiplicative inverse, or reciprocal, of 40.

Solution

40 [1/x] 0.025

Objective 5 *Use the* [y^x] *and* [$\sqrt{\ }$] *Keys.* The product of 3 × 3 can be written as follows.

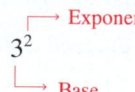

3^2 (2 → Exponent; 3 → Base)

The **exponent** (2 in this case) shows how many times the **base** is multiplied by itself (multiply 3 times itself). The $\boxed{y^x}$ key raises a base to a power; be sure to enter the base first followed by the exponent.

Example 6 Using the $\boxed{y^x}$ Key

Find 5^3.

Solution

$$5 \boxed{y^x} 3 \boxed{=} 125$$

Since $3^2 = 9$, the number 3 is called the **square root** of 9. The symbol $\sqrt{}$ is used to write the square root of a number.

$$\sqrt{9} = 3$$

Example 7 Using the $\boxed{\sqrt{}}$ Key

Find each square root.

(a) $\sqrt{144}$ **(b)** $\sqrt{20}$

Solution

(a) Using the calculator, enter

$$144 \boxed{\sqrt{}}$$

and 12 appears in the display. The square root of 144 is 12.

(b) The square root of 20 is

$$20 \boxed{\sqrt{}} 4.4721360$$

which may be rounded to the desired position.

Objective 6 *Use the* $\boxed{a^b\!/_c}$ *Key.* Many calculators have an $\boxed{a^b\!/_c}$ key that can be used for problems containing fractions and mixed numbers. A mixed number is a number with both a whole number and a fraction such as $7\frac{3}{4}$, which equals $7 + \frac{3}{4}$. The rules for adding, subtracting, multiplying, and dividing both fractions and mixed numbers are given in Chapter 1. Here, we simply show how these operations are done on a calculator.

Example 8 Using the $\boxed{a^b\!/_c}$ Key with Fractions

Solve the following.

(a) $\frac{6}{11} + \frac{3}{4}$

(b) $\frac{3}{8} \div \frac{5}{6}$

Solution

(a) 6 [a b/c] 11 [+] 3 [a b/c] 4 [=] $1\frac{13}{44}$

(b) 3 [a b/c] 8 [÷] 5 [a b/c] 6 [=] $\frac{9}{20}$

NOTE The calculator automatically reduces fractions for you.

Example 9 Using the [a b/c] Key

Solve the following.

(a) $4\frac{7}{8} \div 3\frac{4}{7}$

(b) $\frac{5}{3} \div 27.5$

(c) $65.3 \times 6\frac{3}{4}$

Solution

(a) 4 [a b/c] 7 [a b/c] 8 [÷] 3 [a b/c] 4 [a b/c] 7 [=] $1\frac{73}{200}$

(b) 5 [a b/c] 3 [÷] 27.5 [=] 0.060606061

(c) 65.3 [×] 6 [a b/c] 3 [a b/c] 4 [=] 440.775

Objective 7 *Solve Problems with Negative Numbers.* There are several calculations in business that result in a **negative number** or **deficit amount**.

Example 10 Working with Negative Numbers

The amount in the advertising account last month was $4800, while $5200 was actually spent. Find the balance remaining in the advertising account.

Solution Enter the numbers in the calculator.

4800 [−] 5200 [=] −400

The minus sign in front of the 400 indicates that there is a deficit or negative amount. This value can be written as −$400 or sometimes as ($400), which indicates a negative amount. Some calculators place the minus after the number, or as 400−.

Negative numbers may be entered into the calculator by using the [−] before entering the number. For example, if $3000 is now added to the advertising account in Example 10, the new balance is calculated as follows.

[−] 400 [+] 3000 [=] 2600

The new account balance is $2600.

The +/− key can be used to change the sign of a number that has already been entered. For example, 520 +/− changes +520 to −520.

Objective 8 ***Use the Calculator Memory Function.*** Many calculators feature memory keys, which are a sort of electronic scratch paper. These **memory keys** are used to store intermediate steps in a calculation. On some calculators, a key labeled M or STO is used to store the numbers in the display, with MR or RCL used to recall the numbers from memory. Other calculators have M+ and M− keys. The M+ key adds the number displayed to the number already in memory. For example, if the memory contains the number 0 at the beginning of a problem, and the calculator display contains the number 29.4, then pushing M+ will cause 29.4 to be stored in the memory (the result of adding 0 and 29.4). If 57.8 is then entered into the display, pushing M+ will cause

$$29.4 + 57.8 = 87.2$$

to be stored. If 11.9 is then entered into the display, with M− pushed, the memory will contain

$$87.2 - 11.9 = 75.3$$

The MR key is used to recall the number in memory as needed, with MC used to clear the memory. (Always clear the memory before starting a problem—not doing so is a very common error.)

Scientific calculators typically have one or more storage registers in which to store numbers. These memory keys are usually labeled as STO for store and RCL for recall. For example, 32.5 can be stored in register 1 by

32.5 STO 1

or it can be stored in memory register 2 by 32.5 STO 2 and so forth. Values are retrieved from a particular memory register by using the RCL key followed by the number of the register. For example, RCL 2 recalls the contents of memory register 2.

With a scientific calculator, a number stays in memory until it is replaced by another number or until the memory is cleared. The contents of the memory are saved *even when the calculator is turned off.*

Example 11 Using the Memory Registers

An elevator repairperson counted the number of people entering an elevator and also measured the weight of each group of people. Find the average weight per person.

Number of People	Total Weight
6	839 pounds
8	1184 pounds
4	640 pounds

Solution First, find the weight of all three groups and store in memory register 1.

839 [+] 1184 [+] 640 [=] 2663 [STO] 1

Then find the total number of people.

6 [+] 8 [+] 4 [=] 18

Finally, divide the contents of memory register 1 by the 18 people.

[RCL] 1 [÷] 18 [=] 147.94444 pounds

This value can be rounded as needed.

Objective 9 ***Solve Chain Calculations Using Order of Operations.*** Long calculations involving several operations (adding, subtracting, multiplying, and dividing) must be done in a specific sequence called the **order of operations** and are called **chain calculations**. The logic of the following order of operations is built into most scientific calculators and can help us work problems without having to store a lot of intermediate values.

Solving Chain Calculations

Step 1. Do all operations inside parentheses first.

Step 2. Simplify any expressions with exponents (squares) and find any square roots.

Step 3. Multiply and divide from left to right.

Step 4. Add and subtract from left to right.

Example 12 Using the Order of Operations

Solve the following.

(a) $3 + 7 \times 9\frac{3}{4}$ **(b)** $42.1 \times 5 - 90 \div 4$ **(c)** $6.75^2 \times 9 - 7$

Solution The calculator automatically keeps track of the order of operations for us.

(a) The order of operations tells us to multiply before doing the addition in the problem.

3 [+] 7 [×] 9 [a b/c] 3 [a b/c] 4 [=] $71\frac{1}{4}$

(b) The order of operations tells us to multiply, then divide, and only then subtract.

42.1 [×] 5 [−] 90 [÷] 4 [=] 188

(c) The order of operations tells us to square 6.75 first, then multiply, and finally subtract.

6.75 [x^2] [×] 9 [−] 7 [=] 403.0625

NOTE Scientific calculators keep track of the order of operations for us. All we have to do is enter the problem correctly into the calculator and the calculator does the rest. However, the basic four-function calculator is not programmed to observe the order of operations and can only be used if you enter numbers in the proper order. If you do not get the answer above with your calculator, try using parentheses, as shown next. ■

Objective 10 | ***Use the Parentheses Key.*** The parentheses keys can be used to help establish the order of operations in a more complex chain calculation. For example, $\frac{4}{5 + 7}$ can be written as $\frac{4}{(5 + 7)}$ which can be solved as

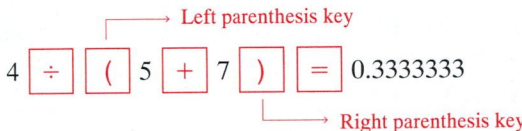

Left parenthesis key

4 [÷] [(] 5 [+] 7 [)] [=] 0.3333333

Right parenthesis key

Example 13 Using Parentheses

Solve the following problem.

$$\frac{16 \div 2.5}{39.2 - 29.8 \times 0.6}$$

Solution Think of this problem as follows.

$$\frac{(16 \div 2.5)}{(39.2 - 29.8 \times 0.6)}$$

Using parentheses to set off the numerator and denominator will help you minimize errors.

[(] 16 [÷] 2.5 [)] [÷] [(] 39.2 [−] 29.8 [×] .6 [)] [=] 0.3001876

Objective 11 | ***Use the Calculator for Problem Solution.***

Example 14 Finding Sale Price

A compact disc player with an original price of $560 is on sale at 10% off. Find the sale price.

Solution If the discount from the original price is 10%, then the sale price is 100% − 10% of the original price.

560 [×] [(] 100 [−] 10 [)] [%] [=] 504

On some calculators the following key strokes will also work:

560 [−] 10 [%] [=] 504

Example 15 Applying Calculator Use to Problem Solving

A home buyer borrows $86,400 at 10% for 30 years. The monthly payment on the loan is $8.78 per $1000 borrowed. Annual taxes are $780, and fire insurance is $453 a year. Find the total monthly payment including taxes and insurance.

Solution The monthly payment is the *sum* of the monthly payment on the loan plus monthly taxes plus monthly fire insurance costs. The monthly payment on the loan is the number of thousands in the loan (86.4) times the monthly payment per \$1000 borrowed (8.78).

86400 [÷] 1000 [×] 8.78 [+] 780 [÷] 12 [+] 453 [÷] 12 [=] 861.342

(780 ÷ 12 → Monthly taxes.)
(453 ÷ 12 → Monthly fire insurance)

To the nearest cent, this amount rounds to \$861.34.

A.1 Exercises

Solve each of the following problems on a calculator. Round each answer to the nearest hundredth.

1. $\begin{array}{r} 384.92 \\ 407.61 \\ 351.14 \\ +\ \ 27.93 \\ \hline \end{array}$

2. $\begin{array}{r} 85.76 \\ 21.94 \\ +\ 39.89 \\ \hline \end{array}$

3. $\begin{array}{r} 6{,}850 \\ 321 \\ +\ 4{,}207 \\ \hline \end{array}$

4. $\begin{array}{r} 781.42 \\ 304.59 \\ +261.35 \\ \hline \end{array}$

5. $\begin{array}{r} 4270.41 \\ -\ \ 365.09 \\ \hline \end{array}$

6. $\begin{array}{r} 3000.07 \\ -\ \ \ 48.12 \\ \hline \end{array}$

7. $\begin{array}{r} 384.96 \\ -\ 129.72 \\ \hline \end{array}$

8. $36.84 - 12.17$

9. $\begin{array}{r} 365 \\ \times\ \ 43 \\ \hline \end{array}$

10. $\begin{array}{r} 27.51 \\ \times\ \ 1.18 \\ \hline \end{array}$

11. 3.7×8.4

12. 62.5×81

13. $\dfrac{375.4}{10.6}$

14. $\dfrac{9625}{400}$

15. $96.7 \div 3.5$

16. $103.7 \div 0.35$

Solve each of the following chain calculations. Round each answer to the nearest hundredth.

17. $\dfrac{9 \times 9}{2 \times 5}$

18. $\dfrac{15 \times 8 \times 3}{11 \times 7 \times 4}$

19. $\dfrac{87 \times 24 \times 47.2}{13.6 \times 12.8}$

20. $\dfrac{2 \times (3 + 4)}{6 + 10}$

21. $\dfrac{2 \times 3 + 4}{6 + 10}$

22. $\dfrac{4200 \times 0.12 \times 90}{365}$

23. $\dfrac{640 - 0.6 \times 12}{17.5 + 3.2}$

24. $\dfrac{16 \times 18 - 0.4 \div 2}{95.4 \times 3 - 0.8}$

25. $\dfrac{14^2 - 3.6 \times 6}{95.2 \div 0.5}$

26. $\dfrac{9^2 + 3.8 \div 2}{14 + 7.5}$

Solve each of the following problems. Reduce any fractions to lowest terms or round to the nearest hundredth.

27. $7\frac{5}{8} \div \left(1 + \frac{3}{8}\right)$

28. $\left(5\frac{1}{4}\right)^2 \times 3.65$

29. $\left(\frac{3}{4} \div \frac{5}{8}\right)^3 \div 3\frac{1}{2}$

30. $\sqrt{6} \times \dfrac{3^2 + 2\frac{1}{2}}{7 \times \frac{5}{6}}$

31. Describe in your own words the order of operations to be used when solving chain calculations. (See Objective 9.)

32. Explain how the parentheses keys are used when solving chain calculations. (See Objective 9.)

Solve each of the following application problems on a calculator. Round each answer to the nearest cent.

33. Bucks County Community College Bookstore bought 397 copies of a computer science book at a net cost of \$23.86 each; 125 copies of an accounting book at \$28.74 each; and 740 copies of a real estate text at \$21.76 each. Find the total paid by the bookstore.

34. Ben Thompson fishes for halibut off the coast of Alaska. His daily catch over the past few days was 263.5 pounds, 122.7 pounds, 82.4 pounds, and 90.8 pounds. Find the average catch per day by summing the values and dividing the sum by 4.

35. Jessica Rodriguez owns a company that installs sprinkling systems. She has three crews made up of 3, 3, and 4 individuals, respectively. One of the workers in each crew is a crew chief who is paid \$9.25 per hour. The others in each crew are paid \$6.80 per hour. Find payroll for a week in which the first two crews worked 32 hours each and the third crew worked 40 hours.

36. Judy Martinez needs to file her expense account claims. She spent 5 nights at the Macon Holiday Inn at \$47.46 per night, 4 nights at the Charlotte Sheraton at \$51.62 per night, and rented a car for 7 days at \$29.95 per day. She drove the car 916 miles with a charge of \$0.24 per mile. Find her total expenses.

37. In Virginia City, the sales tax is 6.5%. Find the tax on each of the following items: (a) a new car costing \$17,908.43 and (b) an office word processor costing \$1463.58.

38. Marja Strutz bought a new commercial fishing boat equipped for sardine fishing at a cost of \$78,250. Additional safety equipment was needed at a cost of \$4820, and sales tax of $7\frac{1}{4}$% was due on the boat and safety equipment. In addition she was charged a licensing fee of \$1135 and a Coast Guard registration fee of \$428. Strutz will pay $\frac{1}{3}$ of the total cost as a down payment and will borrow the balance. How much will she borrow?

39. Ben Fick bought a home for \$80,000. He paid \$8000 down and agreed to make payments of \$528.31 each month for 30 years. By how much does the down payment and the sum of the monthly payments exceed the purchase price?

40. Linda Smelt purchased a 32-unit apartment house for \$620,000. She made a down payment of \$150,000, which she had inherited from her parents, and agreed to make monthly payments of \$5050 for 15 years. By how much does the sum of her down payment and all monthly payments exceed the original purchase price?

41. Ben Hurd wishes to open a small repair shop but only has \$32,400 in cash. He estimates that he will need \$15,000 for equipment, \$2800 for the first month's rent on a building, and about \$28,000 operating expenses until the business is profitable. How much additional funding does he need?

42. Koplan Kitchens wishes to expand their retail store. In order to do so, they must first purchase the \$26,000 lot next door to them; they then anticipate \$120,000 construction costs plus an additional \$28,500 for additional inventory. They have \$50,000 in cash and must borrow the balance from a bank. How much must they borrow?

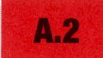

A.2 Financial Calculators

Objectives

1. *Learn the basic conventions used with cash flows.*
2. *Learn the basic financial calculator keys.*
3. *Understand which keys to use for a particular problem.*
4. *Use the financial calculator to solve financial problems.*

Financial calculators are calculators with added features that allow you to make certain compound interest calculations quickly and accurately. These calculators are commonly used by people in business. Financial calculators differ greatly from one another. *Be sure to use the booklet that came with your financial calculator* for specific information about your calculator.

Objective 1 *Learn the Basic Conventions Used with Cash Flows.* Many financial calculators are based on logic that separates cash received by an individual or a company from cash paid out by the same individual or company using + and − signs. *Cash received is given a + sign, and cash paid out is given a − sign.*

Inflows of cash to a person or firm **are positive.** Outflows of cash from the same person or firm are negative. For example, suppose you put \$100 per month into an investment for 10 years and then take out the balance of \$23,000 at the end of 10 years. The \$100 payment each month is shown as a negative number, since cash is leaving you. But the \$23,000 received in 10 years is shown as a positive number, since cash flows to you at that time. We will use this convention throughout this text even though some financial calculators differ.

Objective 2 *Learn the Basic Financial Calculator Keys.* Financial calculators have special functions that allow the user to solve financial problems involving time, interest rates, and money. Many of the compound interest problems presented in this text can be solved using a financial calculator rather than using tables as shown in the text. Most financial calculators have financial keys similar to those shown below.

[*n*] [*i*] [PV] [PMT] [FV]

These keys represent the following functions:

[*n*]	The number of compounding periods
[*i*]	The interest rate *per compounding period*
[PV]	**Present value:** the value in *today's* dollars
[PMT]	The amount of a **level payment** (e.g., \$625 per month); this is used for annuity type problems.
[FV]	**Future value:** the value at *some future date*

NOTE Different financial calculators look and work somewhat differently from one another. You must *look at the instruction book* that came with your calculator to determine how the keys are used with your particular calculator. ■

You will also find that different financial calculators will sometimes give slightly different answers to the same problems because of rounding.

Objective 3 ***Understand Which Keys to Use for a Particular Problem.*** Most simple financial problems require only four of the five financial keys described above. Both the number of compounding periods [n] and the interest rate per compounding period [i] are needed for each financial problem—these two keys will always be used. Which two of the remaining three financial keys ([PV], [PMT], and [FV]) are used depends on the particular problem. Using the convention described above, one of these monetary values will be negative and one will be positive. *The process of solving a financial problem is to enter values for the three variables that are known, then press the key for the unknown, fourth variable.*

For example, if you wish to know the future value of a series of known, equal payments, enter the specific values for [n], [i], and [PMT]. Then press [FV] for the result. Or, if you wish ot know how long it will take for an investment to grow to some specific value at a given interest rate, enter values for [PV], [i], and [FV]. Then press [n] to find the required number of compounding periods.

NOTE Be sure to enter a cash inflow as a positive number or a cash outflow as a negative number. Also be sure to clear all values from the memory of your calculator before working a problem. ■

Objective 4 ***Use the Financial Calculator to Solve Financial Problems.***

Example 1 Given *n*, *i*, and PV, Find FV

Mr. Willis invests \$1000 in an account paying 8% compounded quarterly. Find the future value in $1\frac{1}{2}$ years.

Solution The present value of \$1000 (a cash outflow entered as a negative number) is compounded at 2% per quarter ($8\% \div 4 = 2\%$) for 6 quarters ($1\frac{1}{2} \times 4 = 6$). Enter values for [PV], [i], and [n].

−1000 [PV] 2 [i] 6 [n]

Then press [FV] to find the compound amount at the end of 6 quarters.

[FV] 1126.16

which is the future value.

Example 2 Given *n*, *i*, and PMT, Find FV

Joan Jones plans to invest \$100 at the end of each month in a mutual fund that she believes will grow at 12% per year compounded monthly. Find the future value in 20 years.

Solution She makes 240 payments ($12 \times 20 = 240$) of \$100 each (cash outflows entered as a negative number) into an account earning 1% per month ($12\% \div 12 = 1\%$). Enter values for [*n*], [PMT], and [*i*].

240 [*n*] −100 [PMT] 1 [*i*]

Press [FV] for the result.

[FV] 98925.54

which is the future value.

NOTE The order in which data are entered into the calculator does not matter. Just remember to *press the financial key for the unknown value last.*

Any one of the four values used to solve a particular financial problem can be unknown. Look at the next three examples in which the number of compounding periods [*n*], the payment amount [PMT], and the interest rate per compounding period [*i*], respectively, are unknown.

Example 3 Given *i*, PMT, and FV, Find *n*

Mr. Trebor needs \$140,000 for a new farm tractor. He can invest \$8000 at the end of each month in an account paying 6% per year compounded monthly. How many monthly payments are needed?

Solution The \$8000 monthly payment (cash outflow) will grow at 0.5% per compounding period ($6\% \div 12 = 0.5\%$) until a future value of \$140,000 (cash inflow at a future date) is accumulated. Enter values for [PMT], [*i*], and [FV].

−8000 [PMT] .5 [*i*] 140000 [FV]

Press [*n*] to determine the number of payments.

[*n*] 17 monthly payments of \$8000 each

Actually, 17 payments of \$8000 each in an account earning 0.5% per month will grow to slightly more than \$140,000:

−8000 [PMT] .5 [*i*] 17 [*n*]

Press [FV] to determine the future value.

[FV] 141578.41

which is the future value.

The 17th payment would only need to be

$$\$8000 - (\$141{,}578.41 - \$140{,}000) = \$6421.59$$

in order to accumulate exactly \$140,000.

Example 4 Given *n*, *i*, and FV, Find PMT

Jane Abel wishes to have \$1,000,000 at her retirement in 40 years. Find the payment she must make at the end of each quarter into an account earning 10% compounded quarterly to attain her goal.

Solution Abel makes 160 payments ($40 \times 4 = 160$) into an account earning 2.5% per quarter ($10\% \div 4 = 2.5\%$) until a future value of \$1,000,000 (cash inflow at a future date) is accumulated. Enter values for [*n*], [*i*], and [FV].

160 [*n*] 2.5 [*i*] 1000000 [FV]

Press [PMT] for the quarterly payment:

[PMT] − 490.41

which is the required quarterly payment of cash. Hence, 160 payments of \$490.41 at the end of each quarter in an account earning 10% compounded quarterly will grow to \$1,000,000.

Example 5 Given *n*, PV, and FV, Find *i*

Tom Fernandez bought 200 shares of stock in an oil company at $\$33\frac{1}{2}$ per share. Exactly 3 years later he sold the stock at $\$41\frac{1}{4}$ per share. Find the annual interest rate, rounded to the nearest tenth of a percent, that Fernandez earned on this investment, assuming the company paid no dividends.

Solution In 3 years, the per share price increased from a present value of \$33.50 ($\$33\frac{1}{2}$) to a future value of \$41.25 ($\$41\frac{1}{4}$). The purchase of the stock is a cash outflow and the eventual sale of the stock is a cash inflow. It is not necessary to multiply the stock price times the number of shares: the interest rate indicating the return on the investment is the same whether 1 share or 200 shares are used. Enter values for [*n*], [PV], and [FV].

3 [*n*] −33.50 [PV] 41.25 [FV]

Press [*i*] for the annual interest rate:

[*i*] 7.18%

or about 7.2% per year. Fernandez's return on his original investment compounded at 7.2% per year.

Interest rates can have a great influence on both individuals and businesses. Individuals borrow for homes, cars, and other personal items, whereas firms borrow to buy real estate, expand operations, or cover operating expenses. A small difference in interest rates can make a large difference in costs over time, as shown in the next example.

Example 6 Compare Monthly House Payments

John and Leticia Adams wish to borrow \$62,000 on a 30-year home loan. Find the monthly payment at interest rates of (a) 8% and (b) 9%. Show (c) the monthly savings at the lower rate and (d) the total savings in monthly payments over the 30 years.

Solution

(a) Enter a present value of \$62,000 (cash inflow) with 360 compounding periods ($30 \times 12 = 360$) and a rate of 0.666667% per month ($8\% \div 12 = 0.666667$, rounded) and press [PMT] to find the monthly payment.

62000 [PV] 360 [*n*] .666667 [*i*]

[PMT] −454.93 is the monthly payment at 8% per year, rounded to the nearest cent.

(b) Enter the values again using the new interest rate of 0.75% ($9\% \div 12 = 0.75\%$).

62000 [PV] 360 [*n*] .75 [*i*]

[PMT] −498.87 is the monthly payment at 9% per year, again, rounded to the nearest cent.

(c) The difference in the monthly payments is

$$\$498.87 - \$454.93 = \$43.94$$

(d) The total difference saved over 30 years ($30 \times 12 = 360$ payments) is

$$\$43.94 \times 360 \text{ payments} = \$15{,}818.40$$

The lower interest rate will reduce the Adam's mortgage payments by a total of \$15,818.40 over 30 years.

Example 7 Retirement Planning

Courtney and Nathan Wright plan to retire in 25 years and need \$3500 per month for 20 years.

(a) Find the amount needed at retirement to fund the monthly retirement payments, assuming the annuity earns 9% compounded monthly while payments are being made.

(b) Find the amount of the quarterly payment they must make for the next 25 years to accumulate the necessary funds, assuming earnings of 12% compounded quarterly during the accumulation period.

Solution

(a) The accumulated funds at the end of 25 years is, at their retirement, a present value that must generate a cash inflow to the Wrights of \$3500 per month for 240 months ($20 \times 12 = 240$), assuming earnings of 0.75% per month ($9\% \div 12 = 0.75\%$). Enter values for [*n*], [*i*], and [PMT].

240 [*n*] .75 [*i*] 3500 [PMT]

Press [PV] to find the amount needed at the end of 25 years.

[PV] −389007.34

is the amount they must accumulate.

(b) The Wrights have 25 years of quarterly payments (100 payments that are cash outflows) in an account earning 3% per quarter (12% ÷ 4 = 3%) to accumulate a future value of $389,007.34. The question is one of what quarterly payment is required. Enter values for [n], [i], and [FV].

100 [n] 3 [i] 389007.34 [FV]

Press [PMT] to find the quarterly payment needed:

[PMT] −640.57

is the required quarterly payment.

Thus, the Wrights must make 100 end-of-quarter deposits of $640.57 each into an account earning 3% per quarter. They will then receive 20 years of payments of $3500 per month, assuming 9% per year during the time that payments are made.

A.2 Exercises

Solve the following problems for the missing quantity using a financial calculator. Round dollar answers to the nearest hundredth, interest rates to the nearest hundredth of a percent, and number of compounding periods to the nearest whole number. Assume that any payments are made at the end of the period.

	n	*i*	PV	PMT	FV
1.	10	8%	$3,500	—	______
2.	8	1.5%	$6,400	—	______
3.	10	3%	______	—	$12,000
4.	16	4%	______	—	$8,200
5.	7	8%	—	$300	______
6.	25	2%	—	$1000	______
7.	30	______	—	$319.67	$12,000
8.	50	______	—	$4718.99	$285,000
9.	360	1%	$83,500	______	—
10.	180	0.5%	$125,000	______	—
11.	______	4%	$85,383	$5600	—
12.	______	2%	$3,822	$100	—

Solve each of the following application problems.

13. Tremaine Walker received $2000 as a Christmas bonus. He placed the funds in a 4-year certificate of deposit earning 6% compounded quarterly. Find the future value.

14. Junella Ruiz decides to begin saving at a young age. She has $60 per month taken out of her paycheck and automatically deposited in an account containing technology stocks that she hopes will grow at 1% per month. Find the future value in 15, 30, and 45 years.

15. Mr. and Mrs. Thrash borrowed $86,500 on a 30-year home loan at 9% per year. Find the monthly payment.

16. Terrance Walker wishes to have $20,000 in 10 years when his son begins college. What payment must he make at the end of each quarter in an investment earning 10% compounded quarterly?

17. The *Daily Gazette* needs $340,000 for new printing presses. They can invest $12,000 per month in an account paying 0.8% per month. Find the number of payments that must be paid before reaching their goal. Round to the nearest whole number.

18. Mr. and Mrs. Peters wish to build their dream home and must borrow $110,000 on a 30-year mortgage to do so. Find the highest acceptable annual interest rate, to the nearest tenth of a percent, if they cannot afford a monthly payment above $845.

19. A farmer purchases a tractor costing $345,000, with 25% down and monthly payments at 1% per month for 5 years. Find the monthly payment.

20. A technical school purchases 25 state-of-the-art computers at a cost of $2300 each. They pay 20% down and finance the balance at 0.8% per month for 3 years. Find the monthly payment.

Appendix Terms

Review the following terms to test your understanding of the appendix. For any terms you do not know, refer to the page number found next to the term.

algebraic logic [p. 757]
base [p. 759]
calculator [p. 756]
chain calculations [p. 762]
deficit amount [p. 760]
exponent [p. 759]
financial calculator [p. 766]
floating decimal [p. 757]
future value [p. 766]
graphing calculator [p. 756]
level payment [p. 766]
memory keys [p. 761]
negative number [p. 760]
order of operations [p. 762]
present value [p. 766]
scientific calculator [p. 756]
square root [p. 759]

Appendix B The Metric System

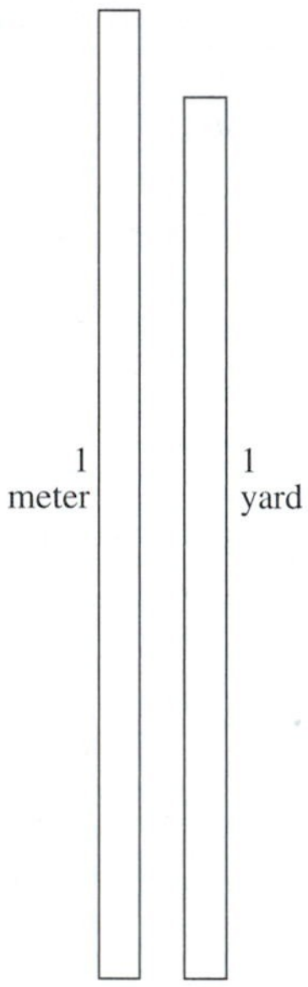

1 yard is 36 inches.
1 meter is about 39.37 inches.

Recall the basic metric units for length, volume, weight, and temperature.

Prefixes in the Metric System
deca = 10 times
kilo = 1000 times
deci = $\frac{1}{10}$ times
centi = $\frac{1}{100}$ times
milli = $\frac{1}{1000}$ times

Today's emphasis on science and technology requires accurate measurements. Early methods of measurement were very crude by modern standards. Parts of the body—the span of the thumb and hand or the length of the foot, for example—were often used as units for measurement. Since the size of body parts differs from one person to another, these measurements are not at all accurate. Many early kings established their own body measures as standard units. Each new king gave a new standard of measure. All these different definitions of standard measure led to confusion.

These problems finally resulted in the formation of standard systems of weights and measures. One of the most useful of these systems is based on decimals. This system is called the **metric system**. It was established by a group of scientists after the French Revolution of 1789, but was not really popularly accepted by the people of France for 50 years.

Today, the metric system is used just about everywhere in the world. In the United States, many industries are switching over to this improved system. The metric system is being taught extensively in elementary schools, and it may eventually replace the current system.

A table on the English system is included here to refresh your memory. Notice that the time relationships are the same in the English and metric systems.

Length		**Weight**	
1 foot	= 12 inches (in.)	1 pound (lb)	= 16 ounces (oz)
1 yard (yd)	= 3 feet (ft)	1 ton (T)	= 2000 pounds (lb)
1 mile (mi)	= 5280 feet (ft)		
Capacity		**Time**	
1 cup (c)	= 8 fluid ounces	1 week (wk)	= 7 days
1 pint (pt)	= 2 cups	1 day	= 24 hours (hr)
1 quart (qt)	= 2 pints (pt)	1 hour (hr)	= 60 minutes (min)
1 gallon (gal)	= 4 quarts (qt)	1 minute (min)	= 60 seconds (sec)

Objectives

1 *Learn the metric system.*
2 *Learn how to convert from one system to the other.*

Objective 1 *Learn the Metric System.* The basic unit of length in the metric system is the **meter**. A meter is a little longer than a yard. For shorter lengths, the units **centimeter** and **millime-**

ter are used. The prefix "centi" means hundredth, so 1 centimeter is one-hundredth of a meter. Thus

$$100 \text{ centimeters} = 1 \text{ meter}$$

The prefix "milli" means thousandth, so 1 millimeter means thousandth of a meter. Thus

$$1000 \text{ millimeters} = 1 \text{ meter}$$

"Meter" is abbreviated m, "centimeter" is cm, and "millimeter" is mm.

Convert from centimeters to millimeters to meters by moving the decimal point, as shown in the following example.

Example 1 Convert the Following Measurements

(a) 6.4 m to cm
(b) 0.98 m to mm
(c) 34 cm to m

Solution

(a) A centimeter is a small unit of measure (a centimeter is about $\frac{1}{2}$ of the diameter of a penny) and a meter is a large unit (a little over 3 feet), so many centimeters make a meter. For this reason, *multiply* by 100 to convert meters to centimeters.

$$6.4 \text{ m} = 6.4 \times 100 = 640 \text{ cm}$$

(b) Multiply by 1000 to convert meters to millimeters.

$$0.98 \text{ m} = 0.98 \times 1000 = 980 \text{ mm}$$

(c) A meter is a large unit of measure, and a centimeter is a smaller unit, so 34 cm are equivalent to a smaller number of meters. Thus, *divide* by 100 to convert centimeters to meters.

$$34 \text{ cm} = \frac{34}{100} = 0.34 \text{ m}$$

Long distances are measured in **kilometer** (km) units. The prefix "kilo" means one thousand. Thus,

$$1 \text{ kilometer} = 1000 \text{ meters}$$

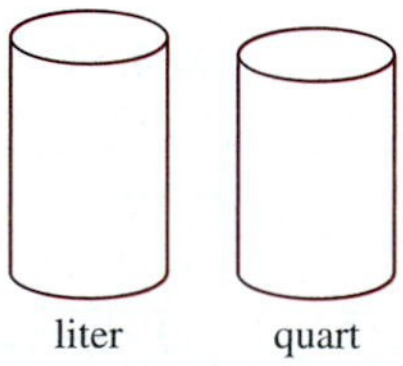

A liter equals about 1.06 quarts.

Since a meter is about a yard, 1000 meters is about 1000 yards, or 3000 feet. Therefore, 1 kilometer is about 3000 feet. One mile is 5280 feet, so 1 kilometer is about 3000/5280 of a mile. Divide 5280 into 3000 to find that 1 kilometer is about 0.6 miles.

The basic unit of volume in the metric system is the **liter** (L) which is a little more than a quart. You may have noticed that Coca Cola is sometimes sold in 2-liter plastic bottles. Again the prefixes "milli" and "centi" are used. Thus,

$$1 \text{ liter} = 100 \text{ centiliters}$$
$$1 \text{ liter} = 1000 \text{ milliliters}$$

Milliliter (mL) and **centiliter** (cL) are such small volumes that they find their main uses in science. In particular, drug dosages are often expressed in milliliters.

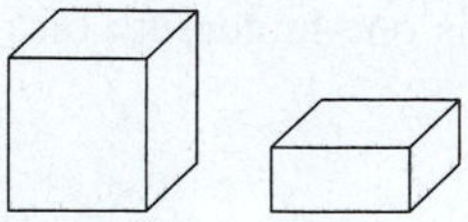

A kilogram equals about 2.2 pounds.

Weight is measured in **grams** (g). A nickel weighs almost exactly 5 grams. **Milligrams** (mg; one-thousandth of a gram) and **centigrams** (cg; one-hundredth of a gram) are so small that they are used mainly in science. A more common measure is the **kilogram** (kg), which is 1000 grams. A kilogram weighs about 2.2 pounds.

$$1000 \text{ grams} = 1 \text{ kilogram}$$

Example 2 Convert the Following Measurements

(a) 650 g to kg
(b) 9.4 L to cL
(c) 4350 mg to g

Solution

(a) A gram is a small unit, and a kilogram is a large unit. Thus, *divide* by 1000 to convert grams to kilograms.

$$650 \text{ g} = \frac{650}{1000} = 0.65 \text{ kg}$$

(b) *Multiply* by 100 to convert liters to centiliters.

$$9.4 \text{ L} = 9.4 \times 100 = 940 \text{ cL}$$

(c) *Divide* by 1000 to convert milligrams to grams.

$$4350 \text{ mg} = \frac{4350}{1000} = 4.35 \text{ g}$$

Objective 2 *Learn How to Convert From One System to the Other.* Eventually, everyone will think in the metric system as easily as in the **English system** of feet, quarts, pounds, and so on. However, during the period of changeover from the English system to the metric system, most people will find it necessary to convert from one system to the other. Approximate conversion can be made with the aid of Table B.1.

Table B.1 ENGLISH-METRIC CONVERSION TABLE

From Metric	To English	Multiply By	From English	To Metric	Multiply By
Meters	Yards	1.09	Yards	Meters	0.914
Meters	Feet	3.28	Feet	Meters	0.305
Meters	Inches	39.37	Inches	Meters	0.0254
Kilometers	Miles	0.62	Miles	Kilometers	1.609
Grams	Pounds	0.00220	Pounds	Grams	454
Kilograms	Pounds	2.20	Pounds	Kilograms	0.454
Liters	Quarts	1.06	Quarts	Liters	0.946
Liters	Gallons	0.264	Gallons	Liters	3.785

Example 3 Convert the Following Measurements

(a) 15 meters to yards
(b) 39 yards to meters
(c) 47 meters to inches
(d) 87 kilometers to miles
(e) 598 miles to kilometers
(f) 12 quarts to liters

Solution

(a) Look at Table B.1 for meters-to-yards, and find the number 1.09. Multiply 15 meters by 1.09.

$$15 \times 1.09 = 16.35 \text{ yards}$$

(b) Read the yards-to-meters row of Table B.1. The number 0.914 appears. Multiply 39 yards by 0.914.

$$39 \times 0.914 = 35.646 \text{ meters}$$

(c) 47 meters $= 47 \times 39.37 = 1850.39$ inches
(d) 87 kilometers $= 87 \times 0.62 = 53.94$ miles
(e) 598 miles $= 598 \times 1.609 = 962.182$ kilometers
(f) 12 quarts $= 12 \times 0.946 = 11.352$ liters

Temperature in the metric system is measured in degrees **Celsius** (abbreviated C). In the celsius scale, water freezes at 0°C and boils at 100°C. This is more sensible than degrees **fahrenheit** (abbreviated F) in use now, in which a mixture of salt and water freezes at 0°F, and 100°F represents the temperature inside the individual Gabriel Fahrenheit's mouth.

Converting From Fahrenheit to Celsius

Step 1 Subtract 32.

Step 2 Multiply by 5.

Step 3 Divide by 9.

These steps can be expressed by the following formula.

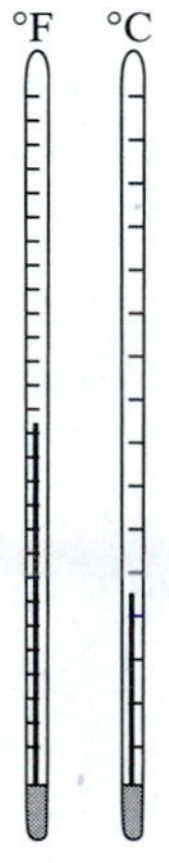

$$C = \frac{5(F - 32)}{9}$$

Example 4 Convert 68°F to Celsius

Solution Use the steps above.

Step 1. Subtract 32. $68 - 32 = 36$

Step 2. Multiply by 5. $36 \times 5 = 180$

Step 3. Divide by 9. $\frac{180}{9} = 20$

Thus, 68°F = 20°C.

Converting From Celsius to Fahrenheit

Step 1. Multiply by 9.

Step 2. Divide by 5.

Step 3. Add 32.

These steps can be expressed by the following formula.

$$F = \frac{9 \times C}{5} + 32$$

Example 5 Convert 11°C to Fahrenheit

Solution Use the steps above.

Step 1. Multiply by 9. $9 \times 11 = 99$

Step 2. Divide by 5. $99 \div 5 = 19.8$

Step 3. Add 32. $19.8 + 32 = 51.8$°F

Thus, 11°C = 51.8°F.

B.1 Exercises

Convert each of the following measurements.

1. 68 cm to m

2. 934 mm to m

3. 4.7 m to mm

4. 7.43 m to cm

5. 8.9 kg to g

6. 4.32 kg to g

7. 39 cL to L

8. 469 cL to L

9. 46,000 g to kg

10. 35,800 g to kg

11. 0.976 kg to g

12. 0.137 kg to g

Convert each of the following measurements. Round to the nearest hundredth.

13. 36 m to yards

14. 76.2 m to yards

15. 55 yards to m

16. 89.3 yards to m

17. 4.7 m to feet

18. 1.92 m to feet

19. 3.6 feet to m

20. 12.8 feet to m

21. 496 km to miles

22. 138 km to miles

23. 768 miles to km

24. 1042 miles to km

25. 683 g to pounds

26. 1792 g to pounds

27. 4.1 pounds to g

28. 12.9 pounds to g

29. 38.9 kg to pounds

30. 40.3 kg to pounds

Work each of the following application problems. Round to the nearest hundredth.

31. One nickel weighs 5 grams. How many nickels are in 1 kilogram of nickels?

32. Seawater contains about 3.5 grams of salt per 1000 milliliters of water. How many grams of salt would 5 liters of seawater contain?

33. Helium weighs about 0.0002 grams per milliliter. A balloon contains 3 liters of helium. How much would the helium weigh?

34. About 1500 grams of sugar can be dissolved in a liter of warm water. How much sugar could be dissolved in 1 milliliter of warm water?

35. Find your height in centimeters.

36. Find your height in meters.

Convert each of the following Fahrenheit temperatures to Celsius. Round to the nearest degree.

37. 104°F	**38.** 86°F	**39.** 536°F
40. 464°F	**41.** 98°F	**42.** 114°F

Convert each of the following Celsius temperatures to Fahrenheit. Round to the nearest degree.

43. 35°C	**44.** 100°C	**45.** 10°C
46. 25°C	**47.** 135°C	**48.** 215°C

In most cases today, medical measurements are given in the metric system. In each of the following problems, a doctor's prescription is given. Some of these are for reasonable amounts of medicine, and some are not. Decide which you think must not be correct.

49. 1940 grams of Kaopectate after each meal

50. 76.8 centiliters of cough syrup every 2 hours

51. 94.3 milliliters of antibiotic every 6 hours

52. 1.4 kilograms of vitamins every 3 hours

53. Apply a bandage 5 centimeters square as needed

54. Soak your feet in 3 milligrams of Epsom salts per 4 liters of water.

Appendix Terms

Review the following terms to test your understanding of the appendix. For any terms you do not know, refer to the page number found next to the term.

Celsius (C) [p. 777]
centiliter [p. 775]
centimeter [p. 774]
centigram [p. 776]
English system [p. 776]
Fahrenheit (F) [p. 777]
gram [p. 776]
kilogram [p. 776]
kilometer [p. 775]
liter [p. 775]
meter [p. 774]
metric system [p. 774]
milligram [p. 776]
milliliter [p. 775]
millimeter [p. 774]

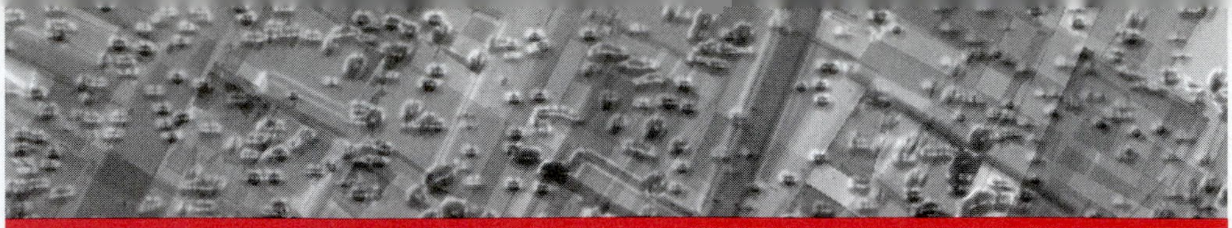

Appendix

C Powers of e

x	e^x	x	e^x	x	e^x	x	e^x	x	e^x
0.00	1.00000000	0.55	1.73325302	1.10	3.00416602	1.65	5.20697983	2.20	9.02501350
0.01	1.01005017	0.56	1.75067250	1.11	3.03435839	1.66	5.25931084	2.21	9.11571639
0.02	1.02020134	0.57	1.76826705	1.12	3.06485420	1.67	5.31216780	2.22	9.20733087
0.03	1.03045453	0.58	1.78603843	1.13	3.09565650	1.68	5.36555597	2.23	9.29986608
0.04	1.04081077	0.59	1.80398842	1.14	3.12676837	1.69	5.41948071	2.24	9.39333129
0.05	1.05127110	0.60	1.82211880	1.15	3.15819291	1.70	5.47394739	2.25	9.48773584
0.06	1.06183655	0.61	1.84043140	1.16	3.18993328	1.71	5.52896148	2.26	9.58308917
0.07	1.07250818	0.62	1.85892804	1.17	3.22199264	1.72	5.58452846	2.27	9.67940081
0.08	1.08328707	0.63	1.87761058	1.18	3.25437420	1.73	5.64065391	2.28	9.77668041
0.09	1.09417428	0.64	1.89648088	1.19	3.28708121	1.74	5.69734342	2.29	9.87493768
0.10	1.10517092	0.65	1.91554083	1.20	3.32011692	1.75	5.75460268	2.30	9.97418245
0.11	1.11627807	0.66	1.93479233	1.21	3.35348465	1.76	5.81243739	2.31	10.07442466
0.12	1.12749685	0.67	1.95423732	1.22	3.38718773	1.77	5.87085336	2.32	10.17567431
0.13	1.13882838	0.68	1.97387773	1.23	3.42122954	1.78	5.92985642	2.33	10.27794153
0.14	1.15027380	0.69	1.99371553	1.24	3.45561346	1.79	5.98945247	2.34	10.38123656
0.15	1.16183424	0.70	2.01375271	1.25	3.49034296	1.80	6.04964746	2.35	10.48556972
0.16	1.17351087	0.71	2.03399126	1.26	3.52542149	1.81	6.11044743	2.36	10.59095145
0.17	1.18530485	0.72	2.05443321	1.27	3.56085256	1.82	6.17185845	2.37	10.69739228
0.18	1.19721736	0.73	2.07508061	1.28	3.59663973	1.83	6.23388666	2.38	10.80490286
0.19	1.20924960	0.74	2.09593551	1.29	3.63278656	1.84	6.29653826	2.39	10.91349394
0.20	1.22140276	0.75	2.11700002	1.30	3.66929667	1.85	6.35981952	2.40	11.02317638
0.21	1.23367806	0.76	2.13827622	1.31	3.70617371	1.86	6.42373677	2.41	11.13396115
0.22	1.24607673	0.77	2.15976625	1.32	3.74342138	1.87	6.48829640	2.42	11.24585931
0.23	1.25860001	0.78	2.18147227	1.33	3.78104339	1.88	6.55350486	2.43	11.35888208
0.24	1.27124915	0.79	2.20339643	1.34	3.81904351	1.89	6.61936868	2.44	11.47304074
0.25	1.28402542	0.80	2.22554093	1.35	3.85742553	1.90	6.68589444	2.45	11.58834672
0.26	1.29693009	0.81	2.24790799	1.36	3.89619330	1.91	6.75308880	2.46	11.70481154
0.27	1.30996445	0.82	2.27049984	1.37	3.93535070	1.92	6.82095847	2.47	11.82244685
0.28	1.32312981	0.83	2.29331874	1.38	3.97490163	1.93	6.88951024	2.48	11.94126442
0.29	1.33642749	0.84	2.31636698	1.39	4.01485005	1.94	6.95875097	2.49	12.06127612
0.30	1.34985881	0.85	2.33964685	1.40	4.05519997	1.95	7.02868758	2.50	12.18249396
0.31	1.36342511	0.86	2.36316069	1.41	4.09595540	1.96	7.09932707	2.51	12.30493006

x	e^x	x	e^x	x	e^x	x	e^x	x	e^x
0.32	1.37712776	0.87	2.38691085	1.42	4.13712044	1.97	7.17067649	2.52	12.42859666
0.33	1.39096813	0.88	2.41089971	1.43	4.17869919	1.98	7.24274299	2.53	12.55350614
0.34	1.40494759	0.89	2.43512965	1.44	4.22069582	1.99	7.31553376	2.54	12.67967097
0.35	1.41906755	0.90	2.45960311	1.45	4.26311452	2.00	7.38905610	2.55	12.80710378
0.36	1.43332941	0.91	2.48432253	1.46	4.30595953	2.01	7.46331735	2.56	12.93581732
0.37	1.44773461	0.92	2.50929039	1.47	4.34923514	2.02	7.53832493	2.57	13.06582444
0.38	1.46228459	0.93	2.53450918	1.48	4.39294568	2.03	7.61408636	2.58	13.19713816
0.39	1.47698079	0.94	2.55998142	1.49	4.43709552	2.04	7.69060920	2.59	13.32977160
0.40	1.49182470	0.95	2.58570966	1.50	4.48168907	2.05	7.76790111	2.60	13.46373804
0.41	1.50681779	0.96	2.61169647	1.51	4.52673079	2.06	7.84596981	2.61	13.59905085
0.42	1.52196156	0.97	2.63794446	1.52	4.57222520	2.07	7.92482312	2.62	13.73572359
0.43	1.53725752	0.98	2.66445624	1.53	4.61817682	2.08	8.00446891	2.63	13.87376990
0.44	1.55270722	0.99	2.69123447	1.54	4.66459027	2.09	8.08491516	2.64	14.01320361
0.45	1.56831219	1.00	2.71828183	1.55	4.71147018	2.10	8.16616991	2.65	14.15403865
0.46	1.58407398	1.01	2.74560102	1.56	4.75882125	2.11	8.24824128	2.66	14.29628910
0.47	1.59999419	1.02	2.77319476	1.57	4.80664819	2.12	8.33113749	2.67	14.43996919
0.48	1.61607440	1.03	2.80106583	1.58	4.85495581	2.13	8.41486681	2.68	14.58509330
0.49	1.63231622	1.04	2.82921701	1.59	4.90374893	2.14	8.49943763	2.69	14.73167592
0.50	1.64872127	1.05	2.85765112	1.60	4.95303242	2.15	8.58485840	2.70	14.87973172
0.51	1.66529119	1.06	2.88637099	1.61	5.00281123	2.16	8.67113766	2.71	15.02927551
0.52	1.68202765	1.07	2.91537950	1.62	5.05309032	2.17	8.75828404	2.72	15.18032224
0.53	1.69893231	1.08	2.94467955	1.63	5.10387472	2.18	8.84630626	2.73	15.33288702
0.54	1.71600686	1.09	2.97427407	1.64	5.15516951	2.19	8.93521311	2.74	15.48698510

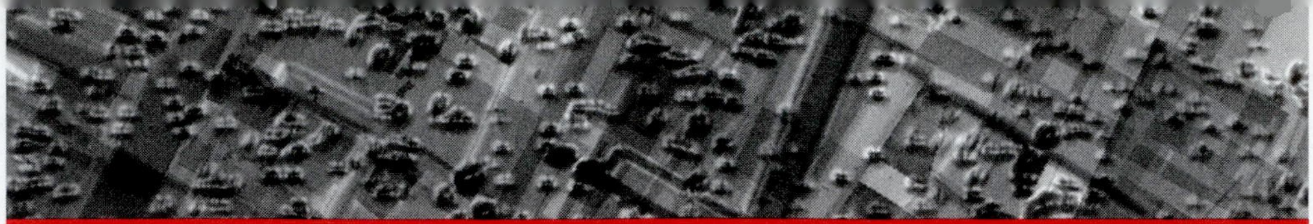

Appendix D

Interest Tables

Rate $\frac{1}{3}\%$	A Compound Interest	B Present Value	C Amount of Annuity	D Present Value of Annuity	E Sinking Fund	F Amortization
n	$(1+i)^n$	$\frac{1}{(1+i)^n}$	$s_{\overline{n}\vert i}$	$a_{\overline{n}\vert i}$	$\frac{1}{s_{\overline{n}\vert i}}$	$\frac{1}{a_{\overline{n}\vert i}}$
1	1.00333333	0.99667774	1.00000000	0.99667774	1.00000000	1.00333333
2	1.00667778	0.99336652	2.00333333	1.99004426	0.49916805	0.50250139
3	1.01003337	0.99006630	3.01001111	2.98011056	0.33222469	0.33555802
4	1.01340015	0.98677704	4.02004448	3.96688760	0.24875347	0.25208680
5	1.01677815	0.98349871	5.03344463	4.95038631	0.19867110	0.20200444
6	1.02016741	0.98023127	6.05022278	5.93061759	0.16528317	0.16861650
7	1.02356797	0.97697469	7.07039019	6.90759228	0.14143491	0.14476824
8	1.02697986	0.97372893	8.09395816	7.88132121	0.12354895	0.12688228
9	1.03040313	0.97049395	9.12093802	8.85181516	0.10963785	0.11297118
10	1.03383780	0.96726972	10.15134114	9.81908487	0.09850915	0.10184248
11	1.03728393	0.96405620	11.18517895	10.78314107	0.08940402	0.09273736
12	1.04074154	0.96085335	12.22246288	11.74399442	0.08181657	0.08514990
13	1.04421068	0.95766115	13.26320442	12.70165557	0.07539656	0.07872989
14	1.04769138	0.95447955	14.30741510	13.65613512	0.06989383	0.07322716
15	1.05118369	0.95130852	15.35510648	14.60744364	0.06512491	0.06845825
16	1.05468763	0.94814803	16.40629017	15.55559167	0.06095223	0.06428557
17	1.05820326	0.94499803	17.46097781	16.50058970	0.05727056	0.06060389
18	1.06173060	0.94185851	18.51918107	17.44244821	0.05399807	0.05733140
19	1.06526971	0.93872941	19.58091167	18.38117762	0.05107015	0.05440348
20	1.06882060	0.93561071	20.64618137	19.31678832	0.04843511	0.05176844
21	1.07238334	0.93250236	21.71500198	20.24929069	0.04605111	0.04938445
22	1.07595795	0.92940435	22.78738532	21.17869504	0.04388393	0.04721726
23	1.07954448	0.92631663	23.86334327	22.10501167	0.04190528	0.04523861
24	1.08314296	0.92323916	24.94288775	23.02825083	0.04009159	0.04342492
25	1.08675344	0.92017192	26.02603071	23.94842275	0.03842307	0.04175640
26	1.09037595	0.91711487	27.11278414	24.86553763	0.03688297	0.04021630
27	1.09401053	0.91406798	28.20316009	25.77960561	0.03545702	0.03879035
28	1.09765724	0.91103121	29.29717062	26.69063682	0.03413299	0.03746632
29	1.10131609	0.90800453	30.39482786	27.59864135	0.03290033	0.03623367
30	1.10498715	0.90498790	31.49614395	28.50362925	0.03174992	0.03508325
31	1.10867044	0.90198130	32.60113110	29.40561055	0.03067378	0.03400712
32	1.11236601	0.89898468	33.70980154	30.30459523	0.02966496	0.03299830
33	1.11607389	0.89599802	34.82216754	31.20059325	0.02871734	0.03205067
34	1.11979414	0.89302128	35.93824143	32.09361454	0.02782551	0.03115885
35	1.12352679	0.89005444	37.05803557	32.98366898	0.02698470	0.03031803
36	1.12727187	0.88709745	38.18156236	33.87076642	0.02619065	0.02952399
37	1.13102945	0.88415028	39.30883423	34.75491670	0.02543957	0.02877291
38	1.13479955	0.88121290	40.43986368	35.63612960	0.02472808	0.02806141
39	1.13858221	0.87828528	41.57466322	36.51441488	0.02405311	0.02738644
40	1.14237748	0.87536739	42.71324543	37.38978228	0.02341194	0.02674527
41	1.14618541	0.87245920	43.85562292	38.26224147	0.02280209	0.02613543
42	1.15000603	0.86956066	45.00180833	39.13180213	0.02222133	0.02555466
43	1.15383938	0.86667175	46.15181436	39.99847389	0.02166762	0.02500095
44	1.15768551	0.86379245	47.30565374	40.86226633	0.02113912	0.02447246
45	1.16154446	0.86092270	48.46333925	41.72318903	0.02063415	0.02396749
46	1.16541628	0.85806249	49.62488371	42.58125153	0.02015118	0.02348451
47	1.16930100	0.85521179	50.79029999	43.43646332	0.01968880	0.02302213
48	1.17319867	0.85237055	51.95960099	44.28883387	0.01924572	0.02257905
49	1.17710933	0.84953876	53.13279966	45.13837263	0.01882077	0.02215410
50	1.18103303	0.84671637	54.30990899	45.98508900	0.01841285	0.02174618

Rate $\frac{5}{12}\%$	A Compound Interest	B Present Value	C Amount of Annuity	D Present Value of Annuity	E Sinking Fund	F Amortization
n	$(1+i)^n$	$\frac{1}{(1+i)^n}$	$s_{\overline{n}\vert i}$	$a_{\overline{n}\vert i}$	$\frac{1}{s_{\overline{n}\vert i}}$	$\frac{1}{a_{\overline{n}\vert i}}$
1	1.00416667	0.99585062	1.00000000	0.99585062	1.00000000	1.00416667
2	1.00835069	0.99171846	2.00416667	1.98756908	0.49896050	0.50312717
3	1.01255216	0.98760345	3.01251736	2.97517253	0.33194829	0.33611496
4	1.01677112	0.98350551	4.02506952	3.95867804	0.24844291	0.25260958
5	1.02100767	0.97942457	5.01484064	4.93810261	0.19834026	0.20250693
6	1.02526187	0.97536057	6.06284831	5.91346318	0.16493898	0.16910564
7	1.02953379	0.97131343	7.08811018	6.88477661	0.14108133	0.14524800
8	1.03382352	0.96728308	8.11764397	7.85205970	0.12318845	0.12735512
9	1.03813111	0.96326946	9.15146749	8.81532916	0.10927209	0.11343876
10	1.04245666	0.95927249	10.18959860	9.77460165	0.09813929	0.10230596
11	1.04680023	0.95529211	11.23205526	10.72989376	0.08903090	0.09319757
12	1.05116190	0.95132824	12.27885549	11.68122200	0.08144082	0.08560748
13	1.05554174	0.94738082	13.33001739	12.62860283	0.07501866	0.07918532
14	1.05993983	0.94344978	14.38555913	13.57205261	0.06951416	0.07368082
15	1.06435625	0.93953505	15.44549896	14.51158766	0.06474378	0.06891045
16	1.06879106	0.93563657	16.50985520	15.44722422	0.06056988	0.06473655
17	1.07324436	0.93175426	17.57864627	16.37897848	0.05688720	0.06105387
18	1.07771621	0.92788806	18.65189063	17.30686654	0.05361387	0.05778053
19	1.08220670	0.92403790	19.72960684	18.23090443	0.05068525	0.05485191
20	1.08671589	0.92020372	20.81181353	19.15110815	0.04804963	0.05221630
21	1.09124387	0.91638544	21.89852942	20.06749359	0.04566517	0.04983183
22	1.09579072	0.91258301	22.98977330	20.98007661	0.04349760	0.04766427
23	1.10035652	0.90879636	24.08556402	21.88887297	0.04151865	0.04568531
24	1.10494134	0.90502542	25.18592053	22.79389839	0.03970472	0.04387139
25	1.10954526	0.90127013	26.29086187	23.69516853	0.03803603	0.04220270
26	1.11416836	0.89753042	27.40040713	24.59269895	0.03649581	0.04066247
27	1.11881073	0.89380623	28.51457549	25.48650517	0.03506978	0.03923645
28	1.12347244	0.89009749	29.63338622	26.37660266	0.03374572	0.03791239
29	1.12815358	0.88640414	30.75685866	27.26300680	0.03251307	0.03667974
30	1.13285422	0.88272611	31.88501224	28.14573291	0.03136270	0.03552936
31	1.13757444	0.87906335	33.01786646	29.02479626	0.03028663	0.03445330
32	1.14231434	0.87541578	34.15544090	29.90021205	0.02927791	0.03344458
33	1.14707398	0.87178335	35.29775524	30.77199540	0.02833041	0.03249708
34	1.15185346	0.86816599	36.44482922	31.64016139	0.02743873	0.03160540
35	1.15665284	0.86456365	37.59668268	32.50472504	0.02659809	0.03076476
36	1.16147223	0.86097624	38.75333552	33.36570128	0.02580423	0.02997090
37	1.16631170	0.85740373	39.91480775	34.22310501	0.02505336	0.02922003
38	1.17117133	0.85384604	41.08111945	35.07695105	0.02434208	0.02850875
39	1.17605121	0.85030311	42.25229078	35.92725416	0.02366736	0.02783402
40	1.18095142	0.84677488	43.42834199	36.77402904	0.02302644	0.02719310
41	1.18587206	0.84326129	44.60929342	37.61729033	0.02241685	0.02658352
42	1.19081319	0.83976228	45.79516547	38.45705261	0.02183637	0.02600303
43	1.19577491	0.83627779	46.98597866	39.29333040	0.02128295	0.02544961
44	1.20075731	0.83280776	48.18175357	40.12613816	0.02075474	0.02492141
45	1.20576046	0.82935212	49.38251088	40.95549028	0.02025008	0.02441675
46	1.21078446	0.82591083	50.58827134	41.78140111	0.01976743	0.02393409
47	1.21582940	0.82248381	51.79905581	42.60388492	0.01930537	0.02347204
48	1.22089536	0.81907102	53.01488521	43.42295594	0.01886263	0.02302929
49	1.22598242	0.81567238	54.23578056	44.23862832	0.01843801	0.02260468
50	1.23109068	0.81228785	55.46176298	45.05091617	0.01803044	0.02219711

Rate $\frac{1}{2}\%$	A Compound Interest	B Present Value	C Amount of Annuity	D Present Value of Annuity	E Sinking Fund	F Amortization
n	$(1 + i)^n$	$\frac{1}{(1 + i)^n}$	$s_{\overline{n}\mid i}$	$a_{\overline{n}\mid i}$	$\frac{1}{s_{\overline{n}\mid i}}$	$\frac{1}{a_{\overline{n}\mid i}}$
1	1.00500000	0.99502488	1.00000000	0.99502488	1.00000000	1.00500000
2	1.01002500	0.99007450	2.00500000	1.98509938	0.49875312	0.50375312
3	1.01507513	0.98514876	3.01502500	2.97024814	0.33167221	0.33667221
4	1.02015050	0.98024752	4.03010012	3.95049566	0.24813279	0.25313279
5	1.02525125	0.97537067	5.05025063	4.92586633	0.19800997	0.20300997
6	1.03037751	0.97051808	6.07550188	5.89638441	0.16459546	0.16959546
7	1.03552940	0.96568963	7.10587939	6.86207404	0.14072854	0.14572854
8	1.04070704	0.96088520	8.14140879	7.82295924	0.12282886	0.12782886
9	1.04591058	0.95610468	9.18211583	8.77906392	0.10890736	0.11390736
10	1.05114013	0.95134794	10.22802641	9.73041186	0.09777057	0.10277057
11	1.05639583	0.94661487	11.27916654	10.67702673	0.08865903	0.09365903
12	1.06167781	0.94190534	12.33556237	11.61893207	0.08106643	0.08606643
13	1.06698620	0.93721924	13.39724018	12.55615131	0.07464224	0.07964224
14	1.07232113	0.93255646	14.46422639	13.48870777	0.06913609	0.07413609
15	1.07768274	0.92791688	15.53654752	14.41662465	0.06436436	0.06936436
16	1.08307115	0.92330037	16.61423026	15.33992502	0.06018937	0.06518937
17	1.08848651	0.91870684	17.69730141	16.25863186	0.05650579	0.06150579
18	1.09392894	0.91413616	18.78578791	17.17276802	0.05323173	0.05823173
19	1.09939858	0.90958822	19.87971685	18.08235624	0.05030253	0.05530253
20	1.10489558	0.90506290	20.97911544	18.98741915	0.04766645	0.05266645
21	1.11042006	0.90056010	22.08401101	19.88797925	0.04528163	0.05028163
22	1.11597216	0.89607971	23.19443107	20.78405896	0.04311380	0.04811380
23	1.12155202	0.89162160	24.31040322	21.67568055	0.04113465	0.04613465
24	1.12715978	0.88718567	25.43195524	22.56286622	0.03932061	0.04432061
25	1.13279558	0.88277181	26.55911502	23.44563803	0.03765186	0.04265186
26	1.13845955	0.87837991	27.69191059	24.32401794	0.03611163	0.04111163
27	1.14415185	0.87400986	28.83037015	25.19802780	0.03468565	0.03968565
28	1.14987261	0.86966155	29.97452200	26.06768936	0.03336167	0.03836167
29	1.15562197	0.86533488	31.12439461	26.93302423	0.03212914	0.03712914
30	1.16140008	0.86102973	32.28001658	27.79405397	0.03097892	0.03597892
31	1.16720708	0.85674600	33.44141666	28.65079997	0.02990304	0.03490304
32	1.17304312	0.85248358	34.60862375	29.50328355	0.02889453	0.03389453
33	1.17890833	0.84824237	35.78166686	30.35152592	0.02794727	0.03294727
34	1.18480288	0.84402226	36.96057520	31.19554818	0.02705586	0.03205586
35	1.19072689	0.83982314	38.14537807	32.03537132	0.02621550	0.03121550
36	1.19668052	0.83564492	39.33610496	32.87101624	0.02542194	0.03042194
37	1.20266393	0.83148748	40.53278549	33.70250372	0.02467133	0.02967139
38	1.20867725	0.82735073	41.73544942	34.52985445	0.02396045	0.02896045
39	1.21472063	0.82323455	42.94412666	35.35308900	0.02328607	0.02828607
40	1.22079424	0.81913886	44.15884730	36.17222786	0.02264552	0.02764552
41	1.22689821	0.81506354	45.37964153	36.98729141	0.02203631	0.02703631
42	1.23303270	0.81100850	46.60653974	37.79829991	0.02145622	0.02645622
43	1.23919786	0.80697363	47.83957244	38.60527354	0.02090320	0.02590320
44	1.24539385	0.80295884	49.07877030	39.40823238	0.02037541	0.02537541
45	1.25162082	0.79896402	50.32416415	40.20719640	0.01987117	0.02487117
46	1.25787892	0.79498907	51.57578497	41.00218547	0.01938894	0.02438894
47	1.26416832	0.79103390	52.83366390	41.79321937	0.01892733	0.02392733
48	1.27048916	0.78709841	54.09783222	42.58031778	0.01848503	0.02348503
49	1.27684161	0.78318250	55.36832138	43.36350028	0.01806087	0.02306087
50	1.28322581	0.77928607	56.64516299	44.14278635	0.01765376	0.02265376

Rate $\frac{3}{4}\%$	A Compound Interest	B Present Value	C Amount of Annuity	D Present Value of Annuity	E Sinking Fund	F Amortization
n	$(1 + i)^n$	$\frac{1}{(1 + i)^n}$	$s_{\overline{n}\rvert i}$	$a_{\overline{n}\rvert i}$	$\frac{1}{s_{\overline{n}\rvert i}}$	$\frac{1}{a_{\overline{n}\rvert i}}$
1	1.00750000	0.99255583	1.00000000	0.99255583	1.00000000	1.00750000
2	1.01505625	0.98516708	2.00750000	1.97772291	0.49813200	0.50563200
3	1.02266917	0.97783333	3.02255625	2.95555624	0.33084579	0.33834579
4	1.03033919	0.97055417	4.04522542	3.92611041	0.24720501	0.25470501
5	1.03806673	0.96332920	5.07556461	4.88943961	0.19702242	0.20452242
6	1.04585224	0.95615802	6.11363135	5.84559763	0.16356891	0.17106891
7	1.05369613	0.94904022	7.15948358	6.79463785	0.13967488	0.14717488
8	1.06159885	0.94197540	8.21317971	7.73661325	0.12175552	0.12925552
9	1.06956084	0.93496318	9.27477856	8.67157642	0.10781929	0.11531929
10	1.07758255	0.92800315	10.34433940	9.59957958	0.09667123	0.10417123
11	1.08566441	0.92109494	11.42192194	10.52067452	0.08755094	0.09505094
12	1.09380690	0.91423815	12.50758636	11.43491267	0.07995148	0.08745148
13	1.10201045	0.90743241	13.60139325	12.34234508	0.07352188	0.08102188
14	1.11027553	0.90067733	14.70340370	13.24302242	0.06801146	0.07551146
15	1.11860259	0.89397254	15.81367923	14.13699495	0.06323639	0.07073639
16	1.12699211	0.88731766	16.93228183	15.02431261	0.05905879	0.06655879
17	1.13544455	0.88071231	18.05927394	15.90502492	0.05537321	0.06287321
18	1.14396039	0.87415614	19.19471849	16.77918107	0.05209766	0.05959766
19	1.15254009	0.86764878	20.33867888	17.64682984	0.04916740	0.05666740
20	1.16118414	0.86118985	21.49121897	18.50801969	0.04653063	0.05403063
21	1.16989302	0.85477901	22.65240312	19.36279870	0.04414543	0.05164543
22	1.17866722	0.84841589	23.82229614	20.21121459	0.04197748	0.04947748
23	1.18750723	0.84210014	25.00096336	21.05331473	0.03999846	0.04749846
24	1.19641353	0.83583140	26.18847059	21.88914614	0.03818474	0.04568474
25	1.20538663	0.82960933	27.38488412	22.71875547	0.03651650	0.04401650
26	1.21442703	0.82343358	28.59027075	23.54218905	0.03497693	0.04247693
27	1.22353523	0.81730380	29.80469778	24.35949286	0.03355176	0.04105176
28	1.23271175	0.81121966	31.02823301	25.17071251	0.03222871	0.03972871
29	1.24195709	0.80518080	32.26094476	25.97589331	0.03099723	0.03849723
30	1.25127176	0.79918690	33.50290184	26.77508021	0.02984816	0.03734816
31	1.26065630	0.79323762	34.75417361	27.56831783	0.02877352	0.03627352
32	1.27011122	0.78733262	36.01482991	28.35565045	0.02776634	0.03526634
33	1.27963706	0.78147158	37.28494113	29.13712203	0.02682048	0.03432048
34	1.28923434	0.77565418	38.56457819	29.91277621	0.02593053	0.03343053
35	1.29890359	0.76988008	39.85381253	30.68265629	0.02509170	0.03259170
36	1.30864537	0.76414896	41.15271612	31.44680525	0.02429973	0.03179973
37	1.31846021	0.75846051	42.46136149	32.20526576	0.02355082	0.03105082
38	1.32834866	0.75281440	43.77982170	32.95808016	0.02284157	0.03034157
39	1.33831128	0.74721032	45.10817037	33.70529048	0.02216893	0.02966893
40	1.34834861	0.74164796	46.44648164	34.44693844	0.02153016	0.02903016
41	1.35846123	0.73612701	47.79483026	35.18306545	0.02092276	0.02842276
42	1.36864969	0.73064716	49.15329148	35.91371260	0.02034452	0.02784452
43	1.37891456	0.72520809	50.52194117	36.63892070	0.01979338	0.02729338
44	1.38925642	0.71980952	51.90085573	37.35873022	0.01926751	0.02676751
45	1.39967584	0.71445114	53.29011215	38.07318136	0.01876521	0.02626521
46	1.41017341	0.70913264	54.68978799	38.78231401	0.01828495	0.02578495
47	1.42074971	0.70385374	56.09996140	39.48616775	0.01782532	0.02532532
48	1.43140533	0.69861414	57.52071111	40.18478189	0.01738504	0.02488504
49	1.44214087	0.69341353	58.95211644	40.87819542	0.01696292	0.02446292
50	1.45295693	0.68825165	60.39425732	41.56644707	0.01655787	0.02405787

Rate $\frac{5}{6}\%$	A Compound Interest	B Present Value	C Amount of Annuity	D Present Value of Annuity	E Sinking Fund	F Amortization
n	$(1 + i)^n$	$\frac{1}{(1 + i)^n}$	$s_{\overline{n}\vert i}$	$a_{\overline{n}\vert i}$	$\frac{1}{s_{\overline{n}\vert i}}$	$\frac{1}{a_{\overline{n}\vert i}}$
1	1.00833333	0.99173554	1.00000000	0.99173554	1.00000000	1.00833333
2	1.01673611	0.98353938	2.00833333	1.97527491	0.49792531	0.50625864
3	1.02520891	0.97541095	3.02506944	2.95068586	0.33057092	0.33890426
4	1.03375232	0.96734970	4.05027836	3.91803557	0.24689661	0.25522994
5	1.04236692	0.95935508	5.08403068	4.87739065	0.19669433	0.20502766
6	1.05105331	0.95142652	6.12639760	5.82881717	0.16322806	0.17156139
7	1.05981209	0.94356349	7.17745091	6.77238066	0.13932523	0.14765856
8	1.06864386	0.93576545	8.23726300	7.70814611	0.12139955	0.12973288
9	1.07754922	0.92803185	9.30590686	8.63617796	0.10745863	0.11579196
10	1.08652880	0.92036217	10.38345608	9.55654013	0.09630705	0.10464038
11	1.09558321	0.91275587	11.46998489	10.46929600	0.08718407	0.09551741
12	1.10471307	0.90521243	12.56556809	11.37450843	0.07958255	0.08791589
13	1.11391901	0.89773134	13.67028116	12.27223976	0.07315138	0.08148472
14	1.12320167	0.89031207	14.78420017	13.16255183	0.06763978	0.07597311
15	1.13256168	0.88295412	15.90740184	14.04550595	0.06286382	0.07119715
16	1.14199970	0.87565698	17.03996352	14.92116292	0.05868557	0.06701890
17	1.15151636	0.86842014	18.18196322	15.78958306	0.05499956	0.06333289
18	1.16111233	0.86124312	19.33347958	16.65082618	0.05172375	0.06005708
19	1.17078827	0.85412540	20.49459191	17.50495158	0.04879336	0.05712669
20	1.18054483	0.84706652	21.66538017	18.35201810	0.04615659	0.05448992
21	1.19038271	0.84006597	22.84592501	19.19208406	0.04377148	0.05210482
22	1.20030256	0.83312327	24.03630772	20.02520734	0.04160373	0.04993706
23	1.21030509	0.82623796	25.23661028	20.85144529	0.03962497	0.04795831
24	1.22039096	0.81940954	26.44691537	21.67085483	0.03781159	0.04614493
25	1.23056089	0.81263756	27.66730633	22.48349240	0.03614374	0.04447708
26	1.24081556	0.80592155	28.89786721	23.28941395	0.03460463	0.04293796
27	1.25115569	0.79926104	30.13868277	24.08867499	0.03317995	0.04151328
28	1.26158199	0.79265558	31.38983846	24.88133057	0.03185744	0.04019078
29	1.27209517	0.78610471	32.65142045	25.66743527	0.03062654	0.03895987
30	1.28269596	0.77960797	33.92351562	26.44704325	0.02947808	0.03781141
31	1.29338510	0.77316493	35.20621158	27.22020818	0.02840408	0.03673741
32	1.30416331	0.76677514	36.49959668	27.98698332	0.02739756	0.03573090
33	1.31503133	0.76043815	37.80375999	28.74742147	0.02645240	0.03478573
34	1.35298993	0.75415354	39.11879132	29.50157501	0.02556316	0.03389650
35	1.33703984	0.74792087	40.44478125	30.24949588	0.02472507	0.03305840
36	1.34818184	0.74173970	41.78182109	30.99123559	0.02393385	0.03226719
37	1.35941669	0.73560962	43.13000293	31.72684521	0.02318572	0.03151905
38	1.37074516	0.72953020	44.48941962	32.45637541	0.02247725	0.03081059
39	1.38216804	0.72350103	45.86016479	33.17987644	0.02180542	0.03013875
40	1.39368611	0.71752168	47.24233283	33.89739813	0.02116746	0.02950079
41	1.40530016	0.71159175	48.63601893	34.60898988	0.02056089	0.02889423
42	1.41701099	0.70571083	50.04131909	35.31470070	0.01998349	0.02831682
43	1.42881942	0.69987851	51.45833008	36.01457921	0.01943320	0.02776653
44	1.44072625	0.69409439	52.88714950	36.70867360	0.01890818	0.02724152
45	1.45273230	0.68835807	54.32787575	37.39703167	0.01840676	0.02674009
46	1.46483840	0.68266916	55.78060805	38.07970083	0.01792738	0.02626071
47	1.47704539	0.67702727	57.24544645	38.75672809	0.01746864	0.02580197
48	1.48935410	0.67143200	58.72249183	39.42816009	0.01702925	0.02536258
49	1.50176538	0.66588297	60.21184593	40.09404307	0.01660803	0.02494136
50	1.51428009	0.66037981	61.71361131	40.75442288	0.01620388	0.02453721

Rate 1%	A Compound Interest	B Present Value	C Amount of Annuity	D Present Value of Annuity	E Sinking Fund	F Amortization
n	$(1+i)^n$	$\frac{1}{(1+i)^n}$	$s_{\overline{n}\|i}$	$a_{\overline{n}\|i}$	$\frac{1}{s_{\overline{n}\|i}}$	$\frac{1}{a_{\overline{n}\|i}}$
1	1.01000000	0.99009901	1.00000000	0.99009901	1.00000000	1.01000000
2	1.02010000	0.98029605	2.01000000	1.97039506	0.49751244	0.50751244
3	1.03030100	0.97059015	3.03010000	2.94098521	0.33002211	0.34002211
4	1.04060401	0.96098034	4.06040100	3.90196555	0.24628109	0.25628109
5	1.05101005	0.95146569	5.10100501	4.85343124	0.19603980	0.20603980
6	1.06152015	0.94204524	6.15201506	5.79547647	0.16254837	0.17254837
7	1.07213535	0.93271805	7.21353521	6.72819453	0.13862828	0.14862828
8	1.08285671	0.92348322	8.28567056	7.65167775	0.12069029	0.13069029
9	1.09368527	0.91433982	9.36852727	8.56601758	0.10674036	0.11674036
10	1.10462213	0.90528695	10.46221254	9.47130453	0.09558208	0.10558208
11	1.11566835	0.89632372	11.56683467	10.36762825	0.08645408	0.09645408
12	1.12682503	0.88744923	12.68250301	11.25507747	0.07884879	0.08884879
13	1.13809328	0.87866260	13.80932804	12.13374007	0.07241482	0.08241482
14	1.14947421	0.86996297	14.94742132	13.00370304	0.06690117	0.07690117
15	1.16096896	0.86134947	16.09689554	13.86505252	0.06212378	0.07212378
16	1.17257864	0.85282126	17.25786449	14.71787378	0.05794460	0.06794460
17	1.18430443	0.84437749	18.43044314	15.56225127	0.05425806	0.06425806
18	1.19614748	0.83601731	19.61474757	16.39826858	0.05098205	0.06098205
19	1.20810895	0.82773992	20.81089504	17.22600850	0.04805175	0.05805175
20	1.22019004	0.81954447	22.01900399	18.04555297	0.04541531	0.05541531
21	1.23239194	0.81143017	23.23919403	18.85698313	0.04303075	0.05303075
22	1.24471586	0.80339621	24.47158598	19.66037934	0.04086372	0.05086372
23	1.25716302	0.79544179	25.71630183	20.45582113	0.03888584	0.04888584
24	1.26973465	0.78756613	26.97346485	21.24338726	0.03707347	0.04707347
25	1.28243200	0.77976844	28.24319950	22.02315570	0.03540675	0.04540675
26	1.29525631	0.77204796	29.52563150	22.79520366	0.03386888	0.04386888
27	1.30820888	0.76440392	30.82088781	23.55960759	0.03244553	0.04244553
28	1.32129097	0.75683557	32.12909669	24.31644316	0.03112444	0.04112444
29	1.33450388	0.74934215	33.45038766	25.06578530	0.02989502	0.03989502
30	1.34784892	0.74192292	34.78489153	25.80770822	0.02874811	0.03874811
31	1.36132740	0.73457715	36.13274045	26.54228537	0.02767573	0.03767573
32	1.37494068	0.72730411	37.49406785	27.26958947	0.02667089	0.03667089
33	1.38869009	0.72010307	38.86900853	27.98969255	0.02575744	0.03572744
34	1.40257699	0.71297334	40.25769862	28.70266589	0.02483997	0.03483997
35	1.41660276	0.70591420	41.66027560	29.40853009	0.02400368	0.03400368
36	1.43076878	0.69892495	43.07687836	30.10750504	0.02321431	0.03321431
37	1.44507647	0.69200490	44.05764714	30.79950994	0.02246805	0.03246805
38	1.45952724	0.68515337	45.95272361	31.48466330	0.02176150	0.03176150
39	1.47412251	0.67836967	47.41225085	32.16303298	0.02109160	0.03109160
40	1.48886373	0.67165314	48.88637336	32.83468611	0.02045560	0.03045560
41	1.50375237	0.66500311	50.37523709	33.49968922	0.01985102	0.02985102
42	1.51878989	0.65841892	51.87898946	34.15810814	0.01927563	0.02927563
43	1.53397779	0.65189992	53.39777936	34.81000806	0.01872737	0.02872737
44	1.54931757	0.64544546	54.93175715	35.45545352	0.01820441	0.02820441
45	1.56481075	0.63905492	56.48107472	36.09450844	0.01770505	0.02770505
46	1.58045885	0.63272764	58.04588547	36.72723608	0.01722775	0.02722775
47	1.59626344	0.62646301	59.62634432	37.35369909	0.01677111	0.02677111
48	1.61222608	0.62026041	61.22260777	37.97395949	0.01633384	0.02633384
49	1.62834834	0.61411921	62.83483385	38.58807871	0.01591474	0.02591474
50	1.64463182	0.60803882	64.46318218	39.19611753	0.01551273	0.02551273

Rate $1\frac{1}{4}\%$	A Compound Interest	B Present Value	C Amount of Annuity	D Present Value of Annuity	E Sinking Fund	F Amortization
n	$(1+i)^n$	$\frac{1}{(1+i)^n}$	$s_{\overline{n}\vert i}$	$a_{\overline{n}\vert i}$	$\frac{1}{s_{\overline{n}\vert i}}$	$\frac{1}{a_{\overline{n}\vert i}}$
1	1.01250000	0.98765432	1.00000000	0.98765432	1.00000000	1.01250000
2	1.02515625	0.97546106	2.01250000	1.96311538	0.49689441	0.50939441
3	1.03797070	0.96341833	3.03765625	2.92653371	0.32920117	0.34170117
4	1.05094534	0.95152428	4.07562695	3.87805798	0.24536102	0.25786102
5	1.06408215	0.93977706	5.12657229	4.81783504	0.19506211	0.20756211
6	1.07738318	0.92817488	6.19065444	5.74600992	0.16153381	0.17403381
7	1.09085047	0.91671593	7.26803762	6.66272585	0.13758872	0.15008872
8	1.10448610	0.90539845	8.35888809	7.56812429	0.11963314	0.13213314
9	1.11829218	0.89422069	9.46337420	8.46234498	0.10567055	0.11817055
10	1.13227083	0.88318093	10.58166637	9.34552591	0.09450307	0.10700307
11	1.14642422	0.87227746	11.71393720	10.21780337	0.08536839	0.09786839
12	1.16075452	0.86150860	12.86036142	11.07931197	0.07775831	0.09025831
13	1.17526395	0.85087269	14.02111594	11.93018466	0.07132100	0.08382100
14	1.18995475	0.84036809	15.19637988	12.77055275	0.06580515	0.07830515
15	1.20482918	0.82999318	16.38633463	13.60054592	0.06102646	0.07352646
16	1.21988955	0.81974635	17.59116382	14.42029227	0.05684672	0.06934672
17	1.23513817	0.80962602	18.81105336	15.22991829	0.05316023	0.06566023
18	1.25057739	0.79963064	20.04619153	16.02954893	0.04988479	0.06238479
19	1.26620961	0.78975866	21.29676893	16.81930759	0.04695548	0.05945548
20	1.28203723	0.78000855	22.56297854	17.59931613	0.04432039	0.05682039
21	1.29806270	0.77037881	23.84501577	18.36969495	0.04193749	0.05443749
22	1.31428848	0.76086796	25.14307847	19.13056291	0.03977238	0.05227238
23	1.33071709	0.75147453	26.45736695	19.88203744	0.03779666	0.05029666
24	1.34735105	0.74219707	27.78808403	20.62423451	0.03598665	0.04848665
25	1.36419294	0.73303414	29.13543508	21.35726865	0.03432247	0.04682247
26	1.38124535	0.72398434	30.49962802	22.08125299	0.03278729	0.04528729
27	1.39851092	0.71504626	31.88087337	22.79629925	0.03136677	0.04386677
28	1.41599230	0.70621853	33.27938429	23.50251778	0.03004863	0.04254863
29	1.43369221	0.69749978	34.69537659	24.20001756	0.02882228	0.04132228
30	1.45161336	0.68888867	36.12906880	24.88890623	0.02767854	0.04017854
31	1.46975853	0.68038387	37.58068216	25.56929010	0.02660942	0.03910942
32	1.48813051	0.67198407	39.05044069	26.24127418	0.02560791	0.03810791
33	1.50673214	0.66368797	40.53857120	26.90496215	0.02466786	0.03716786
34	1.52556629	0.65549429	42.04530334	27.56045644	0.02378387	0.03628387
35	1.54463587	0.64740177	43.57086963	28.20785822	0.02295111	0.03545111
36	1.56394382	0.63940916	45.11550550	28.84726737	0.02216533	0.03466533
37	1.58349312	0.63151522	46.67944932	29.47878259	0.02142270	0.03392270
38	1.60328678	0.62371873	48.26294243	30.10250133	0.02071983	0.03321983
39	1.62332787	0.61601850	49.86622921	30.71851983	0.02005365	0.03255365
40	1.64361946	0.60841334	51.48955708	31.32693316	0.01942141	0.03192141
41	1.66416471	0.60090206	53.13317654	31.92783522	0.01882063	0.03132063
42	1.68496677	0.59348352	54.79734125	32.52131874	0.01824906	0.03074906
43	1.70602885	0.58615656	56.48230801	33.10747530	0.01770466	0.03020466
44	1.72735421	0.57892006	58.18833687	33.68639536	0.01718557	0.02968557
45	1.74894614	0.57177290	59.91569108	34.25816825	0.01669012	0.02919012
46	1.77080797	0.56471397	61.66463721	34.82288222	0.01621675	0.02871675
47	1.79294306	0.55774219	63.43544518	35.38064224	0.01576406	0.02826406
48	1.81535485	0.55085649	65.22838824	35.93148091	0.01533075	0.02783075
49	1.83804679	0.54405579	67.04374310	36.47553670	0.01491563	0.02741563
50	1.86102237	0.53733905	68.88178989	37.01287575	0.01451763	0.02701763

Rate $1\frac{1}{2}\%$	A Compound Interest	B Present Value	C Amount of Annuity	D Present Value of Annuity	E Sinking Fund	F Amortization
n	$(1 + i)^n$	$\frac{1}{(1 + i)^n}$	$s_{\overline{n}\rvert i}$	$a_{\overline{n}\rvert i}$	$\frac{1}{s_{\overline{n}\rvert i}}$	$\frac{1}{a_{\overline{n}\rvert i}}$
1	1.01500000	0.98522167	1.00000000	0.98522167	1.00000000	1.01500000
2	1.03022500	0.97066175	2.01500000	1.95588342	0.49627792	0.51127792
3	1.04567838	0.95631699	3.04522500	2.91220042	0.32838296	0.34338296
4	1.06136355	0.94218423	4.09090337	3.85438465	0.24444479	0.25944479
5	1.07728400	0.92826033	5.15226693	4.78264497	0.19408932	0.20908932
6	1.09344326	0.91454219	6.22955093	5.69718717	0.16052521	0.17552521
7	1.10984491	0.90102679	7.32299419	6.59821396	0.13655616	0.15155616
8	1.12649259	0.88771112	8.43283911	7.48592508	0.11858402	0.13358402
9	1.14338998	0.87459224	9.55933169	8.36051732	0.10460982	0.11960982
10	1.16054083	0.86166723	10.70272167	9.22218455	0.09343418	0.10843418
11	1.17794894	0.84893323	11.86326249	10.07111779	0.08429384	0.09929384
12	1.19561817	0.83638742	13.04121143	10.90750521	0.07667999	0.09167999
13	1.21355244	0.82402702	14.23682960	11.73153222	0.07024036	0.08524036
14	1.23175573	0.81184928	15.45038205	12.54338150	0.06472332	0.07972332
15	1.25023207	0.79985150	16.68213778	13.34323301	0.05994436	0.07494436
16	1.26898555	0.78803104	17.93236984	14.13126405	0.05576508	0.07076508
17	1.28802033	0.77638526	19.20135539	14.90764931	0.05207966	0.06707966
18	1.30734064	0.76491159	20.48937572	15.67256089	0.04880578	0.06380578
19	1.32695075	0.75360747	21.79671636	16.42616837	0.04587847	0.06087847
20	1.34685501	0.74247042	23.12366710	17.16863879	0.04324574	0.05824574
21	1.36705783	0.73149795	24.47052211	17.90013673	0.04086550	0.05586550
22	1.38756370	0.72068763	25.83757994	18.62082437	0.03870332	0.05370332
23	1.40837715	0.71003708	27.22514364	19.33086145	0.03673075	0.05173075
24	1.42950281	0.69954392	28.63352080	20.03040537	0.03492410	0.04992410
25	1.45094535	0.68920583	30.06302361	20.71961120	0.03326345	0.04826345
26	1.47270953	0.67902052	31.51396896	21.39863172	0.03173196	0.04673196
27	1.49480018	0.66898574	32.98667850	22.06761746	0.03031527	0.04531527
28	1.51722218	0.65909925	34.48147867	22.72671671	0.02900108	0.04400108
29	1.53998051	0.64935887	35.99870085	23.37607558	0.02777878	0.04277878
30	1.56308022	0.63976243	37.53868137	24.01583801	0.02663919	0.04163919
31	1.58652642	0.63030781	39.10176159	24.64614582	0.02557430	0.04057430
32	1.61032432	0.62099292	40.68828801	25.26713874	0.02457710	0.03957710
33	1.63447918	0.61181568	42.29861233	25.87895442	0.02364144	0.03864144
34	1.65899637	0.60277407	43.93309152	26.48172849	0.02276189	0.03776189
35	1.68388132	0.59386608	45.59208789	27.07559458	0.02193363	0.03693363
36	1.70913954	0.58508974	47.27596921	27.66068431	0.02115240	0.03615240
37	1.73477663	0.57644309	48.98510874	28.23712740	0.02041437	0.03541437
38	1.76079828	0.56792423	50.71988538	28.80505163	0.01971613	0.03471613
39	1.78721025	0.55953126	52.48068366	29.36458288	0.01905463	0.03405463
40	1.81401841	0.55126232	54.26789391	29.91584520	0.01842710	0.03342710
41	1.84122868	0.54311559	56.08191232	30.45896079	0.01783106	0.03283106
42	1.86884712	0.53508925	57.92314100	30.99405004	0.01726426	0.03226426
43	1.89687982	0.52718153	59.79198812	31.52123157	0.01672465	0.03172465
44	1.92533302	0.51939067	61.68886794	32.04062223	0.01621038	0.03121038
45	1.95421301	0.51171494	63.61420096	32.55233718	0.01571976	0.03071976
46	1.98352621	0.50415265	65.56841398	33.05648983	0.01525125	0.03025125
47	2.01327910	0.49670212	67.55194018	33.55319195	0.01480342	0.02980342
48	2.04347829	0.48936170	69.56521929	34.04255365	0.01437500	0.02937500
49	2.07413046	0.48212975	71.60869758	34.52468339	0.01396478	0.02896478
50	2.10524242	0.47500468	73.68282804	34.99968807	0.01357168	0.02857168

Rate $1\frac{3}{4}\%$	A Compound Interest	B Present Value	C Amount of Annuity	D Present Value of Annuity	E Sinking Fund	F Amortization
n	$(1+i)^n$	$\frac{1}{(1+i)^n}$	$s_{\overline{n}\rvert i}$	$a_{\overline{n}\rvert i}$	$\frac{1}{s_{\overline{n}\rvert i}}$	$\frac{1}{a_{\overline{n}\rvert i}}$
1	1.01750000	0.98280098	1.00000000	0.98280098	1.00000000	1.01750000
2	1.03530625	0.96589777	2.01750000	1.94869875	0.49566295	0.51316295
3	1.05342411	0.94928528	3.05280625	2.89798403	0.32756746	0.34506746
4	1.07185903	0.93295851	4.10623036	3.83094254	0.24353237	0.26103237
5	1.09061656	0.91691254	5.17808939	4.74785508	0.19312142	0.21062142
6	1.10970235	0.90114254	6.26870596	5.64899762	0.15952256	0.17702256
7	1.12912215	0.88564378	7.37840831	6.53464139	0.13553059	0.15303059
8	1.14888178	0.87041157	8.50753045	7.40505297	0.11754292	0.13504292
9	1.16898721	0.85544135	9.65641224	8.26049432	0.10355813	0.12105813
10	1.18944449	0.84072860	10.82539945	9.10122291	0.09237534	0.10987534
11	1.21025977	0.82626889	12.01484394	9.92749181	0.08323038	0.10073038
12	1.23143931	0.81205788	13.22510371	10.73954969	0.07561377	0.09311377
13	1.25298950	0.79809128	14.45654303	11.53764097	0.06917283	0.08667283
14	1.27491682	0.78436490	15.70953253	12.32200587	0.06365562	0.08115562
15	1.29722786	0.77087459	16.98444935	13.09288046	0.05887739	0.07637739
16	1.31992935	0.75761631	18.28167721	13.85049677	0.05469958	0.07219958
17	1.34302811	0.74458605	19.60160656	14.59508282	0.05101623	0.06851623
18	1.36653111	0.73177990	20.94463468	15.32686272	0.04774492	0.06524492
19	1.39044540	0.71919401	22.31116578	16.04605673	0.04482061	0.06232061
20	1.41477820	0.70682458	23.70161119	16.75288130	0.04219122	0.05969122
21	1.43953681	0.69466789	25.11638938	17.44754919	0.03981464	0.05731464
22	1.46472871	0.68272028	26.55592620	18.13026948	0.03765638	0.05515638
23	1.49036146	0.67097817	28.02065490	18.80124764	0.03568796	0.05318796
24	1.51644279	0.65943800	29.51101637	19.46068565	0.03388565	0.05138565
25	1.54298054	0.64809632	31.02745915	20.10878196	0.03222952	0.04972952
26	1.56998269	0.63694970	32.57043969	20.74573166	0.03070269	0.04820269
27	1.59745739	0.62599479	34.14042238	21.37172644	0.02929079	0.04679079
28	1.62541290	0.61522829	35.73787977	21.98695474	0.02798151	0.04548151
29	1.65385762	0.60464697	37.36329267	22.59160171	0.02676424	0.04426424
30	1.68280013	0.59424764	39.01715029	23.18584934	0.02562975	0.04312975
31	1.71224913	0.58402716	40.69995042	23.76987650	0.02457005	0.04207005
32	1.74221349	0.57398247	42.41219955	24.34385897	0.02357812	0.04107812
33	1.77270223	0.56411053	44.15441305	24.90796951	0.02264779	0.04014779
34	1.80372452	0.55440839	45.92711527	25.46237789	0.02177363	0.03927363
35	1.83528970	0.54487311	47.73083979	26.00725100	0.02095082	0.03845082
36	1.86740727	0.53550183	49.56612949	26.54275283	0.02017507	0.03767507
37	1.90008689	0.52629172	51.43353675	27.06904455	0.01944257	0.03694257
38	1.93333841	0.51724002	53.33362365	27.58628457	0.01874990	0.03624990
39	1.96717184	0.50834400	55.26696206	28.09462857	0.01809399	0.03559399
40	2.00159734	0.49960098	57.23413390	28.59422955	0.01747209	0.03497209
41	2.03662530	0.49100834	59.23573124	29.08523789	0.01688170	0.03438170
42	2.07226624	0.48256348	61.27235654	29.56780136	0.01632057	0.03382057
43	2.10853090	0.47426386	63.34462278	30.04206522	0.01578666	0.03328666
44	2.14543019	0.46610699	65.45315367	30.50817221	0.01527810	0.03277810
45	2.18297522	0.45809040	67.59858386	30.96626261	0.01479321	0.03229321
46	2.22117728	0.45021170	69.78155908	31.41647431	0.01433043	0.03183043
47	2.26004789	0.44246850	72.00273637	31.85894281	0.01388836	0.03138836
48	2.29959872	0.43485848	74.26278425	32.29380129	0.01346569	0.03096569
49	2.33984170	0.42737934	76.56238298	32.72118063	0.01306124	0.03056124
50	2.38078893	0.42002883	78.90222468	33.14120946	0.01267391	0.03017391

Rate 2%	A Compound Interest	B Present Value	C Amount of Annuity	D Present Value of Annuity	E Sinking Fund	F Amortization
n	$(1+i)^n$	$\frac{1}{(1+i)^n}$	$s_{\overline{n}\vert i}$	$a_{\overline{n}\vert i}$	$\frac{1}{s_{\overline{n}\vert i}}$	$\frac{1}{a_{\overline{n}\vert i}}$
1	1.02000000	0.98039216	1.00000000	0.98039216	1.00000000	1.02000000
2	1.04040000	0.96116878	2.02000000	1.94156094	0.49504950	0.51504950
3	1.06120800	0.94232233	3.06040000	2.88388327	0.32675467	0.34675467
4	1.08243216	0.92384543	4.12160800	3.80772870	0.24262375	0.26262375
5	1.10408080	0.90573081	5.20404016	4.71345951	0.19215839	0.21215839
6	1.12616242	0.88797138	6.30812096	5.60143089	0.15852581	0.17852581
7	1.14868567	0.87056018	7.43428338	6.47199107	0.13451196	0.15451196
8	1.17165938	0.85349037	8.58296905	7.32548144	0.11650980	0.13650980
9	1.19509257	0.83675527	9.75462843	8.16223671	0.10251544	0.12251544
10	1.21899442	0.82034830	10.94972100	8.98258501	0.09132653	0.11132653
11	1.24337431	0.80426304	12.16871542	9.78684805	0.08217794	0.10217794
12	1.26824179	0.78849318	13.41208973	10.57534122	0.07455960	0.09455960
13	1.29360663	0.77303253	14.68033152	11.34837375	0.06811835	0.08811835
14	1.31947876	0.75787502	15.97393815	12.10624877	0.06260197	0.08260197
15	1.34586834	0.74301473	17.29341692	12.84926350	0.05782547	0.07782547
16	1.37278571	0.72844581	18.63928525	13.57770931	0.05365013	0.07365013
17	1.40024142	0.71416256	20.01207096	14.29187188	0.04996984	0.06996984
18	1.42824625	0.70015937	21.41231238	14.99203125	0.04670210	0.06670210
19	1.45681117	0.68643076	22.84055863	15.67846201	0.04378177	0.06378177
20	1.48594740	0.67297133	24.29736980	16.35143334	0.04115672	0.06115672
21	1.51566634	0.65977582	25.78331719	17.01120916	0.03878477	0.05878477
22	1.54597967	0.64683904	27.29898354	17.65804820	0.03663140	0.05663140
23	1.57689926	0.63415592	28.84496321	18.29220412	0.03466810	0.05466810
24	1.60843725	0.62172149	30.42186247	18.91392560	0.03287110	0.05287110
25	1.64060599	0.60953087	32.03029972	19.52345647	0.03122044	0.05122044
26	1.67341811	0.59757928	33.67090572	20.12103576	0.02969923	0.04969923
27	1.70688648	0.58586204	35.34432383	20.70689780	0.02829309	0.04829309
28	1.74102421	0.57437455	37.05121031	21.28127236	0.02698967	0.04698967
29	1.77584469	0.56311231	38.79223451	21.84438466	0.02577836	0.04577836
30	1.81136158	0.55207089	40.56807921	22.39645555	0.02464992	0.04464992
31	1.84758882	0.54124597	42.37944079	22.93770152	0.02359635	0.04359635
32	1.88454059	0.53063330	44.22702961	23.46833482	0.02261061	0.04261061
33	1.92223140	0.52022873	46.11157020	23.98856355	0.02168653	0.04168653
34	1.96067603	0.51002817	48.03380160	24.49859172	0.02081867	0.04081867
35	1.99988955	0.50002761	49.99447763	24.99861933	0.02000221	0.04000221
36	2.03988734	0.49022315	51.99436719	25.48884248	0.01923285	0.03923285
37	2.08068509	0.48061093	54.03425453	25.96945341	0.01850678	0.03850678
38	2.12229879	0.47118719	56.11493962	26.44064060	0.01782057	0.03782057
39	2.16474477	0.46194822	58.23723841	26.90258883	0.01717114	0.03717114
40	2.20803966	0.45289042	60.40198318	27.35547924	0.01655575	0.03655575
41	2.25220046	0.44401021	62.61002284	27.79948945	0.01597188	0.03597188
42	2.29724447	0.43530413	64.86222330	28.23479358	0.01541729	0.03541729
43	2.34318936	0.42676875	67.15946777	28.66156233	0.01488993	0.03488993
44	2.39005314	0.41840074	69.50265712	29.07996307	0.01438794	0.03438794
45	2.43785421	0.41019680	71.89271027	29.49015987	0.01390962	0.03390962
46	2.48661129	0.40215373	74.33056447	29.89231360	0.01345342	0.03345342
47	2.53634352	0.39426836	76.81717576	30.28658196	0.01301792	0.03301792
48	2.58707039	0.38653761	79.35351927	30.67311957	0.01260184	0.03260184
49	2.63881179	0.37895844	81.94058966	31.05207801	0.01220396	0.03220396
50	2.69158803	0.37152788	84.57940145	31.42360589	0.01182321	0.03182321

Rate $2\frac{1}{2}\%$	A Compound Interest	B Present Value	C Amount of Annuity	D Present Value of Annuity	E Sinking Fund	F Amortization
n	$(1+i)^n$	$\frac{1}{(1+i)^n}$	$s_{\overline{n}\rceil i}$	$a_{\overline{n}\rceil i}$	$\frac{1}{s_{\overline{n}\rceil i}}$	$\frac{1}{a_{\overline{n}\rceil i}}$
1	1.02500000	0.97560976	1.00000000	0.97560976	1.00000000	1.02500000
2	1.05062500	0.95181440	2.02500000	1.92742415	0.49382716	0.51882716
3	1.07689063	0.92859941	3.07562500	2.85602356	0.32513717	0.35013717
4	1.10381289	0.90595064	4.15251562	3.76197421	0.24081788	0.26581788
5	1.13140821	0.88385429	5.25632852	4.64582850	0.19024686	0.21524686
6	1.15969342	0.86229687	6.38773673	5.50812536	0.15654997	0.18154997
7	1.18868575	0.84126524	7.54753015	6.34939060	0.13249543	0.15749543
8	1.21840290	0.82074657	8.73611590	7.17013717	0.11446735	0.13946735
9	1.24886297	0.80072836	9.95451880	7.97086553	0.10045689	0.12545689
10	1.28008454	0.78119840	11.20338177	8.75206393	0.08925876	0.11425876
11	1.31208666	0.76214478	12.48346631	9.51420871	0.08010596	0.10510596
12	1.34488882	0.74355589	13.79555297	10.25776460	0.07248713	0.09748713
13	1.37851104	0.72542038	15.14044179	10.98318497	0.06604827	0.09104827
14	1.41297382	0.70772720	16.51895284	11.69091217	0.06053652	0.08553652
15	1.44829817	0.69046556	17.93192666	12.38137773	0.05576646	0.08076646
16	1.48450562	0.67362493	19.38022483	13.05500266	0.05159899	0.07659899
17	1.52161826	0.65719506	20.86473045	13.71219772	0.04792777	0.07292777
18	1.55965872	0.64116591	22.38634871	14.35336363	0.04467008	0.06967008
19	1.59865019	0.62552772	23.94600743	14.97889134	0.04176062	0.06676062
20	1.63861644	0.61027094	25.54465761	15.58916229	0.03914713	0.06414713
21	1.67958185	0.59538629	27.18327405	16.18454857	0.03678733	0.06178733
22	1.72157140	0.58086467	28.86285590	16.76541324	0.03464661	0.05964661
23	1.76461068	0.56669724	30.58442730	17.33211048	0.03269638	0.05769638
24	1.80872595	0.55287535	32.34903798	17.88498583	0.03091282	0.05591282
25	1.85394410	0.53939059	34.15776393	18.42437642	0.02927592	0.05427592
26	1.90029270	0.52623472	36.01170803	18.95061114	0.02776875	0.05276875
27	1.94780002	0.51339973	37.91200073	19.46401087	0.02637687	0.05137687
28	1.99649502	0.50087778	39.85980075	19.96488866	0.02508793	0.05008793
29	2.04640739	0.48866125	41.85629577	20.45354991	0.02389127	0.04889127
30	2.09756758	0.47674269	43.90270316	20.93029259	0.02277764	0.04777764
31	2.15000677	0.46511481	46.00027074	21.39540741	0.02173900	0.04673900
32	2.20375694	0.45377055	48.15027751	21.84917796	0.02076831	0.04576831
33	2.25885086	0.44270298	50.35403445	22.29188094	0.01985938	0.04485938
34	2.31532213	0.43190534	52.61288531	22.72378628	0.01900675	0.04400675
35	2.37320519	0.42137107	54.92820744	23.14515734	0.01820558	0.04320558
36	2.43253532	0.41109372	57.30141263	23.55625107	0.01745158	0.04245158
37	2.49334870	0.40106705	59.73394794	23.95731812	0.01674090	0.04174090
38	2.55568242	0.39128492	62.22729664	24.34860304	0.01607012	0.04107012
39	2.61957448	0.38174139	64.78297906	24.73034443	0.01543615	0.04043615
40	2.68506384	0.37243062	67.40255354	25.10277505	0.01483623	0.03983623
41	2.75219043	0.36334695	70.08761737	25.46612200	0.01426786	0.03926786
42	2.82099520	0.35448483	72.83980781	25.82060683	0.01372876	0.03872876
43	2.89152008	0.34583886	75.66080300	26.16644569	0.01321688	0.03821688
44	2.96380808	0.33740376	78.55232308	26.50384945	0.01273037	0.03773037
45	3.03790328	0.32917440	81.51613116	26.83302386	0.01226751	0.03726751
46	3.11385086	0.32114576	84.55403443	27.15416962	0.01182676	0.03682676
47	3.19169713	0.31331294	87.66788530	27.46748255	0.01140669	0.03640669
48	3.27148956	0.30567116	90.85958243	27.77315371	0.01100599	0.03600599
49	3.35327680	0.29821576	94.13107199	28.07136947	0.01062348	0.03562348
50	3.43710872	0.29094221	97.48434879	28.36231168	0.01025806	0.03525806

Rate 3%	A Compound Interest	B Present Value	C Amount of Annuity	D Present Value of Annuity	E Sinking Fund	F Amortization
n	$(1+i)^n$	$\frac{1}{(1+i)^n}$	$s_{\overline{n}\vert i}$	$a_{\overline{n}\vert i}$	$\frac{1}{s_{\overline{n}\vert i}}$	$\frac{1}{a_{\overline{n}\vert i}}$
1	1.03000000	0.97087379	1.00000000	0.97087379	1.00000000	1.03000000
2	1.06090000	0.94259591	2.03000000	1.91346970	0.49261084	0.52261084
3	1.09272700	0.91514166	3.09090000	2.82861135	0.32353036	0.35353036
4	1.12550881	0.88848705	4.18362700	3.71709840	0.23902705	0.26902705
5	1.15927407	0.86260878	5.30913581	4.57970719	0.18835457	0.21835457
6	1.19405230	0.83748426	6.46840988	5.41719144	0.15459750	0.18459750
7	1.22987387	0.81309151	7.66246218	6.23028296	0.13050635	0.16050635
8	1.26677008	0.78940923	8.89233605	7.01969219	0.11245639	0.14245639
9	1.30477318	0.76641673	10.15910613	7.78610892	0.09843386	0.12843386
10	1.34391638	0.74409391	11.46387931	8.53020284	0.08723051	0.11723051
11	1.38423387	0.72242128	12.80779569	9.25262411	0.07807745	0.10807745
12	1.42576089	0.70137988	14.19202956	9.95400399	0.07046209	0.10046209
13	1.46853371	0.68095134	15.61779045	10.63495533	0.06402954	0.09402954
14	1.51258972	0.66111781	17.08632416	11.29607314	0.05852634	0.08852634
15	1.55796742	0.64186195	18.59891389	11.93793509	0.05376658	0.08376658
16	1.60470644	0.62316694	20.15688130	12.56110203	0.04961085	0.07961085
17	1.65284763	0.60501645	21.76158774	13.16611847	0.04595253	0.07595253
18	1.70243306	0.58739461	23.41443537	13.75351308	0.04270870	0.07270870
19	1.75350605	0.57028603	25.11686844	14.32379911	0.03981388	0.06981388
20	1.80611123	0.55367575	26.87037449	14.87747486	0.03721571	0.06721571
21	1.86029457	0.53754928	28.67648572	15.41502414	0.03487178	0.06487178
22	1.91610341	0.52189250	30.53678030	15.93691664	0.03274739	0.06274739
23	1.97358651	0.50669175	32.45288370	16.44360839	0.03081390	0.06081390
24	2.03279411	0.49193374	34.42647022	16.93554212	0.02904742	0.05904742
25	2.09377793	0.47760557	36.45926432	17.41314769	0.02742787	0.05742787
26	2.15659127	0.46369473	38.55304225	17.87684242	0.02593829	0.05593829
27	2.22128901	0.45018906	40.70963352	18.32703147	0.02456421	0.05456421
28	2.28792768	0.43707675	42.93092252	18.76410823	0.02329323	0.05329323
29	2.35656551	0.42434636	45.21885020	19.18845459	0.02211467	0.05211467
30	2.42726247	0.41198676	47.57541571	19.60044135	0.02101926	0.05101926
31	2.50008035	0.39998715	50.00267818	20.00042849	0.01999893	0.04999893
32	2.57508276	0.38833703	52.50275852	20.38876553	0.01904662	0.04904662
33	2.65233524	0.37702625	55.07784128	20.76579178	0.01815612	0.04815612
34	2.73190530	0.36604490	57.73017652	21.13183668	0.01732196	0.04732196
35	2.81386245	0.35538340	60.46208181	21.48722007	0.01653929	0.04653929
36	2.89827833	0.34503243	63.27594427	21.83225250	0.01580379	0.04580379
37	2.98522668	0.33498294	66.17422259	22.16723544	0.01511162	0.04511162
38	3.07478348	0.32522615	69.15944927	22.49246159	0.01445934	0.04445934
39	3.16702698	0.31575355	72.23423275	22.80821513	0.01384385	0.04384385
40	3.26203779	0.30655684	75.40125973	23.11477197	0.01326238	0.04326238
41	3.35989893	0.29762800	78.66329753	23.41239997	0.01271241	0.04271241
42	3.46069589	0.28895922	82.02319645	23.70135920	0.01219167	0.04219167
43	3.56451677	0.28054294	85.48389234	23.98190213	0.01169811	0.04169811
44	3.67145227	0.27237178	89.04840911	24.25427392	0.01122985	0.04122985
45	3.78159584	0.26443862	92.71986139	24.51871254	0.01078518	0.04078518
46	3.89504372	0.25673653	96.50145723	24.77544907	0.01036254	0.04036254
47	4.01189503	0.24925876	100.39650095	25.02470783	0.00996051	0.03996051
48	4.13225188	0.24199880	104.40839598	25.26670664	0.00957777	0.03957777
49	4.25621944	0.23495029	108.54064785	25.50165693	0.00921314	0.03921314
50	4.38390602	0.22810708	112.79686729	25.72976401	0.00886549	0.03886549

Rate $3\frac{1}{2}\%$	A Compound Interest	B Present Value	C Amount of Annuity	D Present Value of Annuity	E Sinking Fund	F Amortization
n	$(1+i)^n$	$\frac{1}{(1+i)^n}$	$s_{\overline{n}\rvert i}$	$a_{\overline{n}\rvert i}$	$\frac{1}{s_{\overline{n}\rvert i}}$	$\frac{1}{a_{\overline{n}\rvert i}}$
1	1.03500000	0.96618357	1.00000000	0.96618357	1.00000000	1.03500000
2	1.07122500	0.93351070	2.03500000	1.89969428	0.49140049	0.52640049
3	1.10871788	0.90194271	3.10622500	2.80163698	0.32193418	0.35693418
4	1.14752300	0.87144223	4.21494287	3.67307921	0.23725114	0.27225114
5	1.18768631	0.84197317	5.36246588	4.51505238	0.18648137	0.22148137
6	1.22925533	0.81350064	6.55015218	5.32855302	0.15266821	0.18766821
7	1.27227926	0.78599096	7.77940751	6.11454398	0.12854449	0.16354449
8	1.31680904	0.75941156	9.05168677	6.87395554	0.11047665	0.14547665
9	1.36289735	0.73373097	10.36849581	7.60768651	0.09644601	0.13144601
10	1.41059876	0.70891881	11.73139316	8.31660532	0.08524137	0.12024137
11	1.45996972	0.68494571	13.14199192	9.00155104	0.07609197	0.11109197
12	1.51106866	0.66178330	14.60196164	9.66333433	0.06848395	0.10348395
13	1.56395606	0.63940415	16.11303030	10.30273849	0.06206157	0.09706157
14	1.61869452	0.61778179	17.67698636	10.92052028	0.05657073	0.09157073
15	1.67534883	0.59689062	19.29568088	11.51741090	0.05182507	0.08682507
16	1.73398604	0.57670591	20.97102971	12.09411681	0.04768483	0.08268483
17	1.79467555	0.55720378	22.70501575	12.65132059	0.04404313	0.07904313
18	1.85748920	0.53836114	24.49969130	13.18968173	0.04081684	0.07581684
19	1.92250132	0.52015569	26.35718050	13.70983742	0.03794033	0.07294033
20	1.98978886	0.50256588	28.27968181	14.21240330	0.03536108	0.07036108
21	2.05943147	0.48557090	30.26947068	14.69797420	0.03303659	0.06803659
22	2.13151158	0.46915063	32.32890215	15.16712484	0.03093207	0.06593207
23	2.20611448	0.45328563	34.46041373	15.62041047	0.02901880	0.06401880
24	2.28332849	0.43795713	36.66652821	16.05836760	0.02727283	0.06227283
25	2.36324498	0.42314699	38.94985669	16.48151459	0.02567404	0.06067404
26	2.44595856	0.40883767	41.31310168	16.89035226	0.02420540	0.05920540
27	2.53156711	0.39501224	43.75906024	17.28536451	0.02285241	0.05785241
28	2.62017196	0.38165434	46.29062734	17.66701885	0.02160265	0.05660265
29	2.71187798	0.36874815	48.91079930	18.03576700	0.02044538	0.05544538
30	2.80679370	0.35627841	51.62267728	18.39204541	0.01937133	0.05437133
31	2.90503148	0.34423035	54.42947098	18.73627576	0.01837240	0.05337240
32	3.00670759	0.33258971	57.33450247	19.06886547	0.01744150	0.05244150
33	3.11194235	0.32134271	60.34121005	19.39020818	0.01657242	0.05157242
34	3.22086033	0.31047605	63.45315240	19.70068423	0.01575966	0.05075966
35	3.33359045	0.29997686	66.67401274	20.00066110	0.01499835	0.04999835
36	3.45026611	0.28983272	70.00760318	20.29049381	0.01428416	0.04928416
37	3.57102543	0.28003161	73.45786930	20.57052542	0.01361325	0.04861325
38	3.69601132	0.27056194	77.02889472	20.84108736	0.01298214	0.04798214
39	3.82537171	0.26141250	80.72490604	21.10249987	0.01238775	0.04738775
40	3.95925972	0.25257247	84.55027775	21.35507234	0.01182728	0.04682728
41	4.09783381	0.24403137	88.50953747	21.59910371	0.01129822	0.04629822
42	4.24125799	0.23577910	92.60737128	21.83488281	0.01079828	0.04579828
43	4.38970202	0.22780590	96.84862928	22.06268870	0.01032539	0.04532539
44	4.54334160	0.22010231	101.23833130	22.28279102	0.00987768	0.04487768
45	4.70235855	0.21265924	105.78167290	22.49545026	0.00945343	0.04445343
46	4.86694110	0.20546787	110.48403145	22.70091813	0.00905108	0.04405108
47	5.03728404	0.19851968	115.35097255	22.89943780	0.00866919	0.04366919
48	5.21358898	0.19180645	120.38825659	23.09124425	0.00830646	0.04330646
49	5.39606459	0.18532024	125.60184557	23.27656450	0.00796167	0.04296167
50	5.58492686	0.17905337	130.99791016	23.45561787	0.00763371	0.04263371

Rate $3\frac{3}{4}\%$	A Compound Interest	B Present Value	C Amount of Annuity	D Present Value of Annuity	E Sinking Fund	F Amortization
n	$(1+i)^n$	$\frac{1}{(1+i)^n}$	$s_{\overline{n}\vert i}$	$a_{\overline{n}\vert i}$	$\frac{1}{s_{\overline{n}\vert i}}$	$\frac{1}{a_{\overline{n}\vert i}}$
1	1.03750000	0.96385542	1.00000000	0.96385542	1.00000000	1.03750000
2	1.07640625	0.92901727	2.03750000	1.89287270	0.49079755	0.52829755
3	1.11677148	0.89543834	3.11390625	2.78831103	0.32114005	0.35864005
4	1.15865042	0.86307310	4.23067773	3.65138413	0.23636875	0.27386875
5	1.20209981	0.83187768	5.38932815	4.48326181	0.18555189	0.22305189
6	1.24717855	0.80180981	6.59142796	5.28507162	0.15171219	0.18921219
7	1.29394774	0.77282874	7.83860650	6.05790036	0.12757370	0.16507370
8	1.34247078	0.74489517	9.13255425	6.80279553	0.10949839	0.14699839
9	1.39281344	0.71797125	10.47502503	7.52076677	0.09546517	0.13296517
10	1.44504394	0.69202048	11.86783847	8.21278725	0.08426134	0.12176134
11	1.49923309	0.66700769	13.31288241	8.87979494	0.07511521	0.11261521
12	1.55545433	0.64289898	14.81211550	9.52269392	0.06751230	0.10501230
13	1.61378387	0.61966167	16.36756983	10.14235558	0.06109642	0.09859642
14	1.67430076	0.59726426	17.98135370	10.73961984	0.05561317	0.09311317
15	1.73708704	0.57567639	19.65565447	11.31529623	0.05087595	0.08837595
16	1.80222781	0.55486881	21.39274151	11.87016504	0.04674483	0.08424483
17	1.86981135	0.53481331	23.19496932	12.40497835	0.04311280	0.08061280
18	1.93992927	0.51548271	25.06478067	12.92046106	0.03989662	0.07739662
19	2.01267662	0.49685080	27.00470994	13.41731187	0.03703058	0.07453058
20	2.08815200	0.47889234	29.01738656	13.89620421	0.03446210	0.07196210
21	2.16645770	0.46158298	31.10553856	14.35778719	0.03214862	0.06964862
22	2.24769986	0.44489926	33.27199626	14.80268645	0.03005531	0.06755531
23	2.33198860	0.42881856	35.51969612	15.23150501	0.02815339	0.06565339
24	2.41943818	0.41331910	37.85168472	15.64482411	0.02641890	0.06391890
25	2.51016711	0.39837985	40.27112290	16.04320396	0.02483169	0.06233169
26	2.60429838	0.38398058	42.78129001	16.42718454	0.02337470	0.06087470
27	2.70195956	0.37010176	45.38558838	16.79728630	0.02203343	0.05953343
28	2.80328305	0.35672459	48.08754794	17.15401089	0.02079540	0.05829540
29	2.90840616	0.34383093	50.89083099	17.49784183	0.01964991	0.05714991
30	3.01747139	0.33140331	53.79923715	17.82924513	0.01858762	0.05608762
31	3.13062657	0.31942487	56.81670855	18.14867001	0.01760046	0.05510046
32	3.24802507	0.30787940	59.94733512	18.45654941	0.01668131	0.05418131
33	3.36982601	0.29675123	63.19536019	18.75330063	0.01582395	0.05332395
34	3.49619448	0.28602528	66.56518619	19.03932591	0.01502287	0.05252287
35	3.62730178	0.27568702	70.06138067	19.31501293	0.01427320	0.05177320
36	3.76332559	0.26572242	73.68868245	19.58073535	0.01357060	0.05107060
37	3.90445030	0.25611800	77.45200804	19.83685335	0.01291122	0.05041122
38	4.05086719	0.24686072	81.35645834	20.08371407	0.01229159	0.04979159
39	4.20277471	0.23793805	85.40732553	20.32165212	0.01170860	0.04920860
40	4.36037876	0.22933788	89.61010024	20.55098999	0.01115946	0.04865946
41	4.52389296	0.22104855	93.97047900	20.77203855	0.01064164	0.04814164
42	4.69353895	0.21305885	98.49437196	20.98509739	0.01015286	0.04765286
43	4.86954666	0.20535793	103.18791091	21.19045532	0.00969106	0.04719106
44	5.05215466	0.19793535	108.05745757	21.38839067	0.00925434	0.04675434
45	5.24161046	0.19078106	113.10961223	21.57917173	0.00884098	0.04634098
46	5.43817085	0.18388536	118.35122269	21.76305709	0.00844943	0.04594943
47	5.64210226	0.17723890	123.78939354	21.94029599	0.00807824	0.04557824
48	5.85368109	0.17083268	129.43149579	22.11112866	0.00772609	0.04522609
49	6.07319413	0.16465800	135.28517689	22.27578666	0.00739179	0.04489179
50	6.30093891	0.15870651	141.35837102	22.43449317	0.00707422	0.04457422

Rate 4%	A Compound Interest	B Present Value	C Amount of Annuity	D Present Value of Annuity	E Sinking Fund	F Amortization
n	$(1+i)^n$	$\frac{1}{(1+i)^n}$	$s_{\overline{n}\|i}$	$a_{\overline{n}\|i}$	$\frac{1}{s_{\overline{n}\|i}}$	$\frac{1}{a_{\overline{n}\|i}}$
1	1.04000000	0.96153846	1.00000000	0.96153846	1.00000000	1.04000000
2	1.08160000	0.92455621	2.04000000	1.88609467	0.49019608	0.53019608
3	1.12486400	0.88899636	3.12160000	2.77509103	0.32034854	0.36034854
4	1.16985856	0.85480419	4.24646400	3.62989522	0.23549005	0.27549005
5	1.21665290	0.82192711	5.41632256	4.45182233	0.18462711	0.22462711
6	1.26531902	0.79031453	6.63297546	5.24213686	0.15076190	0.19076190
7	1.31593178	0.75991781	7.89829448	6.00205467	0.12660961	0.16660961
8	1.36856905	0.73069021	9.21422626	6.73274487	0.10852783	0.14852783
9	1.42331181	0.70258674	10.58279531	7.43533161	0.09449299	0.13449299
10	1.48024428	0.67556417	12.00610712	8.11089578	0.08329094	0.12329094
11	1.53945406	0.64958093	13.48635141	8.76047671	0.07414904	0.11414904
12	1.60103222	0.62459705	15.02580546	9.38507376	0.06655217	0.10655217
13	1.66507351	0.60057409	16.62683768	9.98564785	0.06014373	0.10014373
14	1.73167645	0.57747508	18.29191119	10.56312293	0.05466897	0.09466897
15	1.80094351	0.55526450	20.02358764	11.11838743	0.04994110	0.08994110
16	1.87298125	0.53390818	21.82453114	11.65229561	0.04582000	0.08582000
17	1.94790050	0.51337325	23.69751239	12.16566885	0.04219852	0.08219852
18	2.02581652	0.49362812	25.64541288	12.65929697	0.03899333	0.07899333
19	2.10684918	0.47464242	27.67122940	13.13393940	0.03613862	0.07613862
20	2.19112314	0.45638695	29.77807858	13.59032634	0.03358175	0.07358175
21	2.27876807	0.43883360	31.96920172	14.02915995	0.03128011	0.07128011
22	2.36991879	0.42195539	34.24796979	14.45111533	0.02919881	0.06919881
23	2.46471554	0.40572633	36.61788858	14.85684167	0.02730906	0.06730906
24	2.56330416	0.39012147	39.08260412	15.24696314	0.02558683	0.06558683
25	2.66583633	0.37511680	41.64590829	15.62207994	0.02401196	0.06401196
26	2.77246978	0.36068923	44.31174462	15.98276918	0.02256738	0.06256738
27	2.88336858	0.34681657	47.08421440	16.32958575	0.02123854	0.06123854
28	2.99870332	0.33347747	49.96758298	16.66306322	0.02001298	0.06001298
29	3.11865145	0.32065141	52.96628630	16.98371463	0.01887993	0.05887993
30	3.24339751	0.30831867	56.08493775	17.29203330	0.01783010	0.05783010
31	3.37313341	0.29646026	59.32833526	17.58849356	0.01685535	0.05685535
32	3.50805875	0.28505794	62.70146867	17.87355150	0.01594859	0.05594859
33	3.64838110	0.27409417	66.20952742	18.14764567	0.01510357	0.05510357
34	3.79431634	0.26355209	69.85790851	18.41119776	0.01431477	0.05431477
35	3.94608899	0.25341547	73.65222486	18.66461323	0.01357732	0.05357732
36	4.10393255	0.24366872	77.59831385	18.90828195	0.01288688	0.05288688
37	4.26808986	0.23429685	81.70224640	19.14257880	0.01223957	0.05223957
38	4.43881345	0.22528543	85.97033626	19.36786423	0.01163192	0.05163192
39	4.61636599	0.21662061	90.40914971	19.58448484	0.01106083	0.05106083
40	4.80102063	0.20828904	95.02551570	19.79277388	0.01052349	0.05052349
41	4.99306145	0.20027793	99.82653633	19.99305181	0.01001738	0.05001738
42	5.19278391	0.19257493	104.81959778	20.18562674	0.00954020	0.04954020
43	5.40049527	0.18516820	110.01238169	20.37079494	0.00908989	0.04908989
44	5.61651508	0.17804635	115.41287696	20.54884129	0.00866454	0.04866454
45	5.84117568	0.17119841	121.02939204	20.72003970	0.00826246	0.04826246
46	6.07482271	0.16461386	126.87056772	20.88465356	0.00788205	0.04788205
47	6.31781562	0.15828256	132.94539043	21.04293612	0.00752189	0.04752189
48	6.57052824	0.15219476	139.26320604	21.19513088	0.00718065	0.04718065
49	6.83334937	0.14634112	145.83373429	21.34147200	0.00685712	0.04685712
50	7.10668335	0.14071262	152.66708366	21.48218462	0.00655020	0.04655020

Rate $4\frac{1}{2}\%$	A Compound Interest	B Present Value	C Amount of Annuity	D Present Value of Annuity	E Sinking Fund	F Amortization
n	$(1+i)^n$	$\frac{1}{(1+i)^n}$	$s_{\overline{n}\mid i}$	$a_{\overline{n}\mid i}$	$\frac{1}{s_{\overline{n}\mid i}}$	$\frac{1}{a_{\overline{n}\mid i}}$
1	1.04500000	0.95693780	1.00000000	0.95693780	1.00000000	1.04500000
2	1.09202500	0.91572995	2.04500000	1.87266775	0.48899756	0.53399756
3	1.14116613	0.87629660	3.13702500	2.74896435	0.31877336	0.36377336
4	1.19251860	0.83856134	4.27819112	3.58752570	0.23374365	0.27874365
5	1.24618194	0.80245105	5.47070973	4.38997674	0.18279164	0.22779164
6	1.30226012	0.76789574	6.71689166	5.15787248	0.14887839	0.19387839
7	1.36086183	0.73482846	8.01915179	5.89270094	0.12470147	0.16970147
8	1.42210061	0.70318513	9.38001362	6.59588607	0.10660965	0.15160965
9	1.48609514	0.67290443	10.80211423	7.26879050	0.09257447	0.13757447
10	1.55296942	0.64392768	12.28820937	7.91271818	0.08137882	0.12637882
11	1.62285305	0.61619874	13.84117879	8.52891692	0.07224818	0.11724818
12	1.69588143	0.58966386	15.46403184	9.11858078	0.06466619	0.10966619
13	1.77219610	0.56427164	17.15991327	9.68285242	0.05827535	0.10327535
14	1.85194492	0.53997286	18.93210937	10.22282528	0.05282032	0.09782032
15	1.93528244	0.51672044	20.78405429	10.73954573	0.04811381	0.09311381
16	2.02237015	0.49446932	22.71933673	11.23401505	0.04401537	0.08901537
17	2.11337681	0.47317639	24.74170689	11.70719143	0.04041758	0.08541758
18	2.20847877	0.45280037	26.85508370	12.15999180	0.03723690	0.08223690
19	2.30786031	0.43330179	29.06356246	12.59329359	0.03440734	0.07940734
20	2.41171402	0.41464286	31.37142277	13.00793645	0.03187614	0.07687614
21	2.52024116	0.39678743	33.78313680	13.40472388	0.02960057	0.07460057
22	2.63365201	0.37970089	36.30337795	13.78442476	0.02754565	0.07254565
23	2.75216635	0.36335013	38.93702996	14.14777489	0.02568249	0.07068249
24	2.87601383	0.34770347	41.68919631	14.49547837	0.02398703	0.06898703
25	3.00543446	0.33273060	44.56521015	14.82820896	0.02243903	0.06743903
26	3.14067901	0.31840248	47.57064460	15.14661145	0.02102137	0.06602137
27	3.28200956	0.30469137	50.71132361	15.45130282	0.01971946	0.06471946
28	3.42969999	0.29157069	53.99333317	15.74287351	0.01852081	0.06352081
29	3.58403649	0.27901502	57.42303316	16.02188853	0.01741461	0.06241461
30	3.74531813	0.26700002	61.00706966	16.28888854	0.01639154	0.06139154
31	3.91385745	0.25550241	64.75238779	16.54439095	0.01544345	0.06044345
32	4.08998104	0.24449991	68.66624524	16.78889086	0.01456320	0.05956320
33	4.27403018	0.23397121	72.75622628	17.02286207	0.01374453	0.05874453
34	4.46636154	0.22389589	77.03025646	17.24675796	0.01298191	0.05798191
35	4.66734781	0.21425444	81.49661800	17.46101240	0.01227045	0.05727045
36	4.87737846	0.20502817	86.16396581	17.66604058	0.01160578	0.05660578
37	5.09686049	0.19619921	91.04134427	17.86223979	0.01098402	0.05598402
38	5.32621921	0.18775044	96.13820476	18.04999023	0.01040169	0.05540169
39	5.56589908	0.17966549	101.46442398	18.22965572	0.00985567	0.05485567
40	5.81636454	0.17192870	107.03032306	18.40158442	0.00934315	0.05434315
41	6.07810094	0.16452507	112.84668760	18.56610949	0.00886158	0.05386158
42	6.35161548	0.15744026	118.92478854	18.72354975	0.00840868	0.05340868
43	6.63743818	0.15066054	125.27640402	18.87421029	0.00798235	0.05298235
44	6.93612290	0.14417276	131.91384220	19.01838305	0.00758071	0.05258071
45	7.24824843	0.13796437	138.84996510	19.15634742	0.00720202	0.05220202
46	7.57441961	0.13202332	146.09821353	19.28837074	0.00684471	0.05184471
47	7.91526849	0.12633810	153.67263314	19.41470884	0.00650734	0.05150734
48	8.27145557	0.12089771	161.58790163	19.53560654	0.00618858	0.05118858
49	8.64367107	0.11569158	169.85935720	19.65129813	0.00588722	0.05088722
50	9.03263627	0.11070965	178.50302828	19.76200778	0.00560215	0.05060215

Rate 5%	A Compound Interest	B Present Value	C Amount of Annuity	D Present Value of Annuity	E Sinking Fund	F Amortization
n	$(1+i)^n$	$\frac{1}{(1+i)^n}$	$s_{\overline{n}\vert i}$	$a_{\overline{n}\vert i}$	$\frac{1}{s_{\overline{n}\vert i}}$	$\frac{1}{a_{\overline{n}\vert i}}$
1	1.05000000	0.95238095	1.00000000	0.95238095	1.00000000	1.05000000
2	1.10250000	0.90702948	2.05000000	1.85941043	0.48780488	0.53780488
3	1.15762500	0.86383760	3.15250000	2.72324803	0.31720856	0.36720856
4	1.21550625	0.82270247	4.31012500	3.54595050	0.23201183	0.28201183
5	1.27628156	0.78352617	5.52563125	4.32947667	0.18097480	0.23097480
6	1.34009564	0.74621540	6.80191281	5.07569207	0.14701747	0.19701747
7	1.40710042	0.71068133	8.14200845	5.78637340	0.12281982	0.17281982
8	1.47745544	0.67683936	9.54910888	6.46321276	0.10472181	0.15472181
9	1.55132822	0.64460892	11.02656432	7.10782168	0.09069008	0.14069008
10	1.62889463	0.61391325	12.57789254	7.72173493	0.07950457	0.12950457
11	1.71033936	0.58467929	14.20678716	8.30641422	0.07038889	0.12038889
12	1.79585633	0.55683742	15.91712652	8.86325164	0.06282541	0.11282541
13	1.88564914	0.53032135	17.71298285	9.39357299	0.05645577	0.10645577
14	1.97993160	0.50506795	19.59863199	9.89864094	0.05102397	0.10102397
15	2.07892818	0.48101710	21.57856359	10.37965804	0.04634229	0.09634229
16	2.18287459	0.45811152	23.65749177	10.83776956	0.04226991	0.09226991
17	2.29201832	0.43629669	25.84036636	11.27406625	0.03869914	0.08869914
18	2.40661923	0.41552065	28.13238467	11.68958690	0.03554622	0.08554622
19	2.52695020	0.39573396	30.53900391	12.08532086	0.03274501	0.08274501
20	2.65329771	0.37688948	33.06595410	12.46221034	0.03024259	0.08024259
21	2.78596259	0.35894236	35.71925181	12.82115271	0.02799611	0.07799611
22	2.92526072	0.34184987	38.50521440	13.16300258	0.02597051	0.07597051
23	3.07152376	0.32557131	41.43047512	13.48857388	0.02413682	0.07413682
24	3.22509994	0.31006791	44.50199887	13.79864179	0.02247090	0.07247090
25	3.38635494	0.29530277	47.72709882	14.09394457	0.02095246	0.07095246
26	3.55567269	0.28124073	51.11345376	14.37518530	0.01956432	0.06956432
27	3.73345632	0.26784832	54.66912645	14.64303362	0.01829186	0.06829186
28	3.92012914	0.25509364	58.40258277	14.89812726	0.01712253	0.06712253
29	4.11613560	0.24294632	62.32271191	15.14107358	0.01604551	0.06604551
30	4.32194238	0.23137745	66.43884750	15.37245103	0.01505144	0.06505144
31	4.53803949	0.22035947	70.76078988	15.59281050	0.01413212	0.06413212
32	4.76494147	0.20986617	75.29882937	15.80267667	0.01328042	0.06328042
33	5.00318854	0.19978254	80.06377084	16.00254921	0.01249004	0.06249004
34	5.25334797	0.19035480	85.06695938	16.19290401	0.01175545	0.06175545
35	5.51601537	0.18129029	90.32030735	16.37419429	0.01107171	0.06107171
36	5.79181614	0.17265741	95.83632272	16.54685171	0.01043446	0.06043446
37	6.08140694	0.16443563	101.62813886	16.71128734	0.00983979	0.05983979
38	6.38547729	0.15660536	107.70954580	16.86789271	0.00928423	0.05928423
39	6.70475115	0.14914797	114.09502309	17.01704067	0.00876462	0.05876462
40	7.03998871	0.14204568	120.79977424	17.15908635	0.00827816	0.05827816
41	7.39198815	0.13528160	127.83976295	17.29436796	0.00782229	0.05782229
42	7.76158756	0.12883962	135.23175110	17.42320758	0.00739471	0.05739471
43	8.14966693	0.12270440	142.99333866	17.54591198	0.00699333	0.05699333
44	8.55715028	0.11686133	151.14300559	17.66277331	0.00661625	0.05661625
45	8.98500779	0.11129651	159.70015587	17.77406982	0.00626173	0.05626173
46	9.43425818	0.10599668	168.68516366	17.88006650	0.00592820	0.05592820
47	9.90597109	0.10094921	178.11942185	17.98101571	0.00561421	0.05561421
48	10.40126965	0.09614211	188.02539294	18.07715782	0.00531843	0.05531843
49	10.92133313	0.09156391	198.42666259	18.16872173	0.00503965	0.05503965
50	11.46739979	0.08720373	209.34799572	18.25592546	0.00477674	0.05477674

Rate $5\frac{1}{2}\%$	A Compound Interest	B Present Value	C Amount of Annuity	D Present Value of Annuity	E Sinking Fund	F Amortization
n	$(1 + i)^n$	$\frac{1}{(1 + i)^n}$	$s_{\overline{n}\mid i}$	$a_{\overline{n}\mid i}$	$\frac{1}{s_{\overline{n}\mid i}}$	$\frac{1}{a_{\overline{n}\mid i}}$
1	1.05500000	0.94786730	1.00000000	0.94786730	1.00000000	1.05500000
2	1.11302500	0.89845242	2.05500000	1.84631971	0.48661800	0.54161800
3	1.17424138	0.85161366	3.16802500	2.69793338	0.31565407	0.37065407
4	1.23882465	0.80721674	4.34226637	3.50515012	0.23029449	0.28529449
5	1.30696001	0.76513435	5.58109103	4.27028448	0.17917644	0.23417644
6	1.37884281	0.72524583	6.88805103	4.99553031	0.14517895	0.20017895
7	1.45467916	0.68743681	8.26689384	5.68296712	0.12096442	0.17596442
8	1.53468651	0.65159887	9.72157300	6.33456599	0.10286401	0.15786401
9	1.61909427	0.61762926	11.25625951	6.95219525	0.08883946	0.14383946
10	1.70814446	0.58543058	12.87535379	7.53762583	0.07766777	0.13266777
11	1.80209240	0.55491050	14.58349825	8.09253633	0.06857065	0.12357065
12	1.90120749	0.52598152	16.38559065	8.61851785	0.06102923	0.11602923
13	2.00577390	0.49856068	18.28679814	9.11707853	0.05468426	0.10968426
14	2.11609146	0.47256937	20.29257203	9.58964790	0.04927912	0.10427912
15	2.23247649	0.44793305	22.40866350	10.03758094	0.04462560	0.09962560
16	2.35526270	0.42458109	24.64113999	10.46216203	0.04058254	0.09558254
17	2.48480215	0.40244653	26.99640269	10.86460856	0.03704197	0.09204197
18	2.62146627	0.38146590	29.48120483	11.24607447	0.03391992	0.08891992
19	2.76564691	0.36157906	32.10267110	11.60765352	0.03115006	0.08615006
20	2.91775749	0.34272896	34.86831801	11.95038248	0.02867933	0.08367933
21	3.07823415	0.32486158	37.78607550	12.27524406	0.02646478	0.08146478
22	3.24753703	0.30792567	40.86430965	12.58316973	0.02447123	0.07947123
23	3.42615157	0.29187267	44.11184669	12.87504239	0.02266965	0.07766965
24	3.61458990	0.27665656	47.53799825	13.15169895	0.02103580	0.07603580
25	3.81339235	0.26223370	51.15258816	13.41393266	0.01954935	0.07454935
26	4.02312893	0.24856275	54.96598051	13.66249541	0.01819307	0.07319307
27	4.24440102	0.23560450	58.98910943	13.89809991	0.01695228	0.07195228
28	4.47784307	0.22332181	63.23351045	14.12142172	0.01581440	0.07081440
29	4.72412444	0.21167944	67.71135353	14.33310116	0.01476857	0.06976857
30	4.98395129	0.20064402	72.43547797	14.53374517	0.01380539	0.06880539
31	5.25806861	0.19018390	77.41942926	14.72392907	0.01291665	0.06791665
32	5.54726238	0.18026910	82.67749787	14.90419817	0.01209519	0.06709519
33	5.85236181	0.17087119	88.22476025	15.07506936	0.01133469	0.06633469
34	6.17424171	0.16196321	94.07712207	15.23703257	0.01062958	0.06562958
35	6.51382501	0.15351963	100.25136378	15.39055220	0.00997493	0.06497493
36	6.87208538	0.14551624	106.76518879	15.53606843	0.00936635	0.06436635
37	7.25005008	0.13793008	113.63727417	15.67399851	0.00879993	0.06379993
38	7.64880283	0.13073941	120.88732425	15.80473793	0.00827217	0.06327217
39	8.06948699	0.12392362	128.53612708	15.92866154	0.00777991	0.06277991
40	8.51330877	0.11746314	136.60561407	16.04612469	0.00732034	0.06232034
41	8.98154076	0.11133947	145.11892285	16.15746416	0.00689090	0.06189090
42	9.47552550	0.10553504	154.10046360	16.26299920	0.00648927	0.06148927
43	9.99667940	0.10003322	163.57598910	16.36303242	0.00611337	0.06111337
44	10.54649677	0.09481822	173.57266850	16.45785063	0.00576128	0.06076128
45	11.12655409	0.08987509	184.11916527	16.54772572	0.00543127	0.06043127
46	11.73851456	0.08518965	195.24571936	16.63291537	0.00512175	0.06012175
47	12.38413287	0.08074849	206.98423392	16.71366386	0.00483129	0.05983129
48	13.06526017	0.07653885	219.36836679	16.79020271	0.00455854	0.05955854
49	13.78384948	0.07254867	232.43362696	16.86275139	0.00430230	0.05930230
50	14.54196120	0.06876652	246.21747645	16.93151790	0.00406145	0.05906145

Rate 6%	A Compound Interest	B Present Value	C Amount of Annuity	D Present Value of Annuity	E Sinking Fund	F Amortization
n	$(1+i)^n$	$\frac{1}{(1+i)^n}$	$s_{\overline{n}\vert i}$	$a_{\overline{n}\vert i}$	$\frac{1}{s_{\overline{n}\vert i}}$	$\frac{1}{a_{\overline{n}\vert i}}$
1	1.06000000	0.94339623	1.00000000	0.94339623	1.00000000	1.06000000
2	1.12360000	0.88999644	2.06000000	1.83339267	0.48543689	0.54543689
3	1.19101600	0.83961928	3.18360000	2.67301195	0.31410981	0.37410981
4	1.26247696	0.79209366	4.37461600	3.46510561	0.22859149	0.28859149
5	1.33822558	0.74725817	5.63709296	4.21236379	0.17739640	0.23739640
6	1.41851911	0.70496054	6.97531854	4.91732433	0.14336263	0.20336263
7	1.50363026	0.66505711	8.39383765	5.58238144	0.11913502	0.17913502
8	1.59384807	0.62741237	9.89746791	6.20979381	0.10103594	0.16103594
9	1.68947896	0.59189846	11.49131598	6.80169227	0.08702224	0.14702224
10	1.79084770	0.55839478	13.18079494	7.36008705	0.07586796	0.13586796
11	1.89829856	0.52678753	14.97164264	7.88687458	0.06679294	0.12679294
12	2.01219647	0.49696936	16.86994120	8.38384394	0.05927703	0.11927703
13	2.13292826	0.46883902	18.88213767	8.85268296	0.05296011	0.11296011
14	2.26090396	0.44230096	21.01506593	9.29498393	0.04758491	0.10758491
15	2.39655819	0.41726506	23.27596988	9.71224899	0.04296276	0.10296276
16	2.54035168	0.39364628	25.67252808	10.10589527	0.03895214	0.09895214
17	2.69277279	0.37136442	28.21287976	10.47725969	0.03544480	0.09544480
18	2.85433915	0.35034379	30.90565255	10.82760348	0.03235654	0.09235654
19	3.02559950	0.33051301	33.75999170	11.15811649	0.02962086	0.08962086
20	3.20713547	0.31180473	36.78559120	11.46992122	0.02718456	0.08718456
21	3.39956360	0.29415540	39.99272668	11.76407662	0.02500455	0.08500455
22	3.60353742	0.27750510	43.39229028	12.04158172	0.02304557	0.08304557
23	3.81974966	0.26179726	46.99582769	12.30337898	0.02127848	0.08127848
24	4.04893464	0.24697855	50.81557735	12.55035753	0.01967900	0.07967900
25	4.29187072	0.23299863	54.86451200	12.78335616	0.01822672	0.07822672
26	4.54938296	0.21981003	59.15638272	13.00316619	0.01690435	0.07690435
27	4.82234594	0.20736795	63.70576568	13.21053414	0.01569717	0.07569717
28	5.11168670	0.19563014	68.52811162	13.40616428	0.01459255	0.07459255
29	5.41838790	0.18455674	73.63979832	13.59072102	0.01357961	0.07357961
30	5.74349117	0.17411013	79.05818622	13.76483115	0.01264891	0.07264891
31	6.08810064	0.16425484	84.80167739	13.92908599	0.01179222	0.07179222
32	6.45338668	0.15495740	90.88977803	14.08404339	0.01100234	0.07100234
33	6.84058988	0.14618622	97.34316471	14.23022961	0.01027293	0.07027293
34	7.25102528	0.13791153	104.18375460	14.36814114	0.00959843	0.06959843
35	7.68608679	0.13010522	111.43477987	14.49824636	0.00897386	0.06897386
36	8.14725200	0.12274077	119.12086666	14.62098713	0.00839483	0.06839483
37	8.63608712	0.11579318	127.26811866	14.73678031	0.00785743	0.06785743
38	9.15425235	0.10923885	135.90420578	14.84601916	0.00735812	0.06735812
39	9.70350749	0.10305552	145.05845813	14.94907468	0.00689377	0.06689377
40	10.28571794	0.09722219	154.76196562	15.04629687	0.00646154	0.06646154
41	10.90286101	0.09171905	165.04768356	15.13801592	0.00605886	0.06605886
42	11.55703267	0.08652740	175.95054457	15.22454332	0.00568342	0.06568342
43	12.25045463	0.08162962	187.50757724	15.30617294	0.00533312	0.06533312
44	12.98548191	0.07700908	199.75803188	15.38318202	0.00500606	0.06500606
45	13.76461083	0.07265007	212.74351379	15.45583209	0.00470050	0.06470050
46	14.59048748	0.06853781	226.50812462	15.52436990	0.00441485	0.06441485
47	15.46591673	0.06465831	241.09861210	15.58902821	0.00414768	0.06414768
48	16.39387173	0.06099840	256.56452882	15.65002661	0.00389765	0.06389765
49	17.37750403	0.05754566	272.95840055	15.70757227	0.00366356	0.06366356
50	18.42015427	0.05428836	290.33590458	15.76186064	0.00344429	0.06344429

Rate $6\frac{1}{2}\%$	A Compound Interest	B Present Value	C Amount of Annuity	D Present Value of Annuity	E Sinking Fund	F Amortization
n	$(1+i)^n$	$\frac{1}{(1+i)^n}$	$s_{\overline{n}\mid i}$	$a_{\overline{n}\mid i}$	$\frac{1}{s_{\overline{n}\mid i}}$	$\frac{1}{a_{\overline{n}\mid i}}$
1	1.06500000	0.93896714	1.00000000	0.93896714	1.00000000	1.06500000
2	1.13422500	0.88165928	2.06500000	1.82062642	0.48426150	0.54926150
3	1.20794963	0.82784909	3.19922500	2.64847551	0.31257570	0.37757570
4	1.28646635	0.77732309	4.40717462	3.42579860	0.22690274	0.29190274
5	1.37008666	0.72988084	5.69364098	4.15567944	0.17563454	0.24063454
6	1.45914230	0.68533412	7.06372764	4.84101356	0.14156831	0.20656831
7	1.55398655	0.64350621	8.52286994	5.48451977	0.11733137	0.18233137
8	1.65499567	0.60423119	10.07685648	6.08875096	0.09923730	0.16423730
9	1.76257039	0.56735323	11.73185215	6.65610419	0.08523803	0.15023803
10	1.87713747	0.53272604	13.49442254	7.18883022	0.07410469	0.13910469
11	1.99915140	0.50021224	15.37156001	7.68904246	0.06505521	0.13005521
12	2.12909624	0.46968285	17.37071141	8.15872532	0.05756817	0.12256817
13	2.26748750	0.44101676	19.49980765	8.59974208	0.05128256	0.11628256
14	2.41487418	0.41410025	21.76729515	9.01384233	0.04594048	0.11094048
15	2.57184101	0.38882652	24.18216933	9.40266885	0.04135278	0.10635278
16	2.73901067	0.36509533	26.75401034	9.76776418	0.03737757	0.10237757
17	2.91704637	0.34281251	29.49302101	10.11057670	0.03390633	0.09890633
18	3.10655438	0.32188969	32.41006738	10.43246638	0.03085461	0.09585461
19	3.30858691	0.30224384	35.51672176	10.73471022	0.02815575	0.09315575
20	3.52364506	0.28379703	38.82530867	11.01850725	0.02575640	0.09075640
21	3.75268199	0.26647608	42.34895373	11.28498333	0.02361333	0.08861333
22	3.99660632	0.25021228	46.10163573	11.53519562	0.02169120	0.08669120
23	4.25638573	0.23494111	50.09824205	11.77013673	0.01996078	0.08496078
24	4.53305081	0.22060198	54.35462778	11.99073871	0.01839770	0.08339770
25	4.82769911	0.20713801	58.88767859	12.19787673	0.01698148	0.08198148
26	5.14149955	0.19449579	63.71537769	12.39237251	0.01569480	0.08069480
27	5.47569702	0.18262515	68.85687725	12.57499766	0.01452288	0.07952288
28	5.83161733	0.17147902	74.33257427	12.74647668	0.01345305	0.07845305
29	6.21067245	0.16101316	80.16419159	12.90748984	0.01247440	0.07747440
30	6.61436616	0.15118607	86.37486405	13.05867591	0.01157744	0.07657744
31	7.04429996	0.14195875	92.98923021	13.20063465	0.01075393	0.07575393
32	7.50217946	0.13329460	100.03353017	13.33392925	0.00999665	0.07499665
33	7.98982113	0.12515925	107.53570963	13.45908850	0.00929924	0.07429924
34	8.50915950	0.11752042	115.52553076	13.57660892	0.00865610	0.07365610
35	9.06225487	0.11034781	124.03469026	13.68695673	0.00806226	0.07306226
36	9.65130143	0.10361297	133.09694513	13.79056970	0.00751332	0.07251332
37	10.27863603	0.09728917	142.74824656	13.88785887	0.00700534	0.07200534
38	10.94674737	0.09135134	153.02688259	13.97921021	0.00653480	0.07153480
39	11.65828595	0.08577590	163.97362996	14.06498611	0.00609854	0.07109854
40	12.41607453	0.08054075	175.63191590	14.14552687	0.00569373	0.07069373
41	13.22311938	0.07562512	188.04799044	14.22115199	0.00531779	0.07031779
42	14.08262214	0.07100950	201.27110981	14.29216149	0.00496842	0.06996842
43	14.99799258	0.06667559	215.35373195	14.35883708	0.00464352	0.06964352
44	15.97286209	0.06260619	230.35172453	14.42144327	0.00434119	0.06934119
45	17.01109813	0.05878515	246.32458662	14.48022842	0.00405968	0.06905968
46	18.11681951	0.05519733	263.33568475	14.53542575	0.00379743	0.06879743
47	19.29441278	0.05182848	281.45250426	14.58725422	0.00355300	0.06855300
48	20.54854961	0.04866524	300.74691704	14.63591946	0.00332505	0.06832505
49	21.88420533	0.04569506	321.29546665	14.68161451	0.00311240	0.06811240
50	23.30667868	0.04290616	343.17967198	14.72452067	0.00291393	0.06791393

Rate 7%	A Compound Interest	B Present Value	C Amount of Annuity	D Present Value of Annuity	E Sinking Fund	F Amortization
n	$(1+i)^n$	$\frac{1}{(1+i)^n}$	$s_{\overline{n}\rvert i}$	$a_{\overline{n}\rvert i}$	$\frac{1}{s_{\overline{n}\rvert i}}$	$\frac{1}{a_{\overline{n}\rvert i}}$
1	1.07000000	0.93457944	1.00000000	0.93457944	1.00000000	1.07000000
2	1.14490000	0.87343873	2.07000000	1.80801817	0.48309179	0.55309179
3	1.22504300	0.81629788	3.21490000	2.62431604	0.31105167	0.38105167
4	1.31079601	0.76289521	4.43994300	3.38721126	0.22522812	0.29522812
5	1.40255173	0.71298618	5.75073901	4.10019744	0.17389069	0.24389069
6	1.50073035	0.66634222	7.15329074	4.76653966	0.13979580	0.20979580
7	1.60578148	0.62274974	8.65402109	5.38928940	0.11555322	0.18555322
8	1.71818618	0.58200910	10.25980257	5.97129851	0.09746776	0.16746776
9	1.83845921	0.54393374	11.97798875	6.51523225	0.08348647	0.15348647
10	1.96715136	0.50834929	13.81644796	7.02358154	0.07237750	0.14237750
11	2.10485195	0.47509280	15.78359932	7.49867434	0.06335690	0.13335690
12	2.25219159	0.44401196	17.88845127	7.94268630	0.05590199	0.12590199
13	2.40984500	0.41496445	20.14064286	8.35765074	0.04965085	0.11965085
14	2.57853415	0.38781724	22.55048786	8.74546799	0.04434494	0.11434494
15	2.75903154	0.36244602	25.12902201	9.10791401	0.03979462	0.10979462
16	2.95216375	0.33873460	27.88805355	9.44664860	0.03585765	0.10585765
17	3.15881521	0.31657439	30.84021730	9.76322299	0.03242519	0.10242519
18	3.37993228	0.29586392	33.99903251	10.05908691	0.02941260	0.09941260
19	3.61652754	0.27650833	37.37896479	10.33559524	0.02675301	0.09675301
20	3.86968446	0.25841900	40.99549232	10.59401425	0.02439293	0.09439293
21	4.14056237	0.24151309	44.86517678	10.83552733	0.02228900	0.09228900
22	4.43040174	0.22571317	49.00573916	11.06124050	0.02040577	0.09040577
23	4.74052986	0.21094688	53.43614090	11.27218738	0.01871393	0.08871393
24	5.07236695	0.19714662	58.17667076	11.46933400	0.01718902	0.08718902
25	5.42743264	0.18424918	63.24903772	11.65358318	0.01581052	0.08581052
26	5.80735292	0.17219549	68.67647036	11.82577867	0.01456103	0.08456103
27	6.21386763	0.16093037	74.48382328	11.98670904	0.01342573	0.08342573
28	6.64883836	0.15040221	80.69769091	12.13711125	0.01239193	0.08239193
29	7.11425705	0.14056282	87.34652927	12.27767407	0.01144865	0.08144865
30	7.61225504	0.13136712	94.46078632	12.40904118	0.01058640	0.08058640
31	8.14511290	0.12277301	102.07304137	12.53181419	0.00979691	0.07979691
32	8.71527080	0.11474113	110.21815426	12.64655532	0.00907292	0.07907292
33	9.32533975	0.10723470	118.93342506	12.75379002	0.00840807	0.07840807
34	9.97811354	0.10021934	128.25876481	12.85400936	0.00779674	0.07779674
35	10.67658148	0.09366294	138.23687835	12.94767230	0.00723396	0.07723396
36	11.42394219	0.08753546	148.91345984	13.03520776	0.00671531	0.07671531
37	12.22361814	0.08180884	160.33740202	13.11701660	0.00623685	0.07623685
38	13.07927141	0.07645686	172.56102017	13.19347345	0.00579505	0.07579505
39	13.99482041	0.07145501	185.64029158	13.26492846	0.00538676	0.07538676
40	14.97445784	0.06678038	199.63511199	13.33170884	0.00500914	0.07500914
41	16.02266989	0.06241157	214.60956983	13.39412041	0.00465962	0.07465962
42	17.14425678	0.05832857	230.63223972	13.45244898	0.00433591	0.07433591
43	18.34435475	0.05451268	247.77649650	13.50696167	0.00403590	0.07403590
44	19.62845959	0.05094643	266.12085125	13.55790810	0.00375769	0.07375769
45	21.00245176	0.04761349	285.74931084	13.60552159	0.00349957	0.07349957
46	22.47262338	0.04449859	306.75176260	13.65002018	0.00325996	0.07325996
47	24.04570702	0.04158747	329.22438598	13.69160764	0.00303744	0.07303744
48	25.72890651	0.03886679	353.27009300	13.73047443	0.00283070	0.07283070
49	27.52992997	0.03632410	378.99899951	13.76679853	0.00263853	0.07263853
50	29.45702506	0.03394776	406.52892947	13.80074629	0.00245985	0.07245985

Rate $7\frac{1}{2}\%$	A Compound Interest	B Present Value	C Amount of Annuity	D Present Value of Annuity	E Sinking Fund	F Amortization
n	$(1+i)^n$	$\frac{1}{(1+i)^n}$	$s_{\overline{n}\|i}$	$a_{\overline{n}\|i}$	$\frac{1}{s_{\overline{n}\|i}}$	$\frac{1}{a_{\overline{n}\|i}}$
1	1.07500000	0.93023256	1.00000000	0.93023256	1.00000000	1.07500000
2	1.15562500	0.86533261	2.07500000	1.79556517	0.48192771	0.55692771
3	1.24229688	0.80496057	3.23062500	2.60052574	0.30953763	0.38453763
4	1.33546914	0.74880053	4.47292187	3.34932627	0.22356751	0.29856751
5	1.43562933	0.69655863	5.80839102	4.04588490	0.17216472	0.24716472
6	1.54330153	0.64796152	7.24402034	4.69384642	0.13804489	0.21304489
7	1.65904914	0.60275490	8.78732187	5.29660132	0.11380032	0.18880032
8	1.78347783	0.56070223	10.44637101	5.85730355	0.09572702	0.17072702
9	1.91723866	0.52158347	12.22984883	6.37888703	0.08176716	0.15676716
10	2.06103156	0.48519393	14.14708750	6.86408096	0.07068593	0.14568593
11	2.21560893	0.45134319	16.20811906	7.31542415	0.06169747	0.13669747
12	2.38177960	0.41985413	18.42372799	7.73527827	0.05427783	0.12927783
13	2.56041307	0.39056198	20.80550759	8.12584026	0.04806420	0.12306420
14	2.75244405	0.36331347	23.36592066	8.48915373	0.04279737	0.11779737
15	2.95887735	0.33796602	26.11836470	8.82711975	0.03828724	0.11328724
16	3.18079315	0.31438699	29.07724206	9.14150674	0.03439116	0.10939116
17	3.41935264	0.29245302	32.25803521	9.43395976	0.03100003	0.10600003
18	3.67580409	0.27204932	35.67738785	9.70600908	0.02802896	0.10302896
19	3.95148940	0.25306913	39.35319194	9.95907821	0.02541090	0.10041090
20	4.24785110	0.23541315	43.30468134	10.19449136	0.02309219	0.09809219
21	4.56643993	0.21898897	47.55253244	10.41348033	0.02102937	0.09602937
22	4.90892293	0.20371067	52.11897237	10.61719101	0.01918687	0.09418687
23	5.27709215	0.18949830	57.02789530	10.80668931	0.01753528	0.09253528
24	5.67287406	0.17627749	62.30498744	10.98296680	0.01605008	0.09105008
25	6.09833961	0.16397906	67.97786150	11.14694586	0.01471067	0.08971067
26	6.55571508	0.15253866	74.07620112	11.29948452	0.01349961	0.08849961
27	7.04739371	0.14189643	80.63191620	11.44138095	0.01240204	0.08740204
28	7.57594824	0.13199668	87.67930991	11.57337763	0.01140520	0.08640520
29	8.14414436	0.12278761	95.25525816	11.69616524	0.01049811	0.08549811
30	8.75495519	0.11422103	103.39940252	11.81038627	0.00967124	0.08467124
31	9.41157683	0.10625212	112.15435771	11.91663839	0.00891628	0.08391628
32	10.11744509	0.09883918	121.56593454	12.01547757	0.00822599	0.08322599
33	10.87625347	0.09194343	131.68337963	12.10742099	0.00759397	0.08259397
34	11.69197248	0.08552877	142.55963310	12.19294976	0.00701461	0.08201461
35	12.56887042	0.07956164	154.25160558	12.27251141	0.00648291	0.08148291
36	13.51153570	0.07401083	166.82047600	12.34652224	0.00599447	0.08099447
37	14.52490088	0.06884729	180.33201170	12.41536952	0.00554533	0.08054533
38	15.61426844	0.06404399	194.85691258	12.47941351	0.00513197	0.08013197
39	16.78533858	0.05957580	210.47118102	12.53898931	0.00475124	0.07975124
40	18.04423897	0.05541935	227.25651960	12.59440866	0.00440031	0.07940031
41	19.39755689	0.05155288	245.30075857	12.64596155	0.00407663	0.07907663
42	20.85237366	0.04795617	264.69831546	12.69391772	0.00377789	0.07877789
43	22.41630168	0.04461039	285.55068912	12.73852811	0.00350201	0.07850201
44	24.09752431	0.04149804	307.96699080	12.78002615	0.00324710	0.07824710
45	25.90483863	0.03860283	332.06451511	12.81862898	0.00301146	0.07801146
46	27.84770153	0.03590961	357.96935375	12.85453858	0.00279354	0.07779354
47	29.93627915	0.03340428	385.81705528	12.88794287	0.00259190	0.07759190
48	32.18150008	0.03107375	415.75333442	12.91901662	0.00240527	0.07740527
49	34.59511259	0.02890582	447.93483451	12.94792244	0.00223247	0.07723247
50	37.18974603	0.02688913	482.52994709	12.97481157	0.00207241	0.07707241

Rate 8%	A Compound Interest	B Present Value	C Amount of Annuity	D Present Value of Annuity	E Sinking Fund	F Amortization
n	$(1+i)^n$	$\frac{1}{(1+i)^n}$	$s_{\overline{n}\rvert i}$	$a_{\overline{n}\rvert i}$	$\frac{1}{s_{\overline{n}\rvert i}}$	$\frac{1}{a_{\overline{n}\rvert i}}$
1	1.08000000	0.92592593	1.00000000	0.92592593	1.00000000	1.08000000
2	1.16640000	0.85733882	2.08000000	1.78326475	0.48076923	0.56076923
3	1.25971200	0.79383224	3.24640000	2.57709699	0.30803351	0.38803351
4	1.36048896	0.73502985	4.50611200	3.31212684	0.22192080	0.30192080
5	1.46932808	0.68058320	5.86660096	3.99271004	0.17045645	0.25045645
6	1.58687432	0.63016963	7.33592904	4.62287966	0.13631539	0.21631539
7	1.71382427	0.58349040	8.92280336	5.20637006	0.11207240	0.19207240
8	0.85093021	0.54026888	10.63662763	5.74663894	0.09401476	0.17401476
9	1.99900463	0.50024897	12.48755784	6.24688791	0.08007971	0.16007971
10	2.15892500	0.46319349	14.48656247	6.71008140	0.06902949	0.14902949
11	2.33163900	0.42888286	16.64548746	7.13896426	0.06007634	0.14007634
12	2.51817012	0.39711376	18.97712646	7.53607802	0.05269502	0.13269502
13	2.71962373	0.36769792	21.49529658	7.90377594	0.04652181	0.12652181
14	2.93719362	0.34046104	24.21492030	8.24423698	0.04129685	0.12129685
15	3.17216911	0.31524170	27.15211393	8.55947869	0.03682954	0.11682954
16	3.42594264	0.29189047	30.32428304	8.85136916	0.03297687	0.11297687
17	3.70001805	0.27026895	33.75022569	9.12163811	0.02962943	0.10962943
18	3.99601950	0.25024903	37.45024374	9.37188714	0.02670210	0.10670210
19	4.31570106	0.23171206	41.44626324	9.60359920	0.02412763	0.10412763
20	4.66095714	0.21454821	45.76196430	9.81814741	0.02185221	0.10185221
21	5.03383372	0.19865575	50.42292144	10.01680316	0.01983225	0.09983225
22	5.43654041	0.18394051	55.45675516	10.20074366	0.01803207	0.09803207
23	5.87146365	0.17031528	60.89329557	10.37105895	0.01642217	0.09642217
24	6.34118074	0.15769934	66.76475922	10.52875828	0.01497796	0.09497796
25	6.84847520	0.14601790	73.10593995	10.67477619	0.01367878	0.09367878
26	7.39635321	0.13520176	79.95441515	10.80997795	0.01250713	0.09250713
27	7.98806147	0.12518682	87.35076836	10.93516477	0.01144810	0.09144810
28	8.62710639	0.11591372	95.33882983	11.05107849	0.01048891	0.09048891
29	9.31727490	0.10732752	103.96593622	11.15840601	0.00961854	0.08961854
30	10.06265689	0.09937733	113.28321111	11.25778334	0.00882743	0.08882743
31	10.86766944	0.09201605	123.34586800	11.34979939	0.00810728	0.08810728
32	11.73708300	0.08520005	134.21353744	11.43499944	0.00745081	0.08745081
33	12.67604964	0.07888893	145.95062044	11.51388837	0.00685163	0.08685163
34	13.69013361	0.07304531	158.62667007	11.58693367	0.00630411	0.08630411
35	14.78534429	0.06763454	172.31680368	11.65456822	0.00580326	0.08580326
36	15.96817184	0.06262458	187.10214797	11.71719279	0.00534467	0.08534467
37	17.24562558	0.05798572	203.07031981	11.77517851	0.00492440	0.08492440
38	18.62527563	0.05369048	220.31594540	11.82886899	0.00453894	0.08453894
39	20.11529768	0.04971341	238.94122103	11.87858240	0.00418513	0.08418513
40	21.72452150	0.04603093	259.05651871	11.92461333	0.00386016	0.08386016
41	23.46248322	0.04262123	280.78104021	11.96723457	0.00356149	0.08356149
42	25.33948187	0.03946411	304.24352342	12.00669867	0.00328684	0.08328684
43	27.36664042	0.03654084	329.58300530	12.04323951	0.00303414	0.08303414
44	29.55597166	0.03383411	356.94964572	12.07707362	0.00280152	0.08280152
45	31.92044939	0.03132788	386.50561738	12.10840150	0.00258728	0.08258728
46	34.47408534	0.02900730	418.42606677	12.13740880	0.00238991	0.08238991
47	37.23201217	0.02685861	452.90015211	12.16426741	0.00220799	0.08220799
48	40.21057314	0.02486908	490.13216428	12.18913649	0.00204027	0.08204027
49	43.42741899	0.02302693	530.34273742	12.21216341	0.00188557	0.08188557
50	46.90161251	0.02132123	573.77015642	12.23348464	0.00174286	0.08174286

Rate 9%	A Compound Interest	B Present Value	C Amount of Annuity	D Present Value of Annuity	E Sinking Fund	F Amortization
n	$(1 + i)^n$	$\frac{1}{(1 + i)^n}$	$s_{\overline{n}\|i}$	$a_{\overline{n}\|i}$	$\frac{1}{s_{\overline{n}\|i}}$	$\frac{1}{a_{\overline{n}\|i}}$
1	1.09000000	0.91743119	1.00000000	0.91743119	1.00000000	1.09000000
2	1.18810000	0.84167999	2.09000000	1.75911119	0.47846890	0.56846890
3	1.29502900	0.77218348	3.27810000	2.53129467	0.30505476	0.39505476
4	1.41158161	0.70842521	4.57312900	3.23971988	0.21866866	0.30866866
5	1.53862395	0.64993139	5.98471061	3.88965126	0.16709246	0.25709246
6	1.67710011	0.59626733	7.52333456	4.48591859	0.13291978	0.22291978
7	1.82803912	0.54703424	9.20043468	5.03295284	0.10869052	0.19869052
8	1.99256264	0.50186628	11.02847380	5.53481911	0.09067438	0.18067438
9	2.17189328	0.46042778	13.02103644	5.99524689	0.07679880	0.16679880
10	2.36736367	0.42241081	15.19292972	6.41765770	0.06582009	0.15582009
11	2.58042641	0.38753285	17.56029339	6.80519055	0.05694666	0.14694666
12	2.81266478	0.35553473	20.14071980	7.16072528	0.04965066	0.13965066
13	3.06580461	0.32617865	22.95338458	7.48690392	0.04356656	0.13356656
14	3.34172703	0.29924647	26.01918919	7.78615039	0.03843317	0.12843317
15	3.64248246	0.27453804	29.36091622	8.06068843	0.03405888	0.12405888
16	3.97030588	0.25186976	33.00339868	8.31255819	0.03029991	0.12029991
17	4.32763341	0.23107318	36.97370456	8.54363137	0.02704625	0.11704625
18	4.71712042	0.21199374	41.30133797	8.75562511	0.02421229	0.11421229
19	5.14166125	0.19448967	46.01845839	8.95011478	0.02173041	0.11173041
20	5.60441077	0.17843089	51.16011964	9.12854567	0.01954648	0.10954648
21	6.10880774	0.16369806	56.76453041	9.29224373	0.01761663	0.10761663
22	6.65860043	0.15018171	62.87333815	9.44242544	0.01590499	0.10590499
23	7.25787447	0.13778139	69.53193858	9.58020683	0.01438188	0.10438188
24	7.91108317	0.12640494	76.78981305	9.70661177	0.01302256	0.10302256
25	8.62308066	0.11596784	84.70089623	9.82257960	0.01180625	0.10180625
26	9.39915792	0.10639251	93.32397689	9.92897211	0.01071536	0.10071536
27	10.24508213	0.09760781	102.72313481	10.02657992	0.00973491	0.09973491
28	11.16713952	0.08954845	112.96821694	10.11612837	0.00885205	0.09885205
29	12.17218208	0.08215454	124.13535646	10.19828291	0.00805572	0.09805572
30	13.26767847	0.07537114	136.30753855	10.27365404	0.00733635	0.09733635
31	14.46176953	0.06914783	149.57521702	10.34280187	0.00668560	0.09668560
32	15.76332879	0.06343838	164.03698655	10.40624025	0.00609619	0.09609619
33	17.18202838	0.05820035	179.80031534	10.46444060	0.00556173	0.09556173
34	18.72841093	0.05339481	196.98234372	10.51783541	0.00507660	0.09507660
35	20.41396792	0.04898607	215.71075465	10.56682148	0.00463584	0.09463584
36	22.25122503	0.04494135	236.12472257	10.61176282	0.00423505	0.09423505
37	24.25383528	0.04123059	258.37594760	10.65299342	0.00387033	0.09387033
38	26.43668046	0.03782623	282.62978288	10.69081965	0.00353820	0.09353820
39	28.81598170	0.03470296	309.06646334	10.72552261	0.00323555	0.09323555
40	31.40942005	0.03183758	337.88244504	10.75736020	0.00295961	0.09295961
41	34.23626789	0.02920879	369.29186510	10.78656899	0.00270789	0.09270789
42	37.31753197	0.02679706	403.52813296	10.81336604	0.00247814	0.09247814
43	40.67610984	0.02458446	440.84566492	10.83795050	0.00226837	0.09226837
44	44.33695973	0.02255455	481.52177477	10.86050504	0.00207675	0.09207675
45	48.32728610	0.02069224	525.85873450	10.88119729	0.00190165	0.09190165
46	52.67674185	0.01898371	574.18602060	10.90018100	0.00174160	0.09174160
47	57.41764862	0.01741625	626.86276245	10.91759725	0.00159525	0.09159525
48	62.58523700	0.01597821	684.28041107	10.93357546	0.00146139	0.09146139
49	68.21790833	0.01465891	746.86564807	10.94823436	0.00133893	0.09133893
50	74.35752008	0.01344854	815.08355640	10.96168290	0.00122687	0.09122687

Rate 10%	A Compound Interest	B Present Value	C Amount of Annuity	D Present Value of Annuity	E Sinking Fund	F Amortization
n	$(1 + i)^n$	$\frac{1}{(1 + i)^n}$	$s_{\overline{n}\rvert i}$	$a_{\overline{n}\rvert i}$	$\frac{1}{s_{\overline{n}\rvert i}}$	$\frac{1}{a_{\overline{n}\rvert i}}$
1	1.10000000	0.90909091	1.00000000	0.90909091	1.00000000	1.10000000
2	1.21000000	0.82644628	2.10000000	1.73553719	0.47619048	0.57619048
3	1.33100000	0.75131480	3.31000000	2.48685199	0.30211480	0.40211480
4	1.46410000	0.68301346	4.64100000	3.16986545	0.21547080	0.31547080
5	1.61051000	0.62092132	6.10510000	3.79078677	0.16379748	0.26379748
6	1.77156100	0.56447393	7.71561000	4.35526070	0.12960738	0.22960738
7	1.94871710	0.51315812	9.48717100	4.86841882	0.10540550	0.20540550
8	2.14358881	0.46650738	11.43588810	5.33492620	0.08744402	0.18744402
9	2.35794769	0.42409762	13.57947691	5.75902382	0.07364054	0.17364054
10	2.59374246	0.38554329	15.93742460	6.14456711	0.06274539	0.16274539
11	2.85311671	0.35049390	18.53116706	6.49506101	0.05396314	0.15396314
12	3.13842838	0.31863082	21.38428377	6.81369182	0.04676332	0.14676332
13	3.45227121	0.28966438	24.52271214	7.10335620	0.04077852	0.14077852
14	3.79749834	0.26333125	27.97498336	7.36668746	0.03574622	0.13574622
15	4.17724817	0.23939205	31.77248169	7.60607951	0.03147378	0.13147378
16	4.59497299	0.21762914	35.94972986	7.82370864	0.02781662	0.12781662
17	5.05447028	0.19784467	40.54470285	8.02155331	0.02466413	0.12466413
18	5.55991731	0.17985879	45.59917313	8.20141210	0.02193022	0.12193022
19	6.11590904	0.16350799	51.15909045	8.36492009	0.01954687	0.11954687
20	6.72749995	0.14864363	57.27499949	8.51356372	0.01745962	0.11745962
21	7.40024994	0.13513057	64.00249944	8.64869429	0.01562439	0.11562439
22	8.14027494	0.12284597	71.40274939	8.77154026	0.04100506	0.11400506
23	8.95430243	0.11167816	79.54302433	8.88321842	0.01257181	0.11257181
24	9.84973268	0.10152560	88.49732676	8.98474402	0.01129978	0.11129978
25	10.83470594	0.09229600	98.34705943	9.07704002	0.01016807	0.11016807
26	11.91817654	0.08390545	109.18176538	9.16094547	0.00915904	0.10915904
27	13.10999419	0.07627768	121.09994191	9.23722316	0.00825764	0.10825764
28	14.42099361	0.06934335	134.20993611	9.30656651	0.00745101	0.10745101
29	15.86309297	0.06303941	148.63092972	9.36960591	0.00672807	0.10672807
30	17.44940227	0.05730855	164.49402269	9.42691447	0.00607925	0.10607925
31	19.19434250	0.05209868	181.94342496	9.47901315	0.00549621	0.10549621
32	21.11377675	0.04736244	201.13776745	9.52637559	0.00497172	0.10497172
33	23.22515442	0.04305676	222.25154420	9.56943236	0.00449941	0.10449941
34	25.54766986	0.03914251	245.47669862	9.60857487	0.00407371	0.10407371
35	28.10243685	0.03558410	271.02436848	9.64415897	0.00368971	0.10368971
36	30.91268053	0.03234918	299.12680533	9.67650816	0.00334306	0.10334306
37	34.00394859	0.02940835	330.03948586	9.70591651	0.00302994	0.10302994
38	37.40434344	0.02673486	364.04343445	9.73265137	0.00274692	0.10274692
39	41.14477779	0.02430442	401.44777789	9.75695579	0.00249098	0.10249098
40	45.25925557	0.02209493	442.59255568	9.77905072	0.00225941	0.10225941
41	49.78518112	0.02008630	487.85181125	9.79913702	0.00204980	0.10204980
42	54.76369924	0.01826027	537.63699237	9.81739729	0.00185999	0.10185999
43	60.24006916	0.01660025	592.40069161	9.83399753	0.00168805	0.10168805
44	66.26407608	0.01509113	652.64076077	9.84908867	0.00153224	0.10153224
45	72.89048369	0.01371921	718.90483685	9.86280788	0.00139100	0.10139100
46	80.17953205	0.01247201	791.79532054	9.87527989	0.00126295	0.10126295
47	88.19748526	0.01133819	871.97485259	9.88661808	0.00114682	0.10114682
48	97.01723378	0.01030745	960.17233785	9.89692553	0.00104148	0.10104148
49	106.71895716	0.00937041	1057.18957163	9.90629594	0.00094590	0.10094590
50	117.39085288	0.00851855	1163.90852880	9.91481449	0.00085917	0.10085917

Rate 11%	A Compound Interest	B Present Value	C Amount of Annuity	D Present Value of Annuity	E Sinking Fund	F Amortization
n	$(1+i)^n$	$\frac{1}{(1+i)^n}$	$s_{\overline{n}\rvert i}$	$a_{\overline{n}\rvert i}$	$\frac{1}{s_{\overline{n}\rvert i}}$	$\frac{1}{a_{\overline{n}\rvert i}}$
1	1.11000000	0.90090090	1.00000000	0.90090090	1.00000000	1.11000000
2	1.23210000	0.81162243	2.11000000	1.71252333	0.47393365	0.58393365
3	1.36763100	0.73119138	3.34210000	2.44371472	0.29921307	0.40921307
4	1.51807041	0.65873097	4.70973100	3.10244569	0.21232635	0.32232635
5	1.68505816	0.59345133	6.22780141	3.69589702	0.16057031	0.27057031
6	1.87041455	0.53464084	7.91285957	4.23053785	0.12637656	0.23637656
7	2.07616015	0.48165841	9.78327412	4.71219626	0.10221527	0.21221527
8	2.30453777	0.43392650	11.85943427	5.14612276	0.08432105	0.19432105
9	2.55803692	0.39092477	14.16397204	5.53704753	0.07060166	0.18060166
10	2.83942099	0.35218448	16.72200896	5.88923201	0.05980143	0.16980143
11	3.15175729	0.31728331	19.56142995	6.20651533	0.05112101	0.16112101
12	3.49845060	0.28584082	22.71318724	6.49235615	0.04402729	0.15402729
13	3.88328016	0.25751426	26.21163784	6.74987040	0.03815099	0.14815099
14	4.31044098	0.23199482	30.09491800	6.98186523	0.03322820	0.14322820
15	4.78458949	0.20900435	34.40535898	7.19086958	0.02906524	0.13906524
16	5.31089433	0.18829220	39.18994847	7.37916178	0.02551675	0.13551675
17	5.89509271	0.16963262	44.50084281	7.54879440	0.02247148	0.13247148
18	6.54355291	0.15282218	50.39593551	7.70161657	0.01984287	0.12984287
19	7.26334373	0.13767764	56.93948842	7.83929421	0.01756250	0.12756250
20	8.06231154	0.12403391	64.20283215	7.96332812	0.01557564	0.12557564
21	8.94916581	0.11174226	72.26514368	8.07507038	0.01383793	0.12383793
22	9.93357404	0.10066870	81.21430949	8.17573908	0.01231310	0.12231310
23	11.02626719	0.09069252	91.14788353	8.26643160	0.01097118	0.12097118
24	12.23915658	0.08170498	102.17415072	8.34813658	0.00978721	0.11978721
25	13.58546380	0.07360809	114.41330730	8.42174466	0.00874024	0.11874024
26	15.07986482	0.06631359	127.99877110	8.48805826	0.00781258	0.11781258
27	16.73864995	0.05974197	143.07863592	8.54780023	0.00698916	0.11698916
28	18.57990145	0.05382160	159.81728587	8.60162183	0.00625715	0.11625715
29	20.62369061	0.04848793	178.39718732	8.65010976	0.00560547	0.11560547
30	22.89229657	0.04368282	199.02087793	8.69379257	0.00502460	0.11502460
31	25.41044919	0.03935389	221.91317450	8.73314646	0.00450627	0.11450627
32	28.20559861	0.03545395	247.32362369	8.76860042	0.00404329	0.11404329
33	31.30821445	0.03194050	275.52922230	8.80054092	0.00362938	0.11362938
34	34.75211804	0.02877522	306.83743675	8.82931614	0.00325905	0.11325905
35	38.57485103	0.02592363	341.58955480	8.85523977	0.00292749	0.11292749
36	42.81808464	0.02335462	380.16440582	8.87859438	0.00263044	0.11263044
37	47.52807395	0.02104020	422.98249046	8.89963458	0.00236416	0.11236416
38	52.75616209	0.01895513	470.51056441	8.91858971	0.00212535	0.11212535
39	58.55933991	0.01707670	523.26672650	8.93566641	0.00191107	0.11191107
40	65.00086731	0.01538441	581.82606641	8.95105082	0.00171873	0.11171873
41	72.15096271	0.01385983	646.82693372	8.96491065	0.00154601	0.11154601
42	80.08756861	0.01248633	718.97789643	8.97739698	0.00139086	0.11139086
43	88.89720115	0.01124895	799.06546504	8.98864593	0.00125146	0.11125146
44	98.67589328	0.01013419	887.96266619	8.99878011	0.00112617	0.11112617
45	109.53024154	0.00912990	986.63855947	9.00791001	0.00101354	0.11101354
46	121.57856811	0.00822513	1096.16880101	9.01613515	0.00091227	0.11091227
47	134.95221060	0.00741003	1217.74736912	9.02354518	0.00082119	0.11082119
48	149.79695377	0.00667570	1352.69957973	9.03022088	0.00073926	0.11073926
49	166.27461868	0.00601415	1502.49653350	9.03623503	0.00066556	0.11066556
50	184.56482674	0.00541815	1668.77115218	9.04165318	0.00059924	0.11059924

Rate 12	A Compound Interest	B Present Value	C Amount of Annuity	D Present Value of Annuity	E Sinking Fund	F Amortization
n	$(1 + i)^n$	$\frac{1}{(1 + i)^n}$	$s_{\overline{n}\rvert i}$	$a_{\overline{n}\rvert i}$	$\frac{1}{s_{\overline{n}\rvert i}}$	$\frac{1}{a_{\overline{n}\rvert i}}$
1	1.12000000	0.89285714	1.00000000	0.89285714	1.00000000	1.12000000
2	1.25440000	0.79719388	2.12000000	1.69005102	0.47169811	0.59169811
3	1.40492800	0.71178025	3.37440000	2.40183127	0.29634898	0.41634898
4	1.57351936	0.63551808	4.77932800	3.03734935	0.20923444	0.32923444
5	1.76234168	0.56742686	6.35284736	3.60477620	0.15740973	0.27740973
6	1.97382269	0.50663112	8.11518904	4.11140732	0.12322572	0.24322572
7	2.21068141	0.45234922	10.08901173	4.56375654	0.09911774	0.21911774
8	2.47596318	0.40388323	12.29969314	4.96763977	0.08130284	0.20130284
9	2.77307876	0.36061002	14.77565631	5.32824979	0.06767889	0.18767889
10	3.10584821	0.32197324	17.54873507	5.65022303	0.05698416	0.17698416
11	3.47854999	0.28747610	20.65458328	5.93769913	0.04841540	0.16841540
12	3.89597599	0.25667509	24.13313327	6.19437423	0.04143681	0.16143681
13	4.36349311	0.22917419	28.02910926	6.42354842	0.03567720	0.15567720
14	4.88711229	0.20461981	32.39260238	6.62816823	0.03087125	0.15087125
15	5.47356576	0.18269626	37.27971466	6.81086449	0.02682424	0.14682424
16	6.13039365	0.16312166	42.75328042	6.97398615	0.02339002	0.14339002
17	6.86604098	0.14564434	48.88367407	7.11963049	0.02045673	0.14045673
18	7.68996580	0.13003959	55.74971496	7.24967008	0.01793731	0.13793731
19	8.61276169	0.11610678	63.43968075	7.36577686	0.01576300	0.13576300
20	9.64629309	0.10366677	72.05244244	7.46944362	0.01387878	0.13387878
21	10.80384826	0.09255961	81.69873554	7.56200324	0.01224009	0.13224009
22	12.10031006	0.08264251	92.50258380	7.64464575	0.01081051	0.13081051
23	13.55234726	0.07378796	104.60289386	7.71843370	0.00955996	0.12955996
24	15.17862893	0.06588210	118.15524112	7.78431581	0.00846344	0.12846344
25	17.00006441	0.05882331	133.33387006	7.84313911	0.00749997	0.12749997
26	19.04007214	0.05252081	150.33393446	7.89565992	0.00665186	0.12665186
27	21.32488079	0.04689358	169.37400660	7.94255350	0.00590409	0.12590409
28	23.88386649	0.04186927	190.69888739	7.98442277	0.00524387	0.12524387
29	26.74993047	0.03738327	214.58275388	8.02180604	0.00466021	0.12466021
30	29.95992212	0.03337792	241.33268434	8.05518397	0.00414366	0.12414366
31	33.55511278	0.02980172	271.29260646	8.08498569	0.00368606	0.12368606
32	37.58172631	0.02660868	304.84771924	8.11159436	0.00328033	0.12328033
33	42.09153347	0.02375775	342.42944555	8.13535211	0.00292031	0.12292031
34	47.14251748	0.02121227	384.52097901	8.15656438	0.00260064	0.12260064
35	52.79961958	0.01893953	431.66349649	8.17550391	0.00231662	0.12231662
36	59.13557393	0.01691029	484.46311607	8.19241421	0.00206414	0.12206414
37	66.23184280	0.01509848	543.59869000	8.20751269	0.00183959	0.12183959
38	74.17966394	0.01348078	609.83053280	8.22099347	0.00163980	0.12163980
39	83.08122361	0.01203641	684.01019674	8.23302988	0.00146197	0.12146197
40	93.05097044	0.01074680	767.09142034	8.24377668	0.00130363	0.12130363
41	104.21708689	0.00959536	860.14239079	8.25337204	0.00116260	0.12116260
42	116.72313732	0.00856728	964.35947768	8.26193932	0.00103696	0.12103696
43	130.72991380	0.00764936	1081.08261500	8.26958868	0.00092500	0.12092500
44	146.41750346	0.00682978	1211.81252880	8.27641846	0.00082521	0.12082521
45	163.98760387	0.00609802	1358.23003226	8.28251648	0.00073625	0.12073625
46	183.66611634	0.00544466	1522.21763613	8.28796115	0.00065694	0.12065694
47	205.70605030	0.00486131	1705.88375247	8.29282245	0.00058621	0.12058621
48	230.39077633	0.00434045	1911.58980276	8.29716290	0.00052312	0.12052312
49	258.03766949	0.00387540	2142.98057909	8.30103831	0.00046686	0.12046686
50	289.00218983	0.00346018	2400.01824858	8.30449849	0.00041666	0.12041666

Rate 13%	A Compound Interest	B Present Value	C Amount of Annuity	D Present Value of Annuity	E Sinking Fund	F Amortization
n	$(1+i)^n$	$\frac{1}{(1+i)^n}$	$s_{\overline{n}\rvert i}$	$a_{\overline{n}\rvert i}$	$\frac{1}{s_{\overline{n}\rvert i}}$	$\frac{1}{a_{\overline{n}\rvert i}}$
1	1.13000000	0.88495575	1.00000000	0.88495575	1.00000000	1.13000000
2	1.27690000	0.78314668	2.13000000	1.66810244	0.46948357	0.59948357
3	1.44289700	0.69305016	3.40690000	2.36115260	0.29352197	0.42352197
4	1.63047361	0.61331873	4.84979700	2.97447133	0.20619420	0.33619420
5	1.84243518	0.54275994	6.48027061	3.51723126	0.15431454	0.28431454
6	2.08195175	0.48031853	8.32270579	3.99754979	0.12015323	0.25015323
7	2.35260548	0.42506064	10.40465754	4.42261043	0.09611080	0.22611080
8	2.65844419	0.37615986	12.75726302	4.79877029	0.07838672	0.20838672
9	3.00404194	0.33288483	15.41570722	5.13165513	0.06486890	0.19486890
10	3.39456739	0.29458835	18.41974915	5.42624348	0.05428956	0.18428956
11	3.83586115	0.26069765	21.81431654	5.68694113	0.04584145	0.17584145
12	4.33452310	0.23070589	25.65017769	5.91764702	0.03898608	0.16898608
13	4.89801110	0.20416450	29.98470079	6.12181152	0.03335034	0.16335034
14	5.53475255	0.18067655	34.88271190	6.30248807	0.02866750	0.15866750
15	6.25427038	0.15989075	40.41746444	6.46237882	0.02474178	0.15474178
16	7.06732553	0.14149624	46.67173482	6.60387506	0.02142624	0.15142624
17	7.98607785	0.12521791	53.73906035	6.72909298	0.01860844	0.14860844
18	9.02426797	0.11081231	61.72513819	6.83990529	0.01620085	0.14620085
19	10.19742280	0.09806399	70.74940616	6.93796928	0.01413439	0.14413439
20	11.52308776	0.08678229	80.94682896	7.02475158	0.01235379	0.14235379
21	13.02108917	0.07679849	92.46991672	7.10155007	0.01081433	0.14081433
22	14.71383077	0.06796327	105.49100590	7.16951334	0.00947948	0.13947948
23	16.62662877	0.06014448	120.20483667	7.22965782	0.00831913	0.13831913
24	18.78809051	0.05322521	136.83146543	7.28288303	0.00730826	0.13730826
25	21.23054227	0.04710195	155.61955594	7.32998498	0.00642593	0.13642593
26	23.99051277	0.04168314	176.85009821	7.37166812	0.00565451	0.13565451
27	27.10927943	0.03688774	200.84061098	7.40855586	0.00497907	0.13497907
28	30.63348575	0.03264402	227.94989040	7.44119988	0.00438693	0.13438693
29	34.61583890	0.02888851	258.58337616	7.47008839	0.00386722	0.13386722
30	39.11589796	0.02556505	293.19921506	7.49565344	0.00341065	0.13341065
31	44.20096469	0.02262394	332.31511301	7.51827738	0.00300919	0.13300919
32	49.94709010	0.02002119	376.51607771	7.53829857	0.00265593	0.13265593
33	56.44021181	0.01771786	426.46316781	7.55601643	0.00234487	0.13234487
34	63.77743935	0.01567953	482.90337962	7.57169596	0.00207081	0.13207081
35	72.06850647	0.01387569	546.68081897	7.58557164	0.00182922	0.13182922
36	81.43741231	0.01227937	618.74932544	7.59785101	0.00161616	0.13161616
37	92.02427591	0.01086670	700.18673775	7.60871771	0.00142819	0.13142819
38	103.98743178	0.00961655	792.21101365	7.61833426	0.00126229	0.13126229
39	117.50579791	0.00851022	896.19844543	7.62684447	0.00111582	0.13111582
40	132.78155163	0.00753117	1013.70424333	7.63437564	0.00098648	0.13098648
41	150.04315335	0.00666475	1146.48579497	7.64104039	0.00087223	0.13087223
42	169.54876328	0.00589801	1296.52894831	7.64693840	0.00077129	0.13077129
43	191.59010251	0.00521948	1466.07771159	7.65215787	0.00068209	0.13068209
44	216.49681583	0.00461901	1657.66781410	7.65677688	0.00060326	0.13060326
45	244.64140189	0.00408762	1874.16462994	7.66086450	0.00053357	0.13053357
46	276.44478414	0.00361736	2118.80603183	7.66448185	0.00047196	0.13047196
47	312.38260608	0.00320120	2395.25081596	7.66768306	0.00041749	0.13041749
48	352.99234487	0.00283292	2707.63342204	7.67051598	0.00036933	0.13036933
49	398.88134970	0.00250701	3060.62576691	7.67302299	0.00032673	0.13032673
50	450.73592516	0.00221859	3459.50711660	7.67524158	0.00028906	0.13028906

Answers to Selected Exercises

Chapter 1

Section 1.1 (page 7)

1. 382 miles **3.** 130 more crimes **5.** 1200 miles **7.** 2477 pounds **9.** 23,993,000 small and midsize businesses **11.** \$382,325,000 **13.** \$125 **15.** \$20,961 **17.** 20 seats **19.** 35.2 hours **21.** 65.5 million shares **23.** 26 coins **25.** \$35.96 million or \$35,960,000 **27.** (a) 159.1 hours (b) \$14.86 **29.** (a) \$8100 (b) \$17,485

Section 1.2 (page 19)

1. $\frac{11}{8}$ **3.** $\frac{17}{4}$ **5.** $\frac{183}{8}$ **7.** $\frac{101}{8}$ **9.** $\frac{1}{2}$ **11.** $\frac{8}{15}$ **13.** $\frac{5}{8}$ **15.** $\frac{4}{5}$ **17.** $\frac{11}{12}$ **19.** $\frac{8}{15}$ **21.** $3\frac{1}{2}$ **23.** $3\frac{4}{5}$ **25.** $1\frac{3}{11}$ **27.** $1\frac{2}{5}$ **29.** $1\frac{15}{16}$ **31.** $2\frac{17}{32}$ **35.** $\frac{3}{5}$ **37.** $\frac{17}{20}$ **39.** $1\frac{7}{60}$ **41.** $\frac{19}{22}$ **43.** $1\frac{23}{36}$ **45.** $2\frac{5}{24}$ **47.** $97\frac{4}{5}$ **49.** $80\frac{3}{4}$ **51.** $53\frac{17}{24}$ **53.** 187 **55.** $\frac{1}{2}$ **57.** $\frac{1}{2}$ **59.** $\frac{17}{48}$ **61.** $\frac{1}{3}$ **63.** $4\frac{3}{8}$ **65.** $3\frac{11}{24}$ **67.** $9\frac{1}{24}$ **69.** $6\frac{1}{4}$ **75.** $\frac{31}{40}$ inches **77.** $4\frac{3}{8}$ miles **79.** $\frac{3}{16}$ in. **81.** $22\frac{1}{2}$ hours **83.** $342\frac{17}{24}$ acres **85.** 50 cases **87.** $9\frac{1}{6}$ hours **89.** \$$118\frac{5}{8}$

Section 1.3 (page 30)

1. $\frac{5}{12}$ **3.** $\frac{99}{160}$ **5.** $4\frac{1}{2}$ **7.** 69 **9.** $4\frac{7}{12}$ **11.** 90 **13.** $\frac{1}{2}$ **15.** $\frac{3}{4}$ **17.** $1\frac{1}{2}$ **19.** $\frac{2}{3}$ **21.** $3\frac{1}{3}$ **23.** $4\frac{4}{5}$ **27.** \$3170 **29.** \$1794 **31.** \$1828 **33.** $\frac{4}{5}$ **35.** $\frac{6}{25}$ **37.** $\frac{73}{100}$ **39.** $\frac{7}{8}$ **41.** $\frac{3}{80}$ **43.** $\frac{3}{16}$ **45.** 3.5; 3.52 **47.** 0.1; 0.08 **49.** 8.6; 8.64 **51.** 59.0; 58.96 **53.** 0.75 **55.** 0.375 **57.** 0.167 **59.** 0.813 **61.** 0.32 **63.** 0.010 **69.** 36 yards **71.** 600 shares **73.** 12 homes **75.** $21\frac{7}{8}$ oz. **77.** 471 rolls **79.** 60 trips

Chapter 1 Review Exercises (page 36)

1. $\frac{3}{5}$ **2.** $\frac{1}{2}$ **3.** $\frac{1}{3}$ **4.** $\frac{1}{2}$ **5.** $\frac{9}{10}$ **6.** $\frac{7}{11}$ **7.** $\frac{1}{50}$ **8.** $\frac{3}{8}$ **9.** $8\frac{1}{8}$ **10.** $4\frac{2}{3}$ **11.** $1\frac{7}{12}$ **12.** $7\frac{6}{7}$ **13.** $2\frac{2}{3}$ **14.** $8\frac{1}{6}$ **15.** $8\frac{1}{16}$ **16.** $3\frac{1}{32}$ **17.** $1\frac{5}{24}$ **18.** $\frac{7}{8}$ **19.** $\frac{8}{21}$ **20.** $\frac{1}{12}$ **21.** $71\frac{5}{6}$ **22.** $91\frac{5}{6}$ **23.** $4\frac{1}{4}$ **24.** $81\frac{1}{16}$ **25.** \$2470.10 **26.** 14,454 gallons saved **27.** $22\frac{1}{2}$ hours **28.** $35\frac{7}{8}$ gallons **29.** $319\frac{1}{2}$ ft. **30.** $54\frac{11}{24}$ lb. **31.** $\frac{5}{12}$ **32.** $\frac{7}{40}$ **33.** $\frac{1}{2}$ **34.** 16 **35.** $\frac{2}{3}$ **36.** $2\frac{2}{9}$ **37.** $20\frac{5}{6}$ **38.** $6\frac{1}{6}$ **39.** \$3.93 **40.** 3.6 million shares **41.** $42\frac{1}{2}$ acres **42.** \$405.38 **43.** 36 pull cords **44.** $\frac{1}{12}$ of the total profit **45.** $\frac{1}{4}$ **46.** $\frac{5}{8}$ **47.** $\frac{93}{100}$ **48.** $\frac{1}{200}$ **49.** 68.4; 68.43 **50.** 975.5; 975.54 **51.** 0.4; 0.35 **52.** 8.0; 8.03 **53.** 7.0; 6.97 **54.** 0.4; 0.43 **55.** 1.0; 0.96 **56.** 71.2; 71.25 **57.** 0.625 **58.** 0.75 **59.** 0.833 **60.** 0.438

Chapter 1 Summary Exercises (page 37)

(a) \$0.375; \$0.1875; \$0.875; \$0.8125 **(b)** \$12.69 **(c)** \$15.63 **(d)** AT&T; \$35.50 **(e)** \$7030 **(f)** \$5431.25 **(g)** 61 shares **(h)** Coca-Cola \$728.13; Reynolds Metals −\$478.13; \$250 gain

Chapter 1 Net Assets (page 39)

1. $37\frac{9}{16}$ inches **2.** 1 foot $3\frac{5}{8}$ inches **3.** \$$7621\frac{1}{4}$ or \$7621.25 **4.** 272 shares

Chapter 2

Section 2.1 (page 46)

1. 59 **3.** 57 **5.** 13 **7.** 4.2 **9.** 12 **11.** 2 **13.** 3 **15.** 2.2 **17.** 0.8 **19.** 400 **21.** 294 **23.** 7 **25.** 12 **27.** 26 **29.** $\frac{5}{6}$ **31.** $\frac{5}{36}$ **33.** 4 **35.** $11\frac{4}{7}$ **37.** 4.65 **39.** $1\frac{5}{24}$ **41.** $\frac{2}{3}$ **43.** 1 **45.** 3.5 **47.** 13 **49.** 2 **51.** 5 **53.** 7 **55.** 10 **57.** $\frac{16}{19}$ **59.** $\frac{2}{3}$ **61.** $1\frac{1}{6}$ **63.** 2.1 **65.** 0.8

Section 2.2 (page 51)

1. $27 + x$ **3.** $22 + x$ **5.** $x - 4$ **7.** $x - 3\frac{1}{2}$ **9.** $3x$ **11.** $\frac{3}{5}x$ **13.** $\frac{9}{x}$ **15.** $\frac{16}{x}$ **17.** $2.1(4 + x)$ **19.** $7(x - 3)$ **21.** $12y$ **23.** $472 - x$ **25.** $73 - x$ **27.** $\frac{172}{x}$ **29.** $21 - x$ **31.** 15 **33.** 1.5 **35.** 1 **37.** $1\frac{3}{7}$ **39.** 45 **41.** 146 **43.** \$20,500 **45.** 42 deluxe; 63 economy **47.** \$3937.50 announcers; \$6562.50 all other employees **49.** \$15,000; \$52,500 **51.** 28 new; 35 experienced **53.** 81 Atimas; 39 Sentras

Section 2.3 (page 58)

1. \$586.50 **3.** 15.5169 **5.** 0.0625 **7.** 14 **9.** 0.08 **11.** 151.2 **13.** 749.86 **15.** 7.5 **17.** 7 **19.** 24,000 **21.** $\frac{A}{W}$ **23.** $\frac{nRT}{P}$ **25.** $M/(1 + I)^n$ **27.** $(A - P)/P$ **29.** $(M - P)/MT$ **31.** $2A/(b + B)$ **33.** M/e^{ni} **35.** \$47.36 **37.** \$93.80 **39.** (a) \$194.56 (b) \$213.88 (c) \$229.42 **41.** \$236 million **43.** \$48 **45.** \$90,000 **47.** \$390 **49.** 0.13 or 13% **51.** 4 years **53.** \$4000 **55.** \$4500

Section 2.4 (page 65)

1. $\frac{9}{32}$ **3.** $\frac{27}{1}$ **5.** $\frac{4}{3}$ **7.** $\frac{3750}{1}$ **9.** $\frac{225}{1}$ **11.** $\frac{8}{5}$ **13.** $\frac{1}{6}$ **15.** $\frac{4}{15}$ **17.** $\frac{9}{2}$ **19.** T **21.** F **23.** F **25.** F **27.** F **29.** F **31.** T **33.** T **35.** F **37.** T **39.** 7 **41.** 72 **43.** 105 **45.** $3\frac{1}{2}$ **47.** 24 **49.** 8 **53.** 1575 tickets **55.** \$412,800 **57.** \$96 **59.** 36 yards **61.** 1020 miles **63.** \$713,211.20 **65.** \$24,000 **67.** 1475 miles **69.** 3,500,000 cubic meters **71.** U.S. \$193.86

Chapter 2 Review Exercises (page 69)

1. 51 **2.** 50.7 **3.** 16.3 **4.** $5\frac{1}{4}$ **5.** 252 **6.** 136 **7.** 56 **8.** $13\frac{1}{2}$ **9.** 7 **10.** $1\frac{2}{3}$ **11.** 21 **12.** 3.6 **13.** $94x$ **14.** $\frac{1}{2x}$ **15.** $6x + x$ **16.** $5x - 11$ **17.** $3x + 7$ **18.** \$92.35 **19.** \$20,000 **20.** \$108 water; \$432 phone **21.** 76 **22.** 70 child; 30 adult **23.** \$4000 **24.** \$3250 **25.** \$8200 (rounded) **26.** $\frac{I}{PT}$ **27.** $\frac{M - P}{PR}$ **28.** $\frac{R - D}{RD}$ **29.** $\frac{34}{1}$ **30.** $\frac{18}{1}$ **31.** $\frac{20}{1}$ **32.** $\frac{12}{5}$ **33.** $\frac{8}{3}$ **34.** 3 **35.** $6\frac{3}{4}$ **36.** $4\frac{1}{2}$ **37.** 165 **38.** 24 **39.** 4293 bass **40.** 4734 pounds per square inch **41.** \$17.50 **42.** 105 **43.** \$57,000 **44.** \$139.50 **45.** 25

Chapter 2 Summary Exercises (page 70)

(a) \$6100 **(b)** Profit = \$3.18*N* − \$6100 **(c)** 1919 books **(d)** The owner would probably receive less salary **(e)** 2705 books (2704 isn't quite enough)

Chapter 2 Net Assets (page 71)

1. 1336 cars **2.** \$41,904,000,000 **3.** $P = 0.027G$ **4.** \$1,131,408,000

Chapter 3

Section 3.1 (page 77)

1. 20% **3.** 72% **5.** 140% **7.** 37.5% **9.** 462.5% **11.** 0.25% **13.** 0.15% **15.** 345% **17.** 25% **19.** 10% **21.** 2% **23.** 37.5% **25.** 12.5% **27.** 0.5% **29.** 87.5% **31.** 6% **33.** 0.65 **35.** 0.75 **37.** 0.006 **39.** 0.0025 **41.** 3.15 **43.** 2.006 **45.** 5.406 **47.** 0.0007 **53.** 0.5; 50% **55.** $\frac{3}{20}$; 0.15 **57.** $\frac{1}{4}$; 25% **59.** 6.125; 612.5% **61.** $7\frac{1}{4}$; 725% **63.** $\frac{1}{400}$; 0.25% **65.** 0.333 rounded; $33\frac{1}{3}$% **67.** $\frac{3}{400}$; 0.0075 **69.** $\frac{1}{8}$; 0.125 **71.** $2\frac{1}{2}$; 250% **73.** $10\frac{767}{2000}$; 10.3835 **75.** 4.375; 437.5% **77.** $\frac{27}{400}$; 0.0675

Section 3.2 (page 83)

1. 16 guests **3.** \$244.35 **5.** 4.8 feet **7.** 10,185 miles **9.** 182 homes **11.** 148.444 yds **13.** \$5366.65 **15.** \$6.50 **19.** \$52.80 **21.** \$645 **23.** 2024 shoppers **25.** \$92.06 million **27.** (a) 39% female (b) 2.318 million or 2,318,000 male workers **29.** 234 executives **31.** 2156 products **33.** \$135 million **35.** \$51,844.20 **37.** \$510,390 **39.** \$6296.40 **41.** \$6620

Section 3.3 (page 89)

1. 1060 **3.** 187.5 **5.** 1000 **7.** 4800 **9.** 22,000 **11.** 20,000 **13.** \$90,320 **15.** 1750 **17.** 312,500 **19.** 65,400 **21.** 40,000 **25.** 30,000 total workers **27.** 7761 students **29.** \$2800 **31.** 20 million or 20,000,000 adolescents **33.** 130 million returns **35.** \$185,500

Section 3.3 Supplementary Exercises (page 90)

1. \$20,200,000,000 **3.** \$288,150 **5.** 478,175 Mustangs **7.** \$93.9 million **9.** \$39,000 **11.** 836 drivers **13.** \$61.4 million

Section 3.4 (page 95)

1. 10 **3.** 125 **5.** 28.3 **7.** 9.3 **9.** 4.1 **11.** 5.9 **13.** 102.5 **15.** 17.6 **17.** 27.8 **21.** 5.6% **23.** 2% **25.** 8.7% **27.** 10.7% **29.** 22.2%

Section 3.4 Supplementary Exercises (page 97)
1. 0.6 million or 600,000 patients **3.** 14.1% **5.** 16,910 hotels **7.** 41,771 traffic deaths **9.** \$396.05 **11.** (a) 30% (b) 70% **13.** 470,844 workers **15.** 24.4% **17.** 12.5% **19.** 5.5% **21.** 960 candy bars **23.** \$8823 **25.** 23.6% **27.** 5742 deaths **29.** 36%

Section 3.5 (page 105)
1. \$375 **3.** \$27.91 **5.** \$25 **7.** \$854.50 **11.** \$165,500 **13.** \$86.2 million or \$86,200,000 **15.** 160 feet **17.** \$15,161.90 **19.** \$118,080 **21.** \$58.9 billion **23.** \$3864 **25.** \$1.078 billion **27.** 51.2 million **29.** 20,000 students **31.** 69.4 million computers **33.** 6564 homes

Chapter 3 Review Exercises (page 111)
1. 150 members **2.** 24 vans **3.** 1100 shippers **4.** 2.5% **5.** \$3.75 **6.** $\frac{6}{25}$ **7.** 960 loads **8.** $\frac{7}{8}$ **9.** 8.5% **10.** $\frac{1}{200}$ **11.** \$1.66 **12.** 224,000 units **13.** 2.71 million vehicles **14.** 21.43 million people **15.** (a) 11% (b) \$4488 **16.** 68% **17.** \$850 **18.** 1200 backpacks **19.** 2.8% **20.** 7.5 million **21.** 44% **22.** 12.5 ounces **23.** 97,757 copies **24.** \$39,840 **25.** 48.7% **26.** 0.7% **27.** 1.84 billion people **28.** \$1.365 billion **29.** 933.3% **30.** 54.4% **31.** 1.1% **32.** 1,800,000 tourists **33.** 250,000 cars **34.** 148,507 units **35.** \$11.90 per hour **36.** 97 units

Chapter 3 Summary Exercises (page 112)
\$40; −7%; \$4; \$900; \$75; −11%; \$5; \$20; 4900%; −25%

Chapter 3 Net Assets (page 113)
1. \$135,250 **2.** \$56,500 **3.** 8% **4.** 12%

Part 1 Cumulative Review Chapters 1–3 (page 114)
1. \$193 **2.** \$20,961 **3.** \$31,658.27 **4.** 27 months **5.** $\frac{8}{9}$ **6.** $\frac{65}{8}$ **7.** $7\frac{2}{15}$ **8.** $4\frac{5}{12}$ **9.** $13\frac{13}{24}$ **10.** $1\frac{3}{4}$ **11.** 3 **12.** $2\frac{2}{9}$ **13.** $42\frac{1}{2}$ acres **14.** $22\frac{1}{2}$ hours **15.** 130 feet **16.** $\frac{1}{12}$ of the total **17.** $\frac{7}{20}$ **18.** 0.667 **19.** 78.57 **20.** 4732.49 **21.** 62.7 **22.** 215.675 **23.** 26 **24.** 85.4 **25.** 112 **26.** 12 **27.** $\frac{3}{4}x$ **28.** $5x + x$ **29.** $8x - 8$ **30.** $6x + 5$ **31.** $T = 3$ **32.** $P = \$1850$ **33.** $\frac{5}{1}$ **34.** $\frac{7}{5}$ **35.** 9 **36.** $35\frac{1}{5}$ **37.** 744 first-time buyers **38.** \$154,000 **39.** 62.5% **40.** 0.0025 **41.** 450 prospects **42.** \$107.20 **43.** 25% **44.** 150% **45.** 64.5% **46.** 21,888 people **47.** 58,484 students **48.** \$16,760

Chapter 4

Section 4.1 (page 124)
1. \$14.20 **3.** \$20.00 **5.** \$17.10 **7.** \$21.90

9.

857

Mar. 8 20____

Amount \$380.71

To Patty Demko

For Tutoring

Bal. Bro't. For'd.	3971	28
Am't. Deposited	79	26
Total	4050	54
Am't. this Check	380	71
Balance For'd.	3669	83

11.

735

Dec. 4 20____

Amount $37.52

To Paul's Pools

For Chemicals

Bal. Bro't. For'd.	1126	73
Am't. Deposited		
Total	1126	73
Am't. this Check	37	52
Balance For'd.	1089	21

17.

5312

Oct. 10 20____

Amount $39.12

To County Clerk

For License

Bal. Bro't. For'd.	5972	89
Am't. Deposited	752 23	18 32
Total	6748	39
Am't. this Check	39	12
Balance For'd.	6709	27

19. 1629.86; 1379.41; 1230.41; 1348.14; 1278.34; 1608.20; 2026.50; 1916.74; 1302.62; 1270.44; 1791.39 **21.** 3852.48; 3709.32; 3590.92; 3877.24; 3797.24; 2811.02; 2435.52; 3637.34; 2901.66; 2677.72; 3175.73; 3097.49

Section 4.2 (page 131)

1. (a) $2419.76 (b) $203.86 (c) $2215.90 (d) $66.48 (e) $2149.42 **3.** (a) $1591.44 (b) $189.39 (c) $1402.05 (d) $56.08 (e) $1345.97 **5.** (a) $1064.72 (b) $72.83 (c) $991.89 (d) $29.76 (e) $962.13

Section 4.3 (page 139)

1. $5095.47 **3.** $7690.62 **5.** $18,314.72 **7.** $6967.88 **9.** $7498.20 **11.** $4496.01 **17.** $6728.20

Chapter 4 Review Exercises (page 144)

1. $15.90 **2.** $19 **3.** $10.20 **4.** 9517.70 **5.** 9831.34 **6.** 19,415.20 **7.** $1064.72 **8.** $72.83 **9.** $991.89 **10.** $39.68 **11.** $952.21 **12.** $6043.16 **13.** $8992.02 **14.** $1267.21

Chapter 4 Summary Exercises (page 146)

(a) $6101.69 gross deposit **(b)** $5888.13 credit **(c)** $9810.36 total of checks outstanding **(d)** $4882.58 deposits not recorded **(e)** $5188.69 balance

Chapter 4 Net Assets (page 147)

1. $88.60 **2.** (a) $809,197 (b) $793,013.06 **3.** (a) $36,987 (b) $33,213 **4.** $13,512

Chapter 5

Section 5.1 (page 157)
1. 40; 0; $12.15 **3.** 38.75; 0; $11.70 **5.** 40; 5.25; $17.22 **7.** $324; $0; $324 **9.** $302.25; $0; $302.25 **11.** $459.20; $90.41; $549.61 **13.** $13.20; $347.60; $0; $347.60 **15.** $10.80; $288; $48.60; $336.60 **17.** $13.77; $367.20; $58.52; $425.72 **19.** 51; 11; $3.70; $377.40; $40.70; $418.10 **21.** 50.25; 10.25; $4.30; $432.15; $44.08; $476.23 **23.** 50.25; 10.25; $5.10; $512.55; $52.28; $564.83 **25.** 35; 6;$10.05; $234.50; $60.30; $294.80 **27.** 39.5; 3.75; $10.05; $264.65; $37.69; $302.34 **29.** 39.75; 3.5; $15.30; $405.45; $53.55; $459 **33.** $496; $537.33; $1074.67; $12,896 **35.** $426; $923; $1846; $22,152 **37.** $501.92; $1003.85; $2175; $26,100 **39.** $618.46; $1236.92; $1340; $32,160 **41.** $415; $$830; $899.17; $1798.33 **43.** $832 **45.** $487.70 **47.** $703.08 **49.** $384.80 **51.** $335.40 **53.** $793.80 **55.** $925.16 **57.** (a) $1260 biweekly (b) $1365 semimonthly (c) $2730 monthly (d) $32,760 annually **61.** (a) $519.96; $1039.92; $1126.58; $2253.17 (b) $856.21; $1712.42; $1855.13; $3710.25

Section 5.2 (page 165)
1. $208.16 **3.** $421.65 **5.** $1941.75 **7.** $433.37 **9.** $1405 **11.** $688.40 **13.** $748 **15.** $2136 **19.** $5030; $201.20; $491.20 **21.** $4085; $245.10; $245.10 **23.** $9530; $285.90; $285.90 **25.** $2897; $144.85; $354.85 **27.** $438 **29.** $5646 **31.** (a) $2495 (b) $1695 **33.** $412.14

Section 5.3 (page 170)
1. 652; $254.28 **3.** 451; $338.25 **5.** 665; $452.20 **7.** 588; $270.48 **9.** 670; $522.60 **11.** $92.06 **13.** $145.12 **15.** $156 **17.** $260.19 **19.** $284.65 **21.** $297 **23.** $274.80 **25.** $439.52 **27.** $503.80 **29.** $322.57 **31.** $340.73 **35.** $844 **37.** $624.25 **39.** $478.80

Section 5.4 (page 178)
1. $20.13; $4.71 **3.** $28.72; $6.72 **5.** $52.99; $12.39 **7.** $67.35; $15.75 **9.** $131.98 **11.** $265.26 **13.** $11.18 **15.** $368.80; $76.07; $444.87; $27.58; $6.45; $4.45 **17.** $412; $61.80; $473.80; $29.38; $6.87; $4.74 **19.** $327.20; $61.35; $388.55; $24.09; $5.63; $3.89 **21.** $249.60; $63.18; $312.78; $19.39; $4.54; $3.13 **23.** (a) $24.07 (b) $5.63 **25.** (a) $95.67 (b) $22.38 (c) $15.43 **27.** $4569.75; $1068.73 **29.** $4317.33; $1009.70 **31.** $3328.61; $778.46

Section 5.5 (page 187)
1. $241 **3.** $54 **5.** $172 **7.** $74 **9.** $69 **11.** $8.30 **13.** $8.89 **15.** $88.58 **17.** $25.89; $6.05; $12.31; $373.33 **19.** $95.00; $22.22; $162.30; $1252.66 **21.** $120.20; $28.11; $198.90; $1591.55 **23.** $122.21; $28.58; $152.19; $1668.08 **25.** $44.05; $10.30; $43.69; $612.52 **27.** $56.75; $13.27; $165.59; $679.73 **29.** $329.38; $77.03; $699.78; $4206.40 **31.** $26.74; $6.25; $41.17; $357.09 **35.** $506.03 **37.** $522.21 **39.** $3628.43

Section 5.6 (page 194)
1. $2392.04 **3.** $2671.70 **5.** $2162.14 **7.** $9558.80 **9.** $1570.26 **11.** $13,129.72 **13.** $12,831.18 **15.** $17,914.01 **19.** $3355 **21.** $2791.98 **23.** (a) $1120 (b) $69.44 **25.** (a) $1600 (b) $99.20

Chapter 5 Review Exercises (page 200)
1. 40; 8.5; $482.14 **2.** 40; 8; $442 **3.** 38.25; 0; $283.05 **4.** 40; 17.25; $447.95 **5.** $821.60; $890.07; $1780.13; $21,361.60 **6.** $530; $1148.33; $2296.67; $27,560 **7.** $346.15; $692.31; $750; $1500 **8.** $403.85; $807.69; $1750; $21,000 **9.** $760 **10.** $427.50 **11.** $3641.12 **12.** $3374.55 **13.** $242.75 **14.** $134 **15.** $2025 **16.** $465 **17.** $600 **18.** $265.02 **19.** (a) $487.01 (b) $113.90 **20.** (a) $89.90 (b) $113.90 **21.** $18 **22.** $72 **23.** $151 **24.** $93 **25.** $104 **26.** $27 **27.** $1444.30 **28.** $406.12 **29.** $532.60 **30.** $1946.18 **31.** $9386.33 **32.** (a) $31.90 (b) $7.46 (c) $5.14 **33.** (a) $103.84 (b) $28.37 **34.** $5451.25 **35.** (a) $4810.61 (b) $1125.06 **36.** (a) $3424.70 (b) $800.94 **37.** (a) $913 (b) $56.61 **38.** (a) $2140 (b) $132.68

Chapter 5 Summary Exercise (page 202)
(a) $620 **(b)** $279 **(c)** $899 **(d)** $55.74 **(e)** $13.04 **(f)** $161.01 **(g)** $8.99 **(h)** $39.56 **(i)** $393.66

Chapter 5 Net Assets (page 203)
1. $221.40 **2.** $2600 monthly; $1300 semimonthly; $1200 biweekly **3.** (a) $25.22 (b) $5.90 **4.** $4275

Chapter 6

Section 6.1 (page 209)
1. $2.29; $8.38; $86.87 **3.** $2.15; $1.43; $51.28 **5.** $8.68; $14.85; $197.03 **7.** $57.55; $98.66; $978.39 **9.** $1837.50; $504; $31,741.50 **11.** $160; $169.60 **13.** $157.50; $163.80 **15.** $330; $351.45 **17.** $1276.80; $1340.64 **19.** $102.20; $5.11 **21.** $520.30; $31.22 **23.** $19.76; $0.84 **25.** $315; $18.90 **27.** $2753.05; $192.71 **31.** (a) $7.19 (b) $14.86 (c) $141.85 **33.** $380 **35.** $1867.50 **37.** $49.44 **39.** (a) $108 (b) $6.48 **41.** $201.35 **43.** $39,316.75

Section 6.2 (page 216)
1. $25,600 **3.** $60,830 **5.** $325,125 **7.** 12% **9.** 8% **11.** 3% **13.** (a) 2.8% (b) $2.80 (c) $28 **15.** (a) $2.41 (b) $24.10 (c) 24.1 **17.** (a) 7.08% (b) $70.80 (c) 70.8 **21.** $5861.60 **23.** $47,498.22 **25.** $1047.60 **27.** 5.8% **29.** $6273 **31.** $5.25 **33.** $236,800 **35.** $4853.52 **37.** $6295.08 **39.** $200,925 **41.** (a) The second county (b) $75.24 **43.** 8.4% **45.** $2500

Section 6.3 (page 229)
1. $20,600 **3.** $21,710 **5.** $39,031 **7.** $17,150; $2572.50 **9.** $23,301; $3495.15 **11.** $49,199; $8179.22 **13.** $33,300; $5976.50 **15.** $61,524; $13,879.22 **17.** $49,432; $8244.46 **19.** $975.15 tax refund **21.** $305.16 tax refund **23.** $466.62 tax due **27.** $5426.70 **29.** $5037.94 **31.** $665.40 **33.** $6869.94 **35.** $3295.65

Chapter 6 Review Exercises (page 234)
1. $51.13; $85.22; $988.50 **2.** $3.45; $9.48; $99.08 **3.** $825; $1342; $18,667 **4.** $24.22; $52.16; $422.34 **5.** $1134 **6.** $284 **7.** $280 **8.** $350 **9.** $422 **10.** $125 **11.** $279 **12.** $414 **13.** 4.06%; $40.60; 40.6 **14.** 2.7%; $2.70; $27 **15.** $1.27; $12.70; 12.7 **16.** 1.95%; $1.95; 19.5 **17.** $13,632 **18.** $18.50 **19.** $46,500 **20.** 2.8% **21.** $96,200 **22.** $2797.20 **23.** $31,149; $5374.22 **24.** $57,244; $10,431.82 **25.** $38,660; $5799 **26.** $41,702; $8329.06 **27.** 2.2% **28.** $194.88 **29.** $15,701.97 **30.** $5937 **31.** $3106.50 **32.** $8768.34 **33.** $1567.50 tax due **34.** $274.90 tax refund **35.** $900.10 tax refund **36.** $420.60 tax refund

Chapter 6 Summary Exercise (page 236)
(a) Anderson: $3,143,664; Bentonville: $3,488,706 **(b)** Anderson: $785,916; Bentonville: $697,741.20 **(c)** Anderson: $25,149.31; Bentonville: $20,583.37 **(d)** Anderson: $3,395,157.10; Bentonville: $3,694,539.70 **(e)** Anderson

Chapter 6 Net Assets (page 237)
1. $54.69 **2.** $27.80 **3.** $5164.60 **4.** $71,066

Chapter 7

Section 7.1 (page 246)
1. $1851 **3.** $3158 **5.** $9299 **7.** $8973 **9.** $432.40 **11.** $661.74 **13.** $828.85 **15.** $243 **17.** (a) $2010 (b) $670 **19.** (a) $2552.08 (b) $1822.92 **21.** (a) $2654 (b) $2654 **23.** $19,850 **25.** $1134.55 **27.** $36,500 **29.** $12,554.95 **31.** A: $60,000; B $20,000 **33.** 1: $292,500; 2: $260,000; 3: $97,500 **35.** (a) $30,000 (b) A: $17,500; B: $12,500 (c) $6000 **37.** (a) $12,500 (b) 1: $7500; 2: $5000 (c) $7500 **39.** $1661 **41.** $393 **43.** $1057.50 **45.** $412.50 **47.** (a) $1112.50 (b) $1557.50 **49.** (a) $1134 (b) $810 **53.** (a) $19,936.71 (b) $2563.29 **55.** (a) $30,681.82 (b) $14,318.18 **59.** A: $274,000 B: $182,667 C: $91,333 **61.** (a) $375,000 (b) A: $281,250; B: $93,750 **63.** (a) $75,000 (b) 1: $41,667; 2: $20,833; 3: $12,500

Section 7.2 (page 258)
1. $631 **3.** $932 **5.** $535 **7.** $682 **9.** $748 **11.** $356 **15.** $657.80 **17.** $915 **19.** (a) $25,000 (b) $11,500 **21.** (a) $1628 (b) $6936 (c) $100,000 (d) $15,250 **23.** (a) $60,000 (b) $10,250 **27.** (a) life, medical, auto, homeowner's (b) check with employer, avoid specialty insurance, check prices with different agents

Section 7.3 (page 270)
1. $952.50; $485.78; $247.65; $86.49 **3.** $849.10; $433.04; $220.77; $77.10 **5.** $516.80; $263.57; $134.37; $46.93 **7.** $297; $151.47; $77.22; $26.97 **9.** $531; $270.81; $138.06; $48.21 **11.** $1836.10; $936.41; $477.39; $166.72 **13.** $3175; $1619.25; $825.50; $288.29 **17.** $15,500 **19.** $30,000 **21.** $196,800 **23.** 23 yr. 315 days **25.** $289 **27.** 18 years **29.** $160.20 **31.** (a) $384 (b) $50,000 **33.** $190.49 **35.** $221.94 **37.** (a) $444.72 (b) $226.72 (c) $79.18 (d) $889.44; $906.88; $950.16

39. (a) $5660 (b) $11,850 (c) 23 yr. 315 days **41.** $149,500 **43.** (a) $211.50 (b) approx. 20 years (c) $115.75 (d) $109.50
45. (a) $345.50 (b) 18 yr. (c) $297 (d) $286.50

Chapter 7 Review Exercises (page 277)
1. $4630 **2.** $3510 **3.** $434 **4.** $1028 **5.** $1152.45 **6.** $323.40 **7.** $222.90 **8.** $1007.30 **9.** (a) $1312.50 (b) $1837.50 **10.** (a) $1481.25 (b) $493.75 **11.** (a) $1230 (b) $246 **12.** $464 (b) $2320 **13.** $39,473.68 **14.** $72,689.19 **15.** $2731.66 **16.** $35,707.69 **17.** $696 **18.** $1086.55 **19.** $887.60 **20.** $643 **21.** $2281.30 **22.** $246 **23.** $133.50 **24.** $192 **25.** A: $36,000; B: $21,600; C: $14,400 **26.** (a) $125,762 (b) 1: $80,030; 2: $45,732 **27.** (a) $25,000 (b) $9000 **28.** (a) $15,000 (b) $0 **29.** (a) $40,000 (b) $36,800 **30.** (a) $494.70 (b) $252.20 (c) $88.08 (d) $989.40; $1008.80; $1056.96 **31.** $67,500 **32.** (a) $11,320 (b) $23,160 (c) 23 yr. 315 days **33.** (a) $552.80 (b) about 12 yr. (c) $475.20 (d) $392.80 **34.** (a) $231.20 (b) about 18 yr. (c) $185.20 (d) $179.60

Chapter 7 Summary Exercise (page 281)
(a) $39,940.40 **(b)** $370.39 **(c)** $40,310.79 **(d)** $1389.21

Part 2 Cumulative Review Chapters 4–7 (page 282)
1. $1014.40 **2.** $208.15 **3.** $806.25 **4.** $18.14 **5.** $788.11 **6.** $17,984.04 **7.** $595.20 **8.** (a) $2710 (b) $1510 **9.** $679.43 **10.** (a) $102.58 (b) $23.99 (c) $16.54 **11.** $497.49 **12.** $4359.38 **13.** $272 **14.** (a) 2.68% (b) $26.80 (c) 26.8 **15.** (a) $4.62 (b) $46.20 (c) 46.2 **16.** $37,389.93 **17.** $1330.56 **18.** (a) $150,000 (b) A: $83,333; B: $41,667; C: $25,000 **19.** (a) $6090 (b) $25,000 (c) $50,000 (d) $125,050 **20.** (a) $1662.09 (b) $847.34 (c) $295.92 (d) $3324.18; $3389.36; $3551.04

Chapter 7 Net Assets (page 281)
1. $1557 **2.** $2199 **3.** (a) $50,000 (b) $13,800 **4.** (a) $1292 (b) $658.84 (c) $230.09 (d) $2584.68; $2635.36; $2761.08

Chapter 8

Section 8.1 (page 291)
1. foot **3.** sack **5.** great gross **7.** case **9.** drum **11.** cost per thousand **13.** gallon **15.** cash on delivery **17.** $54.00; $57.00; $64.80; $28.40; $297.00; $501.20; $524.95 **21.** 0.72 **23.** 0.512 **25.** 0.54 **27.** 0.4025 **29.** 0.532 **31.** 0.342 **33.** $267.52 **35.** $7.01 **37.** $7.14 **39.** $1254.29 **41.** $11.81 **43.** $640 **49.** $242.99 **51.** (a) 20/15 (b) $1.02 **53.** $16,416 **55.** $33.84 **57.** $189 **59.** (a) $274.02 (b) $18.27

Section 8.2 (page 298)
1. 0.72; 28% **3.** 0.5525; 44.75% **5.** 0.64; 36% **7.** 0.576; 42.4% **9.** 0.675; 32.5% **11.** 0.75; 25% **13.** 0.648; 35.2% **15.** 0.243; 75.7% **17.** 0.45; 55% **19.** 0.7; 30% **21.** 0.5184; 48.16% **23.** 0.5054; 49.46% **27.** $720 **29.** $700 **31.** $1920 **33.** (a) $58.29 (b) $64.76 (c) $6.47 **35.** $740 **37.** (a) $40.11 (b) $44.57 (c) $4.46 **39.** 25.0% **41.** $3857.14 **43.** 5.0%

Section 8.3 (page 305)
1. Oct. 18; Nov. 7 **3.** Mar. 25; Mar. 30 **5.** Sept. 21; Nov. 10 **7.** Jan. 24; Mar. 15 **9.** Jan. 20.; Mar. 6 **11.** $3.03; $160.90 **13.** $0; $101.28 **15.** $14.48; $ 747.66 **17.** $0; $659.50 **19.** $29.15; $607.03 **23.** $1990.59 **25.** $647.42 **27.** $132.59 **29.** $2511.13 **31.** $2021.32 **33.** (a) April 24; May 4; May 14 (b) June 3 **35.** (a) April 25 (b) May 5 **37.** $3547.50 **39.** $3106.60

Section 8.4 (page 311)
1. Mar. 10; Mar. 30 **3.** Dec. 22; Jan. 11 **5.** June 16; July 6 **7.** Aug. 10; Aug. 30 **9.** Aug. 24.; Sept. 13 **11.** $20.47; $661.81 **13.** $0; $194.04 **15.** $59.20; $2900.80 **17.** $168.80; $4051.20 **19.** $0.25; $12.13 **21.** $97.52; $3153.08 **23.** $68.33; $1639.85 **27.** $1900; $1250 **29.** $720; $1030 **31.** $100; $60 **35.** (a) Sept. 15 (b) Oct. 5 **37.** (a) Dec. 23 (b) Jan. 12 **39.** $1509.75 **41.** (a) Dec. 13 (b) $951.27 **43.** $4271.33 **45.** (a) $1000 (b) $920 **47.** $1495.58 **49.** $93.02 **51.** (a) $306.12 (b) $220.68 **53.** (a) $597.94 (b) $522.21

Chapter 8 Review Exercises (page 318)
1. $345.60 **2.** $195.62 **3.** $1828.18 **4.** $826.20 **5.** (a) 0.6375 (b) 36.25% **6.** (a) 0.576 (b) 42.4% **7.** (a) 0.54 (b) 46% **8.** (a) 0.4536 (b) 54.64% **9.** $502.08 **10.** $1545.21 **11.** $537.09 **12.** $2262.11 **13.** Mar. 15; April 4 **14.** May 30; June 19

15. Jan. 10; Jan. 30 **16.** Dec. 19; Jan. 8 **17.** $38.41; $1318.17 **18.** $28.37; $917.23 **19.** $35.02; $907.66 **20.** $44.21; $2166.39 **21.** $306.12; $353.88 **22.** $2597.94; $2712.06 **23.** $505.05; $354.95 **24.** $2113.40; $1736.60 **25.** (a) $394.40 (b) $386.51 (c) $398.06 **26.** (a) Builders Supply (b) $1.91 **27.** $63,295.14 **28.** 10% **29.** $1779.95 **30.** $1798.38 **31.** (a) April 15 (b) $821.24 **32.** (a) $1875 (b) $3405 (c) $75

Chapter 8 Summary Exercise (page 320)
(a) $3844.76 **(b)** June 15 **(c)** July 5 **(d)** $3904.56 **(e)** $2577.32; $1442.58

Chapter 8 Net Assets (page 321)
1. $65.28 **2.** $260.59 **3.** (a) August 20 (b) September 9

Chapter 9

Section 9.1 (page 328)
1. $4.96; $17.36 **3.** 50%; $35.25 **5.** $39.80; 25.1% **7.** $67.50; 20% **9.** $118; 56.2% **11.** $133.65; $628.65 **15.** $102.22 **17.** $13.73 **19.** $14.95 **21.** 17.8% **23.** (a) $95.96 (b) 25% (c) 125% **25.** (a) 132% (b) $97.02 (c) $23.52 **27.** (a) $35 (b) 27.2% (c) $9.52 **29.** $5172.40 **31.** $6.63

Section 9.2 (page 335)
1. $7; $28 **3.** $131.48; $243.48 **5.** $37.20; $55.80 **7.** $46.36; $24.96 **9.** $77.82; 35.2% **11.** $230; $287.50 **13.** $8.46; $22.26; 61.3% **15.** $750; $1050; 28.6% **17.** $357.52; 22% **19.** 25% **21.** 35.1% **23.** 33.3% **25.** 66.7% **29.** $57.50 **31.** $386.75 **33.** $119 **35.** (a) $4.50 (b) $2.88 (c) 64% **37.** 26.5% **39.** (a) $27.72 (b) $10.78 (c) 28% **41.** (a) $0.72 (b) 100% **43.** (a) $13,680 (b) $6080 (c) 44.4% (d) 80% **45.** (a) $13.95 (b) 100% **47.** (a) 43.8% (b) 30.4% **49.** (a) $32.40 (b) 25.9% **51.** (a) $24.90 (b) 12.5% (c) 14.2%

Section 9.3 (page 341)
1. $1.11 **3.** $3.13 **5.** $9 **7.** $1.65 **9.** 24; $8.94 **11.** 18; $22 **13.** 76; $3.75 **15.** 1900; $16.58 **17.** 160; 40; $3.36 **19.** 34; 6; $78.53 **21.** 108; 36; $4.17 **23.** 700; 300; $4.71 **27.** $4.22 **29.** $10.50 **31.** $15 **33.** $308.75 **35.** $177.78 **37.** $24.13

Chapter 9 Review Exercises (page 346)
1. $6.40; $38.40 **2.** $24.60; $30.75 **3.** $147; 50% **4.** $45.90; 42.5% **5.** $18.08; $90.40 **6.** $105; 25% **7.** $34.70; $52.05 **8.** 38.5%; $460.20 **9.** $1920; $2400 **10.** $64.50; 50% **11.** $8; 46%; 31.5% **12.** $474.28; $948.56; 50% **13.** 25% **14.** 50% **15.** 18.1% **16.** $16\frac{2}{3}$% **17.** $1.25 **18.** $32.50 **19.** $13.59 **20.** $17.04 **21.** 135; 15; $6.67 **22.** 72; 18; $2.38 **23.** 216; 72; $8.27 **24.** 700; 300; $4.71 **25.** $12.50 **26.** 18% **27.** 40% **28.** (a) $37.99 (b) 19.0% (c) 23.5% **29.** (a) $19,050.25 (b) 47% **30.** $4 **31.** (a) 20% (b) 25% **32.** $4.12

Chapter 9 Summary Exercise (page 348)
(a) $8058.20 **(b)** $10,063.50 **(c)** $2005.30 **(d)** 20% **(e)** 25%

Chapter 9 Net Assets (page 349)
1. $56 **2.** $105.30 **3.** (a) $3.75 (b) 30% (c) 42.9% **4.** $74

Chapter 10

Section 10.1 (page 357)
1. 25%; $645 **3.** 45%; $13.86 **5.** $6.50; $1.30 **7.** $2.70; $1.62 **9.** $21.75; $21.75 **11.** $857; 20% **13.** $60; $10; none **15.** $16; $22; $6 **17.** $385; $250; $60 **19.** $29; $10; $39 **23.** $239 **25.** 15% **27.** $298 opeating loss **29.** (a) $63.62 operating loss (b) $40.75 (c) 27% **31.** (a) $256.50 (b) 11%

Section 10.2 (page 362)
1. $11,680 **3.** $22,673 **5.** $24,500 **7.** 2.81; 2.83 **9.** 7.94; 7.98 **11.** 10.25; 10.25 **13.** 4.66; 4.69 **17.** $54,496.15 **19.** 5.43 **21.** 6.32 **23.** 11.09 **25.** 4.17

Section 10.3 (page 370)
1. $827 **3.** $560 **5.** (a) $182 (b) $195 (c) $170 **7.** (a) $2352 (b) $2385 (c) $2313 **9.** $1770 **11.** (a) $48 (b) $50 (c) $51 **13.** (a) $1251 (b) $1430 (c) $1040 **15.** (a) $6144 (b) $6784 (c) $5500 **17.** $130,600 **19.** $30,641

Chapter 10 Review Exercises (page 376)
1. $28.80; $67.20 **2.** $15; $5 **3.** $5.40; $2.70 **4.** 25%; $585 **5.** $48; $66 **6.** $100; $7; none **7.** $100; $70; $8 **8.** $6.25; $0.75; none **9.** $41,051 **10.** $348,468 **11.** $76,411.80 **12.** $32,859.80 **13.** 5.73; 5.76 **14.** 7.94; 7.98 **15.** 4.66; 4.69 **16.** 4.56; 4.63 **17.** $566 **18.** $1242 **19.** $3640 **20.** $6017 **21.** 28% **22.** (a) $78.40 (b) 7% **23.** $32,568 **24.** 10.77 **25.** 4.57 **26.** (a) $6489 (b) $5887.50 (c) $6675 **27.** $115,000 **28.** $75,600

Chapter 10 Summary Exercise (page 378)
(a) $125 **(b)** $2062.50 **(c)** $375 **(d)** none

Chapter 10 Net Assets (page 379)
1. 12% **2.** (a) $1531.25 (b) 8% **3.** 27.65 at retail **4.** (a) $6063 (b) $5990 (c) $6040

Part 3 Cumulative Review Chapters 8–10 (page 380)
1. $201.60 **2.** $240.47 **3.** 0.72; 28% **4.** 0.64; 36% **5.** 0.532; 46.8% **6.** 0.27; 73% **7.** $762.57 **8.** $839.25 **9.** $3745.73 **10.** $4598.65 **11.** $288.14; 100%; 50% **12.** $114.66; $152.88; 25% **13.** 700 pr.; 300 pr.; $2.36 **14.** 135; 15; $20 **15.** $48; $66; $18 **16.** $50; $35; $4 **17.** $64,877 **18.** $211,981 **19.** 5.73 at retail; 5.76 at cost **20.** 4.66 at retail; 4.69 at cost **21.** (a) $1533.44 (b) $1916.80 (c) $383.36 **22.** $3506.44 **23.** $4200 **24.** 20% **25.** (a) $124.99 (b) 25% (c) $33\frac{1}{3}$% **26.** 28% **27.** (a) $131.10 operating loss (b) $45.60 absolute loss **28.** $65,139 average inventory **29.** $1125 weighted average method **30.** (a) $1157 FIFO (b) $1100 LIFO

Chapter 11

Section 11.1 (page 386)
1. $850 **3.** $785.33 **5.** $1340.63 **7.** $644.90 **9.** $3821.17 **11.** $500 **13.** $4062.50 **15.** 7% **17.** 14 **19.** 1.5 years **21.** $4200 **23.** 11 months **25.** 8% **27.** $146.25 **29.** $1,525,000 **31.** 8 months **33.** 4% **37.** $170.67 **39.** 14% **41.** $12.72 **43.** $830; $25,730

Section 11.2 (page 393)
1. 209 days **3.** 36 days **5.** 196 days **7.** October 12 **9.** March 10 **11.** April 11 **13.** $45 **15.** $162.50 **17.** $41.25 **19.** $67.13 **21.** $104.11 **23.** $98.30 **25.** $114.68 **27.** $157.04 **31.** $118.36 **33.** $2632.77 **35.** $3234 **37.** $2400 **39.** (a) $532.60 (b) $540 (c) $7.40 **41.** (a) $9057.53 (b) $9183.33 (c) $125.80 **43.** $55.67 **45.** First Bank; $9.34

Section 11.3 (page 400)
1. $637.60; $9137.50 **3.** $191.72; $5991.72 **5.** $1080; $9720 **7.** $133.33; $4933.33 **9.** 70; $8733.78 **11.** $10\frac{1}{2}$%; $14,490 **13.** $16,400; $717.50 **15.** $14,800; 180 **17.** $126; 180 **19.** $42,459.44; 15% **21.** $490; $12,490 **23.** 140 days; $512.36 **25.** 18%; $70 **27.** $820,000; $34,166.67

Section 11.4 (page 408)
1. $4932 **3.** $11,462.50 **5.** $4529.22 **7.** $2400 **9.** $8400 **11.** $8700 **15.** $6,755,576.39 **17.** $58,437.50 **19.** $1211.76 **21.** $7104.85 **23.** $3489.33 **25.** $397.50 loss **27.** $5800 in 150 days

Chapter 11 Review Exercises (page 415)
1. $1512 **2.** $840 **3.** $8710 **4.** $2400 **5.** 9% **6.** 7.5% **7.** 9 months **8.** 30 months **9.** 114 days **10.** 115 days **11.** $258.90 **12.** $336.58 **13.** $222.51 **14.** $131.34 **15.** $43.17 **16.** $3241.33 **17.** $12,199.80 **18.** $6181.82 **19.** 7% **20.** $8\frac{1}{4}$% **21.** 40 days **22.** 90 days **23.** $6270 **24.** $7272.60 **25.** $12,000 **26.** $6500 **27.** 9% **28.** 12% **29.** $6418.15 **30.** $11,612.90 **31.** $8093.75 **32.** $19,550.56 **33.** $809.69 **34.** $15,598.34 **35.** $84.38 **36.** $19,297.50 **37.** 250 days **38.** $3600 **39.** 9.5% **40.** 17.3% **41.** 90 days **42.** $9127.31 **43.** $19,528.16 **44.** $11,948.33 **45.** $852 gain **46.** $2811 loss

Chapter 11 Summary Exercises (page 418)

September 3, \$600, \$12,600; July 16, \$186, \$6386; January 25 of the following year, \$971.11, \$19,371.11; Totals, \$1757.11, \$38,357.11

Chapter 11 Net Assets (page 419)

1. 8300 homes; \$1,535,500,000 **2.** I = \$1,680,000; M = \$29,680,000 **3.** \$315,000

Chapter 12

Section 12.1 (page 426)

1. Helen Spence **3.** Leta Clendenen **5.** 90 days **7.** January 25 **9.** December 20 **11.** June 30 **13.** March 17 **15.** April 23 **17.** November 9; \$6225 **19.** June 10; \$5354.17 **21.** \$8400 **23.** 120 days **25.** 11% **27.** \$12,954 **31.** (a) January 9 (b) \$1150.78 **33.** (a) December 13 (b) \$22,732.50 **35.** \$42,500 **37.** (a) April 7 (b) \$28,400 **39.** 90 **41.** 8% **43.** \$2004

Section 12.2 (page 435)

1. \$103.50; \$4496.50 **3.** \$441.75; \$5758.25 **5.** \$66.50; \$8333.50 **7.** April 13; \$8143.33 **9.** February 17; \$11,430 **11.** 8/3; \$1140; \$22,860 **13.** 11%; 8/7; \$13,740 **15.** \$7200; 6%; 2/10 **17.** \$220; \$5780 **19.** 120 days **21.** $12\frac{1}{2}$% **23.** \$3329.48 **25.** \$3815; 12.1% **27.** \$167,088.61; 15.2% **29.** (a) \$24,625,000 (b) \$25,000,000 (c) \$375,000 (d) 6.09%

Section 12.3 (page 440)

1. 9.28% **3.** 15.48% **5.** 11.18% **7.** 9.73% **11.** 13% simple interest **13.** \$17.23

Section 12.4 (page 446)

1. 83 days **3.** 59 days **5.** \$4698 **7.** \$14,100 **9.** 61 days; \$73.30; \$3531.70 **11.** 68 days; \$159.98; \$6897.85 **13.** 81 days; \$286.50; \$10,324.61 **15.** (a) 113 days (b) \$475.86 (c) \$12,157.47 **17.** (a) \$218.75 (b) \$4468.75 **19.** (a) \$22,950 (b) \$23,340 (c) \$390 **21.** (a) \$33,093.33 (b) \$35,202.22 (c) \$2108.89 **23.** (a) \$81,900 (b) \$79,688.70

Chapter 12 Review Exercises (page 451)

1. \$462.78; \$10,262.78 **2.** $10\frac{1}{2}$%; \$3078.75 **3.** 240 days; \$8640 **4.** \$12,300; \$12,915 **5.** May 8; \$12,360 **6.** January 5; \$6416.67 **7.** \$480; \$17,520 **8.** \$1365; \$24,635 **9.** (a) 99 days (b) \$410.85 (c) \$12,039.15 **10.** (a) 50 days (b) \$146.92 (c) \$8668.29 **11.** 11.53% **12.** 9.40% **13.** 11.32% **14.** \$39.488.33 **15.** \$891.23 **16.** 9% **17.** 180 days **18.** 200 days **19.** 14% **20.** 11%; 10.71% **21.** 13.04% **22.** \$13,874.39 **23.** 150 days **24.** \$2216.67 **25.** 15.9% **26.** \$18,400 **27.** \$8318.53 **28.** 270 days **29.** \$30,212 **30.** \$866.67; \$25,133.33 **31.** \$24,366.67 **32.** \$15,750.67 **33.** \$85,420.83 **34.** \$7951.05 **35.** \$6377.09 **36.** 13.48%

Chapter 12 Summary Exercise (page 453)

(a) \$1,120,000; Interest to be paid to the Japanese investment house:

$I = \$80{,}000{,}000 \times 0.09 \times \frac{180}{360} = \$3{,}600{,}000$

Interest from the Canadian firm:

$I = \$38{,}000{,}000 \times 0.11 \times \frac{180}{360} = \$2{,}090{,}000$

Interest from the French contractor:

$B = \$27{,}500{,}000 \times 0.128 \times \frac{180}{360} = \$1{,}760{,}000$

Interest from the Louisiana company:

$B = \$14{,}500{,}000 \times 0.12 \times \frac{180}{360} = \$870{,}000$

Interest received by Bank of America:	\$2,090,000
	1,760,000
	+ 870,000
	\$4,720,000
Interest received by Bank of America:	\$4,720,000
Interest paid by Bank of America:	− 3,600,000
Difference:	\$1,120,000

(b) $77,370,000

Loan		Amount loaned out
First		$38,000,000
Second	$27,500,000 − $1,760,000 =	$25,740,000
Third	$14,500,000 − $870,000 =	$13,630,000
	Total loaned out	$77,370,000

Chapter 12 Net Assets (page 455)
1. 31.5 million pairs **2.** $63,421.05 of debt per employee **3.** $866,875; $37,366,875; Feb. 9 **4.** $400,000; $11,600,000

Chapter 13

Section 13.1 (page 465)
1. $15,116.54; $3116.54 **3.** $7600.62; $1600.62 **5.** $21,724.52 **7.** $1341.54 **9.** $6032.76 **11.** $5755.82 **13.** $300; $338.23; $38.23 **15.** $3163.37; $3775.92; $612.55 **17.** 8.24% **19.** 16.08% **21.** (a) $3762.97 (b) $3773.98 **23.** (a) $2624.77 (b) $2667.70 (c) $2689.86 (d) $2704.89 (e) $2400 **25.** $12,043.41 **27.** (a) $11,248.64; $11,268.25 (b) $19.61 **29.** $10,538.50 **31.** 6.09%; 5.12%; the first option **33.** $18,127.97 **35.** $12,250.08 **39.** $439.20 **41.** $85,280.20

Section 13.2 (page 475)
1. $38.16 **3.** $35.62 **5.** $7267.81 **7.** $2992 **9.** $5637.43 **11.** $17,271.15 **13.** $4673.06 **15.** $1432.90 **19.** $5845.41 **21.** (a) $118.49 (b) $11,638.49 **23.** $0 **25.** $175.61 **27.** $9019.97; $1019.97 **29.** $4354.27; $253.57 **31.** (a) $901,988.58 (b) $101,988.58 **33.** 10% compounded semiannually; $76.03

Section 13.3 (page 480)
1. 3 **3.** 3 **5.** $6400 **7.** 6% **9.** 8% **13.** (a) $14,347.42 (b) $16,162.26 **15.** 10% **17.** $1\frac{1}{2}$ years **19.** 28 years **21.** 20 years **23.** 35 years **25.** $40,224.31 **27.** $7185.73 **29.** $21,247.04

Section 13.4 (page 485)
1. $3810.39; $989.61 **3.** $9077.95; $3122.05 **5.** $7543.32; $956.68 **9.** $3564.93; $1435.07 **11.** $23,368.76 **13.** $7096.44 **15.** $3800 in 5 years **17.** $38,288.45; $31,409.86 **19.** $23,520; $18,534.49 **21.** (a) $26,620 (b) $20,989.69

Chapter 13 Review Exercises (page 490)
1. $18,093.36; $5693.36 **2.** $8867.39; $1867.39 **3.** $6455.47; $1655.47 **4.** $23,758.73; $5758.73 **5.** $11,261.45; $2261.45 **6.** $14,576.65; $2576.65 **7.** 7.19% **8.** 8.24% **9.** 7.12% **10.** 9.38% **11.** $24.29 **12.** $24.21 **13.** $22.68 **14.** $3523.23 **15.** $7248.84 **16.** $9648.39 **17.** $4934.61 **18.** $7328.66 **19.** $12,118.30 **20.** $20,058.47; $9458.47 **21.** $7095.34; $2095.34 **22.** 6% **23.** 10% **24.** 8% **25.** 3 **26.** $3\frac{1}{2}$ **27.** $12,178.08; $2121.92 **28.** $3071.58; $928.42 **29.** $3661.63; $2338.37 **30.** $2361.30; $638.70 **31.** $4120.01 **32.** $5238.54 **33.** $40,056.23; $22,056.23 **34.** $258.14 **35.** $29,462.37 **36.** (a) $15,173.40 (b) $13,481.37 **37.** 4 **38.** approx. 12 years

Chapter 13 Summary Exercise (page 492)
(a) $4,053,386; $3,016,100 **(b)** $2,539,386; $1,889,542 **(c)** $2,798,302; $2,082,199 **(d)** $2,539,386; $1,889,542; $2,798,302; $2,082,199; $4,053,386; $3,016,100

Chapter 13 Net Assets (page 493)
1. 9,620,000,000 checks per year **2.** $4,094,400 **3.** $4834.03 **4.** $145,600,000

Chapter 14

Section 14.1 (page 504)
1. 15.19292972 **3.** 18.59891389 **5.** $58,248.89; $34,448.89 **7.** $73,105.94; $48,105.94 **9.** $67,410.39; $22,610.39 **11.** $46,016.57; $7616.57 **13.** $13,475.82; $3875.82 **15.** $138,997.66; $33,733.66 **17.** $11,320.10 **19.** $13,673.64 **21.** $16,577.77 **23.** $36,895.94 **25.** $57,782.55 **27.** $59,794.96 **29.** $122,676.78 **31.** $263,879.29 **35.** $29,117.87;

$9117.87 **37.** $5303.43; $1703.43 **39.** $1413.72 **41.** $952.33 **43.** 28 quarterly payments; $15,516.67 **45.** 45 monthly payments; $40,745.91 **47.** $7188.27 **49.** (a) $18,366.78 (b) $3966.78 **51.** $34,186.78; $14,706.78 **53.** 31 months; $4075.14 **55.** (a) $15,100.50 (b) $13,566.97

Section 14.2 (page 512)
1. 6.81086449 **3.** 8.82711975 **5.** $14,992.53 **7.** $11,901.98 **9.** $6540.57 **13.** $839,255.24 **15.** $239,326.76 **17.** $12,675.75; $4124.25 **19.** (a) $160,121.64 (b) $117,647.09 **21.** (a) $48,879.75 (b) no **23.** $79,636.35 **25.** $51,000 down and $8000 per month **27.** $420,000 today **29.** $28,847.04 **31.** $27,050.13

Section 14.3 (page 517)
1. 0.05427783 **3.** 0.00295961 **5.** $1957.50 **7.** $576.19 **11.** (a) $106,747.76 (b) $699.22 **13.** (a) $305,866.11 (b) $2003.48 **15.** (a) $2087.67 (b) $2947.96 **17.** $9572.45 **19.** $404,143.76; $766,849.92 **21.** $17,047.28; $443,229.18; $443,229.18; $213,090.95; $35,458.33; $691,778.46; $691,778.46; $213,090.98; $55,342.28; $960,211.72 **23.** (a) $53,392.91 (b) $37,567.83

Chapter 14 Review Exercises (page 522)
1. $69,152.45 **2.** $34,426.47 **3.** $113,736 **4.** $49,153.29 **5.** $163,383.42 **6.** $97,457.37 **7.** $14,943.67; $2891.62 **8.** $134,489.45; $21,989.45 **9.** $38,253.24 **10.** $5948.27 **11.** $8055.06 **12.** $5094.30 **13.** $11,298.18 **14.** $1332.74 **15.** $8702.22 **16.** $623.89 **17.** $53,922.04 **18.** 31 years **19.** $734.85 **20.** $5596.62 **21.** $25,723.79 **22.** $2676.72 **23.** $32,625.58 **24.** $98,713.33 **25.** $179,999.95; $125,806.46 **26.** $21,866.87; $21,866.87; $1968.02; $45,701.76; $45,701.76; $21,866.87; $4113.16; $71,681.79; $71,681.79; $21,866.85; $6451.36; $100,000.00

Chapter 14 Summary Exercises (page 524)
(a) $439,542.06 **(b)** $53,046.70 **(c)** $520,820.32 **(d)** No, short by $81,278.26

Chapter 14 Net Assets (page 525)
1. $58,484.75 **2.** $104,000,000 **3.** $434,762.08 **4.** No, they will be short by $24,272,142.95

Chaper 15

Section 15.1 (page 534)
1. $10.03 **3.** $4.87 **5.** $13.99; $853.80; $853.80; $14.51; $1002.06; $1002.06; $17.04; $985.40; $985.40; $16.75; $958.58 **7.** $691.16; $10.37; $720.44 **9.** $312.91; $4.69; $285.94 **11.** 681.52; $10.22; $769.01 **13.** (a) $42.01 (b) $28 (c) $14.01 **15.** (a) $72.76 (b) $38.80 (c) $33.96

Section 15.2 (page 544)
1. $20,896.40; $3496.40 **3.** $180; $30 **5.** $100; $10 **7.** 12.50% **9.** 10.25% **11.** 11.25% **15.** 11% **17.** 12.75% **19.** (a) 732.212 pesos (b) 13.75% **21.** (a) $209,239.80 (b) 12%

Section 15.3 (page 550)
1. $5130.57; $130.57 **3.** $3754.46; $254.46 **5.** $4333.78; $333.78 **7.** $385.52 **9.** $21.90 **11.** $139.81 **17.** (a) $6569.44 (b) $131,569.44 **19.** $83.24 **21.** (a) $51.82 (b) $490.58 **23.** $41,176.11; $1876.11

Section 15.4 (page 556)
1. 0.11328724 **3.** 0.10334306 **5.** $1030.32 **7.** $404.73 **9.** $4185.25 **11.** $161.95 **13.** $152.64; $695.04 **15.** $321.92; $3452.16 **17.** $15,987.97; $147,663.16

19.

Payment Number	Amount of Payment	Interest for Period	Portion to Principal	Principal at End of Period
0	—	—	—	$4000.00
1	$1207.68	$320.00	$887.68	3112.32
2	1207.68	248.99	958.69	2153.63
3	1207.68	172.29	1035.39	1118.24
4	1207.70	89.46	1118.24	0

21. Payment Number	**Amount of Payment**	**Interest for Period**	**Portion to Principal**	**Principal at End of Period**
0	—	—	—	$14,500.00
1	$374.77	$132.92	$241.83	14,258.15
2	374.77	130.70	244.07	14,014.08
3	374.77	128.46	246.31	13,767.77
4	374.77	126.20	248.57	13,519.20
5	374.77	123.93	250.84	13,268.36

25. Payment Number	**Amount of Payment**	**Interest for Period**	**Portion to Principal**	**Principal at End of Period**
0	—	—	—	$8000.00
1	$576.99	$80.00	$496.99	7503.01
2	576.99	75.03	501.96	7001.05
3	576.99	70.01	506.98	6494.07
⋮	⋮	⋮	⋮	⋮
14	576.99	11.37	565.62	571.28
15	576.99	5.71	571.28	0

Section 15.5 (page 564)
1. $532.74 **3.** $1111.08 **5.** $1022.90 **9.** $651.21 **11.** $523.87 **13.** $873.82 **15.** Yes, they are qualified

17. Payment Number	**Amount of Payment**	**Interest for Period**	**Portion to Principal**	**Principal at End of Period**
0	—	—	—	$122,500.00
1	$1135.58	$765.63	$369.95	122,130.05
2	1135.58	763.31	372.27	121,757.78

19. $763.84; $95,321.60 **21.** $81,000

Chapter 15 Review Exercises (page 570)
1. $3.65 **2.** $40.51 **3.** $14.18 **4.** $552.56; $8.29; $551.80 **5.** $216.51; $3.25; $203.40 **6.** $2700; $2915.14; $215.14 **7.** $4600; $5040; $440 **8.** $8000; $9600; $1600 **9.** 13% **10.** 13.75% **11.** 12.75% **12.** 11.5% **13.** $10,876.33; $476.33 **14.** $7309.87; $309.87 **15.** $915.66; $115.66 **16.** $4641.89; $141.89 **17.** $3638.63 **18.** $2395.83 **19.** $795.42 **20.** $392.89 **21.** $442.51; $1220.24 **22.** $234.38; $2343.96 **23.** $245.94; $2805.12 **24.** $574.16; $2224.80 **25.** $582.38 **26.** $523.90 **27.** $956 **28.** $819.60 **29.** (a) $2788.22 (b) $173.22 **30.** $57.28; $0.86 **31.** $18,756.32; 10.50% **32.** $2822.98 **33.** $629.48 **34.** (a) $1398.25 (b) $360,580

Chapter 15 Summary Exercise (page 572)
(a) $495.08; $420.06; $950.53; $149.46; $2015.13 **(b)** $2015.13; $215.00; $120.00; $210.83; $2560.96 **(c)** $392.36; $285.64; $695.83; $149.46; $187.00; $120.00, $210.83, $2041.12 **(d)** $519.84

Chapter 15 Net Assets (page 574)
1. $369 **2.** $559.44; $676.20 **3.** $375.32; $535.03

Part 4 Cumulative Review Chapters 11–15 (page 575)
1. $150 **2.** $417.64 **3.** $2400.17 **4.** $12,449.83 **5.** 9.7% **6.** 13.5% **7.** 80 days **8.** 150 days **9.** $270; $8730 **10.** $33.18; $841.82 **11.** 10.3% **12.** 11.3% **13.** $1947.90 **14.** $7599.42 **15.** $12,686.07; $86.07 **16.** $7562.10; $62.10 **17.** $583.49;

$416.51 **18.** $12,182.15; $6817.85 **19.** $9214.23 **20.** $13,950.64 **21.** $13,367.28 **22.** $639,160.10 **23.** $403.46 **24.** $112.90 **25.** $10,129.08 **26.** $2173.88 **27.** (a) $12,106.67 (b) $106.67 **28.** (a) $416.67 (b) $14,583.33 **29.** $7993.14 **30.** $94 **31.** (a) $94.91 (b) $50.62 (c) $44.29 **32.** (a) $12,340 (b) $1840 (c) $9500 (d) 12% **33.** $3906.25 **34.** $9.828 million **35.** (a) $11,957.04 (b) $27,257.87 **36.** $469.22 **37.** $20.01 **38.** $881.44

Chapter 16

Section 16.1 (page 583)

1. 20% **3.** 12.5% **5.** 5% **7.** $6\frac{2}{3}$% **9.** 1.25% **11.** 2% **13.** $450 **15.** $800 **17.** $840 **19.** $2850 **21.** $5000 **23.** $2775 **25.** $73,000

29.

Year	Computation	Amount of Depreciation	Accumulated Depreciation	Book Value
0	—	—	—	$12,000
1	($33\frac{1}{3}$% × $9000)	$3000	$3000	$9,000
2	($33\frac{1}{3}$% × $9000)	$3000	$6000	$6,000
3	($33\frac{1}{3}$% × $9000)	$3000	$9000	$3,000

31. Book values: $8200; $7000; $5800; $4600; $3400; $2200 **33.** (a) $55,000 depreciation (b) $1,025,000 book value **35.** (a) $2527 depreciation (b) $20,211 book value **37.** (a) $12\frac{1}{2}$% (b) $90 (c) $790

Section 16.2 (page 589)

1. 40% **3.** 25% **5.** $13\frac{1}{3}$% **7.** 20% **9.** $33\frac{1}{3}$% **11.** 4% **13.** $3600 **15.** $4200 **17.** $1900 **19.** $3360 **21.** $1215 **23.** $6834 **25.** $750

29.

Year	Computation	Amount of Depreciation	Accumulated Depreciation	Book Value
0	—	—	—	$14,400
1	(50% × $14,400)	$7200	$7,200	$7,200
2	(50% × $7200)	$3600	$10,800	$3,600
3	(50% × $3600)	$1800	$12,600	$1,800
4	—	$1800*	$14,400	$0

*To depreciate to 0 scrap value

31. Book values: $19,125; $14,344; $10,758; $8068; $6051; $4538; $3500; $3500 **33.** $1153 **35.** $235 **37.** (a) 25% (b) $1450 (c) $4425 (d) $1375

Section 16.3 (page 596)

1. $\frac{4}{10}$ **3.** $\frac{6}{21}$ **5.** $\frac{7}{28}$ **7.** $\frac{10}{55}$ **9.** $1640 **11.** $10,000 **13.** $600 **15.** $7700 **17.** $2700 **19.** $890 **21.** $2400

25.

Year	Amount of Computation	Accumulated Depreciation	Depreciation	Book Value
0	—	—	—	$10,800
1	($\frac{6}{21}$ × $8400)	$2400	$2400	$8,400
2	($\frac{5}{21}$ × $8400)	$2000	$4400	$6,400
3	($\frac{4}{21}$ × $8400)	$1600	$6000	$4,800
4	($\frac{3}{21}$ × $8400)	$1200	$7200	$3,600
5	($\frac{2}{21}$ × $8400)	$800	$8000	$2,800
6	($\frac{1}{21}$ × $8400)	$400	$8400	$2,400

27. Book values: \$2100; \$1600; \$1200; \$900; \$700; \$600 **29.** \$6000 **31.** \$24,584 **33.** (a) \$2120 (b) \$1696 **35.** (a) $\frac{8}{36}$ (b) \$2360 (c) \$10,620 (d) \$4750

Section 16.4 (page 603)

1. \$0.35 **3.** \$0.029 **5.** \$68.75 **7.** \$30 **9.** \$17,940 **11.** \$17,280 **13.** \$2775 **15.** \$6960 **19.** \$1313; \$2250 **21.** \$1667; \$1833 **23.** \$450; \$825 **25.** \$900; \$1088

27.

Year	Computation	Amount of Depreciation	Accumulated Depreciation	Book Value
0	—	—	—	\$6800
1	(1350 × \$1.26)	\$1701	\$1701	\$5099
2	(1820 × \$1.26)	\$2293	\$3994	\$2806
3	(730 × \$1.26)	\$920	\$4914	\$1886
4	(1100 × \$1.26)	\$1386	\$6300	\$500

29. \$26,350 **31.** \$3650; \$14,600 **33.** \$450; \$1620 **35.** \$2775; \$5365

Section 16.5 (page 610)

1. 19.2% **3.** 11.52% **5.** 20% **7.** 3.636% **9.** 5.76% **11.** 2.564% **13.** \$1751 **15.** \$3226 **17.** \$4800 **19.** \$6254 **21.** \$16,920 **23.** \$2756 **25.** \$79,364

29.

Year	Computation	Amount of Depreciation	Accumulated Depreciation	Book Value
0	—	—	—	\$10,980
1	(33.33% × \$10,980)	\$3660	\$3,660	\$7,320
2	(44.45% × \$10,980)	\$4881	\$8,541	\$2,439
3	(14.81% × \$10,980)	\$1626	\$10,167	\$813
4	(7.41% × \$10,980)	\$813*	\$10,980	\$0

*Due to rounding in prior years

31. Book values: \$110,430; \$88,344; \$70,675; \$56,540; \$45,227; \$36,184; \$28,147; \$20,110; \$12,061; \$4024; \$0 **33.** \$1066 **35.** Year 1: \$12,326; years 2–5: \$12,307

Chapter 16 Review Exercises (page 616)

1. 20%; 40%; $\frac{5}{15}$ **2.** $16\frac{2}{3}\%$; $33\frac{1}{3}\%$; $\frac{6}{21}$ **3.** 25%; 50%; $\frac{4}{10}$ **4.** 10%; 20%; $\frac{10}{55}$ **5.** 5%; 10%; $\frac{20}{210}$ **6.** $12\frac{1}{2}\%$; 25%; $\frac{8}{36}$ **7.** 11.52% **8.** 33.33% **9.** 4.522% **10.** 17.49% **11.** 2.564% **12.** 3.174% **13.** \$940 **14.** \$43,000 **15.** \$21,375 **16.** \$7400 **17.** year 1: \$2700; year 2: \$2025; year 3: \$1350; year 4: \$675 **18.** \$4000 **19.** \$1530 **20.** (a) \$4836; \$2666; \$3007; \$4712 (b) \$15,264; \$12,598; \$9591; \$4879 **21.** \$15,120 **22.** \$24,000; \$19,200; \$14,400; \$9600; \$4800 **23.** \$2900 **24.** \$1260; \$2003 **25.** \$2840; \$5112 **26.** \$5702 **27.** \$4746 **28.** \$37,575 **29.** \$6222 **30.** \$4788 **31.** \$1157; \$1984; \$1417; \$1012; \$723 **32.** \$2,440,928

Chapter 16 Summary Exercise (page 618)

(a) \$114,000 **(b)** \$61,560 **(c)** \$228,000 **(d)** \$57,000 straight-line; \$24,624 double-declining balance; \$38,000 sum-of-the-years'-digits

Chapter 16 Net Assets (page 619)

1. \$11,700 **2.** \$4860 **3.** (a) \$0.18 (b) \$9000 **4.** \$6480

Chapter 17

Section 17.1 (page 625)

1. (a) \$316,350 (b) \$88,050 (c) \$65,350 **3.** \$324,200

5. FUTURE TECH COMPUTING
INCOME STATEMENT
YEAR ENDING DECEMBER 31

Gross sales			\$284,000
Returns			\$6,000
Net sales			\$278,000
Inventory, January 1		\$58,000	
Cost of goods purchased	\$232,000		
Freight	\$3,000		
Total cost of goods purchased		\$235,000	
Total of goods available for sale		\$293,000	
Inventory, December 31		\$69,000	
Cost of goods sold			\$224,000
Gross profit			\$54,000
Expenses			
Salaries and wages		\$15,000	
Rent		\$6,000	
Advertising		\$2,000	
Utilities		\$1,000	
Taxes on inventory, payroll		\$3,000	
Miscellaneous expenses		\$4,000	
Total expenses			\$31,000
Net Income before Taxes			\$23,000
Income taxes			\$2,400
Net Income			\$20,600

7. KATHY GILMORE, CONSULTANT
INCOME STATEMENT FOR THE
YEAR ENDING DECEMBER 31

Gross sales			\$170,500
Returns			0
Net sales			\$170,500
Inventory, January 1		0	
Cost of goods purchased	0		
Freight	0		
Total cost of goods purchased		0	
Total of goods available for sale		0	
Inventory, December 31		0	
Cost of goods sold			0
Gross profit			\$170,500

Expenses		
Salaries and wages	$63,000	
Rent	$28,000	
Advertising	$12,000	
Utilities	$4,000	
Taxes on inventory, payroll	$3,800	
Miscellaneous expenses	$9,400	
Total expenses		$120,200
Net Income before Taxes		$50,300
Income taxes		$6,800
Net Income		$43,500

Section 17.2 (page 633)

1. 50.7%; 29.2%

3. CAPITAL APPLIANCE CENTER

	Amount	Percent	Percent from Table 17.1
Net sales	$900,000	100.0%	100.0%
Cost of goods sold	617,000	68.6%	66.9%
Gross profit	283,000	31.4%	33.1%
Wages	108,900	12.1%	11.9%
Rent	20.700	2.3%	2.4%
Advertising	27,000	3%	2.5%
Total expenses	216,000	24%	26%
Net income before taxes	67,000	7.4%	7.2%

5. BEST TIRES, INC.
COMPARATIVE INCOME STATEMENT

	This Year		Last Year	
	Amount	Percent	Amount	Percent
Gross sales	$1,856,000	100.3%	$1,692,000	100.7%
Returns	6,000	0.3%	12,000	0.7%
Net sales	$1,050,000	100.0%	$1,680,000	100.0%
Cost of goods sold	1,202,000	65.0%	1,050,000	62.5%
Gross profit	648,000	35.0%	630,000	37.5%
Wages	152,000	8.2%	148,000	8.8%
Rent	82,000	4.4%	78,000	4.6%
Advertising	111,000	6.0%	122,000	7.3%
Utilities	32,000	1.7%	17,000	1.0%
Taxes on inv., payroll	17,000	0.9%	18,000	1.1%
Miscellaneous expenses	62,000	3.4%	58,000	3.5%
Total expenses	456,000	24.6%	441,000	26.3%
Net income	$192,000	10.4%	$189,000	11.3%

7. 82.7%; 17.3%; 13.9%; 3.4%; 6.5%; 0.8%; 1.0% **9.** 67.9%; 32.1%; 23.5%; 8.6%; 12.3%; 2.4%; 1.4%

Section 17.3 (page 639)

1. APPLE CONSTRUCTION
BALANCE SHEET FOR DECEMBER 31 (IN MILLIONS)

Assets			
Current assets			
Cash	$273		
Notes receivable	$312		
Accounts receivable	$264		
Inventory	$180		
Total current assets		$1029	
Plants assets			
Land	$466		
Buildings	$290		
Fixtures	$28		
Total plant assets		$784	
Total Assets			$1813
Liabilities			
Current liabilities			
Notes payable	$312		
Accounts payable	$63		
Total current liabilities		$375	
Long-Term liabilities			
Mortgages payable	$212		
Long-term notes payable	$55		
Total long-term liabilities		$267	
Total Liabilities			$642
Owner's Equity			
Owner's equity	$1171		
Total Liabilities and Owner's Equity			$1813

Section 17.4 (page 647)

1. INTERSTATE RUBBER SUPPLY
COMPARATIVE BALANCE SHEET

	This Year		Last Year	
	Amount	Percent	Amount	Percent
Assets				
Current assets				
Cash	$52,000	13%	$42,000	13.1%
Notes receivable	$8,000	2%	$6,000	1.9%
Accounts receivable	$148,000	37%	$120,000	37.5%
Inventory	$153,000	38.3%	$120,000	37.5%
Total current assets	$361,000	90.3%	$288,000	90%

	This Year		Last Year	
	Amount	**Percent**	**Amount**	**Percent**
Plant assets				
Land	$10,000	2.5%	$8,000	2.5%
Buildings	$14,000	3.5%	$11,000	3.4%
Fixtures	$15,000	3.8%	$13,000	4.1%
Total plant assets	$39,000	9.8%	$32,000	10%
Total Assets	$400,000	100.0%	$320,000	100.0%
Liabilities				
Current liabilities				
Accounts payable	$3,000	0.8%	$4,000	1.3%
Notes payable	$201,000	50.3%	$152,000	47.5%
Total current liabilities	$204,000	51%	$156,000	48.8%
Long-term liabilities				
Mortgages payable	$20,000	5%	$16,000	5%
Long-term notes payable	$58,000	14.5%	$42,000	13.1%
Total long-term liabilities	$78,000	19.5%	$58,000	18.1%
Total Liabilities	$282,000	70.5%	$214,000	66.9%
Owner's Equity	$118,000	29.5%	$106,000	33.1%
Total Liabilities and Owner's Equity	$400,000	100.0%	$320,000	100.0%

3. (a) 1.77:1 (b) 1.02:1 (c) No, current ratio is low. **5.** (a) 1.79:1 (b) 0.62:1 (c) No, both ratios are low. **7.** 2.13:1; 1.27:1 **9.** 2.39 or 239% **11.** 1.47 or 147% **15.** 6.1% **17.** 16.9 times; 21.6 days

Chapter 17 Review Exercises (page 654)

1. $187,400; $98,200 **2.** $263,730; $138,930 **3.** $141,600; $43,200 **4.** $235,880; $23,580 **5.** $416,600 **6.** $126,500 **7.** $42,700 **8.** $282,100 **9.** $188,800; 0.56 **10.** $90,600; 0.23

11. LORI'S BOUTIQUE
INCOME STATEMENT
FOR THE YEAR ENDING DECEMBER 31

Gross sales			$175,000
Returns			$8,000
Net sales			$167,000
Inventory, January 1		$44,000	
Cost of goods purchased	$126,000		
Freight	$2,000		
Total cost of goods purchased		$128,000	
Total of goods available for sale		$172,000	
Inventory, December 31		$52,000	
Cost of goods sold			$120,000
Gross profit			$47,000
Expenses			
Salaries and wages		$9,000	

Rent	$4,000	
Advertising	$1,500	
Utilities	$1,000	
Taxes on inventory, payroll	$2,000	
Miscellaneous expenses	$3,000	
Total expenses		$20,500
Net Income before Taxes		$26,500

12. THE GUITAR WAREHOUSE
INCOME STATEMENT
FOR THE YEAR ENDING DECEMBER 31

Gross sales			$2,215,000
Returns			$26,000
Net sales			$2,189,000
Inventory, January 1		$215,000	
Cost of goods purchased	$1,123,000		
Freight	4,000		
Total cost of goods purchased		$1,127,000	
Total of goods available for sale		$1,342,000	
Inventory, December 31		$265,000	
Cost of goods sold			$1,077,000
Gross profit			$1,112,000
Expenses			
Salaries and wages		$154,000	
Rent		$59,000	
Advertising		$11,000	
Utilities		$12,000	
Taxes on inventory, payroll		$10,000	
Miscellaneous expenses		$9,000	
Total expenses			$255,000
Net Income before Taxes			$857,000
Income taxes			$242,300
Net Income			$614,700

13. 59.8%; 12.8% **14.** 61.2%; 29.1%

15. ANDY'S STEAK HOUSE

	Amount	**Percent**	**Percent from Table 17.1**
Net sales	$300,000	100%	100%
Cost of goods sold	$125,000	41.7%	48.4%
Gross profit	$175,000	58.3%	51.6%
Wages	$72,000	24.0%	26.4%
Rent	$12,000	4.0%	2.8%
Advertising	$5,700	1.9%	1.4%
Total expenses	$123,000	41.0%	43.7%
Net income	$52,000	17.3%	7.9%

16. GASKETS, INC.
BALANCE SHEET
FOR DECEMBER 31

	Assets		
Current assets			
Cash	$240,000		
Notes receivable	$180,000		
Accounts receivable	$460,000		
Inventory	$225,000		
Total current assets		$1,105,000	
Plant assets			
Land	$180,000		
Buildings	$260,000		
Fixtures	$48,000		
Total plant assets		$488,000	
Total Assets			$1,593,000
	Liabilities		
Current liabilities			
Notes payable	$410,000		
Accounts payable	$882,000		
Total current liabilities		$1,292,000	
Long-term liabilities			
Mortgages payable	$220,000		
Long-term notes payable	$194,000		
Total long-term liabilities		$414,000	
Total liabilities			$1,706,000
	Owner's Equity		
Owner's equity	($113,000)		
Total Liabilities and Owner's Equity			$1,593,000

17. 1.32:1; 0.46:1; 1.78:1 **18.** 0.83:1; 0.04:1; 4.17:1 **19.** 0.78:1; 0.71:1; 1.11:1 **20.** 5.0; 73 days **21.** 8.6; 42.4 days

22.

	This Year		Last Year	
	Amount	Percent	Amount	Percent
Current assets				
Cash	$28,000	18.8%	$22,000	21.1%
Notes receivable	$12,000	8.1%	$15,000	14.4%
Accounts receivable	$39,000	26.2%	$31,500	30.1%
Inventory	$22,000	14.8%	$20,000	19.1%
Total current assets	$101,000	67.8%	$88,500	84.7%
Total plant and equipment	$48,000	32.2%	$16,000	15.3%
Total Assets	$149,000	100.0%	$104,500	100.0%
Total current liabilities	$38,000	25.5%	$36,000	34.4%

Chapter 17 Summary Exercise (page 658)

(a) WALKER BICYCLE SHOP
INCOME STATEMENT
YEAR ENDING DECEMBER 31

Gross sales			$212,000
Returns			$12,500
Net sales			$199,500
Inventory, January 1		$44,000	
Cost of goods purchased	$75,000		
Freight	$8,000		
Total cost of goods purchased		$83,000	
Total of goods available for sale		$127,000	
Inventory, December 31		$26,000	
Cost of goods sold			$101,000
Gross profit			$98,500
Expenses			
Salaries and wages		$37,000	
Rent		$12,000	
Advertising		$2,000	
Utilities		$3,000	
Taxes on inventory, payroll		$7,000	
Miscellaneous expenses		$4,500	
Total expenses			$65,500
Net Income before Taxes			$33,000

(b) 106.3%; 18.5%; 6.3%; 6%; 50.6%; 1.5%

(c) WALKER BICYCLE SHOP
BALANCE SHEET
DECEMBER 31

	Assets		
Current assets			
Cash	$62,000		
Notes receivable	$2,500		
Accounts receivable	$8,200		
Inventory	$26,000		
Total current assets		$98,700	
Plant assets			
Land	$7,600		
Buildings	$28,000		
Fixtures	$13,500		
Total plant assets		$49,100	
Total assets			$147,800

	Liabilities		
Current liabilities			
Notes payable	$4,500		
Accounts payable	$27,000		
Total current liabilities		$31,500	
Long-term liabilities			
Mortgages payable	$15,000		
Long-term notes payable	$8,000		
Total long-term liabilities		$23,000	
Total liabilities			$54,500
	Owner's Equity		
Owner's equity	$93,300		
Total Liability and Owner's Equity			$147,800

(d) 3.13:1; 2.31:1

Chapter 17 Net Assets (page 661)
1. $489,795.92; Answers will vary **2.** 1.506 **3.** $15 billion; Answers will vary

Chapter 18

Section 18.1 (page 667)
1. $0; $0 **3.** $0.30; $0.14 **5.** $20; $0 **7.** $0.10 **9.** $8; $2.56 **11.** $1.60; $0.86 **13.** $20; $5.06 **15.** $5.83 **17.** $4.38 **19.** (a) $0.77 (b) $0.90 (c) 14.2% **21.** $26,666.67

Section 18.2 (page 678)
1. $57\frac{7}{16}$ **3.** $59\frac{5}{8}$ **5.** $30\frac{7}{8}$ **7.** 1.44 **9.** 808,300 **11.** $41,562.50 **13.** $4500 **15.** $2687.50 **17.** $5607.88 **19.** $45,979.50 **21.** $2177.06 **23.** $37,960.93 **25.** $9012.44 **27.** $4013.73 **29.** $43,120.55 **31.** 0.9% **33.** 3.6% **35.** 1.4% **37.** 19 **39.** 9 **41.** $48,707.25 **43.** $2330.96 **47.** (a) $47,512.76 (b) $77,380.98 (c) $29,868.22

Section 18.3 (page 684)
1. $71\frac{1}{2}$ **3.** $9\frac{3}{4}$; $97.50 **5.** −2 **7.** $40,050 **9.** $94,500 **11.** $45,437.50 **13.** $26,000 **17.** $18,800 **19.** $18,237.50 **21.** (a) $18,087.50 (b) $1187.50 **23.** $4500 **25.** $93,333.33

Section 18.4 (page 689)
1. $29,000 each **3.** $36,800 for 1; $9200 for 2 **5.** $16,920 for 1; $8080 for 2 **7.** $10,000 for 1; $12,000 for 2; 0 for 3 **9.** $42,250 each **11.** $3120; $6240; $6240 **13.** $22,500; $37,500; $52,500; $67,500 **15.** $23,400 for Finch; $33,600 for Renz **17.** (a) $21,000 for Coker; $39,000 for Toms (b) $7500 for Coker; $1500 by Toms to Coker **19.** $22,179 to 1; $46,464 to 2; $41,357 to 3

Section 18.5 (page 694)
1. $60,000; $120,000; $150,000 **3.** $18,000; $27,000; $30,000; $45,000 **5.** $64,615; $37,692; $37,692 **7.** $9739; $20,870; $2087; $15,304 **9.** $286,000; $156,000; $208,000 **11.** $190,476; $228,571; $266,667; $114,286 **13.** $24,000; $96,000; $72,000; $108,000; $12,000; $48,000 **15.** $8079; $21,545; $16,158; $10,099; $12,119 **17.** $720; $960; $3360; $960

Chapter 18 Review Exercises (page 702)
1. (a) $3.25 (b) $0.65 **2.** (a) $12 (b) $7.23 **3.** (a) $8.75 (b) $2.69 **4.** $1.55 **5.** $2.73 **6.** $3.70 **7.** $56\frac{1}{8}$ **8.** $34\frac{1}{2}$ **9.** $12\frac{5}{8}$ **10.** $+\frac{3}{8}$ **11.** 0.06 **12.** 413,100 **13.** 0.5% **14.** $21\frac{13}{16}$ **15.** $10\frac{3}{4}$ **16.** $1500 **17.** $5456.25 **18.** $55,550 **19.** $8449.88 **20.** $25,240.44 **21.** $3016.45 **22.** $8367.13 **23.** 0.6% **24.** 0.2% **25.** 59 **26.** 38 **27.** $91\frac{3}{4}$ **28.** 1 **29.** 2033 **30.** No change **31.** $27,675 **32.** $6\frac{3}{4}$%; $67.50 **33.** $45,500 **34.** $98,625 **35.** $16,000 to each **36.** $48,000 to 1; $72,000 to 2

37. \$38,571 to 1; \$51,429 to 2 **38.** \$25,000; \$41,667; \$33,333 **39.** \$23,256; \$58,140; \$43,605 **40.** \$25,870; \$16,630; \$22,174 **41.** \$18,562.50 **42.** \$24,156

Chapter 18 Summary Exercise (page 704)

(a) 30,000; 20,000; 50,000 **(b)** \$4 preferred; \$0.73 common **(c)** \$53,900; \$54,600; \$84,500 **(d)** \$34,375; \$4125 **(e)** Math—\$62,125; English—\$44,375; Reading—\$35,500 **(f)** Yes; probably not

Chapter 18 Net Assets (page 705)

1. \$29,067.06 **2.** \$3,250,000; \$7250 per year **3.** \$4,222,395.83; \$154,687.50

Chapter 19

Section 19.1 (page 712)

1. \$65.08 **3.** 97% **5.** 51.4% **7.** 4 **9.** 6 **11.** 5

13.

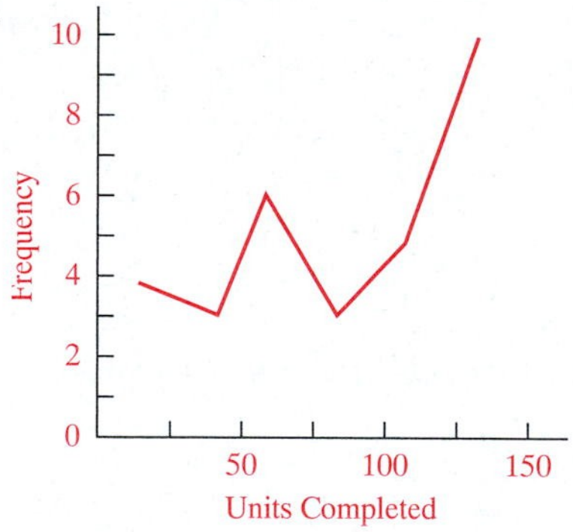

15. 23 **17.** 7 **21.** (a) almost 1 acre (b) almost $3\frac{7}{8}$ acres (c) 2 mph (d) 99 inches

23.

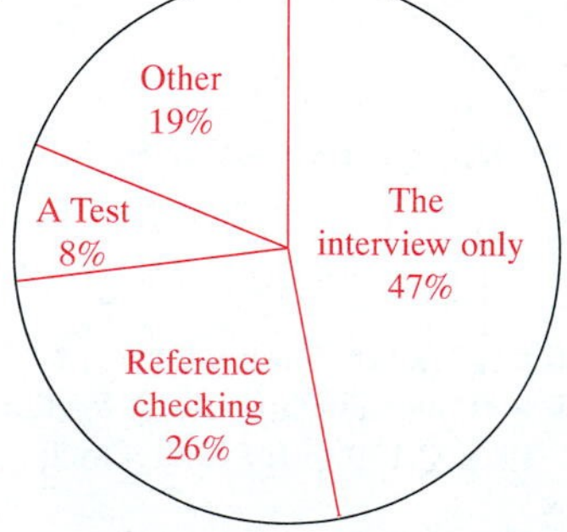

25. (a) 3.5% (b) 2% to 5% (c) 4% to 13% (d) 10%

Section 19.2 (page 721)

1. 228 **3.** 3689.6 **5.** 3945 pounds **7.** \$1300 **9.** \$657,125 **11.** 39.4 **13.** 167.2 **17.** \$31,000 **19.** 1.62 **21.** 76.3 **23.** 105.8 **25.** 68%

Section 19.3 (page 727)

1. 32 **3.** 88 **5.** 0.83 **7.** 60 **9.** none **11.** 4 and 6 **15.** 81.2 feet; 82 feet **19.** 115.1; 114.5; 114.5 **21.** 3.01 lb. (rounded); probably so

Section 19.4 (page 734)

1. 18; 1.9 **3.** 21; 1.6 **5.** 55.9; 3.1 **7.** 43 **9.** 388 **13.** 100 **15.** 136 **17.** 27 **19.** 190 **21.** 2.5% **23.** 95% **25.** 50% **27.** 16% **29.** 95% **31.** $\frac{1}{2}$% **33.** 84% **35.** 97.5% **37.** 84% **39.** 97.5% **41.** 68% **43.** 1 claim

Section 19.5 (page 737)

1. 244.4 **3.** 60 **5.** 168 **7.** \$154.30 **9.** \$244.50 **11.** \$151.40 **13.** \$309,000 **15.** \$11,300 **17.** medical costs

Chapter 19 Review Exercises (page 744)

1. 1; 3; 10; 3; 2; 1 **2.** 6

3.

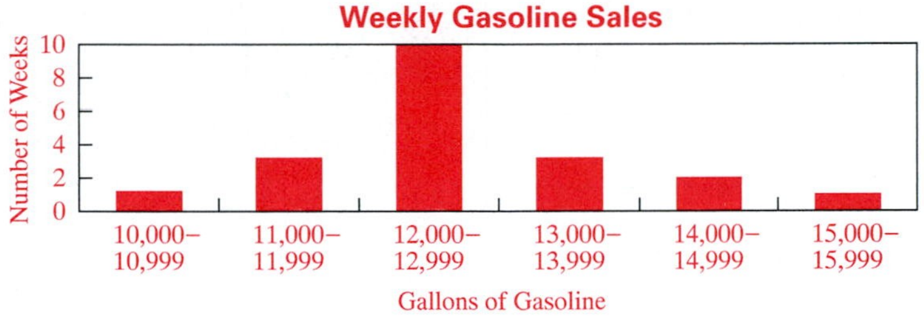

4. 72°; 10%; 108°; 108°; 36°

5.

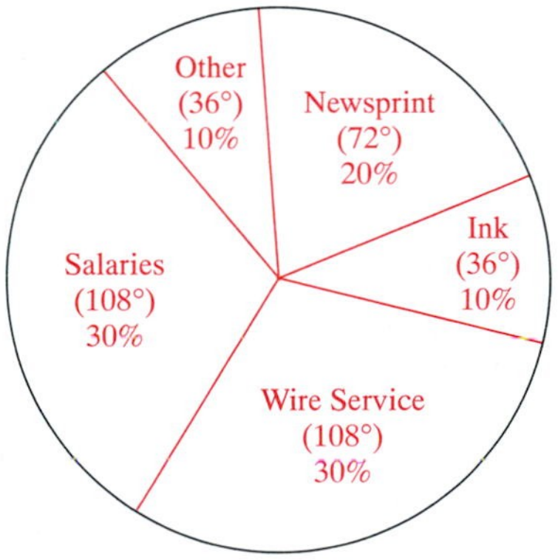

6. 60% **7.** 23.4; 20; no mode **8.** 85.6; 85; 82 **9.** 20.3; 20.5; 21 **10.** 43.3; 44; 44 **11.** 6.9; 7; 7 **12.** 2.4; 2.4; 2.4 **13.** 26.1 **14.** 34.8 **15.** 8; 3 **16.** 74.5; 74.5 **17.** 67; 22.77 **18.** 13; 4.05 **19.** 21; 7.21 **20.** 42; 14.88 **21.** 143.6 **22.** 164.0 **23.** 52.8 **24.** 162.2 **25.** \$50,700 **26.** 84%

27.

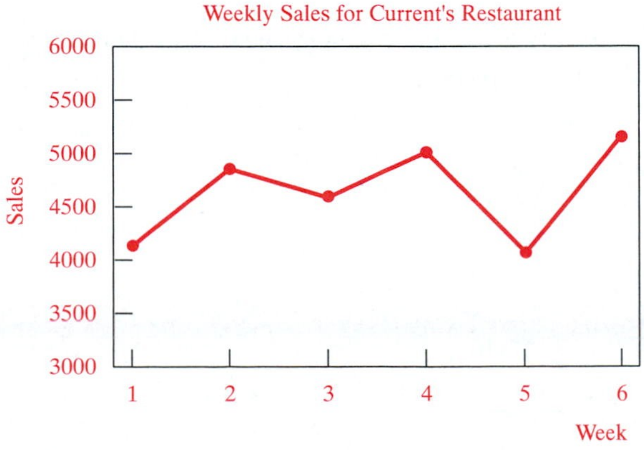

28. 2.54; 2.31 **29.** 177.3; \$159,570

30.

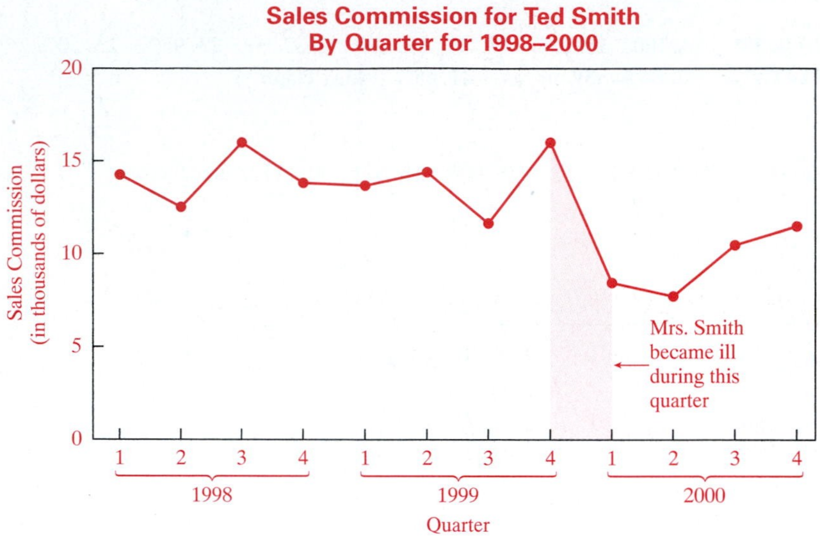

Chapter 19 Summary Exercise (page 748)

(a) Store 1: \$7.5, \$7.4, no mode; Store 2: \$8.2, \$8.2, \$8.2

(b)

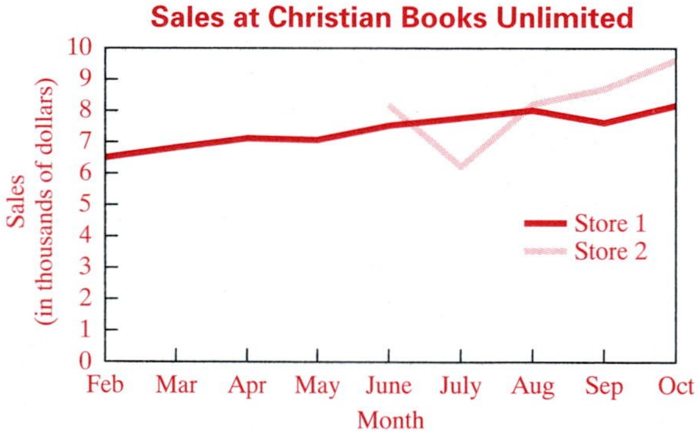

(c) Sales at Store 2 seem to be growing faster than at Store 1

Part 5 Cumulative Review Chapters 16–19 (page 750)

1. (a) 25% (b) \$1160 (c) \$2520 **2.** (a) \$3280 (b) \$2624 **3.** \$2760; \$2415; \$7645 **4.** \$15,330; \$26,920 **5.** (a) \$16,800 (b) \$33,600 (c) \$28,000 **6.** (a) 10 years (b) 5 years (c) 3 years (d) 27.5 years (e) 7 years

7.

Year	Computation	Amount of Depreciation	Accumulated Depreciation	Book Value
0	—	—	—	\$96,000
1	(33.33% × \$96,000)	\$32,000	\$32,000	\$64,000
2	(44.45% × \$96,000)	\$42,672	\$74,672	\$21,328
3	(14.81% × \$96,000)	\$14,218	\$88,890	\$7110
4	(7.41% × \$96,000)	\$7,110*	\$96,000	0

*to depreciate to 0

8. THE FASHION SHOPPE
INCOME STATEMENT
YEAR ENDING DECEMBER 31

Gross sales			$240,800
Returns			$4,300
Net sales			$236,500
Inventory, January		$48,300	
Cost of Goods Purchased	$102,000		
Freight	$2,900		
Total cost of goods purchased		$153,200	
Total of goods available for sale		$104,900	
Inventory, December 31		$41,500	
Cost of goods sold			$111,700
Gross profit			$124,800
Expenses			
Salaries and wages		$32,400	
Rent		$15,000	
Advertising		$2,200	
Utilities		$3,100	
Taxes on inventory, payroll		$6,100	
Miscellaneous expenses		$8,900	
Total expenses			$67,700
Net Income Before Taxes			$57,100
Income taxes			$11,400
Net Income			$45,700

9. 47.2%; 52.8%; 24.1%; 19.3%

10. THE FASHION SHOPPE
BALANCE SHEET FOR DECEMBER 31

	Assets		
Current assets			
Cash	$28,400		
Notes receivable	$8,400		
Accounts receivable	$3,800		
Inventory	$41,500		
Total current assets		$82,100	
Plant assets			
Land	$0		
Buildings	$0		
Fixtures	$12,200		
Total plant assets		$12,200	
Total Assets			$94,300
	Liabilities		
Current liabilities			
Notes payable	$4,800		
Accounts payable	$32,500		
Total current liabilities		$37,300	

Long-term liabilities			
Mortgages payable	$0		
Long-term notes payable	$0		
Total long-term liabilities		$0	
Total Liabilities			$37,300
Owner's Equity			
Owner's equity		$57,000	
Total Liabilities and Owner's Equity			$94,300

11. 2.20; 1.09; 0.92 **12.** $200 **13.** $0.35 **14.** $0.385 per share **15.** $8675.80 **16.** (a) 0.6% (rounded) (b) 30 (rounded) **17.** $8509.71 **18.** $20, 0.3%, $24\frac{1}{8}$, or $24.125 **19.** 6.0%, 5 bonds, $90\frac{3}{8}$, or $903.75 **20.** $30.82; 27.8% **21.** $24,437.50 **22.** $36,750 **23.** (a) $24,000 to each (b) $36,000 to partner 1; $12,000 to partner 2 (c) $19,200 to partner 1; $28,000 to partner 2 **24.** Padgett—$42,750; Harden—$42,250 **25.** $127,500; $63,750; $148,750 **26.** $72,000; $43,200; $64,800 **27.** $78,734.62; $52,407.73; $8857.64 **28.** $560; $580; 3.6% increase **29.** 50% decrease

30.

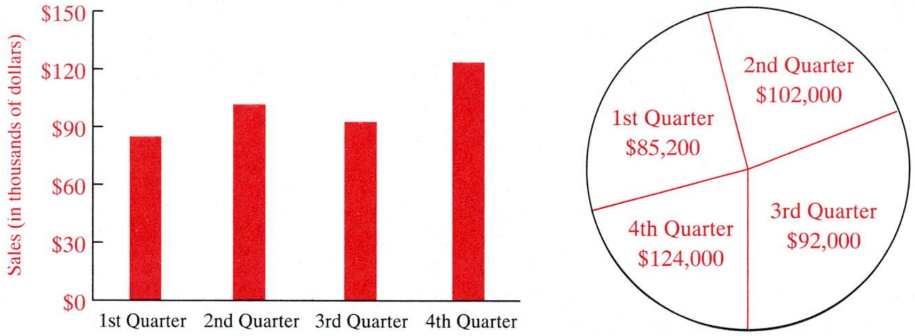

31. (a) 16.7; 16.5; 16 (b) 27.3; 27; 24 **32.** 2.69 **33.** 75; 13; 5.2 **34.** 47.5% **35.** 57 **36.** $109,100

Chapter 19 Net Assets (page 749)

1. $1522.57; $1475 **2.** $617; $210 **3.** $1102.57; $1942.57

Appendix A

Section A.1 (page 764)

1. 1771.60 **3.** 11,378 **5.** 3905.32 **7.** 255.24 **9.** 15,695 **11.** 31.08 **13.** 35.42 **15.** 27.63 **17.** 8.1 **19.** 566.14 **21.** 0.63 **23.** 30.57 **25.** 0.92 **27.** $5\frac{6}{11}$ **29.** 0.49 **33.** $29,167.32 **35.** $2648.40 **37.** (a) $1164.05 (b) $95.13 **39.** $118,191.60 **41.** $13,400

Section A.2 (page 771)

1. $7556.24 **3.** $8929.13 **5.** $2676.84 **7.** 1.5% **9.** $858.89 **11.** 24 **13.** $2537.97 **15.** $696 **17.** 26 **19.** $5755.75

Appendix B

Section B.1 (page 778)

1. 0.68 m **3.** 4700 mm **5.** 8900 g **7.** 0.39 L **9.** 46 kg **11.** 976 kg **13.** 39.24 yards **15.** 50.27 m **17.** 15.42 feet **19.** 1.10 m **21.** 307.52 miles **23.** 1235.71 km **25.** 1.50 pounds **27.** 1861.4 g **29.** 85.58 pounds **31.** 200 **33.** 0.6 g **35.** Answers will vary **37.** 40°C **39.** 280°C **41.** 37°C **43.** 95°F **45.** 50°F **47.** 275°F **49.** Unreasonable **51.** Reasonable **53.** Reasonable

Glossary

For further information on any of these terms, see the index.

1099 forms These forms show miscellaneous income received such as stock dividends or interest income.

1040EZ This form is used by some single individuals when filing their income tax returns.

absolute or **gross loss** The loss resulting when the selling price is less than the cost.

accelerated cost recovery system (ACRS) The method of depreciation required on all federal income tax returns for property acquired after January 1, 1981 and before 1986.

accelerated depreciation Depreciation which has a rate greater than a straight-line rate of depreciation. For example, declining balance, sum-of-the-years'-digits, or ACRS.

accelerated mortgages Home mortgages that are paid off faster than the typical 30 years for most home mortgages.

accidental death benefit Coverage which pays an additional death benefit if the insured dies as the result of an accident.

accounts payable Funds owed by a firm that must be paid within a short period of time such as a year or less.

accounts receivable Funds owed to a firm, by customers, that must be paid within a short period of time such as a year or less.

accounts receivable turnover Net sales divided by average accounts receivable.

accumulated depreciation A running balance or total of the depreciation to date on an asset.

acid-test ratio The ratio of current assets and current liabilities. Also called quick ratio.

actual selling price *See* reduced price.

actuary A person who determines insurance premiums.

addition rule The same value can be added to both sides of an equation.

adjustable rate mortgage *See* variable rate loan.

adjusted bank balance This number represents the current checking account balance.

adjusted gross income (AGI) The sum total of all income received from wages, salaries, interest, and dividends less any adjustments to income such as sick pay and moving expenses.

adult operator A driver of a motor vehicle over a certain age, usually 25 years of age or older.

agreed ratio Partners of a business divide the profits using some given rule. These profit divisions are sometimes given as a ratio.

algebraic logic The logic used by most electronic calculators.

allocation of overhead Dividing the overhead among the various products or lines of business of a company.

allowances See withholding allowances.

amortization schedule A table showing the equal payment necessary to pay off a loan for a specific amount of money including interest, for a specific amount of time. Also called a repayment schedule.

amortization table See amortization schedule.

amortized When principal and interest are paid by a sequence of equal payments (on a loan).

amount financed The difference between the cash price of an item and the down payment.

amount of an annuity The sum of the compound amounts of all the periodic payments into an annuity, compounded to the end of the term.

annual meeting A meeting open to all owners of stock where the management of the firm is open to questions from stockholders, and where the board of directors is elected.

annual percentage rate A rate of interest that must be stated for each loan, by federal regulation. This annual percentage rate is designed to help consumers compare interest rates. This is the true or effective rate of interest.

annuity Periodic payments of a given, fixed amount of money.

annuity certain An annuity with a fixed beginning date and a fixed ending date.

annuity due An annuity with payments made at the beginning of a time period.

anticipation The seller's offer of interest for early payment of an invoice, in addition to a cash discount.

APR *See* annual percentage rate.

AS OF An invoice which is postdated "AS OF" a future date.

assessed The procedure whereby a local official, called the assessor, makes an estimate of the fair market value of property.

assessed valuation The value for property tax purposes set by the tax assessor on a piece of property.

assessment rate A certain percent used in an area to determine assessed valuation.

assets Items of value owned by a firm.

ATM cards Many people now carry automatic teller machine cards which results in a debit to their account when they obtain cash from an ATM machine or when they make a purchase.

attached A seizure of property by court order.

automated teller machine (ATM) A machine allowing 24-hour banking.

average *See* Mean.

average age of accounts receivable One year (365 days) divided by accounts receivable turnover rate.

average daily balance A method of calculating the balance owed on a revolving charge account. With this method, the balance on the account is found at the end of each day in the month and the total is divided by the number of days in the month.

average daily balance method A method of calculating the finance charge on a revolving charge account by using the average balance at the end of each day during a month.

average inventory Determined by dividing the sum of all inventories taken by the number of times inventory was taken.

bad check A check that is not honored because there are insufficient funds in the checking account.

balanced Agreement reached between the bank statement amount and the depositor's checkbook balance.

balance brought forward The amount left in a checking account after previous checks written have been subtracted; the current balance.

balance forward *See* balance brought forward.

balance sheet A summary of the financial condition of a firm at one point in time.

bank discount *See* Discount.

bank statement A list of all charges and deposits made against and to a checking account, usually sent out monthly by the bank.

bank statement balance This is the checking account balance appearing on the front of the bank statement.

banker's interest *See* Ordinary interest.

banker's ratio *See* Current ratio.

bar graph A graph that uses bars to show data.

base In the number 3^2, the number 3 is the base. *See* Exponent. The starting point or reference point or that to which something is being compared.

base year A previous year against which something is being compared.

basic percent equation Rate × Base = Part, or $R \times B = P$.

beneficiary The person receiving insurance benefits upon the occurrence of a certain event.

bimodal A set of data in which two numbers occur equally often.

blank endorsement The endorsement of a check with a signature alone.

board of directors A group of people who represent the stockholders.

bodily injury insurance Another name for liability insurance.

bond A promise by a corporation or government to pay a certain fixed amount of money at a certain time in the future.

book value The cost of an asset minus any depreciation to date.

borrow Taking a number from one column of numbers in a problem in order to make a larger number in the column to which the borrowed number is added.

breakeven point The cost of an item plus the operating expenses associated with the item. Above this amount a profit is made, below it, a loss is incurred.

breaking even This occurs when the reduced price at which at item is sold just covers cost plus operating expenses.

business checking account The type of checking account used by businesses.

business owner's package policy A business insurance policy insuring many additional perils beyond fire.

cancellation A process used to simplify multiplication and division of fractions using a modification of the method of writing fractions in lowest terms.

cancellation rate Another name for short-term rate.

canceled check A check is canceled after credit has been received for the amount of the check by the depositor's bank.

capital The amount of money originally invested in a firm.

carrier The insurance company. Also known as the insurer.

cash discount A discount offered by the seller allowing the buyer to take a discount if payment is made within a specified period of time.

cash settlement option Life insurance benefits which are paid in cash.

cash value The value in cash remaining after a policy holder has canceled or borrowed against a life insurance policy.

cashier's check A check written by the financial institution itself and having the full faith and backing of the institution.

certificate of deposit Money placed in a time deposit account.

chain calculation A long sequence of calculations.

chain discount Involves two or more individual discounts.

charge accounts An account that allows an individual to charge purchases at a specific store.

chargeback A fee charged to a production employee for a rejected item of production.

check register A single page record of checks written and deposits made to a checking account.

checks outstanding Checks written that have not reached and cleared the bank as of the statement date.

check stub A stub attached to the check and retained for keeping a record of checks written.

circle graph A circle broken up into various parts, based on percentages of 360°.

classes The intervals of a frequency distribution.

class frequency The frequency of values for each class.

class mark The midpoint of an interval.

C.O.D. A shipping term meaning *cash on delivery*.

coefficient A number used to multiply a variable.

coinsurance clause A fire insurance clause which places part of the risk of fire loss upon the insured.

collateral Goods pledged as security for a loan—in the event that a loan is not paid off, the collateral can be seized by the lender and sold to pay the debt.

collision insurance A form of automobile insurance that pays for repairs to the insured's car in case of an accident.

commission A fee paid to an employee for transacting a piece of business or performing a service. A charge for buying or selling stock.

common denominator A number that all the denominators of a fraction problem divide into evenly.

common stock Ordinary capital stock not sharing the privileges of preferred stock.

companion or **spouse insurance** This lets the insured add a companion or spouse to a policy, resulting in both being insured.

comparative balance sheet A balance sheet used to compare the assets and liabilities of a firm at two different periods.

comparative income statement Preparation of a vertical analysis for two or more years in order to compare income or expense items for each year analyzed.

comparative line graph One graph which shows how several different things relate.

compensatory time (comp time) Time off given an employee to compensate for overtime previously worked.

complement The number which must be added to a given discount to get 1 or 100%.

compound amount The final amount, both principal and interest, after money is deposited at compound interest.

compound interest Interest computed on both principal and interest.

compounded daily Interest is paid for every day that money is on deposit in a savings account.

comprehensive insurance A form of automobile insurance that pays for damage to the insured's car caused by fire, theft, vandalism, and so on.

consolidate loans Several smaller loans can sometimes be consolidated, or brought together into one larger loan, in order to reduce payments or gain more favorable terms.

consolidated A consolidated statement is one that includes the financial results of all subsidiaries of a company.

consumer The ultimate user of a product or service, the public.

Consumer Price Index (CPI) A monthly publication by the federal government showing the change in the cost of living.

contingent annuity An annuity with a variable beginning or ending date.

continuous compounding An account offering continuous compounding features interest compounded every instant.

conversion formulas These formulas allow one to convert from percent markup on cost to percent markup on selling price or from percent markup on selling price to percent markup on cost.

convertible preferred stock One share is convertible into a stated number of shares of common stock at some date.

corporate bond A bond issued by a corporation.

corporation A form of business organization offering limited liability to shareholders—no more money may be lost than has been invested.

cost The price paid to a supplier after trade and cash discounts have been taken. This price includes transportation and insurance charges. The amount paid for a depreciable asset. This is the amount used to determine depreciation.

cost accounting A detailed look at the costs of material, labor and overhead underlying a product or a production process.

cost of goods sold The amount paid by a firm for the goods it sold during the time period covered by an income statement.

country club billing A type of billing received on revolving charge plans in which actual carbon copies of charges are returned.

credit cards Cards that allow an individual to make charges at several different businesses such as retail stores and restaurants.

credit limit The limit on the amount an individual can charge.

creditor *See* lender.

credit union share draft accounts A credit union account that may be used as a checking account.

cross-products In the proportion $\frac{a}{b} = \frac{c}{d}$, the cross-products are *ad* and *bc*.

cumulative preferred stock Stock requiring that any dividends not paid in the past must be paid before common stockholders receive any money.

current assets Cash or items that can be converted into cash within a given period of time, such as a year.

current balance *See* adjusted bank balance.

current (short-term) liabilities Those items which must be paid by a firm within a given period of time, such as a year.

current ratio The quotient of current assets and current liabilities. Also called banker's ratio.

current yield The annual dividend per share of stock divided by the current price per share.

daily overtime Some companies pay overtime on all hours over 8 that are worked in one day.

debit card A bank card used at a point of sale terminal. The amount of the purchase is instantly subtracted from the customer's account and credit is given to the seller's account.

debtor The person or firm borrowing money.

debt-to-equity ratio All liabilities divided by owner's equity.

decimal A number written with a decimal point, such as 4.3 or 7.22.

decimal equivalent of a fraction The decimal fraction that is equal to a proper fraction. For example, $\frac{1}{2} = 0.5$, $\frac{3}{8} = 0.375$.

decimal number A fraction with a denominator that is a power of ten, which is written with a decimal point, such as 4.3 or 7.22.

decimal point The starting point in the decimal system (.).

declining-balance depreciation An accelerated depreciation method.

declining-balance method A method of depreciation using a declining balance rate.

(200%) declining-balance method An accelerated method of depreciation using twice or 200% of the straight-line rate. Also called double-declining-balance method.

(150%) declining-balance method An accelerated method of depreciation using one and one-half or 150% of the straight-line rate.

(125%) declining-balance method An accelerated method of depreciation using one and one-fourth or 125% of the straight-line rate.

decrease problem Often called a difference problem; the part equals the base minus some portion of the base. Usually the base must be found.

decreasing term insurance A form of life insurance in which the insured pays a fixed premium until age 60 or 65 with the amount of life insurance decreasing periodically.

deductible The amount of the deductible is paid by the insured with the balance of the loss being paid by the insurance company.

deductions Amounts that are subtracted from the gross earnings of an employee to arrive at the amount of money the employee actually receives, net pay.

deferred payment price The cash price of an item, plus any finance charge.

deficit number *See* negative number.

denominator The number below the line in a fraction. For example, in the fraction $\frac{7}{9}$, 9 is the denominator.

deposit slip (deposit ticket) The form used for making a bank savings or checking account deposit.

depreciable amount The amount to be depreciated over the life of the asset.

depreciation A method used to spread the value of an asset over the several years of its life.

depreciation fraction The fraction used with sum-of-the-years'-digits depreciation. The numerator is the year and the denominator is the sum of the years of life of the asset.

depreciation schedule A schedule or table showing the depreciation rate, amount of depreciation, book value, and accumulated depreciation for each year of an asset's life.

destination The city or town where goods or merchandise are being shipped.

differential-piece rate A piece rate designed to pay a greater amount for each unit of production as the number of units produced is increased.

discount An amount subtracted from the price of a product or service which helps the buyer purchase at a lower cost and increase profits. The amount of interest charged on a note. Also called a bank discount.

discount broker A stockbroker who charges less than full price for buying and selling stocks.

discount fee A bank collects a discount fee on credit card deposits from the merchant.

discount note A note where interest is deducted in advance.

discount period The period of time in the discounting process.

discounting a note The holder of a note sometimes sells the note to a bank before its maturity date. This gives cash to the holder earlier than otherwise.

dispersion Spread of data.

distributive property A property of algebra that says a number on the outside of the parentheses should be multiplied times each term inside the parentheses.

dividends Money paid by a company to the holders of a stock.

divisibility rules These rules allow us to quickly determine if a number is evenly divisible by 2, 3, 4, 5, 6, 8, 9 and/or 10.

docking Same meaning as chargeback.

double-declining-balance method *See* (200%) declining-balance method.

double-time Twice the regular hourly rate. A premium often paid for working holidays and Sunday.

Dow Jones Industrial Average A commonly quoted average of the stock prices of 30 large industrial, publicly held corporations.

down payment An amount paid when an item is bought.

draw An amount paid by an employer to salespeople at regular intervals. This is often paid against future income.

drawing account An account from which a salesperson can receive payment against future commissions.

duplicate statement A duplicate of a checking account statement issued by the bank.

early withdrawal penalty A fine or amount of money charged by the lending institution for withdrawal of money earlier than the time agreed upon by the depositor and the institution. These penalties apply only to interest, not to principal.

earnings per share The difference of the net income of a corporation and any dividends on preferred shares, divided by the number of common shares outstanding.

effective annual yield *See* effective rate of interest.

effective rate of interest The simple interest rate corresponding to a given discount rate. Also called true rate of interest. The actual percent of interest earned during a year.

electronic banking The use of electronic technology in banking. Such services as direct deposits, ATM cards, debit cards, and home and business banking are all a part of electronic banking.

electronic commerce People and companies can now purchase products and pay bills electronically using the Internet.

electronic funds transfer Many companies choose to move money electronically using computer networks, rather than using checks that are mailed.

Employer's Quarterly Federal Tax Return The form (Form 941) is sent to the IRS along with quarterly payments of FICA and withholding tax.

ending inventory The value at cost of inventory at the end of a period.

end-of-month dating (EOM) In cash discounts, the time period beginning at the end of the month the invoice is dated. Proximo and prox. have the same meaning.

endowment policy A form of life insurance guaranteeing the payment of a fixed amount of money to a given individual whether or not the insured person lives.

equal shares Partners in a business share all profits equally.

equation A statement that says two expressions are equal.

equivalent cash price A single amount today equal to the present value of an annuity.

escrow account An account maintained by real estate lenders and used to pay taxes and insurance. Also called an impound account.

exact interest Simple interest calculated using 365 days in a year.

excise tax A tax charged on specific items which are purchased. Tobacco, alcoholic beverages, and gasoline have an excise tax.

executive officers President, vice president, and so on, of a corporation.

exponent In the number 3^2 the small number 2 is the exponent. It says to multiply 3 by itself. A number that tells how many times a number is used in a product. For example, $3^2 = 3 \times 3 = 9$.

extended term insurance The nonforfeiture option which gives the insured term insurance for a fixed number of years and days.

extension total On an invoice, the product of the number of items times the unit price.

extra dating (extra, ex., or x) Extra time allowed in determining the net payment date of a cash discount.

face value The amount shown on the face of a note. *See* Par value.

face value of the policy The amount of an insurance policy.

factor A person who buys accounts receivable of a firm (accounts receivable represent money owed to the firm).

factoring The term for when a business sells part of its accounts receivable to a financial institution.

Fair Labor Standards Act A federal law setting work conditions and standards of employee treatment.

fair market value The price for which a piece of property could reasonably be expected to be sold in the market.

FAS (free alongside ship) Free alongside the ship on the loading dock with all freight charges to that point paid by the shipper.

Federal Insurance Contributions Act (FICA) A federal act requiring that a specified amount of money be collected from the paycheck of almost all nongovernmental employees, which is used by the federal government to pay pensions, survivors benefits, and disability.

Federal Tax Deposit Coupon (Form 8109) Employers use this form when depositing taxes due to the IRS.

Federal Truth in Lending Act An act passed in 1969 which requires that all interest rates be given as comparable percents.

Federal Truth-in-Lending Law in 1969 (Regulation Z) A federal law passed in 1969 that establishes uniform methods of disclosing information on finance charges and interest rates.

Federal Unemployment Tax Act (FUTA) A federal act covering unemployment insurance.

financial calculator A calculator that includes common business applications such as present value, future value, payments and interest rates.

finance charges Charges paid to obtain credit.

first-in, first-out method (FIFO) Inventory valuation method following the flow of goods, first-in, first-out.

fixed amount annuity A settlement option which pays a fixed amount per month to life insurance beneficiaries.

fixed assets *See* plant and equipment assets.

fixed period annuity A settlement option for life insurance beneficiaries paying a sum of money for a fixed period of time.

fixed rate loan A home loan made at a fixed rate of interest for a fixed period of time.

flat fee checking account Some banks offer checking accounts that charge a flat fee each month for a checking account and perhaps an ATM card, debit card and credit card.

floating decimal point A feature on electronic calculators that locates the decimal point in the answer.

floor space Overhead is sometimes allocated among lines of business based on the amount of floor space used by each line.

FOB (free on board) Free on board shipping point means that the buyer pays for shipping. Free on board destination means that the seller pays for shipping.

formula A rule showing how quantities are related.

formula for base In a percent problem, the base is the part divided by the rate.

formula for rate In a percent problem, the rate is the part divided by the base.

fraction Used to indicate a part of a whole. For example, $\frac{3}{4}$ means that the whole is divided into 4 parts and we are considering 3 of them.

frequency distribution table A table showing the number of times one or more events occur.

fringe benefits Companies often provide benefits other than a paycheck, including child care or medical insurance.

future value The amount an investment grows to at some future date.

graph A visual presentation of numeric data.

graphing calculator A calculator that allows functions to be graphed.

gross earnings The total amount of money earned by an employee before any deductions are taken.

gross loss *See* absolute loss.

gross profit The difference between the amount received from customers for goods and what the firm paid for the goods.

gross profit method A method used to estimate inventory value at cost which utilizes cost amounts.

gross sales The total amount of money received from customers for the goods or services sold by the firm.

group insurance plans An insurance plan which includes a group of people employed by the same company or belonging to the same organization.

grouped data Items combined into groups (taken from a table) to simplify information for more immediate comprehension.

guaranteed conversion privilege This provision allows the insured to convert term insurance to ordinary or variable life insurance without physical examination.

half-year convention Under MACRS, property placed in service or disposed of is allowed one-half year of depreciation.

home banking A system that allows the customer to do banking from the home or business using the telephone and computer.

homeowner's policy A policy for homeowners providing fire, theft, and vandalism protection.

horizontal analysis Prepared by finding the amount of any change from last year to current year, both in dollars and as a percent.

hourly wage A rate of pay expressed as so much per hour. *See* Time rate.

hundredths Refers to the number of parts out of one hundred.

IMMA *See* insured money market accounts.

impound account *See* escrow account.

improper fraction A fraction with a numerator larger than or equal to the denominator. For example, $\frac{7}{5}$ is an improper fraction; $\frac{1}{9}$ is not.

incentive rate A payment system based on the amount of work completed.

income statement A summary of all the income and expenses involved in running a business for a given period of time.

income tax method *See* accelerated cost recovery system (MACRS).

income tax withholding Federal income tax withheld from gross earnings by the employer.

increase problem Often called an amount problem; the part equals the base plus some portion of the base, resulting in a new value. Usually the base must be found.

indicator words These words help indicate to us whether addition, subtraction, multiplication or division is used in solving the problem.

Individual Retirement Account (IRA) An account that permits an individual to establish a retirement plan and to deduct any contributions to the account.

inflation The increase over time in the price levels of goods and services.

initial inventory The value at cost of inventory at the beginning of a period.

inspection Sometimes we can quickly find a common denominator by inspecting the denominators of two or three fractions.

installment loan A loan paid off in a series of equal payments made at equal periods of time. Car loans are examples of installment loans.

insufficient funds (NSF) Not enough funds in a checking account for the bank to honor the check.

insured A person or business that has purchased insurance. Also known as the policyholder.

insured money market accounts (IMMAs) Accounts which are insured up to a certain maximum by the federal government, and which offer a higher rate than passbook accounts.

insurer The insurance company.

intangible assets These are assets of a company that are not tangible and include something like the good name and reputation of a company.

interest A charge paid for borrowing money or a fee received for lending money.

interest-in-advance note *See* simple discount note.

interest on investment and agreed ratio Sometimes one partner will put up a large share of money to start a firm, while other partners operate it. The partner putting up the larger share of money gets interest on the investment before any further division of profits.

interest paid accounts Some checking accounts receive interest based on the amount in the account.

interest rate spread The difference between the interest rate charged on loans and interest rate paid on deposits.

Internal Revenue Service (IRS) The IRS is the branch of the United States Government that is responsible for collecting the taxes due to the government.

inventory The value of all goods on hand for sale.

inventory turnover The number of times each year that the average inventory is sold, also called stock turnover.

inventory valuation Determining the value of merchandise in stock. Four common methods are specific identification, weighted average cost, FIFO, and LIFO.

invoice A document which helps businesses keep track of sales and purchases.

invoice total The total amount owed on an invoice.

IRA See Individual Retirement Account.

irregulars Items that are blemished or have flaws and must be sold at a reduced price.

itemized billing A method of credit card billing in which purchases are listed, along with payments, with no actual receipts returned to the user.

itemized deductions A taxpayer can itemize certain expenses, such as interest and taxes on a home and cash donations among others, and deduct from income before calculating taxes.

junk bonds Bonds of bankrupt or troubled companies.

last-in, first-out method (LIFO) Inventory valuation method following the flow of goods, last-in, first-out.

late fees Fees charged by lenders for late payments.

least common denominator The smallest whole number that all the denominators of two or more fractions evenly divide into.

left side In the equation $4x = 28$, the left side is $4x$.

lender The person or firm making the loan.

liabilities Expenses which must be paid by a firm.

liability or **bodily injury insurance** Coverage which provides protection from suit by an injured party.

like fractions Two fractions that have the same denominator.

like terms Algebraic terms that have the same variables raised to the same power.

limited liability No more money can be lost than has been invested.

limited payment life insurance A form of life insurance in which premiums are paid for only a certain fixed number of years.

line graph A graph that uses a line to show data.

liquid assets Cash or items which can be converted to cash quickly.

liquidity A firm is liquid if it has the ability to pay its bills as they come due.

list price The suggested retail price or final consumer price given by the manufacturer or supplier.

loan payoff table A table used to decide on the payment that will amortize a loan.

long-term liabilities Those items which will be paid after one year.

lowest terms A fraction is written in lowest terms when no number except the number 1 divides evenly into both the numerator and denominator of the fraction.

lump sum Benefits from a retirement plan can be paid all at once in a lump sum.

luxury tax A name sometimes given to excise tax.

maintenance charge per month A flat charge for maintaining a checking account.

maker The person borrowing the money on a note.

manufacturer The assembler of component parts or finished products.

margin *See* gross profit.

marital status A married individual pays less income tax than does a single person with the same income.

markdown A reduction from the original selling price. It may be expressed as a dollar amount or as a percent of the original selling price.

marketable securities Stocks, bonds, and other securities that can be converted quickly to cash.

marketing channels The path or steps that goods take from manufacturer to consumer.

markup (margin or gross profit) The difference between the cost and the selling price.

markup conversion formula A formula used to convert markup from one base to the other base.

markup equivalents A markup of 20% on selling price is equivalent to a markup of 25% on cost.

markup formula The formula used when working with markup. Cost + Markup = Selling price.

markup on cost Markup that is calculated as a percent of cost.

markup on selling price Markup that is calculated as a percent of selling price.

markup with spoilage The calculation of markup including deduction for spoiled or unsaleable merchandise.

MasterCard A credit card plan.

mathematics of buying The mathematics involving trade and cash discounts.

maturity date The date a loan is due.

maturity value The total amount, principal, and interest, that must be repaid when a loan is paid off. It equals face value plus interest.

mean The sum of all the numbers divided by the number of numbers.

measure of central tendency A number that tries to estimate the middle of a set of data. Measures of central tendency include the mean, median, and mode.

median The middle number in an ordered array.

medical and dental expenses Medical and dental expenses in excess of a certain percent of an individual's adjusted gross income may be deducted.

medical insurance Insurance providing medical protection in the event of accident or injury.

medicare tax Part of the social security tax (FICA) until 1991. Since 1991 medicare tax has been collected separately.

memory keys A feature on electronic calculators which allows answers to be stored for future use and recalled.

merchant batch header ticket A form used to deposit credit card sales in a business checking account.

method of prime numbers The least common denominator of two or more fractions can be found using this method.

metric system A system of measurement established by a group of scientists after the French Revolution of 1789.

middlemen Those along the marketing channels, such as wholesalers, brokers, and retailers.

mill A mill is one-tenth of a cent or one-thousandth of a dollar.

miscellaneous deductions In some cases, deductions such as union dues, qualified education expenses, income tax preparation fees, etc. may be deducted from income before calculating income taxes.

mixed number The sum of a fraction and a whole number. For example, $1\frac{1}{5}$ or $2\frac{5}{9}$ are mixed numbers.

mode The most common number in a list of numbers.

modified accelerated cost recovery system (MACRS) The Tax Reform Act of 1986 replaces the ACRS with the MACRS.

money order An instrument which is purchased and used in place of cash. It is usually preferred over a personal or business check.

mortality table A table showing statistics on life expectancy, survival, and death rates.

multiple carriers More than one insurance company sharing an insurable risk.

multiplication rule Both sides of an equation can be multiplied by the same number.

municipal bond A bond issued by a municipality such as a city or school district.

mutual company An insurance company owned by the policyholders who receive a dividend.

mutual fund Typically receives money from many different small investors and reinvests the funds in stocks and/or bonds.

negative number (deficit number) A number which is less than zero; a negative balance or deficit. For example, −$800 or ($800).

net cost The cost or price after allowable discounts have been taken.

net cost equivalent or **percent paid** The decimal number derived from the product of the complements of the trade discounts. This number may be multiplied by the list price to find the net cost (price).

net income Net income before taxes minus taxes.

net income after taxes *See* net income.

net income before taxes Gross sales − Returns − Cost of goods sold − Operating Expenses

net pay The amount of money actually received by an employee after deductions are taken from gross pay.

net price The net price is the list price minus the trade discount minus the cash discount.

net profit (or **net earnings**) The difference between gross margin and expenses. After the cost of goods and operating expenses are subtracted from total sales, the remainder is net profit.

net sales The value of goods bought by customers after subtracting goods returned.

net worth Same as owner's equity.

no-fault insurance Motor vehicle insurance which pays directly to the insured no matter who causes the accident.

nominal rate *See* stated rate.

noncustomer check cashing A service that allows an individual who is not a bank customer to cash a check upon payment of a fee.

nonforfeiture options Options available to the insured when canceling the insurance policy.

nonparticipating A form of stock which will never pay dividends above the stated rate.

nonparticipating policy The type of life insurance policy issued by a stock insurance company.

nonsmoker's discount A discount given to nonsmokers because they are better insurance risks.

nonsufficient funds (NSF) Not enough funds in a checking account for the bank to honor the check.

normal curve The bell-shaped curve commonly used in statistics.

notary service A service that provides notarization, which is required on certain business documents and transfers.

NOW account This account uses a "Negotiable Order of Withdrawal," which works and looks like a check.

numerator The number above the line in a fraction. For example, in the fraction $\frac{5}{8}$, 5 is the numerator.

odd lot Fewer than 100 shares of stock.

odd-lot differential An additional charge for buying or selling stocks when the number of shares is not a multiple of 100.

open-end credit An account that is not paid off in a fixed period of time; MasterCard and VISA accounts are examples of open-end credit.

operating expenses (or **overhead**) Expenses of operating a business. Wages, salaries, rent, utilities, and advertising are examples of operating expenses.

operating loss The loss resulting when the selling price is less than the breakeven point.

order of operations The rules determining which calculations must be done first in chain calculations.

ordered array An arrangement of a list of numbers from smallest to largest.

ordinary annuity An annuity with payments made at the end of a given period of time.

ordinary dating method A method for calculating the discount date and the net payment date. Days are counted from the date of the invoice.

ordinary interest Simple interest calculated assuming 360 days in a year. Also called banker's interest.

ordinary life insurance (Whole life insurance, straight life insurance) A form of life insurance in which the insured pays a constant premium until death or retirement, whichever occurs sooner. Upon retirement, monthly payments are made by the company to the insured until the death of the insured.

original investment Partners divide profits of a business on the basis of original investments by each partner.

overdraft This occurs when a customer writes a check for which there are insufficient funds in the account.

overdraft protection The bank service of honoring checks written on an account which has insufficient funds.

overhead *See* operating expenses.

override A commission received by a sales supervisor or department head based on total sales of the sales group or department.

over the limit fees Many lenders charge a fee in the event the debt exceeds the amount authorized by the issuer of the card.

overtime The number of hours worked by an employee in excess of 40 hours per week, or 8 hours per day.

overtime premium method Payment of overtime as a premium. All hours worked are paid at regular rate. Overtime hours are paid at $\frac{1}{2}$ rate. Gross earnings are the sum of these.

owner's equity The difference between assets and liabilities. Also called proprietorship or net worth.

paid-up insurance A nonforfeiture option which provides paid-up insurance of a certain amount.

par value The amount printed on a stock certificate; usually the price at which a share of stock is first offered to the public.

part The result of multiplying the base times the rate.

partial payment A payment which is less than the total owed on an invoice; a cash discount may be earned.

partial-year depreciation The amount of depreciation that is determined for the asset during a period less than one year.

participating A type of stock that could be affected by an increase in the dividend.

participating policy The type of life insurance policy issued by a mutual insurance company.

partnership A business formed by two or more people.

passbook account A bank account used for day-in and day-out deposits of money. These accounts usually have the lowest interest rates of any accounts, but have no penalties when money is withdrawn.

pay period The time period for which an employee is paid.

payee The person who loans the money and will receive payment on a note.

payer *See* maker.

payment period The time between the payments of annuity.

payments for life A life insurance settlement option which pays an annuity for life.

payments for life with a guaranteed number of years A life insurance settlement option which pays a certain amount per month for the life of the insured but guarantees a certain length of time in the event that the insured dies before this guaranteed time period.

payroll ledger A chart showing all payroll information.

PE ratio *See* price-earnings ratio.

per debit charge A charge per check. Usually continues regardless of the number of checks written.

percent A percent is one hundredth. For example, 2 percent means 2 parts of a hundred.

percent or **rate of markdown** The markdown expressed as a percent of original price.

percent or **rate of markup** The markup expressed as a percent of original price.

percentage method of withholding Used to determine federal withholding tax. This method does not require several pages of tables needed with the wage bracket method.

percent formula The basic percent formula is $P = B \times R$, or Part = Base × Rate.

percent key The electronic calculator key [%] which moves the decimal point two places to the left when used following multiplication or division.

peril insurance Insurance which pays upon a loss by the insured.

period of compounding Amount of time between the addition of interest to a deposit or loan.

periodic inventory A physical inventory taken at regular intervals.

periodic payments A series of payments made at regular intervals in time.

perpetual inventory A continuous inventory system normally utilizing a computer.

personal checking account The type of checking account used by individuals.

personal exemption Each taxpayer currently gets a deduction for each dependent, including the taxpayer.

personal identification number (PIN) A special code that must be entered when using an ATM card or a debit card.

personal income tax A type of tax charged by states and the federal government to individuals. The tax is based on income.

personal property Property other than real estate such as furnishings, appliances, cars, trucks, clothing, boats, and money.

physical inventory An actual physical count of each item in stock at a given time.

piecework rate A method of pay by which an employee receives so much money per item completed.

PIN A personal identification number is needed in addition to an ATM card in order to make a cash withdrawal.

plant assets *See* plant and equipment assets.

plant and equipment assets Items owned by a firm which will not be converted to cash within a year. Also called fixed assets or plant assets.

point of sale terminal This refers to the point in a store where the sale is actually made—commonly at a cash register where scanners are used to record the items purchased.

policy A contract between an insured and an insurance company.

policy fee An annual fee charged by insurance companies to cover the cost of processing the policy.

policyholder A person or business that has purchased insurance. Also known as the insured.

population The entire group being studied.

postdating Dating in the future; on an invoice **"AS OF"** dating.

preferred stock Stock which pays dividends before common stockholders receive any dividends.

premium The amount of money charged for an insurance policy.

premium factor A factor used to convert annual premiums to either semiannual, quarterly, or monthly premiums of an insurance policy.

premium payment An additional payment for extra service.

premium rate A higher amount of pay given for additional hours worked or additional units produced.

present value An amount that can be invested today to produce a given amount in the future.

present value of an annuity (1) The lump sum that can be deposited today that will amount to the same final total as would the periodic payment of an annuity. (2) A lump sum that could be deposited today so that equal periodic withdrawals could be made.

price-earnings ratio (PE ratio) The price per share of stock divided by the annual earnings per share of the stock.

price relative The quotient of the current price and the price in some past year with the quotient multiplied by 100.

prime number A number divisible without remainder only by itself or 1 (such as 7 or 13).

prime rate The interest rate at which large, financially secure corporations borrow money.

principal An amount of money either borrowed, loaned, or deposited. The initial amount of money deposited.

proceeds The amount of money the borrower receives after subtracting the discount from the face value of a note.

processing (check) A check given to a merchant is processed by being deposited with the merchant's bank where it is then routed to a Federal Reserve bank and then to the payer's bank.

promissory note A document in which one person agrees to pay money to another person, a certain amount of time in the future, and at a certain rate of interest.

proper fraction A fraction in which the numerator is smaller than the denominator. For example, $\frac{2}{3}$ is a proper fraction; $\frac{9}{5}$ is not.

property damage insurance A type of automobile insurance that pays for damages caused to the property of others.

property tax rate The tax rate applied to the assessed value of property.

proportion A proportion says that two ratios are equal.

proprietorship *See* owner's equity.

prox. *See* proximo dating.

proximo dating In cash discounts, the time period beginning at the end of the month the invoice is dated. End-of-month dating (EOM) and "prox." have the same meaning.

Publication 534 The Internal Revenue Service publication which gives a complete coverage of depreciation.

purchase invoice The invoice or document received by the purchaser of goods or services from the seller.

quick ratio *See* acid-test ratio.

quota bonus A plan which pays a bonus to an employee after reaching a quota.

range The difference between the largest value and the smallest value in a set of numbers.

rate A number followed by "%" or "percent."

rate of commission Many people in sales are paid based on a commission or percent of sales.

rate of interest The percent of interest charged for one year.

ratio A quotient of two quantities.

ratio of net income after taxes to average owner's equity Net income divided by average owner's equity.

raw data A set of data before analysis.

real estate Real property such as land and buildings.

real property All land, buildings, and other improvements attached to the land.

receipt of goods dating (ROG) In cash discounts, time is counted from the date that goods are received.

reciprocal The result of interchanging the numerator and denominator of a fraction.

reconciliation The process of checking a bank statement against the depositor's own personal records.

recourse Merchants sometimes sell debts that are owed them. If the person owing the money is unavailable and the bank buying the debt has recourse to the merchant—the merchant is liable for the debt.

recovery class The class into which property is placed under MACRS (3-, 5-, 7- 10-, 15-, 20-, 27.5-, 31.5-, or 39-year class).

recovery period The number of years over which the cost of an asset is recovered using the MACRS.

rediscounting The process in which one financial institution discounts a note at a second institution.

reduced net profit This occurs when a markdown decreases the selling price to a point which is still above the breakeven point.

reduced price The selling price after subtracting the markdown, also called sale price and actual selling price.

Regulation DD A federal law requiring that interest paid on funds in savings accounts be paid based on the exact number of days.

Regulation Z *See* Federal Truth in Lending Act.

repayment schedule *See* amortization schedule.

repeating decimal A decimal which repeats one or more digits without ending. A bar is often placed over the repeating digit(s). For example, $.333\overline{3}$ and $.16\overline{16}$ are both repeating decimals.

repossess The act by a lender of taking back ownership of an item when payments have not been made.

Request for Earnings and Benefit Estimate Statement This form is used by an individual when checking with the Social Security Administration to see if they have the correct information on the individual's previous social security contributions.

residual value *See* salvage value.

restricted endorsement Endorsement of a check so that only the person or company given the check may cash it.

retail method A method used to estimate inventory value at cost which utilizes both cost and retail amounts.

retail price The price at which merchandise is offered for sale to the public.

retailer A firm that sells directly to the consumer.

returned check A check that has been deposited but returned to the bank due to nonsufficient funds (NSF) is said to be returned.

returned deposit item The return to the bank of an item which has been deposited, due to any number of irregularities.

returned goods Merchandise returned due to incorrect shipment or damage.

returns The total value of all goods returned by customers.

revolving charge account *See* open-end credit.

right side In the equation $4x = 28$, the right side is 28.

ROG *See* Receipt-of-Goods Dating.

Roth IRA Contributions to this type of IRA are not deductible when made, but funds in the IRA grow tax-free and eventual withdrawals are also tax-free.

round lot Multiple of 100 shares of stock.

rounded decimals Decimals reduced to a number with fewer decimals.

Rule of 78 A method of calculating interest charges that need not be paid because the loan was paid off earlier than planned.

rules for divisibility Rules which help determine whether a number is evenly divisible by another number.

salary A fixed amount of money per pay period.

salary and agreed ratio Same as agreed ratio, except that a salary may be allowed to one partner or the other in addition to the profit division.

salary plus commission A commission is paid as a premium in addition to salary.

sale price *See* reduced price.

sales invoice The invoice or document retained by the seller of goods or services, a copy of which is sent to the purchaser.

sales quota An expected level of production. A premium may be paid for surpassing quota.

sales tax A tax placed on sales to the final consumer. The tax is collected by the state, county, or local government.

sales value Value of sales for each department of a company.

salvage value or scrap value The value of an asset at the end of its useful life. For depreciation purposes, this is often an estimate.

sample A portion of the entire population being studied.

savings account *See* passbook account.

Schedule 1 (Form 1040) This is the basic form used for reporting interest dividend income.

scientific calculator A calculator that includes common math operations used in science including the grouping of operations using parentheses.

SDI (State Disability Insurance) deduction The deduction for a state disability insurance program.

self-employed individuals Individuals who work for themselves instead of for the government or a company owned by someone else.

selling price The price at which merchandise is offered for sale to the public. The cost of an item plus its markup.

series discount *See* chain discount.

settlement options Methods of receiving life insurance benefits in addition to a cash payment.

shift-differential A premium paid for working a less desirable shift, such as swing shift or graveyard shift.

shipping point The location from which merchandise is shipped by the seller to the buyer.

short-term or **cancellation rate** A rate used when charging for short-term policies and the refunds given when policies are canceled by the policyholder.

silent partner A partner who invests in a partnership, but takes no part in running it.

similar terms *See* like terms.

simple annuity An annuity with payment dates matching the compounding period.

simple discount note A note whose interest is deducted in advance from the face value, with only the difference given to the borrower.

simple interest Interest computed only on the principal.

simple interest notes Notes on which interest is found by formulas for simple interest.

single discount A discount expressed as a single percent and not as a series or chain discount.

single discount equivalent to a series discount A series or chain discount which is expressed as a single discount.

sinking fund A fund set up to receive equal periodic payments in order to pay off an obligation at some fixed time in the future.

sliding-scale commission A graduated commission plan giving a higher rate to top producing salespeople.

SMP The abbreviation for a special multi-perils policy. This policy gives additional insurance coverage to businesses. *See* Business owner's package policy.

Social Security *See* Federal Insurance Contributions Act (FICA).

sole proprietorship A business owned by one person.

solution A number that can replace a variable in an equation and result in a true statement.

special endorsement An endorsement to a specific payee.

specific identification method Inventory valuation method which identifies the cost of each individual item.

split-shift premium A premium paid for working a split shift. For example, an employee who is on 4 hours, off 4 hours, and then on 4 hours.

spoilage Merchandise which becomes unsaleable. Usually considered when calculating markup.

square root The square root ($\sqrt{\ }$) of a number is a number when multiplied by itself equals that number. The square root of 9, $\sqrt{9}$, is 3, since $3 \times 3 = 9$.

square root key The electronic calculator key $\boxed{\sqrt{\ }}$ which calculates the square root of the number on the calculator.

Stafford loan Government loans that can be subsidized—they are frequently used by students.

standard deduction A deduction used to reduce taxable income for taxpayers who do not itemize their deductions.

standard deviation A measurement of the dispersion of a set of data.

state income tax Some states levy a state income tax on income.

state withholding State income tax withheld from gross earnings by the employer.

stated rate The rate of interest quoted by a bank, also called the nominal rate.

statistics Data and/or the analysis of data.

stock A share of stock represents partial ownership of a corporation.

stock certificates Documentation of stock ownership.

stock company An insurance company owned by stockholders. No dividend is paid to policyholders.

stock exchange A place or mechanism at or through which stocks can be bought and sold.

stock fund There are many mutual funds that contain stocks.

stock ratio Numbers used to compare stocks—typically the current yield and the PE ratio.

stock turnover *See* inventory turnover.

stockbroker A person who buys and sells stock for the public.

stockholders The owners of a corporation.

stockholders' equity The difference between a corporation's assets and liabilities.

stop payment order A request that the bank not pay on a check previously written.

straight commission A fixed amount or percent for each unit of work. Earnings are based on performance alone.

straight life insurance Another name for ordinary or whole life insurance.

straight-line depreciation A depreciation method where depreciation is spread evenly over the life of the asset.

substitution Replacing the variable in an equation by the solution; substitution is used to check a solution.

suicide clause A clause which excludes suicide as an insurable cause of death (usually for the first two years of the policy).

sum of the balances method See Rule of 78.

sum-of-the-years' digits method An accelerated depreciation method using a depreciation fraction.

T-account form This is a method for reconciling a bank statement in which the bank statement balance is on the left and the checkbook balance is on the right side.

T-bills *See* U.S. Treasury bills.

tangible assets These are physical assets such as machinery, trucks, and buildings.

tax deduction Any expense that the IRS will allow a taxpayer to subtract from adjusted gross income.

tax rate schedule A schedule which shows the individual tax rates for tax filing status.

taxable income Adjusted gross income, minus exemptions, minus deductions.

taxes Individuals and companies pay many types of taxes including sales tax, income tax, and property tax among others.

telephone transfer The transfer of funds with a verbal request over the telephone.

term A single letter, a single number, or the product of a number and a letter.

term insurance A form of insurance providing protection for a fixed length of time.

term of an annuity The time from the beginning of the first payment into an annuity until the end of the last payment.

term of the note The length of time until a note is due.

territorial ratings Insurance companies use territorial ratings to adjust for the fact that automobile accidents occur more often in certain territories than in others.

time The number of years or fraction of a year for which the loan is made.

time-and-a-half rate Many employees are paid $1\frac{1}{2}$ times the normal rate of pay for any hours worked in excess of 40 hours per week, or 8 hours per day.

time card A card that is helpful in preparing the payroll. The time card includes such information as the dates of the pay period, the employee's name, and the number of hours worked.

time deposit account A savings account in which the depositor agrees to leave money for a certain period of time.

time rate Earnings based on hours worked, not work accomplished.

time value of money, or **value of money** The average interest rate for which money is loaned at a given time.

total installment cost Find the total installment cost by multiplying the amount of each payment on a loan and the number of payments; then add any down payment.

total invoice amount The sum of all the extension totals on an invoice.

trade discount The discount offered to businesses. This discount is expressed either as a single discount (such as 25%) or a series discount (such as 20/10) and is subtracted from the list price.

transfer tax A tax charged by some cities and states on the purchase or sale of stock.

Treasury bills *See* U.S. Treasury bills.

Treasury bill fund Some funds invest most of their assets in Treasury bills.

true rate of interest *See* effective rate of interest.

turnover at cost Found by the following formula.

$$\frac{\text{Cost of goods sold}}{\text{Amount of inventory at cost}}$$

turnover at retail Found by the following formula.

$$\frac{\text{Retail sales}}{\text{Average inventory at retail}}$$

underinsured motorist insurance Insurance coverage which covers the insured when involved in an accident with a driver who is underinsured. Coverage for bodily injury above the amounts of insurance carried by the underinsured driver.

underwriter An insurance company employee who determines the risk factors involved in the occurrence of various insurable losses. This helps determine the insurance premium.

unearned interest The amount of interest not owed when a loan is paid off early.

unemployment insurance tax A tax paid by employers. The money is used to pay unemployment benefits to qualified unemployed workers.

uniform product codes (UPC) Bar codes found on each product in most stores; used for efficient inventory control by stores. It also provides greater accuracy and perhaps faster service to the customer.

uninsured motorist insurance Insurance coverage which covers the insured when involved in an accident with a driver who is not insured.

unit price The cost of each unit on an invoice.

units-of-production method A depreciation method using the units produced to determine depreciation allowance.

units shipped The number of units shipped.

United States Rule A method of handling partial loan payoffs; any payment is first applied to the interest owed on the loan, with any balance then used to reduce the principal amount of the loan.

universal life insurance Allows the insured to vary the amount of premium and type of protection depending on changing insurance needs.

unlike fractions Fractions having different denominators.

unpaid balance The balance outstanding on a revolving charge account at the end of a billing period.

unpaid balance method A method of calculating the finance charge on a revolving charge account by using the balance at the end of the previous month.

unsaleable items Merchandise which cannot be sold. Usually considered when calculating markup.

useful life The estimated life of an asset. The IRS gives guidelines of useful life for depreciation purposes.

U.S. Treasury bills A loan of money to the United States government. Treasury bills (or T-bills) are a very safe way to invest money.

variable A letter that represents a number.

variable commission A rate of commission that depends on the total amount of the sales, with the rate increasing as sales increase.

variable life insurance Provides life insurance protection and allows the insured to select investment funds to invest the balance of the premium.

variable rate loan A home loan made at an interest rate that varies with market conditions.

variance The square of the standard deviation.

vertical analysis The process of listing each of the important items on an income statement as a percent of total net sales or each item on a balance sheet as a percent of total assets.

Visa A credit card plan.

wage bracket method of withholding Used to determine federal withholding tax. This method requires several pages of tables.

waiver of premium clause Allows insurance to continue without payment of premium when insured becomes disabled.

weighted average method Inventory valuation method where the cost of all purchases during a time period is divided by the number of units purchased.

weighted mean A mean calculated by using weights, so that each number is multiplied by its frequency.

whole life insurance (Ordinary or straight life insurance) A form of life insurance in which the insured pays a constant premium until death or retirement, whichever occurs sooner. Upon retirement, monthly payments are made by the company to the insured until the death of the insured.

wholesaler The middleman; purchases from manufacturers or other wholesalers and sells to retailers.

wire transfer The instant electronic transfer of funds from one account to another.

withholding allowance These allowances, for employees, their spouses and dependents, determine the amount of withholding tax taken from gross earnings.

worker's compensation insurance Insurance which provides payments to an employee who is unable to work due to a job related injury or illness.

W-2 form The wage and tax statement given to the employee each year by the employer.

youthful operator A driver of a motor vehicle under a certain age, usually 25 years of age or younger.

zero coupon bond A bond that does not pay annual interest, rather it only pays the face value of the bond at maturity.

Index of Applications

Business

Business Equipment

Construction

Family

Food Service Industry

Foreign Affairs

General Interest

Sales & Marketing

Science & Technology

Stocks/Investments

Taxes

Index

The Number of Each of the Days of the Year*

Number of Days	Jan.	Feb.	Mar.	Apr.	May	June	July	Aug.	Sept.	Oct.	Nov.	Dec.	Number of days
1	1	32	60	91	121	152	182	213	244	274	305	335	1
2	2	33	61	92	122	153	183	214	245	275	306	336	2
3	3	34	62	93	123	154	184	215	246	276	307	337	3
4	4	35	63	94	124	155	185	216	247	277	308	338	4
5	5	36	64	95	125	156	186	217	248	278	309	339	5
6	6	37	65	96	126	157	187	218	249	279	310	340	6
7	7	38	66	97	127	158	188	219	250	280	311	341	7
8	8	39	67	98	128	159	189	220	251	281	312	342	8
9	9	40	68	99	129	160	190	221	252	282	313	343	9
10	10	41	69	100	130	161	191	222	253	283	314	344	10
11	11	42	70	101	131	162	192	223	254	284	315	345	11
12	12	43	71	102	132	163	193	224	255	285	316	346	12
13	13	44	72	103	133	164	194	225	256	286	317	347	13
14	14	45	73	104	134	165	195	226	257	287	318	348	14
15	15	46	74	105	135	166	196	227	258	288	319	349	15
16	16	47	75	106	136	167	197	228	259	289	320	350	16
17	17	48	76	107	137	168	198	229	260	290	321	351	17
18	18	49	77	108	138	169	199	230	261	291	322	352	18
19	19	50	78	109	139	170	200	231	262	292	323	353	19
20	20	51	79	110	140	171	201	232	263	293	324	354	20
21	21	52	80	111	141	172	202	233	264	294	325	355	21
22	22	53	81	112	142	173	203	234	265	295	326	356	22
23	23	54	82	113	143	174	204	235	266	296	327	357	23
24	24	55	83	114	144	175	205	236	267	297	328	358	24
25	25	56	84	115	145	176	206	237	268	298	329	359	25
26	26	57	85	116	146	177	207	238	269	299	330	360	26
27	27	58	86	117	147	178	208	239	270	300	331	361	27
28	28	59	87	118	148	179	209	240	271	301	332	362	28
29	29		88	119	149	180	210	241	272	302	333	363	29
30	30		89	120	150	181	211	242	273	303	334	364	30
31	31		90		151		212	243		304		365	31

Comparing Simple Interest and Simple Discount Rates

The simple interest rate R and the simple discount rate D are calculated by the formulas

$$R = \frac{D}{1 - DT} \quad \text{and} \quad D = \frac{R}{1 + RT},$$

where T is time in years.

Compound Interest

If P dollars are deposited at a rate of interest i per period for n periods, then the *compound amount* M, or the final amount on deposit, is

$$M = P(1 + i)^n$$

The interest earned I is

$$I = M - P$$

(Use column A of the interest table.)

Present Value at Compound Interest

The *present value* P of the future amount M at an interest rate of i per period for n periods is

$$P = \frac{M}{(1 + i)^n} \qquad \text{(Use column B of the interest table.)}$$

Unearned Interest

The *unearned interest* is given by $U = F\left(\frac{N}{P}\right)\left(\frac{1 + N}{1 + P}\right)$

where: U = Unearned interest; N = Number of payments remaining; F = Finance charge; P = Total number of payments

Annuities, Sinking Funds, and Amortization

Lump Sums. A lump sum is deposited today; to find the *compound amount* in the future, use the formula $M = P(1 + i)^n$ and column A of the interest table in Appendix D.

To find the lump sum which is the *present value* today of a known amount in the future, use $P = \frac{M}{(1 + i)^n}$ and column B.

Making Periodic Payments. To find the *amount of an annuity* when periodic payments are made for a fixed period of time, use $S = R \cdot s_{\overline{n}|i}$ and column C.

To find the amount that could be deposited today that would be equivalent to a series of periodic payments, find the *present value of an annuity* by using the formula $A = R \cdot a_{\overline{n}|i}$ and column D.

Find Periodic Payments. To find the periodic payment that must be made to produce some fixed total in the future, use the formula for a *sinking fund*,

$$R = S\left(\frac{1}{s_{\overline{n}|i}}\right), \text{ and column E.}$$

The periodic payment that will pay off, or amortize, a loan is given by

$$R = A\left(\frac{1}{a_{\overline{n}|i}}\right) \text{ and column F.}$$